MOLECULAR BIOLOGY
INTELLIGENCE
UNIT

NF-κB/Rel Transcription Factor Family

Hsiou-Chi Liou, Ph.D.
Division of Immunology
Department of Medicine
Weill Medical College of Cornell University
New York, New York, U.S.A.

LANDES BIOSCIENCE / EUREKAH.COM
GEORGETOWN, TEXAS
U.S.A.

SPRINGER SCIENCE+BUSINESS MEDIA
NEW YORK, NEW YORK
U.S.A.

NF-κB/Rel Transcription Factor Family

Molecular Biology Intelligence Unit

Landes Bioscience / Eurekah.com
Springer Science+Business Media, LLC

ISBN: 0-387-33572-2   Printed on acid-free paper.

Copyright ©2006 Landes Bioscience and Springer Science+Business Media, LLC

All rights reserved. This work may not be translated or copied in whole or in part without the written permission of the publisher, except for brief excerpts in connection with reviews or scholarly analysis. Use in connection with any form of information storage and retrieval, electronic adaptation, computer software, or by similar or dissimilar methodology now known or hereafter developed is forbidden.
The use in the publication of trade names, trademarks, service marks and similar terms even if they are not identified as such, is not to be taken as an expression of opinion as to whether or not they are subject to proprietary rights.
While the authors, editors and publisher believe that drug selection and dosage and the specifications and usage of equipment and devices, as set forth in this book, are in accord with current recommendations and practice at the time of publication, they make no warranty, expressed or implied, with respect to material described in this book. In view of the ongoing research, equipment development, changes in governmental regulations and the rapid accumulation of information relating to the biomedical sciences, the reader is urged to carefully review and evaluate the information provided herein.

Springer Science+Business Media, LLC, 233 Spring Street, New York, New York 10013, U.S.A.
http://www.springer.com

Please address all inquiries to the Publishers:
Landes Bioscience / Eurekah.com, 810 South Church Street, Georgetown, Texas 78626, U.S.A.
Phone: 512/ 863 7762; FAX: 512/ 863 0081
http://www.eurekah.com
http://www.landesbioscience.com

Printed in the United States of America.

9 8 7 6 5 4 3 2 1

Library of Congress Cataloging-in-Publication Data

Liou, Hsiou-Chi.
 NF-kB/Rel transcription factor family / Hsiou-Chi Liou.
   p. ; cm. -- (Molecular biology intelligence unit)
 Includes index.
 ISBN 0-387-33572-2
 1. NF-kappa B (DNA-binding protein) 2. Transcription factors. I. Title. II. Series: Molecular biology intelligence unit (Unnumbered)
 [DNLM: 1. NF-kappa B. 2. Transcription Factors. QU 475 L763n 2006]
 QP552.N46L56 2006
 611'.018166--dc22
                          2006013944

Dedication

This book is dedicated to Dr. David Baltimore
for his pioneer work in the NF-κB field
and leadership in several frontiers of biomedical sciences.

CONTENTS

Foreword xiii

Preface xix

1. **Structural Analysis of NF-κB and IκB Proteins** 1
   Tom Huxford and Gourisankar Ghosh
   NF-κB Structure 4
   NF-κB/DNA Complex 6
   NF-κB/IκB Complex 7
   NF-κB p50 Homodimer/RNA Aptamer Complex 9

2. **NF-κB Signal Transduction by IKK Complexes** 12
   Zhi-Wei Li and Michael Karin
   NF-κB, IκB, IKK and the Canonical Pathway
     for NF-κB Activation 12
   IKKα and Noncanonical Pathway for NF-κB Activation 15
   Modification of NF-κB and Its Signaling Molecules 17
   Molecules Involved in Multiple NF-κB Activation
     Signaling Pathway 19

3. **Receptors and Adaptors for NF-κB Signaling** 26
   Shao-Cong Sun and Edward W. Harhaj
   Antigen Receptors 26
   Receptor-Proximal Signaling Events 26
   Toll-Like Receptors 30
   TNF Receptor Superfamily 32
   Other Receptors Mediating NF-κB Activation 36

4. **Cellular Dynamics of NF-κB Associated Proteins** 41
   Daliya Banerjee and Ranjan Sen
   Nucleocytoplasmic Shuttling of IκB Proteins 41
   Signal Induced Degradation of IκB Proteins 44
   Shuttling of IκB Kinase Components 46
   Shuttling of Rel Proteins 47

5. **NF-κB in Lymphopoiesis** 51
   Estefania Claudio, Keith Brown and Ulrich Siebenlist
   NF-κB in Early Stages of Lymphocyte Development 51
   Double Negative, PreTCR+ Thymocytes 54
   Double Positive Thymocytes 54
   Single-Positive Thymocytes and Peripheral T Cells 56

6. **NF-κB and Immune Cell Effector Functions** 70
   *Hsiou-Chi Liou, Biao Feng, Wenzhi Tian, Shuhua Cheng
   and Constance Y. Hsia*
   Brief Overview of the Immune System 70
   NF-κB and Innate Immunity 71
   NF-κB in Dendritic Cells 72
   NF-κB in T Cell Development and Selection 74
   NF-κB in T Cell Effector Function 75
   NF-κB in B Lymphocyte Clonal Expansion
     and Cell Fate Determination 77
   NF-κB in the Germinal Center Immune Response 78

7. **Roles of NF-κB in Autoimmunity** 84
   Stacey Garrett and Youhai H. Chen
   Roles of NF-κB in T Cell Development 84
   Roles of NF-κB in Mature Lymphocytes 85
   Roles of NF-κB in Apoptosis, Proliferation and Autoimmunity 86
   Roles of NF-κB in Autoimmune Encephalomyelitis and Diabetes 87

8. **The Central Role of NF-κB in the Regulation of Immunity
   to Infection** 91
   Cristina M. Tato and Christopher A. Hunter
   NF-κB: An Evolutionarily Conserved System Associated
     with Innate Immunity 93
   Toll-Like Receptors and Innate Immunity 93
   The Role of NF-κB in Resistance to Infection 94
   Pathogens That Interfere with NF-κB 100

9. **Molecular Basis of Oncogenesis by NF-κB:
   From a Bird's Eye View to a RELevant Role in Cancer** 112
   Yongjun Fan, Jui Dutta, Nupur Gupta and Céline Gélinas
   Constitutive Rel/NF-κB Activity Is a Hallmark
     of Many Human Cancers 113
   Molecular Basis for Oncogenesis by Rel/NF-κB 117
   Functional Consequences of Rel/NF-κB-Mediated
     Gene Activation in Oncogenesis 119
   Other Means for NF-κB to Participate in Oncogenesis 121
   Conclusions and Perspectives for Therapy 122

10. **NF-κB in Human Cancers** 131
    Elaine J. Schattner, Richard R. Furman and Alejandro Bernal
    NF-κB Controls Cellular Proliferation, Adhesion and Survival 132
    Tumors in Which NF-κB Is Implicated in Pathogenesis 133
    Virus-Associated Tumors 135
    Inhibition of NF-κB in Cancer Treatment 136

11. **NF-κB in Neurons: Behavioral and Physiologic Roles in Nervous System Function** 147
    Jonathan M. Levenson, Marina Pizzi and J. David Sweatt
    A Synaptic Messenger 147
    Signaling Pathways That Regulate Neuronal NF-κB 149
    Glutamate 150
    Growth Factors 150
    Calcium 150
    Reactive Oxygen Species 150
    Protein Kinases 150
    A Model for Activation of NF-κB in Neurons 151
    Synaptic Plasticity 152
    Memory Formation 154
    Fear Motivated Learning: An Explanation 154
    Fear Potentiated Startle 155
    Fear Conditioning 155
    Radial Arm Maze 157

12. **Inhibitors of NF-κB Activity: Tools for Treatment of Human Ailments** 162
    Vinay Tergaonkar, Qiutang Li and Inder M. Verma
    Inhibition of NF-κB Can Be Achieved at Multiple Points in the Pathway 163
    Biological Inhibitors 163
    Synthetic Inhibitors 168
    Future Directions 170

Index 179

EDITOR

Hsiou-Chi Liou
Division of Immunology
Department of Medicine
Weill Medical College of Cornell University
New York, New York, U.S.A.
Email: hcliou@med.cornell.edu
Preface, Chapter 6

CONTRIBUTORS

Daliya Banerjee
Department of Biology
Brandeis University
Waltham, Massachusetts, U.S.A.
Chapter 4

Alejandro Bernal
Division of Hematology
  and Medical Oncology
Weill Medical College
  of Cornell University
New York, New York, U.S.A.
Chapter 10

Keith Brown
Laboratory of Immunoregulation
NIAID
National Institutes of Health
Bethesda, Maryland, U.S.A.
Chapter 5

Youhai H. Chen
Department of Pathology
  and Laboratory Medicine
School of Medicine
University of Pennsylvania
Philadelphia, Pennsylvania, U.S.A.
Email: yhc@mail.med.upenn.edu
Chapter 7

Shuhua Cheng
Division of Immunology
Department of Medicine
Weill Medical College
  of Cornell University
New York, New York, U.S.A.
Chapter 6

Estefania Claudio
Laboratory of Immunoregulation
NIAID
National Institutes of Health
Bethesda, Maryland, U.S.A.
Chapter 5

Jui Dutta
Center for Advanced Biotechnology
  and Medicine
Graduate Program in Biochemistry
  and Molecular Biology
University of Medicine and Dentistry
  of New Jersey
Piscataway, New Jersey, U.S.A.
Chapter 9

Yongjun Fan
Center for Advanced Biotechnology
  and Medicine
University of Medicine and Dentistry
  of New Jersey
Piscataway, New Jersey, U.S.A.
Chapter 9

Biao Feng
Division of Immunology
Department of Medicine
Weill Medical College
  of Cornell University
New York, New York, U.S.A.
Chapter 6

Richard R. Furman
Division of Hematology
  and Medical Oncology
Weill Medical College
  of Cornell University
New York, New York, U.S.A.
Chapter 10

Stacey Garrett
Department of Pathology
  and Laboratory Medicine
School of Medicine
University of Pennsylvania
Philadelphia, Pennsylvania, U.S.A.
Chapter 7

Céline Gélinas
Center for Advanced Biotechnology
  and Medicine
Department of Biochemistry
Robert Wood Johnson Medical School
University of Medicine and Dentistry
  of New Jersey
Piscataway, New Jersey, U.S.A.
Email: gelinas@cabm.rutgers.edu
Chapter 9

Gourisankar Ghosh
Department of Chemistry
  and Biochemistry
University of California, San Diego
La Jolla, California, U.S.A.
Email: gghosh@ucsd.edu
Chapter 1

Nupur Gupta
Center for Advanced Biotechnology
  and Medicine
Graduate Program in Biochemistry
  and Molecular Biology
University of Medicine and Dentistry
  of New Jersey
Piscataway, New Jersey, U.S.A.
Chapter 9

Edward W. Harhaj
Department of Microbiology
  and Immunology
and
Sylvester Comprehensive Cancer Center
University of Miami School of Medicine
Miami, Florida, U.S.A.
Chapter 3

Constance Y. Hsia
SG Cowen & Co.
New York, New York, U.S.A.
Chapter 6

Christopher A. Hunter
Department of Pathobiology
School of Veterinary Medicine
University of Pennsylvania
Philadelphia, Pennsylvania, U.S.A.
Email: chunter@vet.upenn.edu
Chapter 8

Tom Huxford
Department of Chemistry
  and Biochemistry
University of California, San Diego
La Jolla, California, U.S.A.
Email: thuxford@chem.ucsd.edu
Chapter 1

Michael Karin
Department of Pharmacology
School of Medicine
University of California, San Diego
La Jolla, California, U.S.A.
Email: karinoffice@ucsd.edu
Chapter 2

Jonathan M. Levenson
Department of Neuroscience
Baylor College of Medicine
Houston, Texas, U.S.A.
and
Department of Pharmacology
The Waisman Center
  for Human Development
University of Wisconsin Medical School
Madison, Wisconsin, U.S.A.
Email: jlevenson@wisc.edu
Chapter 11

Qiutang Li
Laboratory of Genetics
The Salk Institute for Biological Studies
La Jolla, California, U.S.A.
Chapter 12

Zhi-Wei Li
H. Lee Moffitt Cancer Center
  and Research Institute
Department of Interdisciplinary
  Oncology
University of South Florida
Tampa, Florida, U.S.A.
Email: liz@moffitt.usf.edu
Chapter 2

Marina Pizzi
Division of Pharmacology
  and Experimental Therapeutics
Department of Biomedical Sciences
  and Biotechnologies
School of Medicine
University of Brescia
Brescia, Italy
Email: pizzi@med.unibs.it
Chapter 11

Elaine J. Schattner
Department of Medicine
Weill Medical College
  of Cornell University
New York, New York, U.S.A.
Email: ejsch@med.cornell.edu
Chapter 10

Ranjan Sen
Laboratory of Cellular
  and Molecular Biology
National Institute on Aging
Baltimore, Maryland, U.S.A.
Email: rs465z@nih.gov
Foreword, Chapter 4

Ulrich Siebenlist
Laboratory of Immunoregulation
NIAID
National Institutes of Health
Bethesda, Maryland, U.S.A.
Email: usiebenlist@niaid.nih.gov
Chapter 5

Shao-Cong Sun
Department of Microbiology
  and Immunology
Pennsylvania State University
  College of Medicine
Hershey, Pennsylvania, U.S.A.
Email: sxs70@psu.edu
Chapter 3

J. David Sweatt
Department of Neuroscience
Baylor College of Medicine
Houston, Texas, U.S.A.
Email: jsweatt@bcm.tmc.edu
Chapter 11

Cristina M. Tato
Department of Pathobiology
School of Veterinary Medicine
University of Pennsylvania
Philadelphia, Pennsylvania, U.S.A.
Chapter 8

Vinay Tergaonkar
Laboratory of Genetics
The Salk Institute for Biological Studies
La Jolla, California, U.S.A.
Email: tergaonkar@smtp.salk.edu
Chapter 12

Wenzhi Tian
Division of Immunology
Department of Medicine
Weill Medical College
  of Cornell University
New York, New York, U.S.A.
Chapter 6

Inder M. Verma
Laboratory of Genetics
The Salk Institute for Biological Studies
La Jolla, California, U.S.A.
Email: verma@salk.edu
Chapter 12

FOREWORD

The First Few Years of NF-κB

When Hsiou-Chi Liou first asked me to provide a personal view of the "early years of NF-κB research" I hesitated, in large part because retrospection seemed to be of little value at a time when we should be looking forward, not backward. However, after mulling over the idea for some time, I decided to do it because the evolution of NF-κB, from the quest to understand B cell-specific transcription of immunoglobulin (Ig) κ genes to its present industry-sized scope, is a fascinating example of biological multi-tasking by a limited gene family. In this context, I felt that the early development of the project may be of some interest. What follows, thus, is a personal view of the lead-up to the identification of NF-κB and the years immediately thereafter.

Towards the end of my Ph.D. in organic chemistry I became interested in biology and approached David Baltimore for a post-doctoral position. In the early 80s the Baltimore lab had three major research directions: molecular immunology, retrovirology and the biology of poliovirus. My closest encounter with the life sciences prior to this had been graduate courses in biophysical chemistry and biochemistry, which had not prepared me sufficiently to favor one of these topics over the others. Therefore, to apply for post-doctoral fellowships I wrote three short blurbs in each of these areas for David's consideration. He picked one on the control of κ gene recombination by DNA methylation for further elaboration into a proposal. The choice landed me in the immunology sub-group and, when I reached M.I.T. a year later, I started as planned with chromatin structural assays and DNA methylation studies in Abelson virus transformed cell lines. One of the earliest observations we made was that the κ intron enhancer coincided with a DNase 1 hypersensitive site in cell lines that actively recombine Ig κ genes. The altered chromatin structure reflected in the hypersensitive site, with the underlying sequence of the enhancer as its apparent cause, piqued my interest as a chemist. Thus, I initiated studies to "reproduce enhancer function in vitro." David, a biochemist at heart, was enthusiastically supportive.

Phil Sharp's laboratory was next door to ours. Phil, also a Ph.D. in chemistry, was enthusiastic about another chemist's conversion to molecular biology and lent me his copy of *Bacteria and Their Viruses* as a rite of initiation. He put me in touch with Andy Fire when I spoke to him about studying enhancer-dependent transcription in vitro. With Andy's expertise and cooperation, we developed the first in vitro transcription extracts from B lymphocyte cell lines and used them to study the Vk promoter and the κ enhancer. Though the effects of promoter sequences were evident in transcription assays, we did not observe κ enhancer-dependent elevation of transcription in vitro. Months of disappointing results provided the time and the incentive to think, and rethink, the strategy. While these early studies were in progress I had spoken to the two people at M.I.T who had experience with enhancers. Alex Varshavsky was the first to describe that the SV40 enhancer was located in a nucleosome-free region of the mini-chromosome, and Alex Rich had ideas about enhancers and Z-DNA. We even used anti-Z-DNA antibodies obtained from Rich in some experiments. These conversations forcefully emphasized how little was known about these regulatory sequences and indicated that going for a functional readout in vitro may have been somewhat premature. In retrospect, good examples of promoter/enhancer communication over significant distances remain elusive to this day. Therefore, we decided to step back from the functional approach and began to "simply look" for components of the enhancer.

The electrophoretic mobility shift assay (EMSA) had been developed to study the interactions of purified bacterial DNA binding proteins with their cognate sites. This was very different from the objective we had in mind, which was to identify new DNA binding proteins; however, its intrinsic simplicity strongly recommended it as a method to work with. As a source of putative enhancer binding proteins I settled on a new transcription-competent nuclear extract procedure that had recently been developed in Bob Roeder's laboratory, and set about putting the two together using immunoglobulin-related regulatory sequences as targets. Shortly into this new line of experimentation I was joined by Harinder Singh, then newly arrived in Phil Sharp's laboratory.

Together we made two modifications that proved to be invaluable in making EMSA a widely-used technique in the eukaryotic gene regulation community. First, the use of small, functionally inactive fragments of enhancer DNA allowed greater sensitivity and significantly increased the signal-to-noise ratio in binding gels. Though at the time it was difficult to decide whether a functionally intact enhancer sequence was essential to recruit the necessary proteins to DNA, the advantages noted above encouraged us to continue with small fragments. Indeed, studies with the origin recognition complex since then have shown that intact sequences are sometimes necessary for DNA binding. Second, the use of synthetic nucleic acid polymers with low sequence complexity also increased the sensitivity of the assay. Finally, with additional modifications to co-opt existing DNase 1 footprinting and methylation interference assays for use with EMSA, we generated the first data sets using Vk promoter, κ enhancer and m heavy chain gene enhancer sequences. Amongst the proteins identified in this screen were octamer binding proteins which have since been shown to be important for stem cell biology, immunoglobulin expression and cell-cycle regulated genes, basic helix-loop-helix (bHLH) proteins that play important roles in lymphocyte development, leucine zipper-containing bHLH proteins that are related to c-myc and NF-κB.

NF-κB was identified as a mobility shifted band using the k3 fragment of the κ intron enhancer, and interference assays revealed an 11 base pair binding site towards one end of the fragment; we referred to it as the κB site. The first version of our manuscript continued to refer to the protein(s) that generated the EMSA complex as k3 binding protein(s). Upon reading that draft David suggested we give it a more user-friendly name, which led us to call the protein(s) NF-κB (for nuclear factor that binds the κB site). Unlike all other proteins identified in the initial screen, NF-κB DNA binding activity was detected only in extracts from cells that expressed immunoglobulin light chain genes, including mature B cell lines of human and murine origin as well as plasmacytomas, but not pre-B cell lines that did not express Ig κ or λ. The close correspondence between DNA binding and light chain gene expression suggested that NF-κB was an important κ gene regulatory factor.

To further strengthen the idea, we used the pre-B cell line, 70Z, that had been shown to activate κ gene transcription in response to lipopolysaccharide (LPS) or phorbol esters (PMA). Moreover, κ transcription in response to these agents occurred in the absence of new protein synthesis, providing an especially stringent constraint to test the validity of the hypothesis. We showed that NF-κB induction by these agents indeed occurred in the presence of translation inhibitors cycloheximide and anisomycin. These crucial experiments substantiated the link of NF-κB expression to κ gene transcription and provided the first evidence for a post-translational mechanism of NF-κB activation. Soon thereafter Mike Lenardo and Jackie Pierce mutated the κB site of the κ enhancer and showed that it was essential for enhancer activity.

NF-κB as an important κ enhancer activating protein may have been the end of the story, but for two other observations. First, control cells treated with translation inhibitors in the absence of LPS, or PMA, showed low but consistent activation of NF-κB DNA binding. This led us to postulate the existence of a "short-lived inhibitor" that suppressed NF-κB DNA binding in unactivated cells. Second, the induction of NF-κB in pre-B cells by a general activator such as PMA prompted us to question whether NF-κB activation and function might extend to cell types that had no connection to immunoglobulin expression. In the original studies we found that NF-κB DNA binding was induced in HeLa (epithelial) cells and Jurkat (T lymphocyte) cells by PMA in the absence of protein synthesis. These observations provided the first glimpse that NF-κB may be more than a B lineage-specific κ gene activating protein. Identification of additional tissue-unrestricted NF-κB activating signals soon followed, the earliest ones being double-stranded RNA, TNFα and serum.

A series of biochemical analyses, primarily in David Baltimore's laboratory, fleshed out the characters implicated from this early work. Patrick

Baeuerle showed that the nonDNA-binding form of NF-κB resided in the cell cytosol and could be converted to the DNA binding form in elegant experiments that used a combination of strong anionic detergent followed by its removal in a mixed micelle. The inhibitory activity, which was biochemically separable from NF-κB, was called IκB, and ultimately proved to be the predicted "short-lived inhibitor." Baeuerle and Baltimore also showed that NF-κB is a heterodimer composed of 50 and 65 kD subunits, of which the latter was essential for IκB-mediated inhibition. The detergent release assay greatly aided subsequent purification of NF-κB, first by allowing a simple scan of a variety of nonlymphoid tissue for the most abundant source of NF-κB, and also by providing the means to reveal DNA binding activity for capture on a sequence-specific DNA affinity matrix.

Sankar Ghosh used this strategy to purify NF-κB from rabbit lung extracts. Peptide sequencing and molecular cloning of the 50 kD component revealed the relationship of NF-κB to the avian oncogene *v-rel*, as well as the need for proteolytic processing to convert the precursor p105 to a DNA binding form. Sankar's purification scheme also yielded p65 and IκBα polypeptides, the former of which was cloned by Gary Nolan to reveal another *v-rel* homologous gene. The NF-κB/*v-rel* connection, manifest in a 300 amino acid DNA binding Rel homology domain (RHD), brought those working on *v-rel*, and its cellular counterpart *c-Rel*, into the rapidly expanding NF-κB community. One of the early positive outcomes of this fusion was the identification of v-Rel-associated pp40 protein as the ortholog of IκBα purified from rabbit lung. While these studies were ongoing in the Baltimore lab, Alain Israel and colleagues independently cloned p105 in their search for DNA binding proteins that regulated MHC Class I and b2 microglobulin genes. In parallel, a human cDNA encoding IκBα was isolated by Steve Haskill and Al Baldwin.

As requested by the editor, I have highlighted only the earliest years of NF-κB research here. Obviously, much has happened since then as exemplified by the contributions to this book. Some highlights include the identification of two new Rel family members, the IκB kinases and unique signaling pathways to NF-κB downstream of various cell surface receptors. Additionally, the phenotypes of genetic deletions of NF-κB and NF-κB-associated molecules have revealed the biological scope of this family of proteins. More than 16,000 publications attest to a role for NF-κB in virtually all cell types and all vertebrate species examined to date, with contributions to early embryogenesis, defense against pathogens and the regulation of cell viability. The function of the κ enhancer NF-κB binding site, however, remains enigmatic since the demonstration by Yang Xu and colleagues that its mutation in the endogenous locus does not affect κ gene recombination or expression. Amongst the variety of phenomena that are modulated by NF-κB, its role in inflammatory processes has attracted the greatest attention. Both as a factor that regulates expression of pro-inflammatory cytokines and a factor that is activated by pro-inflammatory cytokines, NF-κB is a hot target of the pharmaceutical industry. Ironically, the very ubiquitousness of NF-κB that attracted its vast following may also complicate its use as a therapeutic by increasing the likelihood of nonspecific effects. The objective for the future will be to find, or better still, to design, small molecule regulators that maximize positive effects while minimizing the negative.

Ranjan Sen, Ph.D.
Laboratory of Cellular and Molecular Biology
National Institute on Aging
Baltimore, Maryland, U.S.A.

PREFACE

Since its discovery by Ranjan Sen and David Baltimore in the late 80s, NF-κB has drawn the attention of experimental biologists, medical professionals, and the biotech/pharmaceutical industry for its broad and diverse role in all aspects of human biology and disease. Inspired by the wealth of knowledge in this field, I have had the privilege of editing this book so that I could share some of these recent and exciting findings.

This book is meant to provide the most current information gathered on the NF-κB transcription factor family. Written by experts of the subject, the book covers such topics as the structural basis and molecular mechanism of NF-κB signal transduction, transcriptional regulation, and target gene expression. Multiple roles for NF-κB are also explored for a growing number of NF-κB dependent biological and pathological processes. In particular, the function of NF-κB in modulating the immune response to pathogenic infection, as well as its involvement in autoimmunity, cancer, and neuronal development, are each emphasized in different chapters. Current strategies to intervene in the NF-κB signaling pathway are further discussed as a potential means of therapy.

Although the book covers many aspects of NF-κB biology, one can anticipate that new functional roles for NF-κB in other biological or medical disciplines will continue to emerge with the increasing availability of new reagents, disease models, and technology. The role of NF-κB in muscle physiology, ischemia, neuronal degenerative diseases, or cardiovascular disease, for instance, remain uncharted areas that warrant investigation based upon a growing body of evidence implicating NF-κB activity in these processes. Certainly the ability to manipulate NF-κB through inactivation, deletion, regulated expression, or gain-of-function mutation, may extend current research beyond inflammation, autoimmunity, and tumorigenesis into novel applications for vaccine development, tumor immunotherapy, and the control of infectious disease.

Finally, I would like to express my appreciation for the significant contribution from each of the authors on the NF-κB field and the book.

Hsiou-Chi Liou, Ph.D.

CHAPTER 1

Structural Analysis of NF-κB and IκB Proteins

Tom Huxford* and Gourisankar Ghosh

Abstract

By binding specifically to DNA sequences found within their enhancer elements, transcription factors of the NF-κB family activate the expression of genes involved in cellular immunity, inflammation, development, and apoptosis. The maintenance of proper cellular function requires the tight control of NF-κB levels. Regulation of NF-κB is accomplished primarily through its association with members of the IκB family of transcription factor inhibitor proteins. Structural characterization of various NF-κB dimers in complex with target site DNA, IκB inhibitor proteins, and an in vitro selected RNA aptamer reveals how conformational rearrangements of the versatile NF-κB molecule accommodate high affinity binding to different partners. The relative mobility of three structural and functional units permits large conformational changes in NF-κB without changing the DNA binding surfaces. The structures of NF-κB bound to IκBα and IκBβ further illustrate how some of the disordered elements of these proteins become ordered and participate in the formation of these complexes.

Introduction

The NF-κB transcription factor system is notable for the vast array of diverse stimuli that induce its activity, the rapidity with which it converts from an inactive to active state, and the numerous genes whose transcription is directly influenced by NF-κB activity.[1,2] Though originally discovered and characterized as a nuclear factor with binding specificity toward an element within the immunoglobulin kappa light chain gene enhancer of B lymphocytes, transcription factors of the NF-κB family are now recognized as being present in virtually all resting cell types as stable cytoplasmic complexes with a member of the IκB family of inhibitor proteins.

NF-κB Activation Pathway

Inflammatory cytokines (TNF-α, interleukin-1), growth factors and hormones (EGF, insulin), bacterial products (lipopolysaccharide, lipoteichoic acid), viruses (HIV-1, HTLV-1), and their products (double-stranded RNA, Tax protein) initiate signal transduction pathways that converge upon the multisubunit IκB kinase complex (IKK).[3] IKK phosphorylates IκB associated in complex with inactive NF-κB. This leads to the rapid ubiquitinylation and degradation of complex-associated IκB via the 26 S proteasome.[4] Removal of IκB potentiates NF-κB nuclear translocation by unmasking the type I nuclear localization signal (NLS) present within every

---

*Corresponding Author: Tom Huxford—Department of Chemistry and Biochemistry, University of California, San Diego, La Jolla, California, U.S.A. Email: thuxford@chem.ucsd.edu

NF-κB/Rel Transcription Factor Family, edited by Hsiou-Chi Liou.
©2006 Landes Bioscience and Springer Science+Business Media.

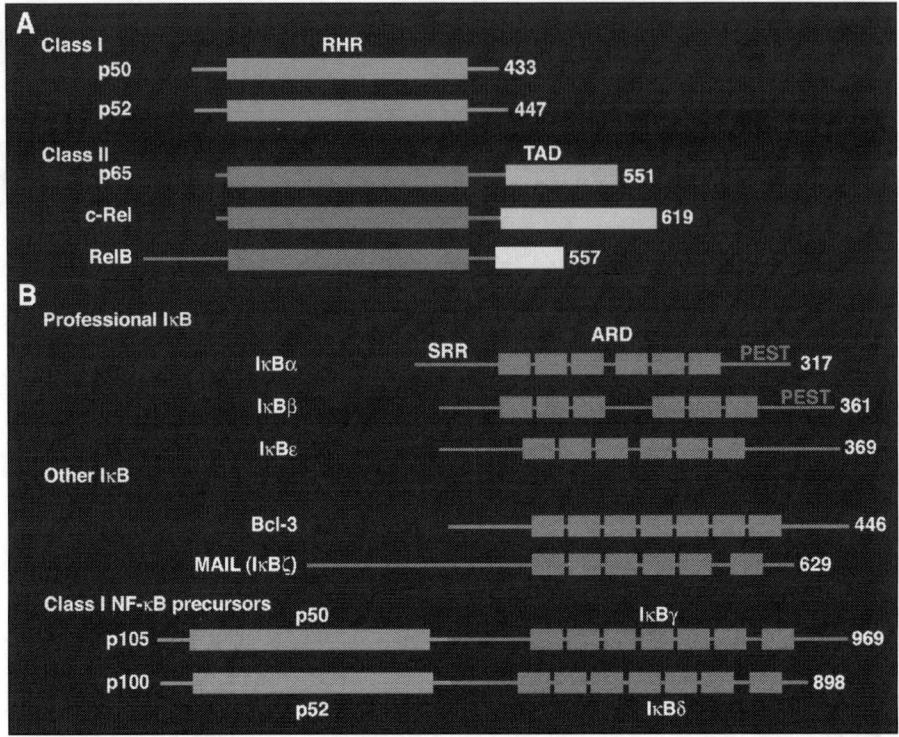

Figure 1. Domain organization of mammalian NF-κB and IκB family proteins. A) Subunits of the mammalian NF-κB transcription factor family all contain the amino-terminal Rel homology region (RHR). The polypeptides are classified based on the absence (class I) or presence (class II) of a carboxy-terminal transcriptional activation domain (TAD). Amino acid numbering corresponds to the size of the human gene products. B) Mammalian IκB family proteins each contain the ankyrin repeat domain (ARD). Additionally, the "professional" IκB proteins (IκBα, IκBβ, and IκBε) contain the amino-terminal signal response region (SRR). IκBα and IκBβ exhibit a carboxy-terminal PEST region. Bcl-3 and MAIL (IκBζ) are classified separately as they function in the nucleus. The p105 and p100 proteins function both as cytosolic IκB proteins an as precursors of the class I NF-κB subunits p50 and p52, respectively.

NF-κB subunit.[5,6] Free NF-κB dimers rapidly accumulate in the nucleus where they bind with specificity to 10 base pair DNA elements present within the promoters of target genes. These elements, collectively referred to as κB DNA, share the consensus sequence 5'-GGGRNNYYCC-3' (where R, N, and Y represent purine, any, and pyrimidine nucleotide bases, respectively).[7] Active nuclear NF-κB upregulates the expression of hundreds of genes including cytokines (interleukins-1, -2, and -6), immunoreceptors (immunoglobin kappa light chain, MHC class I, T-cell receptor β chain), cellular adhesion molecules (ICAM-1, ELAM-1), and many others.[3] Among these is included the gene that encodes IκBα. Within minutes of NF-κB induction, newly translated IκBα appears in the nucleus where it is capable of removing NF-κB from κB DNA and facilitating transport of the inactive complex back to the cytoplasm thus reestablishing the preinduction state.[8-10]

Mammalian NF-κB Family Proteins

The mammalian NF-κB family of inducible, dimeric transcription factors is composed of five subunits, p50, p52, p65 (RelA), c-Rel, and RelB (Fig. 1A). These associate one with another to form active homo- and heterodimers. The five polypeptide subunits are related through a highly conserved amino-terminal sequence of approximately 300 amino acids in length known as the Rel homology region (RHR). Contained within the RHR are all of the amino acid sequences necessary for dimerization, site specific DNA binding, binding to IκB inhibitor proteins, and nuclear localization.

The type I NF-κB subunits p50 and p52 are generated by proteolytic processing of the precursor proteins p105 and p100, respectively. Each of the p105 and p100 precursors contains an IκB-like ankyrin repeat domain within its carboxy-terminal half. These ankyrin repeat domains are sometimes referred to as IκBγ and IκBδ, respectively. As a result of this domain organization, p105 and p100 function in the cell as IκB family inhibitors of NF-κB activity. The proteolytic removal of the carboxy-terminal ankyrin repeat domains occurs either in a constitutive co-translational manner in the case of p50 or by a stimulus-dependent mechanism in p52.[11-13] Mature p50 and p52 subunits are integral components of strong transactivating NF-κB dimers, most notably the p50/p65 NF-κB heterodimer. They lack inherent transcriptional potential, however, and NF-κB dimers consisting only of p50 and p52 subunits fail to activate transcription of reporter genes.[14]

The p65, c-Rel, and RelB subunits each contain unique transcriptional activation domains carboxy-terminal to their respective RHR. Consequently, NF-κB dimers that contain at least one of these subunits, classified as type II, act as strong activators of transcription.

Mammalian IκB Family Proteins

The mammalian IκB family of transcription factor inhibitor proteins consists of IκBα, IκBβ, IκBε, Bcl-3, and MAIL (IκBζ), as well as the aforementioned p105 and p100 NF-κB precursors (Fig. 1B).[15-17] The primary distinguishing characteristic of the IκB family proteins is a centrally-located ankyrin repeat domain (ARD) consisting of six or seven ankyrin repeats. Ankyrin repeats are helical tandem repeating units of approximately 33 amino acids in length related in structure to leucine rich repeats (LRR), HEAT repeats, armadillo repeats (ARM), and tetratricopeptide repeats.[18,19] Tandem repeats stack sequentially to form an elongated domain devoid of a classical hydrophobic core. Several hundred ARD-containing proteins have been identified exhibiting as many as twenty-four consecutive ankyrin repeats. The function of ankyrin repeat proteins is varied though many participate in protein-protein interactions.

Of the IκB family proteins, IκBα, IκBβ, and IκBε serve to bind and sequester NF-κB dimers in the cytosol and respond to NF-κB inducing stimuli by releasing NF-κB. These three inhibitors are sometimes referred to as the "professional" IκBs. In contrast, Bcl-3 functions within the nucleus where it binds to NF-κB on DNA and alters the transcriptional signal, while p105 and p100 serve as inactive NF-κB subunit precursors. The precise function of MAIL (IκBζ) remains unknown, though it appears to reside primarily in the nucleus, prefers binding NF-κB p50 subunits, and is involved in interleukin-6 production.

Amino-terminal to the ARD in the "professional" IκB proteins is a region that contains the conserved serine sites of phosphorylation by IKK and lysine sites of polyubiquitinylation. Both of these features are required for stimulus-dependent removal of IκB and NF-κB activation. This amino-terminal element, known as the signal response region (SRR), lacks ordered three-dimensional structure in solution. Carboxy-terminal to the ARD of IκBα and IκBβ lies a polypeptide region rich in the amino acids proline, glutamic acid, serine, and threonine. This appropriately named PEST region is common to many proteins that, like IκBα, show a high rate of turnover in resting cells.[20,21] It has been shown that the PEST region of IκBα also functions in binding NF-κB and inhibiting DNA binding.[22,23]

NF-κB Structural Biology

Structural biology attempts to address cell biological questions by determining structures of key molecules or molecular complexes. Successful elucidation of the structure will often suggest mechanisms of molecular action that could not otherwise be addressed. The study of

NF-κB signaling has benefited in recent years from a number of structural studies. In this chapter, we will review these structural studies of NF-κB and IκB proteins within the context of the above outlined NF-κB activation pathway.

NF-κB Structure

The RHR structure from various NF-κB homo- and heterodimers has been determined by X-ray crystallography bound to different κB DNA sequences.[24-32] These structures provided the first atomic resolution views of NF-κB. They reveal the subunit domain organization and they suggest mechanisms for dimerization selectivity, and DNA binding specificity.

The NF-κB RHR is composed of three structural elements, which can be separated by treatment of the molecules with the appropriate proteases. Beginning from the amino-terminus, these are the amino-terminal domain, the dimerization domain, and the NLS polypeptide (Fig. 2).

Amino-Terminal Domain Structure

The NF-κB amino-terminal domain exhibits a variation of the immunoglobulin (Ig) fold. This domain, present in numerous proteins, is especially prevalent among immune receptors. The secondary structure of the protein fold is entirely beta strand in nature. It is formed as two antiparallel beta sheets of three and four strands sandwich one upon the other. The NF-κB amino-terminal Ig-like domain contains a noncanonical insertion prior to its final two beta strands. This insertion is largest in p50 where it forms two long alpha helices separated by an extended coil. In p65 the insertion consists of only one alpha helix.

The amino-terminal domain is sometimes referred to as the DNA binding domain. It earns this nickname because all of the amino acids involved in direct contact and readout of target DNA bases originate from this domain. The majority of these contacts are mediated by amino acids on the loop linking the first and second beta strands. This loop, referred to as L1, is highly conserved in both sequence and structure among all NF-κB subunits.

Dimerization Domain Structure

The NF-κB dimerization domain, or carboxy-terminal domain as it is sometimes referred to, also assumes an Ig-like fold with two stacked antiparallel beta sheets of three and four strands each. Amino acid side chains from the first, second, and fifth beta strands of two NF-κB dimerization domains contact each other symmetrically to form the dimer interface. Each subunit contributes 14 dimer forming amino acid side chains. The dimer interface is mainly hydrophobic in nature and buries approximately 1500 $Å^2$ solvent accessible surface area.

The RelB homodimer is unique among NF-κB family dimers. It is not observed in vivo and fails to form heterodimers with either p65 or c-Rel. The recently determined X-ray crystal structure of the RelB dimerization domain in the absence of binding partner shows that it does form at high concentration. Interestingly, the structure reveals that two RelB subunits contact one another with 8 Å greater separation and a rotation of 30° relative to the dimer interface of other NF-κB dimers. Furthermore, the individual subunits cross over after the fourth beta strand so that the resulting protein is an intertwined dimer. This result is surprising as relatively few amino acid substitutions exist between RelB and the classical p50 homodimer (Vu and Huang, unpublished result). Currently, efforts are underway to determine the source of this significant structural difference and its possible biological implications.

Prior to the last beta strand of the NF-κB dimerization domain resides an amino acid sequence containing a consensus serine phosphorylation site target for cyclic-AMP-dependent protein kinase. Phosphorylation at this p65 subunit serine residue number 276 has been shown to be a requirement for maximum activation of gene transcription.[33]

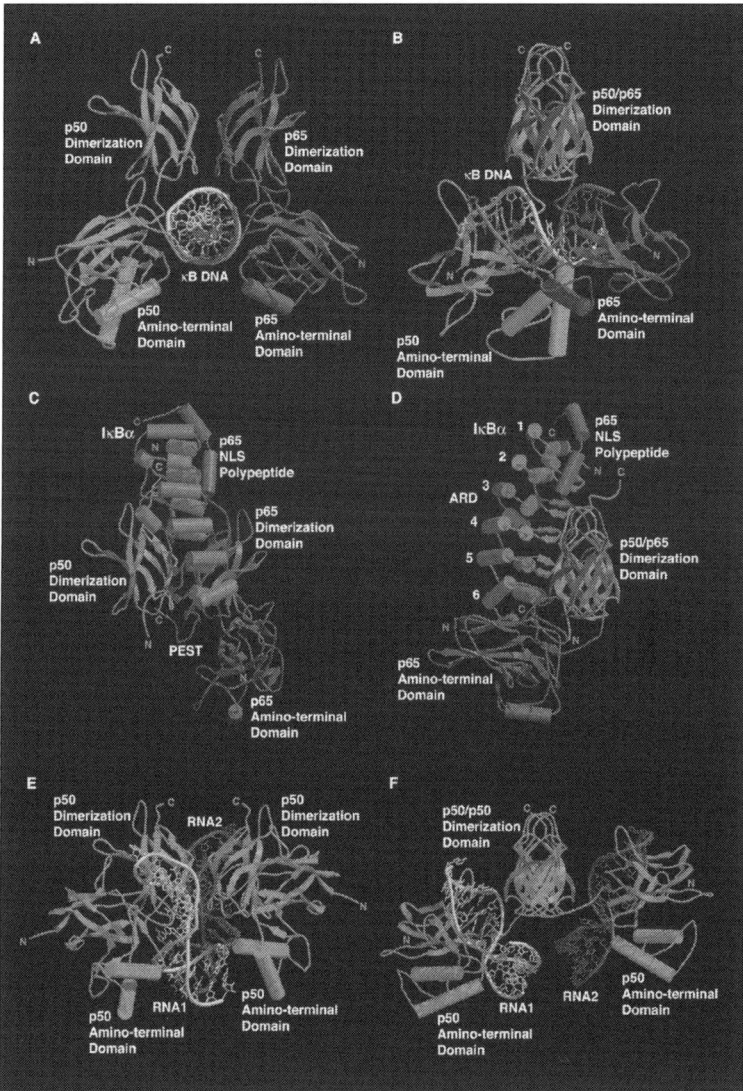

Figure 2. Ribbon diagram representations of NF-κB RHR bound to κB DNA, IκBα, and an RNA aptamer. A) The NF-κB p50/p65 heterodimer bound to κB DNA. The p50 subunit is shown in green, the p65 subunit is red, and the DNA double helix is depicted in two-tone grey. The structure is viewed down the pseudosymmetrical dimer interface and the separable structural elements of the complex are labeled. Note that the NLS polypeptide segment is present but not visible in the NF-κB heterodimer bound to DNA. B) Same complex rotated 90° with respect to the vertical axis in A). C) The NF-κB p50/p65 heterodimer in complex with IκBa. NF-κB subunits are oriented and colored as in A). IκBα is colored purple. D) NF-κB/IκBα complex rotated 90° with respect to C). Ankyrin repeat domain (ARD) is labeled and individual ankyrin repeats are numbered. Note the difference in the position of the NF-κB p65 subunit amino-terminal domain relative to it DNA bound conformation. E) NF-κB p50 homomdimer bound to a selected RNA aptamer and aligned along the dimerization domain symmetry axis as in A) and C). The p50 subunits are shown in green and the two RNA molecules are depicted in two shades of grey. F) The p50 homodimer/RNA complex rotated 90° about the vertical axis. Again, note the drastic change in the position of the amino-terminal domains in complex with RNA.

NLS Polypeptide Structure

A small segment of approximately 30 amino acids carboxy-terminal to the dimerization domain constitutes the third structurally independent component of the NF-κB RHR. This region, within which is contained the type I nuclear localization sequence, is present but unobserved in a number of the NF-κB/DNA complex crystal structures. It is also absent from the high resolution structures of the NF-κB dimerization domains solved in the absence of DNA.[34,35] The systematic lack of observed electron density in maps calculated from X-ray diffraction data is indicative of disordered regions within the protein. Furthermore, it has been shown that the removal of this segment leads to improved crystal quality and extended X-ray diffraction resolution.[36] These observations suggest that, in its DNA bound conformation, this NF-κB NLS-containing polypeptide represents a flexible segment that lacks an ordered structure. As discussed later in this chapter, such is not the case when NF-κB binds to IκB inhibitor proteins.

Structurally Related Proteins

Since the original determination of the NF-κB RHR structure two additional transcription factor families, NFAT and STAT, have been found to use a similar mode of binding to DNA with Ig-like domains. The nuclear factor of activated T-cells or NFAT proteins share modest sequence homology with the NF-κB RHR. NFAT1-4 function as monomers and require additional regulators of transcription for stable DNA complex formation.[37] NFAT5, on the other hand, is dimeric both in solution and bound to DNA. The recently determined X-ray crystal structure of an NFAT1 bound to a pseudopalindromic target DNA sequence reveals how two monomeric NFAT proteins can cooperate in binding DNA. The structure shows striking similarity to the NF-κB/DNA complex crystal structures, though the two proteins contact one another asymmetrically as opposed to the symmetrical interactions displayed by the NF-κB dimerization domain.[38] The signal transducer and activator of transcription or STAT proteins bind DNA through an amino-terminal domain with structural similarity to the amino-terminal Ig-like domain of NF-κB.[39,40] Like NF-κB, NFAT and STAT are involved in lymphocyte activation and the immune response.

NF-κB/DNA Complex

In complex with κB DNA the NF-κB RHR resembles the wings of a butterfly (Fig. 2A). The NF-κB RHR contacts the DNA major groove along one entire turn of the double helix (Fig. 2B). The most striking observation from NF-κB/DNA complex structural determination is that it does so entirely through ordered loops. This is a stark contrast to the canonical sequence specific DNA binding domains that employ secondary structure elements to "read" the DNA sequence. The NF-κB DNA contacting loops connect the beta strands of the amino-terminal and dimerization domains and are, therefore, held in place due to the domain stability. In all, ten loops make DNA contacts. Five of these are contributed by the amino-terminal domain while the five others emanate from the dimerization domain. This arrangement serves to explain, in part, why such a large protein is required for binding to a relatively small DNA half site of five base pairs.

In general, DNA binding specificity is derived from DNA base-contacting amino acid side chains. Nearly all of the base-specific contacts in NF-κB/DNA complexes are mediated by amino acids within the first loop of the amino-terminal domain, the so-called loop L1. The p50 subunit contributes side chains Arg54, Arg56, and Glu60 that contact the second and third guanine bases within the κB consensus DNA sequence. His64, also from the loop L1 of the p50 amino-terminal domain, contacts the first guanine. A similar set of side chains are contributed by the homologous loop from the NF-κB p65 subunit, with the notable exception that the His64 of p50 is replaced by Ala43 in p65. As a consequence, a 5'-G is not required within the κB DNA half sites contacted by p65 subunits.

The modular organization of the NF-κB RHR allows for significant movement of the amino-terminal domains relative to the dimerization domains. When the first NF-κB p65 homodimer/DNA complex was determined, it was noted that one subunit contacts its κB DNA half site in a manner similar to that observed in the p50/p65 heterodimer while the second p65 subunit amino-terminal domain rotates 18° in order to maximize nonspecific contacts with the DNA backbone.[27] This rearrangement was later determined to result from the choice of a pseudopalindromic κB DNA target with one too many A:T base pairs. When two additional κB DNA targets were used for crystallization, it was found that by varying amino-terminal domain placement, the NF-κB p65 homodimer could accommodate binding to diverse κB DNA sites without sacrificing overall binding affinity.[30] The discovery that the amino-terminal domain is capable of significant global movement relative to the dimerization domain in order to accommodate various binding partners has become a recurring theme in many of the subsequently determined NF-κB complex crystal structures.

NF-κB/IκB Complex

The determination of two different NF-κB p50/p65 heterodimer/IκBα complex cocrystal structures allows for analysis of the structure of IκBα as well as comparison of the NF-κB RHR with its DNA-bound conformation (Fig. 2C,D).[41,42] Comparison of this structure with that of IκBβ, solved in complex with the NF-κB p65 homodimer dimerization domain, and Bcl-3, determined as a free ARD, allow for general and specific rules of IκB structure and point to their differences in function.[43,44]

IκB Ankyrin Repeat Structure

As revealed in the NF-κB/IκB complex structures, the ankyrin repeats of IκBα and IκBβ exhibit the typical fold. This consists of a closely packed helix-turn-helix motif followed by a loop that extends perpendicularly from the end of the second helix and closes in a tight beta-turn. Consecutive repeats stack with roughly 11 Å spacing. The conservation of small amino acid side chains at the turn in the helix pair allows for a slight curvature in the ARD. Both IκBα and IκBβ contain nonconsensus insertions between ankyrin repeats 3 and 4. The insert is only six amino acids in IκBα. IκBβ, however, contains a sequence of 47 amino acids that bears little homology to any known proteins. The placement of the inserts is on the back of the ARD at some distance from the NF-κB binding partner, and all but ten of the IκBβ insert residues are disordered in the NF-κB p65 homodimer/IκBβ complex structure.[43] Therefore, the insert region does not appear to affect NF-κB binding directly, though the large insert between ankyrin repeats 3 and 4 of IκBβ has been shown to mediate interaction with the small Ras-like GTPase κB-Ras.[45]

One interesting aspect of the IκB ankyrin repeat domains is their instability relative to other ARD-containing proteins (C.A. Hughes, unpublished results). One way in which this is manifest is that the IκBα ARD has resisted crystallization in the absence of NF-κB binding partner despite considerable efforts on the part of one of the authors. By contrast, the structures of a number of proteins containing either actual or idealized ARDs have been determined by X-ray crystallography.[44,46-49] One of these is the IκB family member Bcl-3, which contains seven ankyrin repeats.[44] The relative instability of IκBα and IκBβ are indicative of their role as substrates for rapid proteolysis by the 26 S proteasome.

IκB PEST Region

The length of the IκBα PEST region included within the crystal differs for the two NF-κB/IκBα complex crystal structures. The structure that contains the longer PEST reveals that it forms a sigmoidal polypeptide loop devoid of secondary structure. The poorer quality of the electron density maps and higher temperature factors associated with the amino acids in this region indicate that a significant amount of thermal motion is exhibited by the IκBα PEST region in the NF-κB/IκBα complex crystals. The corresponding region is present but not observed in the crystal structure of the NF-κB p65 homodimer/IκBβ complex.

NF-κB/IκB Complex Interface

As previously mentioned, the two NF-κB/IκBα complex crystal structures contain unique elements by virtue of the design of protein constructs employed in the preparation of the complex crystals.[50,51] As a result of these differences, one structure reveals a protein-protein interaction surface of 4300 Å^2 while the second contains 3800 Å^2 of buried surface area. When the portions common to the two structures are superimposed, the resulting complex buries 4800 Å^2 solvent accessible surface area. A comparison of the elements common to both structures reveals a nearly identical contact area, however, indicating the correctness of the two structures and the precision of the experimental technique. Both structures reveal that NF-κB and IκBα contact one another through a discontinuous interface. The contribution of three discrete, discontinuous patches to the overall protein-protein interface explains its tight binding affinity with dissociation constants measured in solution on the order of 1 nM.

The NF-κB p65 subunit NLS polypeptide makes extensive contacts with ankyrin repeats 1 and 2 of IκBα. Within this interaction, the type I NLS of the p65 subunit adopts an alpha helical structure. In this conformation, amino acids Lys301, Arg302, and Arg304 participate in direct ion pairing interactions with amino acid side chains in IκBα. Additional hydrophobic interactions tether amino acids carboxy-terminal to the basic NLS sequence and the top of the first ankyrin repeat of IκBα. As mentioned previously, this same NLS polypeptide region is disordered in several NF-κB/DNA structures. When bound by its cognate receptor, importin-α, the type I NLS mediates translocation of NF-κB from the cytoplasm to the nucleus. X-ray crystallographic studies on ARM repeat containing importin-α proteins bound to various NLS peptides have revealed that these bind in an extended polypeptide conformation.[52,53] Taken together, these results indicate that the NLS polypeptide element of the NF-κB p65 subunit is capable of making a transition from alpha helical structure in its IκB bound state to extended beta-like structure upon binding to importin-α and, finally, to unstructured polypeptide upon binding DNA.

Next within the discontinuous NF-κB/IκBα binding interface is the interaction of IκBα ankyrin repeats 3, 4, and 5 with the dimerization domains of the NF-κB p50/p65 heterodimer. One surprising observation is that the p50 subunit dimerization domain mediates the vast majority of contacts within this interaction surface. This is counterintuitive in light of the cellular and biochemical data that clearly show a preference of IκBα toward binding NF-κB dimers containing at least one p65 subunit.[54] A closer inspection of the contact surfaces of IκBα and NF-κB throughout this region of the complex interface reveals few specific side chain interactions. It has been shown that, although this binding surface contributes significantly to overall NF-κB/IκBα binding affinity, the p65 NLS polypeptide is chiefly responsible for directing the specificity of IκBα toward p65-containing NF-κB homo- and heterodimers.[35]

The third and final component of the discontinuous NF-κB/IκBα complex interface involves the amino-terminal domain of the NF-κB p65 subunit and the carboxy-terminal PEST region of IκBα. This portion of the structure exhibits significantly higher thermal motion when compared to the rest of both NF-κB/IκBα complex X-ray crystal structures. In the complex, the acidic IκBα PEST juxtaposes with an acidic patch unique to the bottom of the NF-κB p65 dimerization domain. The result is the formation of an extended composite acidic surface measuring some 40 Å by 20 Å. Relative to its DNA bound conformation, the p65 amino-terminal domain rotates by 170° and translates nearly 40 Å. Aside from its new position, the p65 subunit amino-terminal domain remains virtually unchanged. The consequence of this *en bloc* movement is to place the basic DNA contacting surface of p65 in a position to oppose the newly formed acidic surface. This long-range electrostatic interaction is thought to play a major role in the DNA-inhibitory binding activity of IκBα. In support of this hypothesis, increasing the strength of the acidic surface through hyperphosphorylation of the IκBα

PEST by protein kinase CK2 (casein kinase II) in vitro has been shown to increase NF-κB binding affinity and DNA-inhibitory binding activity of IκBα.[54]

NF-κB p50 Homodimer/RNA Aptamer Complex

The recent determination of the X-ray crystal structure of the NF-κB p50 homodimer bound to an RNA aptamer reveals a new binding conformation available to the NF-κB RHR.[55] Whereas one NF-κB dimer binds to one decameric κB DNA sequence with each subunit binding symmetrically to one five base pair half site, an RNA aptamer selected both in vitro and in vivo in yeast for high affinity binding to NF-κB p50 binds only one subunit of the homodimer.[56-58] The complex crystal structure reveals how this is possible (Fig. 2E, F).

The RNA aptamer forms a hairpin structure that is bound by the DNA contacting amino acid residues on the p50 amino-terminal domain loop L1. As in the case of the NF-κB p65 subunit in complex with IκBα, the NF-κB p50 subunit amino-terminal domain changes its relative orientation in order to accommodate binding the RNA aptamer. Relative to its DNA bound conformation, the domain swings 40° and translates by 10 Å to bind one RNA molecule away from the dimer interface and classical location of DNA binding. What is surprising about this domain movement is that it results in an orientation in the opposite extreme position relative to the dimerization domain platform from that exhibited in the NF-κB/IκBα complex structures. Even more interesting, the DNA binding loop L1 and its DNA-contacting amino acid side chains remain virtually unchanged from their positions in any of the other NF-κB structures. The high affinity binding exhibited by the p50 homodimer/RNA complex results from the evolution of an arrangement of chemical groups on the selected RNA aptamer that perfectly complements the DNA-binding properties of the loop L1 side chains and, therefore, mimics the natural κB DNA target site.[59]

Conclusions

The NF-κB RHR is a versatile binding factor by virtue of the relative flexibility of its three structural components: amino-terminal domain, dimerization domain, and NLS polypeptide. The observation that all of the DNA binding loops of NF-κB are preformed and most of the DNA-contacting amino acid side chains approximate their DNA bound conformations even in the absence of any binding partner indicates that the hinge points that join the three structured elements dictate the NF-κB binding conformations (C.B. Phelps, unpublished data). In particular, a flexible linker of approximately 10 amino acids in length joining the amino-terminal and dimerization domains is chiefly responsible for allowing the relative motion of the amino-terminal domain. Other multidomain transcription factors, STAT for example, lack this flexible linker. One manifestation of this difference is the recent successful crystallization and X-ray crystal structure determination of *Dictyostelium discoideum* STATa homodimer in the absence of DNA.[60] By contrast, the dynamic NF-κB RHR has continually resisted crystallization and structural determination in the absence of some binding partner.

The IκBα and IκBβ proteins represent a subset of ARD-containing proteins that display a low degree of folding stability. It is not hard to imagine that the selective pressure exerted by nature to engineer an NF-κB binding partner that responds to stimuli by being rapidly degraded would result in a molecule with the characteristics of the IκB proteins. It is interesting, however, that both NF-κB and IκB proteins exhibit such flexibility and yet bind with such high affinity and stability.

References

1. Ghosh S, May MJ, Kopp EB. NF-κB and Rel proteins: Evolutionarily conserved mediators of immune responses. Annu Rev Immunol 1998; 16:225-260.
2. Silverman N, Maniatis T. NF-κB signaling pathways in mammalian and insect innate immunity. Genes Dev 2001; 15(18):2321-2342.
3. Pahl HL. Activators and target genes of Rel/NF-κB transcription factors. Oncogene 1999; 18(49):6853-6866.

4. Karin M, Ben-Neriah Y. Phosphorylation meets ubiquitination: The control of NF-κB activity. Annu Rev Immunol 2000; 18:621-663.
5. Beg AA, Ruben SM, Scheinman RI et al. IκB interacts with the nuclear localization sequences of the subunits of NF-κB: A mechanism for cytoplasmic retention. Genes Dev 1992; 6(10):1899-1913.
6. Zabel U, Henkel T, Silva MS et al. Nuclear uptake control of NF-κB by MAD-3, an I k B protein present in the nucleus. EMBO J 1993; 12(1):201-211.
7. Grimm S, Baeuerle PA. The inducible transcription factor NF-κB: Structurefunction relationship of its protein subunits. Biochem J 1993; 290(Pt 2):297-308.
8. Sun SC, Ganchi PA, Ballard DW et al. NF-κB controls expression of inhibitor IκBα: Evidence for an inducible autoregulatory pathway. Science 1993; 259(5103):1912-1915.
9. Arenzana-Seisdedos F, Thompson J, Rodriguez MS et al. Inducible nuclear expression of newly synthesized IκBα negatively regulates DNA-binding and transcriptional activities of NF-κB. Mol Cell Biol 1995; 15(5):2689-2696.
10. Hoffmann A, Levchenko A, Scott ML et al. The IκB-NF-κB signaling module: Temporal control and selective gene activation. Science 2002; 298(5596):1241-1245.
11. Lin L, DeMartino GN, Greene WC. Cotranslational biogenesis of NF-κB p50 by the 26S proteasome. Cell 1998; 92(6):819-828.
12. Lin L, DeMartino GN, Greene WC. Cotranslational dimerization of the Rel homology domain of NF-κB1 generates p50-p105 heterodimers and is required for effective p50 production. EMBO J 2000; 19(17):4712-4722.
13. Senftleben U, Cao Y, Xiao G et al. Activation by IKKα of a second, evolutionary conserved, NF-κB signaling pathway. Science 2001; 293(5534):1495-1499.
14. Franzoso G, Bours V, Park S et al. The candidate oncoprotein Bcl-3 is an antagonist of p50/NF-κB-mediated inhibition. Nature 1992; 359(6393):339-342.
15. Whiteside ST, Israël A. IκB proteins: Structure, function and regulation. Semin Cancer Biol 1997; 8(2):75-82.
16. Kitamura H, Kanehira K, Okita K et al. MAIL, a novel nuclear IκB protein that potentiates LPS-induced IL-6 production. FEBS Lett 2000; 485(1):53-56.
17. Yamazaki S, Muta T, Takeshige K. A novel IκB protein, IκB-ζ, induced by proinflammatory stimuli, negatively regulates nuclear factor-κB in the nuclei. J Biol Chem 2001; 276(29):27657-27662.
18. Sedgwick SG, Smerdon SJ. The ankyrin repeat: A diversity of interactions on a common structural framework. Trends Biochem Sci 1999; 24(8):311-316.
19. Groves MR, Barford D. Topological characteristics of helical repeat proteins. Curr Opin Struct Biol 1999; 9(3):383-389.
20. Rogers S, Wells R, Rechsteiner M. Amino acid sequences common to rapidly degraded proteins: The PEST hypothesis. Science 1986; 234(4774):364-368.
21. Pando MP, Verma IM. Signal-dependent and -independent degradation of free and NF-κB-bound IκBα. J Biol Chem 2000; 275(28):21278-21286.
22. Ernst MK, Dunn LL, Rice NR. The PEST-like sequence of IκBα is responsible for inhibition of DNA binding but not for cytoplasmic retention of c-Rel or RelA homodimers. Mol Cell Biol 1995; 15(2):872-882.
23. Whiteside ST, Ernst MK, LeBail O et al. N- and C-terminal sequences control degradation of MAD3/IκBα in response to inducers of NF-κB activity. Mol Cell Biol 1995; 15(10):5339-5345.
24. Ghosh G, van Duyne G, Ghosh S et al. Structure of NF-κB p50 homodimer bound to a κB site. Nature 1995; 373(6512):303-310.
25. Müller CW, Rey FA, Sodeoka M et al. Structure of the NF-κB p50 homodimer bound to DNA. Nature 1995; 373(6512):311-317.
26. Cramer P, Larson CJ, Verdine GL et al. Structure of the human NF-κB p52 homodimer-DNA complex at 2.1 Å resolution. EMBO J 1997; 16(23):7078-7090.
27. Chen YQ, Ghosh S, Ghosh G. A novel DNA recognition mode by the NF-kB p65 homodimer. Nat Struct Biol 1998; 5(1):67-73.
28. Chen FE, Huang DB, Chen YQ et al. Crystal structure of p50/p65 heterodimer of transcription factor NF-κB bound to DNA. Nature 1998; 391(6665):410-413.
29. Cramer P, Varrot A, Barillas-Mury C et al. Structure of the specificity domain of the Dorsal homologue Gambif1 bound to DNA. Structure Fold Des 1999; 7(7):841-852.
30. Chen YQ, Sengchanthalangsy LL, Hackett A et al. NF-κB p65 (RelA) homodimer uses distinct mechanisms to recognize DNA targets. Structure Fold Des 2000; 8(4):419-428.
31. Escalante CR, Shen L, Thanos D et al. Structure of NF-κB p50/p65 heterodimer bound to the PRDII DNA element from the interferon-β promoter. Structure (Camb) 2002; 10(3):383-391.
32. Berkowitz B, Huang DB, Chen-Park FE et al. The x-ray crystal structure of the NF-κB p50.p65 heterodimer bound to the interferon beta-κB site. J Biol Chem 2002; 277(27):24694-24700.

33. Zhong H, Voll RE, Ghosh S. Phosphorylation of NF-κB p65 by PKA stimulates transcriptional activity by promoting a novel bivalent interaction with the coactivator CBP/p300. Mol Cell 1998; 1(5):661-671.
34. Huang DB, Huxford T, Chen YQ et al. The role of DNA in the mechanism of NF-κB dimer formation: Crystal structures of the dimerization domains of the p50 and p65 subunits. Structure 1997; 5(11):1427-1436.
35. Huxford T, Mishler D, Phelps CB et al. Solvent exposed noncontacting amino acids play a critical role in NF-κB/IκBα complex formation. J Mol Biol 2002; 324(4):587-597.
36. Cramer P, Müller CW. Engineering of diffraction-quality crystals of the NF-κB p52 homodimer: DNA complex. FEBS Lett 1997; 405(3):373-377.
37. Chen L, Glover JN, Hogan PG et al. Structure of the DNA-binding domains from NFAT, Fos and Jun bound specifically to DNA. Nature 1998; 392(6671):42-48.
38. Jin L, Sliz P, Chen L et al. An asymmetric NFAT1 dimer on a pseudo-palindromic κB-like DNA site. Nat Struct Biol 2003; 10(10):807-811.
39. Becker S, Groner B, Müller CW. Three-dimensional structure of the Stat3β homodimer bound to DNA. Nature 1998; 394(6689):145-151.
40. Chen X, Vinkemeier U, Zhao Y et al. Crystal structure of a tyrosine phosphorylated STAT-1 dimer bound to DNA. Cell 1998; 93(5):827-839.
41. Huxford T, Huang DB, Malek S et al. The crystal structure of the IκBα/NF-κB complex reveals mechanisms of NF-κB inactivation. Cell 1998; 95(6):759-770.
42. Jacobs MD, Harrison SC. Structure of an IκBα/NF-κB complex. Cell 1998; 95(6):749-758.
43. Malek S, Huang DB, Huxford T et al. X-ray crystal structure of an IκBβ•NF-κB p65 homodimer complex. J Biol Chem 2003; 278(25):23094-23100.
44. Michel F, Soler-Lopez M, Petosa C et al. Crystal structure of the ankyrin repeat domain of Bcl-3: A unique member of the IκB protein family. EMBO J 2001; 20(22):6180-6190.
45. Chen Y, Wu J, Ghosh G. κB-Ras binds to the unique insert within the ankyrin repeat domain of IκBβ and regulates cytoplasmic retention of IκBβ•NF-κB complexes. J Biol Chem 2003; 278(25):23101-23106.
46. Venkataramani R, Swaminathan K, Marmorstein R. Crystal structure of the CDK4/6 inhibitory protein p18INK4c provides insights into ankyrin-like repeat structure/function and tumor-derived p16INK4 mutations. Nat Struct Biol 1998; 5(1):74-81.
47. Kohl A, Binz HK, Forrer P et al. Designed to be stable: Crystal structure of a consensus ankyrin repeat protein. Proc Natl Acad Sci USA 2003; 100(4):1700-1705.
48. Binz HK, Stumpp MT, Forrer P et al. Designing repeat proteins: Well-expressed, soluble and stable proteins from combinatorial libraries of consensus ankyrin repeat proteins. J Mol Biol 2003; 332(2):489-503.
49. Binz HK, Amstutz P, Kohl A et al. High-affinity binders selected from designed ankyrin repeat protein libraries. Nat Biotechnol 2004; 22(5):575-582.
50. Baeuerle PA. IκB-NF-κB structures: At the interface of inflammation control. Cell 1998; 95(6):729-731.
51. Cramer P, Müller CW. A firm hand on NFκB: Structures of the IκBα-NF-κB complex. Structure Fold Des 1999; 7(1):R1-6.
52. Conti E, Uy M, Leighton L et al. Crystallographic analysis of the recognition of a nuclear localization signal by the nuclear import factor karyopherin α. Cell 1998; 94(2):193-204.
53. Conti E, Kuriyan J. Crystallographic analysis of the specific yet versatile recognition of distinct nuclear localization signals by karyopherin α. Structure Fold Des 2000; 8(3):329-338.
54. Phelps CB, Sengchanthalangsy LL, Huxford T et al. Mechanism of IκB alpha binding to NF-κB dimers. J Biol Chem 2000; 275(38):29840-29846.
55. Huang DB, Vu D, Cassiday LA et al. Crystal structure of NF-κB $(p50)_2$ complexed to a high-affinity RNA aptamer. Proc Natl Acad Sci USA 2003; 100(16):9268-9273.
56. Lebruska LL, Maher IIIrd LJ. Selection and characterization of an RNA decoy for transcription factor NF-κB. Biochemistry 1999; 38(10):3168-3174.
57. Cassiday LA, Maher IIIrd LJ. In vivo recognition of an RNA aptamer by its transcription factor target. Biochemistry 2001; 40(8):2433-2438.
58. Cassiday LA, Maher IIIrd LJ. Yeast genetic selections to optimize RNA decoys for transcription factor NF-k B. Proc Natl Acad Sci USA 2003; 100(7):3930-3935.
59. Ghosh G, Huang DB, Huxford T. Molecular mimicry of the NF-κB DNA target site by a selected RNA aptamer. Curr Opin Struct Biol 2004; 14(1):21-27.
60. Soler-Lopez M, Petosa C, Fukuzawa M et al. Structure of an activated Dictyostelium STAT in its DNA-unbound form. Mol Cell 2004; 13(6):791-804.

CHAPTER 2

NF-κB Signal Transduction by IKK Complexes

Zhi-Wei Li* and Michael Karin

Abstract

Transcription factor NF-κB plays a major role in many physiological and pathological processes while its regulation is best understood in inflammatory and immune system. The central event in NF-κB signaling pathway is the activation of IKK complex, the convergent point of diverse NF-κB activation signaling. This review addresses the cell signaling of IKK and NF-κB activation in response to various immune and inflammatory stimuli as revealed by the analysis of mice and cells lacking specific signaling transducers.

NF-κB, IκB, IKK and the Canonical Pathway for NF-κB Activation

NF-κB is a master transcription factor that plays a major role in inflammatory and immune response.[1] It was originally found in nuclei of B cells and named for its ability binding the κ-chain enhancer of immunoglobulin in B cells. NF-κB was later found in the cytoplasm of all cell types, where it enters the nucleus upon stimulation. NF-κB transcription factors are evolutionarily conserved from insects to mammals. In mammals, the NF-κB family consists of five members (p65/RelA, RelB, c-Rel, p50/NF-κB1 and p52/NF-κB2). These proteins share an N-terminal domain of about 300 amino acids, which bears homology to the product of the *v-rel* oncogene, the Rel homology domain (RHD), and includes regions for DNA binding, dimerization and nuclear translocation (Fig. 1). DNA binding by NF-κB requires dimerization and most members of this family form both homo- and heterodimers except for RelB, which forms only heterodimers with p50 or p52. Mammalian NF-κB proteins can be classified into two groups; the first group, consisting of p65 (RelA), RelB and c-Rel, are expressed as mature proteins and possess a transcriptional activation domain at their C-termini. NF-κB dimers containing any one of these subunits can activate target gene transcription upon induction by certain stimuli. The second group consists of p50 (NF-κB1) and p52 (NF-κB2), which are first expressed as large precursors p105 and p100, respectively. NF-κB1 precursor p105 is constitutively processed to produce p50, whereas p52 is proteolytically released from p100 only upon stimulation. Both p50 and p52 lack a potent transcriptional activation domain and therefore cannot activate transcription as homodimers, or as p50/p52 heterodimers. In fact, p50/p52 dimers may suppress expression of NF-κB target genes.

The C-termini of p105 and p100 contain multiple ankyrin repeats, which are required for association with NF-κB and are the distinguishing structural feature of the IκBs, the specific inhibitors of NF-κB (Fig. 1). Therefore, p105 and p100 can serve an IκB-like function by

*Zhi-Wei Li—H. Lee Moffitt Cancer Center and Research Institute; Department of Interdisciplinary Oncology, University of South Florida, 3011 West Holly Drive, SRB-22344, Tampa, Florida 33612, U.S.A. Email: liz@moffitt.esf.edu.

NF-κB/Rel Transcription Factor Family, edited by Hsiou-Chi Liou.
©2006 Landes Bioscience and Springer Science+Business Media.

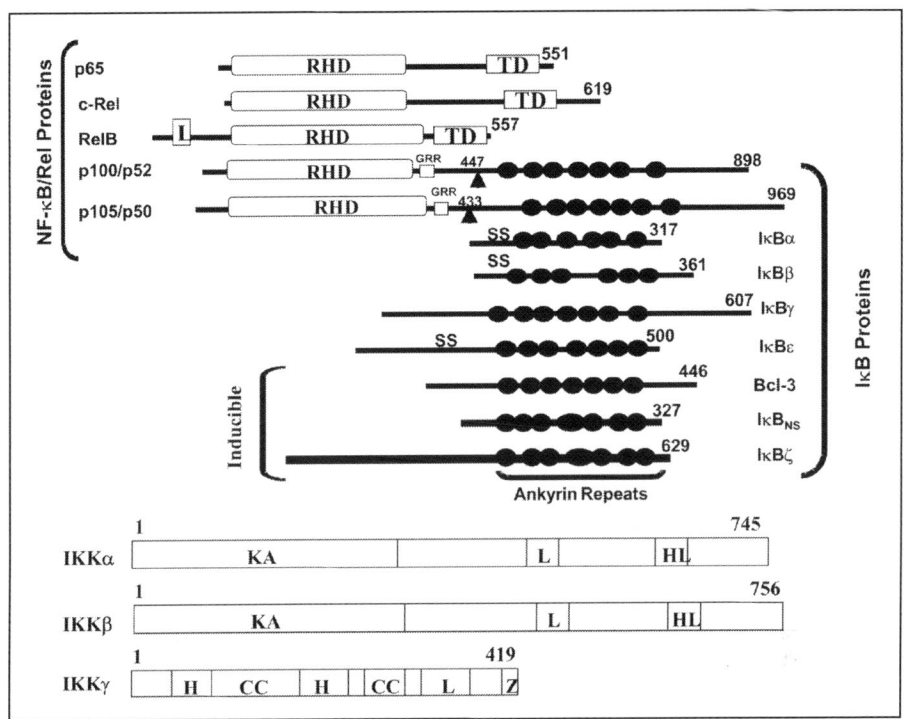

Figure 1. Mammalian NF-κB, IκB and IKK proteins. CC: coiled coil; GRR: glycine rich repeat; H: α-helix; HLH: helix-loop-helix; L: leucine zipper; RHD: Rel homology domain; SS: serine phosphorylation sites; TD: transactivation domain; Z: zinc finger. Modified from Li ZW, Rickert RC and Karin M. Genetic dissection of antigen receptor induced-NF-kappaB activation. Mol Immunol 2004; 41(6-7):701-714.

retaining RelA, RelB or c-Rel in the cytoplasm. Three major mammalian IκB proteins, IκBα, IκBβ and IκBε, have been identified.[1] These IκBs have overlapping yet distinct inhibitory specificity and thus can differentially inhibit NF-κB dimers. In addition, the C-terminal portion of p105 can be expressed as an independent transcript that encodes IκBγ, which is expressed only in the lymphoid cells. Another mammalian IκB family member is the nuclear protein Bcl-3.[1] Although it contains ankyrin repeats, Bcl-3 functions as a transcriptional activator with p50 or p52 homodimers, rather than an inhibitor of NF-κB. This activity may be caused by Bcl-3-mediated displacement of p50 or p52 homodimers from NF-κB binding site to allow binding of NF-κB molecules with transactivation domains, such as p65, c-Rel and RelB. Alternatively, Bcl-3 may also activate gene transcription by its own transactivation domain.[2] Bcl-3 production is inducible and is required for humoral immune response. Other two inducible IκB family members are IκBζ (also called MAIL or INAP) and IκB$_{NS}$.[3-6] IκBζ is required for Toll-like receptor (TLR) and interleukin 1 (IL-1) receptor (IL-1R) activation induced production of IL-6,[7] and IκB$_{NS}$ is induced by TCR (T cell receptor) activation,[6] suggesting that other inducible IκBs may exist and respond to diverse stimulation. However, how these inducible IκBs function has yet to be determined.

In addition to the ankyrin repeats, the C-terminal acidic region of IκBs is necessary for their inhibitory activity. The PEST motif in the C-terminal acidic region of IκB is the target site of IκB phosphorylation that is responsible for the basal turnover of these proteins and their induced degradation in response to UV irradiation.[8] The IκBs inhibit NF-κB activity by masking the nuclear localization signal (NLS) of NF-κB, thereby retaining NF-κB in the cytoplasm

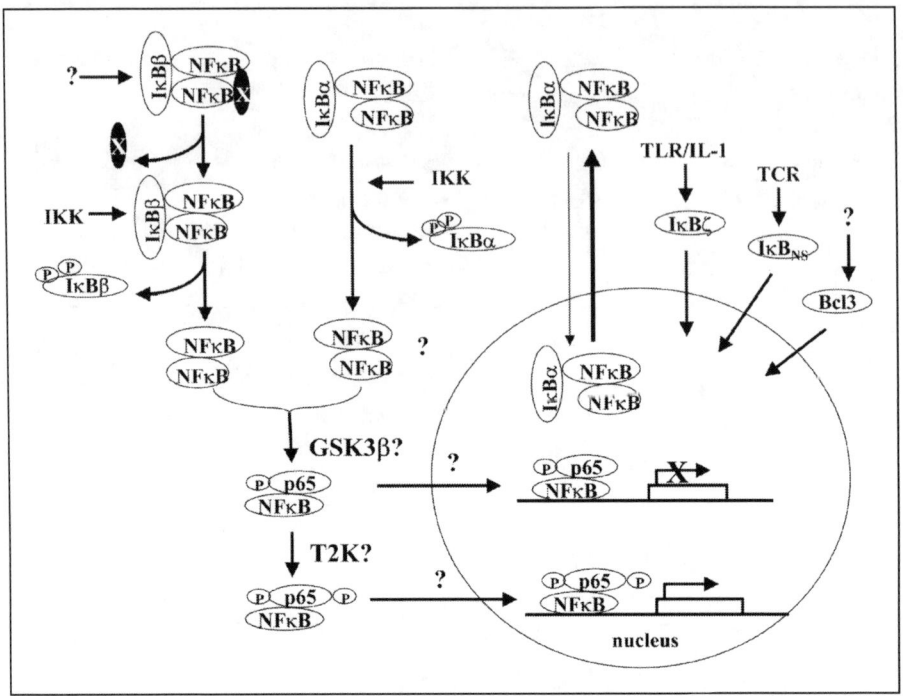

Figure 2. Regulation of NF-κB activity by IκBs and protein kinases. IκBα:NFκB complex shuttles between the cytoplasm and nucleus although majority of the complex is in the cytoplasm. The IκBβ:NF-κB complex is retained in cytoplasm by IκBβ and protein X such as κB-Ras. Upon stimulation, unidentified molecule removes X from IκBβ:NF-κB complex. IKK phosphorylation of IκBs leads to the ubiquitination and degradation. Released NF-κB is phosphorylated and then induces gene expression. Nuclear import of inducible IκBs (Bcl3, IκBζ, IκB$_{NS}$) occurs only upon certain stimuli. For p65 containing NF-κB, GSK3β phosphorylation might enhance DNA binding, and further T2K phosphorylation might lead to gene transcription. Question markers indicate unidentified or unconfirmed molecule(s). Modified from Li ZW, Rickert RC and Karin M. Genetic dissection of antigen receptor induced-NF-kappaB activation. Mol Immunol 2004; 41(6-7):701-714.

(Fig. 2). There are two variations of this model.[9] One mechanism is used by IκBβ and causes cytoplasmic retention of NF-κB due to the masking of two NLSs on NF-κB dimers. Interaction between the NF-κB:IκBβ complex and the small guanosine triphosphatases κB-Ras-1, -2 also contribute to NF-κB activation. When binding to κB-Ras, IκBβ cannot be phosphorylated by IKK, thus blocking the NF-κB activation signal from IKK.[10] IκBα and IκBε, which both mask one NLS of NF-κB, utilize the other mechanism in which the 2nd NLS of NF-κB and the nuclear export signal (NES) of IκBα or IκBε causes the IκB:NF-κB complex to shuttle between the nucleus and cytoplasm. Recently, Moorthy et al suggested that IκBs, including IκBγ (the c-terminus of p105), utilize the same binding mechanism, and the localization of IκB:NF-κB complex might be controlled by other protein(s).[11] It was suggested that κB-Ras is a protein that interacts only with IκBβ but not other IκB family members. The interaction of κB-Ras with the NF-κB:IκBβ complex causes the binding of two NF-κB NLS motifs by IκBβ. Removal of κB-Ras release one NLS and leads to nuclear import of the NF-κB:IκBβ complex.[12] If this is true, free κB-Ras should be detected upon NF-κB activation stimuli. The differential control between IκBα and IκBβ may lead to biphasic activation of NF-κB. As a target gene of NF-κB, IκBα is promptly upregulated upon NF-κB activation and therefore controls the fast transient activation of NF-κB, whereas IκBβ, whose transcription is not controlled by NF-κB, controls the persistent activation of NF-κB.[13] Two MAP3Ks, MEKK3 and MEKK2 were suggested to regulate the biphasic activation of NF-κB upon TNFα (tumor necrosis factor alpha) and IL-1α by participating in assembling of IκBα:NF-κB/IKK and IκBβ:NF-κB/IKK complex, respectively.[14]

In response to extracellular stimuli, the IκBs are rapidly phosphorylated due to activation of a protein kinase complex called the IκB kinase (IKK).[15] This phosphorylation event targets the IκBs for polyubiquitination and degradation by the 26S proteasome and thereafter the release of NF-κB dimers that translocate to the nucleus to regulate gene transcription. IKK (Fig. 1) contains two closely related catalytic subunits, IKKα and IKKβ, which contain a protein kinase domain at their N-terminal portion, whereas their C-terminal portion contains protein interaction motifs such as a leucine zipper (LZ) and a helix-loop-helix (HLH) domain. IKKα and IKKβ can both directly phosphorylate IκBs and their activity depends on dimerization through their leucine zipper motifs. In addition, IKK activity also depends on the HLH motif that may act as an intramolecular activator of the kinase domain. While the major native IKK complex is based on IKKα:IKKβ heterodimers, in vitro, both IKKα and IKKβ can also form functional homodimers. IKKβ is the major kinase controlling canonical pathway of NF-κB activation, in which phosphorylation of IκB by IKK release NF-κB to enter nuclear and regulate gene expression. The native IKK complex also contains a regulatory subunit, IKKγ that can form homodimers and is necessary for assembly of the IKK complex and recruitment of upstream activators to the IKK complex. The C-terminal domain of IKKγ is essential for IKK kinase activity,[16] while the N-terminus is required for binding of IKKγ to the catalytic subunits and therefore is also important for IKK activity.[17] Recently, it was found that the C-terminal oligomerization domain of IKKγ is required for dimerization whereas tetramerization, which enhances IKK kinase activity, needs the N-terminal domain.[18] Consistent with this finding, Weil et al found that the IKKγ N-terminus is sufficient for IKK and NF-κB activation when recruited to the plasma membrane.[19] Although reports identifying signaling events that only require IKKα or IKKβ are beginning to emerge, the similar phenotypes of IKKβ-null and IKKγ-null mice strongly suggest that IKKγ is required for activation of IKKβ.[20] IKKα$^{-/-}$, IKKβ$^{-/-}$ double knockout mice showed the same phenotype as IKKγ knockout mice, suggesting that IKKγ may also control IKKα activation in the context of the tri-subunit IKK complex. IKKγ-deficient mice die earlier than IKKα- or IKKβ- deficient mice, making it difficult to precisely determine the function of IKKγ in a variety of physiological processes. Hopefully, the generation of conditional IKKγ knockout mice will circumvent this difficulty. Most recently, another protein, ELKS (for the relative abundance of its constitutive amino acids: glutamic acid (E), leucine (L), lysine (K), and serine (S)), was suggested to be an IKK regulatory subunit.[21] Knocking down ELKS by RNA interfering leads to defect in NF-κB activation, including reduction in IKK kinase activity, IκB phosphorylation and degradation, NF-κB DNA binding activity, NF-κB targeting gene expression and the protection of cell death induced by TNFα. It was suggested that ELKS functions probably by recruiting IκBα to the IKK complex. However, the physiological function of ELKS remains to be explored.

IKKα and Noncanonical Pathway for NF-κB Activation

In addition to the canonical NF-κB activation pathway that is mostly dependent upon IKKγ-regulated IKKβ activation, IκB phosphorylation and degradation, and then NF-κB activation,[15] there are at least two situations in which NF-κB activation was reported to depend only on IKKα (Fig. 3). One is IKKα-dependent p100 processing, which is believed to be the mechanism for LTβ (lymphotoxin beta) and Blys (B lymphocyte stimulator, also called BAFF, TALL-1, zTNF4 or THANK) induced NF-κB activation.[9] Ligation of the LTβR (LTβ receptor) or Blys receptor (BR3) leads to NIK (NF-κB-inducing kinase) activation which in turn phosphorylates IKKα and thereby activates NF-κB by phosphorylating p100 and causing the release of p50:RelB dimers. Indeed, NIK-deficient mice are defective in LTβ induced NF-κB

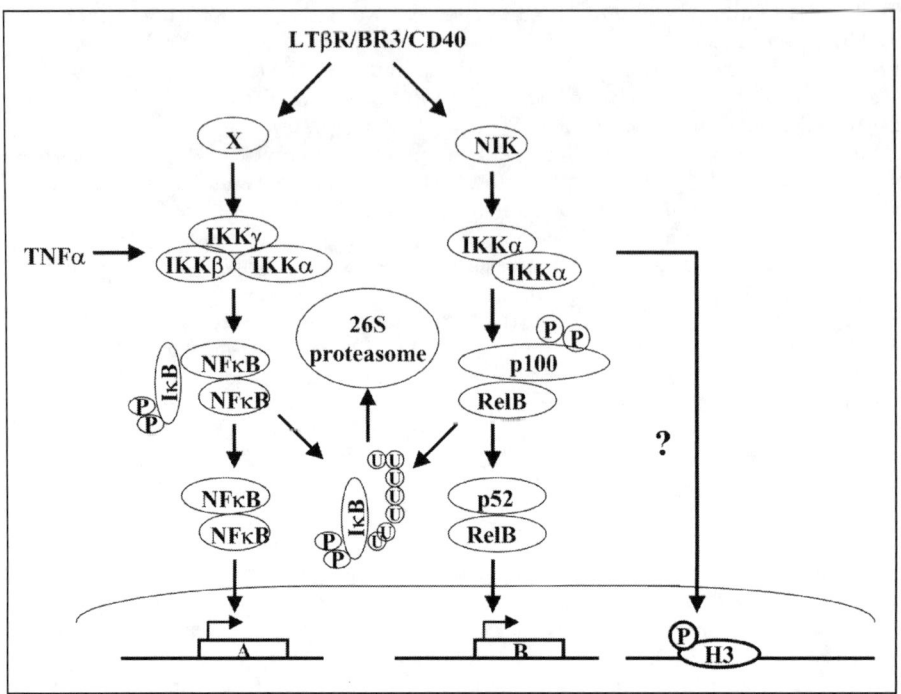

Figure 3. NF-κB activation is differentially regulated by IKKα and IKKβ. Upon LTβ/Blys/CD40L stimulation, their receptor LTβR/BR3/CD40 activates NIK and thereafter IKKα homodimers, which in turn phosphorylate p100 and lead to p100 processing, and release p52/RelB heterodimers. An unidentified X signal pathway (it contains TRAF6 for CD40 pathway) activates the IKK holoenzyme composed of all three subunits. In most cases, the activity of this complex depends on IKKβ. Modified from Li ZW, Rickert RC and Karin M. Genetic dissection of antigen receptor induced-NF-kappaB activation. Mol Immunol 2004; 41(6-7):701-714.

activation.[22] IKKα-deficient mice show defective p100 processing in B cells and this might be due to a defect in the phosphorylation of p100 by IKKα.[23] In the case of Blys signaling, in vivo data also support this NIK-IKKα-p100 model. Blys knockout mice showed reduced generation of mature follicular B cells.[24] This phenotype is also observed in IKKα$^{-/-}$ fetal liver transplanted mice and in NF-κB2$^{-/-}$ mice.[23,25-27] Blys/BR3 signaling promotes p100 processing and thereafter prevents apoptotic B cell death. A BR3-Fc fusion protein, which blocks BR3 signaling, inhibits p100 processing and attenuates the development of autoimmune disease in a mouse model.[28] Blys-induced p100 processing and NF-κB activation are impaired in NIK mutant B cells, and p100 processing is independent of IKKγ, the regulator of the canonical pathway of NF-κB activation.[29] The involvement of IKKα in Blys signaling, however, has not been determined using IKKα-deficient B cells. In NF-κB2$^{-/-}$ B cells, other NF-κBs still respond to Blys stimulation.[29] This seems to be not simply due to the compensation of other NF-κB components since Blys-mediated cell survival is impaired in IKKβ-deficient B cells (Z.W. Li, unpublished result). Together, these findings suggested that IKKα is not the only signaling molecule that regulates Blys-mediated B cell survival. With respect to LTβR stimulation, induction of some NF-κB target genes requires NIK and IKKα, whereas expression of other genes depends only on IKKβ and, presumably, IKKγ. RelB upregulation and p100 processing as well as translocation of p52:RelB dimers into the nucleus depend only on NIK and IKKα. However, p100 expression is controlled by RelA and IKKβ.[30] Consistent with these findings, LTβR stimulation leads to IκBα degradation and a shift of DNA binding molecules from RelA- to RelB-containing NF-κB dimers. Prior activation of the IKKβ-dependent pathways results in upregulation of p100:RelB dimers available for IKKα-induced processing.[31] Saccani et al also found that upon proinflammatory stimulation, NF-κB mediated gene transcription in monocyte-derived dendritic cells is fine-tuned by exchange of NF-κB dimmers.[32] However, the mechanism controlling this dimer exchange is still unclear. CD40L is another NF-κB activator that utilizes noncanonical pathway,[33] although it also utilizes canonical pathway.[34] CD40 was known to recruit TRAF6 (TNF receptor associated factor 6) for NF-κB activation as justified by analysis of TRAF6 knockout mice.[35] It would be interesting to determine how TRAF6 differentially affects both canonical and noncanonical pathways that activate NF-κB through recruitment of IKKβ and IKKα, respectively.

The second mechanism proposed for IKKα-regulated NF-κB activation is the phosphorylation of histone H3 residue serine 10 by IKKα.[36,37] Although both groups suggested that IKKα and not IKKβ, is the kinase phosphorylating this residue in vitro, more work is needed to demonstrate this in vivo and resolve a number of questions raised from the discrepancy between the in vitro evidence for H3 phosphorylation by IKKα and contrasting in vivo findings.[38] Most prominently, it is well established that IKKα-deficiency in mice results in postnatal death but normal TNFα signaling, whereas IKKβ-deficiency in mice results in liver apoptosis due to defective TNFα-depended NF-κB activation.[9] A crucial question therefore is why the phenotype of IKKα$^{-/-}$ mice is so different from that of IKKβ$^{-/-}$ mice if IKKα phosphorylation of histone H3 is critical for TNFα induced NF-κB activation. It is most likely that IKKα-dependent H3 phosphorylation is of little physiological relevance in vivo. Recently, it was found that the skeletal morphological defect in IKKα null mice was attributed to failed epidermal differentiation that is regulated by kinase-independent functions of IKKα.[39] Although this defect was related to increased FGF8 (fibroblast growth factor 8) expression, how IKKα inhibits FGF8 expression is still a myth.[39] Interestingly, nuclear translocation of IKKα is required by this inhibitory function. This finding suggested that IKKα may do play a role by functioning in nucleus.

Modification of NF-κB and Its Signaling Molecules

Phosphorylation of NF-κB subunits also contributes to the regulation of NF-κB activation by facilitate the recruitment of various transcription cofactors.[9,40] Several kinases, including PKAc (catalytic subunit of protein kinase A), MSK1 (mitogen- and stress-activated kinase-1), RSK1 (ribosomal subunit kinase-1), PI3K/AKT, CKII (casein kinase II), GSK3β (glucose synthase kinase 3β), T2K (TRAF2-associated kinase, also called TBK or NAK), IKK, PKCζ (protein kinase C) and NIK, were suggested to phosphorylate NF-κB subunits,[9,40] and GSK3β, T2K, IKK, PKCζ and NIK were confirmed by gene targeting to be essential for NF-κB-regulated gene transcription by particular stimuli.[20] While additional evidence suggests that IKK, PKCζ and NIK act upstream of IκB, GSK3β and T2K are the more likely candidates to act downstream of IKK and phosphorylate p65.[9] In response to a variety of stimuli, cells deficient in either GSK3β or T2K exhibit normal IκB degradation and NF-κB nuclear translocation, but are defective in NF-κB target gene transcription. Furthermore, either GSK3β or T2K knockout mice die during mid-gestation due to massive liver apoptosis,[41,42] which is the same phenotype observed in IKKβ$^{-/-}$ or p65$^{-/-}$ mice.[43-46] However, NF-κB DNA binding activity is only affected in the GS3Kβ knockout, and not in the T2K knockout (Fig. 2), suggesting that these kinases differentially control p65 activity downstream of IκB degradation. To provide further evidence for the phosphorylation of p65 by GSK3β or T2K, it will be important to compare p65 phosphorylation in wild type and GSK3β- or T2K-deficient cells in response to particular stimuli. Further in vivo data to support this hypothesis could be generated by attempting to rescue the GSK3β- or T2K-deficiencies by overexpression of constitutively activated p65.

Acetylation of RelA is also reported to affect NF-κB activation, perhaps by regulating the interaction between newly synthesized IκB and RelA, This acetylation is reversible and

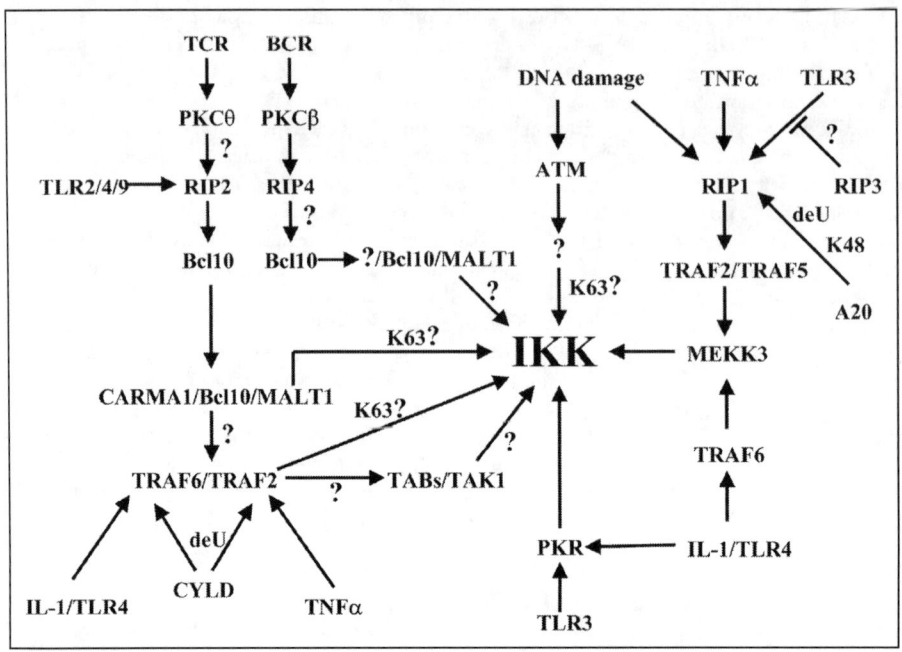

Figure 4. Activation of IKK complex. Different isoforms of PKCs and RIPs are utilized by different signaling pathway to activate IKK, whereas TAK1 and MEKK3 are involved in several IKK activation pathway. K48, K63 or deU represent ubiquitination that target recipient for degradation, activation, or de-ubiquitination, respectively. The K63 ubiquitination target in IKK complex is IKKγ. TRAF2/TRAF5 was confirmed to be involved only in TNFα, not in DNA damage and TLR3 induced IKK activation. PKR is involved in TLR3, not TLR4 induced IKK activation. Question markers indicate unidentified molecule(s) or unconfirmed pathway(s).

therefore controls the strength and duration of NF-κB activation.[47] The acetyltransferases and deacetylase in this reversible event were suggested to be the p300/CBP (cAMP-responsive element-binding protein) and p300/CBP-associated factor, and HDAC3 (histone deacetylase 3), respectively.[47,48] In addition to RelA acetylation, it was also reported that p50 acetylation upon stimulation affects NF-κB activation, perhaps by enhancing the DNA binding activity of p50.[47,48] The enzymes responsible for p50 acetylation/deacetylation have not been described, neither the mechanism that coordinates acetylation/deacetylation and activation signaling pathways. Several questions should be answered to justify the in vivo role of NF-κB acetylation, the most important one is whether NF-κB acetylation/decetylation is a physiologically relevant event. This could be done by analyzing NF-κB activation in mice or cells deficient in p300/CBP or other putative NF-κB modification enzymes.

Ubiquitination might be the hottest field in NF-κB community recently. IκB Ubiquitination plays an essential role in IκB degradation and NF-κB activation induced by various extracellular stimuli.[15] Recent works suggested that several other NF-κB signaling molecules could also been ubiquitylated and the ubiquitination of these molecules is required for transducing NF-κB activation signal (Fig. 4). The type of ubiquitin linkage controls the fate of an ubiquitylated protein. K48-linked polyubiquitination generally targets proteins for degradation in the proteasome, whereas K63-linked ubiquitination can regulate protein function.[49] TRAF6 is the first NF-κB signal transducer reported to be an ubiquitin ligase to mediate IKK activation.[50,51] Later, TRAF2 was also found to be an ubiquitin E3 ligase,[52] although its role as an ubiquitin ligase in NF-κB activation has yet to be explored. Another NF-κB signaling regulator A20 was reported to be a de-ubiquitination/polyubiquitination enzyme downregulating NF-κB signaling by de-ubiquitinating RIP (receptor interacting protein) and target it for degradation by polyubination.[53] The most interesting topic might be the ubiquitination of IKKγ. Upon nuclear export inhibitor Leptomycin B treatment, it was found that IKKγ competes with p65 and IKKα for binding to the N terminus of CBP and therefore leads to transcriptional repression of the NF-κB pathway.[54] Whether this is a truly physiological situation remains to be clarified since IKKγ-deficient mice show the phenotype similar to IKKβ- or p65-deficient mice, although it is possible that IKKγ has other function in addition to the activation of IKKβ and p65 upon TNFα stimulation during liver development. The existence of IKK-unbound free IKKγ has been reported by others.[55] These free IKKγ molecules shuttle between cytoplasma and nucleus. Upon genotoxic stress, free IKKγ is sumoylated in an ATM (ataxia telangiectasia mutant) independent manner. The sumoylation leads to nuclear localization of IKKγ. Although IKKγ is later desumoylated, this modification is required by the sequential ubiquitylation in an ATM-dependent manner and ultimately activation of IKK in the cytoplasma as evidenced by the ATM RNA interfering knocking down analysis. This work proposed a novel mechanism for NF-κB activation upon genotoxic stress. The reality of this novel mechanism remains to be testified by analyzing the sumoylation modification enzymes.

Inhibition of the family cylindromatosis tumor suppressor gene (CYLD) enhances the activation of NF-κB. CYLD binds to IKKγ and regulate IKK activity by de-ubiquitination of TRAF2, and to a less extent, of TRAF6.[56-58] CYLD may function through binding to TRIP (TRAF-interacting protein) and stabilize TRIP by removing the ubiquitins and thereby blocking NF-κB activation (Fig. 4).[59] The ubiquitination of IKKγ was also suggested to be regulated by c-IAP1 in TNFα activation of NF-κB.[60] Using purified protein and a cell-free system, it was found that Bcl-10 activates NF-κB through the intrinsic ubiquitin ligase activity of paracaspase MALT1 (mucosa associated lymphoid tissue),[61] or through MALT1 to induce oligomerization and activation of TRAF6 (Fig. 4).[62] TAK1 (transforming growth factor β-activated kinase 1) is further required to phosphorylate IKKβ in response to TCR activation.[62] Although different groups proposed unidentical model regarding IKKγ ubiquitination in the same Bcl-10 signaling pathway, the above work established signal flow chart from Bcl-10, MALT1, TRAF6, TAK1 and IKK to NF-κB activation, and emphasized the significance of IKKγ ubiquitination. Since the essential role of TRAF6 in TCR activation of NF-κB does not match the phenotype of TRAF6 knockout T cells, any significant functional defect of that has not been reported, there might be other missing piece(s) in this NF-κB activation signaling pathway, or compensation from other molecule such as TRAF2 as suggested.[62] Analysis of NF-κB activation in TRAF2-/-, TRAF6-/- double knockout T cells would be very helpful to elucidate this signaling pathway. It is also interesting to see how TRAF2 and TRAF6 function in TCR activation of NF-κB without affecting TNFα and IL-1 signaling wherein these two molecules were required for NF-κB activation.

Molecules Involved in Multiple NF-κB Activation Signaling Pathway

NF-κB activation signaling pathway is best understood in the immune and inflammatory system. Genetics dissection of antigen receptor induced-NF-κB activation was recently reviewed.[63] Receptors and adaptors for NF-κB were also reviewed in this book (see Chapter 3). Recent studies suggested that certain signaling molecules could be utilized by different pathways to activate IKK (Fig. 4). Following receptor-proximal signaling events upon TCR or BCR activation, different PKC isoforms are utilized to induce NF-κB activation during T cell and B cell development. In pro-B cells, preBCR activation of NF-κB is regulated by PKCλ through both IKKα and IKKβ.[64] In mature B cells, BCR activation of NF-κB may be regulated by PKCβ through IKKα.[65,66] It is important to note, however, that an additional IKKα-mediated mechanism other than increased NF-κB DNA binding activity must account for the PKCβ-associated defect since DNA binding by NF-κB is not the major consequence of PKCβ inactivation.[64] This is also different from PKCθ-regulated NF-κB activation in T cells, which utilizes IKKβ instead of IKKα.[67]

Downstream signaling from PKC to NF-κB activation is believed to be mediated by CARMA1 (CARD carrying member of the MAGUK family proteins 1), Bcl10 and MALT1,[68] although how PKC activates CARMA1 and how MALT1 activates IKK remain to be determined in vivo. Of notice, TNFα- and IL-1- mediated NF-κB activation is not affected by CARMA1-, Bcl10- or MALT1-knockout, suggesting that these three signaling molecules are unique for a PKC-dependent pathway. Interestingly, although all of these three molecules are required for NF-κB activation induced by antigen receptor ligation, their functions in lymphocyte development are different. Only B cell and not T cell development is defective in mice deficient in CARMA1.[69] However, both B and T cell development are defective in Bcl10 and MALT1 knockout mice.[70-72] These findings suggest that in addition to CARMA1, other molecules might be involved downstream of PKC in T cells, which may also recruit Bcl10 and MALT1 to induce NF-κB activation. Ruland et al also suggested that except for MALT1, other molecule might exist in B cells to relay BCR activation signal to NF-κB activation since the B cell development is only affected moderately in MALT1 deficient mice.[72]

Recent progress suggested that some known NF-κB signaling transducers might be involved in several signaling pathway, although further in vivo studies are needed to verify it. The essential role of RIP for NF-κB activation has been established long time ago. RIP1, the NF-κB activator in TNFα signaling pathway,[73] was recently reported to be essential for TLR3-mediated NF-κB activation.[74] It would be interesting to determine how the RIP1 downstream signaling molecules function in TLR3 signaling without affecting TNFα signaling pathway. It was also reported that RIP1 is essential for DNA-damage-induced NF-κB activation by inducing IκBα degradation, suggesting that this activation may go through IKK.[75] Indeed, upon DNA damage, it was confirmed that RIP forms a complex with IKK, and IκBα degradation requires IKKβ. Interestingly, the kinase activity of RIP is not required and IKK activation was not confirmed by kinase assay. Although RIP is involved in both TNFα and DNA damage-induced NF-κB activation, other signaling molecules, such as TNFR1 (TNF receptor 1), TRAF2, TRAF5 and FADD (Fas-associated death domain protein) are involved only in TNFα induced NF-κB activation. Although ATM was found to be required for the formation of RIP-IKK complex upon DNA damage, how does RIP activate IKK remains to be solved.[75]

RIP2, another RIP family protein, is involved in TLR-mediated NF-κB activation.[76,77] RIP2$^{-/-}$ cells and mice exhibit impaired responses, including defective NF-κB activation, cytokine production and are resistant to endotoxic shock in response to LPS, dsRNA and peptidoglycan stimulation. It was suggested that RIP2 plays a role in signaling by TLR2/4/9 and Nod1.[76,77] RIP2 is also involved in TCR signaling to NF-κB activation. RIP2-deficient T cells show severely reduced NF-κB activation upon TCR engagement, as well as some other defect in TCR signaling, T cell differentiation and function,[76,77] whereas RIP1 is important for TNFR2 signaling in thymocyte development and apoptosis, but is not required for thymocyte proliferation.[78] How RIP1 and RIP2 are involved in TCR signaling differentially is not yet clear. Ruefli-Brasse et al reported that upon TCR activation, RIP2 associates and phosphorylates Bcl-10 and therefore involved in TCR mediated NF-κB activation.[79]

Different from other RIP family members, RIP3 is dispensable for NF-κB activation by several NF-κB activators, including the engagement of TCR, BCR, TNFR1, TLR2 and TLR4.[80] In contrast, RIP3 negatively regulates the RIP1-induced NF-κB activation in TLR3 signaling pathway.[74] Analysis of transgenic mice expressing a kinase dead version of another RIP family member, RIP4, suggested that RIP4 may be required for BCR signaling to NF-κB.[81] RIP4 was cloned based on its association with PKCβ,[82] suggesting that the involvement of RIP1 or RIP2 in TCR signaling may allow for the interaction with other PKC isoforms. Biochemical experiments suggested that RIP4 is involved in PKC activation of NF-κB that is independent of Bcl10.[83] However, additional in vivo data is needed to confirm the existence of this

Bcl10-independent signaling pathway in NF-κB activation. A detailed understanding of RIP function awaits further investigation.

Downstream of RIP, TRAF2 and TRAF5 are the mediators of IKK activation in TNFα induced NF-κB activation signaling pathway. TRAF2 regulates TNFα induced NF-κB activation in cooperation with TRAF5. TRAF2 and TRAF5 single-knockouts show a mild phenotype, but TRAF2/TRAF5 double knockout are severely impaired in TNFα-induced NF-κB activation, and are therefore very sensitive to TNFα-induced apoptosis.[84-86] How TRAF2/5 and RIP activate IKK and NF-κB is still controversial.[9] It was reported that MEKK3 is involved in IKK activation downstream of TRAF2 and RIP in response to TNFα stimulation.[87] Interestingly, MEKK3 appears to be a multiple edge sword, it is also required for IL-1 and LPS induced activation of NF-κB, as well as JNK and p38, but not ERK,[88] perhaps due to the existence of TRAF6, another member of TRAF family in the IL-1 and TLR signaling pathway.

NIK, which is required for activation of the noncanonical NF-κB pathway by Blys in B cells as discussed above, also plays a role in TCR induced NF-κB activation, which is distinct from the PKC mediated signaling pathway.[89] Since NF-κB activation is only slightly attenuated in NIK mutant thymocytes as well as mature T cells, it appears that NIK may not play a major role in NF-κB activation in T cells. PKR, the sensor kinase for virus infection and perhaps also involved in LPS stimulated NF-κB activation,[90] was reported to directly activate IKK by protein-protein interaction.[91,92] Using Bone marrow or fetal liver derived macrophages of various knockout mice, Hsu et al validated the TLR 4 pathway in macrophage.[93] Consistent with the previous finding, in addition to the role in antivirus infection, they found that PKR is required for macrophage apoptosis after activation of TLR4. However, perhaps due to cell type specificity, PKR is not required for the activation of MAPK p38, JNK, ERK or activation of IKK in response to LPS stimulation in macrophage. This is different from that in embryonic fibroblast where PKR is required for p38 activation in response to LPS and other proinflammatory stimuli.[90] Hence, the full physiological function of PKR in NF-κB activation remains to be determined in a variety of cell types.

TAK1 might be the most potential candidate as IKK kinase in multiple signaling pathways, perhaps due to its interaction with TRAF6. Interaction of TAK1 and TRAF6 involves TAK1 binding proteins TAB1 and TAB2. Association of TAK1 with TRAF6 leads to activation of TAK1, and activated TAK1 in turn activates IKK.[50] However, TAB1 was suggested to be important for heart development.[94] Analysis of TAB2 knockout mice indicated that the TAK1:TAB complex is not essential for IL-1 and TNFα signaling, but is required for preventing liver apoptosis,[95] suggesting the existence of other signaling molecules that link TRAF6 and IKK. One candidate of these signaling molecules is TAB3.[96] TAB3 and TAB2 bind cooperatively, but not competitively, to TRAF6, and TAB3 associates with both TRAF2 and TRAF6 and therefore links them with TAK1. RNA interfering experiment demonstrated that TAB2 and TAB3 play a redundant but critical role in the IL-1- and TNF-induced activation of TAK1.[96] These finding explained the involvement of TAK1 in both IL-1 and TNF signaling pathway.[97] If this were the truly physiological situation, it would be interesting to explore the differential role of TAK1 and MEKK3, another MAP3K involved in both IL-1 and TNF induced NF-κB activation.[87,88] Much convincible data should be the analysis of TAK1 knockout mice and cells.

The critical role of IKK and many of its upstream signaling molecules in NF-κB activation has been confirmed by gene targeting experiments. IKK upstream signaling molecules are diverse. A better understanding of NF-κB activation may provide helpful information for the design of drugs targeting NF-κB in various diseases such as cancer, inflammation, autoimmune diseases and infectious disease. However, due to the general involvement of NF-κB in developmental and functional aspects of various cells, cell type-specific inhibition of NF-κB is needed for meaningful NF-κB targeted therapies. Attractive targets for such drugs include specific inhibitors for PKCθ in T lymphocytes, PKCβ in B lymphocytes, or Bcl10 in T and B cells. On the other hand, transcriptional targets of NF-κB could be more specific target, and might also represent interesting modalities for drug intervention.

References

1. Ghosh S, May MJ, Kopp EB. NF-kappa B and Rel proteins: Evolutionarily conserved mediators of immune responses. Annu Rev Immunol 1998; 16:225-260.
2. Viatour P, Bentires-Alj M, Chariot A et al. NF- kappa B2/p100 induces Bcl-2 expression. Leukemia 2003; 17(7):1349-1356.
3. Yamazaki S, Muta T, Takeshige K. A novel IkappaB protein, IkappaB-zeta, induced by proinflammatory stimuli, negatively regulates nuclear factor-kappaB in the nuclei. J Biol Chem 2001; 276(29):27657-27662.
4. Haruta H, Kato A, Todokoro K. Isolation of a novel interleukin-1-inducible nuclear protein bearing ankyrin-repeat motifs. J Biol Chem 2001; 276(16):12485-12488.
5. Kitamura H, Kanehira K, Okita K et al. MAIL, a novel nuclear I kappa B protein that potentiates LPS-induced IL-6 production. FEBS Lett 2000; 485(1):53-56.
6. Fiorini E, Schmitz I, Marissen WE et al. Peptide-induced negative selection of thymocytes activates transcription of an NF-kappa B inhibitor. Mol Cell 2002; 9(3):637-648.
7. Yamamoto M, Yamazaki S, Uematsu S et al. Regulation of Toll/IL-1-receptor-mediated gene expression by the inducible nuclear protein IkappaBzeta. Nature 2004; 430(6996):218-222.
8. Kato Jr T, Delhase M, Hoffmann A et al. CK2 is a C-terminal IkappaB kinase responsible for NF-kappaB activation during the UV response. Mol Cell 2003; 12(4):829-839.
9. Ghosh S, Karin M. Missing pieces in the NF-kappaB puzzle. Cell 2002; 109(Suppl):S81-96.
10. Fenwick C, Na SY, Voll RE et al. A subclass of Ras proteins that regulate the degradation of IkappaB. Science 2000; 287(5454):869-873.
11. Moorthy AK, Ghosh G. p105.Ikappa Bgamma and prototypical Ikappa Bs use a similar mechanism to bind but a different mechanism to regulate the subcellular localization of NF-kappa B. J Biol Chem 2003; 278(1):556-566.
12. Chen Y, Wu J, Ghosh G. KappaB-Ras binds to the unique insert within the ankyrin repeat domain of IkappaBbeta and regulates cytoplasmic retention of IkappaBbeta x NF-kappaB complexes. J Biol Chem 2003; 278(25):23101-23106.
13. Ladner KJ, Caligiuri MA, Guttridge DC. Tumor necrosis factor-regulated biphasic activation of NF-kappa B is required for cytokine-induced loss of skeletal muscle gene products. J Biol Chem 2003; 278(4):2294-2303.
14. Schmidt C, Peng B, Li Z et al. Mechanisms of proinflammatory cytokine-induced biphasic NF-kappaB activation. Mol Cell 2003; 12(5):1287-1300.
15. Karin M, Ben-Neriah Y. Phosphorylation meets ubiquitination: The control of NF-[kappa]B activity. Annu Rev Immunol 2000; 18:621-663.
16. Rothwarf DM, Zandi E, Natoli G et al. IKK-gamma is an essential regulatory subunit of the IkappaB kinase complex. Nature 1998; 395(6699):297-300.
17. May MJ, D'Acquisto F, Madge LA et al. Selective inhibition of NF-kappaB activation by a peptide that blocks the interaction of NEMO with the IkappaB kinase complex. Science 2000; 289(5484):1550-1554.
18. Tegethoff S, Behlke J, Scheidereit C. Tetrameric oligomerization of IkappaB kinase gamma (IKKgamma) is obligatory for IKK complex activity and NF-kappaB activation. Mol Cell Biol 2003; 23(6):2029-2041.
19. Weil R, Schwamborn K, Alcover A et al. Induction of the NF-kappaB cascade by recruitment of the scaffold molecule NEMO to the T cell receptor. Immunity 2003; 18(1):13-26.
20. Li Q, Verma IM. NF-kappaB regulation in the immune system. Nat Rev Immunol 2002; 2(10):725-734.
21. Sigala JL, Bottero V, Young DB et al. Activation of transcription factor NF-kappaB requires ELKS, an IkappaB kinase regulatory subunit. Science 2004; 304(5679):1963-1967.
22. Yin L, Wu L, Wesche H et al. Defective lymphotoxin-beta receptor-induced NF-kappaB transcriptional activity in NIK-deficient mice. Science 2001; 291(5511):2162-2165.
23. Senftleben U, Cao Y, Xiao G et al. Activation by IKKalpha of a second, evolutionary conserved, NF-kappa B signaling pathway. Science 2001; 293(5534):1495-1499.
24. Schiemann B, Gommerman JL, Vora K et al. An essential role for BAFF in the normal development of B cells through a BCMA-independent pathway. Science 2001; 293(5537):2111-2114.
25. Caamano JH, Rizzo CA, Durham SK et al. Nuclear factor (NF)-kappa B2 (p100/p52) is required for normal splenic microarchitecture and B cell-mediated immune responses. J Exp Med 1998; 187(2):185-196.
26. Franzoso G, Carlson L, Poljak L et al. Mice deficient in nuclear factor (NF)-kappa B/p52 present with defects in humoral responses, germinal center reactions, and splenic microarchitecture. J Exp Med 1998; 187(2):147-159.

27. Kaisho T, Takeda K, Tsujimura T et al. IkappaB kinase alpha is essential for mature B cell development and function. J Exp Med 2001; 193(4):417-426.
28. Kayagaki N, Yan M, Seshasayee D et al. BAFF/BLyS receptor 3 binds the B cell survival factor BAFF ligand through a discrete surface loop and promotes processing of NF-kappaB2. Immunity 2002; 17(4):515-524.
29. Claudio E, Brown K, Park S et al. BAFF-induced NEMO-independent processing of NF-kappa B2 in maturing B cells. Nat Immunol 2002; 3(10):958-965.
30. Dejardin E, Droin NM, Delhase M et al. The lymphotoxin-beta receptor induces different patterns of gene expression via two NF-kappaB pathways. Immunity 2002; 17(4):525-535.
31. Muller JR, Siebenlist U. Lymphotoxin beta receptor induces sequential activation of distinct NF-kappa B factors via separate signaling pathways. J Biol Chem 2003; 278(14):12006-12012.
32. Saccani S, Pantano S, Natoli G. Modulation of NF-kappaB activity by exchange of dimers. Mol Cell 2003; 11(6):1563-1574.
33. Coope HJ, Atkinson PG, Huhse B et al. CD40 regulates the processing of NF-kappaB2 p100 to p52. EMBO J 2002; 21(20):5375-5385.
34. Zarnegar B, He JQ, Oganesyan G et al. Unique CD40-mediated biological program in B cell activation requires both type 1 and type 2 NF-kappaB activation pathways. Proc Natl Acad Sci USA 2004; 101(21):8108-8113.
35. Lomaga MA, Yeh WC, Sarosi I et al. TRAF6 deficiency results in osteopetrosis and defective interleukin-1, CD40, and LPS signaling. Genes Dev 1999; 13(8):1015-1024.
36. Yamamoto Y, Verma UN, Prajapati S et al. Histone H3 phosphorylation by IKK-alpha is critical for cytokine-induced gene expression. Nature 2003; 423(6940):655-659.
37. Anest V, Hanson JL, Cogswell PC et al. A nucleosomal function for IkappaB kinase-alpha in NF-kappaB-dependent gene expression. Nature 2003; 423(6940):659-663.
38. Israel A. Signal transduction: A regulator branches out. Nature 2003; 423(6940):596-597.
39. Sil AK, Maeda S, Sano Y et al. IkappaB kinase-alpha acts in the epidermis to control skeletal and craniofacial morphogenesis. Nature 2004; 428(6983):660-664.
40. Chen LF, Greene WC. Shaping the nuclear action of NF-kappaB. Nat Rev Mol Cell Biol 2004; 5(5):392-401.
41. Bonnard M, Mirtsos C, Suzuki S et al. Deficiency of T2K leads to apoptotic liver degeneration and impaired NF-kappaB-dependent gene transcription. EMBO J 2000; 19(18):4976-4985.
42. Hoeflich KP, Luo J, Rubie EA et al. Requirement for glycogen synthase kinase-3beta in cell survival and NF-kappaB activation. Nature 2000; 406(6791):86-90.
43. Beg AA, Sha WC, Bronson RT et al. Embryonic lethality and liver degeneration in mice lacking the RelA component of NF-kappa B. Nature 1995; 376(6536):167-170.
44. Li ZW, Chu W, Hu Y et al. The IKKbeta subunit of IkappaB kinase (IKK) is essential for nuclear factor kappaB activation and prevention of apoptosis. J Exp Med 1999; 189(11):1839-1845.
45. Li Q, Van Antwerp D, Mercurio F et al. Severe liver degeneration in mice lacking the IkappaB kinase 2 gene. Science 1999; 284(5412):321-325.
46. Tanaka M, Fuentes ME, Yamaguchi K et al. Embryonic lethality, liver degeneration, and impaired NF-kappa B activation in IKK-beta-deficient mice. Immunity 1999; 10(4):421-429.
47. Chen L, Fischle W, Verdin E et al. Duration of nuclear NF-kappaB action regulated by reversible acetylation. Science 2001; 293(5535):1653-1657.
48. Kiernan R, Bres V, Ng RW et al. Post-activation turn-off of NF-kappa B-dependent transcription is regulated by acetylation of p65. J Biol Chem 2003; 278(4):2758-2766.
49. Weissman AM. Themes and variations on ubiquitylation. Nat Rev Mol Cell Biol 2001; 2(3):169-178.
50. Wang C, Deng L, Hong M et al. TAK1 is a ubiquitin-dependent kinase of MKK and IKK. Nature 2001; 412(6844):346-351.
51. Deng L, Wang C, Spencer E et al. Activation of the IkappaB kinase complex by TRAF6 requires a dimeric ubiquitin-conjugating enzyme complex and a unique polyubiquitin chain. Cell 2000; 103(2):351-361.
52. Shi CS, Kehrl JH. Tumor necrosis factor (TNF)-induced germinal center kinase-related (GCKR) and stress-activated protein kinase (SAPK) activation depends upon the E2/E3 complex Ubc13-Uev1A/TNF receptor-associated factor 2 (TRAF2). J Biol Chem 2003; 278(17):15429-15434.
53. Wertz IE, O'Rourke KM, Zhou H et al. De-ubiquitination and ubiquitin ligase domains of A20 downregulate NF-kappaB signalling. Nature 2004.
54. Verma UN, Yamamoto Y, Prajapati S et al. Nuclear role of I kappa B Kinase-gamma/NF-kappa B essential modulator (IKK gamma/NEMO) in NF-kappa B-dependent gene expression. J Biol Chem 2004; 279(5):3509-3515.
55. Huang TT, Wuerzberger-Davis SM, Wu ZH et al. Sequential modification of NEMO/IKKgamma by SUMO-1 and ubiquitin mediates NF-kappaB activation by genotoxic stress. Cell 2003; 115(5):565-576.

56. Brummelkamp TR, Nijman SM, Dirac AM et al. Loss of the cylindromatosis tumour suppressor inhibits apoptosis by activating NF-kappaB. Nature 2003; 424(6950):797-801.
57. Kovalenko A, Chable-Bessia C, Cantarella G et al. The tumour suppressor CYLD negatively regulates NF-kappaB signalling by deubiquitination. Nature 2003; 424(6950):801-805.
58. Trompouki E, Hatzivassiliou E, Tsichritzis T et al. CYLD is a deubiquitinating enzyme that negatively regulates NF-kappaB activation by TNFR family members. Nature 2003; 424(6950):793-796.
59. Regamey A, Hohl D, Liu JW et al. The tumor suppressor CYLD interacts with TRIP and regulates negatively nuclear factor kappaB activation by tumor necrosis factor. J Exp Med 2003; 198(12):1959-1964.
60. Tang ED, Wang CY, Xiong Y et al. A role for NF-kappaB essential modifier/IkappaB kinase-gamma (NEMO/IKKgamma) ubiquitination in the activation of the IkappaB kinase complex by tumor necrosis factor-alpha. J Biol Chem 2003; 278(39):37297-37305.
61. Zhou H, Wertz I, O'Rourke K et al. Bcl10 activates the NF-kappaB pathway through ubiquitination of NEMO. Nature 2004; 427(6970):167-171.
62. Sun L, Deng L, Ea CK et al. The TRAF6 ubiquitin ligase and TAK1 kinase mediate IKK activation by BCL10 and MALT1 in T lymphocytes. Mol Cell 2004; 14(3):289-301.
63. Li ZW, Rickert RC, Karin M. Genetic dissection of antigen receptor induced-NF-kappaB activation. Mol Immunol 2004; 41(6-7):701-714.
64. Saijo K, Schmedt C, Su IH et al. Essential role of Src-family protein tyrosine kinases in NF-kappaB activation during B cell development. Nat Immunol 2003; 4(3):274-279.
65. Saijo K, Mecklenbrauker I, Santana A et al. Protein kinase C beta controls nuclear factor kappaB activation in B cells through selective regulation of the IkappaB kinase alpha. J Exp Med 2002; 195(12):1647-1652.
66. Su TT, Guo B, Kawakami Y et al. PKC-beta controls I kappa B kinase lipid raft recruitment and activation in response to BCR signaling. Nat Immunol 2002; 3(8):780-786.
67. Lin X, O'Mahony A, Mu Y et al. Protein kinase C-theta participates in NF-kappaB activation induced by CD3-CD28 costimulation through selective activation of IkappaB kinase beta. Mol Cell Biol 2000; 20(8):2933-2940.
68. Ruland J, Mak TW. From antigen to activation: Specific signal transduction pathways linking antigen receptors to NF-kappaB. Semin Immunol 2003; 15(3):177-183.
69. Hara H, Wada T, Bakal C et al. The MAGUK family protein CARD11 is essential for lymphocyte activation. Immunity 2003; 18(6):763-775.
70. Ruefli-Brasse AA, French DM, Dixit VM. Regulation of NF-kappaB-dependent lymphocyte activation and development by paracaspase. Science 2003; 302(5650):1581-1584.
71. Ruland J, Duncan GS, Elia A et al. Bcl10 is a positive regulator of antigen receptor-induced activation of NF-kappaB and neural tube closure. Cell 2001; 104(1):33-42.
72. Ruland J, Duncan GS, Wakeham A et al. Differential requirement for Malt1 in T and B cell antigen receptor signaling. Immunity 2003; 19(5):749-758.
73. Kelliher MA, Grimm S, Ishida Y et al. The death domain kinase RIP mediates the TNF-induced NF-kappaB signal. Immunity 1998; 8(3):297-303.
74. Meylan E, Burns K, Hofmann K et al. RIP1 is an essential mediator of Toll-like receptor 3-induced NF-kappa B activation. Nat Immunol 2004; 5(5):503-507.
75. Hur GM, Lewis J, Yang Q et al. The death domain kinase RIP has an essential role in DNA damage-induced NF-kappa B activation. Genes Dev 2003; 17(7):873-882.
76. Kobayashi K, Inohara N, Hernandez LD et al. RICK/Rip2/CARDIAK mediates signalling for receptors of the innate and adaptive immune systems. Nature 2002; 416(6877):194-199.
77. Chin AI, Dempsey PW, Bruhn K et al. Involvement of receptor-interacting protein 2 in innate and adaptive immune responses. Nature 2002; 416(6877):190-194.
78. Cusson N, Oikemus S, Kilpatrick ED et al. The death domain kinase RIP protects thymocytes from tumor necrosis factor receptor type 2-induced cell death. J Exp Med 2002; 196(1):15-26.
79. Ruefli-Brasse AA, Lee WP, Hurst S et al. Rip2 participates in Bcl10 signaling and T-cell receptor-mediated NF-kappaB activation. J Biol Chem 2004; 279(2):1570-1574.
80. Newton K, Sun X, Dixit VM. Kinase RIP3 is dispensable for normal NF-kappa Bs, signaling by the B-cell and T-cell receptors, tumor necrosis factor receptor 1, and Toll-like receptors 2 and 4. Mol Cell Biol 2004; 24(4):1464-1469.
81. Cariappa A, Chen L, Haider K et al. A catalytically inactive form of protein kinase C-associated kinase/receptor interacting protein 4, a protein kinase Cbeta-associated kinase that mediates NF-kappaB activation, interferes with early B cell development. J Immunol 2003; 171(4):1875-1880.
82. Chen L, Haider K, Ponda M et al. Protein kinase C-associated kinase (PKK), a novel membrane-associated, ankyrin repeat-containing protein kinase. J Biol Chem 2001; 276(24):21737-21744.

83. Muto A, Ruland J, McAllister-Lucas LM et al. Protein kinase C-associated kinase (PKK) mediates Bcl10-independent NF-kappa B activation induced by phorbol ester. J Biol Chem 2002; 277(35):31871-31876.
84. Tada K, Okazaki T, Sakon S et al. Critical roles of TRAF2 and TRAF5 in tumor necrosis factor-induced NF-kappa B activation and protection from cell death. J Biol Chem 2001; 276(39):36530-36534.
85. Yeh WC, Shahinian A, Speiser D et al. Early lethality, functional NF-kappaB activation, and increased sensitivity to TNF-induced cell death in TRAF2-deficient mice. Immunity 1997; 7(5):715-725.
86. Nakano H, Sakon S, Koseki H et al. Targeted disruption of Traf5 gene causes defects in CD40- and CD27-mediated lymphocyte activation. Proc Natl Acad Sci USA 1999; 96(17):9803-9808.
87. Yang J, Lin Y, Guo Z et al. The essential role of MEKK3 in TNF-induced NF-kappaB activation. Nat Immunol 2001; 2(7):620-624.
88. Huang Q, Yang J, Lin Y et al. Differential regulation of interleukin 1 receptor and Toll-like receptor signaling by MEKK3. Nat Immunol 2004; 5(1):98-103.
89. Matsumoto M, Yamada T, Yoshinaga SK et al. Essential role of NF-kappa B-inducing kinase in T cell activation through the TCR/CD3 pathway. J Immunol 2002; 169(3):1151-1158.
90. Goh KC, deVeer MJ, Williams BR. The protein kinase PKR is required for p38 MAPK activation and the innate immune response to bacterial endotoxin. EMBO J 2000; 19(16):4292-4297.
91. Chu WM, Ostertag D, Li ZW et al. JNK2 and IKKbeta are required for activating the innate response to viral infection. Immunity 1999; 11(6):721-731.
92. Bonnet MC, Weil R, Dam E et al. PKR stimulates NF-kappaB irrespective of its kinase function by interacting with the IkappaB kinase complex. Mol Cell Biol 2000; 20(13):4532-4542.
93. Hsu LC, Park JM, Zhang K et al. The protein kinase PKR is required for macrophage apoptosis after activation of Toll-like receptor 4. Nature 2004; 428(6980):341-345.
94. Komatsu Y, Shibuya H, Takeda N et al. Targeted disruption of the Tab1 gene causes embryonic lethality and defects in cardiovascular and lung morphogenesis. Mech Dev 2002; 119(2):239-249.
95. Sanjo H, Takeda K, Tsujimura T et al. TAB2 is essential for prevention of apoptosis in fetal liver but not for interleukin-1 signaling. Mol Cell Biol 2003; 23(4):1231-1238.
96. Ishitani T, Takaesu G, Ninomiya-Tsuji J et al. Role of the TAB2-related protein TAB3 in IL-1 and TNF signaling. EMBO J 2003; 22(23):6277-6288.
97. Takaesu G, Ninomiya-Tsuji J, Kishida S et al. Interleukin-1 (IL-1) receptor-associated kinase leads to activation of TAK1 by inducing TAB2 translocation in the IL-1 signaling pathway. Mol Cell Biol 2001; 21(7):2475-2484.

CHAPTER 3

Receptors and Adaptors for NF-κB Signaling

Shao-Cong Sun* and Edward W. Harhaj

Cells communicate with their environment via various surface receptors that recognize specific molecules (ligands) present in the extracellular environment. Upon binding to a specific ligand, the receptors transduce signals into the cell, triggering multiple intracellular signaling cascades that lead to gene expression and other biochemical events involved in specific cellular functions. The signaling pathway that leads to activation of the transcription factor NF-κB has drawn much attention, since it regulates diverse biological processes such as immune and inflammatory responses, cell growth and survival, and tumorigenesis.[1-4] In this chapter, we summarize the receptors and their intracellular signaling molecules that target NF-κB activation. Consistent with its pleiotropic biological functions, the NF-κB signaling pathway responds to signals initiated by a large variety of receptors (Table 1). The focus of this chapter will be on three major families of the NF-κB-inducing receptors: the antigen receptors that mediate antigen recognition by lymphocytes, the toll-like receptors (TLR) involved in innate immune responses and the connection between innate and adaptive immune responses, and members of the tumor necrosis factor receptor (TNFR) superfamily that mediate diverse biological functions.

Antigen Receptors

Lymphocytes of the adaptive immune system have evolved a mechanism to recognize a great diversity of antigens in order to detect and combat a wide range of pathogens. This unique function of lymphocytes is mediated by their surface antigen receptors, the B-cell receptors (BCR) and T-cell receptors (TCR), multiprotein complexes composed of clonally variable antigen-binding subunits and invariant signaling chains.[5] TCR recognizes antigenic peptides bound to major histocompatibility complex (MHC) molecules on antigen presenting cells (APC), whereas BCR recognizes epitopes associated intact antigens. However, the intracellular signaling network leading to NF-κB activation is highly similar for these two types of antigen receptors. In both cases, the network involves receptor-proximal signaling events characterized by the activation of protein tyrosine kinases (PTKs), signal amplification involving phosphorylation and membrane-recruitment of adaptor molecules and other factors, and activation of the NF-κB effector kinase, IKK (IκB kinase), by a distal signaling complex composed of several recently characterized adaptor proteins.[6]

Receptor-Proximal Signaling Events

Following antigen binding, both BCR and TCR form clusters and are recruited to membrane compartments, enriched in cholesterol and sphingolipid, known as lipid rafts.[7] Many other signaling factors, including Src family of PTKs, the protein tyrosine phosphatase CD45,

---

*Shao-Cong Sun—Department of Microbiology and Immunology, Rm. C6708, Pennsylvania State University College of Medicine, 500 University Drive, Hershey, Pennsylvania 17033, U.S.A. Email: sxs70@psu.edu

NF-κB/Rel Transcription Factor Family, edited by Hsiou-Chi Liou.
©2006 Landes Bioscience and Springer Science+Business Media.

Table 1. NF-κB-inducing receptors

Receptors	Major Function
Antigen receptors	
BCR	B-cell development and activation
TCR	T-cell development and activation
Pattern recognition receptors	
TLR1-TLR11	Activation of macrophages, neutrophils, DCs, B cells, etc
NOD1 and 2	Intracellular pathogen recognition
Scavenger receptors	Activation of phagocytes
TNFR superfamily	
TNFR 1 and 2	Activation and apoptosis of diverse cell types
4-1BB	T-cell costimulation
Baff-R	B-cell maturation
CD27	T-cell costimulation
CD30	T-cell activation, apoptosis
CD40	B-cell differentiation, DC maturation
Fas	Cell death, lymphocyte homeostasis
DR4, 5	Apoptosis of transformed cells, thymocytes
EDAR	Ectodermal differentiation
XEDAR	Skeletal muscle homeostasis?
LTβR	Secondary lymphoid tissue development
OX-40	T-cell memory
RANK	Osteoclast and DC maturation
RELT	T-cell costimulation?
TIR domain-containing cytokine receptors	
IL-1R	Host defense and inflammation
IL-18R	T helper 1 (T_H1) cell development and function
Cell adhesion molecules	
α5β1 integrin	Cell adhesion and angiogenesis
α5β3 integrin	Survival of endothelial cells
α6β4 integrin	Survival of mammary epithelial cells
β2 integrins	Activation of neutrophils and monocytes
G protein-coupled receptors	
KSHV-GPCR	Transformation of vascular endothelial cells
C3a, C5a receptors	Proinflammatory in monocytes
CXCR1, 2, 6	Chemoattractant and proinflammatory
Bradykinin receptor (B2)	Proinflammatory peptide
Proteinase-activated receptors	Proinflammatory in endothelial cells and microglia
Lysophosphatidic acid receptors	Growth factor for fibroblasts and endothelial cells
Growth factor receptors	
GM-CSF-R	Proliferation and differentiation of hematopoetic cells
NGF-R (p75, TrkA)	Survival of neurons
PDGF-R	Survival, proliferation and migration of mesenchymal cells
EGF receptors	Proliferation, differentiation and survival

and adaptor molecules, are also concentrated in the lipid rafts. As a result of their aggregation within the lipid rafts, several Src family of PTKs are activated, which constitutes the initial step of intracellular signaling mediated by the antigen receptors. The primary members of the Src PTKs include Lck and Fyn in T cells, and Fyn, Lyn, and Blk in B cells. Upon activation, these kinases phosphorylate specific tyrosine residues in the immunoreceptor tyrosine-based activation motifs (ITAMs) of the cytoplasmic tails of BCR and TCR invariant chains. Phosphorylated ITAMs serve as docking sites for recruiting PTKs of the Syk family, which includes Syk in B cells and ZAP-70 in T cells. Once activated in the receptor complex, Syk and ZAP-70 amplify the receptor signals by phosphorylating various downstream targets (Fig. 1).

Adaptors Are Central Components Involved in Signal Amplification by ZAP-70 and Syk

Among the targets of ZAP70 and Syk are a number of adaptor molecules that play a central role in transducing the receptor-proximal signal to downstream signaling cascades.[8,9] One such adaptor that primarily functions in T cells is LAT (linker of activation in T cells), which is identified as a protein associated with the plasma membrane via palmitoylated cysteine residues. Upon phosphorylation by ZAP-70, LAT interacts with phospholipase C-γ (PLC-γ) and recruits this lipid enzyme to the inner face of the plasma membrane, where it is activated via phosphorylation by PTKs from the Src, Syk, and Tec families. Activated PLC-γ digests the membrane phospholipids phosphotidylinositol bisphosphate (PIP_2) into two important second messengers, diacyl glycerol (DAG) and inositol 1,4,5-triphosphate (IP_3), which mediate protein kinase C (PKC) activation and calcium mobilization, respectively (Fig. 1). In B cells, a functional homolog of LAT, termed BLNK (B-cell linker, also named SLP-65) plays a similar role. BLNK is a cytoplasmic protein, but can be recruited to the membrane and phosphorylated by Syk upon BCR stimulation. The phosphorylated BLNK interacts with PLC-γ2 and a Tec PTK, Bruton's tyrosine kinase (Btk), and the complex formation allows PLC-γ2 to be phosphorylated and activated, resulting in generation of DAG and IP3.

Another adaptor protein phosphorylated by ZAP-70 in T cells is SLP-76. Like the B-cell adaptor BLNK, SLP-76 is a cytoplasmic protein but is recruited to the membrane upon phosphorylation by ZAP-70. This adaptor interacts with a guanine nucleotide exchange factor (GEF), Vav, and serves to recruit Vav to the plasma membrane for activation. This activation process requires the cooperative action of the TCR signal and a costimulatory signal, which in naïve T cells is primarily mediated by the CD28 molecule. As a result, both SLP-76 and Vav are required for NF-κB activation by the TCR/CD28 signals.[10,11] Vav appears to exert its NF-κB-inducing function in different ways. For example, Vav regulates the recruitment of PKCθ to lipid rafts,[12] an essential step in the activation of this key signaling component of the NF-κB pathway.[13] This function of Vav also requires its canonical target, the small GTP-binding protein Rac. Additionally, Vav has been suggested to function in a more upstream signaling step involving activation of PLC-γ.[14] More recently, Vav has been shown to directly interact with a component of IKK, IKKα, which is required for Vav-induced NF-κB activation.[15]

The adaptor molecule SHC (SH2 domain-containing transforming protein) is another factor that connects the receptor-proximal signal to downstream pathways leading to NF-κB activation. Upon TCR ligation, SHC is rapidly recruited to the TCR complex, where it is phosphorylated by ZAP-70 and the Src family of PTKs.[16] Interestingly, SHC is required for activation of a specific member of the NF-κB family, c-Rel, that plays a critical role in the production of the T-cell growth factor interleukin-2 (IL-2).[17] How SHC regulates this specific axis of NF-κB signaling is not clear.

Connecting the Antigen Receptor Signals to NF-κB Activation by Carma/Bcl10/MALT1

Activation of PKC isoforms is an important consequence of the receptor-proximal signal transduction and signal amplification in B and T cells. In T cells, PKCθ plays a critical role in

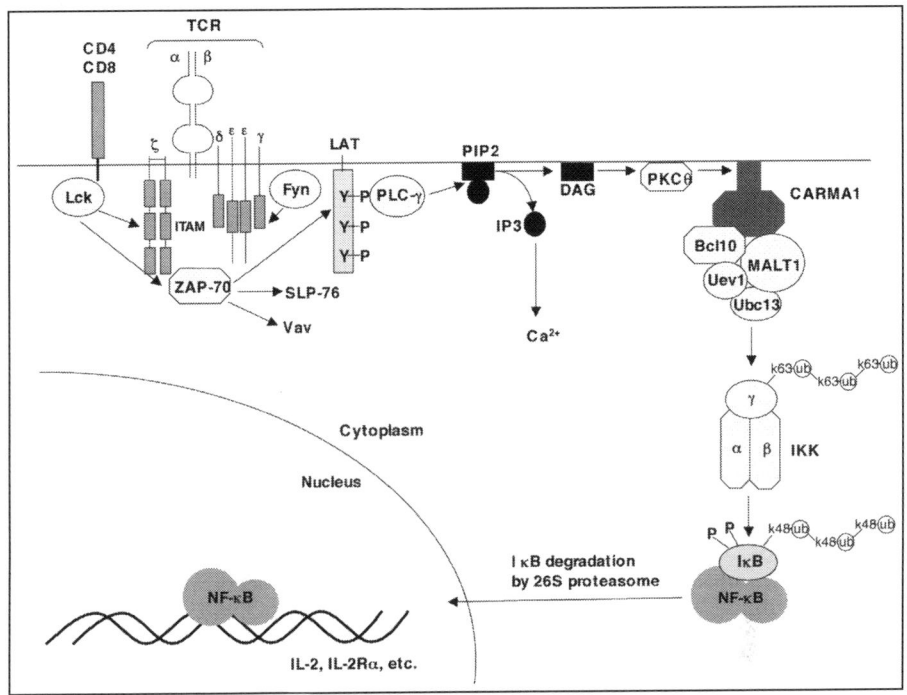

Figure 1. NF-κB signaling pathway initiated through the TCR. TCR ligation triggers the activation of Src PTKs, Lck and Fyn, which phosphorylate immunoreceptor tyrosine-based activation motifs (ITAMs) located in the cytoplasmic domains of the TCR invariant chains. The phosphorylated ITAMs serve as an anchor to recruit ZAP-70 to the TCR complex, where it is activated through phosphorylation by Lck. The activated ZAP-70 amplifies the TCR signal by phosphorylating a number of target proteins, including LAT, SLP-76, and Vav, all of which are involved in NF-κB activation. Upon membrane recruitment, LAT activates PLC-γ, which in turn cleaves the membrane lipid PIP2 to DAG and IP3, leading to activation PKC and calcium mobilization, respectively. PKCθ induces the formation of an IKK-activating signalsome that activates IKK through a mechanism that involves Lys63-linked IKKγ ubiquitination.

NF-κB activation by the TCR signal.[18] PKCθ-deficient T cells are defective in the activation of both NF-κB and AP-1, which is associated with a blockade in IL-2 production and cell proliferation in response to TCR ligation. A functional homologue of PKCθ in B cells is a conventional PKC isoform, PKCβ, which is required for NF-κB activation by the BCR signal.[19,20]

Significant progress has been made recently to understand the signaling events that link PKCs to NF-κB activation. Central to these events is the coordinated action of a protein complex composed of CARMA1, Bcl10, and MALT1[18] (Fig. 1). Genetic deficiency in any one of these molecules abolishes NF-κB activation by antigen receptor ligation.[18] CARMA1 (also known as CARD11 and Bimp3) is a member of the CARD-containing membrane-associated guanylate kinase (MAGUK) family and is predominantly expressed in lymphoid tissues. CARMA1 interacts with Bcl10 via the CARD domain present in both proteins. Additionally, CARMA1 may also interact with MALT1,[21] a member of the paracaspase family, although MALT1 is generally thought to associate with the CARMA1/Bcl10 complex through binding to Bcl10.[18] Within this trimeric complex, CARMA1 provides a functional link with the upstream signals. CARMA1 is normally distributed diffusely in the plasma membrane but is rapidly recruited into the lipid rafts upon TCR crosslinking;[22] this relocalization appears to require PKCθ, although the underlying mechanism remains unclear. Via physical interaction with CARMA1, both Bcl10 and

MALT1 are also recruited to the lipid rafts, where they undergo oligomerization and trigger the assembly of a ubiquitin-dependent IKK-activating signalsome.[23,24]

According to two recent studies,[23,24] Bcl10 and MALT1 associate with two ubiquitin-conjugating enzymes, Ubc13 and Uev1A (also known as MMS2), and catalyze Lys63-linked ubiquitination of the IKK regulatory subunit, IKKγ (also named NEMO), which in turn is required for IKK activation by the TCR signal (Fig. 1). What remains controversial is the identity of the E3 ubiquitin ligase. Although one study suggests that MALT1 may function as a ubiquitin ligase,[24] the other study challenges this idea and provides in vitro evidence for the involvement of a known ubiquitin ligase, TNF receptor-associated factor 6 (TRAF6), in the ubiquitination of IKKγ and activation of IKK downstream of TCR.[23] Notwithstanding, both studies establish IKKγ ubiquitination as a mechanism of IKK activation by the Bcl10/MALT1 complex.

Toll-Like Receptors

Toll-like receptors (TLRs) form a major family of pattern recognition receptors (PRRs) that mediate detection of microbes based on general structural features, known as pathogen-associated molecular patterns (PAMPs).[25] TLRs not only play a critical role in the early-phase host defense against infections but also regulate the nature and magnitude of the adaptive immune response.[26,27] To date, eleven TLRs have been identified in mammals, which recognize a broad range of microbial components, such as peptidoglycans, double-stranded RNA, lipopolysaccharide, and CpG DNA motifs of bacterial origin.[28] The TLRs contain an extracellular domain with leucine-rich repeats (LRR) and a cytoplasmic region sharing homology with the IL-1 receptor (IL-1R), termed Toll/IL-1R homology (TIR) domain. The LRR mediates ligand recognition, while the TIR domain is required for initiating intracellular signaling. Largely through activation of NF-κB and MAP kinases (MAPKs), TLRs induce the expression of various proinflammatory cytokines, including TNF-α, IL-1, IL-6, and IL-12. Additionally, certain TLR members, TLR3 and TLR4, also induce expression of interferon-responsive genes through activation of both NF-κB and transcription factors belonging to the interferon-regulated factor (IRF) family (Fig. 2).

Adaptors Involved in TLR-Proximal Signaling

Initiation of intracellular signaling by TLRs requires a family of TIR domain-containing adaptor proteins; these include myeloid differentiation protein 88 (MyD88), TIR domain-containing adaptor protein (TIRAP, also named MAL for MyD88-adaptor like), TRIF (also named TICAM-1), and TRAM.[28] Upon ligand binding, TLRs recruit the adaptors to their cytoplasmic tails via TIR/TIR interactions. Although most of the adaptors are differentially used by the different TLRs, MyD88 serves as a common adaptor for all TLR members. In addition to the TIR domain, MyD88 contains a death domain that mediates interaction with a family of death domain-containing serine/threonine kinases, IRAKs (interleukin-1-receptor-associated kinases). MyD88 functions to recruit IRAKs to the TLR complex for their activation, which in turn is required for initiation of downstream signaling events, including activation of IKK and its target NF-κB. Germline inactivation of the MyD88 gene abolishes NF-κB activation by all the TLRs, except for TLR3 and TLR4 that retain partial and delayed NF-κB signaling activity.[28] Moreover, induction of proinflammatory cytokine genes by all the TLRs is blocked in MyD88-deficient mice, although the ability of TLR3 and TLR4 to induce interferon-responsive genes is retained. Thus, TLR3 and TLR4 have both MyD88-dependent and independent signaling pathways, whereas the signaling function of the other TLR members is completely dependent on MyD88.

In contrast to the universal role of MyD88, TIRAP is selectively involved in the signaling function of TLR2 and TLR4 (Fig. 2). Gene targeting studies suggest that TIRAP functions in the MyD88-dependent signaling pathway but is not involved in the MyD88-independent TLR signaling.[29,30] The MyD88-independent signaling function of TLR3 and TLR4 is me-

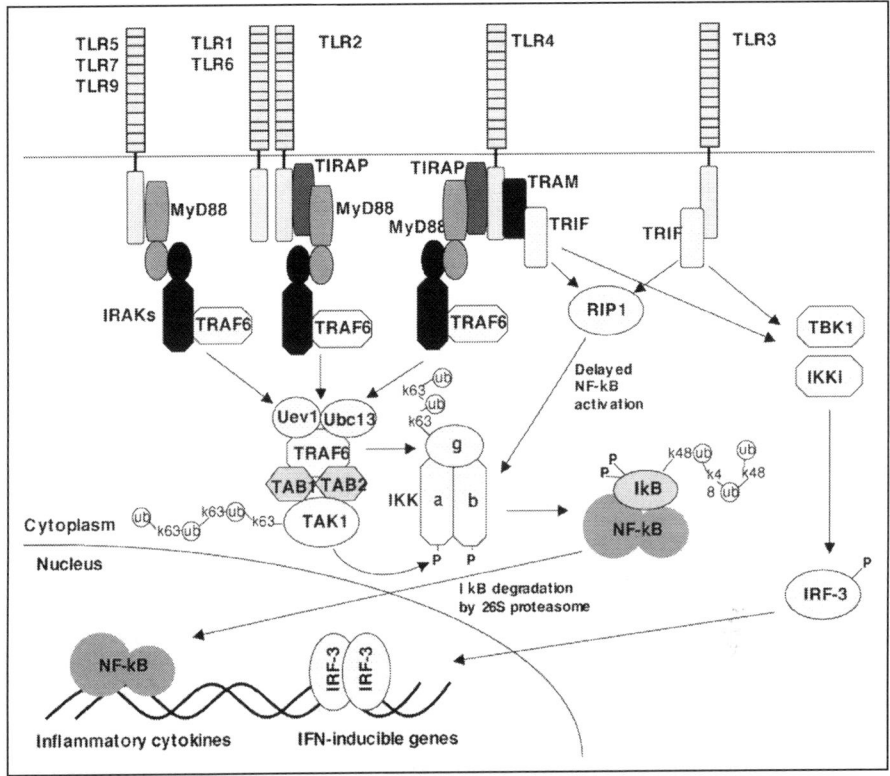

Figure 2. NF-κB signaling pathway initiated through TLRs. TLRs interact with microbial components via their extracellular leucine repeat domain and intracellular signaling adaptors via their cytoplasmic TIR domain. A common TLR adaptor is MyD88, which is involved in NF-κB activation by all the TLRs. MyD88 recruits IRAKs and TRAF6 to the receptor complex, triggering the assembly of a signalsome composed of TRAF6 and several other proteins. TRAF6 functions as a ubiquitin ligase inducing the Lys63-linked ubiquitination of both IKKγ and an IKK-activating kinase TAK1. Both IKKγ ubiquitination and TAK1-mediated IKK phosphorylation may contribute to IKK activation. The adaptor TRIF mediates delayed activation of NF-κB via the RIP1 kinase. Additionally, TRIF is required for activation of the IRF-3 by TLR3 and TLR4.

diated by the more recently identified adaptor TRIF. TRIF deficiency blocks the delayed activation of NF-κB and the induction of interferon-inducible genes by TLR3 and TLR4, although it has no effect on NF-κB activation by other TLRs or the MyD88-dependent early-phase NF-κB activation by TLR4.[31,32] TRIF directly interacts with TLR3 and binds to TLR4 via another adaptor, TRAM.[33,34] As a result, TRAM is specifically required for the TRIF-dependent and MyD88-independent signaling function of TLR4.[28]

Connecting the TLR Signals to NF-κB Activation by IRAKs and TRAF6

IRAK family of kinases plays a critical role in transducing the MyD88-dependent TLR signals. Gene targeting studies demonstrate that IRAK-4 is a key member of this family that is essential for NF-κB activation by both TLRs and IL-1R.[35] IRAK-4 appears to function as an activator of another IRAK member, IRAK-1. Upon ligand binding, both IRAK4 and IRAK1 are recruited to the receptor complex, where IRAK-1 is phosphorylated, likely by IRAK-4.[36] IRAK-1 in turn recruits the adaptor molecule TRAF6 for activation. The IRAK-1/TRAF6 complex is then released to the cytoplasm, resulting in the formation of a signalsome composed of TRAF6, an IKK-activating kinase-TGF-β activating kinase (TAK1), two adaptor proteins, TAB1 and TAB2, as well as the ubiquitin-conjugating enzymes Ubc13 and Uev1A[37-40] (Fig. 2). Within this protein complex, TRAF6 functions as an E3 ubiquitin ligase to catalyze Lys63-linked polyubiquitination, which appears to trigger the activation of TAK1 and subsequent activation of IKK.[40] Recent studies suggest that the TLR-mediated NF-κB activation in macrophages and certain subset of B cells also requires Bcl10,[41] an adaptor known to be involved in IKK activation downstream of antigen receptors. Interestingly, as discussed above, Bcl10 is a component of the TRAF6 signaling complex in T cells.[23] It is thus conceivable that Bcl10 may also participate in TRAF6-induced IKK activation in the TLR signaling pathway.

MyD88-Independent Pathway of NF-κB Activation

As noted above, both TLR3 and TLR4 induce delayed NF-κB activation in MyD88-deficient cells, and this MyD88-independent signaling pathway is mediated by the adaptor TRIF (Fig. 2). Unlike the MyD88-dependent signaling pathway, the TRIF-dependent pathway of NF-κB activation does not seem to go through IRAKs but instead involves direct interactions with downstream signaling factors. A recent study suggests that TRIF directly interacts with TRAF6 and TAK1, resulting in activation of both NF-κB and IRF-3.[42] The N-terminal region of TRIF contains three TRAF6-binding motifs that are required for the TRIF/TRAF6 interaction. Interestingly, the C-terminal region of TRIF induces NF-κB signaling via another mechanism that involves interaction with a serine/threonine kinase termed receptor-interacting protein 1 (RIP1).[43] This region of TRIF contains a RIP homotypic interaction motif (RHIM) that mediates the interaction of TRIF with RIP1. The critical role of RIP1 in TRIF-mediated NF-κB activation is supported by studies using RIP1-deficient mouse embryonic fibroblasts.[43]

IKK, a Kinase Controlling Two Signaling Pathways Downstream of TLR4

Recent studies revealed a surprising function of IKK downstream of the TLR4. In addition to its role in NF-κB activation, the canonical IKK plays an essential role in the activation of Tpl2 (also named Cot),[44] a MAPK kinase kinase (MAP3K) specifically regulating activation of ERK MAPK by the TLR4 ligand LPS.[45] This novel function of IKK is mediated through phosphorylation of the nfκb1 gene product p105,[44] which functions as both a stabilizer and inhibitor of Tpl2.[46,47] In all the cell types so far analyzed, Tpl2 and p105 exist as a stable complex,[46-48] which appears to contain at least one other protein, ABIN-2.[49] In NFκB1-deficient macrophages and other cell types, Tpl2 is rapidly degraded due to the lack of its binding protein p105, which is associated with a defect in LPS-stimulated activation of ERK.[46] Interestingly, activation of the Tpl2 complex requires phosphorylation and degradation of p105, a process that requires the action of IKKβ and IKKγ.[44] By regulating the activation of Tpl2, IKK targets another TLR-mediated signaling pathway involved in ERK MAPK activation.

TNF Receptor Superfamily

The tumor necrosis factor receptor (TNFR) superfamily consists of at least 30 members, which regulate cell proliferation, apoptosis, and/or differentiation. TNFRs contain cytoplasmic regions of varying lengths, some of which contain death domains that dictate the specificity of the desired signal response. All of the receptors appear to utilize common mechanisms to activate the transcription factors NF-κB and AP-1. Upon ligand binding, TRAF family members are recruited to the receptor and coordinate signaling cascades consisting of adaptor proteins, kinases, and other enzymatic components. Recent studies reveal an emerging theme of complexity in TNFR signaling mechanisms, including usage of multiple receptors by single ligands, cross-talk between NF-κB and AP-1 signaling networks, and ubiquitination of signaling proteins as regulatory mechanisms.

TNFR1/TNFR2

Tumor necrosis factor-α (TNF-α) is a proinflammatory cytokine and a key regulator of cell death, proliferation, inflammation, and immunity. TNF-α binds to two distinct receptors, TNFR1 and TNFR2. TNFR1 binds to soluble TNF-α whereas TNFR2 binds to membrane-bound TNF-α.[50] Genetic deletion of either receptor ablates the majority of TNF signaling, suggesting cooperativity between the two receptors.[51] A trimeric form of TNF-α binds to TNFR1 and activates NF-κB and AP-1 transcription factors, which in turn regulate the expression of genes involved in apoptosis and inflammation.[52] Inhibition of NF-κB signaling, or treatment of cells with protein synthesis inhibitors, sensitizes cells to TNF-mediated apoptosis, indicating a critical role for NF-κB in the transcription of genes involved in the protection from cell death.[50]

TNFR1 ligation elicits a signaling cascade that leads to activation of NF-κB and c-jun NH$_2$-terminal kinase (JNK). Binding of TNF-α to TNFR1 triggers the trimerization and translocation of TNFR1 to lipid rafts. TNFR1 recruits the adaptor protein TRADD that serves as a platform for the binding of other signaling proteins, including RIP1 and TRAF2, known to mediate activation of IKK and NF-κB.[53] In the absence of TRAF2, TNF-α still activates NF-κB, however the combined deletion of TRAF2 and TRAF5 abrogates TNF-mediated NF-κB activation.[54] In addition to the TNFR1, TRADD, RIP1, and TRAF complex in the plasma membrane, a distinct complex comprised of TRADD, RIP1, FADD, and caspase-8 assembles in the cytoplasm and mediates cell death.[55] However, simultaneous activation of NF-κB protects against cell death by inducing the expression of the caspase-8 inhibitor, FLIP$_L$, which assembles into the cytoplasmic death-inducing complex.[55] TNFR2 does not directly trigger cell death, since FADD is not recruited to TNFR2. However, TNFR2 may influence cell death by regulating the activity of JNK which may be pro-apoptotic.[56] Thus, the relative levels of TNFR1 and TNFR2 expression, as well as IKK and JNK activation, are important factors for TNF-mediated cell death.

There is considerable evidence for extensive cross-talk between the NF-κB and AP-1 pathways during TNF-α signaling. Murine embryonic fibroblasts (MEFs) lacking IKKβ or the NF-κB subunit RelA/p65 exhibit prolonged TNF-α-mediated JNK activation and increased apoptosis.[57,58] It has been proposed that NF-κB-induced genes such as GADD45β or X-linked-inhibitor of apoptosis (XIAP) may inhibit cell death by suppressing JNK activation.[57,58] GADD45β directly binds and inhibits the catalytic activity of MKK7, a MAP kinase kinase that is upstream of JNK, as a mechanism to inhibit TNF-α-mediated JNK activation.[59]

IKK is recruited to the TNFR via a direct interaction of the IKK regulatory subunit IKKγ with RIP1.[60] In addition, the chaperone proteins, Cdc37 and Hsp90, interact with IKK and mediate their shuttling from the cytoplasm to the membrane.[61] TNF-α is a potent activator of IKKβ, a central component of the canonical NF-κB pathway. IKKβ phosphorylates IκBs leading to their ubiquitination and degradation by the proteasome, thus allowing the liberation and nuclear translocation of NF-κB heterodimers (see Fig. 3). In contrast, TNF-α does not stimulate p100 processing to p52 in the noncanonical NF-κB pathway that is dependent on the MAP3K, NF-κB inducing kinase (NIK), and IKKα.[62,63]

The mechanism of IKK activation within the TNFR complex in lipid rafts has remained elusive. Since ubiquitination of signaling proteins by the assembly of Lys63 (K63)-linked ubiquitin chains is critical for the activation of IKK by T cell receptor stimulation of T cells,[23] it is possible that a similar mechanism exists for TNFR1 signaling. Indeed, a recent study suggests that A20, a negative regulator of TNF-mediated NF-κB activation, inactivates RIP1 by a unique mechanism involving deubiquitination of K63-linked ubiquitin chains as well as the coordinated catalysis of K48-linked ubiquitin chains, leading to RIP1 inactivation and subsequent degradation by the proteasome.[64] Moreover, IKKγ is also a target of ubiquitination by TNF-α stimulation.[65] Thus, it is tempting to speculate that ubiquitination of signaling proteins plays an important regulatory role for the activation of IKK by TNFR superfamily members.

The role of MAP3Ks in activation of NF-κB by TNF-α has been extensively studied. Although many MAP3Ks, such as NIK or MEKK1, activate NF-κB when overexpressed, genetic studies have revealed that only MEKK3 and TAK1 are critical for TNF-mediated NF-κB activation. MEKK3 interacts with RIP, and links RIP to activation of IKK by directly phosphorylat-

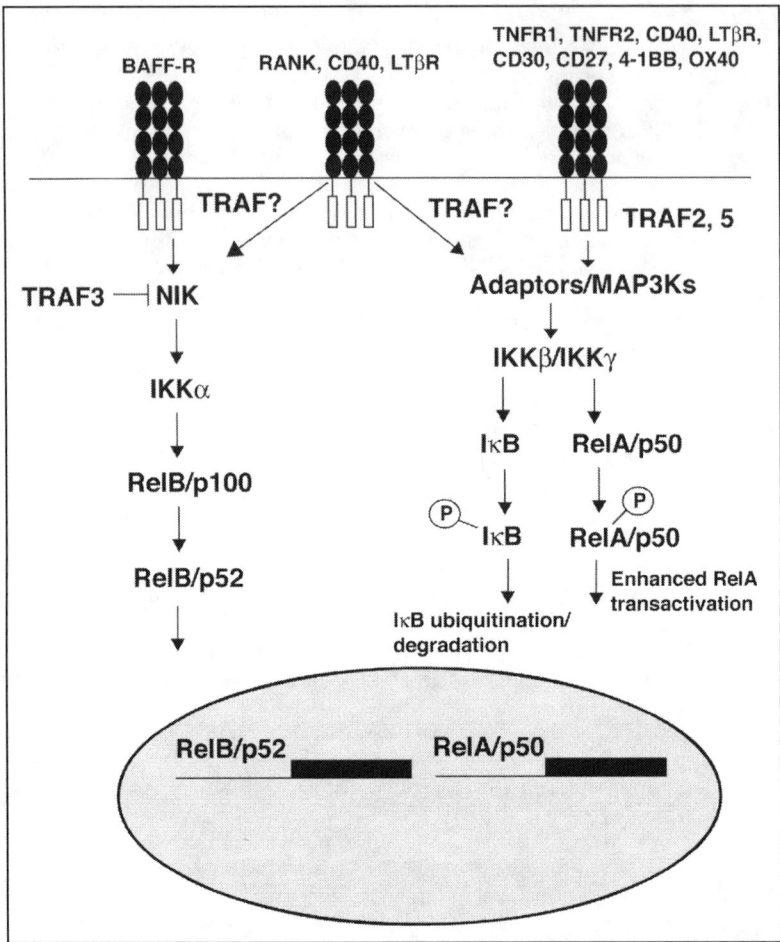

Figure 3. NF-κB signaling pathways activated by TNFR family members. The majority of TNFR proteins activate the canonical NF-κB pathway that consists of IκB degradation and nuclear translocation of NF-κB heterodimers. TNF-α, and possibly other TNF family members, mediates the phosphorylation of RelA that potentiates RelA-mediated transactivation. A single TNFR family member, BAFF-R, activates only the noncanonical NF-κB pathway that results in processing of p100 to p52. A subset of TNFR proteins, RANK, CD40 and LTβR, activate both the canonical and noncanonical pathways.

ing IKKβ.[66] Knockdown of TAK1 expression by siRNA transfection reduced TNF-α-mediated NF-κB activation, suggesting that TAK1 may be a critical component of TNF signaling.[67]

TNF signaling also enhances NF-κB activation by mediating site-specific phosphorylation of multiple serine residues within RelA, all of which appear to enhance NF-κB transactivation. For example, casein kinase II mediates the phosphorylation of RelA on serine 529 in response to TNF.[68] In addition, serine 536 of RelA is phosphorylated by IKK during TNF-α stimulation.[69] TNF-α stimulation also induces RelA phosphorylation on serine 311 that is dependent on the atypical protein kinase C family member, PKCζ.[70] Finally, RelA is phosphorylated on serine 276 by the mitogen- and stress-activated protein kinase-1 (MSK1) within the nucleus in response to TNF-α stimulation.[71] It is unclear which of the identified phosphoacceptor sites within RelA are critical for TNF-α-mediated NF-κB transactivation. Reconstitution of RelA[-/-] MEFs with RelA point mutants revealed that only mutation of serine 276 abrogated TNF-α-mediated NF-κB activation.[72] However, different approaches, such as the generation of RelA mutant knock-in mice, may be required to appreciate the role of specific RelA serines in response to TNF-α signaling.

CD40

CD40 is a member of the TNFR superfamily and is expressed in a wide variety of cell types such as B cells, dendritic cells (DCs), macrophages, endothelial and epithelial cells, and neurons.[73] The ligand for CD40, CD154 (also known as CD40L), is mainly expressed in activated T cells, but it may also be expressed in monocytes. Binding of CD40L to CD40 regulates cell survival, proliferation, germinal center formation, and immunoglobulin class switching in B cells as well as antigen presentation, maturation, and survival of dendritic cells.[73] Binding of CD40 to CD40L triggers clustering of CD40 within lipid rafts where signaling complexes are assembled. The intracellular domain of CD40 contains two binding sites for TRAF proteins and has been shown to interact with TRAF 1, 2, 3, 5, and 6.[74] The TRAFs then couple CD40 signaling to the activation of NF-κB, phosphoinositide 3-kinase (PI3K), and Map kinases ERK, p38, and JNK.[74]

CD40 activates the canonical NF-κB pathway in B cells and dendritic cells.[75] CD40 ligation of B cells also triggers the noncanonical NF-κB pathway resulting in NIK/IKKα-dependent p100 processing and p52/RelB nuclear translocation.[76] The contributions of the canonical and noncanonical NF-κB pathways to the CD40 gene program have been recently evaluated. CD40 signaling through the canonical NF-κB pathway regulates the proliferation of B cells whereas both pathways are required for B cell aggregation.[77] Furthermore, the gene program elicited by CD40 in B cells requires specific contributions from both the canonical and noncanonical NF-κB pathways.[77]

The signaling intermediates downstream of CD40 that mediate IKK activation have not been well characterized. Act1 (also known as CIKS) was identified as an IKKγ binding protein that is recruited to CD40 in response to stimulation with CD40L.[78] In addition, Act1 interacts with TRAF3 and TRAF6, and may play a role in CD40-mediated NF-κB activation.[79] It has been difficult to ascertain the relative contributions of individual TRAF proteins in CD40-mediated NF-κB activation due to issues of redundancy. A recent study using a B cell line lacking TRAF2 revealed that both TRAF2 and TRAF6 are required for CD40-mediated NF-κB activation.[80] Conversely, TRAF3 may play a negative regulatory role in NF-κB activation. Specifically, TRAF3 may be a negative regulator of CD40-mediated p100 processing by inducing the degradation of NIK.[81] Future studies will require the use of cells deficient in multiple TRAF family members to circumvent issues with redundancy.

Lymphotoxin Beta Receptor

The lymphotoxin beta receptor (LT-βR) is critical for the development and organization of lymphoid tissue. LTβ is expressed in stromal tissue and binds to LTα1β2 and the closely related ligand LIGHT that is expressed in lymphocytes.[82] The LTβR utilizes TRAF proteins to propagate signals leading to the activation of NF-κB and AP-1. The cytoplasmic domain of LTβR binds to TRAF2, 3, and 5, however only TRAF2 and TRAF5 are required for LTβR-mediated activation of NF-κB.[83] LTβR signaling activates both the canonical and noncanonical NF-κB pathways. LTβR mediates the processing of p100, resulting in nuclear translocation of p52/RelB dimmers.[84] A distinct panel of genes can be attributed to the canonical and noncanonical NF-κB pathways induced by LTβR-mediated NF-κB activation. Whereas the expression of the adhesion molecule VCAM-1 was dependent on the canonical pathway, the expression of chemokine and cytokine genes such as SLC, ELC, BLC, SDF-1-α, and BAFF were regulated by the noncanonical pathway.[84]

BAFF Receptor

B cell activating factor (BAFF) plays an important role in the development and survival of peripheral B cells.[85] BAFF binds to three distinct receptors: BAFF-R (also known as BR-3), B cell maturation antigen (BCMA), and transmembrane activator and CAML interactor (TACI).

However, only BAFF-R is specific for BAFF since BCMA and TACI also interact with the related ligand APRIL. BAFF-R is a potent activator of the noncanonical NF-κB pathway that involves NIK-dependent p100 processing in B cells.[86,87] BAFF-mediated NF-κB activation appears to selectively activate the noncanonical pathway since IκBα degradation was not observed after BAFF stimulation of B cells.[86] BAFF activation of p100 processing contributes to the survival of splenic transitional B cells, likely through the induction of anti-apoptotic factors bcl-2 and bcl-x.[86] BAFF-mediated p100 processing requires de novo protein synthesis, which is also common to CD40- and LTβ-mediated p100 processing. It is not clear if this reflects the synthesis of an unknown protein required for p100 processing or continued synthesis of NIK and/or p100. Since induction of p100 processing is associated with NIK accumulation,[81] it is likely that the de novo synthesis of NIK is an important step in noncanonical NF-κB signaling. The intracellular signaling proteins downstream of BAFF-R that lead to NIK activation are unknown, although it has been reported that TRAF3 binds to BAFF-R and negatively regulates NF-κB activation.[88]

RANK

RANK and its ligand, RANKL, are critical regulators of bone remodeling, differentiation of osteoclasts, dendritic cell survival, and lymph node formation.[89] RANK is expressed in osteoclast precursors, myeloid precursors, activated B cells, and dendritic cells. RANK signaling promotes cell survival and differentiation. The RANK protein has a long cytoplasmic region (383 amino acids) which binds to TRAF 1, 2, 3, 5 and 6 and mediates activation of NF-κB, JNK, and AKT.[89] RANK signaling activates the canonical NF-κB pathway in osteoclasts and dendritic cells.[90] RANKL also triggers NIK-dependent p100 processing and promotes osteoclastogenesis.[91] In addition to its roles in osteoclastogenesis and immune regulation, RANK is also important for mammary gland development. RANK signaling requires IKKα for regulation of the expression of cyclin D1, a gene that is critical for mammary epithelial cell proliferation.[92] However, it is unclear if both the canonical and noncanonical NF-κB pathways are required for RANK-mediated mammary epithelial cell proliferation.

OX40, CD27, CD30, and 4-1BB

A host of other TNFR superfamily members such as OX40, CD27, CD30 and 4-1BB activate NF-κB signaling, however little is known regarding the mechanisms of NF-κB activation by these receptors.[51] Each of the receptors has been reported to activate the canonical NF-κB signaling pathway via binding to specific TRAFs. OX40 has been shown to bind to TRAF2, 3, and 5 with TRAF2 and 5 playing positive regulatory roles.[93] CD27 binds to TRAF2 and 3, and while TRAF2 appears to be an activator, TRAF3 may inhibit NF-κB activation.[94] CD30 interacts with TRAF2 and 5.[95] 4-1BB binds to TRAF1, 2, and 3, although only a dominant negative TRAF2 inhibits 4-1BB-mediated NF-κB activation.[96] It must be noted that many of the conclusions derived from these studies relied on overexpression of dominant negative TRAF mutants. Thus, it is important to confirm these findings with cells lacking individual TRAF proteins.

Other Receptors Mediating NF-κB Activation

In addition to the three families of immune receptors discussed above, a large number of other receptors are known to induce NF-κB activation (Table 1). IL-1R and IL-18R contain the TIR domain and activate NF-κB using the MyD88-dependent pathway of the TLRs.[97] Both IL-1R and IL-18R play an important role in mediating inflammatory signals.[98] G-protein-coupled receptors (GPCRs) form a large family of cell surface receptors that signal through the heterotrimeric G proteins.[99] The GPCRs are involved in diverse biological functions, such as leukocyte chemotaxis and activation, platelet aggregation, and hormone action. One of the major downstream signaling events of the GPCRs is activation of NF-κB. Although the pathways leading to NF-κB activation by the different GPCRs are not well defined and appear to be divergent, it seems clear that they converge to the activation of IKK.[99] Another type of NF-κB-inducing receptors are the integrins, a family of cell adhesion molecules involved in cell-cell interaction, adhesion, angiogenesis, cell trafficking and survival.[100] A number of integrins, such as the β2, α5β1, α5β3 and α6β4 integrins, also induce NF-κB signaling in multiple cell types, although the pathways have not been well defined.[101] In addition, growth factor receptors, including those for epidermal growth factor (EGF) and nerve growth factor (NGF), mediate activation of NF-κB in numerous cell types. Many growth factors act as potent cellular mitogens and have diverse roles in normal biological processes and pathological states such as cancer.[102] Little is known regarding the mechanisms of NF-κB activation by growth factors, although a recent report indicates that EGF stimulation requires IKKα and RelA for optimal induction of the *c-fos* gene.[103] In addition to cell surface receptors, certain NF-κB-inducing receptors are located within cells. One example is NOD2, a CARD domain-containing protein that specifically recognizes peptidoglycans, and acts as a cytoplasmic sensor for microorganisms.[104]

Acknowledgement

The research projects in the authors' laboratories are supported by grants from the National Institutes of Health: 1 R01 CA94922 and 1 R01 AI057555 to S.-C. S. and 1 R01 CA99926 to E. W. H.

References

1. Li Q, Verma IM. NF-kappaB regulation in the immune system. Nat Rev Immunol 2002; 2:725-734.
2. Kucharczak J, SM J, Fan Y et al. To be, or not to be: NF-kappaB is the answer—role of Rel/NF-kappaB in the regulation of apoptosis. Oncogene 2003; 22:8961-8982.
3. Sun S-C, Xiao G. Deregulation of NF-kB and its upstream kinases in cancer. Cancer and Metastasis Reviews 2003; 22:405-422.
4. Greten FR, Karin M. The IKK/NF-kappaB activation pathway-a target for prevention and treatment of cancer. Cancer Lett 2003; 206:193-199.
5. Janeway CA, Travers P, Walport M et al. Immunobiology 5 ed. New York: Garland, 2001.
6. Ruland J, Mak TW. Transducing signals from antigen receptors to nuclear factor kappaB. Rev Immunol 2003; 193:93-100.
7. Jun JE, Goodnow CC. Scaffolding of antigen receptors for immunogenic versus tolerogenic signaling. Nat Immunol 2003; 4:1057-1064.
8. Tomlinson MG, Lin J, Weiss A. Lymphocytes with a complex: Adapter proteins in antigen receptor signaling. Today Immunol 2000; 21:584-591.
9. Wilkinson B, Wang H, Rudd CE. Positive and negative adaptors in T-cell signalling. Immunology 2004; 111:368-374.
10. Herndon TM, Shan XC, Tsokos GC et al. ZAP-70 and SLP-76 regulate protein kinase C-theta and NF-kappa B activation in response to engagement of CD3 and CD28. J Immunol 2001; 166:654-664.
11. Costello PS, Walters AE, Mee PJ et al. The Rho-family GTP exchange factor Vav is a critical transducer of T cell receptor signals to the calcium, ERK, and NF-kappaB pathways. Proc Natl Acad Sci USA 1999; 96:3035-3040.
12. Villalba M, Coudronniere N, Deckert M et al. A novel functional interaction between Vav and PKCtheta is required for TCR-induced T cell activation. Immunity 2000; 12:151-160.
13. Bi K, Tanaka Y, Coudronniere N et al. Antigen-induced translocation of PKC-theta to membrane rafts is required for T cell activation. Nat Immunol 2001; 2:556-563.
14. Reynolds LF, Smyth LA, Norton T et al. Vav1 transduces T cell receptor signals to the activation of phospholipase C-gamma1 via phosphoinositide 3-kinase-dependent and -independent pathways. J Exp Med 2002; 195(9):1103-1114.
15. Piccolella E, Spadaro F, Ramoni C et al. Vav-1 and the IKK alpha subunit of I kappa B kinase functionally associate to induce NF-kappa B activation in response to CD28 engagement. J Immunol 2003; 170(6):2895-2903.
16. Simeoni L, Kliche S, Lindquist J et al. Adaptors and linkers in T and B cells. Curr Opin Immunol 2004; 16(3):304-313.
17. Iwashima M, Takamatsu M, Yamagishi H et al. Genetic evidence for Shc requirement in TCR-induced c-Rel nuclear translocation and IL-2 expression. Proc Natl Acad Sci USA 2002; 99(7):4544-4549.

18. Ruland J, Duncan GS, Wakeham A et al. Differential requirement for Malt1 in T and B cell antigen receptor signaling. Immunity 2003; 19(5):749-758.
19. Su TT, Guo B, Kawakami Y et al. PKC-beta controls I kappa B kinase lipid raft recruitment and activation in response to BCR signaling. Nat Immunol 2002; 3:780-786.
20. Saijo K, Mecklenbrauker I, Santana A et al. Protein kinase C beta controls nuclear factor kappaB activation in B cells through selective regulation of the IkappaB kinase alpha. J Exp Med 2002; 195:1647-1652.
21. Che T, You Y, Wang D et al. MALT1/paracaspase is a signaling component downstream of CARMA1 and mediates T cell receptor-induced NF-kappaB activation. J Biol Chem 2004; 279(16):15870-15876.
22. Gaide O, Favier B, Legler DF et al. CARMA1 is a critical lipid raft-associated regulator of TCR-induced NF-kappa B activation. Nat Immunol 2002; 3:836-843.
23. Sun L, Deng L, Ea C-K et al. The TRAF6 ubiquitin ligase and TAK1 kinase mediate IKK activation by BCL10 and MALT1 in T lymphocytes. Mol Cell 2004; 14:289-301.
24. Zhou H, Wertz I, O'Rourke K et al. Bcl10 activates the NF-kappaB pathway through ubiquitination of NEMO. Nature 2004; 427:167-171.
25. Takeda K, Kaisho T, Akira S. Toll-like receptors. Annu Rev Immunol 2003; 21:335-376.
26. Medzhitov R Toll-like receptors and innate immunity. Nat Rev Immunol 2001; 1:135-145.
27. Schnare M, Barton GM, Holt AC et al. Toll-like receptors control activation of adaptive immune responses. Nat Immunol 2001; 2:947-950.
28. Akira S, Takeda K. Toll-like receptor signalling. Nat Rev Immunol 2004; 4:499-511.
29. Fitzgerald KA, Palsson-McDermott EM, Bowie AG et al. Mal (MyD88-adapter-like) is required for Toll-like receptor-4 signal transduction. Nature 2001; 413:78-83.
30. Yamamoto M, Sato S, Hemmi H et al. Essential role for TIRAP in activation of the signaling cascade shared by TLR2 and TLR4. Nature 2002; 420:324-329.
31. Hoebe K, Du X, Georgel P et al. Identification of Lps2 as a key transducer of MyD88-independent TIR signalling. Nature 2003; 424(6950):743-748.
32. Yamamoto M, Sato S, Hemmi H et al. Role of adaptor TRIF in the MyD88-independent toll-like receptor signaling pathway. Science 2003; 301(5633):640-643.
33. Oshiumi H, Sasai M, Shida K et al. TIR-containing adapter molecule (TICAM)-2, a bridging adapter recruiting to toll-like receptor 4 TICAM-1 that induces interferon-beta. J Biol Chem 2003; 278(50):49751-49762.
34. Yamamoto M, Sato S, Hemmi H et al. TRAM is specifically involved in the Toll-like receptor 4-mediated MyD88-independent signaling pathway. Nat Immunol 2003; 4(11):1144-1150.
35. Suzuki N, Suzuki S, Duncan GS et al. Severe impairment of interleukin-1 and Toll-like receptor signaling in mice lacking IRAK-4. Nature 2002; 416:750-756.
36. Suzuki N, Suzuki S, Yeh W-C. IRAK-4 as the central TIR signaling mediator in innate immunity. TRENDS Immunol 2002; 23:503-506.
37. Takaesu G, Kishida S, Hiyama A et al. TAB2, a novel adaptor protein, mediates activation of TAK1 MAPKKK by linking TAK1 to TRAF6 in the IL-1 signal transduction pathway. Mol Cell 2000; 5(4):649-658.
38. Ninomiya-Tsuji J, Kishimoto K, Hiyama A et al. The kinase TAK1 can activate the NIK-I kappaB as well as the MAP kinase cascade in the IL-1 signalling pathway. Nature 1999; 398:252-256.
39. Deng L, Wang C, Spencer E et al. Activation of the IkB kinase complex by TRAF6 requires a dimeric ubiqutitin-conjugating enzyme complex and a unique polyubiquitin chain. Cell 2000; 103:351-361.
40. Wang C, Deng L, Hong M et al. TAK1 is a ubiquitin-dependent kinase of MKK and IKK. Nature 2001; 412:346-351.
41. Fischer K-D, Tedford K, Wirth T. New roles for Bcl10 in B-cell development and LPS response. TRENDS Immunol 2004; 25:113-116.
42. Sato S, Sugiyama M, Yamamoto M et al. Toll/IL-1 receptor domain-containing adaptor inducing IFN-beta (TRIF) associates with TNF receptor-associated factor 6 and TANK-binding kinase 1, and activates two distinct transcription factors, NF-kappa B and IFN-regulatory factor-3, in the Toll-like receptor signaling. J Immunol 2003; 171(8):4304-4310.
43. Meylan E, Burns K, Hofmann K et al. RIP1 is an essential mediator of Toll-like receptor 3-induced NF-kappa B activation. Nat Immunol 2004; 5(5):503-507.
44. Waterfield M, Jin W, Reiley W et al. IkappaB is an essential component of the Tpl2 signaling pathway. Mol Cell Biol 2004; 24(13):6040-6048.
45. Dumitru CD, Ceci JD, Tsatsanis C et al. TNF-alpha induction by LPS is regulated posttranscriptionally via a Tpl2/ERK-dependent pathway. Cell 2000; 103:1071-1083.

46. Waterfield MR, Zhang M, Norman LP et al. NF-kappaB1/p105 regulates lipopolysaccharide-stimulated MAP kinase signaling by governing the stability and function of the Tpl2 kinase. Mol Cell 2003; 11:685-694.
47. Beinke S, Deka J, Lang V et al. NF-kappaB1 p105 negatively regulates TPL-2 MEK kinase activity. Mol Cell Biol 2003; 23(14):4739-4752.
48. Belich MP, Salmeron A, Johnston LH et al. TPL-2 kinase regulates the proteolysis of the NF-kappaB-inhibitory protein NF-kappaB1 p105. Nature 1999; 397:363-368.
49. Lang V, Symons A, Watton SJ et al. ABIN-2 forms a ternary complex with TPL-2 and NF-kappa B1 p105 and is essential for TPL-2 protein stability. Mol Cell Biol 2004; 24(12):5235-5248.
50. Baud V, Karin M. Signal transduction by tumor necrosis factor and its relatives. Trends Cell Biol 2001; 11(9):372-377.
51. Aggarwal BB. Signalling pathways of the TNF superfamily: A double-edged sword. Nat Rev Immunol 2003; 3(9):745-756.
52. Chen G, Goeddel DV. TNF-R1 signaling: A beautiful pathway. Science.2002; 296(5573):1634-1635.
53. Legler DF, Micheau O, Doucey MA et al. Recruitment of TNF receptor 1 to lipid rafts is essential for TNFalpha-mediated NF-kappaB activation. Immunity 2003; 18:655-664.
54. Tada K, Okazaki T, Sakon S et al. Critical roles of TRAF2 and TRAF5 in tumor necrosis factor-induced NF-kappa B activation and protection from cell death. J Biol Chem 2001; 276(39):36530-36534.
55. Micheau O, Tschopp J. Induction of TNF receptor I-mediated apoptosis via two sequential signaling complexes. Cell 2003; 114(2):181-190.
56. Maeda S, Chang L, Li ZW et al. IKKbeta is required for prevention of apoptosis mediated by cell-bound but not by circulating TNFalpha. Immunigy 2003; 19:725-737.
57. De Smaele E, Zazzeroni F, Papa S et al. Induction of gadd45β by NF-kB downregulates pro-apoptotic JNK signaling. Nature 2001; 414:308-313.
58. Tang G, Minemoto Y, Dibling B et al. Inhibition of JNK activation through NF-kappaB target genes. Nature 2001; 414(6861):313-317.
59. Papa S, Zazzeroni F, Bubici C et al. Gadd45 beta mediates the NF-kappa B suppression of JNK signalling by targeting MKK7/JNKK2. Nat Cell Biol 2004; 6:146-153.
60. Zhang SQ, Kovalenko A, Cantarella G et al. Recruitment of the IKK signalosome to the p55 TNF receptor: RIP and A20 bind to NEMO (IKKgamma) upon receptor stimulation. Immunity 2000; 12(3):301-311.
61. Chen G, Cao P, Goeddel DV. TNF-induced recruitment and activation of the IKK complex require Cdc37 and Hsp90. Mol Cell 2002; 9(2):401-410.
62. Senftleben U, Cao Y, Xiao G et al. Activation of IKKα of a second, evolutionary conserved, NF-κB signaling pathway. Science 2001; 293:1495-1499.
63. Xiao G, Harhaj EW, Sun SC. NF-kappaB-inducing kinase regulates the processing of NF-kappaB2 p100. Mol Cell 2001; 7:401-409.
64. Wertz IE, O'Rourke KM, Zhou H et al. De-ubiquitination and ubiquitin ligase domains of A20 downregulate NF-kappaB signalling. Nature 2004.
65. Tang ED, Wang CY, Xiong Y et al. A role for NF-kappaB essential modifier/IkappaB kinase-gamma (NEMO/IKKgamma) ubiquitination in the activation of the IkappaB kinase complex by tumor necrosis factor-alpha. J Biol Chem 2003; 278:37297-37305.
66. Yang J, Lin Y, Guo Z et al. The essential role of MEKK3 in TNF-induced NF-kappaB activation. Nat Immunol 2001; 2:620-624.
67. Takaesu G, Surabhi RM, Park KJ et al. TAK1 is critical for IkappaB kinase-mediated activation of the NF-kappaB pathway. J Mol Biol 2003; 326(1):105-115.
68. Wang D, Westerheide SD, Hanson JL et al. Tumor necrosis factor α-induced phosphorylation of RelA/p65 on Ser529 is controlled by casein kinase II. J Biol Chem 2000; 275:32592-32597.
69. Sakurai H, Chiba H, Miyoshi H et al. IκB kinases phosphorylate NF-κB p65 subunit on serine 536 in the transactivation domain. J Biol Chem 1999; 274:30353-30356.
70. Duran A, Diaz-Meco MT, Moscat J. Essential role of RelA Ser311 phosphorylation by zetaPKC in NF-kappaB transcriptional activation. EMBO J 2003; 22(15):3910-3918.
71. Vermeulen L, De Wilde G, Van Damme P et al. Transcriptional activation of the NF-kappaB p65 subunit by mitogen- and stress-activated protein kinase-1 (MSK1). EMBO J 2003; 22:1313-1324.
72. Okazaki T, Sakon S, Sasazuki T et al. Phosphorylation of serine 276 is essential for p65 NF-kappaB subunit-dependent cellular responses. Biochem Biophys Res Commun 2003; 300(4):807-812.
73. Bishop GA, Hostager BS. The CD40-CD154 interaction in B cell-T cell liaisons. Cytokine Growth Factor Rev 2003; 14:297-309.
74. Harnett MM. CD40: A growing cytoplasmic tale. Sci STKE. 2004; 2004(237):25.
75. Kosaka Y, Calderhead DM, Manning EM et al. Activation and regulation of the IkappaB kinase in human B cells by CD40 signaling. Eur J Immunol 1999; 29:1353-1362.

76. Coope HJ, Atkinson PG, Huhse B et al. CD40 regulates the processing of NF-kappaB2 p100 to p52. EMBO J 2002; 15:5375-5385.
77. Zarnegar B, He JQ, Oganesyan G et al. Unique CD40-mediated biological program in B cell activation requires both type 1 and type 2 NF-kappaB activation pathways. Proc Natl Acad Sci USA 2004; 101(21):8108-8113.
78. Qian Y, Zhao Z, Jiang Z et al. Role of NF kappa B activator Act1 in CD40-mediated signaling in epithelial cells. Proc Natl Acad Sci USA 2002; 99(14):9386-9391.
79. Kanamori M, Kai C, Hayashizaki Y et al. NF-kappaB activator Act1 associates with IL-1/Toll pathway adaptor molecule TRAF6. FEBS Lett 2002; 532:241-246.
80. Hostager BS, Haxhinasto SA, Rowland SL et al. Tumor necrosis factor receptor-associated factor 2 (TRAF2)-deficient B lymphocytes reveal novel roles for TRAF2 in CD40 signaling. J Biol Chem 2003; 278:45382-45390.
81. Liao G, Zhang M, Harhaj EW et al. Regulation of the NF-kappaB-inducing kinase by tumor necrosis factor receptor-associated factor 3-induced degradation. J Biol Chem 2004; 279(25):26243-26250.
82. Tumanov AV, Grivennikov SI, Shakhov AN et al. Dissecting the role of lymphotoxin in lymphoid organs by conditional targeting. Immunol Rev 2003; 195:106-116.
83. Saitoh T, Nakano H, Yamamoto N et al. Lymphotoxin-beta receptor mediates NEMO-independent NF-kappaB activation. FEBS Lett 2002; 532(1-2):45-51.
84. Dejardin E, Droin NM, Delhase M et al. The Lymphotoxin-beta receptor induces different patterns of gene expression via two NF-kappaB pathways. Immunity 2002; 17:525-535.
85. Batten M, Groom J, Cachero TG et al. BAFF mediates survival of peripheral immature B lymphocytes. J Exp Med 2000; 192(10):1453-1466.
86. Claudio E, Brown K, Park S et al. BAFF-induced NEMO-independent processing of NF-kappaB2 in maturing B cells. Nat Immunol 2002; 3:958-965.
87. Kayagaki N, Yan M, Seshasayee D et al. BAFF/BLyS receptor 3 binds the B Cell survival factor BAFF ligand through a discrete surface loop and promotes processing of NF-kappaB2. Immunity 2002; 17:515-524.
88. Xu L-G, Shu H-B. TNFR-associated factor 3 is associated with BAFF-R and negatively regulates BAFF-R-mediated NF-κB activation and IL-10 production. J Immunol 2002; 169:6883-6889.
89. Theill LE, Boyle WJ, Penninger JM. RANK-L and RANK: T cells, bone loss, and mammalian evolution. Annu Rev Immunol 2002; 20:795-823.
90. Wong BR, Josien R, Choi Y. TRANCE is a TNF family member that regulates dendritic cell and osteoclast function. J Leukoc Biol 1999; 65(6):715-724.
91. Novack DV, Yin L, Hagen-Stapleton A et al. The IkappaB function of NF-kappaB2 p100 controls stimulated osteoclastogenesis. J Exp Med 2003; 198(5):771-781.
92. Cao Y, Bonizzi G, Seagroves TN et al. IKKalpha provides an essential link between RANK signaling and cyclin D1 expression during mammary gland development. Cell 2001; 107(6):763-775.
93. Kawamata S, Hori T, Imura A et al. Activation of OX40 signal transduction pathways leads to tumor necrosis factor receptor-associated factor (TRAF) 2- and TRAF5-mediated NF-kappaB activation. J Biol Chem 1998; 273(10):5808-5814.
94. Yamamoto H, Kishimoto T, Minamoto S. NF-kappaB activation in CD27 signaling: Involvement of TNF receptor-associated factors in its signaling and identification of functional region of CD27. J Immunol 1998; 161(9):4753-4759.
95. Horie R, Watanabe T, Ito K et al. Cytoplasmic aggregation of TRAF2 and TRAF5 proteins in the Hodgkin-Reed-Sternberg cells. Am J Pathol 2002; 160:1647-1654.
96. Jang IK, Lee ZH, Kim YJ et al. Human 4-1BB (CD137) signals are mediated by TRAF2 and activate nuclear factor-kappa B. Biochem Biophys Res Commun 1998; 242(3):613-620.
97. O'Neill LA. The interleukin-1 receptor/Toll-like receptor superfamily: Signal transduction during inflammation and host defense. Science's stke 2000; 44:1-11.
98. Dunne A, O'Neill LA. The interleukin-1 receptor/Toll-like receptor superfamily: Signal transduction during inflammation and host defense. Sci STKE 2003; 2003(171):re3.
99. Ye R. Regulation of nuclear factor κB by G-protein-coupled receptors. J Leukocyte Biol 2001; 70:839-848.
100. Pribila JT, Quale AC, Mueller KL et al. Integrins and T cell-mediated immunity. Annu Rev Immunol 2004; 22:157-180.
101. Reyes-Reyes M, Mora N, Zentella A et al. Phosphatidylinositol 3-kinase mediates integrin-dependent NF-kappaB and MAPK activation through separate signaling pathways. J Cell Sci 2001; 114(Pt 8):1579-1589.
102. Jorissen RN, Walker F, Pouliot N et al. Epidermal growth factor receptor: Mechanisms of activation and signalling. Exp Cell Res 2003; 284(1):31-53.
103. Anest V, Cogswell PC, Baldwin Jr AS. IκB Kinase α and p65/RelA contribute to optimal epidermal growth factor-induced c-fos Gene expression independent of IκBα degradation. J Biol Chem 2004; 279(30):31183-31189.

CHAPTER 4

Cellular Dynamics of NF-κB Associated Proteins

Daliya Banerjee and Ranjan Sen*

Introduction

NF-κB DNA binding activity, and thus NF-κB-dependent gene expression, is regulated by the IκB family of inhibitory proteins. In unstimulated cells NF-κB proteins are complexed to IκB proteins and do not bind DNA. The observation that NF-κB/IκB complexes are located in the cytoplasm led to the model that the IκB proteins serve as cytoplasmic "tethers" that keep NF-κB out of the nucleus.[1,2] This was proposed to occur by masking the nuclear localization sequence (NLS) of the p65/RelA component of NF-κB.[3] Cell stimulation leads to IκB degradation and release of DNA binding NF-κB that activates gene transcription. However, even the earliest studies with IκB proteins indicated a level of complexity beyond a simple sequestration model. For example, IκBα was shown to efficiently disrupt NF-κB/DNA complexes.[4] Because such complexes only formed in the nucleus, this in vitro property was not easily reconciled with its function as a cytoplasmic tether. Furthermore, the model did not satisfactorily explain why the NLS of p50, which did not interact with IκBα,[3] did not direct nuclear translocation of the p50/p65/IκB heterotrimer even if the p65/NLS was blocked by IκB. Indeed, when the X-ray crystal structures of NF-κB/ IκB complexes became available, it was also apparent that the p50 NLS was "exposed".[5,6] This chapter summarizes developments in the past five years that lead to a more dynamic view of NF-κB cell biology.

Nucleocytoplasmic Shuttling of IκB Proteins

The family of small IκB proteins includes IkBα, β and ε (Fig. 1); in addition, the C-termini of the precursors p100 and p105 regulate sub-cellular distribution and DNA binding of Rel proteins. Finally, the ankyrin domain-containing proteins Bcl-3 and IκBζ resemble IκBs, but are functionally quite different because they do not inhibit NF-κB DNA binding. The properties of even the inhibitory IκBs differ considerably between each other. For example, unlike IκBα, IκBβ binds NF-κB/DNA complexes without disrupting them[7] and also interacts with a unique cytosolic component κB-Ras.[8] Early indications that IκBα was not "simply" a cytoplasmic tether came from the detection of this protein in nuclear extracts. Kerr et al first showed that anti-IκBα antibody super-shifted a small proportion of the NF-κB/DNA complex detected by electrophoretic mobility shift assays using WEHI 231 (B cell) nuclear extracts.[9] Thereafter, IκBα was transiently detected in nuclear extracts derived from HeLa cells treated with TNFα.[10] These authors proposed that newly synthesized IκBα migrated to the nucleus to inhibit NF-κB activity and terminate the NF-κB response. The ability of IκBα to restore

---

*Corresponding Author: Ranjan Sen—Laboratory of Cellular and Molecular Biology, National Institute on Aging, 5600 Nathan Shock Drive, Baltimore, Maryland 21224 U.S.A. Email: rs465z@nih.gov

NF-κB/Rel Transcription Factor Family, edited by Hsiou-Chi Liou.
©2006 Landes Bioscience and Springer Science+Business Media.

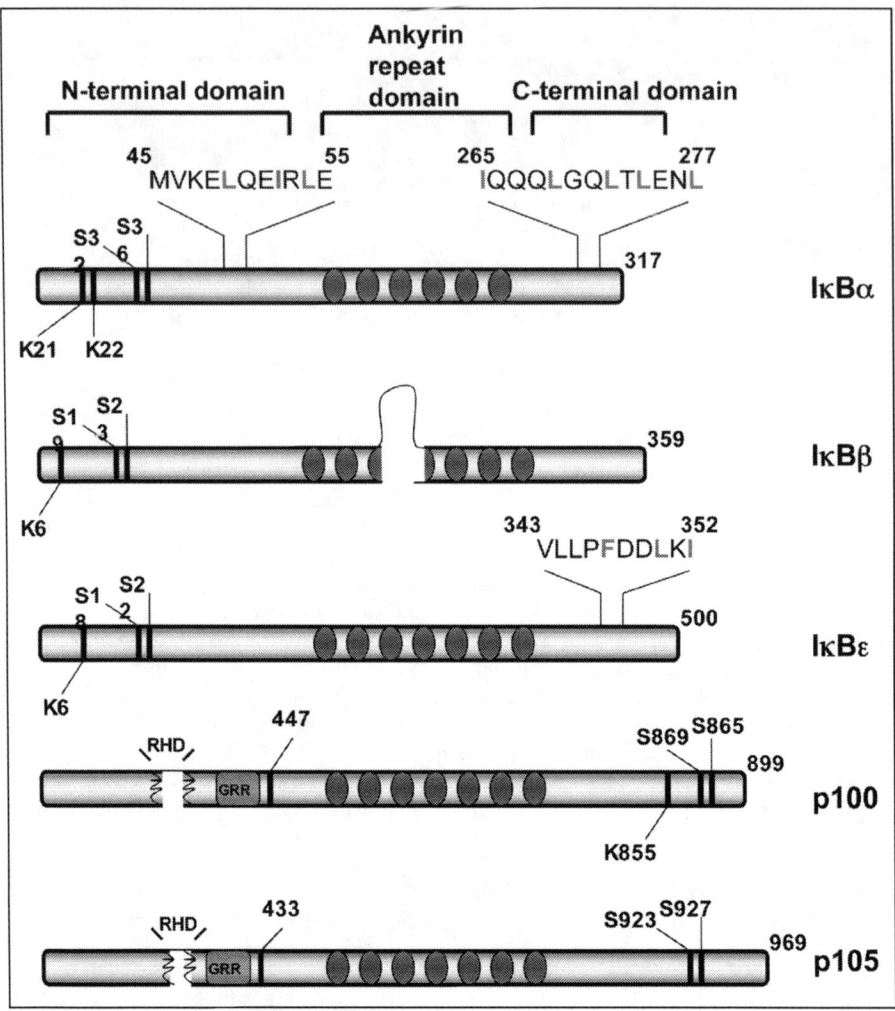

Figure 1. The family of IκB proteins. Schematic of IκB family members highlighting features discussed in this chapter. The ankyrin repeats are designated as red ovals, with the third repeat of IκBβ shown as interrupted by a loop. The amino acid sequences shown in single letter code represent the nuclear export sequences (NES) identified in IκBα and IκBε; the hydrophobic residues that are important for function are indicated in red. Serine residues indicated in the N-terminal domains of IκBα, β and ε are the ones that are phosphorylated by IκB kinases and recognized by β-TRCP1. The lysine residues indicated in these IκBs are the sites of ubiquitination that mark the proteins for proteasomal degradation. For p100 and p105, the full structure comprising the Rel homology domain (RHD) at the N-terminus is not shown, to emphasize the IκB-like part of these molecules. Phosphorylated serines in these proteins are located C-terminal to the ankyrin repeats as indicated. The blue box labeled GRR show the position of a glycine-rich repeat, located between the DNA binding RHD and the inhibitory ankyrin repeat domains, that is involved in constitutive processing.

cytoplasmic NF-κB was demonstrated by microinjection experiments into Xenopus oocytes. In these studies nuclear injection of recombinant IκBα resulted in partially restoring NF-κB to the cytosol.[11]

An important role for nuclear IκBα was firmly established by the identification of two nuclear export sequences (NES) in this protein.[12-14] Both sequences (Fig. 1) are leucine/isoleucine-rich motifs that interact with the exportin receptor CRM1 in the presence of GTP-bound Ran protein.[15] Proteins that contain such motifs bind to CRM1 in the GTP-rich nuclear environment and the complex is brought to the nuclear pore for translocation to the cytoplasm. Upon hydrolysis of the Ran-bound GTP to GDP the complex falls apart and the cargo protein is delivered to the cytoplasm. This vectorial transport process critically depends upon the Ran (GTP)/Ran(GDP) gradient between the nucleus and the cytoplasm. The CRM1 exportin is conserved from yeast to metazoa, though the metazoan species is selectively inhibited by the drug leptomycin B (LMB),[16] which is widely used to study CRM1-dependent nuclear export.

The IκBα NESs determine sub-cellular location of IκBα itself, as well as IκBα-associated Rel proteins, in yeast and mammalian cells. A green fluorescent protein (GFP) IκBα fusion protein is located mainly in the cytoplasm of transfected mammalian cells or transformed yeast strains.[14] However, inhibiting mammalian CRM1 with LMB, or attenuating yeast CRM1p genetically, leads to exclusively nuclear localization in both situations.[13,14] These observations suggest that that GFP- IκBα is in a state of dynamic flux between the nucleus and the cytoplasm. The appearance of most of the protein in the cytosol, either by direct visualization or in cellular extracts, thus reflects a snapshot of the state at which these measurements are made. Continuous movement in and out of the nucleus requires a nuclear localization sequence, to direct the protein to the nucleus, in addition to the NES that directs it out of the nucleus. A classical lysine-rich NLS has not been identified in IκBα. Rather, nuclear localization function is attributed to the second ankyrin domain of IκBα.[17] Consistent with the suggestion of a nonclassical NLS, nuclear import in vitro has been shown to be independent of importin α, the major classical NLS-recognizing protein.[18] Of the two NESs present in IκBα, mutation of the N-terminal NES, but not the C-terminal NES, is sufficient to enforce nuclear localization.[13,14] These observations suggest that the N-terminal NES is the stronger determinant of sub-cellular localization of IκBα.

Coexpression of p65/RelA and IκBα results in cytosolic localization of both proteins. However, as in the case of IκBα alone, this is also a momentary representation of a fluxional state. LMB-mediated inhibition of CRM1 export in mammalian cells, or genetic mutation of the endogenous *crm1* gene in yeast, leads to a complete shift of the complex to the nucleus indicative of ongoing shuttling.[13,14] Two possible NLSs may mediate nuclear entry of the complex. First, the classical NLS of p65, though attenuated by interaction with IκBα, may still be available to interact with importin α. Accessibility of the p65 NLS in complex with IκBα is indicated in biochemical[19] as well as X-ray crystallographic studies.[5,6] Alternatively, the second ankyrin domain of IκBα may still be available for import duty. There is no direct experimental support for either model at present. Presumably, the two pathways can be distinguished by determining whether nuclear import is dependent, or independent, of importin α. Nuclear export of the complex is mediated by the IκBα NESs as well as an NES in the C-terminal domain of p65/RelA (see below).[20] In coexpression studies, mutation of either element does not significantly alter cytoplasmic localization, however, the p65/ IκBα complex is predominantly nuclear when both IκBα and p65/RelA NESs are simultaneously mutated.[21] Thus, homodimers of p65 are not sequestered in the cytoplasm but shuttle continuously. While identification and characterization of sequences that regulate subcellular distribution of Rel complexes were carried out in transfection studies, the principles apply for endogenous Rel/IκBα complexes as well. For example, IκBα relocates to a primarily nuclear location in LMB-treated mammalian cells, and p65/RelA levels in the nucleus are also elevated under these conditions.

IκBα has also been shown to associate with hnRNP1,[22] a nucleocytoplasmic shuttling protein that is involved in mRNA export from the nucleus. The interaction requires an RNA binding domain in hnRNP1 and C-terminal sequences in IκBα. A functional role for this interaction was assessed in an erythroleukemia cell line that lacks hnRNP1. NF-κB activation induced by the latent membrane protein (LMP1) of Epstein-Barr virus was reduced in these cells, and restored upon reconstitution with hnRNP1. At present it is unclear whether other NF-κB inducing stimuli also utilize this pathway, or the mechanism by which it works.

IκBε is also a nucleo-cytoplasmic shuttling protein.[23] Like IκBα, IκBε nuclear import is mediated by a nonclassical NLS in the ankyrin domains. Hannink and colleagues have identified a functional NES in the C-terminal domain of IκBε that interacts with CRM1 and mediates cytoplasmic retention and post-induction repression of Rel proteins. In vitro binding assays show that the association of IκBε to CRM1 via its NES is comparable to the strength of IκBα/CRM1 binding, suggesting that the export potential of both these proteins are similar. Thus, IκBα and ε share many common features of sub-cellular regulation of Rel proteins. One important difference between these two is the kinetics with which they are reexpressed after signal-induced degradation,[23] and thus their efficacy in terminating the NF-κB response. The rapid re-expression of IκBα suits it for rapid shut-down of NF-κB-dependent gene transcription, while the slower kinetics of IκBε re-expression make it more suitable for terminating long-lived NF-κB responses. How NF-κB-dependent promoters are distinguished for early, or late, shut-down by IκBα, or IκBε, respectively, remains an important question for future studies.

The properties of IκBβ differ considerably from those of IκBα and IκBε. Most importantly, from the perspective of this chapter, IκBβ-associated Rel proteins do not shuttle between the nucleus and cytoplasm; rather, these complexes are sequestered in the cytosol as envisaged in the classical pathway of NF-κB/IκB function.[19] One of the mechanisms postulated for the lack of nuclear entry of IκBβ-associated p65 is a closer association of IκBβ with the p65 NLS. More effective blockade of the NLS may be mediated by κB-Ras, a small Ras-like G protein, that interacts with IκBβ via the 40 amino-acid loop located in the third ankyrin repeat of IκBβ. Ghosh and colleagues have shown that deletion of this loop makes IκBβ a shuttling protein.[8] While nuclear entry in the absence of κB-Ras association may be explained by exposure of an NLS, it is less clear how an IκBβ-associated complex is exported from the nucleus since an export sequence has not yet been identified in this protein. One possibility is that IκBβ/p65 complexes may be exported using the NES located at the C-terminus of p65. This mechanism would not apply to c-Rel associated complexes because c-Rel does not contain an NES. Alternatively, IκBβ may associate with other proteins that direct it out of the nucleus.

Another difference between IκBα and IκBβ is that the latter is more refractory to signal-induced degradation compared to IκBα.[7,24,25] This has also been attributed to association with κB-Ras which, in addition to contacting the IκBβ loop, also requires an intact N-terminus of IκBβ for efficient binding.[26] Biochemical experiments indicate that κB-Ras-associated IκBβ is not efficiently phosphorylated by IκB kinase β(IKKβ) in vitro; conversely, IκBβ degradation in response to TNFα is enhanced in 293T cells in which endogenous κB-Ras is down-modulated by siRNA. Because only a proportion of cellular IκBβ is associated with κB-Ras, a testable prediction of this model is that the portion of IκBβ not degraded after TNFα treatment should be enriched in κB-Ras associated IκBβ. Another possibility discussed below is that the absence of nucleo-cytoplasmic shuttling of IκBβ plays a role in its reduced responsiveness to NF-κB inducing signals.

Signal Induced Degradation of IκB Proteins

Cellular stimulation leads to phosphorylation of IκB proteins and their degradation by the 26S proteasome.[1,27,28] The signal transduction pathway that activates IκB kinases is discussed in Chapters 2 and 3 of this book and will not be further considered here. Phospho-IκBs are targeted to the proteasome via poly-ubiquitination of a specific lysine residue at the N-terminus of each IκB (Fig. 1). The ubiquitination process is initiated by the β-TrCP protein that is the recognition component of the multi-subunit ubiquitin-conjugating enzyme SCF (Skpsl-culin-1-F-box).[29-31] By binding to phosphorylated IκB, SCF$^{\beta TrCP}$ brings to it a particular ubiquitin conjugating enzyme (UbcH5) that ligates a "string" of ubiquitin moieties to IκB, thus making it a substrate for proteasomal degradation.[27] Though some components of the

IκB kinase play a role in the nucleus (see below), most of the evidence suggests that IκB phosphorylation is a cytoplasmic event.[32-35] For IκBα-associated complexes this means that phosphorylation must occur as the shuttling complex passes through the cytoplasm. Indeed, the best evidence that phosphorylation must be a cytosolic event comes from the observation that nuclear IκBα/NF-κB complexes, produced in LMB-treated HeLa cells, are refractory to signal-induced degradation.[33] Since IκBβ-associated complexes reside in the cytoplasm, sub-cellular compartmentalization of the kinase and the IκB is less of an issue in this case.

Recognition of the phosphorylated IκB is the next essential step towards NF-κB activation. A quandary arises from the observation by Ben-Neriah and colleagues that β-TrCP1, the major phospho- IκBα recognizing protein, is located primarily in the nucleus.[36] The phospho-protein recognition domain of nuclear β-TrCP1 is complexed to a ubiquitous phospho-ribonucleoprotein hnRNP-U; this interaction, however, does not lead to ubiquitination and degradation of hnRNP-U (thus, it is referred to as a pseudo-substrate). Davis et al showed that a phospho-peptide comprising a part of the N-terminal domain of IκBa binds better to β-TRCP1 than the corresponding phosphopeptide from hnRNP-U. They suggested that the higher affinity of the IκBα phosphopeptide displaces the hnRNP-U peptide in activated cells. The shuttling property of IκBα complexes provides a ready mechanism by which phospho-IκBα/NF-κB complexes bind to nuclear β-TrCP1. Specifically, once NF-κB/IκB complexes are phosphorylated in the cytoplasm they re-enter the nucleus and bind to β-TrCP1. However, it has not been experimentally verified that phospho-IκBα/NF-κB complexes continue to shuttle. Studies with β-TrCP1-deficient lymphocytes show that IκBα degradation is reduced but not eliminated, indicating that the closely related β-TrCP2 protein may partially compensate for the lack of β-TrCP1.[37] Phosphorylated IκBβ is also recognized by β-TRCP1 en route to being ubiquitinated and degraded.[30] How cytosolic IκBβ-containing complexes meet up with nuclear β-TRCP1 remains unclear. One possibility is that β-TRCP2, which is mainly located in the cytoplasm, may serve as the major IκBβ recognizing protein. A prediction of this model is that IκBβ degradation would be less affected in β-TRCP1-deficient cells. Though β-TrCP2 has been shown to bind phospho- IκBβ, IκBβ degradation is affected more severely than IκBα in β-TrCP1-deficient thymocytes.[37] The relative contributions of the two β-TRCPs for the degradation of shuttling, and nonshuttling, IκBs thus remains murky.

Several possibilities may be considered for what happens subsequent to nuclear recognition of phospho-IκBα/NF-κB by β-TrCP1. First, the β-TrCP1-associated SCF may lead to poly-ubiquitination in the nucleus, and degradation via nuclear proteasomes giving rise to DNA binding NF-κB in the nucleus. Alternatively, the β-TrCP1 associated complex may be exported to the cytoplasm to undergo the more typical poly-ubiquitination and proteasomal degradation; the DNA binding NF-κB would then translocate to the nucleus to activate genes. We favor the hypothesis that the complex is exported out to the cytoplasm to be degraded, in part because of conflicting data regarding nuclear NF-κB activation. A possible complication is that the NES of IκBα, which is located close to the phosphorylation sites, may be obscured by the associated SCF$^{\beta TrCP}$ complex. We suggest that an alternate export mechanism, such as that mediated via hnRNP-1 associated with the C-terminus of IκBα, may predominate under these circumstances to bring the complex to the cytosol. Once in the cytosol, poly-ubiquitination and IκBα degradation would release DNA binding NF-κB.

An interesting aspect of this model is that the shuttling cycle predicts an increase of β-TrCP1 in the cell cytosol, which transiently puts it in the same sub-cellular compartment as IκBβ. This cytosolic β-TrCP1 can bind to phosphorylated IκBβ leading to its poly-ubiquitination and degradation. The need for cytosolic β-TrCP1, generated by IκBα degradation, may in part be the reason that IκBβ is degraded more slowly, and less efficiently, compared to IκBα in response to most signals. Ghosh and colleagues have previously proposed that the reduced responsiveness of IκBβ is due to association of κB-Ras protein with the loop of IκBβ, which inhibits IκBβ phosphorylation by the IKK complex.[26] Perhaps reduced phosphorylation, as well as reduced recognition of phospho-IκBβ, contribute to the relative stability of IκBβ in stimulated cells.

Shuttling of IκB Kinase Components

The first indication that signaling components in the NF-κB activation pathway are also in a state of dynamic flux between the nucleus and cytoplasm came from the observations of Birbach et al.[38] They used leptomycin B to block nuclear export and examined the subcellular distribution of several "upstream" activators of NF-κB. Of the various proteins examined, the NF-κB inducing kinase (NIK) rapidly shifted to the nucleus in response to LMB, as did the IκB kinase α subunit (IKKα), though with slower kinetics. More recently the adapter subunit IKKγ (NEMO) has been shown to accumulate in the nucleus of HeLa cells in response to LMB.[39] However, IKKβ, the dominant kinase involved in NF-κB activation in response to cytokines such as TNFα and IL-1, appears to be strictly located in the cytoplasm.

The relevance of NIK nucleo-cytoplasmic shuttling for NF-κB induction is unclear at present. This kinase has been implicated in the nonclassical NF-κB activation pathway which leads to the processing of p100 in a IKKα-dependent manner and release of p52/Rel B heterodimers for gene transcription.[40,41] The proposed mechanism involves NIK-induced phosphorylation of p100 at a C-terminal domain, which recruits IKKβ resulting in additional p100 phosphorylation. This form is recognized by β-TrCP and eventually results in degradation of the C-terminal ankyrin repeat domain by the proteasome. The need for nucleocytoplasmic shuttling is not apparent in this series of steps. However, it is intriguing to note that NIK, p100 and IKKα can all be located in the nucleus under appropriate conditions, raising the possibility that shuttling between compartments may play a role in coordinating this outcome. In analogy with the model proposed above for nuclear association of phosphorylated IκBα and β-TrCP, phospho-p100 may also interact with β-TrCP in the nucleus.

Sub-cellular dynamics of NEMO provide a satisfactory mechanism for the activation of NF-κB in response to DNA damage. Miyamoto and colleagues showed that a small fraction of cellular NEMO is modified by covalent attachment of the ubiquitin-like molecule, SUMO, in response to camptothecin, an inhibitor of DNA topoisomerase I.[42] Because sumoylation is believed to be a nuclear phenomenon,[43] the shuttling of NEMO in and out of the nucleus allows the substrate to be present in the right compartment, without the need to invoke the additional step of DNA-damage induced NEMO translocation. SUMO-modified NEMO is a target of the ATM kinase which, by an as yet undetermined mechanism, leads to replacement of SUMO with ubiquitin. Lysines 277 and 309 on NEMO have been identified as possible sites of these modifications. The nuclear accumulation of sumoylated NEMO may be due to retention in the nucleus or retardation of export from the nucleus. In contrast to sumoylated NEMO, the ubiquitin-modified form is exported to the cytoplasm where it presumably associates with and activates IκB kinases to induce IκB phosphorylation.

Nuclear NEMO has also been proposed to play a transcription regulatory role.[39] The work of Gaynor and colleagues shows that NEMO interacts with an N-terminal domain of the transcriptional coactivator CBP. This part of CBP also interacts with p65/RelA and IKKα, leading to the idea that NEMO may sequester CBP and prevent it from interacting with, and thereby activating, p65-dependent gene expression. Taken together these observations strongly support an important role for NEMO in the nucleus.

Immunofluorescence studies demonstrate that IKKα is located in the nucleus and cytoplasm of cells, and LMB treatment slowly shifts the balance towards nuclear localization. One of the functions of IKKα in the nucleus is the activation of NF-κB-dependent gene transcription. This occurs by recruitment of IKKα to the promoters of these genes via an association with the CBP coactivator and, presumably, other promoter-bound transcription factors. At the promoter IKKα likely phosphorylates histone H3 on Ser 10, causing transient increase of phosphorylated H3.[44,45] Promoter phosphorylation then triggers histone acetylation by histone acetyl transferases such as CBP. The model is well substantiated by the loss of promoter phosphorylation, and reduced histone acetylation, in IKKα-deficient cells. These observations are consistent with the idea that nucleocytoplasmic shuttling of IKKα ensures that some of this protein is in the right cellular compartment to be recruited to appropriate promoters in response to cell stimulation. The mechanism of recruitment, however, remains unclear. Because IKKα directly associates with the N-terminus of CBP, one possibility is that a IKKα/CBP complex is recruited to the promoter. Alternatively, IKKα may interact with one or more promoter-bound transcription factors, though it does not interact with p65/RelA directly. It is also not clear whether IKKα is required at all or only a subset of NF-κB-dependent promoters; conversely, it is likely that IKKα may also activate NF-κB-independent promoters via its interaction with CBP, and perhaps other HATs. It is noteworthy that several other histone H3 kinases have been previously identified that phosphorylate Ser 10. The rules that govern the circumstances under which each kinase activates a set of genes remains to be determined.

Shuttling of Rel Proteins

Features of the Rel proteins themselves also contribute to their sub-cellular location and, importantly, distinguish between the two closely related family members p65/Rel A and c-Rel. The RHDs of these two proteins are sufficiently similar that it has been difficult to identify DNA sequences that specifically bind one, or the other, protein. There is less similarity in the C-terminal domains and it is well established that the transcriptional activation potential of p65/Rel A is substantially higher than that of c-Rel (at least when assayed with multimerized κB sites-containing promoters). In addition to the differences in transactivation potential, p65/Rel A contains a nuclear export sequence located in the C-terminal domain,[20] but c-Rel does not.[21] Location of an NES close to the transactivation domain has also been observed in NF-AT family members,[46] though the functional relevance of this juxtaposition in either protein is unclear.

The subcellular distribution of NF-κB/IκB complexes is determined by the cumulative action of NLSs and NESs present in a particular multi-protein complex. For example, ectopically expressed p65 protein, that is present as a homodimer, is located primarily in the nucleus, presumably because the two NLSs dominate over the two NESs. Coexpression of IκBα together with p65 drives the homodimer to the cytoplasm, presumably as the result of attenuating p65 NLSs via association with IκBα and the provision of a strong IκBα NES to the complex (this heterotrimer contains two attenuated NLSs and three NESs). Interestingly, coexpression of p65 with an NES-mutated IκBα also leads to cytoplasmic localization of the complex.[21] The observation indicates that the weaker (than IκBα) NESs of p65 are sufficient to skew the balance towards the cytoplasm when the p65 NLSs are attenuated by interaction with IκBα. In a heterotrimer that contains no NESs, where p65 and the IκBα NESs have been mutated, nuclear expression of the p65/IκBα complex increases. Studies with p65 deletion mutants have also confirmed that the NLS of p65 is functionally attenuated in complex with IκBα, as suggested by the X-ray crystallographic studies. Because c-Rel does not contain an NES, its subcellular distribution closely parallels that of an NES-mutated p65. Two things change when p65 is complexed to p50. First, the complex contains one less NES (from p65) and secondly, the p50 NLS, that is not in contact with IκBα, is stronger than the attenuated NLS of p65; both changes favor nuclear localization of the Rel heterodimer. In the p50/p65/IκBα heterotrimer, however, the additional strong NES of IκBα results in net cytosolic expression based on nuclear export prevailing over nuclear import.

Though the relative importance of various import and export sequences to sub-cellular distribution were dissected in ectopic expression assays, the behavior of endogenous cellular complexes corroborates the observed patterns. Thus, cytoplasmic p65/IκBα complexes in lymphocyte cell lines redistribute to the nucleus when nuclear export is blocked by leptomycin B.[21] Not all of the p65 goes to the nucleus however, presumably because a part of it is associated with IκBβ and therefore does not shuttle. Consistent with this observation, LMB-induced c-Rel redistribution is only observed in cells where c-Rel is associated with IκBα, but not in cells where it is predominantly associated with IκBβ.[21]

Greene and coworkers have recently identified an additional novel mechanism that regulates nuclear localization of p65/Rel A.[47-49] Following up on the observation that the histone de-acetylase inhibitor trichostatin A increases NF-κB-dependent transcription, they found that p65/Rel A is acetylated at several lysine residues (K122, K123, K218, K221 and K310) in the RHD, by the coactivator CBP. This presumably occurs in the nucleus during p65-dependent gene transcription. Acetyl modification of the RHD prevents interaction with IκBα and, as a consequence, p65/Rel A remains in a transcription competent state in the nucleus for a longer time. Thus, down-regulation of p65/Rel A from the nucleus requires removal of the acetyl group as well as re-expression of IκBα following cell stimulation.

The complex pattern of NLSs, NESs and post-translational modifications of Rel proteins that regulates cellular location is likely to be important for appropriate function of these proteins. Based simply on the numbers of NLSs and NESs it is likely that there is graded nuclear propensity amongst Rel family homo-and hetero-dimers. The homodimer of p65 is the least likely to be localized to the nucleus in the presence, or absence, of IκBα, with heterodimerization increasing the probability of nuclear location. This idea provides a plausible explanation for the subunit composition (p50/c-Rel) of constitutive NF-κB in the nuclei of mature B cells. We have previously proposed that nuclear p50/c-Rel in B cells may be the result of more efficient p65 export from the nucleus after generation of free pools of both p65 and c-Rel as a consequence of enhanced IκBα degradation.[21] It is also likely that this hierarchy dictates the order in which Rel proteins are removed from the nucleus by IκBα (or IκBε) after cellular activation. That is, p65 homodimers would be removed first (2 NESs and attenuated NLSs), followed by c-Rel homodimers (only attenuated NLSs). Heterodimers with p50 would tend to be more nuclear compared to the homodimers because of the unattenuated p50 NLS; but, the additional NES of p65 would increase its tendency to be cytoplasmic compared to p50/c-Rel. In this context it is interesting to note that p65, with the strongest transcription activation domain amongst Rel family members, may be the one most likely to lead to unwanted gene expression if present in the nucleus for longer than "required".

In summary, the central theme of this review is to highlight the dynamic nature of NF-κB and associated proteins in the cell. In the unstimulated cell continuous nucleo-cytoplasmic shuttling of several components maintains a nuclear environment that is free of functional transcription factor, yet provides the means for cytosolic or nuclear post-translational modifications that are essential for a variety of stimulus-induced NF-κB activation. In cells responding to stimulus, the same features of intracellular movement provides a means to terminate the response. While the duration of NF-κB-dependent gene expression may vary depending on the cell-type, or the activating stimulus, the underlying principles outlined here represent a framework in which to analyze cellular responses.

References

1. Karin M, Ben-Neriah Y. Phosphorylation meets ubiquitination: The control of NF-[kappa]B activity. Annu Rev Immunol 2000; 18:621-663.
2. Rothwarf DM, Karin M. The NF-kappa B activation pathway: A paradigm in information transfer from membrane to nucleus. Sci STKE 1999; 1999(5):RE1.
3. Zabel U, Henkel T, Silva MS et al. Nuclear uptake control of NF-kappa B by MAD-3, an I kappa B protein present in the nucleus. EMBO J 1993; 12(1):201-211.
4. Zabel U, Baeuerle PA. Purified human I kappa B can rapidly dissociate the complex of the NF-kappa B transcription factor with its cognate DNA. Cell 1990; 61(2):255-265.
5. Huxford T, Huang DB, Malek S et al. The crystal structure of the IkappaBalpha/NF-kappaB complex reveals mechanisms of NF-kappaB inactivation. Cell 1998; 95(6):759-770.
6. Jacobs MD, Harrison SC. Structure of an IkappaBalpha/NF-kappaB complex. Cell 1998; 95(6):749-758.
7. Suyang H, Phillips R, Douglas I et al. Role of unphosphorylated, newly synthesized I kappa B beta in persistent activation of NF-kappa B. Mol Cell Biol 1996; 16(10):5444-5449.
8. Chen Y, Wu J, Ghosh G. KappaB-Ras binds to the unique insert within the ankyrin repeat domain of IkappaBbeta and regulates cytoplasmic retention of IkappaBbeta x NF-kappaB complexes. J Biol Chem 2003; 278(25):23101-23106.

9. Kerr LD, Inoue J, Davis N et al. The rel-associated pp40 protein prevents DNA binding of Rel and NF-kappa B: Relationship with I kappa B beta and regulation by phosphorylation. Genes Dev 1991; 5(8):1464-1476.
10. Arenzana-Seisdedos F, Thompson J, Rodriguez MS et al. Inducible nuclear expression of newly synthesized I kappa B alpha negatively regulates DNA-binding and transcriptional activities of NF-kappa B. Mol Cell Biol 1995; 15(5):2689-2696.
11. Arenzana-Seisdedos F, Turpin P, Rodriguez M et al. Nuclear localization of I kappa B alpha promotes active transport of NF-kappa B from the nucleus to the cytoplasm. J Cell Sci 1997; 110(Pt 3):369-378.
12. Johnson C, Van Antwerp D, Hope TJ. An N-terminal nuclear export signal is required for the nucleocytoplasmic shuttling of IkappaBalpha. EMBO J 1999; 18(23):6682-6693.
13. Huang TT, Kudo N, Yoshida M et al. A nuclear export signal in the N-terminal regulatory domain of IkappaBalpha controls cytoplasmic localization of inactive NF-kappaB/IkappaBalpha complexes. Proc Natl Acad Sci USA 2000; 97(3):1014-1019.
14. Tam WF, Lee LH, Davis L et al. Cytoplasmic sequestration of rel proteins by IkappaBalpha requires CRM1-dependent nuclear export. Mol Cell Biol 2000; 20(6):2269-2284.
15. Gorlich D, Kutay U. Transport between the cell nucleus and the cytoplasm. Annu Rev Cell Dev Biol 1999; 15:607-660.
16. Fukuda M, Asano S, Nakamura T et al. CRM1 is responsible for intracellular transport mediated by the nuclear export signal. Nature 1997; 390(6657):308-311.
17. Sachdev S, Hoffmann A, Hannink M. Nuclear localization of IkappaB alpha is mediated by the second ankyrin repeat: The IkappaB alpha ankyrin repeats define a novel class of cis-acting nuclear import sequences. Mol Cell Biol 1998; 18(5):2524-2534.
18. Turpin P, Hay RT, Dargemont C. Characterization of IkappaBalpha nuclear import pathway. J Biol Chem 1999; 274(10):6804-6812.
19. Tam WF, Sen R. IkappaB family members function by different mechanisms. J Biol Chem 2001; 276(11):7701-7704.
20. Harhaj EW, Sun SC. Regulation of RelA subcellular localization by a putative nuclear export signal and p50. Mol Cell Biol 1999; 19(10):7088-7095.
21. Tam WF, Wang W, Sen R. Cell-specific association and shuttling of IkappaBalpha provides a mechanism for nuclear NF-kappaB in B lymphocytes. Mol Cell Biol 2001; 21(14):4837-4846.
22. Hay DC, Kemp GD, Dargemont C et al. Interaction between hnRNPA1 and IkappaBalpha is required for maximal activation of NF-kappaB-dependent transcription. Mol Cell Biol 2001; 21(10):3482-3490.
23. Lee SH, Hannink M. Characterization of the nuclear import and export functions of Ikappa B(epsilon). J Biol Chem 2002; 277(26):23358-23366.
24. Thompson JE, Phillips RJ, Erdjument-Bromage H et al. I kappa B-beta regulates the persistent response in a biphasic activation of NF-kappa B. Cell 1995; 80(4):573-582.
25. Tran K, Merika M, Thanos D. Distinct functional properties of IkappaB alpha and IkappaB beta. Mol Cell Biol 1997; 17(9):5386-5399.
26. Chen Y, Vallee S, Wu J et al. Inhibition of NF-kappaB activity by IkappaBbeta in association with kappaB-Ras. Mol Cell Biol 2004; 24(7):3048-3056.
27. Ben-Neriah Y. Regulatory functions of ubiquitination in the immune system. Nat Immunol 2002; 3(1):20-26.
28. Hayden MS, Ghosh S. Signaling to NF-kappaB. Genes Dev 2004; 18(18):2195-2224.
29. Yaron A, Hatzubai A, Davis M et al. Identification of the receptor component of the IkappaBalpha-ubiquitin ligase. Nature 1998; 396(6711):590-594.
30. Wu C, Ghosh S. beta-TrCP mediates the signal-induced ubiquitination of IkappaBbeta. J Biol Chem 1999; 274(42):29591-29594.
31. Suzuki H, Chiba T, Suzuki T et al. Homodimer of two F-box proteins betaTrCP1 or betaTrCP2 binds to IkappaBalpha for signal-dependent ubiquitination. J Biol Chem 2000; 275(4):2877-2884.
32. Rodriguez MS, Thompson J, Hay RT et al. Nuclear retention of IkappaBalpha protects it from signal-induced degradation and inhibits nuclear factor kappaB transcriptional activation. J Biol Chem 1999; 274(13):9108-9115.
33. Huang TT, Miyamoto S. Postrepression activation of NF-kappaB requires the amino-terminal nuclear export signal specific to IkappaBalpha. Mol Cell Biol 2001; 21(14):4737-4747.
34. Renard P, Percherancier Y, Kroll M et al. Inducible NF-kappaB activation is permitted by simultaneous degradation of nuclear IkappaBalpha. J Biol Chem 2000; 275(20):15193-15199.
35. Nelson DE, Ihekwaba AE, Elliott M et al. Oscillations in NF-kappaB signaling control the dynamics of gene expression. Science 2004; 306(5696):704-708.

36. Davis M, Hatzubai A, Andersen JS et al. Pseudosubstrate regulation of the SCF(beta-TrCP) ubiquitin ligase by hnRNP-U. Genes Dev 2002; 16(4):439-451.
37. Nakayama K, Hatakeyama S, Maruyama S et al. Impaired degradation of inhibitory subunit of NF-kappa B (I kappa B) and beta-catenin as a result of targeted disruption of the beta-TrCP1 gene. Proc Natl Acad Sci USA 2003; 100(15):8752-8757.
38. Birbach A, Gold P, Binder BR et al. Signaling molecules of the NF-kappa B pathway shuttle constitutively between cytoplasm and nucleus. J Biol Chem 2002; 277(13):10842-10851.
39. Verma UN, Yamamoto Y, Prajapati S et al. Nuclear role of I kappa B Kinase-gamma/NF-kappa B essential modulator (IKK gamma/NEMO) in NF-kappa B-dependent gene expression. J Biol Chem 2004; 279(5):3509-3515.
40. Bonizzi G, Karin M. The two NF-kappaB activation pathways and their role in innate and adaptive immunity. Trends Immunol 2004; 25(6):280-288.
41. Pomerantz JL, Baltimore D. Two pathways to NF-kappaB. Mol Cell 2002; 10(4):693-695.
42. Huang TT, Wuerzberger-Davis SM, Wu ZH et al. Sequential modification of NEMO/IKKgamma by SUMO-1 and ubiquitin mediates NF-kappaB activation by genotoxic stress. Cell 2003; 115(5):565-576.
43. Hay RT. SUMO: A history of modification. Mol Cell 2005; 18(1):1-12.
44. Anest V, Hanson JL, Cogswell PC et al. A nucleosomal function for IkappaB kinase-alpha in NF-kappaB-dependent gene expression. Nature 2003; 423(6940):659-663.
45. Yamamoto Y, Verma UN, Prajapati S et al. Histone H3 phosphorylation by IKK-alpha is critical for cytokine-induced gene expression. Nature 2003; 423(6940):655-659.
46. Zhu J, McKeon F. NF-AT activation requires suppression of Crm1-dependent export by calcineurin. Nature 1999; 398(6724):256-260.
47. Chen L, Fischle W, Verdin E et al. Duration of nuclear NF-kappaB action regulated by reversible acetylation. Science 2001; 293(5535):1653-1657.
48. Kiernan R, Bres V, Ng RW et al. Post-activation turn-off of NF-kappa B-dependent transcription is regulated by acetylation of p65. J Biol Chem 2003; 278(4):2758-2766.
49. Chen LF, Mu Y, Greene WC. Acetylation of RelA at discrete sites regulates distinct nuclear functions of NF-kappaB. EMBO J 2002; 21(23):6539-6548.

CHAPTER 5

NF-κB in Lymphopoiesis

Estefania Claudio, Keith Brown and Ulrich Siebenlist*

NF-κB has long been recognized as a critical mediator of acute immune and stress responses, poised to coordinate the defensive response of the host to pathogenic threats. Beyond these roles, NF-κB is increasingly recognized for its roles in lymphoid organogenesis and development of hematopoietic cell lineages. These functions have been discovered primarily through analyses of mouse models deficient in NF-κB factors, inhibited for NF-κB activity or lacking key components required to signal to NF-κB.

Here we review NF-κB contributions in developing T and B lymphocytes (summarized in Tables 1 and 2). For the purpose of this review we will consider development to be completed once lymphocytes have matured into naïve, peripheral, long-lived recirculating cells. Discussed elsewhere in this book are NF-κB's roles in responses of mature lymphocytes, such as in their antigen-driven activation, proliferation and differentiation. As will become evident in this review, NF-κB is particularly important for the survival of developing lymphocytes, helping to rescue these cells from default death pathways at various stages of their development. However, NF-κB also contributes to differentiation and may even have a role in apoptosis in some situations. Thus NF-κB exhibits context-dependent functions that differ depending on the developmental stage. The structure, regulation and signaling pathways for activation of NF-κB are discussed in other chapters of this book.

NF-κB in Early Stages of Lymphocyte Development

Hematopoietic precursor cells lacking individual NF-κB or IκB proteins can generate mature B and T cells, indicating that no single subunit or inhibitor is essential for development of mature lymphocytes.[1-4] The resulting cells are, however, functionally deficient. Mice lacking RelA died about 7 days before birth from acute TNF-induced liver apoptosis, so lymphopoiesis could only be studied in radiation chimeras made by transfer of fetal liver cells (containing hematopoietic precursors) from RelA-deficient embryos into lethally irradiated wild-type mice.[5] The chimeric mice survived with hematopoietic cells derived from transferred precursors. They developed peripheral lymphocytes, albeit significantly diminished in numbers and impaired in function, when compared to chimeras with wild-type fetal liver cells (to be discussed later below).

In contrast to the RelA single knockout (KO), chimeras generated from mice deficient in both RelA and NF-κB1 (p50/p105)[5] essentially failed to develop lymphocytes. Similarly, lymphopoiesis was absent in radiation chimeras generated with hematopoietic precursors from mice lacking IKKβ.[6] IKKβ is the primary catalytic component of the IKK complex for most NF-κB activating signals and specifically for inflammatory cytokine signals.[3,4,7] Like RelA KO mice, IKKβ KO mice died in utero due to massive apoptosis of hepatocytes.[6] Finally, chimeras generated with fetal liver cells deficient in both RelA and c-Rel appeared to be largely, though

---

*Corresponding Author: Ulrich Siebenlist—Laboratory of Immunoregulation, NIAID, National Institutes of Health, Bethesda, Maryland 20892-1876, U.S.A. Email: US3N@nih.gov

NF-κB/Rel Transcription Factor Family, edited by Hsiou-Chi Liou.
©2006 Landes Bioscience and Springer Science+Business Media.

Table 1. Cell-autonomous roles of NF-κB in developing T lymphocytes

Stage	Role
T Lymphocyte Precursors	Probably needed to protect precursors from death induced by high levels of TNFα.
Double Negative CD4⁻ CD8⁻ thymocytes.	pre-TCR induced activation of NF-κB important for survival during progression of DN thymocytes from stage III/L to IV and to DP Inhibition of NF-κB by IκBα superrepressor triggers apoptosis.
Double Positive CD4⁺ CD8⁺TCR⁺	Self-antigen/TCR-induced NF-κB activity may help set the threshold for signal strength during positive and negative selection of DP thymocytes. Inhibition of NF-κB by IκBα superrepressor interferes with negative selection (conversion of a strong to a weaker signal, leading to survival?), but also with positive selection (further reduction of a weak signal, leading to death by neglect?). May explain pro- and anti-apoptotic roles of NF-κB, depending on the in vivo assay. Different hypothesis suggests negative selection mediated by TCR-induced expression of $I\kappa B_{NS}$. $I\kappa B_{NS}$ may act like a supperrepressor to inhibit NF-κB in strongly self-reactive DP thymocytes, thus inducing death by eliminating the protection from apoptosis by NF-κB.
Single Positive CD4⁺, CD8⁺ Thymocytes and Peripheral T cells	Required for survival of T cells as they progress from SP thymocytes to long-lived peripheral T cells. Conditional knockout of NEMO or knockin of dominant negative IKKβ induces apoptosis of SP thymocytes (especially CD8⁺) and leads to complete loss of peripheral T cells. Less severe loss of NF-κB activity in IKKβ knockouts and IκBα superrepressor transgenic mice results in partial reduction of peripheral T cells (CD8⁻). IKKβ-mediated activation of NF-κB required for development of T_r and NKT cell subsets. IKKβ knockouts lack T_r and NKT cells in thymus. T_r also reduced in NF-κB1 and c-Rel dKO and thymic NKTs reduced in IκBα superrepressor transgenic.

not completely, impaired in generation of peripheral lymphocytes.[8] While these data imply an intrinsic role for NF-κB in developing hematopoietic cells, this role was not cell-autonomous. Cotransfer of wild-type hematopoietic precursors together with those from the above double knockouts (dKOs) or the IKKβ KO mice significantly rescued lymphopoiesis of mutant cells.[5,6,8,9]

Subsequent work has shown that NF-κB's critical function in early lymphopoiesis may be to prevent excessive formation of cytotoxic TNFα. If the IKKβ knockout was placed on a TNFR1-deficient background, this not only rescued the resulting embryos from early liver apoptosis and death, it also relieved the block in lymphopoiesis.[6] Lack of IKKβ was associated with increased granulopoiesis, the likely cause of excessive TNFα production. In addition though, TNFα may have been particularly effective in inducing apoptosis in the developing mutant lymphocytes, since these lymphocytes lacked the anti-apoptotic protection normally afforded to them by NF-κB. So, NF-κB probably also has a cell-autonomous, anti-apoptotic role during lymphopoiesis. Excessive granulopoiesis was seen in radiation chimeras generated with precursors from mice deficient in IKKβ, or RelA and NF-κB1 or RelA and c-Rel.[5,6,8,9] Most

Table 2. Cell-autonomous roles of NF-κB in developing B lymphocytes

Stage	Role
B Lymphocyte Precursors	Probably needed to protect precursors from death induced by high levels of TNFα.
Pre-B	pre-BCR-induced NF-κB aids survival of pre-B cells during progression from large to small pre-B and immature B cells.
	B cells deficient in NF-κB1 and NF-κB2 or containing an IκBα superrepressor generate fewer small pre-B and immature B cells (rescued with Bcl-x_L).
	NF-κB may also contribute to light chain locus demethylation/ germline transcription in small pre-B cells.
Immature	(Self-antigen induced signaling via BCR in immature B cells may induce apoptosis during negative selection due to limited signaling to NF-κB by immature BCR complexes). Possible role in homing to spleen.
Transitional (T1, T2)	NF-κB essential for survival and for full phenotypic maturation. Activation of the alternative pathway by BAFF/BAFFR and the classical pathway by unknown signal important for survival of T1 and subsequent stages.
	Complete development block at T1 in NF-κB1 and NF-κB2 dKO and at T1/T2 in RelA and c-Rel dKO due to apoptosis. Survival and partial maturation rescued with Bcl-2 anti-apoptotic transgene, but not full maturation.
	Reduction of transitional B cells in conditional NEMO KO or dominant negative IKKβ knockin mice (incomplete loss of wild-type gene/protein at this stage).
Marginal Zone (MZ)	Needed for generation of MZ B cells, probably for survival and maturation. Signals likely to include BCR, BAFFR.
	Single KOs of NF-κB1, NF-κB2 or RelB lack MZ B cells, and RelA or c-Rel KOs have reduced numbers.
Mature (follicular, recirculating)	NF-κB needed for survival through BCR and BAFFR activation and other signals.
	Reduction of mature B cells in mice deficient in BCR signaling components dedicated to activation of NF-κB.
	Loss of NEMO or IKKβ in B cells in conditional knockouts leads to complete elimination of mature B cells.
	Partial reduction of mature B cells harboring an IκBα superrepressor transgene (incomplete inhibition), or lacking IKKα (required for activation via the alternative pathway).

likely, NF-κB-dependent extracellular factors are involved in limiting granulopoiesis and thus excessive TNFα production.

The NEMO (IKKγ) subunit is required for IKKβ activation in the classical pathway of NF-κB activation.[1,3,4,7] In contrast to the above factors, NEMO appeared to be required for the generation of peripheral lymphocytes even when wild-type hematopoietic cells were present.[10,11] In female mice in which one X chromosome carried a NEMO mutation, random lyonization produced hematopoietic precursors both with and without NEMO. Despite this mixture, which mimics a mutant/wild-type chimera, only NEMO-expressing circulating lymphocytes were observed. Similar skewing occurred in female human patients with Incontinentia

Pigmenti, where one NEMO allele is mutated.[12] It is not known at what stage of their development NEMO-deficient lymphocytes failed to thrive. Since loss of NEMO more completely blocks NF-κB activation than either loss of IKKβ or two NF-κB factors,[7] the absence of circulating NEMO-deficient lymphocytes may reflect a need for NF-κB in early lymphopoiesis. However, NEMO-deficient ES cells were able to generate some IgM$^+$ immature B cells in an in vitro differentiation system, suggesting that early development of B cells per se is not completely blocked.[13] On the other hand, the in vitro generated cells survived poorly. The true extent of this defect can only be assessed in vivo, however, under conditions where mutant cells are forced to compete with wild-type cells and where they encounter many other cells and signals.

Double Negative, PreTCR$^+$ Thymocytes

T lymphocyte development occurs largely in the thymus. The most immature thymocytes (distinguished by cell surface markers) are termed CD4$^-$CD8$^-$ double negative (DN), since they lack the coreceptors, CD4 and CD8. These cells mature to become CD4$^+$CD8$^+$ double positives (DP), then CD4$^+$ or CD8$^+$ single positive (SP) thymocytes, which finally migrate to the periphery.[14] DN thymocytes bearing the heat-stable HSA or CD24 antigen are subdivided into four stages by their expression of CD44 and IL-2Rα (CD25): (I) CD44$^+$CD25$^-$, (II) CD44$^+$CD25$^+$, (III) CD44$^-$CD25$^+$, and (IV) CD44$^-$CD25$^-$.[14]

T cell receptor (TCR) β chain rearrangement is necessary for stage III to IV transition.[15] In cells with rearranged TCR β chains, β associates with the pre-Tα chain to form the pre-TCR, which is thought to signal in the absence of any known ligands. TCRβ$^+$ stage III cells are referred to as "L" cells (about 15% of stage III cells) and are larger than the still unrearranged stage III, "E" cells. NF-κB was constitutively activated in thymocytes, with highest activity in stages IIIL and IV.[16,17] A further increase in NF-κB activity in IKKβ-transgenic mice generated more stage IV cells. Moreover, IKKβ expression in RAG1$^{-/-}$ thymocytes, which are blocked in pre-TCR assembly, still enabled some differentiation through stage IV to DP cells. Finally, mice with an IκBα superrepressor transgene had fewer stage IV thymocytes and NF-κB inhibition in isolated stage III L and IV cells triggered apoptosis.[17] Thus, pre-TCR signaling appears to activate NF-κB, providing a survival signal for stage III L and IV thymocytes, although involvement of other receptors has not been rigorously excluded. The pre-B cell receptor (BCR) may play an analogous role in B cell development (see below).

Expression of the IκBα superrepressor blocked development only partially at this stage,[17] possibly due to incomplete inhibition and/or alternative pre-TCR-generated survival signals. Thus, NF-κB certainly contributes to, but may not be essential for T cell development at this stage. Cells escaping the block eventually populate later stages, and continued impairment of NF-κB activity may actually obscure the earlier defect. This compensatory effect appears to be due to a pro-apoptotic role for NF-κB at the DP stage, such that repression of NF-κB may actually increase cell numbers at the later stage (see below).

Double Positive Thymocytes

DP thymocytes express both CD4 and CD8 chains, as well as TCRα and β chains (a mature TCR).[14] They are subject to both positive and negative selection. DP thymocytes with TCRs that strongly recognize self-peptides (in association with MHCs) are generally eliminated by apoptosis, as are those thymocytes that fail to see any peptides (death by neglect), while those with weak but measurable recognition of MHC-associated peptides are positively selected and progress to the SP stage. Multiple reports imply critical roles for NF-κB in DP thymocytes and even specifically in positive and negative selection, but to date no consensus has emerged as to exactly which biologic processes NF-κB regulates and by what mechanisms. This may reflect differences in the biologic assays employed in the various studies, as well as differences in the quality and quantity of NF-κB inhibition achieved. The situation may be complicated if NF-κB makes several distinct contributions that influence numbers and progression of thymocytes.

DP thymocytes underwent massive apoptosis in mice treated with dexamethasone, which also functions as an inhibitor of NF-κB, suggesting a need for the transcription factor to protect DP cells against what are likely endogenous apoptotic insults.[18] c-myc may have been the anti-apoptotic target of NF-κB in this instance, since ectopic c-myc partially rescued DP cells after dexamethasone treatment.[19]

Early studies showed that inhibition of NF-κB with IκB transgenes in T cells led to increased, rather than decreased numbers of DP thymocytes.[20,21] While this could have been due to a block in progression/maturation to SP cells, which were somewhat reduced in numbers,[21,22] it could also reflect a pro-apoptotic role of NF-κB in DP thymocytes. Indeed, IκBα superrepressor-expressing DP thymocytes resisted apoptosis normally induced by anti-CD3 administration in vivo,[23] as did DP thymocytes expressing dominant negative IKKβ.[24] However, this does not prove a cell-autonomous, anti-apoptotic activity of NF-κB. It is possible that the resistance to apoptosis was the result of defective expression of cell-extrinsic factors, normally dependent on NF-κB. T cells and thymocytes carrying the superrepressor[23,25] or dominant negative forms of both IKKα and IKKβ[24] are known to be significantly impaired in induced expression of extrinsically-acting cytokines, cytokines which might indirectly affect thymocyte numbers.

While in vivo administration of anti-CD3 is not likely to faithfully model negative selection, another report directly implicates NF-κB in this process.[26] Male mice expressing a TCR transgene directed against an MHC class I restricted H-Y (male) self-antigen represent a model for negative selection, since self-reactive DP thymocytes are largely eliminated, preventing the development of SP self-reactive T cells. However, if such negatively selected T cells also expressed an IκB superrepressor transgene (partially inhibiting NF-κB activity), then negative selection was blunted and many more DP thymocytes were observed. Thus NF-κB was concluded to have a pro-apoptotic effect. The aforementioned study also suggested a role for NF-κB during positive selection, however.[26] Expression of the TCR and IκB superrepressor transgenes in positively selecting mouse background models resulted in fewer than normal positively selected TCR⁺ thymocytes.

It is conceivable that inhibition of NF-κB with the superrepressor might have interfered with both negative and positive selection by reducing TCR signaling strength. TCR-induced phosphorylation of the receptor-proximal ZAP-70 kinase was reduced when NF-κB was inhibited, suggesting that NF-κB normally enhances this critical early signaling step in some unknown way.[26] Inhibiting NF-κB therefore might have dampened the strong TCR signal, thereby also preventing apoptosis, while the lesser signal normally resulting in positive selection might have become so weakened that the result was death by neglect. Thus, NF-κB might normally contribute to both negative and positive selection of DP cells.

Given such potentially complex scenarios, this transcription factor might well appear to have different functions, depending on the experimental conditions employed (quality and strength of NF-κB inhibition; the cell types in which NF-κB is inhibited/reduced; the biologic assay). Not surprisingly then, some studies come to different conclusions than the ones cited above. One study failed to note any effects of an IκB superrepressor on negative selection, while positive selection, especially of CD8⁺ SP cells was demonstrated to be significantly impaired.[21] That latter result is also consistent with the loss of peripheral T cells and of CD8⁺ cells in particular, which has been noted in a number of studies involving IκB transgenes.[20-23,27] Two studies failed to note any significant perturbations of T cell development upon inhibition of NF-κB activity, one involving an IκB superrepressor[25] and another dominant negative-acting IKKα and/or IKKβ transgenes.[24]

Finally it has been reported that negative selection of DP thymocytes may be mediated by specific inhibition of NF-κB. In vivo administration of a negatively selecting peptide into mice bearing the matched TCR transgene (VSV8 peptide into N15 TCR transgenic RAG2⁻/⁻, H-2^b mice) induced the expression of IκB$_{NS}$ in DP thymocytes, but, interestingly, not in mature T cells.[28] This novel IκB-like protein inhibited NF-κB activity and acted like a superrepressor, since it lacks the phosphorylation and ubiquitination sites found in the signal-responsive IκBα, β and ε proteins. Ectopic expression of this protein led to a partial reduction of DP and SP thymocytes as well as a marked increase in apoptosis of thymocytes in response to anti-CD3 stimulation. It remains to be confirmed that this IκB-like protein is critical to negative selection, but if so, it suggests that negative selection induces death via inhibition of NF-κB. While this seems at odds with the reported pro-apoptotic roles of NF-κB in negative selection discussed above, it still remains possible that expression of IκB$_{NS}$ is in some way controlled by NF-κB. If so, this transcription factor would again have pro- and anti-apoptotic activities, depending on the timing and context. NF-κB may have multiple, complex inputs into DP thymocyte selection, where small differences in the level and/or quality of NF-κB activity could make the difference between life and death.

As discussed above, transgenic expression of dominant negative IKKβ prevented the massive apoptosis of DP thymocytes upon in vivo administration of anti-CD3 antibodies.[24] However, the opposite effect was observed with a dominant negative IKKα transgene, as this apparently made DP thymocytes even more susceptible to induced apoptosis.[24] This suggests the possibility that IKKα is part of an anti-apoptotic pathway, while IKKβ, the primary signaling kinase downstream of the TCR, may communicate an apoptotic signal, even if only indirectly. RelB complexes may be involved in mediating the anti-apoptotic role of IKKα. Firstly, IKKα regulates RelB activation via the so-called alternative or nonclassical/noncanonical pathway of activation.[3,4,7,29] IKKα, but not IKKβ or NEMO, is necessary for processing of the IκB-like inhibitor of RelB, p100, to p52. This results in nuclear translocation of the RelB/p52 dimer. Secondly, RelB-deficiency in mice resulted in fewer mature SP thymocytes, and this defect was cell-autonomous.[30] When cultured, the mutant thymocytes were more prone to apoptose, and when stimulated via the TCR, it was the DP and early SP mutant thymocyte populations in particular that exhibited higher rates of apoptosis when compared to their wild-type counterparts.

NF-κB's participation in DN thymocyte survival and DP apoptosis could also be inferred indirectly from studies of the HLH proteins, E2A and HEB. Transgenic mice, whose thymocytes expressed E2A/HEB inhibitors Id1 and/or Tal1 underwent apoptosis at the DP stage.[31,32] The late DN thymocytes of these mice exhibited very high NF-κB activity, which may have been activated by a strong pre-TCR-derived signal, particularly strong due to the absence of negative modulation exerted by E2A/HEB. Even without TCR rearrangements these DN thymocytes can progress to the DP stage, presumably due to a higher basal signaling in these cells. E2A/HEB may normally set a higher threshold for pre-TCR derived signals, although the mechanisms for this are unknown. Once at the DP stage, however, the abnormally high TCR activation signal may have been interpreted as stimulation by self-antigens, triggering apoptosis (equivalent to negative selection). The involvement of NF-κB in apoptosis in the Id1/Tal1 transgenic DP thymocytes is shown more directly by rescue from apoptosis by transgenic IκB superrepressor. During normal DN differentiation, pre-TCR signaling may induce some expression of the HLH inhibitor, Id3, allowing NF-κB activation and developmental progression. TCR stimulation at the DP stage by self-antigens may cause higher levels of Id3, leading to even more NF-κB activation and thus triggering apoptosis,[32] either directly or indirectly (see above).

Single-Positive Thymocytes and Peripheral T Cells

Several early reports noted that expression of IκB superrepressor transgenes in developing thymocytes resulted in decreases in peripheral SP cells, in particular of TCRαβ^{high} CD8αβ$^+$ peripheral T cells.[20-23,27,33,34] Loss of IKKβ (on a TNFRI deficient background, so mice survived, see above) decreased numbers of thymocytes and peripheral T cells, and this was speculated to be due to a defect in TCR-induced proliferation based on in vitro observations.[6] Loss of IKKα, on the other hand, did not effect numbers of developing or mature T cells.[35,36] Finally, radiation chimeras generated with c-Rel and RelA double-deficient hematopoietic precursors contained fewer peripheral T cells, a defect not rescued by a Bcl-2 transgene.[8,9] These early observations supported a role(s) for NF-κB in SP thymocytes/peripheral T cells.

More recently the roles of NF-κB in development and maintenance of T cells have also been explored in CD4-promoter driven, cre-mediated conditional T cell knockouts of IKKβ and NEMO/IKKγ, and in knockins of an activation-deficient IKKβ (which functions as an IKK dominant-negative).[37,38] While this approach would seem well suited to test the roles of IKKs during DP thymocyte selection, it was useful for assessing the need for NF-κB only after DP selection has largely been completed. This is so because developing thymocytes in this conditional knockout model retain at least some IKK activity until they have progressed to the early SP stage, as only then the vast majority of cells have completed cre-mediated recombination and the wild-type protein has been lost.[37]

Mere deletion of IKKβ did not grossly affect the generation and maintenance of naïve peripheral T cells, although peripheral T, especially CD8$^+$ T cells were somewhat reduced,[37] in line with results obtained with transgenic IκB superrepressor mice (see above).

However, the absence of a more pronounced effect is most certainly due to compensation by IKKα, since conditional deletion of NEMO or knockin of the activation-deficient IKKβ completely prevented the appearance of any peripheral mutant T cells.[37] Furthermore, even SP thymocytes, especially CD8$^+$ ones, were markedly reduced in the conditional NEMO knockouts, and this was accompanied by increased apoptosis of these thymocytes.[37] Therefore, a cell-autonomous IKK activity, mediated by either IKKα or IKKβ is critical for maturation/survival of SP T cells. Interestingly, elimination of TCR-induced activation of NF-κB in Bcl-10 knockouts did not lead to loss of mature T cells, suggesting that it was not the TCR, but an as yet unknown signal that was responsible for NF-κB activation and thus generation/survival of mature T cells.[37] This signal may be analogous to the BAFF/BAFFR survival signal that is required in maturing and mature B cells (see below).

While loss of IKKβ activity was largely compensated by IKKα during development of the bulk population of T cells, regulatory T cells (T_r)[37] and NKT[38] cells were exquisitely sensitive to loss of IKKβ, as were memory T cells in the periphery.[37] Therefore, IKKβ appears to have a unique function in these subpopulations that cannot be substituted for by IKKα. Support for the notion that these T cell subtypes may require greater NF-κB activity also comes from the observed reduction of regulatory (and memory) T cells in NF-κB1 and c-Rel double-deficient mice,[39] and of NKT cells in mice transgenic for the IκB superrepressor.[39] In the latter case thymic as opposed to peripheral NKT cells were preferentially reduced, probably the result of incomplete inhibition by the superrepressor in the thymus, allowing some cells to escape into the periphery and eventually fill this niche. The generation of NKT cell thymic precursors is also dependent on expression of RelB in stromal cells.[40,41]

NKT, T_r and memory cells exhibit a more activated phenotype.[38] Quite possibly these cells have an increased need for NF-κB activity, such that loss of IKKβ cannot be sufficiently compensated by IKKα. DP thymocytes serve as precursors for T_r cells. Unlike regular DP thymocytes, which are eliminated by recognition of self-antigens, T_r cells instead are positively selected by self-antigens, probably due to the identity and context of the antigen-presenting cells. CD4$^+$CD25$^+$ T_r cells act as supressors of autoreactive effector T cells in the periphery and thus are critical to maintain tolerance and prevent autoimmunity. NKT cells too are positively selected by recognition of self-antigens in the thymus, in this case self-lipids presented by the nonclassical MHC class I-like molecule CD1d.

Although development of T_r, NKT and memory cells depends on antigen, only T_r and CD4$^+$ memory cells could be shown to depend on TCR-driven NF-κB activation.[38] This conclusion is based on the absence of these latter cells in mice deficient in components needed for TCR signal-induced NF-κB activation, such as Bcl-10.[38] T_r cells are missing in both thymus and periphery in Bcl-10 knockouts. In contrast, NKT cells showed no such dependence on these signaling components. This suggests that TCR-independent signals are responsible for NF-κB activation in developing and peripheral NKT cells, although it remains theoretically possible that NKT TCRs activate NF-κB via an unknown signaling pathway. While NKT cells were normal in CARMA1/Card11, PKCθ or MALT1 knockouts (all are required for TCR-induced NF-κB activation), peripheral though not thymic NKT cells were reduced in Bcl-10 knockouts.[38] Therefore, surprisingly, an unknown peripheral, nonTCR signaling pathway may involve Bcl-10. (The main contributions of NF-κB to T cell development are summarized in Table 1.)

Pre-B Cells

After birth, early B cell development takes place in the bone marrow. The earliest recognized precursors are termed pre-pro-B[42] (for nomenclature used in this review see Hardy et al[43]). In late pro-B cells, immunoglobulin μ heavy chains undergo rearrangement and association with surrogate light chains (Vpre-B and λ5) to form the pre-B cell receptor (pre-BCR). Cells advance to the large pre-B stage and expand. Large pre-B cells eventually progress to become noncycling small pre-B cells that no longer express the pre-BCR, but which begin to rearrange κ or λ light chains. Once successfully rearranged, a κ or λ chain combines with the μ heavy chain to form a B cell receptor (BCR). Expression of a BCR (IgM) on the cell surface marks the beginning of so-called immature B cells, which eventually leave the bone marrow to complete their full maturation in the spleen.[43]

Several lines of evidence suggest that NF-κB participates in both κ and λ rearrangement, although proof for an essential role remains elusive. The Igκ locus is regulated by a distal (3' Eκ) and by a proximal, intronic enhancer (iEκ).[44-46] The latter enhancer contains the first κB element identified.[47] These regulatory elements function as transcriptional enhancers in transfection experiments, although their in vivo role may be more complex[46] (see below). Homozygous disruption of either the iEκ[47] or the 3' Eκ partially impaired rearrangement of this locus, but did not completely abolish it.[46] However, disruption of both enhancers resulted in a near complete block of V to Jκ rearrangements.[46] This suggests that the two enhancers encode somewhat redundant activities in controlling the rearrangement process.

Work by Bergman and colleagues[48-50] specifically implicated the κB element within the iEκ enhancer in demethylation of the Igκ locus in B cells, a necessary step in the rearrangement process that occurs in small pre-B cells once chromatin has become 'accessible'. These investigators also proposed that 'opening' of chromatin and demethylation only occurred on one of the alleles, which would explain the allelic exclusion phenomenon. The rearranging allele may be marked early on by epigenetic mechanisms associated with the order of replication of the two chromosomes. The conclusion that the κB site controlled rearrangement via demethylation was based largely on the demethylation of a transfected, in vitro methylated part of the Igκ locus that included the iEκ enhancer. Demethylation occurred only on open chromatin,[50] and only with an intact κB element.[48-50] Furthermore, demethylation was dependent on activated NF-κB. S107 plasmacytoma cells are blocked in NF-κB activation in response to many signals and they lack constitutive activity. Accordingly, demethylation of the transfected piece of the Igκ locus was blocked in these cells, but could be induced by cotransfection of RelB. RelA failed to do so, probably because it was inhibited and remained in the cytoplasm in S107 cells. Thus RelB appeared to control the demethylation process, at least in these cells.[50]

Data obtained with Abelson murine leukemia virus transformed pre-B cells in culture has implicated NF-κB in control of Igκ locus germline transcription and rearrangement.[46] Germline transcripts precede and appear to be necessary for rearrangement, perhaps signifying the accessibility of chromatin. The Abelson virus encoded v-abl oncogene may impair NF-κB and other transcription factors. The transformed cells are frozen in a large pre-B-like state, since they cycle and fail to rearrange their Igκ locus. Stimulation with LPS activated NF-κB, increased binding to the iEκ κB site and induced κ germline transcription and rearrangement. Similarly, inactivation of the v-abl protein (using drugs or a temperature-sensitive (ts)-v-abl) resulted in germline transcription and rearrangement.[46,51] If the v-abl transformed cells were engineered to also express the IκBα mutant superrepressor, then LPS-induced or temperature (ts-v-abl)-induced germline transcription and rearrangement of κ were blocked.[52] Thus rearrangement of Igκ induced by relief from the v-abl block did require NF-κB activity.

Not all observations appear to be entirely consistent with the hypothesis that activation of NF-κB in pre-B cells might regulate germline transcription and demethylation and thus rearrangement of the Igκ locus. Early in vivo footprinting studies demonstrated that the κB site in the iEκ enhancer was already occupied at the pro-B cell stage.[46] While this does not rule out a role for this site (or NF-κB) in germline transcription and/or demethylation, and thus in rearrangement, it appears to rule out that binding of newly activated NF-κB to the iEκ is an initiating event as cells transit from large to small preB cells. Instead, progression was correlated with changes in occupancy of various transcription factor sites within the 3'Eκ[46] (and not the iEκ). Therefore NF-κB and the κB element may contribute to making the Igκ locus more accessible or assist in the assembly of a demethylation complex, but they appear not to initiate these processes. Because proteins were bound to the iEκ κB (and surrounding sites) even prior to pre-BCR formation, occupancy of this site must be independent of the pre-BCR. However, it is also possible that a pre-BCR-derived signal modified or even exchanged the proteins that interacted with the iEκ κB site, given that in vivo footprinting could not identify the proteins bound.

NF-κB has been suggested to regulate λ light chain germline expression/rearrangement.[53] Three synergistic NF-κB sites were identified in the human λ enhancer[54] while the mouse λ enhancer was reported to contain only a mutated NF-κB site, which may explain its low activity.[55] Nonetheless, inhibition of NF-κB activity via the IκB superrepressor prevented not only κ but also λ rearrangement in response to temperatureinduced inactivation of ts-v-abl in transformed mouse pre-B cells.[53]

Another study failed to implicate NF-κB in κ or λ rearrangement or expression. Mice with IκB superrepressor under the control of the μ heavy chain enhancer and thus expressed in B lineage cells produced near normal levels of $κ^+$ and $λ^+$ cells.[56] However, the superrepressor may not have been very effective in this case since both precursor and mature B cells from these mice still exhibited significant constitutive NF-κB activity. In addition, the strong LPS signal still induced NF-κB in the superrepressor-bearing splenic B cells ex vivo, although the presumably weaker BCR-induced NF-κB activation was impaired.[56] Rearrangement may require only minimal NF-κB activity and given the plasticity of the developmental process, any decrease in the rate of production of $κ^+$ or $λ^+$ cells could have easily been covered up as precursors continue to be generated and as $κ^+$ or $λ^+$ cells may expand to fill later developmental stages. Mice deficient in both NF-κB1 and NF-κB2 generated κ+ or $λ^+$ cells as well, although at somewhat reduced rates[57] (see below). Once generated, these B cells had normal levels of light chain expression on their surface,[57] suggesting that these NF-κB factors did not regulate final light chain expression.

Our results and others, suggest that NF-κB may act as a survival factor during generation of small pre-B cells from pro-B cells[58] (Claudio and Siebenlist, unpublished). In adoptive transfer experiments, precursors lacking NF-κB1 and NF-κB2 produced reduced numbers of small pre-B and immature B cells in bone marrow as compared to the numbers of earlier-staged large pre-B/pro-B cells and this effect was greatly exacerbated in the presence of competing wild-type cells (Claudio and Siebenlist unpublished). Therefore, while the defect in mutant developing B cells was relatively mild, it became dramatic when these mutant cells were forced to compete with wild-type cells, presumably because the wild-type cells fill developmental niches more effectively. The transition of mutant pro-B to small pre-B was also partially blocked in bone marrow cultures ex vivo (Claudio and Siebenlist, unpublished). In this assay bone marrow cells are stimulated with IL-7 for several days, during which time pro-B cells selectively expand to make up nearly the entire culture. Withdrawal of IL-7 then stops expansion and reveals B cells advancing towards later developmental stages. In cultures generated with NF-κB1 and NF-κB2 double-deficient bone marrow we observed relatively fewer small pre-B and immature B cells. Those that were generated appeared to be more prone to apoptosis. Feng et al[58] have also recently noted a reduction in small pre-B and immature B cells in radiation chimeras generated with bone marrow cells transduced ex vivo with retroviruses encoding the IκB superrepressor (dominant-negative). A Bcl-X transgene expressed in these same B cells rescued the loss of cells during the pro-B to small pre-B/immature B transition, confirming the role of NF-κB as a survival factor during development of B cells in bone marrow and specifically in generation of small pre-B cells. These data suggest but do not prove that the pre-BCR signal may be responsible for activation of NF-κB at this stage.

NF-κB's role in generating small pre-B and immature B cells may be analogous to that in T cells, where it is needed for optimal survival at the preTCR stage (see above).[17] Supporting this notion, btk[-/-] mice (lacking Bruton's tyrosine kinase (BTK), a member of the BCR signal pathway involved in activation of NF-κB) have fewer small pre-B cells.[59] In humans the block in small pre-B cell generation is complete in the absence of a functional BTK.[60] NF-κB may enhance survival of small pre-B cells or their immediate precursors, with the pre-BCR providing the initial activating signal.

Further evidence for a role for NF-κB in pre-BCR signaling comes from Blk-, Lyn- and Fyn- triple-deficient mice.[61] pre-B cells from these mutant mice were impaired in anti-Igβ-stimulated pre-BCR-mediated NF-κB signaling (apparently via PKCλ), while signaling via tyrosine-phosphorylation of Syk remained intact. The triple-deficient mice were blocked in pro- to pre-B transition, generating few small pre-B and immature cells and scarcely any mature B cells. The remaining triple knockout pre-B cells exhibited increased apoptotic rates.

Additional evidence points to activation of NF-κB in pre-B cells or their immediate precursors. Analysis of B cells during their progression from pro-B to small pre-B and immature B cells in bone marrow cultures (see description of culture above) revealed activation of NF-κB in response to withdrawal of IL-7, which stopped proliferation of pro-B cells. The NF-κB activity consisted mostly of p50/RelA and RelA/cRel dimers.[62] Pre-BCR formation is the only known positive signaling in this system and may therefore be the trigger for NF-κB activation.

Immature B Cells

IgM first appears on immature B cells. IgM[+] B cells with strong self-antigen reactivity may apoptose or undergo further light chain rearrangement to change their IgM specificity. Immature/Transitional B cells then migrate to the spleen where they mature.[43] BCR stimulation in immature B cells induces apoptosis, while it induces proliferation in mature B cells,[63] suggesting that these cells differ in their linkage from the BCR to downstream targets.

The WEHI 231 tissue culture cell line resembles immature B cells based on cell surface markers and thus provides a possible model for negative selection at the immature/transitional B cell stage. BCR stimulation of WEHI 231 transiently induced NF-κB activity beyond already high basal levels, but this was then followed by a significant decrease in activity to below basal levels and the cells apoptosed.[64] Lowered NF-κB activity correlated with down-regulation of c-myc and cyclin D2 and with increased cyclin-dependent kinase inhibitor (CKI) p27 (Kip1) synthesis.[65] This caused growth arrest and apoptosis.[65] Apoptosis could be blocked by costimulation with CD40 ligand, which induced NF-κB and one of its targets, c-myc.[66] This suggests that BCR signaling in WEHI 231 immature B cells ultimately down-regulates NF-κB to effect negative selection, which contrasts with up-regulation of NF-κB in mature B cells.

Whether the transformed cell line WEHI 231 with constitutively high NF-κB activity can be a model for normal physiologic negative selection of immature B cells in vivo is questionable. Nevertheless, it has been suggested that ex vivo immature B cells may not induce NF-κB to the same extent as mature cells (unpublished observations reported in review by Monroe[63]). Although calcium influx and tyrosine-phosphorylation events in immature B cells were largely similar to those seen in mature B cells, only mature B cells recruited the BCR into lipid rafts after stimulation, while immature cells failed to do so. This is likely the reason why BCR stimulation in immature B cells occurred in the relative absence of phosphatidyl inositol biphosphate hydrolysis and thus generation of diacylglycerol (DAG). DAG is essential for activation of many Protein Kinase C (PKC) family members. Importantly, PMA-induced activation of PKCs rescued immature B cells from BCR-induced apoptosis.[63]

These data do not prove that lack of PKC activation and consequently impaired NF-κB activation are the reasons why BCR-stimulated immature B cell apoptosed. However, in BCR-stimulated mature B cells activation of the DAG-dependent PKCβ isoform appeared to assure survival via activation of NF-κB. BCR-stimulated, mature PKCβ$^{-/-}$ B cells exhibited normal tyrosine-phosphorylation events (including upregulation of early cell cycle entry proteins), but did not proliferate well due to poor survival.[67-69] Activation of NF-κB in response to BCR stimulation was specifically impaired in these mutant cells, but not in response to CD40.[67-69] In consequence, BCR-induced expression of NF-κB controlled survival genes, such as Bcl-x, was abrogated. Further analysis showed that PKCβ was required to recruit the adaptors CARMA1 and Bcl-10 into lipid rafts, which ultimately recruit and activate the IKKs and NF-κB.[70] PKCβ may play a role analogous to that of PKCθ in T cells (see above). PKCβ-deficient B cells resemble B cells from XID mice (mutated in BTK) in their response to BCR stimulation, reportedly because BTK, among other functions is responsible for sustained BCR-induced activation of PKCβ and thus NF-κB.[70] In any case, it remains to be shown whether BCR-induced activation of NF-κB and its targets is significantly impaired in immature B cells when compared to mature B cells.

Generation and Maintenance of Mature B Cells

Immature B cells leave the bone marrow and migrate to the spleen. These cells are so-called transitional B cells. Final development occurs in the spleen where transitional B cells evolve through as many as three distinguishable stages (transitional 1 through 3) before they become mature, naïve B cells.[71,72] Transitional B cells also give rise to marginal zone B cells, which do not recirculate. Transitional 2 B cells move into and accumulate in B cell follicles, a stage when B cells first respond positively to stimulation via their BCR, so the transition from stage 1 to 2 is functionally highly significant.[43,71,72] After reaching full maturity in the B cell follicles, these cells recirculate through the entire periphery, including lymph nodes and bone marrow.[43] Development of transitional B cells in the spleen requires NF-κB, since mice deficient in NF-κB1 + NF-κB2[57] or in RelA + c-Rel[8,9] are blocked in B cell development just before/at the transitional 2 stage, when transitional B cells become more like mature cells; these mutant mice are totally devoid of any mature follicular/recirculating B cells. NF-κB1 + NF-κB2 dKO mice also lack peritoneal B1 B cells (Claudio and Siebenlist, unpublished results). The mechanisms behind these developmental blocks will be discussed below.

Immature/transitional B cells migrating from the bone marrow to the spleen in NF-κB1, NF-κB2 double knockout (dKO) mice may already be somewhat at a disadvantage before they reach their final block in the spleen (see above and below). Relatively fewer transitional 1 B cells were observed in the spleen than expected, based on the number of immature B cells generated in dKO bone marrow[73] (Park and Siebenlist, unpublished results). This relatively mild loss became more apparent in radiation chimeras, when mutant immature/transitional B cells were forced to compete with adoptively cotransferred (and differentially marked) wild-type B cells. Based on these results we speculated that B cells might have a defect in homing to spleen, a defect intrinsic to B cells, since the adoptively transferred host mice produced normal levels of stroma-derived chemokines in the wild-type spleen. Immature/transitional B cells from NF-κB1, NF-κB2 dKO mice were previously noted to lack expression of the chemokine receptor, CXCR5 (BLR1).[74] In addition, the chemokine receptor CCR7 was impaired in its expression in mutant immature/transitional B cells (Park and Siebenlist, unpublished observations), consistent with prior data linking high CCR7 expression with high NF-κB activity in Hodgkins Lymphomas.[75] Both receptors have been suggested to contribute to entry of B cells into the spleen, even though this is not their primary function.[76] Therefore, although NF-κB1, NF-κB2 dKO B cells were still able to migrate into the spleen, lack of both CXCR5 and CCR7 may have been responsible for the lower efficiency, which was much more noticeable in the presence of wild-type cells.

The developing NF-κB1, NF-κB2 dKO B cells that did migrate into the spleen were severely impaired in survival; they were especially prone to undergo apoptosis when placed in culture, even without any stimulation.[73] RelA, c-Rel dKO developing B cells were also impaired in survival, based on their very rapid turnover in spleens in vivo.[9] Furthermore, both NF-κB1, NF-κB2 dKO and RelA, c-Rel dKO B cells were largely lacking in expression of the anti-apoptotic proteins A1 and Bcl2.[9,73] Together these findings suggested that impaired survival in both compound knockout B cells was responsible for the complete developmental block encountered during the transitional phases. The block in both mutants is intrinsic and autonomous to their B cells, based on adoptive cotransfer experiments of mutant together with wild-type bone marrow.

The notion that impaired survival is responsible for the block in the compound mice is supported also by the rescue of developmental progression in the presence of a Bcl-2 transgene. Exogenous expression of the anti-apoptotic Bcl-2 protein in NF-κB1, NF-κB2 dKO(Claudio and Siebenlist, unpublished) and RelA, c-Rel dKO[9] mice allowed mutant B cells to accumulate in spleen and express mature B cell markers. It did not, however, promote full maturation nor did it restore the ability of these B cells to proliferate after BCR stimulation[9] (Claudio and Siebenlist, unpublished). Expression of the mature markers CD21 and CD62L, for example, was not fully restored in NF-κB1, NF-κB2 dKO B cells by a Bcl-2 transgene and these cells remained functionally impaired in basal (and antigen-specific) immunoglobulin production (Claudio and Siebenlist, unpublished). NF-κB is therefore critical not only for survival of transitional B cells, but also for their full functional maturation.

What are the relevant targets of NF-κB that assure survival of developing B cells? Bcl-2 is likely to be one of these targets, based on the data discussed above and additional data presented below. Furthermore, several reports have concluded that that the Bcl-2 gene can be directly regulated by κB elements that may bind several different NF-κB complexes.[77-79] Bcl-2 expression normally rises as B cells progress through the transitional B cell phase. Bcl-2 is not essential for development, however, since Bcl-2 knockout mice were able to generate mature B cells, albeit at significantly reduced levels, coupled with loss of the remaining cells' long-term survival.[80,81] A1 is a known direct target of in particular c-Rel/NF-κB, but c-Rel is not essential for development and A1 is unlikely to be essential. Mice deficient in either A1 or c-Rel had normal B cell development.[80,81] However, only one of three potentially expressed isotypes of the A1 gene was eliminated, so questions regarding A1 remain. Nevertheless, the most likely hypothesis is that various NF-κB factors and their anti-apoptotic targets have at least partially redundant activities to insure optimal survival, such that elimination of only one factor or one anti-apoptotic target does not obviously block development of B cells, especially when there is a continuous influx of newly formed transitional B cells into the spleen.

Although single knockouts of NF-κB factors showed no significant defects in development of regular follicular B cells, all single knockouts were nearly completely (NF-κB1, NF-κB2, RelB) or partially (RelA, c-Rel) devoid of marginal zone (MZ) B cells, indicating that the generation/maintenance of this mature B cell population is highly dependent on NF-κB activity[82,83] (Claudio and Siebenlist unpublished). In addition, even though mature follicular B cells were generated in the single KOs, these cells were partially impaired in survival in at least some situations, revealing the importance of all NF-κB factors in overall B cell survival. For example, while B cells matured normally in NF-κB1 single KOs, the mature cells turned over more rapidly in vivo and showed decreased survival in unstimulated cultures ex vivo.[84] B cells also matured normally in c-Rel-deficient mice, but the mature cells were impaired in proliferation and survival ex vivo when stimulated via the BCR.[84] Loss of RelA resulted in a significant decrease of Bcl-2 and cFLIP, making these mutant B cells highly sensitive to TNF-induced apoptosis; in the presence of high levels of TNF few RelA-deficient B cells were generated.[85] B cell development was also largely normal in NF-κB2-deficient mice, save for a small reduction in numbers of mature cells, but the survival of mature and, even more surprising, of transitional B cells was impaired ex vivo.[73] The life of wild-type immature/transitional B cells in ex vivo cultures could be significantly extended by the addition of BAFF, a member of the TNF family, but this factor had no effect on NF-κB2-deficient immature/transitional B cells (see below).

The latter results suggested that NF-κB2 was already important quite early in the life of transitional B cells, at least ex vivo, which may explain why NF-κB1, NF-κB2 dKO B cells were blocked at the transitional 1 stage and thus slightly earlier than B cells deficient in RelA and c-Rel.[81] The latter mutant cells were reported to progress to the transitional 2 stage, albeit with much reduced numbers. Transitional B cells may be exposed to several different survival signals as they progress through the different stages, signals that appear to be partially redundant, but may mediate their effects in part via distinct NF-κB complexes.

A striking advance in our understanding of the factors underlying B cell development was the elucidation of the role of the B cell Activating Factor (BAFF). BAFF was found to be critical for B cell survival at the transitional stage, as well as for long-term survival of mature cells.[86] This ligand engages three receptors, TACI, BCMA and BAFF receptor (BAFFR). The first two receptors are also engaged by a BAFF-related ligand, APRIL and their deletion in knockout mice did not block B cell generation. However, B cell development in BAFF$^{-/-}$, BAFFR$^{-/-}$ and TACI-Ig transgenic mice (which produce a decoy receptor for BAFF and APRIL) was blocked during progression from transitional 1 to transitional 2 B cells (T1 to T2). There were few T2 B cells and a complete absence of mature B cells.[86-90] Consistent with the survival function of BAFF signaling in B cell development, transgenic over-expression of anti-apoptotic Bcl-2 restored the mature follicular B cell population in mice containing a TACI transgene (in which BAFF function is inhibited) and in BAFFR$^{-/-}$ mice.[88,90] However, expression of some of the mature B cell markers was still somewhat impaired and marginal zone B cells were not restored at all, suggesting roles in addition to survival for BAFF.[88,90]

As discussed, BAFF-promoted survival of transitional B cells ex vivo was dependent on NF-κB2. BAFF not only failed to extend the life of transitional B cells derived from NF-κB2-deficient and BAFFR mutant (A/WySnJ) mice, but also from *aly/aly* mice (mutated in the NF-κB-inducing kinase, NIK).[73] This indicated that BAFF/BAFFR signaling was mediated, at least in part, by the alternative (non-classical, non-canonical) pathway of activation, in which the NIK and IKKα kinases cooperate to induce the processing of p100 to the p52 form of NF-κB2. BAFF indeed induced p100 processing in mature and in developing B cells, starting with immature/transitional 1 B cells, which led to nuclear accumulation of p52/RelB and possibly other complexes. BAFF-induced processing and NF-κB activation was completely independent of NEMO (IKKγ), which together with IKKβ is a necessary component of the classical, IKK-complex-mediated pathway.[73]

Loss of BAFF or BAFFR led to a nearly complete block in B cell development from the T1 to T2 transition onward, although the functionally inactivating mutation of the BAFFR in A/WySnJ mice was slightly less restrictive.[89] As discussed, a BAFF/BAFFR-induced survival signal for transitional (and mature) cells is transmitted via the alternative activation pathway (NIK - IKKα - NF-κB2). However, functional loss of NIK (*aly/aly* mutation) or IKKα or NF-κB2 did not cause as severe a block in B cell development as loss of BAFF or BAFFR.[35,36,73,89,91] Radiation chimeras with NIK-mutant or IKKα-deficient (or IKKα-mutant) bone marrow reduced, but did not eliminate the mature B cell population, and chimeras with NF-κB2-deficient bone marrow contained near-normal numbers of mature B cells.[35,36,73,91] (One report also noted a mild reduction of transitional B cells in IKKα-deficient chimera[36]). Therefore, signaling via BAFF/BAFFR may contribute to survival (and/or differentiation) via a second pathway, independent of the alternative activation pathway. The reason why specifically those B cells lacking NF-κB2 appeared less impaired in vivo than those lacking NIK or IKKα may be due in part to the loss of the inhibitory p100 precursor protein in the former, but not the latter knockouts (p100 is the main inhibitor of RelB).

It is not known what other pathways may be activated by the BAFF receptor. One report suggests a very weak, but functionally significant stimulation of the classical NF-κB activation pathway by BAFF in mature B cells, a pathway that reportedly involved NF-κB1.[92] If such occurred also in transitional B cells, it might explain why only loss of both NF-κB2 and NF-κB1 led to a block in B cell development during the transitional stage. However, the block in these doubly deficient mutant B cells occurred slightly earlier than that seen in BAFF- or BAFFR-deficient mice since the latter contained some T2 cells. In addition, BAFF failed to promote survival of transitional B cells lacking only NF-κB2, but containing NF-κB1, while those lacking NF-κB1 responded well to BAFF.[73] Regardless of whether BAFF can transmit some signal via the classical pathway, this NEMO/IKKβ-dependent pathway is likely to contribute to survival of transitional B cells in vivo (presumably in response to some signal). CD19 promoter-Cre recombination enzyme-driven conditional knockouts of NEMO or conditional knockins of an IKKβ mutant in B cells resulted in noticeably fewer transitional B cells in the spleen (in addition to loss of mature B cells, see below).[93] As discussed previously here, such conditional knockouts/knockins are more difficult to interpret: although loss of the targeted DNA segment increases as B cells progress, it is never complete, and some cells escape. Aside from that, even cells in which the DNA was recently deleted will continue to express the protein for some period of time thereafter while cells progress. It is therefore difficult to pinpoint exactly when loss of classical pathway (NEMO/IKKβ) first interferes with developmental progression. However in the case of the conditional NEMO knockout and IKKβ mutant knockin the first problem must have occurred during the transitional stage (or even earlier), since B cells present at this developmental stage were apparently already somewhat counter-selected for those that had escaped recombination.

In transitional B cells ex vivo, BAFF led to increased expression of Bcl-2 via an NF-κB2-dependent (alternative) pathway of activation.[73] In mature B cells BAFF has been reported to increase expression of in particular A1 and Bcl-x_L.[94] It is thus possible that transitional and mature B cells differ in their response to BAFF, possibly due to differences in expression of other receptors for BAFF. Recently an additional new mechanism of action for BAFF in mature B cell survival has been suggested by a report that BAFF in an unknown way blocks the (spontaneous) nuclear localization of PKCδ, thereby preventing the pro-apoptotic function of this protein in the nucleus.[95]

Apart from BAFF, to what other signals do transitional B cells respond? The MHC class II chaperone, invariant chain Ii, has been reported as essential for B cell maturation via a mechanism in which a proteolytically released cytoplasmic fragment of Ii promotes RelA-mediated transactivation.[96] These findings were based on initial studies with Ii-deficient mice in which B cell maturation appeared to be largely blocked during in the final transitional stages. However, new studies suggest that while Ii-deficient mice do have reduced numbers of mature follicular B cells, they actually have normal or even increased numbers of marginal zone B cells; furthermore the reduction of mature follicular B cells is due to a shortened B cell life span rather than a developmental block per se. More importantly, Ii-deficient compound knockouts in which expression of MHC class II molecules is also severely reduced or completely eliminated exhibited normal B cell development, suggesting that Ii-free MHC class II molecules specifically interfered with B cell survival through an unknown mechanism and that neither Ii nor MCH class II are required for normal B cell development.[97,98]

The B cell receptor (BCR) is important for B cell development and it can signal activation of NF-κB, but does it play a role in transitional B cells? The BCR is certainly essential for maintenance of mature B cells,[99] as the CD23 promoter-Cre recombinase-driven conditional loss of the BCR leads to disappearance of mature B cells within days. The importance of BCR signaling during at least the final transitional maturation can be inferred from the phenotype of mice deficient in Bruton's tyrosine kinase, a component of the BCR pathway in mature B cells which mediates its functions in part via NF-κB.[100] These mice were impaired in the last stage of B cell maturation (transitional stage 2 to mature B cells) and T2 cells accumulated; they also were partially impaired in generation of B1 B cells.[100] Further evidence for the importance of the BCR in late (transitional) B cell development comes from mice deficient in CARMA-1/CARD 11, Bcl-10 or MALT-1. These adaptor proteins act downstream of the BCR (and TCR) in the signal pathway to NF-κB. Mice deficient in CARMA-1, Bcl-10 or MALT-1 generated few peritoneal B1 and MZ B cells and reduced numbers of mature B cells.[101,102] There is however presently no evidence for a role of BCR or this pathway during the earlier transitional phase of B cells (T1).

The maintenance or long-term survival of mature B cells depends on signaling from the BCR and BAFF receptors (see above) and possibly other receptors, mediated at least in part by the classical and alternative pathways of NF-κB activation. Conditional knockouts of the main components of the classical pathway, namely NEMO[93] and IKKβ[103] revealed that as cells progressed and ongoing deletion of targeted alleles should have neared completion, the small population of mature and in particular of marginal zone B cells still had a high percentage of wild-type NEMO/IKKβ alleles, as these 'escapees' were they only ones that had survived. Furthermore, mature B cells were impaired in survival ex vivo and exhibited increased turnover in vivo as even escaping cells eventually lost the wild-type alleles. Finally, if generation of newly maturing cells was blocked by administration of IL-7 antibodies, then all previously generated mature B cells disappeared, long before such cells would normally expire, as the Cre-mediated deletion of NEMO or IKKβ alleles approached completion in the remaining cells. Thus loss of the classical NF-κB activation pathway is incompatible with long-term survival and thus maintenance of all mature B cells, including B2 and B1 cells.[93,103] This conclusion is consistent with other data. A superrepressor IκBα transgene expressed in B cells resulted in dose-dependent reduction of the mature, recirculating B cell pool.[56] c-Rel, NF-κB1 double knockout mice had slightly fewer B2 B cells and a more sizeable reduction of B1 B cells, although the authors suggest that this may have been due to impaired proliferation on top of a defect in survival of mature B cells already noted in single NF-κB1 knockout mice.[104] A dominant-negative IKKβ mutant transgene expressed in B cells appeared to have no dramatic effect on development, most likely due to insufficient blocking of the wild-type activity.[105] But proliferation and antibody responses were impaired, which was also noted across the board not only for KOs of members of the classical NF-κB activation pathway, but also for single KOs of the various NF-κB factors, suggesting that functions of mature cells are even more sensitive to loss of the classical pathway or of NF-κB factors. Radiation chimeras generated with fetal liver cells from IKKα knockouts[36] or mutant IKKα knockins[35] (IKKα KOs die in utero due to non-hematopoietic defects) were reported to have a reduction in the mature, recirculating pool of B cells, with those cells present exhibiting increased turnover, reduced expression of anti-apoptotic A1, reduced survival ex vivo and impaired processing of p100 NF-κB2 to p52. Together with results showing a requirement for BAFF in survival of mature B cells in the periphery, the BAFF-induced processing of p100 mediated by IKKα appears to be essential for the proper maintenance of the peripheral B cell pool. (The main contributions of NF-κB to B cell development are summarized in Table 2.)

Perspectives

NF-κB is a critical contributor to the development and maintenance of lymphocytes. It does so primarily via intrinsic and cell-autonomous functions in lymphocytes, but also via non-cell autonomous mechanisms and activities in other cells. Activation of NF-κB transcription factors in response to signals encountered by developing lymphocytes leads to their survival; without such signals these lymphocytes undergo apoptosis by default. As lymphocytes mature, NF-κB's intrinsic and cell-autonomous roles appear to increase and to include contributions to differentiation and eventually proliferation. Upon maturation the maintenance of these cells also requires NF-κB. While both T and B cells critically rely on this transcription factor for their maturation and differentiation into distinct mature subsets, B cells appear to depend on a broader range and quantity of NF-κB activity.

Many important questions remain unanswered. Particularly intriguing are the likely complex functions of NF-κB in positive and negative selection of T lymphocytes, the yet-unknown activation signals that guide B cell development in the spleen, the possible direct involvement of NF-κB in demethylation and the specific gene targets of NF-κB at developmental junctions.

Throughout most of development individual NF-κB factors appear able to substitute for each other's cell-autonomous survival functions. Only certain compound knockouts or loss of entire NF-κB activation pathways result in noticeable deficits of mutant lymphocytes, and even then in some cases mature lymphocytes may be formed. However, the effects of the loss of a given factor or factors have usually been observed in the context of the original mutant mouse or in radiation chimeras. This tends to minimize the degree to which mutant lymphocytes are impaired, since they do not have to compete with their wild-type counterparts. When mutant cells are forced to compete, important contributions of NF-κB are readily revealed; wild-type cells increasingly fill available niches during developmental progression and ultimately mutant cells are almost entirely lost from the periphery. This mechanism may have evolved to ensure that only cells with properly functioning NF-κB systems reach or persist in the mature pool, given the importance of this transcription factor system in host defense. While only minimal NF-κB activity may be sufficient to enable progress to full maturity per se, mutant lymphocytes in which NF-κB activity was only partially impaired may be eliminated in competition with wild-type cells, since their developmental progression is comparatively inefficient.

References

1. Brown K, Claudio E, Siebenlist U. NF-κB. In: Smolen JS, Lipsky PE, eds. Targeted Therapies in Rheumatology. London: Publ Martin Dunitz, 2002:381-401.
2. Gerondakis S, Grossmann M, Nakamura Y et al. Genetic approaches in mice to understand Rel/NF-kappaB and IkappaB function: Transgenics and knockouts. Oncogene 1999; 18:6888-6895.
3. Li Q, Verma IM. NF-kappaB regulation in the immune system. Nat Rev Immunol 2002; 2:725-734.
4. Ghosh S, Karin M. Missing pieces in the NF-kappaB puzzle. Cell 2002; (Suppl):S81-96.
5. Horwitz BH, Scott ML, Cherry SR et al. Failure of lymphopoiesis after adoptive transfer of NF-kappaB-deficient fetal liver cells. Immunity 1997; 6:765-772.
6. Senftleben U, Li ZW, Baud V et al. IKKbeta is essential for protecting T cells from TNFalpha-induced apoptosis. Immunity 2001; 14:217-230.
7. Hayden MS, Ghosh S. Signaling to NF-kappaB. Genes Dev 2004; 18:2195-2224.
8. Grossmann M, Metcalf D, Merryfull J et al. The combined absence of the transcription factors Rel and RelA leads to multiple hemopoietic cell defects. Proc Natl Acad Sci USA 1999; 96:11848-1153.
9. Grossmann M, O'Reilly LA, Gugasyan R et al.The anti-apoptotic activities of Rel and RelA required during B-cell maturation involve the regulation of Bcl-2 expression. EMBO J 2000; 19:6351-6360.
10. Schmidt-Supprian M, Bloch W, Courtois G et al. NEMO/IKK gamma-deficient mice model Incontinentia pigmenti. Mol Cell 2000; 5:981-989.
11. Makris C, Godfrey VL, Krahn-Senftleben G et al. Female mice heterozygous for IKK gamma/NEMO deficiencies develop a dermatopathy similar to the human X-linked disorder Incontinentia pigmenti. Mol Cell 2000; 5(6):969-979.
12. Smahi A, Courtois G, Rabia Sh et al. The NF-kappaB signalling pathway in human diseases: From Incontinentia pigmenti to ectodermal dysplasias and immune-deficiency syndromes. Hum Mol Genet 2002; 11:2371-2375.
13. Kim S, La Motte-Mohs RN, Rudolph D et al. The role of nuclear factor-kappaB essential modulator (NEMO) in B cell development and survival. Proc Natl Acad Sci USA 2003; 100:1203-1208.
14. Germain RN. T-cell development and the CD4-CD8 lineage decision. Nature Reviews Immunology 2002; 2:309-322.
15. Hoffman ES, Passoni L, Crompton T et al. Productive T-cell receptor beta-chain gene rearrangement: Coincident regulation of cell cycle and clonality during development in vivo. Genes Dev 1996; 10:948-962.
16. Sen J, Venkataraman L, Shinkai Y et al. Expression and induction of nuclear factor-kappa B-related proteins in thymocytes. J Immunol 1995; 154:3213-3221.
17. Voll RE, Jimi E, Phillips RJ et al. NF-kappa B activation by the preT cell receptor serves as a selective survival signal in T lymphocyte development. Immunity 2000; 13:677-689.
18. Ivanov VN, Nikolic-Zugic J. Biochemical and kinetic characterization of the glucocorticoid-induced apoptosis of immature $CD4^+CD8^+$ thymocytes. Int Immunol 1998; 10:1807-1817.
19. Wang W, Wykrzykowska J, Johnson T et al. A NF-kappa B/c-myc-dependent survival pathway is targeted by corticosteroids in immature thymocytes. J Immunol 1999; 162:314-322.
20. Esslinger CW, Wilson A, Sordat B et al. Abnormal T lymphocyte development induced by targeted overexpression of IkappaB alpha. J Immunol 1997; 158:5075-5078.
21. Hettmann T, Leiden JM. NF-kappa B is required for the positive selection of $CD8^+$ thymocytes. J Immunol 2000; 165:5004-5010.

22. Esslinger CW, Jongeneel CV, MacDonald HR. Survival-independent function of NF-kappaB/Rel during late stages of thymocyte differentiation. Mol Immunol 1998; 35:847-852.
23. Hettmann T, DiDonato J, Karin M et al. An essential role for nuclear factor kappaB in promoting double positive thymocyte apoptosis. J Exp Med 1999; 189:145-158.
24. Ren H, Schmalstieg A, van Oers NS et al. I-kappa B kinases alpha and beta have distinct roles in regulating murine T cell function. J Immunol 2002; 168:3721-3731.
25. Ferreira V, Sidenius N, Tarantino N et al. In vivo inhibition of NF-kappa B in T-lineage cells leads to a dramatic decrease in cell proliferation and cytokine production and to increased cell apoptosis in response to mitogenic stimuli, but not to abnormal thymopoiesis. J Immunol 1999; 162:6442-6450.
26. Mora AL, Stanley S, Armistead W et al. Inefficient ZAP-70 phosphorylation and decreased thymic selection in vivo result from inhibition of NF-kappaB/Rel. J Immunol 2001; 167:5628-5635.
27. Boothby MR, Mora AL, Scherer DC et al. Perturbation of the T lymphocyte lineage in transgenic mice expressing a constitutive repressor of nuclear factor (NF)-kappaB. J Exp Med 1997; 185:1897-1907.
28. Fiorini E, Schmitz I, Marissen WE et al. Peptide-induced negative selection of thymocytes activates transcription of an NF-kappa B inhibitor. Mol Cell 2002; 9:637-648.
29. Mueller JR, Siebenlist U. Lymphotoxin beta receptor induces sequential activation of distinct NF-kappa B factors via separate signaling pathways. J Biol Chem 2003; 278:12006-12012.
30. Guerin S, Baron ML, Valero R et al. RelB reduces thymocyte apoptosis and regulates terminal thymocyte maturation. Eur J Immunol 2002; 32:1-9.
31. Kim D, Peng XC, Sun XH. Massive apoptosis of thymocytes in T-cell-deficient Id1 transgenic mice. Mol Cell Biol 1999; 19:8240-8253.
32. Kim D, Xu M, Nie L et al. Helix-loop-helix proteins regulate preTCR and TCR signaling through modulation of Rel/NF-kappaB activities. Immunity 2002; 16:9-21.
33. Mora AL, Chen D, Boothby M et al. Lineage-specific differences among $CD8^+$ T cells in their dependence of NF-kappa B/Rel signaling. Eur J Immunol 1999; 29:2968-2980.
34. Attar RM, Macdonald-Bravo H, Raventos-Suarez C et al. Expression of constitutively active IkappaB beta in T cells of transgenic mice: Persistent NF-kappaB activity is required for T-cell immune responses. Mol Cell Biol 1998; 18:477-487.
35. Senftleben U, Cao Y, Xiao G et al. Activation by IKKalpha of a second, evolutionary conserved, NF-kappa B signaling pathway. Science 2001; 293:1495-1499.
36. Kaisho T, Takeda K, Tsujimura T et al. IkappaB kinase alpha is essential for mature B cell development and function. J Exp Med 2001; 193:417-426.
37. Schmidt-Supprian M, Courtois G, Tian J et al. Mature T cells depend on signaling through the IKK complex. Immunity 2003; 19:377-389.
38. Schmidt-Supprian M, Tian J, Grant EP et al. Differential dependence of $CD4^+CD25^+$ regulatory and natural killer-like T cells on signals leading to NF-kappaB activation. Proc Natl Acad Sci USA 2004; 101:4566-4571.
39. Zheng Y, Vig M, Lyons J et al. Combined deficiency of p50 and cRel in $CD4^+$ T cells reveals an essential requirement for nuclear factor kappaB in regulating mature T cell survival and in vivo function. J Exp Med 2003; 197:861-874.
40. Sivakumar V, Hammond KJ, Howells N et al. Differential requirement for Rel/nuclear factor kappa B family members in natural killer T cell development. J Exp Med 2003; 197:1613-1621.
41. Elewaut D, Shaikh RB, Hammond KJ et al. NIK-dependent RelB activation defines a unique signaling pathway for the development of V alpha 14i NKT cells. J Exp Med 2003; 197:1623-1633.
42. Li Y, Hayakawa K, Hardy RR. The regulated expression of B lineage associated genes during B cell differentiation in bone marrow and fetal liver. J Exp Med 1993; 178:951-960.
43. Hardy RR, Hayakawa K. B cell development pathways. Annu Rev Immunol 2001; 19:595-62.
44. Queen C, Baltimore D. Immunoglobulin gene transcription is activated by downstream sequences. Cell 1983; 33:741-748.
45. Sen R, Baltimore D. Inducibility of kappa immunoglobulin enhancer-binding protein NF-kappa B by a posttranslational mechanism. Cell 1986; 47:921-8.
46. Schlissel MS. Regulation of activation and recombination of murine Igκ locus. Immunol Rev 2004; 200:215-223.
47. Xu Y, Davidson L, Alt FW et al. Deletion of the Ig kappa light chain intronic enhancer/matrix attachment region impairs but does not abolish V kappa J kappa rearrangement. Immunity 1996; 4:377-85.
48. Kirillov A, Kistler B, Mostoslavsky R et al. A role of nuclear NF-kappaB in B-cell-specific demethylation of the Igkappa lovud. Nat Genet 1996; 4:435-41.
49. Mostoslavsky R, Kirillov A, Ji Y-H et al. Demethylation and the establishment of κ allelic exclusion. Cold Spring Harb Symp Quant Biol.

50. Goldmit M, Bergman Y. Monoallelic gene expression: A repertoire of recurrent themes. Immunolog Rev 2004; 200:197-214.
51. Muljo SA, Schlissel MS. A small molecule Abl kinase inhibitor induces differentiation of Abelson virus-transformed preB cell lines. Nat Immunol 2003; 4:31-37.
52. Scherer DC, Brockman JA, Bendall HH et al. Corepression of RelA and c-rel inhibits immunoglobulin kappa gene transcription and rearrangement in precursor B lymphocytes. Immunity 1996; 4:435-41.
53. Bendall HH, Sikes ML, Oltz EM. Transcription factor NF-kappa B regulates Ig lambda light chain gene rearrangement. J Immunol 2001; 167(1):264-9.
54. Blomberg BB, Rudin CM, Storb U. Identification and localization of an enhancer for the human lambda L chain Ig gene complex. J Immunol 1991; 147(7):2354-8.
55. Combriato G, Klobeck H-G. Regulation of Human Igλ light chain gene expression by NF-κB. J Immunol 2002; 168:1259-1266.
56. Bendall HH, Sikes ML, Ballard DW et al. An intact NF-kappa B signaling pathway is required for maintenance of mature B cell subsets. Mol Immunol 1999; 36:187-95.
57. Franzoso G, Carlson L, Xing L et al. Requirement for NF-kappaB in osteoclast and B-cell development. Genes Dev 1997; 11:3482-96.
58. Feng B, Cheng S, Pear WS et al. NF-κB inhibitor blocks B cell development at two checkpoints. Medical Immunology 2004; 3:1.
59. Middendorp S, Dingjan GM, Hendriks RW. Impaired precursor B cell differentiation in Bruton's tyrosine kinase-deficient mice. J Immunol 2002; 168:2695-2703.
60. Conley ME, Rohrer J, Rapalus L et al. Defects in early B-cell development: Comparing the consequences of abnormalities in preBCR signaling in the human and the mouse. Immunol Rev 2000; 178:75-90.
61. Saijo K, Schmedt C, Su IH et al. Essential role of Src-family protein tyrosine kinases in NF-kappaB activation during B cell development. Nat Immunol 2003; 4:274-279.
62. Kistler B, Rolink A, Marienfeld R et al. Induction of nuclear factor-kappa B during primary B cell differentiation. J Immunol 1998; 160:2308-17.
63. King LB, Monroe JG. Immunobiology of the immature B cell: Plasticity in the B-cell antigen receptor-induced response fine tunes negative selection. Immunolog Rev 2000; 176:86-104.
64. Wu M, Lee H, Bellas RE et al. Inhibition of NF-kappaB/Rel induces apoptosis of murine B cells. EMBO J 1996; 15:4682-90.
65. Banerji L, Glassford J, Lea NC et al. BCR signals target p27(Kip1) and cyclin D2 via the PI3-K signalling pathway to mediate cell cycle arrest and apoptosis of WEHI 231 B cells. Oncogene 2001; 20:7352-67.
66. Donjerkovic D, Scott DW. Activation-induced death in B lymphocytes. Cell Research 2000; 10:179-192.
67. Leitges M, Schmedt C, Guinamard R et al. Immunodeficiency in protein kinase C beta-deficient mice. Science 1996; 273:788-91.
68. Saijo K, Mecklenbrauker I, Santana A et al. Protein kinase C beta controls nuclear factor kappaB activation in B cells through selective regulation of the IkappaB kinase alpha. J Exp Med 2002; 195:1647-52.
69. Guo B, Su TT, Rawlings DJ. Protein Kinase C family functions in B-cell activation. Curr Op Immunol 2004; 16:367-373.
70. Su TT, Guo B, Kawakami Y et al. PKC-beta controls I kappa B kinase lipid raft recruitment and activation in response to BCR signaling. Nat Immunol 2002; 3:780-6.
71. Cancro MP. Peripheral B-cell maturation: The intersection of selection annd homeostasis. Immunolog Rev 2004; 197:89-101.
72. Rathmell JC. B cell Homeostasis: Digital survival or analog growth? Immunolog Rev 2002; 197:116-128.
73. Claudio E, Brown K, Park S et al. BAFF-induced NEMO-independent processing of NF-kappaB2 in maturing B cells. Nat Immunol 2002; 10:958-65.
74. Wolf I, Pevzner V, Kaiser E et al. Downstream activation of a TATA-less promoter by Oct-2, Bob1, and NF-kappaB directs expression of the homing receptor BLR1 to mature B cells. J Biol Chem 1998; 273:28831-6.
75. Mathias S, Hinz M, Anagnostopoulos I et al. Aberrantly expressed c-Jun and JunB are a hallmark of Hodgkin lymphoma cells, stimulate proliferation and synergize with NF-kappaB. EMBO J 2002; 15:4104-13.
76. Muller G, Hopken U, Lipp M. The impact of CCR7 and CXCR5 on lymphoid organ development and systemic immunity. Immunolog Rev 2003; 195:117-135.

77. Catz SD, Johnson JL. Transcriptional regulation of Bcl2 by nuclear factor kappa B and its significance in prostate cancer. Oncogene 2001; 20:7342-51.
78. Kurland JF, Kodym R, Story MD et al. NF-kappaB1 (p50) homodimers contribute to transcription of the bcl-2 oncogene. J Biol Chem 2001; 276:45380-6.
79. Viatour P, Bentires-Alj M, Chariot A et al. NF-kappaB2/p100 induces Bcl2 expression. Leukemia 2003; 17:1349-56.
80. Gugasyan R, Grumont R, Grossmann M et al. Rel/NF-kappaB transcription factors: Key mediators of B-cell activation. Immunol Rev 2000; 176:134-40.
81. Gerondakis S, Strasser A. The role of Rel/NF-κB transcription factor in B lymphocyte survival. Sem Immunol 2003; 15:159-166.
82. Cariappa A, Liou HC, Horwitz BH et al. Nuclear factor kappa B is required for the development of marginal zone B lymphocytes. J Exp Med 2000; 192:1175-82.
83. Weih DS, Yilmaz ZB, Weih F. Essential role of relB in germinal center and marginal zone formation and proper expression of homing chemokines. J Immunol 2001; 167:1909-19.
84. Grumont RJ, Rourke IJ, O'Reilly LA et al. B lymphocytes differentially use the Rel and nuclear factor kappaB1 (NF-kappaB1) transcription factors to regulate cell cycle progression and apoptosis in quiescent and mitogen-activated cells. J Exp Med 1998; 187:663-74.
85. Prendes M, Zheng Y, Beg AA. Regulation of developing B cell survival by RelA-containing NF-kappaB complexes. J Immunol 2003; 171:3963-9.
86. Mackay F, Browning JL. BAFF: A fundamental survival factor for B cells. Nat Rev Immunol 2002; 7:465-75.
87. Gross JA, Dillon SR, Mudri S et al. TACI-Ig neutralizes molecules critical for B cell development and autoimmune disease. impaired B cell maturation in mice lacking BLyS. Immunity 2001; 15:289-302.
88. Sasaki Y, Casola S, Kutok JL et al. TNF family member B cell-activating factor (BAFF) receptor-dependent roles for BAFF in B cell Physiology. J Immunol 2004; 173:2245-2252.
89. Shulga-Morskaya S, Dobles M, Walsh ME et al. B cell activating factor belonging to the TNF family acts through separate receptors to support B cell survival and T cell-independent Antibody formation. J Immunol 2004; 173:348-59.
90. Tardivel A, Tinel A, Lens S et al. The anti-apoptotic factor Bcl-2 can functionally substitute for the B cell survival but not for the marginal zone B cell differentiation activity of BAFF. Eur J immunol 2004; 34:509-18.
91. Yamada T, Mitani T, Yorita K et al. Abnormal immune function of hemopoietic cells from alymphoplasia (aly) mice, a natural strain with mutant NF-kappa B-inducing kinase. J Immunol 2000; 165:804-12.
92. Hatada EN, Do RK, Ortofsky A et al. NF-kappaB1 p50 is required for BlyS attenuation of apoptosis but dispensable for processing of NF-kappaB p100 to p52 in quiescent mature B cells. J Immunol 2003; 171:761-8.
93. Pasparakis M, Schmidt-Supprian M, Rajewsky K. IκB Kinase signaling is essential for maintenance of mature B cells. J Exp Med 2002; 196:743-752.
94. Hsu BL, Harless SM, Lindsley RC et al. Cutting edge: BlyS enables survival of transitional and mature B cells through distinct mediators. J Immunol 2002; 168(12):5993-6.
95. Mecklenbrauler I, Kalled SL, Leitges M et al. Regulation of B-cell survival by BAFF-dependent PKCδ-mediated nuclear signaling. Nature 2004; 431:456-61.
96. Matza D, Kerem A, Shachar I. Invariant chain, a chain of command. Trends Immunolog 2003; 24:264-268.
97. Benlagha K, Park S-H, Guinamard R et al. Mechanisms governing B cell developmental defects in invariant chain-deficient mice. J Immunol 2004; 172:2076-83.
98. Maehr R, Kraus M, Ploegh HL. Mice deficient in invariant-chain and MHC class II exhibit a normal mature B2 cell compartment. Eur J Immunol 2004; 34:2230-6.
99. Kraus M, Alimzhanov MB, Rajewsky N et al. Survival of resting mature B lymphocytes depends on BCR signaling via the Ig alpha/beta heterodimer. Cell 2004; 117:787-800.
100. Khan WN. Regulation of B lymphocyte development and activation by Bruton's tyrosine kinase. Immunol Res 2001; 23:147-56.
101. Xue L, Morris SW, Orihuela C et al. Defective development and function of Bcl10-deficient follicular, marginal zone and B1 B cells. Nat Immunol 2003; 4:857-864.
102. Thome M. CARMA-1, Bcl-10 and MALT1 in lymphocyte development and activation. Nature Rev Immunol 2004; 4:348-59.
103. Li Z-W, Omori SA, Labuda T et al. IKKβ is required for peripheral B cell survival and proliferation. J Immunol 2003; 170:4630-7.
104. Pohl T, Gugasyan R, Grumont RJ et al. The combined absence of NF-kappa B1 and c-Rel reveals that overlapping roles for these transcription factors in the B cell lineage are restricted to the activation and function of mature cells. Proc Natl Acad Sci 2002; 99:4514-9.

CHAPTER 6

NF-κB and Immune Cell Effector Functions

Hsiou-Chi Liou,* Biao Feng, Wenzhi Tian, Shuhua Cheng and Constance Y. Hsia

Abstract

Initially identified as a constitutive nuclear factor of kappa light chain immunoglobulin in B lymphocytes (NF-κB), the NF-κB transcription factor family now consists of 5 mammalian members (p50/NF-κB1, p52/NF-κB2, p65/RelA, c-Rel, and RelB). Individual knockouts of each NF-κB subunit in mice have shown that NF-κB is functionally expressed in nearly every tissue and cell type. The best documented roles for NF-κB lie in their ability to modulate the development, activation, and effector functions of immune cells. NF-κB participates in both innate and adaptive immunity through regulation of target genes including anti-apoptotic molecules, cell cycle regulators, cytokines, surface receptors, and various other immune modulators. This chapter will focus on the contribution of each NF-κB member to immune cell differentiation and effector function, particularly with regard to macrophages, dendritic cells, T lymphocytes, and B lymphocytes.

Brief Overview of the Immune System

The primary purpose of the vertebrate immune system is to protect the host from infection by foreign pathogens such as bacteria, viruses, and parasites. Distinct immune cell types carry out two main branches of host immunity, namely the innate and adaptive arms of immune regulation. However components of the innate system also play an important part in shaping the adaptive system. Here we review key elements of the immune system that will be relevant to this chapter.

Immune responses are initiated by host cells which recognize foreign antigens as potentially harmful substances in the body. Innate responses are typically generated against microbial antigens through recognition of invariable patterns expressed on molecules produced by pathogens. To identify these antigens, phagocytes such as macrophages and neutrophils bear pattern-recognition receptors (PRR) on their cell surface. The breadth of antigenic responses induced upon recognition by PRRs is restricted however, as only a limited number of fixed structures are effectively detected. As a consequence, the adaptive immune system has evolved to allow more specific recognition of pathogenic antigens, and entails use of antigen-specific receptors to identify these unique epitopes. Antigen-specific receptors are found only on the surface of lymphocytes, and are generated through genetic recombination to produce receptors bearing variable regions that can recognize a potentially infinite number of epitopes. This complex strategy invites a perplexing dilemma as to how the immune system distinguishes "foreign" versus "self" antigens. Mechanisms must therefore be in place to spare the host from

---

*Corresponding Author: Hsiou-Chi Liou—Division of Immunology, Department of Medicine, Weill Medical College of Cornell University, 515 East 71 Street, S-210, New York, New York 10021, U.S.A. Email: hcliou@med.cornell.edu.

NF-κB/Rel Transcription Factor Family, edited by Hsiou-Chi Liou.
©2006 Landes Bioscience and Springer Science+Business Media.

attacking its own tissues. To avoid harmful autoimmune responses, nature has evolved a fail-safe method to remove self-reactive lymphocytes via a process termed negative selection. Negative selection ensures that only nonself reactive lymphocytes are allowed to populate peripheral lymphocyte pools in the body.

As it stands, the immune system is prepared to guard against foreign invasion and has developed similar methods for the surveillance of abnormal cell growth (i.e., tumors). Modern medicine, not surprisingly, is developing approaches to manipulate these mechanisms for combating cancer, or conversely to enable organ graft tolerance by suppressing undesirable immune reactions. Failure or dysregulation of the immune system therefore has multiple health implications including the development of immunodeficiency, tumorigenesis, hypersensitivity, graft rejection, or autoimmune disease. An understanding of how the immune system operates holds tremendous potential for the treatment of these and many other serious conditions.

NF-κB and Innate Immunity

Innate and adaptive immunity comprise two important aspects of host defense against microbial infection. Innate immunity occurs through phagocytosis of pathogenic material and secretion of anti-microbial or inflammatory mediators by cells such as macrophages and neutrophils. This response typically occurs within minutes or hours of infection. Phagocytes recognize pathogens through several forms of pattern-recognition receptors, including mannose receptors, scavenger receptors, and the more recently identified family of Toll-like receptors (TLRs). Recognition by these receptors leads to (1) enhanced phagocytic activity, (2) production of anti-microbial products such as nitric oxide (NO), defensins, and proteolytic enzymes, (3) production of anti-inflammatory cytokines such as TNF-α, IL-1, and IL-6, (4) production of chemokines or trafficking and adhesion molecules, and (5) expression of costimulatory molecules.[1]

Notably, signaling through all members of the TLR family converge upon the activation of NF-κB through shared signaling pathways (see chapter on Receptor and Adaptor Signaling). Multiple studies support the role of NF-κB in the production of inflammatory mediators that control innate immune responses against bacterial infection. NF-κB induced expression of NO synthase (iNOS) for instance, has been well-documented and highlights one of the key pathways involved in respiratory NO production.[2] Subsequent studies in NF-κB knockout mice have shown that deletion of p65 and c-Rel decreases production of iNOS, as well as the expression of TNF-α, IL-1, and IL-6 in macrophages.[3,4] These reports implicate NF-κB in the regulation of inflammatory cytokines and the respiratory burst in phagocytes. Other studies in mice lacking p50 illustrate the physiological significance of NF-κB-mediated host defense where an inability to clear L. monocytogenes and greater susceptibility to infection with S. pneumoniae is seen.[5] The observed immunodeficiencies are attributed to a hyporesponsive reaction to bacterial lipopolysaccharide (LPS) through a TLR4-dependent recognition pathway. These data suggest that intact NF-κB protein is essential for TLR signaling during activation of macrophages and neutrophils in the clearance of bacteria.

Corroborating these findings is a recent report on a human subject experiencing a recurrent bacterial infection. The study showed that leukocytes derived from the patient displayed profound hyporesponsiveness to LPS and IL-1 concurrent with diminished downstream NF-κB activation.[6] Evidence suggests that the patient carried specific mutations in the IRAK-4 molecule, a kinase downstream of the TLR4 and IL-1R pathways that is required for NF-κB activation (see chapter on Receptors and Adaptors for NF-κB Signaling). Thus, studies such as this support the notion that regulation of inflammatory cytokines, chemokines, and costimulatory molecules by NF-κB is critical to phagocyte activation, maturation, trafficking to inflamed tissues, as well as the ability to present pathogenic antigens to infiltrating T lymphocytes. NF-κB therefore constitutes an important aspect of frontline defense during infection.

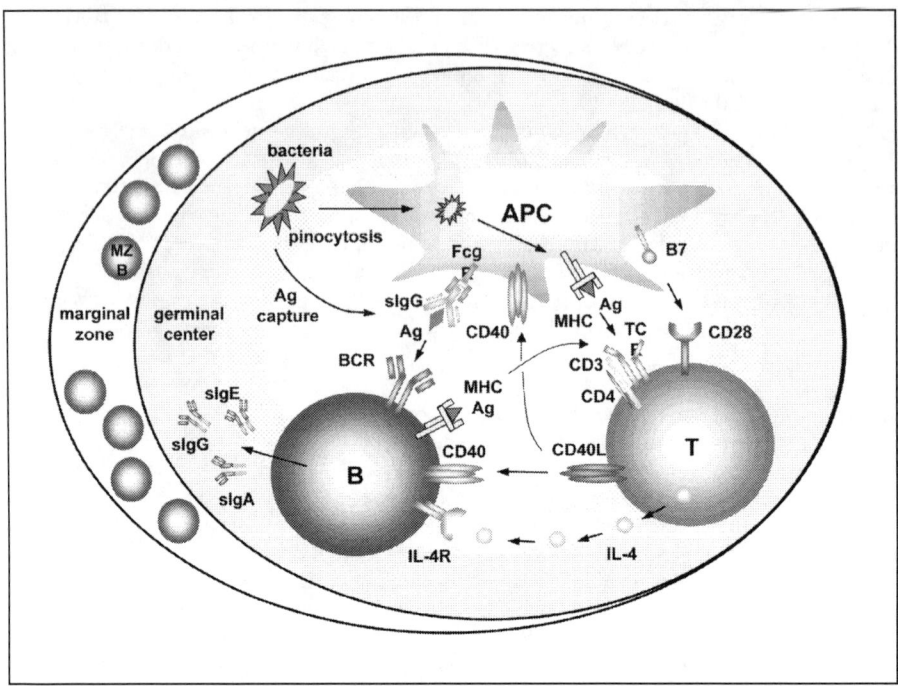

Figure 1. Germinal center response in secondary lymphoid tissues.

NF-κB in Dendritic Cells

Whereas innate immunity can provide initial protection from pathogenic infection, many bacteria and viruses have evolved intricate mechanisms to conceal their identity. Gram-negative bacteria, for instance, are encapsulated by a thick sugar coat that prevents the contents of the cell from being recognized by immune cells. Viruses, on the other hand, may hide inside cells by integrating into host DNA. Mechanisms to accommodate the recognition of disguised pathogens in what is known as the adaptive immune response have subsequently evolved in most higher organisms.[7] The adaptive response involves processing of pathogenic antigens into small peptides that are then presented by MHC molecules to reactive T cells (Fig. 1). MHC class I is expressed on the surface of all cells, but enhanced expression of MHC I and II is found on antigen presenting cells (APC). Antigen receptors on T cells, known as the T cell receptor (TCR), recognize these peptides presented on MHC molecules through APC-T cell interactions. Importantly, this differs from the recognition of native or whole antigens by the B cell receptor (BCR; see below). Soluble BCR is also known as antibody or immunoglobulin, and is produced by B lymphocyte lineages to allow direct recognition of antigens in their unprocessed forms.

The most common type of APCs are dendritic cells (DC) which serve to bridge innate and adaptive immune responses by capturing and processing intracellular antigens for presentation to T lymphocytes in local lymph nodes.[8,9] DCs originate from a common myeloid and lymphoid hematopoietic precursor in the bone marrow and migrate to the periphery as "immature" dendritic cells. Recent studies have devoted much effort to describe the signals that lead to DC maturation, as it appears that the quality or status of DC maturation can determine whether a T cell response is primed or suppressed. During pathogenic infection, numerous signals effectively induce DC maturation including microbial products that bind TLRs on DCs, pro-inflammatory cytokines (such as TNFα and IL-1) produced by infected tissues and

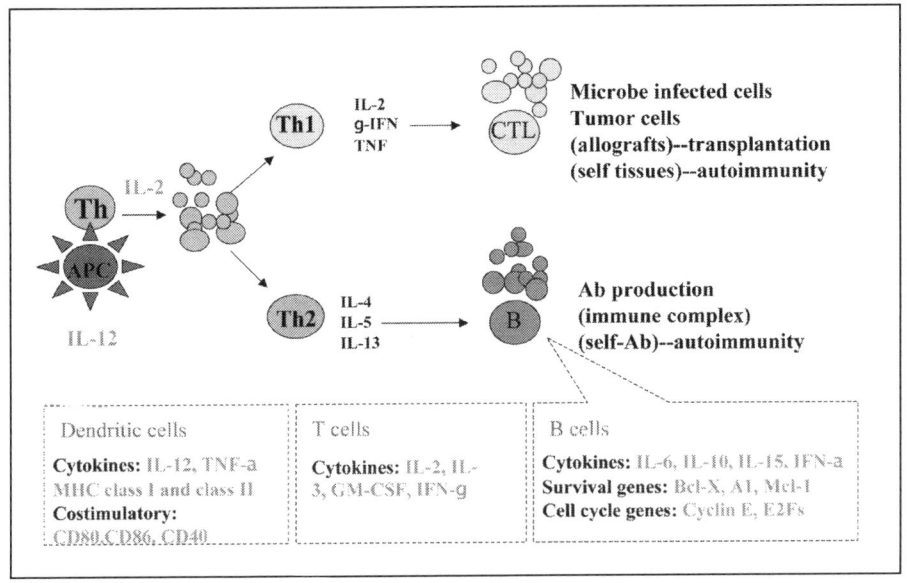

Figure 2. NF-κB/Rel and immune cell effector function.

activated macrophages, and other necrotic cellular components.[8,9] Mature DCs are characterized by enhanced expression of MHC class I and II molecules, costimulatory receptors such as B7-1 (CD80), B7-2 (CD86), and CD40, as well as adhesion molecules such as ICAM-1. Each of these proteins is critical to priming and maintaining strong interactions with T cells via complementary receptors on the lymphocyte (TCR, CD28, CD40L, and LFA-1 respectively).[8-10] DCs may also produce IL-12 and IL-23 to polarize differentiation of T helper cells into T_H1 cells that are crucial for activating cytotoxic T lymphocytes during attack against intracellular pathogens[11] (Fig. 2).

At the intracellular level, NF-κB plays a pivotal role in mediating both DC development and maturation. It was initially observed that RelB knockout mice exhibit defective development of thymic DC and myeloid-derived DEC205⁻ CD8⁻ DCs.[12-14] Subsequent studies on other NF-κB knockout mice show that loss of both p50 and p65 abolishes the formation of lymphoid and myeloid DCs.[15] Interestingly however, deletion of c-Rel or both p50/c-Rel does not affect DC development, but does impair the activation of mature DCs. These studies therefore suggest that different NF-κB members serve distinct roles in DC function and development i.e., p50, p65, and RelB are required for DC development, whereas c-Rel and p50 are important for DC survival and costimulatory function.

Since NF-κB is one of the key mediators of TLR and CD40 signaling (critical DC maturation factors), deletion of NF-κB would appear to compromise DC function (Fig. 2). As mentioned above, DC maturation is a crucial aspect of antigen presenting ability via upregulation of MHC and costimulatory molecules. The promoter regions of B7.1, B7.2, and MHC class II genes all contain NF-κB binding motifs. It is believed that NF-κB controls transcriptional regulation of each of these molecules, and blocking NF-κB activity by IkBα overexpression or using pharmacological inhibitors significantly reduces surface expression of each of these receptors on DCs.[16,17] Yet interestingly, individual knockouts of p50, p65, or c-Rel only minimally reduces expression of B7.1, B7.2, and MHC class II receptors on DCs,[15,18] suggesting that NF-κB members may compensate for each other during transcription of these genes, or alternatively that other transcription factors can participate in this capacity.

Studies also implicate NF-κB in the regulation of DC survival and IL-12 cytokine expression. It appears that during DC-T cell interaction, DCs receive survival signals from T cells via CD40L and TRANCE (Fig. 1). DCs lacking both p50 and c-Rel are particularly vulnerable to apoptosis, an effect attributed to diminished Bcl-X anti-apoptotic gene expression.[15] Other studies indicate that c-Rel is an important regulator of IL-12 expression by controlling the transcription of IL-12p40 and IL-12p35 subunits in macrophages and dendritic cells upon TLR signaling.[19-23] Deficient IL-12 production in c-Rel knockout DCs may explain why c-Rel deficient DCs produce insufficient amounts of T_H1 and T_H2 cytokines in vitro. Furthermore, these DCs fail to stimulate both allogeneic and antigen-specific T cell responses.[18] The defect in IL-12 production is consistent with the finding that c-Rel knockout mice are protected from autoimmune encephalomyelitis, a condition primarily mediated by T_H1 responses. The protection afforded through c-Rel deletion is manifested at two levels: reduced production of IL-12 by DCs and decreased production of IFN-γ by T cells (see chapter on Roles of NF-κB in Autoimmunity).[20-26]

NF-κB in T Cell Development and Selection

Multiple studies support the role of NF-κB in shaping the T cell compartment by influencing both thymocyte development and peripheral T cell effector functions. During T cell development in the thymus, $CD4^-CD8^-$ double negative (DN) thymocytes transit through a $CD4^+CD8^+$ double positive (DP) stage before maturing into $CD4^+$ or $CD8^+$ single positive (SP) cells. Selection begins in the DN stage (the first checkpoint) where signaling through a precursor TCR (preTCR) consisting of a genetically-rearranged TCRβ chain and an invariant surrogate α chain (pTα), is necessary for the subsequent rearrangement and expression of the TCRα chain. Once expressed, the TCRα chain pairs with the TCRβ chain to form a functional TCR receptor on the cell surface. Only cells expressing functional TCRs at the surface are allowed to progress further for development.

In vitro and in vivo experiments demonstrate that NF-κB is critical to the survival and proliferation of immature thymocytes.[27,28] Studies reveal that crosslinking the preTCR can activate molecules such as p56-lck, ZAP70, ras, and PLC-γ, resulting in Ca^{+2} flux and downstream activation of NF-AT and NF-κB. It is hypothesized that proliferation-inducing cell cycle regulators (e.g., cyclins) and anti-apoptotic survival genes (e.g., Bcl-2 family) are activated by NF-κB in DN thymocytes, and may occur through mechanisms similar to those previously described for mature lymphocytes.[29-32,33,34] For instance, in T cell specific IkBα-overexpressing mice (IκB broadly inhibits most NF-κB isoforms), a reduction in the total number of DP and SP thymocytes as well peripheral $CD4^+$ and $CD8^+$ T cells is observed.[35-38] Individual and combinatorial NF-κB knockout studies would help confirm these findings, however only c-Rel and p65 knockout studies have been performed thus far. Deletion of either or both of these genes appears to have little effect on early thymocyte development, and may reflect redundant contributions by other NF-κB family members that potentially compensate for c-Rel or p65 deletion in early thymocytes.[39,40]

The second checkpoint takes place during the DP stage where thymocytes failing to express TCR die by neglect while those expressing a functional TCR undergo positive and negative selections.[41,42] T cells expressing different TCRs are initially circulated through thymic epithelium where they encounter peptides presented on MHC molecules of thymic epithelia. Together this MHC-peptide complex binds to TCRs with varying degrees of avidity. Evidence suggests that positive selection involves weak interaction (low avidity) of the TCR with polymorphic residues on the MHC coreceptor.[41] In contrast, TCRs exhibiting strong avidity for MHC-peptides are signaled to die via apoptosis (negative selection), in what is thought to be a mechanism for deleting potentially self-reactive T cells. Newer studies, however, suggest that some T cells may instead become regulatory T cells which exhibit suppressive activity.[41]

The strength and threshold of TCR signaling has been researched to some extent and data from these studies indicate that these factors may be important in determining the fate of DP and SP thymocytes.[41,43,44] The manner in which cells interpret TCR signal strength is not known however. It is unclear, for example, whether all TCR signals activate a defined set of signaling pathways such as Ca^{+2} mobilization or activation of NF-κB and NF-AT transcription factors. Furthermore, the role of NF-κB in positive and negative selection can only be speculated at present, but emerging evidence suggests that NF-κB proteins do participate to some extent.

A clue to NF-κB's involvement in positive selection emerged from studies using IκBα transgenic mice where broad inhibition of NF-κB in T cell lineages led to a severe reduction of DP and SP thymocytes as well peripheral $CD4^+$ and $CD8^+$ T cells.[35-38] Moreover, deletion of c-Rel and/or p65 has also been associated with decreased peripheral T cell numbers.[39,40] These studies indicate that NF-κB is required for survival of thymocytes, and thereby constitutes an essential requirement for positive selection. With regard to negative selection, it is well documented that thymic epithelial cells present abundant self-peptides, presumably to aid the negative selection of self-reactive thymocytes. Interestingly, both cortical and medullary thymic epithelial cells not only express common self-antigens but also tissue-specific self-antigens that are often expressed only in restricted tissues (i.e., insulin in pancreatic β-cell). This rather promiscuous form of gene expression in the thymic environment is thought to provide a means for eliminating harmful self-reactive thymocytes through either cell death or the generation of regulatory T cells.

A connection between NF-κB and negative selection was first reported in the study of aly/aly mice that harbor a mutation in the NF-κB Inducing Kinase (NIK) gene.[45] The data showed that $NIK^{aly/aly}$ mutant mice develop autoimmune diseases and multi-organ inflammation with disorganized thymic structure, much like RelB knockout mice.[46] More recent reports describe a reduction in $CD4^+CD25^+$ regulatory T cells in these mice, and that this deficiency is associated with impaired processing of the p52 precursor protein p100, as well as low levels of RelB production.[45] Autoimmune conditions in $NIK^{aly/aly}$ mice are rescuable however, by transferring $CD4^+CD25^+$ regulatory T cells into $NIK^{aly/aly}$ hosts, thereby implying that the suppressive function of these cells is relevant to the prevention of self-reactive T cell activation. Further data from the same study suggest that NIK activation (e.g., via RelB and p52) may contribute to the generation of $CD4^+CD25^+$ regulatory T cells through promiscuous expression of temporally controlled tissue-specific genes in thymic stromal cells. Validation of the hypothesis that NF-κB is involved in negative selection or immune tolerance could ideally be tested in p52/RelB knockout mice.

Finally, a recent study suggests that p50 and c-Rel may also contribute to the generation of $CD4^+CD25^+$ T-regulatory cells.[47] c-Rel is a known transcriptional activator of IL-2 in activated T cells and both c-Rel and p50 have been detected at the CD28 responsive element in the IL-2 promoter.[48,49] Considering that numerous reports now suggest a role for IL-2 and STAT5 in the generation of T-regulatory cells,[43,50-55] NF-κB dependent activation of IL-2 my serve as the potential mechanism by which this occurs. Further work to determine whether individual NF-κB knockout mice develop autoimmune disease, as evidenced for IL-2 knockout, IL-2 receptor knockout, and STAT5 knockout mice, will be critical to establishing this relationship.[52-55]

NF-κB in T Cell Effector Function

Key features of adaptive immunity include (1) the ability to recognize an almost infinite number of antigens, (2) clonal selection and expansion of antigen-specific lymphocytes, and (3) development of lymphocyte memory to ensure a more rapid and efficient response upon reencounter with the same antigen. Antigen-specific $CD4^+$ and $CD8^+$ T lymphocytes exist at low frequency in naïve hosts. Upon infection or immunization, clonal expansion of antigen-specific T cells is necessary to generate large numbers of effector cells. A small portion of effector cells will also differentiate into memory cells to ensure rapid recall response. Evidence suggests that NF-κB controls lymphocyte clonal expansion and effector function through the regulation of cell survival proteins, cell cycle regulators, cytokines, chemokines, surface receptors, and adhesion molecules.[56] Studies also suggest that development of memory and T-regulatory cells may involve NF-κB.[47]

Cytokines serve unequivocal functions in shaping T effector cell development. $CD4^+$ T cells exist primarily to aid the expulsion of pathogens through cytokine-dependent activation of phagocytes, cytotoxic T cells (CTL), natural killer cells (NK) and B cells (e.g., antibody-secreting B cells) (Fig. 2). Upon TCR ligation, naïve $CD4^+$ T cells differentiate into two distinct subsets: T helper 1 (T_H1) and T helper 2 (T_H2). Each of these cells is characterized by the production of a specific set of cytokines. T_H1 cells express IFN-γ and TNF-α which promote phagocyte, NK, and CTL responses, while T_H2 cells express IL-4, IL-5, IL-9, and IL-13 that largely aid the activation of B cells.[57,58] Mice deficient in T_H1 cells demonstrate that T_H1 dependent responses are essential to protect the host against intracellular pathogenic infection,[59,60] whereas T_H2 deficiency results in loss of protection against extracellular pathogens.[61,62] Hence the action of $CD4^+$ T helper cells is primarily supportive in nature. In contrast, $CD8^+$ T cells mediate active cytotoxic killing through the production of cytolytic enzymes such as perforin and granzyme B and/or the production of IFN-γ and TNF-α.[63]

During the presentation of antigen to T cells, only antigen-specific cells undergo clonal expansion into effector cell populations. This initial wave of proliferation is mediated by the expression of IL-2 upon TCR and CD28 signaling. IL-2 is perhaps the most influential T cell cytokine during this phase that not only drives the replication of T cell clones, but also promotes downstream differentiation of $CD4^+$ T cells into helper cells. When combined with IL-12 produced by DCs, IL-2 promotes differentiation of $CD8^+$ T cells into cytotoxic T lymphocytes and production of IFN-γ.[64-67] The importance of IL-2 and IFN-γ is demonstrated by the absence of T_H1 and CTL mediated anti-viral immune responses in IL-2 knockout mice.[63,68] Studies show that upon LCMV infection, viral specific $CD8^+$ T cells fail to proliferate in IL-2 knockouts, correlating with an observed decrease in IFN-γ production and a reduction in CTL mediated killing of infected cells.[69,70]

IL-2 expression in T cells is coordinately regulated by NF-κB, NF-AT and AP-1 transcription factors.[49,71-73] In particular, studies show that the CD28 responsive element in the IL-2 promoter possesses high affinity for c-Rel homodimer binding,[49] whereby c-Rel deficient T cells exhibit impaired expression of IL-2, proliferate poorly in response to TCR signals, and are unable to generate $CD8^+$ effector T cells.[74-76] c-Rel is not only essential to expression of IL-2, but regulates the expression of other cytokines such as IL-3, GM-CSF, and IFN-γ as well.[74-76] Given that c-Rel mediates IL-2 and IFN-γ expression during T_H1 responses, there is precedence that loss of anti-viral responses and decreased memory cell production occur in the absence of c-Rel or other NF-κB members (also see chapter on Regulation of Immunity to Infection). Certainly such hypothesis could be tested in relevant NF-κB knockout systems.

Recent studies have identified several transcription factors that are regulated through antigen receptor signaling, as well as cytokines, which direct T_H1/T_H2 differentiation. For example, T-bet appears to be required for T_H1 polarization, whereas GATA-3 is required for generation of T_H2 cells.[57,58,77] Reports indicate that different NF-κB members may contribute to either T_H1 or T_H2 development as well. Initially, researchers thought that NF-κB participated only in the development of T_H1 responses as studies in c-Rel deficient mice showed defective production of T_H1 cytokines (IL-12 and IFN-γ),[19-21,23] and IκBα transgenic mice displayed impaired T_H1 responses correlating with defective survival, clonal expansion, and loss of T_H1-mediated IFN-γ production.[78,79] Further examination however, reveals that p50 deficient T cells exhibit impaired production of T_H2 cytokines associated with decreased GATA3 expression as well.[80] These studies suggest that p50 acts upstream of GATA3 gene expression.[80] More recent studies with Bcl-3 knockout mice (an IκB member) further demonstrate that Bcl-3, possibly in complex with p50, drives GATA-3 expression and participates in T_H2 differentiation. By comparison, RelB appears to control T_H1 differentiation by regulating T-bet gene expression (Mark Boothby, manuscript in preparation). Hence the contribution of different NF-κB or IκB members to either T_H1 or T_H2 responses is more complex than originally believed.

NF-κB in B Lymphocyte Clonal Expansion and Cell Fate Determination

Individual NF-κB factors are implicated in various stages of B cell differentiation ranging from immature cell to effector cell development (for discussion of early B cell development, see chapter on NF-κB and Lymphopoiesis). This section will therefore focus on the signaling consequences of antigen-activated B cells and how interactions among various B cells, T cells, and DCs contribute to germinal center formation and terminal differentiation of B cells into antibody-producing plasma cells and memory B cells (Fig. 1).

There are several means by which mature B cells can encounter foreign antigen. Antigens captured by DCs in tissue can be passed to B cells after DC migration to the lymph node whereas antigens circulating in peripheral blood can be trapped in lymphoid organs through high endothelial venules (HEV). Upon antigen binding in lymphoid follicles, mature B cells undergo clonal expansion in a manner similar to T cells. Notably, the BCR and TCR share analogous signaling pathways including homologous tyrosine kinases and adaptor molecules that induce activation of NF-κB, NF-AT, and AP-1 transcription factors (see chapter on Receptors and Adaptors for NF-κB Signaling). Among these factors, NF-κB dimers (particularly c-Rel/p50 heterodimers) appear to be absolutely required for lymphocyte survival and cell cycle progression.[56] Numerous reports conclude that c-Rel expression in B cells is critical to expression of pro-survival molecules (e.g., Bcl-X, Bfl-1, Mcl-1),[29-32] cell cycle regulators (e.g., cyclin E, E2F3a),[33,34] and cytokines (e.g., IL-6, IL-10, IL-15)[32,56] (Fig. 2).

More and more evidence suggests that the level of NF-κB activation may dictate the fate of lymphocytes upon antigen encounter, i.e., clonal expansion vs. anergy or tolerance. In T lymphocytes, chronic exposure to antigenic signals in the absence of costimulation can lead to anergy (state of unresponsiveness). Anergy of T cells is associated with activation of NF-AT without coactivation of NF-κB and AP-1.[81] Subsequent studies show that NF-AT dimers, in the absence of NF-κB and AP-1, can induce expression of molecules that negatively regulate TCR signaling (and potentially BCR signaling), including Cbl-b and Itch (both E3 ubiquitin ligases) and a ubiquitin-binding component which cooperatively target PLC-γ and PKC-θ for degradation, thus leading to termination of TCR signaling.[82,83] These studies suggest that the primary function of costimulatory signals is to perhaps facilitate activation of NF-κB and AP-1, which together with NF-AT enable the induction of a different set of target genes such as cytokines and chemokines that collectively lead to immune reactivity rather than immune tolerance.[83,84] A similar phenomenon is described in B lymphocytes whereby B cells anergic to self-antigen exhibit specific blockade of c-Rel, p65, and JNK activation.[85]

It has been proposed that differential recruitment of CARMA1 or Cbl adaptor proteins to the intracellular BCR signaling tail (and potentially the TCR) promotes formation of an immunosome or tolerosome, respectively, and ultimately leads to opposite outcomes, i.e., immunogenic response vs tolerogenic response.[86] A notable feature of the immunosome is the ability of CARMA1 to recruit various NF-κB signaling complexes, Bcl-10, MALT1, PKC, and IKK to the intracellular proximal receptor chains where signaling is amplified and sustained within lipid rafts (see chapter on Receptors and Adaptors for NF-κB Signaling). By comparison, the tolerosome is composed of Cbl ubiquitin ligase, E2-conjugating enzyme, and endophilin complexes that target proteins into endosomal compartments for degradation. Hence, an antigenic signal may result in an immunogenic response if NF-κB signaling components are recruited to the BCR-lipid raft, while a tolerogenic response may result in the absence of recruitment.

Cell fate determination has also been investigated in immature B cell development. While mature B cells are signaled to proliferate upon BCR ligation, it was observed that immature B cells undergo growth arrest and apoptosis instead.[87,88] The propensity for BCR-induced death in this manner potentially ensures that immature B cells exhibiting self-reactivity are deleted prior to their appearance in the periphery. Molecules expressed in high concentration in the bone marrow, such as self-antigens, will thus bind the BCR with high affinity and result in apoptosis before the cells have a chance to emerge from the bone marrow. Conversely, those that are not deleted presumably do not recognize self-antigens and are therefore less prone to induce autoimmunity.

Only until recently it became clear that immature B cells are defective in their ability to mobilize several key signaling components necessary for NF-κB activation, including PKC, PI3K, and Akt,[89] (Cheng S, Feng B, Hsia C, and Liou H-C, manuscript in preparation). It is proposed that immature B cells destined to die fail to express critical survival genes (e.g., Bcl-X) and cell cycle progression factors (e.g., E2F, cyclin E) lying downstream of these signaling pathways. Indeed new evidence now suggests that both Bcl-X and E2F3a are regulated by NF-κB in mature B lymphocytes.[33,34] In attempt to better understand the decision to respond to antigenic signals or else undergo tolerance, comparison of NF-κB induced target genes under certain conditions or states of development will be necessary. Recent insights have been offered through microarray profiling of c-Rel dependent targets which provide a glimpse of relevant genes including transcription factors, adhesion molecules, intracellular signaling proteins, C-type lectin-like receptors, and metabolic enzymes (Hsia, C, Owyang, A, Tian, W, Hsu, J, Xiang, J, and Liou manuscript in preparation). These findings indicate that the role NF-κB plays in controlling lymphocyte fate extends beyond the regulation of proliferation or survival and may include differentiation, trafficking, metabolism, and cell-cell communication.

NF-κB in the Germinal Center Immune Response

During microbial infection, dendritic cells capture antigens and migrate to the T-cell zone in lymphoid tissue where they activate antigen-specific T cells (Fig. 1). Antigen-specific B cells are similarly trapped in the T-cell zone upon encounter with BCR-specific antigens. Numerous studies document the importance of $CD4^+$ T cell help in promoting B cell proliferation, differentiation, and the formation of germinal centers.[90] Germinal centers are specialized microenvironments within lymphoid tissue where interaction among follicular DCs, T cells, and B cells is highly concentrated. The result of these interactions may lead to vigorous B cell proliferation, Ig somatic hypermutation, and Ig class switching. Importantly, the T cell-B cell interface involves multiple pairs of receptor-ligand interactions including CD40L-CD40, CD28-B7, and CD30L-CD30. These pairings effectively determine the outcome of an immune response, controlled in large part by cytokines produced within the interaction environment. T_H2 cytokines such as IL-4, IL-5, IL-6, IL-13, and TGFβ for example, facilitate Ig class switching to IgG, IgE, and IgA subclasses.[91]

Not surprisingly, certain NF-κB knockout mice display defective germinal center and antibody phenotypes. Poor germinal center formation is observed in c-Rel deficient mice for instance, correlating with defective Ig class switching and failure to develop memory B cells upon immunization.[30-34,74,92] While this defect appears to be cell-intrinsic, other NF-κB knockout studies show that p52 and RelB deficiency results in both intrinsic and extrinsic defects of B cell follicles including disorganized formation of follicular dendritic cell networks. Interestingly, the extrinsic effect is attributed to loss of NF-κB dependent BLC chemokine expression,[14,93,94] a gene expressed on lymphoid stromal tissue which normally provides chemotractant signals for B cell retention in the lymphoid follicles.

The ultimate purpose of B cell activation is to generate antibodies that will bind and neutralize microbial antigens through various means. This can be achieved through immunoglobulin class switching from IgM/IgD to soluble IgG, IgA, or IgE isotypes. Depending on the isotype, each immunoglobulin carries out different effector functions such as neutralization, opsonization, or activation of the complement system. Class switching to a particular isotype depends upon the accessibility of transcription factors to enhancer regions within the heavy chain (CH) constant locus. In particular, NF-κB binding sites have been identified within the Cγ3, Cγ1 and Cε chain enhancer, and thus implicated in the germline transcription of IgG3, IgG1, and IgE isotypes.[95,96] Cells lacking c-Rel for example, exhibit loss of IgG1 and IgE production resulting from impaired Cγ1 and Cε germline transcription.[97,98] Similarly, impaired germline transcription of Cγ3 and Cε in p50 or p65 deleted B cells results in defective

Table 1. NF-κB function in immune cells

Immune Cell Type	NF-κB Function
Phagocytes (macrophages, neutrophils)	Phagocytosis iNOS, TNFα, IL-1, IL-6 Chemokines Costimulatory molecules
Dendritic cells	Development (p50, p65, RelB) Maturation (p50, c-Rel) Survival (p50, c-Rel) Cytokines (IL-12, IL-23) Costimulatory molecules (B7.1, B7.2) MHC class II
NK cells	Unknown
T lymphocytes	DN thymocyte expansion DP, SP thymocyte survival Positive selection Negative selection (unknown) Survival and proliferation Cytokines (IL-2, IL-3, IFN-γ) CD4+ T_H1 differentiation (c-Rel, RelB) CD4+ T_H2 differentiation (p50, Bcl-3) CD8+ CTL effector function Memory T cell (unknown) T-regulatory cell (unknown)
B lymphocytes	Development and maturation (p50, p65, p52) Survival (Bcl-x, Bfl-1, Mcl-1) (c-Rel) Proliferation (cyclin E, E2F3a) (c-Rel) Cytokines (IL-6, IL-10, IL-15) Germinal center formation Ig class switching Plasma cells (unknown) Memory B cells (unknown)

Note: Listed in the parentheses are NF-κB members required for carrying out described functions, not excluding the potential participation of or cooperation with other NF-κB

switching to IgG3 and IgE.[99,100] These selective defects demonstrate specific NF-κB requirements for individual isotypes, some of which overlap.

Concluding Remarks

The NF-κB transcription factor family plays an important role in innate and adaptive immunity by regulating various aspects of immunological development and effector cell function (Table 1). Although NF-κB members display significant overlapping activities, each member is also unique with regard to differential responses to receptor signals and distinct target gene profiles. Future challenges include systematic analysis of NF-κB activity on downstream targets in various immunomodulatory capacities including inflammatory diseases, autoimmune diseases, transplant rejection, as well as host defense against pathogenic infection. As such, these research efforts hold significant promise toward the discovery and development of therapeutic targets for the prevention of immunopathological conditions via manipulation of NF-κB signaling.

References

1. Takeda K, Kaisho T, Akira S. Toll-like receptors. Annu Rev Immunol 2003; 21:335-376.
2. Xie QW, Kashiwabara Y, Nathan C. Role of transcription factor NF-kappa B/Rel in induction of nitric oxide synthase. J Biol Chem 1994; 269(7):4705-4708.
3. Ouaaz F, Li M, Beg AA. A critical role for the RelA subunit of nuclear factor kappaB in regulation of multiple immune-response genes and in Fas-induced cell death. J Exp Med 1999; 189(6):999-1004.
4. Grigoriadis G, Zhan Y, Grumont RJ et al. The Rel subunit of NF-kappaB-like transcription factors is a positive and negative regulator of macrophage gene expression: Distinct roles for Rel in different macrophage populations. EMBO J 1996; 15(24):7099-7107.
5. Sha WC, Liou H-C, Tuomanen EI et al. Targeted disruption of the p50 subunit of NF-kB leads to multifocal defects in immune responses. Cell 1995; 80:321-330.
6. Medvedev AE, Lentschat A, Kuhns DB et al. Distinct mutations in IRAK-4 confer hyporesponsiveness to lipopolysaccharide and interleukin-1 in a patient with recurrent bacterial infections. J Exp Med 2003; 198(4):521-531.
7. Janeway Jr CA, Medzhitov R. Innate immune recognition. Annu Rev Immunol 2002; 20:197-216.
8. Banchereau J, Steinman RM. Dendritic cells and the control of immunity. Nature 1998; 392(6673):245-252.
9. Mellman I, Steinman RM. Dendritic cells: Specialized and regulated antigen processing machines. Cell 2001; 106(3):255-258.
10. Huppa JB, Davis MM. T-cell-antigen recognition and the immunological synapse. Nat Rev Immunol 2003; 3(12):973-983.
11. Trinchieri G, Pflanz S, Kastelein RA. The IL-12 family of heterodimeric cytokines: New players in the regulation of T cell responses. Immunity 2003; 19(5):641-644.
12. Burkly L, Hession C, Ogata L et al. Expression of relB is required for the development of thymic medulla and dendritic cells. Nature 1995; 373(6514):531-536.
13. Wu L, D'Amico A, Winkel KD et al. RelB is essential for the development of myeloid-related CD8alpha- dendritic cells but not of lymphoid-related CD8alpha+ dendritic cells. Immunity 1998; 9(6):839-847.
14. Weih DS, Yilmaz ZB, Weih F. Essential role of RelB in germinal center and marginal zone formation and proper expression of homing chemokines. J Immunol 2001; 167(4):1909-1919.
15. Ouaaz F, Arron J, Zheng Y et al. Dendritic cell development and survival require distinct NF-kappaB subunits. Immunity 2002; 16(2):257-270.
16. Yoshimura S, Bondeson J, Brennan FM et al. Role of NFkappaB in antigen presentation and development of regulatory T cells elucidated by treatment of dendritic cells with the proteasome inhibitor PSI. Eur J Immunol 2001; 31(6):1883-1893.
17. Yoshimura S, Bondeson J, Brennan FM et al. Antigen presentation by murine dendritic cells is nuclear factor-kappa B dependent both in vitro and in vivo. Scand J Immunol 2003; 58(2):165-172.
18. Boffa DJ, Feng B, Sharma V et al. Selective loss of c-Rel compromises dendritic cell activation of T lymphocytes. Cell Immunol 2003; 222(2):105-115.
19. Sanjabi S, Hoffmann A, Liou HC et al. Selective requirement for c-Rel during IL-12 P40 gene induction in macrophages. Proc Natl Acad Sci USA 2000; 97(23):12705-12710.
20. Weinmann AS, Mitchell DM, Sanjabi S et al. Nucleosome remodeling at the IL-12 p40 promoter is a TLR-dependent, Rel- independent event. Nat Immunol 2001; 2(1):51-57.
21. Hilliard BA, Mason N, Xu L et al. Critical roles of c-Rel in autoimmune inflammation and helper T cell differentiation. J Clin Invest 2002; 110(6):843-850.
22. Grumont R, Hochrein H, O'Keeffe M et al. c-Rel regulates interleukin 12 p70 expression in CD8(+) dendritic cells by specifically inducing p35 gene transcription. J Exp Med 2001; 194(8):1021-1032.
23. Mason N, Aliberti J, Caamano JC et al. Cutting edge: Identification of c-Rel-dependent and -independent pathways of IL-12 production during infectious and inflammatory stimuli. J Immunol 2002; 168(6):2590-2594.
24. Boffa D, Feng B, Sharma V et al. Selective loss of c-Rel compromises dendritic cell activation of T lymphocytes. Cell Immunol 2003; 222(2):105-15.
25. Lamhamedi-Cherradi SE, Zheng S, Hilliard BA et al. Transcriptional regulation of type I diabetes by NF-kappa B. J Immunol 2003; 171(9):4886-4892.
26. Mason NJ, Liou HC, Hunter CA. T cell-intrinsic expression of c-Rel regulates Th1 cell responses essential for resistance to Toxoplasma gondii. J Immunol 2004; 172(6):3704-3711.
27. Voll RE, Jimi E, Phillips RJ et al. NF-kappa B activation by the preT cell receptor serves as a selective survival signal in T lymphocyte development. Immunity 2000; 13(5):677-689.
28. Kim D, Xu M, Nie L et al. Helix-loop-helix proteins regulate preTCR and TCR signaling through modulation of Rel/NF-kappaB activities. Immunity 2002; 16(1):9-21.

29. Andjelic S, Hsia C, Suzuki H et al. Phosphatidylinositol 3-kinase and NF-kappa B/Rel are at the divergence of CD40-mediated proliferation and survival pathways. J Immunol 2000; 165(7):3860-3867.
30. Owyang AM, Tumang JR, Schram BR et al. c-Rel is required for the protection of B cells from antigen receptor- mediated, but not Fas-mediated, apoptosis. J Immunol 2001; 167(9):4948-4956.
31. Grossmann M, O'Reilly LA, Gugasyan R et al. The anti-apoptotic activities of Rel and RelA required during B-cell maturation involve the regulation of Bcl-2 expression. EMBO J 2000; 19(23):6351-6360.
32. Tumang JR, Hsia CY, Tian W et al. IL-6 rescues the hyporesponsiveness of c-Rel deficient B cells independent of Bcl-xL, Mcl-1, and Bcl-2. Cell Immunol 2002; 217(1-2):47-57.
33. Hsia CY, Cheng S, Owyang AM et al. c-Rel regulation of the cell cycle in primary mouse B lymphocytes. Int Immunol 2002; 14(8):905-916.
34. Cheng S, Hsia CY, Leone G et al. Cyclin E and Bcl-xL cooperatively induce cell cycle progression in c-Rel-/- B cells. Oncogene 2003; 22(52):847
35. Esslinger CW, Wilson A, Sordat B et al. Abnormal T lymphocyte development induced by targeted overexpression of IkappaB alpha. J Immunol 1997; 158(11):5075-5078.
36. Boothby MR, Mora AL, Scherer DC et al. Perturbation of the T lymphocyte lineage in transgenic mice expressing a constitutive repressor of nuclear factor (NF)-kappaB. J Exp Med 1997; 185(11):1897-1907.
37. Esslinger CW, Jongeneel CV, MacDonald HR. Survival-independent function of NF-kappaB/Rel during late stages of thymocyte differentiation. Mol Immunol 1998; 35(13):847-852.
38. Hettmann T, DiDonato J, Karin M et al. An essential role for nuclear factor kappaB in promoting double positive thymocyte apoptosis. J Exp Med 1999; 189(1):145-158.
39. Strasser A, Grumont RJ, Stanley ML et al. The transcriptional regulator Rel is essential for antigen receptor-mediated stimulation of mature T cells but dispensable for positive and negative selection of thymocytes and T cell apoptosis. Eur J Immunol 1999; 29(3):928-935.
40. Grossmann M, O'Reilly LA, Gugasyan R et al. The anti-apoptotic activities of Rel and RelA required during B-cell maturation involve the regulation of Bcl-2 expression. EMBO J 2000; 19(23):6351-6360.
41. von Boehmer H, Aifantis I, Gounari F et al. Thymic selection revisited: How essential is it? Immunol Rev 2003; 191:62-78.
42. Germain RN. Ligand-dependent regulation of T cell development and activation. Immunol Res 2003; 27(2-3):277-286.
43. Smith KA. The threshold hypothesis: How the immune system uses the IL-2 molecule to discriminate beteween self and nonself. Med Immunololgy 2004.
44. Gett AV, Sallusto F, Lanzavecchia A et al. T cell fitness determined by signal strength. Nat Immunol 2003; 4(4):355-360.
45. Kajiura F, Sun S, Nomura T et al. NF-kappa B-inducing kinase establishes self-tolerance in a thymic stroma-dependent manner. J Immunol 2004; 172(4):2067-2075.
46. Weih F, Durham SK, Barton DS et al. p50-NF-kappaB complexes partially compensate for the absence of RelB: Severely increased pathology in p50(-/-)relB(-/-) double-knockout mice. J Exp Med 1997; 185(7):1359-1370.
47. Zheng Y, Vig M, Lyons J et al. Combined deficiency of p50 and cRel in CD4+ T cells reveals an essential requirement for nuclear factor kappaB in regulating mature T cell survival and in vivo function. J Exp Med 2003; 197(7):861-874.
48. Liou HC, Jin Z, Tumang J et al. c-Rel is crucial for lymphocyte proliferation but dispensable for T cell effector function. Int Immunol 1999; 11(3):361-371.
49. Huang DB, Chen YQ, Ruetsche M et al. X-ray crystal structure of proto-oncogene product c-Rel bound to the CD28 response element of IL-2. Structure (Camb) 2001; 9(8):669-678.
50. Thornton AM, Donovan EE, Piccirillo CA et al. Cutting edge: IL-2 is critically required for the in vitro activation of CD4+CD25+ T cell suppressor function. J Immunol 2004; 172(11):6519-6523.
51. Almeida AR, Legrand N, Papiernik M et al. Homeostasis of peripheral CD4+ T cells: IL-2R alpha and IL-2 shape a population of regulatory cells that controls CD4+ T cell numbers. J Immunol 2002; 169(9):4850-4860.
52. Furtado GC, Curotto de Lafaille MA, Kutchukhidze N et al. Interleukin 2 signaling is required for CD4(+) regulatory T cell function. J Exp Med 2002; 196(6):851-857.
53. Barmeyer C, Horak I, Zeitz M et al. The interleukin-2-deficient mouse model. Pathobiology 2002; 70(3):139-142.
54. Snow JW, Abraham N, Ma MC et al. Loss of tolerance and autoimmunity affecting multiple organs in STAT5A/5B-deficient mice. J Immunol 2003; 171(10):5042-5050.
55. Antov A, Yang L, Vig M et al. Essential role for STAT5 signaling in CD25+CD4+ regulatory T cell homeostasis and the maintenance of self-tolerance. J Immunol 2003; 171(7):3435-3441.
56. Liou HC, Hsia CY. Distinctions between c-Rel and other NF-kappaB proteins in immunity and disease. Bioessays 2003; 25(8):767-780.

57. Szabo SJ, Sullivan BM, Peng SL et al. Molecular mechanisms regulating Th1 immune responses. Annu Rev Immunol 2003; 21:713-758.
58. Murphy KM, Reiner SL. The lineage decisions of helper T cells. Nat Rev Immunol 2002; 2(12):933-944.
59. Thierfelder WE, van Deursen JM, Yamamoto K et al. Requirement for Stat4 in interleukin-12-mediated responses of natural killer and T cells. Nature 1996; 382(6587):171-174.
60. Kaplan MH, Sun YL, Hoey T et al. Impaired IL-12 responses and enhanced development of Th2 cells in Stat4-deficient mice. Nature 1996; 382(6587):174-177.
61. Takeda K, Tanaka T, Shi W et al. Essential role of Stat6 in IL-4 signalling. Nature 1996; 380(6575):627-630.
62. Shimoda K, van Deursen J, Sangster MY et al. Lack of IL-4-induced Th2 response and IgE class switching in mice with disrupted Stat6 gene. Nature 1996; 380(6575):630-633.
63. Seder RA, Ahmed R. Similarities and differences in CD4+ and CD8+ effector and memory T cell generation. Nat Immunol 2003; 4(9):835-842.
64. Smith KA. Interleukin-2: Inception, impact, and implications. Science 1986; 240:1169-1176.
65. Cantrell DA, Smith KA. The interleukin-2 T-cell system: A new cell growth model. Science 1984; 224:1312-1316.
66. Ehlers S, Smith KA. Differentiation of T cell lymphokine gene expression: The in vitro acquisition of T cell memory. J Exp Med 1991; 173(1):25-36.
67. Blattman JN, Grayson JM, Wherry EJ et al. Therapeutic use of IL-2 to enhance antiviral T-cell responses in vivo. Nat Med 2003; 9(5):540-547.
68. Wherry EJ, Ahmed R. Memory CD8 T-cell differentiation during viral infection. J Virol 2004; 78(11):5535-5545.
69. Cousens LP, Orange JS, Biron CA. Endogenous IL-2 contributes to T cell expansion and IFN-gamma production during lymphocytic choriomeningitis virus infection. J Immunol 1995; 155(12):5690-5699.
70. Su HC, Cousens LP, Fast LD et al. CD4+ and CD8+ T cell interactions in IFN-gamma and IL-4 responses to viral infections: Requirements for IL-2. J Immunol 1998; 160(10):5007-5017.
71. Granelli-Piperno A, Nolan P. Nuclear transcription factors that bind to elements of the IL-2 promoter. Induction requirements in primary human T cells. J Immunol 1991; 147(8):2734-2739.
72. Chen D, Rothenberg EV. Molecular basis for developmental changes in interleukin-2 gene inducibility. Mol Cell Biol 1993; 13(1):228-237.
73. Rooney JW, Hoey T, Glimcher LH. Coordinate and cooperative roles for NF-AT and AP-1 in the regulation of the murine IL-4 gene. Immunity 1995; 2(5):473-483.
74. Kontgen F, Grumont RJ, Strasser A et al. Mice lacking the c-rel proto-oncogene exhibit defects in lymphocyte proliferation, humoral immunity, and interleukin-2 expression. Genes & Dev 1995; 9:1965-1977.
75. Gerondakis S, Strasser A, Metcalf D et al. Rel-deficient T cells exhibit defects in production of interleukin 3 and granulocyte-macrophage colony-stimulating factor. Proc Natl Acad Sci USA 1996; 93(8):3405-3409.
76. Liou HC, Jin Z, Tumang J et al. c-Rel is crucial for lymphocyte proliferation but dispensable for T cell effector function. Int Immunol 1999; 11(3):361-371.
77. Ho IC, Glimcher LH. Transcription: Tantalizing times for T cells. Cell 2002; 109(Suppl):S109-120.
78. Aronica MA, Mora AL, Mitchell DB et al. Preferential role for NF-kappa B/Rel signaling in the type 1 but not type 2 T cell-dependent immune response in vivo. J Immunol 1999; 163(9):5116-5124.
79. Corn RA, Aronica MA, Zhang F et al. T cell-intrinsic requirement for NF-kappa B induction in postdifferentiation IFN-gamma production and clonal expansion in a Th1 response. J Immunol 2003; 171(4):1816-1824.
80. Das J, Chen CH, Yang L et al. A critical role for NF-kappa B in GATA3 expression and TH2 differentiation in allergic airway inflammation. Nat Immunol 2001; 2(1):45-50.
81. Macian F, Garcia-Cozar F, Im SH et al. Transcriptional mechanisms underlying lymphocyte tolerance. Cell 2002; 109(6):719-731.
82. Heissmeyer V, Macian F, Im SH et al. Calcineurin imposes T cell unresponsiveness through targeted proteolysis of signaling proteins. Nat Immunol 2004; 5(3):255-265.
83. Macian F, Im SH, Garcia-Cozar FJ et al. T-cell anergy. Curr Opin Immunol 2004; 16(2):209-216.
84. Heissmeyer V, Rao A. E3 ligases in T cell anergy—turning immune responses into tolerance. Sci STKE 2004; 2004; 29:241.
85. Healy JI, Dolmetsch RE, Lewis RS et al. Quantitative and qualitative control of antigen receptor signalling in tolerant B lymphocytes. Novartis Found Symp 1998; 215:137-144.
86. Jun JE, Goodnow CC. Scaffolding of antigen receptors for immunogenic versus tolerogenic signaling. Nat Immunol 2003; 4(11):1057-1064.

87. Monroe JG. B-cell antigen receptor signaling in immature-stage B cells: Integrating intrinsic and extrinsic signals. Curr Top Microbiol Immunol 2000; 245(2):1-29.
88. Fuentes-Panana EM, Bannish G, Monroe JG. Basal B-cell receptor signaling in B lymphocytes: Mechanisms of regulation and role in positive selection, differentiation, and peripheral survival. Immunol Rev 2004; 197:26-40.
89. King LB, Norvell A, Monroe JG. Antigen receptor-induced signal transduction imbalances associated with the negative selection of immature B cells. J Immunol 1999; 162(5):2655-2662.
90. Liu YJ, de Bouteiller O, Fugier-Vivier I. Mechanisms of selection and differentiation in germinal centers. Curr Opin Immunol 1997; 9(2):256-262.
91. Cerutti A, Zan H, Schaffer A et al. CD40 ligand and appropriate cytokines induce switching to IgG, IgA, and IgE and coordinated germinal center and plasmacytoid phenotypic differentiation in a human monoclonal IgM+IgD+ B cell line. J Immunol 1998; 160(5):2145-2157.
92. Grumont JR, Rourke IJ, Gerondakis S. Rel-dependent induction of A1 transcription is required to protect B cells from antigen receptor ligation-induced apoptosis. Gene & Dev 1999; 13:400-411.
93. Caamano JH, Rizzo CA, Durham SK et al. Nuclear factor (NF)-kappa B2 (p100/p52) is required for normal splenic microarchitecture and B cell-mediated immune responses. J Exp Med 1998; 187(2):185-196.
94. Franzoso G, Carlson L, Poljak L et al. Mice deficient in nuclear factor (NF)-kappa B/p52 present with defects in humoral responses, germinal center reactions, and splenic microarchitecture. J Exp Med 1998; 187(2):147-159.
95. Cerutti A, Schaffer A, Shah S et al. CD30 is a CD40-inducible molecule that negatively regulates CD40-mediated immunoglobulin class switching in nonantigen-selected human B cells. Immunity 1998; 9(2):247-256.
96. Schaffer A, Cerutti A, Shah S et al. The evolutionarily conserved sequence upstream of the human Ig heavy chain S gamma 3 region is an inducible promoter: Synergistic activation by CD40 ligand and IL-4 via cooperative NF-kappa B and STAT-6 binding sites. J Immunol 1999; 162(9):5327-5336.
97. Zelazowski P, Carrasco D, Rosas FR et al. B cells genetically deficient in the c-Rel transactivation domain have selective defects in germline CH transcription and Ig class switching. J Immunol 1997; 159(7):3133-3139.
98. Zelazowski P, Shen Y, Snapper CM. NF-kappaB/p50 and NF-kappaB/c-Rel differentially regulate the activity of the 3'alphaE-hsl,2 enhancer in normal murine B cells in an activation-dependent manner. Int Immunol 2000; 12(8):1167-1172.
99. Snapper CM, Zelazowski P, Rosas FR et al. B cells from p50/NF-kappa B knockout mice have selective defects in proliferation, differentiation, germ-line CH transcription, and Ig class switching. J Immunol 1996; 156(1):183-191.
100. Horwitz BH, Zelazowski P, Shen Y et al. The p65 subunit of NF-kappa B is redundant with p50 during B cell proliferative responses, and is required for germline CH transcription and class switching to IgG3. J Immunol 1999; 162(4):1941-1946.

CHAPTER 7

Roles of NF-κB in Autoimmunity

Stacey Garrett and Youhai H. Chen*

Abstract

Autoimmune diseases are the result of improper and uncontrolled immune responses against self-antigens. The NF-κB/Rel family of transcription factors is known to play important roles in the initiation and regulation of immunity against pathogens and foreign antigens. Over time, evidence has accumulated implicating NF-κB as a mediator of autoimmunity and potential therapeutic target for treating autoimmune diseases. In this chapter, we discuss the role of NF-κB in apoptosis and cell proliferation as well as tolerance and autoimmunity.

Introduction

Once described as "horror autotoxicus", autoimmunity has become a well-recognized phenomenon.[1] At its surface, autoimmunity may be described simply as the incomplete tolerance of the host's immune system to self antigens. Studies showing that autoantibodies and autoreactive T cells can be detected in healthy mice and humans render the previous definition of autoimmunity overly simplistic.[2-5] Recent studies indicate that autoimmunity does not always lead to autoimmune disease leading researchers to conclude that genetic and environmental factors are also required for the development of autoimmune diseases.

Autoimmune diseases occur on two scales: systemic and organ-specific. Additionally, different diseases are driven by different cell types, primarily T or B cells. Autoimmunity is essentially an immune response, albeit an uncontrolled and improper one. The NF-κB family of transcription factors has been implicated as one of the master controllers of the immune response. As detailed in other chapters of this book, these molecules have complex effects on the processes of lymphocyte proliferation, cytokine production, and apoptosis. Each of these cellular processes is intricately involved in immunity and autoimmunity.

Roles of NF-κB in T Cell Development

Proper thymic development is critical for normal immune function. The thymus is responsible for positively and negatively selecting T cells as they progress from the CD4+CD8+ (double positive) stage to the single positive stage. Through the years, the role of NF-κB in this process has been hotly debated.

Early studies involving deletion of single NF-κB family members or blockage of c-Rel and Rel A activation via transgenic expression of mutated IκBα indicated that NF-κB played no role in thymocyte development or selection.[6-9] By contrast, the absence of a functional IκB kinase causes a dramatic reduction in the total numbers of thymocytes, presumably due to its

*Corresponding Author: Youhai H. Chen—Department of Pathology and Laboratory Medicine, School of Medicine, University of Pennsylvania, Philadelphia, Pennsylvania 19104, U.S.A. Email: yhc@mail.med.upenn.edu.

NF-κB/Rel Transcription Factor Family, edited by Hsiou-Chi Liou.
©2006 Landes Bioscience and Springer Science+Business Media.

role in protecting cells from TNF-α induced apoptosis.[10] Recently these findings have been questioned in models employing TCR transgenic mice.

By crossing the IκBα(ΔN) transgenic mouse (which blocks c-Rel and RelA activation in the T cell lineage by out competing endogenous IκBα) to the DO11.10 TCR transgenic mouse, Mark Boothby and colleagues found that positive and negative selection were inhibited. This inhibition was characterized by diminished ZAP-70 phosphorylation in response to stimulation. This study differs from others in that it examined thymocyte development on two different genetic backgrounds – H-2d x H-2d allowed for examination of positive selection while H-2d x H-2b allowed for study of negative selection. The use of IκBα(ΔN) may be superior to that of other dominant negative transgenic IκBα molecules in that it is expressed at a higher level and therefore achieves a more complete blockade of c-Rel and RelA activation. While IκBα(ΔN) protects thymocytes in vivo from negative selection, this study supports signal strength-based theories of negative selection by showing that high levels of negatively-selecting peptide stimulation in vitro can bypass the protective effects of IκBα(ΔN).[11]

Additional evidence for a role of NF-κB in mediating central tolerance has been shown through studies of the aly/aly strain of mice. The aly mutation renders the NF-κB inducing kinase (NIK) incapable of binding to IKK-α. As a result, the development of lymphoid organs is abnormal and these mice experience autoimmunity characterized by chronic inflammation of several organs.[12-14]

Fumiko Kajiura and colleagues performed crucial studies to illustrate the mechanism by which this mutation leads to autoimmunity. These researchers showed that disease identical to that of aly/aly mice could be transferred to nude mice via thymic engraftment. The phenotype can also be rescued by transfer of normal CD4+ T cells and the inflammation is not completely resolved if only CD4+CD25- T cells are transferred. They went on to demonstrate by FACS that the total number of CD25+ regulatory T cells are decreased in the periphery and thymi of aly/aly mice, although these cells are functional in vitro at suppressing T cell proliferation. To support the argument that regulatory T cell production was affected by this NIK defect, they showed that the total levels of FoxP3, a protein currently believed to be specific to regulatory T cells, was decreased in both CD4+ T cells and the thymi of aly/aly mice.[15] Additionally, PCR analysis of AIRE, a family of transcription factors responsible for the expression of peripheral antigens in the thymus, demonstrated that defects in central tolerance were directly related to defects of negative selection in the aly/aly thymi. AIRE message was decreased and the expression of specific peripheral antigens, like salivary protein 1, was completely absent in the thymi of aly/aly mice.[16] As a result, negative selection to peripheral antigens was incomplete.

This study provides critical information about the role of NF-κB in the processes by which the thymic microenvironment establishes central tolerance. A critical question remains as to the role of NIK in the production and maintenance of regulatory T cells though. This study does not answer the question as to whether the decrease in regulatory T cell numbers is due to dysregulation of AIRE and peripheral antigen expression in the thymus or to the role of NIK in the generation of regulatory T cells themselves.

These studies demonstrate that NF-κB does have a role in the development of T cells in the thymus. However, only in the case of the aly/aly strain of mice do these defects actually lead to autoimmune disease. Yet the connection between NF-κB and autoimmunity is supported by the fact that single gene knockouts of NF-κB members affect lymphocyte function to the point that these mutations render mice resistant to the development of autoimmunity. One exception to this statement is the RelB-/- mice which develop diseases similar to the aly/aly mice due to the effects of TNF-α.

Roles of NF-κB in Mature Lymphocytes

Several studies have shown increased NF-κB activity in autoimmune diseases like myasthenia gravis, rheumatoid arthritis, and multiple sclerosis while systemic lupus erythematosus patients have been described to have decreased NF-κB activity in lymphocytes.[17-19] Schmid and Pahan demonstrated that activated NF-κB dimers could be found in the spinal cords of rats with experimental allergic encephalomyelitis (EAE).[20] These observations as well as the critical role of NF-κB in the initiation of effector immune responses and lymphocyte survival have made NF-κB an intriguing target of autoimmune research in recent years. Additionally, our laboratory has shown that deficiencies in NF-κB1 and c-Rel render mice resistant to the development of autoimmunity.[21-23]

One of the key studies providing a link between NF-κB dysregulation and autoimmunity demonstrated aberrant cytokine regulation in autoimmune mice. It had long been observed that MRL and NZB mouse strains, both of which are prone to the development of autoimmunity, produced reduced levels of IL-1 in vitro. This observation was similar to that found for monocytes of lupus patients (rev in 24). Hartwell and colleagues showed that the expression of inflammatory cytokine transcripts like IL-1 and TNF-α occurred with a fatser kinetics and disappeared earlier than in control Balb/c cells. This correlated with a more rapid repression of IL-1 transcriptional activity in MRL/+ macrophages. This rapid repression was accompanied by accelerated reduction of nuclear NF-κB as shown by EMSA while the faster loss of cytokine mRNAs was not associated with decreased mRNA stability in the MRL/+ mice. Thus, researchers concluded that differences in transcription were directly due to the amount of NF-κB in the nucleus resulting in a decrease of IL-1 protein expressed by MRL/+ macrophages in vitro.

Beller's laboratory followed up these findings by showing that the dysregulation of NF-κB was the cause of the abnormal IL-12 production observed in NOD and NZB/W mice. Interestingly, these two strains show different patterns of IL-12 production for different reasons. NOD mice display elevated levels of IL-12 as is generally associated with Th1 mediated organ specific diseases while NZB/W show a reduction in IL-12 before the development of pathology similar to systemic lupus.

Using thioglycolate elicited peritoneal macrophages, Liu and Beller showed that these defects were due to unique binding patterns of Rel dimers to the IL-12p40 promoter. In NOD mice, there was a preferential association of c-Rel with the p40 κB site due to hyperphosphorylation of c-Rel dimers. In contrast, p50 was found bound to the p40 κB site in NZB/W mice. This difference appeared to be due to elevated levels of IκB specifically sequestering c-Rel dimers in the cytoplasm. These findings demonstrated that different patterns of cytokine production can be directly explained by the effects of dysregulation of NF-κB mediated transcription.[25]

Roles of NF-κB in Apoptosis, Proliferation and Autoimmunity

Perhaps one of the strongest links between NF-κB and autoimmunity is the role of this family of transcription factors in regulating both proliferation and apoptosis. One of the most relied upon treatments for autoimmunity involves the use of prednisone to decrease inflammation as well as to prime lymphocytes for apoptosis. As a result, there has been much study of the link of Rel family members to both of these processes over the years.

Several studies have shown a link between NF-κB and apoptosis.[8,26,27] Cells deficient in RelA/p65 show an increased sensitivity to TNF-α and other mediators of apoptosis like radiation (rev in 28). Additionally, c-Rel deficiency results in enhanced B cell apoptosis following mitogen stimulation unless both the BCR and CD40 are signaled.[29] NF-κB also controls many pro-survival genes like c-IAP, A20, and c-Myc.[28]

This led Vallabhapuvapu and colleagues to determine if blocking NF-κB activity with a mutant IκBα molecule targeted to the T cell lineage would alleviate disease in gld/gld mice. Gld mice lack functional FasL, develop lymphadenopathy and autoimmunity characterized by increased serum immunoglobulin and autoantibodies (reviewed in ref. 30). Use of this transgenic mutant IκBα resulted in a dramatic reduction of lymphadenopathy and a complete elimination of the abnormal thy1+B220+CD4-CD8- T cells. Additionally, reduced proliferative responses and an increase in apoptosis of peripheral T cells were shown resulting in partial correction of B cell abnormalities. Nuclear extracts revealed increased amounts of p50 homodimers in unstimulated and stimulated nuclear extracts of the mutant IκBα gld/gld mice.[31] This study provides intriguing possibilities of targeting NF-κB activity in specific lymphocyte compartments for the treatment of autoimmune diseases.

Another interesting relationship between NF-κB, apoptosis and autoimmunity has been demonstrated recently by Mei Wu's laboratory. They generated a transgenic (Tg) mouse in which immediate early response gene X-1 (IEX-1) was constitutively overexpressed in the lymphoid lineage which resulted in susceptibility to a lupus-like autoimmune disease due to accumulation of activated T cells. Additionally, they showed that IEX-1 Tg T cells are resistant to apoptosis mediated by ligation of Fas and TCR.[32] Furthermore, the IEX-1 promoter is specifically regulated by NF-κB in coordination with p53 and c-Myc. The IEX-1 promoter can be activated by multiple Rel dimers including p65/c-Rel, p65/p50, and p50 homodimers.[33]

This data combined with previous studies, which showed that IEX-1 is upregulated in response to mitogens and NF-κB activating cytokines, argues that IEX-1 is an important NF-κB target which functions to promote T cell survival during an immune response. Further studies need to be performed to determine if it possesses similar functions in other cell types since there is evidence indicating that IEX-1 can promote apoptosis in cultured cells under certain conditions (reviewed in ref. 32). Given the lupus-like phenotype of IEX-1 Tg mice, this gene may be a promising therapeutic target for autoimmune diseases.

Roles of NF-κB in Autoimmune Encephalomyelitis and Diabetes

The link between c-Rel transcriptional control of IL-12p35 and p40 led our laboratory to examine the role of c-Rel and NF-κB1 in both experimental autoimmune encephalomyelitis (EAE) and type I diabetes.[34-36] We found that mice deficient in NF-κB1 experienced significantly less EAE due to reduced proliferation and effector function of myelin oligodendrocyte glycoprotein (MOG)-specific T cells.[21] Similarly, c-Rel deficient animals experienced significantly less disease than controls as well as a fourteen day delay in disease onset following MOG immunization. The c-Rel knockout group also showed only a 16 percent incidence of disease. Splenocytes from MOG-immunized c-Rel deficient mice showed deficiencies in proliferation, IL-2 and IFN-γ production as have been previously described in response to other stimulations.[22,37-39] Interestingly, we also observed a significant increase in IL-4 production as well.

This led us to compare the levels of interferon-γ versus IL-4 production of NF-κB1 deficient B6:129 mice, c-Rel deficient C57BL/6 mice, and their respective controls to determine if the skewing of Th cytokines was a result of the strain background of the different knockouts. We found that control B6:129 mice produced slightly more IL-4 than control C57BL/6 mice, but both produced similar amounts of IFN-γ. In contrast, the c-Rel knockout B6 mice generated as much IL-4 as the B6:129 strain control while producing almost no IFN-γ. The NF-κB1 knockout mice produced a similar level of IFN-γ as its control while making almost ten fold less IL-4. This data indicates that NF-κB1 and c-Rel may influence the Th differentiation of T cells. Mechanistically, it is unclear if this Th effect is T cell intrinsic or the result of NF-κB transcriptional regulation of the microenvironment in which the T cell is primed.

An argument for the effect of NF-κB on the microenvironment of the T cell was made by showing that deficiency in c-Rel reduces the amount of IL-12p40 produced by bone marrow derived dendritic cells, microglia and astrocytes. Also, lack of c-Rel decreases the amount of IL-23 p19 message produced by antigen presenting cells stimulated with LPS while the level of T-bet expression by T cells is unchanged compared to wild type cells. IL-23p19 has been implicated as necessary and sufficient for induction of EAE suggesting that the defects in antigen-presenting cell function in c-Rel deficient mice may be more influential than the T cell defects.[40-42] This is supported by data demonstrating that wild type T cells stimulated with anti-CD3 in the presence of c-Rel deficient DCs produce 5 fold less IFN-γ than controls. Despite this finding, supplementation of IL-12 to c-Rel-/- CD4+ T cells stimulated with anti-CD3 and anti-CD28 does not increase IFN-γ production indicating that the role of c-Rel in Th differentiation is not limited to its affects on the priming DCs.

Similarly, using low dose streptozotocin to induce type I diabetes, we showed that both c-Rel and NF-κB1 deficient mice experienced significantly reduced incidence of diabetes as well as a delay in disease onset. Splenocytes from c-Rel deficient mice displayed increased IL-4 production as well as increased IL-10 production in response to anti-CD3 stimulation. Additionally, both c-Rel-/- and NF-κB1-/- bone marrow derived dendritic cells stimulated with LPS produced significantly reduced TNF-α. To further investigate the mechanism of diabetes resistance in these mice, apoptosis of monocytes was studied. In the absence of GM-CSF and IL-4, NF-κB1 deficiency resulted in a two fold increase in apoptosis of bone marrow derived DCs while apoptosis was increased in c-Rel deficient granulocytes and macrophages. Stat-1 expression in NF-κB1-/- macrophages was also decreased. This study provides further evidence for a significant role for NF-κB family members in shaping the innate immune response and its impact on the development of autoimmune diseases.[22]

In the streptozotocin-induced diabetes model, the loss of c-Rel and NF-κB1 have multiple effects on factors affecting disease development. Beyond the well-known effects on T cells, it is significant that NF-κB1 deficiency also affected the survival of antigen presenting cells. This effect could very well decrease the amount of T cell priming in vivo due to decreased antigen presentation in lymphoid organs. The decreased TNF-α production could also slow disease progression by decreasing the number of lymphocytes trafficking to both lymphoid organs and the pancreas. This effect could also explain the decreased disease in the c-Rel deficient animals. More implications of this study may become clear as the exact mechanism of streptozotocin-induced diabetes is determined.

Conclusion

The role of NF-κB in cell proliferation and survival has long been known. NF-κB dimers influence the transcription not only of growth factors and cytokines, but also regulators of cell cycle (reviewed in ref. 43). Multiple studies of NF-κB knockout mice have demonstrated decreased lymphocyte proliferation in response to mitogens as well as defects in Th differentiation. These results combined with evidence implicating NF-κB's role in immune tolerance establish NF-κB as a potential mediator of autoimmunity and therapeutic target. However, evidence is also accumulating suggesting that some of the NF-κB members or regulators can prevent autoimmune diseases under certain circumstances. Thus, both RelB and NIK may protect against autoimmunity through their control of TNF-α signaling. Depending on the cell types involved in the initial stages of autoimmune diseases and the types of the NF-κB activated, the roles of the NF-κB family may vary. Environmental and genetic factors which result in hyper-responsiveness or a shift in NF-κB species activation can tip the fine balance from normal immune function to autoimmunity. Further studies are needed to better understand the mechanisms of action of different members of NF-κB in promoting and inhibiting autoimmunity.

Acknowledgements

This work was supported by grants from the National Institutes of Health (AI50059, DK065848, AI053052), U.S.A.

References

1. Ehrlich P, Morgenroth J. On haemolysis: Third communication. The Collected Papers of Paul Ehrlich. London: Pergamon, 1957; 2:178-206.
2. Kakkanaiah VN, Seth A, Nagarkatti M et al. Autoreactive T cell clones isolated from normal and autoimmune-susceptible mice exhibit lymphokine secretory and functional properties of both Th1 and Th2 cells. Clin Immunol Immunopathol 1990; 57:148-162.
3. Grabar P. Autoantibodies and the physiological role of immunoglobulins. Immunol Today 1983; 4:337-339.
4. Hooper B, Whittingham S, Mathews JD et al. Autoimmunity in a rural community. Clin Exp Immunol 1972; 12:79-87.

5. Hawkins BR, O'Connor KJ, Dawkins RL et al. Autoantibodies in an Australian population. I. Prevalence and persistence. J Clin Lab Immunol 1979; 2:211-215.
6. Palmer E. Negative selection-clearing out the bad apples from the T-cell repertoire. Nat Rev Immunol 2003; 3:383-91.
7. Strasser A, Grumont RJ, Stanley ML et al. The transcriptional regulator Rel is essential for antigen receptor-mediated stimulation of mature T cells but dispensable for positive and negative selection of thymocytes and T cell apoptosis. Eur J Immunol 1999; 29:928-35.
8. Ferreira V, Sidenius N, Tarantino N et al. In vivo inhibition of NF-kB in T lineage cells leads to a dramatic decrease in cell proliferation and cytokine production and to increased cell apoptosis in response to mitogenic stimuli, but not to abnormal thymopoiesis. J Immunol 1999; 162:6442-50.
9. Gerondakis S, Grossman M, Nakamura Y et al. Genetic approaches to study Rel/NF-kB/Ikb function in mice. Oncogene 1999; 18(49):6888-95, (rev).
10. Senftleben UZ, Li ZW, Baud V et al. IKK beta is essential for protecting T cells from TNFalpha-induced apoptosis. Immunity 2001; 14:217.
11. Mora AL, Stanley S, Armistead W et al. Inefficient ZAP-70 phosphorylation and decreased thymic selection in vivo result from inhibition of NF-kB/Rel. J Immunol 2001; 167:5628-35.
12. Miyawaki S, Nakamura Y, Suzuka H et al. A new mutation, aly, that induces a generalized lack of lymph nodes accompanied by immunodeficiency in mice. Eur J Immunol 1994; 24:429-92.
13. Shinkura R, Kitada K, Matsuda F et al. Alymphoplasia is caused by a point mutation in the mouse gene encoding NF-kB-inducing kinase. Nat Genet 1999; 22:74.
14. Tsubata R, Tsubata T, Hiai H et al. Autoimmune disease of exocrine organs in immunodeficient alymphoplasia mice: A spontaneous modle for Sjögren's syndrome. Eur J Immunol 1996; 26:2472.
15. Hori S, Nomura T, Sakaguchi S. Control of regulatory T cell development by the transcription factor FOXP3. Science 2003; 299:1057.
16. Kajiura F, Sun S, Nomura T et al. NF-kB-inducing kinase establishes self-tolerance in a thymic stroma-dependent manner. J Immunol 2004; 172:2067-75.
17. Nagata Y, Onodera H, Ohuchi M et al. Decreased expression of c-myc family genes in thymuses from myasthenia gravis patients. J Neuroimmunol 2001; 115:199-202.
18. Collantes E, Blázquez MV, Mazorra V et al. Nuclear factor-kB activity in T cells from patients with rheumatic diseases: A preliminary report Ann Rheum Dis 1998; 57:738-41.
19. Tsokos GC, Wong HK, Enyedy EJ et al. Immune cell signaling in lupus. Curr Opin Rheumatol 2000; 12:355-63.
20. Pahan K, Schmid M. Activation of nuclear factor-kB in the spinal cord of experimental allergic encephalomyelitis. Neuroscience Letters 2000; 287:17-20.
21. Hilliard B, Samoilova EB, Liu TST et al. Experimental autoimmune encephalomyelitis in NF-kB-deficient mice: Roles of NF-kB in the activation and differentiation of autoreactive T cells. J Immunol 1999; 163:2938-43.
22. Hilliard B, Mason N, Xu L et al. Critical roles of c-Rel in autoimmune inflammation and helper T cell differentiation. J Clin Invest 2002; 110:843-50.
23. Lamhamedi-Cherradi S, Zheng S, Hilliard BA et al. Transcriptional regulation of type I diabetes by NF-kB. J Immunol 2003; 171:4886-92.
24. Hartwell DW, Fenton MJ, Levine JS et al. Aberrant cytokine regulation in macrophages from young autoimmune prone mice: Evidence that the intrinsic defect in MRL macrophage IL-1 expression is transcriptionally controlled. Mol Immunol 1995; 32(10):743-51.
25. Liu J, Beller DI. Distinct pathways for NF-kB regulation are associated with aberrant macrophage IL-12 production in lupus- and diabetes-prone mouse strains. J Immunol 2003; 1709:4489-96.
26. Boothy MR, Mora AL, Scherer DC et al. Perturbation of the T lymphocyte lineage in transgenic mice expressing a constitutive repressor of nuclear factor (NF-kB. J Exp Med 1997; 185:1897-1907.
27. Hettmann T, DiDonato J, Karin M et al. An essential role for nuclear factor kB in promoting double positive thymocyte apoptosis. J Exp Med 1999; 189:145-58.
28. Grossmann M, Nakamura Y, Grumont R et al. New insights into the roles of Rel/NF-kB transcription factors in immune function, hemopoiesis and human disease. Int J Biochem & Cell Bio 1999; 31:1209-19.
29. Grumont RJ, Rourke IJ, O'Reilly LA et al. B lymphocytes differentially use the Rel and nuclear factor kB1 (NF-kB1) transcription factors to regulate cell cycle progression and apoptosis in quiescent and mitogen activated cells. J Exp Med 1998; 187:663-74.
30. Cohen PL, Eisenberg RA. Lpr and gld: Single gene models of systemic autoimmunity and lymphoproliferative disease. Annu Rev Immunol 1991; 9:243-69.
31. Vallabhapurapu S, Ryseck RP, Malewicz M et al. Inhibition of NF-kB in T cells blocks lymphoproliferation and partially resuces autoimmune disease in gld/gld mice. Eur J Immunol 2001; 31:2612-22.

32. Zhang Y, Schlossman SF, Edwards RA et al. Impaired apoptosis, extended duration of immune responses, and a lupus like autoimmune disease in IEX-1 transgenic mice. PNAS 2002; 99(2):878-33.
33. Huang YH, Wu JY, Zhang Y et al. Synergistic and opposing regulation of the stress-responsive gene IEX-1 by p53, c-Myc, and multiple NF-kB/rel complexes. Oncogene 2002; 21:6819-28.
34. Grumont R, Hochrein H, O'Keeffe M et al. c-Rel regulates interleukin 12p70 expression in CD8+ dendritic cells by specifically inducing p35 gene transcription. J Exp Med 2001; 194:1021-1032.
35. Grigoriadis G, Zhan Y, Grumont RJ et al. The Rel subunit of NF-kB-like transcription factors is a positive and negative regulator of macrophage gene expression: Distinct roles for Rel in different macrophage populations. EMBO J 1996; 15:7099.
36. Sanjabi S, Hoffmann A, Liou HC et al. Selective drequirements for c-Rel during IL-12p40 gene inductions in macrophages. Proc Natl Acad Sci USA 2000; 97:12705-10.
37. Kontgen F, Grumont RJ, Strasser A et al. Mice lacking the c-Rel proto-oncogene exhibit defects in lymphocyte proliferation, humoral immunity, and interleukin-2 expression. Genes and Dev 1995; 9:1965-77.
38. Harling-McNabb L, Deliyannis G, Jackson DC et al. Mice lacking the transcription factor subunit Rel can clear an influenza infection and have functional anti-viral cytotoxic T cells but do not develop an optimal antibody response. Int Immunol 1999; 11:1431.
39. Liou HC, Jin Z, Tumang J et al. c-Rel is crucial for lymphocyte proliferation but dispensable for T cell effector function. Int Immunol 1999; 11:361.
40. Wiekowski MT, Leach MW, Evans Ew et al. Ubitiquous transgenic expression of the Il-23 subunit p19 induces multiorgan inflammation, runting, infertility, and premature death. J Immunol 2001; 166:7653-70.
41. Zhang GX et al. Role of Il-12 receptor B1 in regulation of T cell response by APC in experimental autoimmune encephalomyelitis. J Immunol 2003; 171:4485-92.
42. Murphy CA, Langrish CL, Chen Y et al. Divergent pro- and anti-inflammatory roles for IL-23 and IL-12 in joint autoimmune inflammation. J Exp Med 2003; 198(12):1951-7.
43. Baeuerle PA, Baltimore D. NF-kB: Ten years after. Cell 1996; 87:13-20.

CHAPTER 8

The Central Role of NF-κB in the Regulation of Immunity to Infection

Cristina M. Tato and Christopher A. Hunter*

Introduction

The NF-κB family of transcription factors is a group of evolutionarily conserved proteins involved in lymphoid organogenesis, the development of immune cells as well as the coordination of many aspect of innate and adaptive immunity to infection.[1-10] Although the events involved in the activation of these transcription factors is covered in greater detail in other chapters of this book, a brief overview here will highlight some key points. In particular it is important to note that numerous stimuli which act through a variety of receptors initiate the activation of distinct pathways that recruit unique combinations of scaffolding and signaling proteins that ultimately converge on the IκB kinase (IKK) complex. IκB is a protein that is bound to dimers of NF-κB and which retains these transcription factors in the cytoplasm. Activation of IKK leads to the phosphorylation of IκB and its degradation that allow nuclear localization of NF-κB.[3,11-13] Depending on the stimulus and the cell type, different combinations of homo- and hetero-dimers of these transcription factors are activated. Once in the nucleus, NF-κB is involved in the regulation of numerous genes involved in immune function including the production of IκB proteins that provide a feedback mechanism to limit NF-κB activity.[14,15] In part, because of the central role of these transcription factors in the regulation of the immune system this pathway has become one of the best studied signaling pathways and become a focus of drug discovery for the treatment of inflammatory diseases and cancer. Nevertheless, the aim of this chapter is to focus on the principal role of NF-κB in the development of protective immunity to infection.

Though numerous stimuli lead to the activation of NF-κB, some of the best characterized are associated with pathogens. The binding of diverse microbial products (LPS, bacterial DNA, peptidoglycans, and parasite mucins) that contain pathogen associated molecular patterns (PAMPs) to pattern recognition receptors (PRRs) such as Toll-like receptors (TLR) or NOD proteins, results in the activation of NF-κB which initiates distinct profiles of gene expression associated with innate responses to pathogens.[16-21] Some of the best examples of genes that are regulated by these events are cytokines, such as IL-12 and TNF-α, that have a prominent role in innate immunity. Other events that fall into this category include the expression of chemokines and adhesion molecules which are necessary for the migration of immune cells into sites of inflammation.[22-27] NF-κB signaling also regulates the production of effector molecules such as nitric oxide and perforin, which are directly involved in the control of pathogens.[28-30]

*Corresponding Author: Christopher A. Hunter—Department of Pathobiology, School of Veterinary Medicine, University of Pennsylvania, Philadelphia, U.S.A. Email: chunter@vet.upenn.edu

NF-κB/Rel Transcription Factor Family, edited by Hsiou-Chi Liou.
©2006 Landes Bioscience and Springer Science+Business Media.

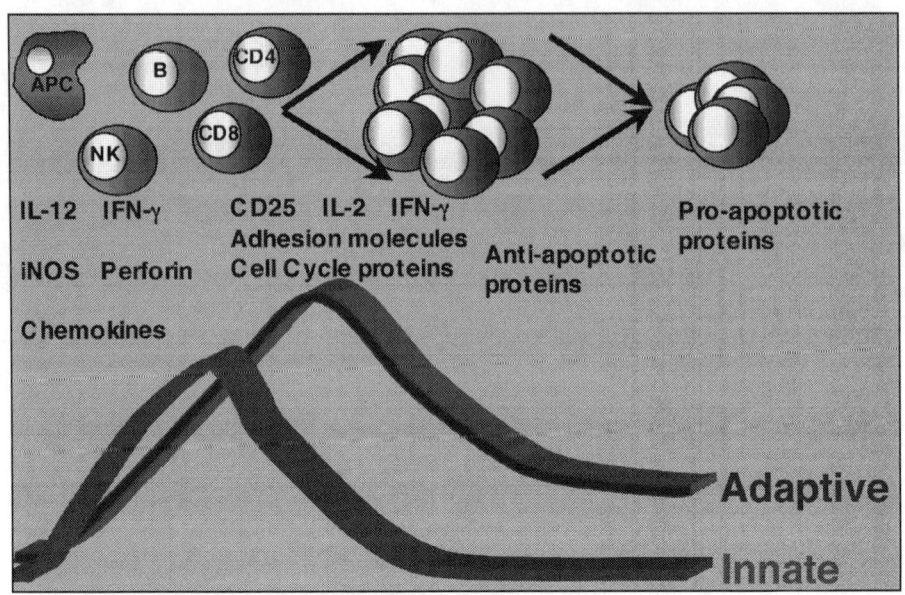

Figure 1. NF-κB target genes. As the immune response transitions from innate to adaptive immunity, NF-κB activation leading to the transcription of target genes is necessary for almost every aspect of resistance to a variety of pathogens. NF-κB promotes activation of innate immune cells such as macrophages and NK cells, and the production of key cytokines, such as IL-12 and IFN-γ, important for differentiation into a Th1 response. Development and expansion of a T cell effector population, as well as the proliferation and isotype switching of plasma cells for specific antibody production are all dependent on signaling pathways leading to NF-κB activation. Finally, the ability of effector cells to contract and down-regulate the immune response, along with the ability to maintain an antigen-specific memory population also involve NF-κB mediated gene transcription.

The activation of NF-κB is not restricted to microbial stimuli or innate responses and many receptors that activate NF-κB initiate events that directly influence adaptive functions of the immune system. Historically, NF-κB activity was first described in B cells[31,32] and it is clear that these transcription factors regulate many aspects of humoral immunity. Specifically BLyS signaling through NF-κB$_2$ regulates B cell homeostasis while CD40-mediated activation of NF-κB plays an essential role in plasma cell class switching required for production of high affinity antibody.[33-43] It is also clear that the other major arm of adaptive immunity, the T cell response, is dependent on NF-κB. For example, the ability of microbial stimuli to induce dendritic cell maturation and upregulate Class II expression required for efficient antigen presentation to T cells is dependent on NF-κB.[44-46] T cell responses can also be divided broadly into type 1 and 2 responses. Type 1 immunity is associated with cell mediated responses dominated by the production of IFN-γ and resistance to intracellular pathogens. In contrast, type 2 immunity is more commonly associated with T cell production of cytokines such as IL-4, 5 and 13 which influence humoral responses and resistance to helminth parasites. The development of these events depends on many cytokines and their receptors, as well as proteins involved in the regulation of T cell proliferation. The events which influence the polarity of a T cell response are closely linked to the innate NF-κB-dependent production of pro-inflammatory cytokines, in particular IL-12. Many of the genes for these factors have NF-κB binding sites in their promoters, and molecular studies have correlated the binding of specific NF-κB members to these sites and with gene expression,[22,23,36,47-50] (summarized in Fig. 1). In addition, stimulation through the T cell receptor, costimulatory molecules as well as many cytokine receptors, such as those for TNF-α and IL-1, result in NF-κB activity which affects the response of T cells.[50-60] Many of these effects of NF-κB on lymphocyte function can be attributed to its role in the regulation of life and death of these cells. Thus, NF-κB regulates expression of anti-apoptotic (Bcl-x_L) and cell cycle proteins (cyclin D1) that have a prominent role in the expansion and contraction of the immune response.[61-69] Since these events are known to influence the development of memory responses, it seems likely that these transcription factors have a role in directing the development and maintenance of immunological memory, but there are few studies which directly address this issue.[70-72]

The previous paragraph associates NF-κB with the innate recognition of invading microorganisms as well as the regulation of many facets of the adaptive immune response. However, in order to appreciate the complexity of this system in resistance to infection, it is necessary to draw from a variety of experimental systems and studies. In this chapter we will briefly review several aspects of the evolutionary history of NF-κB signaling and innate immunity, and highlight how studies with NF-κB deficient mice have helped to define the role of individual family members in resistance to different pathogens. The last two sections deal with case studies of natural infection in humans and the ability of pathogens to subvert NF-κB signaling in order to promote their own survival.

NF-κB: An Evolutionarily Conserved System Associated with Innate Immunity

The Toll receptor in *Drosophila melanogaster* was first identified as having an important role in the development of dorsoventral patterning in embryos,[73,74] but the recognition that Toll was homologous to the human IL-1 receptor implicated it as having an additional role in immunity. This was illustrated by studies in which signaling through Toll was shown to promote production of anti-microbial peptides such as defensins, essential for resistance to bacterial and fungal infections in flies.[73,75,76] One of the strengths of Drosophila as an experimental system was the availability of forward and reverse genetic approaches which helped to identify many of the events downstream of Toll. The use of these techniques revealed that stimulation of Toll leads to activation of the kinase Pelle, the subsequent degradation of the inhibitor Cactus and the activation of the transcription factor Dif. Each of these proteins is evolutionarily conserved with mammalian homologues in the NF-κB pathway: Pelle is homologous to the IL-1 receptor associated kinase (IRAK); Cactus is homologous to IκB; and the transcription factors NF-κB1 and RelA are equivalent to Dif.[77-80] The strong similarities between the signaling induced by Toll and IL-1 was important in developing the concept that NF-κB represented an evolutionarily conserved system associated with innate immunity.

Although Toll had a role in resistance to fungal and gram positive bacterial infections it was not required for resistance to gram negative bacteria. Subsequent studies led to the recognition that the immune deficiency gene, *imd*, was required for resistance to gram negative bacteria but not the protective responses to fungi and gram positive bacteria.[81,82] Furthermore, the identification of a death domain within the adaptor molecule encoded by *imd* revealed a striking similarity of this pathway to mammalian TNF signaling.[83] Moreover, similar to the Toll pathway, other genes identified in the imd pathway also had mammalian homologues involved in the NF-κB pathway. The drosophila kinases dIKK-β and dIKK-γ were homologous to IKK-β (IκB kinase-β) and IKK-γ whereas Relish was equivalent to the precursors of NF-$κB_1$ and NF-$κB_2$.[3,84-88] Thus, for Drosophila, two receptors that lead to the activation of pathways homologous to NF-κB were found to have a central role in distinguishing between different classes of pathogens and directing the development of appropriate anti-microbial responses.

Toll-Like Receptors and Innate Immunity

In the early 90's Charles Janeway proposed that receptors encoded in evolutionarily conserved germ-line were involved in the recognition of microbial products and that these affected subsequent adaptive responses.[89] Given the role of Toll receptors in the innate recognition of pathogens, and their ability to activate conserved signaling pathways associated with immunity, these molecules represented candidates that could form the basis for the recognition of pathogens in mammals. This hypothesis led to the identification of a human Toll-like receptor (TLR) based on homology with Toll and studies which established that signaling through this receptor led to the activation of NF-κB and was associated with the innate activation of genes (IL-6 and B7) that influenced adaptive responses.[90,91] To date 11 human TLRs have been cloned which recognize a wide variety of microbial products from bacteria, parasites and yeast (TLR1,2,4,5,6,9,11),[92-101] to heat shock proteins and viral and fungal products (TLR4,7,8).[102-108] In contrast, only one or two ligands have been identified for others (TLR3,5,7,8,9) including double- and single-stranded RNA or unmethylated CpG DNA.[105-110] Moreover, for some TLRs (TLR2,4) there are coreceptors required for recognition of PAMPs.[93,96,98] There are additional complexities to this system that include the formation of heterodimers by several TLRs (TLR1,2,6) that are required for recognition of ligands[95] and major differences in cell and tissue distribution of certain receptors. For example, these receptors can be expressed not only on lymphoid cells (TLR4,7,10) such as B cells, macrophages and dendritic cells[90,105,107,108,111,112] as well as T and NK cells,[113] but also epithelial cells of the intestine, liver, kidney and bladder (TLR5,11).[99,101] Nevertheless, while the distribution of these pattern recognition receptors contributes to the development of pathogen–specific responses, downstream of these receptors there are also complex signaling pathways that converge on the generic activation of NF-κB. However, downstream of NF-κB activity there is evidence that different TLRs promote distinct patterns of gene expression associated with different pathogens.[114] The molecular events that tailor this response to different classes of pathogens is poorly understood. Indeed, it is unclear whether different PAMPs lead to the activation of different NF-κB homo- or hetero-dimers and affect the strength and duration of this signaling cascade, or whether these latter events influence the response to different PAMPs. In addition, another level of regulation that is likely to affect the events downstream of TLR signaling are epigenetic changes that alter gene accessibility in different cell types. Clearly, additional studies are needed to address these questions, but the major point of these studies is that the immune system has evolved a variety of germ line encoded receptors linked to NF-κB, that distinguish invading microorganisms and direct the development of appropriate immune responses.

The Role of NF-κB in Resistance to Infection

During infection, the coordination of innate and adaptive immunity is a complex process and, as discussed earlier, NF-κB has been implicated in many of these processes. However, the most striking evidence for the importance of NF-κB in immune function is provided by the generation of mice deficient in various components of this pathway. In the last decade there has been a remarkable growth in the number of studies that have been performed to assess how mice that lack TLRs, cytokine receptors, adaptor molecules used in these pathways or individual NF-κB family members respond to infection, and some of these are summarized in Table 1. While these reports establish the critical role for NF-κB in the development of immunity to infection, several studies have provided unexpected insights into the biology of NF-κB.

Initial studies that linked TLR4 with the recognition of LPS provided an important conceptual breakthrough that linked Toll with the best characterized microbial product that induced inflammation.[115] Although there are multiple TLRs and several TLR deficient mice, it has with notable exceptions[101] been difficult to link a particular TLR with susceptibility to a particular micro-organism or class of pathogens. A partial explanation for this may be that pathogens tend to express more than one PAMP and there seems to be some overlap in what different TLRs can recognize. Nevertheless, there has been much more success in linking signaling machinery downstream of TLR ligation with specific pathogens. One example is provided by the TLR/IL-1R (TIR) signaling pathway which recruits the kinases IRAK-1/4 which are necessary for IKK and TRAF-6 activation which in turn lead to NF-κB dependent transcription of pro-inflammatory cytokines.[116] Consistent with this central role in the TIR pathway, studies using IRAK$^{-/-}$ mice revealed that they were more susceptible to infection with

Table 1. Murine gene KO infection phenotypes

Genotype	Increased Susceptibility	Immune Phenotype	
NF-κB1	L. monocytogenes S. pneumoniae L. major T. muris C. jejuni* H. hepaticus*	Decreased: Antibody production; Proliferation IFN-γ; Th1/Th2 response	Increased: Apoptosis of infected cells
NF-κB2	L. monocytogenes T. gondii L. major T. muris	Decreased: Antibody; Proliferation Macrophage function IL-12; IFN-γ Th2 response	Increased: IL-2 GM-CSF Apoptosis DC activation
c-Rel	L. monocytogenes S. pneumoniae T. gondii L. major T. muris Influenza	Decreased: Antibody response Proliferation IL-2, IL-12, IFN-γ APC response	
RelB	Lymphocytic choriomeningitis virus L. monocytogenes T. gondii	Decreased: Antibody production IFN-γ	Increased: Inflammation Chemokine production
Bcl-3	L. monocytogenes S. pneumoniae T. gondii Influenza	Decreased: Antibody response; Macrophage function; Th1 response	Increased:
IκBα(DN)	T. gondii Reovirus	Decreased: NK activation; Proliferation; IFN-γ; Cytotoxicity	
IRAK	CMV P. acnes	Decreased: NK function; Th1 response; IFN-γ	
MyD88	C. albicans Gram(+) bacteria Gram(-) bacteria M. tuberculosis; M. avium B. burgdorfei T. gondii	Decreased: Macrophage function; Cytokine production: IL-12, IFN-γ and TNF-α	
NIK	T. spiralis M. leprae vesicular stomatitis virus	Decreased: Antibody production; Th2 response Mast cell function; IL-2; IL-6; IL-18; IFN-γ	

murine cytomegalovirus (CMV) or *Propionibacterium acnes*.[117] Similarly, MyD88, an adaptor protein associated with IRAK signaling, is required for optimal resistance to a wide variety of pathogens including bacteria such as *Staphylococcus aureus* and *Listeria monocytogenes*, as well as the intracellular parasite *Toxoplasma gondii* and the spirochete *Borrelia burgdorferi*.[118-127] However, interpretation of these studies can be difficult as IL-1, IL-18 and multiple TLRs use

MyD88, which can make it hard to distinguish which specific pathway is required for protective responses. Nevertheless, a common finding of these studies was that the susceptibility of MyD88$^{-/-}$ mice was associated with a profound defect in macrophage and dendritic cell function, consistent with an important role for MyD88 in TLR-induced signaling.

Although NF-κB plays an important role in innate recognition of infection, immune cells involved in the development of adaptive immunity employ a variety of receptors that recruit specific kinases that lead to NF-κB activity. For example, stimulation through the T cell receptor recruits NIK (NF-κB inducing kinase) which in turn initiates NF-κB signaling.[128] Mutations in this kinase result in a significant immune deficiency that leads to increased susceptibility to diverse pathogens such as the helminth *Trichinella spiralis* and the bacterium *Mycobacteria leprae*.[129,130] During these infections these mice are unable to mount an appropriate cell mediated response, consistent with an impaired antigen specific response by T cells and typically exhibit a lack of antibody production. Other studies have shown that the NIK mutation causes a specific defect in B cell class switching during infection with vesicular stomatitis virus (VSV), but that the susceptibility of these mice to lymphocytic choriomeningitis virus (LCMV) is actually due to structural defects in secondary lymph node formation and is not cell intrinsic.[131] This latter study illustrates the difficulty in using genetically manipulated mice that may exhibit defects that are not intrinsic to the NF-κB signaling pathways, but are a secondary cause of abnormal tissue development and structure.

While the section above illustrates how defects in the NF-κB signaling pathway can lead to increased susceptibility to infection, many of these kinases and adaptor molecules are also involved in the activation of other signaling pathways. As a consequence, some of the most definitive data on the role of NF-κB in resistance to infection have been dependent on mice that lack individual family members. However, many of these studies are associated with their own set of problems. For example, the prototypical NF-κB dimer is composed of RelA (p65) and NF-κB$_1$ (p50) but the ability to study the role of RelA in resistance to infection is compromised because the loss of RelA results in embryonic lethality. In contrast, deletion of NF-κB$_1$ does not affect viability, and this has allowed examination of the role of NF-κB$_1$ in resistance to infection. A combination of studies revealed that although NF-κB$_1^{-/-}$ mice have normal responses to *Escherichia coli* and *Haemophilus influenzae* they are more susceptible to infection with *L. monocytogenes*, *Streptococcus pneumoniae* or *Leishmania major*.[132,133] For some of these studies, how the absence of NF-κB$_1$ leads to increased susceptibility has not been defined, but this transcription factor has been implicated in the development of type 1 (dominated by the production of IFN-γ) and type 2 (characterized by the production of IL-4) T cell responses. In one case, susceptibility to *L. major* was found to be the result of a defect in the production of IFN-γ, associated with reduced proliferation of antigen specific CD4$^+$ T cells.[132] Similarly, NF-κB$_1^{-/-}$ mice infected with the gut dwelling helminth *Trichuris muris* failed to develop the type 2 responses required to expel these worms.[134] Interestingly, chronic infection with *T. muris* does not normally lead to the development of inflammation, but in the absence of NF-κB$_1$, infection results in a severe colitis. This particular phenotype does not appear to be unique to challenge with *T. muris* as NF-κB$_1^{-/-}$ mice heterozygous for RelA expression (p65$^{+/-}$) develop gastroenteritis when infected with *Camplyobacter jejuni* or *Helicobacter hepaticus*.[135-137] The basis for these inflammatory phenotypes remains an open question, but they do indicate a role for NF-κB$_1$ as a negative regulator of inflammation in the gut, and these observations are pertinent to studies in human patients with colitis that are discussed in a later section.

A comparison of the canonical and non canonical pathways of NF-κB activation is outlined in detail in other chapters. However, one of the family members most closely associated with the noncanonical pathway is NF-κB$_2$ (p100/p52). Several immune stimuli, such as CD40L, BLyS and lymphotoxin-α which play a prominent role in B cell function activate this pathway and this is consistent with homeostatic defects in the B cell populations of mice that lack NF-κB$_2$.[33,35,42] In terms of infection, these mice display increased susceptibility to several intracellular organisms including *L. monocytogenes*, *T. gondii* and *L. major*.[35,138,139] Interestingly, mice deficient in Bcl-3, an IκBα family member that is involved in the processing of NF-κB$_2$, are also susceptible to some of these same infections, but the basis for these phenotypes remains unclear.[140,141] In contrast, more is known about how the absence of NF-κB$_2$ affects the response to these pathogens. For example, NF-κB$_2$$^{-/-}$ mice infected with *T. gondii* develop appropriate innate and adaptive responses during the early phase of infection, but as the infection progresses there is a loss of T cells that are required for long term resistance to this persistent parasite. As a consequence, there is reactivation of the infection and the development of severe disease in the brain.[138] Similarly, NF-κB$_2$$^{-/-}$ mice infected with *L. major* develop a chronic disease characterized by nonhealing lesions associated with reduced production of IFN-γ. Resistance to *L. major* is dependent on a complex feedback loop in which the CD40L/CD40 interaction is required for the production of IL-12 which in turn stimulates T cell production of IFN-γ.[142] In the absence of NF-κB$_2$, macrophages stimulated through CD40 have reduced IL-12 responses and this provides a likely basis for the susceptibility of these mice to *L. major*.[139] However, NF-κB$_2$ is not just associated with the development of cell mediated immunity as NF-κB$_2$ deficient mice infected with *T. muris* fail to develop the type 2 responses required for resistance to this helminth.[134] Interestingly, in contrast to NF-κB$_1$$^{-/-}$ mice, persistence of *T. muris* in the absence of NF-κB$_2$ is not associated with the development of colitis. The opportunity to directly compare the response of the NF-κB$_1$$^{-/-}$ and NF-κB$_2$$^{-/-}$ mice to this pathogen indicates distinct roles for these transcription factors in the development of type 2 responses and emphasize the unique role of NF-κB$_1$ in the regulation of inflammation in the gut.[134]

Another NF-κB member that is activated via the noncanonical pathway, is RelB. The observation that RelB$^{-/-}$ mice spontaneously develop a lethal inflammatory disease mediated by T cells[143] indicates a critical role for RelB in immune homeostasis, but has restricted research on their response to infection. Nevertheless, these mice have been used in a small number of studies and have been shown to be more susceptible to LCMV as well as *L. monocytogenes*.[144] Furthermore, susceptibility of RelB$^{-/-}$ mice to *T. gondii* is associated with an inability of T and NK cells to produce IFN-γ.[145] These studies suggest a prominent role for this family member in the development of Th1 responses to intracellular pathogens, but underlying defects in the immune system of RelB$^{-/-}$ mice make some of these results difficult to interpret.

The NF-κB family member, c-Rel, is widely expressed in lymphoid cells and has been implicated in the regulation of macrophage as well as T and B cell functions. However, c-Rel$^{-/-}$ mice do not have any major developmental defects in their immune system, although they lack marginal zone B cells[146] and are surprisingly immune competent. Nevertheless, there are several reports that establish a role for c-Rel in the regulation of multiple macrophage functions, in particular production of IL-12 in response to a large number of microbial and inflammatory stimuli.[53,147] Consistent with these reports are studies which revealed that in the absence of c-Rel, mice infected with *L. major* developed more severe lesions than wild type mice.[28] While c-Rel is important in the production of IL-12 there are c-Rel independent pathways to produce IL-12 and this is illustrated by studies with *T. gondii*.[53,148] However, mice deficient for c-Rel also failed to recover from infection with *T. gondii*, due to functional defects in both accessory cell and T cell populations.[148] Several studies have also indicated a role for c-Rel in B cell function[50,149] and while many aspects of acquired immunity are still intact in c-Rel$^{-/-}$ mice, during infection with the influenza virus, these mice have a defect in their ability to produce neutralizing antibodies.[150]

Together, these in vivo studies continue to highlight a critical role for NF-κB in the development and maintenance of the immune response to many pathogens. Yet, in all of these studies, the molecular basis for many of the observed defects remains unclear and few studies have managed to truly understand how defects in specific innate or adaptive compartments contribute to the phenotypes observed. One problem with the use of conventional gene knockout mice is that since the gene of interest is absent from all lymphoid cells, the ability to distinguish direct versus indirect effects of the missing gene can be extremely difficult. One way to address this issue is to generate transgenic mice in which a degradation deficient form of IκBα, (IκBα(ΔN)) which functions as a global inhibitor of NF-κB, is selectively expressed in

Table 2. Human genetic mutations

Gene	Mutation Type	Susceptibility	Prognosis	Immune Characteristics
NEMO	Hypomorphic Recessive	Pyogenic bacteria Gram(+) and gram(-) Fungal; Viral	Acute and recurrent 50% of cases develop disseminated disease.	XL-EDA-ID Hyper IgM; hypo IgG; Decreased NK activity; Impaired cell mediated responses.
IkBa	Hypermorphic Dominant	Pyogenic bacteria; Gram(+) and gram(-)	Chronic and recurrent infections	AD-EDA-ID hyper IgM; Lymphocytosis; Severe T cell immunodeficiency
IRAK-4	Amorphic Recessive	Pyogenic bacteria; Fungal; Opportunistic	Acute in early childhood. Severity decreases as child matures.	Complete TLR signaling deficiency. Decreased inflammatory cytokines.
p50	Single nucleotide polymorphism	Gram(+) and gram(-) bacteria	Chronic ulcerative colitis; IBD?	Increased local inflammatory response; mechanism unknown
Nod2	?Single nucleotide polymorphism	Gram(+) and gram(-) bacteria	Chron's Disease	Tissue specific, decreased innate immune function

different cell types. This approach has emphasized the global role of this transcription family in a tissue specific way for the development of innate and adaptive immune responses to a variety of pathogens.[151-153] Thus, mice in which the IκBα(ΔN) transgene is expressed in hepatocytes are more susceptible to *L. monocytogenes*. Similarly, mice in which the NK and T cells express this transgene are highly susceptible to infection with *T. gondii* as a consequence of decreased NK cell activation and antigen specific CD4$^+$ T cell expansion and subsequent failure to produce IFN-γ.[153]

Human Gene Deficiencies

The studies described in the previous section illustrate the key the role of NF-κB signaling in the coordination of the immune response required for resistance to different pathogens in diverse model systems. Nevertheless, the importance of these transcription factors during natural infection in humans is underscored by the identification of individuals with genetic defects in this pathway. With regard to infectious disease, there are currently four known categories of genetic mutations associated with NF-κB which result in mild to severe forms of immunodeficiency.[154] Table 2, contains a summary of these mutations, their phenotype and disease morbidity.

IRAK-4 Deficiency

As mentioned earlier, the kinase IRAK-4 is part of the TIR signaling pathway and plays a prominent role in the activation of NF-κB in response to many stimuli. To date, four individuals have been described with amorphic mutations in IRAK-4 that are associated with immunodeficiency.[155-158] Importantly, although NF-κB has been associated with development[33,159-162] and haematopoietic cell ontogeny,[8,9,151,163] patients with this mutation have no overt physical developmental defects or deficiencies in lymphocyte populations. However, affected individuals have severe infections with pyogenic bacteria, such as *S. pneumoniae* and *S. aureus* in early childhood associated with sepsis and cellulitis.[155,158] These individuals are also susceptible to recurrent infections with fungi (*Candida albicans*) and other opportunistic infections.[164] In these patients susceptibility to gram(+) infections is characterized by a poor inflammatory response including reduced IL-6 and IFN-γ production.[158] Two of the four patients also exhibited impaired responses to LPS, or stimulation through TLRs 1-6 and TLR-9.[155-157,164] Moreover, consistent with a role for IRAK-4 in B cell function, antibody responses after vaccination to polysaccharide and protein antigens were below normal range and were shown to diminish after the last booster.[164] The reduced production of antibody by B cells could be the result of either an intrinsic defect in B cell TLR signaling, or due to a lack of T cell help. Interestingly, as these patients reach adolescence the severity and frequency of bacterial infection decreases significantly which may indicate the development of sufficient adaptive responses to cope with these challenges.[158]

Mutations in NEMO

As discussed at the start of this chapter, the IKK complex has an important role in the activation of NF-κB. NEMO is a central component of this complex and is required for NF-κB signaling in response to many developmental and inflammatory stimuli. The *NEMO* gene is located on the X chromosome, and similar to NEMO deficient mice, amorphic mutations primarily affect males, causing a lethal form of incontinentia pigmenti (IP) in utero.[165,166] Subsequently identified hypomorphic mutations were found to lead to a collection of developmental abnormalities associated with anhidrotic ectodermal dysplasia (EDA) which includes partial or total absence of teeth, conical teeth, sparcity of hair and an absence of sweat glands leading to dry skin.[167] Furthermore, immunodeficiency (ID) characterized by increased susceptibility to a variety of infectious diseases is a hallmark of this X-linked mutation.[168] The immunological phenotype of these patients is complex, in that most individuals have normal lymphocyte distribution and in vitro proliferation of PBMCs in response to mitogens, but impaired expansion in response to tetanus or diphtheria antigens.[169] Additionally, patients typically present with decreased IgG, but hyper-IgM in the peripheral blood, suggestive of a defect in CD40-mediated class switching in B cells.[168-174] Consistent with this observation, activation of B cells via CD40 ligation was found to be defective in most of the individuals studied.[173-175]

Given the central role of NEMO in the activation of NF-κB, it is not surprising that these patients are highly susceptible to a range of pathogens. Typically, boys with this mutation present with multiple, recurrent pyogenic bacterial infections early in life, including *S. pneumoniae*, *S. aureus* and *H. influenza*, with about 50% of identified cases succumbing to disseminated infection.[168,169,174] Mycobacterial, fungal and viral infections are less frequent in these patients, but are often severe when they occur.[169,173,174,176-178] NK cell cytotoxicity is also significantly impaired in the majority of these individuals, along with a defect in the production of the NK and T cell growth factor IL-2.[169,176] Since NK cells play a central role in resistance to many herpes viruses,[179] it is not surprising that these patients are also susceptible to this group of pathogens, but treatment with IL-2 was able to induce NF-κB activation and partially restore NK cell activity.[176] Overall inflammatory responses during infection were significantly impaired in all cases described[154,172] and characterized by a lack of responsiveness to LPS, IL-1β, IL-18, TNFα and CD40L.[174] Together, the identification of multiple immune defects in these patients and their susceptibility to multiple pathogens indicates the central role of NEMO for the activation of NF-κB.

Mutations in IκBα

The ubiquitination and proteasomal degradation of IκBα represents a major checkpoint in the activation of NF-κB. Hypermorphic mutations of the *IKBA* gene are associated with a dominant form of EDA-ID that results in a specific T cell immunodeficiency and increased susceptibility to infectious disease.[180] This mutation, which was not inherited, prevented IκBα degradation while IκBβ and IκBε degradation remained normal.[180] The patient described had a failure to thrive and presented with chronic diarrhea and recurrent bronchopneumonitis from 2 months of age.[180] This child also suffered from numerous gram(+) and (-) pyogenic bacterial infections, the details of which have not yet been published. The profound T cell immunodeficiency associated with this mutation was characterized by a complete absence of γ/δ T cells and memory α/β T cells, and a lack of responsiveness of the naïve T cell population, indicating a role in development and maintenance of human lymphocyte populations as well as activation and effector function in vivo. In this patient, reduced NF-κB DNA binding activity in response to TIR and TNF-α stimulation was observed, which correlated with decreased production of IL-6, IL-2 and IFN-γ. Thus, the increased severity of disease in this patient is likely related to the ablation of multiple NF-κB signaling pathways in T cells.[180] Despite an overall lymphocytosis, there was a complete absence of serum antibody for specific antigens, indicating a defect in CD40 signaling in B cells, a common theme observed in the various groups of patients described so far.

Colitis and NF-κB1

The defects described above are the result of either deletions or missense mutations which lead to the production of mutated proteins with reduced or altered function. In contrast, recent studies have associated a single nucleotide insertion/deletion polymorphism (SNP) in the 5'-promoter region of the *NF-κB1* gene with chronic ulcerative colitis.[181] This unique polymorphism is linked to decreased binding of nuclear proteins to the *NF-κB1* promoter.[181] This result coincides with studies in NF-κB$_1$$^{-/-}$ mice that were shown to have an increased susceptibility to infection-induced colitis.[134,136,137] Because of the dual role of NF-κB$_1$ in transcriptional activation as well as suppression, the basis for this susceptibility remains elusive. One receptor which may be directly related to NF-κB$_1$ activation and surveillance of mucosal pathogens, is Nod2. This protein, along with the related molecule Nod1, works as a cytosolic receptor that recognizes peptidoglycans from invading bacteria and induces the activation of NF-κB.[55,182] Recently, a mutation in the gene encoding Nod2 has been linked to the pathogenesis of colitis in human and mouse studies[54,183,184] and supports a role for this bacterial "sensor" in regulation of NF-κB$_1$ activation. In contrast to the NF-κB$_1$ SNP that is associated with an increase in inflammatory cell activation, the Nod2 defect leads to a decrease in activation in response to bacterial LPS and peptidoglycan.[183,184] Clearly more studies are required to address the role of NF-κB in the development of this enteric disease.

Pathogens That Interfere with NF-κB

The sections above have focused on the role of NF-κB in the coordination of the immune responses required for resistance to infection. However, as a consequence of the critical role these transcription factors play in innate and adaptive immunity, this pathway acts as a strong selective pressure for pathogens. It is now recognized that microorganisms have developed specific strategies to block or enhance this intracellular signaling pathway in order to promote their own replication, survival and dissemination within the host. Indeed, almost every aspect of the NF-κB pathway has been targeted by pathogens and the following section along with (Table 3) highlight what is currently known about different pathogens that interact with NF-κB and the specific point in the pathway that each target.

One general strategy that has been developed by pathogens to reduce the ability of cells to become activated and/or respond appropriately to stimuli is to alter the surface expression of activating receptors on target cells. Certain viruses, such as CMV reduce NF-κB activity by suppressing surface expression of cytokine receptors, thus preventing infected cells from becoming activated, migrating and interacting with T cells.[185,186] Another example is provided by the bacterium *Ehrlichia chaffeensis* that is able to directly down regulate specific TLRs and their coreceptors on the surface of infected human monocytes.[187] These events would presumably lead to a reduced capacity of cells to sense the presence of these organisms and direct the development of protective immunity.

Table 3. Pathogen inhibition of NF-κB

Pathway	Pathogen	Proposed Mechanism	Target
Receptor Expression	MCMV[185]	NK	TNFR1/2
	huCMV[186]	NK	CCR7
	Ehrlichia chaffeenis[187]	NK	TLR2/4'CD14
	HIV[101]	NK	Mannose receptor
Adaptor Protein Complex Formation	Vaccinia virus	Dominant negative, A52R competes w/ host protein	MyD88
Kinase Activation	Yersinia pestis Y. pseudotuberculosis Y. enterocolitica[189-194]	YopJ, YopH, YopP and others interrupt kinase and ubiquitan function	MAPK; SUMO-1; IKKs
	Uropathogenic E. coli[195]	Soluble factors	MAPK
IκBα Degradation	Measles virus[196]	NK	IκBα phosphorylation
	Pox virus[197]	NK	IκBα degradation
	HIV[198-199]	Viral protein, Vpu competes w/ host protein	β-TrCP
	Francisella tularensis[200]	23kDa protein	IκBα degradation
	Salmonella[201]	NK	IκBα ubiquitination
Nuclear Translocation	African swine fever virus[202]	A238L acts as non-degradable IκB homolog	NF-κB dimers
	M. ulcerans	Soluble toxin	RelA phosphorylation
DNA Binding	Epstein-Barr virus[204]	ZEBRA	RelA
	T. gondii[205-207]	NK	NF-κB dimers
	Schistosoma masoni[208]	Parasite factor	DNA binding complex
Unknown	L. donovani[209]	Ceramide production	NK

Another group of gram negative bacteria that are particularly impressive in their ability to interfere with NF-κB signaling are the extracellular bacteria *Yersinia*. These pathogens use a type III secretion system to inject virulence factors, known as *Yersinia* outer proteins (Yop), into target host cells such as macrophages, epithelial cells, fibroblasts and lymphocytes, and directly inhibit kinase activation. Thus, YopJ of *Y. pseudotuberculosis* targets MAPK kinases, which are upstream of IκB phosphorylation.[189] In addition, these proteins also have proteolytic activity and can target ubiquitin-like molecules and thereby disrupt the degradation of regulatory proteins, including IκB.[190] Similarly, YopP of *Y. enterocolitica* can bind to IKKβ and thereby prevent activation of NF-κB and cause apoptosis in macrophages.[192,193] *Y. pseudotuberculosis* uses YopH, a tyrosine phosphatase, that has been shown to block all antigen specific receptor signaling in T and B cells by inhibiting phosphorylation of proteins early in the pathway[194] and this virulence factor is likely able to inhibit TCR-mediated activation of NF-κB. It is thought that the inhibition of NF-κB activation not only limits the ability of these lymphocytes to produce cytokines, but may also promote their apoptosis and so inhibit activation of the immune system.[210] Together, these studies highlight a common theme among these pathogens: events upstream of IκB degradation represent good targets to inhibit NF-κB activation which emphasizes the importance of IκB as a major checkpoint in this pathway.

Epstein-Barr virus can differentially affect the activation of NF-κB, depending on the cell type in question. In infected T cells, a viral protein is able to bind to RelA and inhibit NF-κB activity most likely blocking transcription of anti-apoptotic proteins and rendering infected T cells susceptible to apoptosis.[204] In contrast, the viral transformation of B cells and the subsequent development of lymphoproliferative disease is associated with sustained NF-κB activation.[211,212] This strategy of transforming cells through sustained NF-κB activation which promotes proliferation and resistance to apoptosis has also been used by other viruses such as human T-cell leukemia virus (HTLV) and the parasite *Theileria parva*.[213-215] This process is analogous to the elevated NF-κB activity that is thought to contribute to the transformation of certain cancer cells. Nevertheless, regardless of the mechanism, it is striking to note the number of viruses, bacteria and parasites which have evolved ways to activate or inhibit the NF-κB system for their own benefit. This is likely a reflection of the long term evolutionary relationship between these micro-organisms and this conserved signaling pathway involved in the recognition and elimination of these pathogens.

Conclusions and Future Directions

In the last 20 years there have been many important advances in our understanding of the receptors and signaling molecules involved in the development of protective immunity to infection. The experimental and clinical studies discussed in this chapter provide several illustrations of the importance of NF-κB in resistance to infection. In particular, the identification of numerous germ line encoded receptors that activate NF-κB provides an insight into how the innate system distinguishes different classes of pathogens. Interestingly, pathogens have targeted almost every aspect of the NF-κB signaling pathway to promote their own replication. It seems possible that as we explore the specifics of how individual pathogens alter intracellular signaling, these studies may provide new insights into aspects of NF-κB signaling that are not well understood. Similarly, with the identification of a growing cohort of patients with defects in NF-κB signaling, it seems likely that the study of these natural mutants will provide a better understanding of the role of some of the individual molecules involved in this intricate pathway.

From an experimental perspective, attempts to delineate the role of individual NF-κB members in resistance to infection has been difficult for several reasons. With few exceptions the response of each knockout strain has yet to be evaluated for the same pathogens[134] and more complete comparative studies are still needed. Moreover, because all of the NF-κB family members are involved in so many aspects of the immune system in many different cell types that cross-regulate each other, it has been difficult to study mice deficient in single NF-κB family members and fully understand their phenotypes. The availability of inducible approaches to ablate specific elements of NF-κB signaling in a cell and tissue specific fashion[216-219] will lead to the development of more sophisticated tools to address specific questions and overcome some of the intrinsic difficulties with the approaches that have been widely used to date Lastly, it is important to recognize that because of its role in inflammatory processes and the development of cancer, the NF-κB system has represented an attractive target for the development of anti-inflammatory and anti-tumor drugs. However, it seems likely that the same knowledge that drives the development of new pharmaceuticals could be used to develop ways to promote and direct the development of protective immunity, either at the level of vaccine development or for the treatment of chronic infections.

Acknowledgements

This work is supported by the NIH Grant AI 46288, Parasitology Training Grant AI07532 and the State of Pennsylvania.

References

1. Chen F, Castranova V, Shi X. New insights into the role of nuclear factor-kappaB in cell growth regulation. Am J Pathol 2001; 159(2):387-397.
2. Ghosh S, May MJ, Kopp EB. NF-κB and Rel proteins: Evolutionarily conserved mediators of immune responses. Annu Rev Immunol 1998; 16:225-260.
3. Karin M, Delhase M. The IκB kinase (IKK) and NF-κB: Key elements of proinflammatory signalling. Semin Immunol 2000; 12(1):85-98.
4. Karin M, Ben-Neriah Y. Phosphorylation meets ubiquitination: The control of NF-κB activity. Annu Rev Immunol 2000; 18:621-663.
5. Medzhitov R, Janeway Jr CA. Innate immune induction of the adaptive immune response. Cold Spring Harb Symp Quant Biol 1999; 64:429-435.
6. Naumann M. Nuclear factor-kappa B activation and innate immune response in microbial pathogen infection. Biochem Pharmacol 2000; 60(8):1109-1114.
7. Silverman N, Maniatis T. NF-κB signaling pathways in mammalian and insect innate immunity. Genes Dev 2001; 15(18):2321-2342.
8. Stanic AK, Bezbradica JS, Park JJ et al. NF-κB controls cell fate specification, survival, and molecular differentiation of immunoregulatory natural T lymphocytes. J Immunol 2004; 172(4):2265-2273.
9. Stanic AK, Bezbradica JS, Park JJ et al. Cutting edge: The ontogeny and function of Vα14Jα18 natural T lymphocytes require signal processing by protein kinase C theta and NF-κB. J Immunol 2004; 172(8):4667-4671.
10. Li ZW, Rickert RC, Karin M. Genetic dissection of antigen receptor induced-NF-κB activation. Mol Immunol 2004; 41(6-7):701-714.
11. Kopp EB, Medzhitov R. The Toll-receptor family and control of innate immunity. Curr Opin Immunol 1999; 11(1):13-18.
12. Yamamoto Y, Gaynor RB. IκB kinases: Key regulators of the NF-κB pathway. Trends Biochem Sci 2004; 29(2):72-79.
13. Sakurai H, Suzuki S, Kawasaki N et al. Tumor necrosis factor-α-induced IKK phosphorylation of NF-κB p65 on serine 536 is mediated through the TRAF2, TRAF5, and TAK1 signaling pathway. J Biol Chem 2003; 278(38):36916-36923.
14. Chiao PJ, Miyamoto S, Verma IM. Autoregulation of IκBα activity. Proc Natl Acad Sci USA 1994; 91(1):28-32.
15. Finco TS, Baldwin AS. Mechanistic aspects of NF-κB regulation: The emerging role of phosphorylation and proteolysis. Immunity 1995; 3(3):263-272.
16. Imler JL, Hoffmann JA. Toll and Toll-like proteins: An ancient family of receptors signaling infection. Rev Immunogenet 2000; 2(3):294-304.
17. MacDonald TT, Pettersson S. Bacterial regulation of intestinal immune responses. Inflamm Bowel Dis 2000; 6(2):116-122.
18. Ropert C, Gazzinelli RT. Signaling of immune system cells by glycosylphosphatidylinositol (GPI) anchor and related structures derived from parasitic protozoa. Curr Opin Microbiol 2000; 3(4):395-403.
19. Ropert C, Almeida IC, Closel M et al. Requirement of mitogen-activated protein kinases and IκB phosphorylation for induction of proinflammatory cytokines synthesis by macrophages indicates functional similarity of receptors triggered by glycosylphosphatidylinositol anchors from parasitic protozoa and bacterial lipopolysaccharide. J Immunol 2001; 166(5):3423-3431.
20. Takeuchi O, Akira S. Toll-like receptors; Their physiological role and signal transduction system. Int Immunopharmacol 2001; 1(4):625-635.
21. Zhang G, Ghosh S. Toll-like receptor-mediated NF-κB activation: A phylogenetically conserved paradigm in innate immunity. J Clin Invest 2001; 107(1):13-19.
22. Plevy SE, Gemberling JH, Hsu S et al. Multiple control elements mediate activation of the murine and human interleukin 12 p40 promoters: Evidence of functional synergy between C/EBP and Rel proteins. Mol Cell Biol 1997; 17(8):4572-4588.
23. Yoshimoto T, Nagase H, Ishida T et al. Induction of interleukin-12 p40 transcript by CD40 ligation via activation of nuclear factor-kappaB. Eur J Immunol 1997; 27(12):3461-3470.
24. Collart MA, Baeuerle P, Vassalli P. Regulation of tumor necrosis factor alpha transcription in macrophages: Involvement of four kappa B-like motifs and of constitutive and inducible forms of NF-κB. Mol Cell Biol 1990; 10(4):1498-1506.
25. Ledebur HC, Parks TP. Transcriptional regulation of the intercellular adhesion molecule-1 gene by inflammatory cytokines in human endothelial cells. Essential roles of a variant NF-κB site and p65 homodimers. J Biol Chem 1995; 270(2):933-943.

26. Collins T, Read MA, Neish AS et al. Transcriptional regulation of endothelial cell adhesion molecules: NF-κB and cytokine-inducible enhancers. FASEB J 1995; 9(10):899-909.
27. Muller S, Kammerbauer C, Simons U et al. Transcriptional regulation of intercellular adhesion molecule-1: PMA- induction is mediated by NF-κB. J Invest Dermatol 1995; 104(6):970-975.
28. Grigoriadis G, Zhan Y, Grumont RJ et al. The Rel subunit of NF-κB-like transcription factors is a positive and negative regulator of macrophage gene expression: Distinct roles for Rel in different macrophage populations. EMBO J 1996; 15(24):7099-7107.
29. Xie QW, Kashiwabara Y, Nathan C. Role of transcription factor NF-κB/Rel in induction of nitric oxide synthase. J Biol Chem 1994; 269(7):4705-4708.
30. Zhou J, Zhang J, Lichtenheld MG et al. A role for NF-κB activation in perforin expression of NK cells upon IL-2 receptor signaling. J Immunol 2002; 169(3):1319-1325.
31. Sen R, Baltimore D. Inducibility of kappa immunoglobulin enhancer-binding protein NF-κB by a posttranslational mechanism. Cell 1986; 47(6):921-928.
32. Sen R, Baltimore D. Multiple nuclear factors interact with the immunoglobulin enhancer sequences. Cell 1986; 46(5):705-716.
33. Caamano JH, Rizzo CA, Durham SK et al. Nuclear factor (NF)-κB2 (p100/p52) is required for normal splenic microarchitecture and B cell-mediated immune responses. J Exp Med 1998; 187(2):185-196.
34. Doi TS, Takahashi T, Taguchi O et al. NF-κB RelA-deficient lymphocytes: Normal development of T cells and B cells, impaired production of IgA and IgG1 and reduced proliferative responses. J Exp Med 1997; 185(5):953-961.
35. Franzoso G, Carlson L, Poljak L et al. Mice deficient in nuclear factor (NF)-κB/p52 present with defects in humoral responses, germinal center reactions, and splenic microarchitecture. J Exp Med 1998; 187(2):147-159.
36. Kontgen F, Grumont RJ, Strasser A et al. Mice lacking the c-rel proto-oncogene exhibit defects in lymphocyte proliferation, humoral immunity, and interleukin-2 expression. Genes Dev 1995; 9(16):1965-1977.
37. Snapper CM, Zelazowski P, Rosas FR et al. B cells from p50/NF-κB knockout mice have selective defects in proliferation, differentiation, germ-line CH transcription, and Ig class switching. J Immunol 1996; 156(1):183-191.
38. Tumang JR, Owyang A, Andjelic S et al. c-Rel is essential for B lymphocyte survival and cell cycle progression. Eur J Immunol 1998; 28(12):4299-4312.
39. Do RK, Hatada E, Lee H et al. Attenuation of apoptosis underlies B lymphocyte stimulator enhancement of humoral immune response. J Exp Med 2000; 192(7):953-964.
40. Hatada EN, Do RK, Orlofsky A et al. NF-κB1 p50 is required for BLyS attenuation of apoptosis but dispensable for processing of NF-κB2 p100 to p52 in quiescent mature B cells. J Immunol 2003; 171(2):761-768.
41. Laabi Y, Strasser A. Immunology. Lymphocyte survival—ignorance is BLys. Science 2000; 289(5481):883-884.
42. Kayagaki N, Yan M, Seshasayee D et al. BAFF/BLyS receptor 3 binds the B cell survival factor BAFF ligand through a discrete surface loop and promotes processing of NF-κB2. Immunity 2002; 17(4):515-524.
43. Kanakaraj P, Migone TS, Nardelli B et al. BLyS binds to B cells with high affinity and induces activation of the transcription factors NF-κB and ELF-1. Cytokine 2001; 13(1):25-31.
44. Speirs K, Lieberman L, Caamano J et al. Cutting edge: NF-κB2 is a negative regulator of dendritic cell function. J Immunol 2004; 172(2):752-756.
45. King NJ, Kesson AM. Interaction of flaviviruses with cells of the vertebrate host and decoy of the immune response. Immunol Cell Biol 2003; 81(3):207-216.
46. Gobin SJ, Montagne L, Van Zutphen M et al. Upregulation of transcription factors controlling MHC expression in multiple sclerosis lesions. Glia 2001; 36(1):68-77.
47. Sica A, Dorman L, Viggiano V et al. Interaction of NF-κB and NFAT with the interferon-γ promoter. J Biol Chem 1997; 272(48):30412-30420.
48. Hofer S, Rescigno M, Granucci F et al. Differential activation of NF-κB subunits in dendritic cells in response to Gram-negative bacteria and to lipopolysaccharide. Microbes Infect 2001; 3(4):259-265.
49. Gerondakis S, Grumont R, Rourke I et al. The regulation and roles of Rel/NF-κB transcription factors during lymphocyte activation. Curr Opin Immunol 1998; 10(3):353-359.
50. Liou HC, Jin Z, Tumang J et al. c-Rel is crucial for lymphocyte proliferation but dispensable for T cell effector function. Int Immunol 1999; 11(3):361-371.
51. Medzhitov R, Preston-Hurlburt P, Kopp E et al. MyD88 is an adaptor protein in the hToll/IL-1 receptor family signaling pathways. Mol Cell 1998; 2(2):253-258.

52. Aune TM, Mora AL, Kim S et al. Costimulation reverses the defect in IL-2 but not effector cytokine production by T cells with impaired IκBα degradation. J Immunol 1999; 162(10):5805-5812.
53. Mason N, Aliberti J, Caamano JC et al. Cutting edge: Identification of c-Rel-dependent and -independent pathways of IL-12 production during infectious and inflammatory stimuli. J Immunol 2002; 168(6):2590-2594.
54. Inohara N, Ogura Y, Chen FF et al. Human Nod1 confers responsiveness to bacterial lipopolysaccharides. J Biol Chem 2001; 276(4):2551-2554.
55. Girardin SE, Philpott DJ. Mini-review: The role of peptidoglycan recognition in innate immunity. Eur J Immunol 2004; 34(7):1777-1782.
56. Kim JG, Lee SJ, Kagnoff MF. Nod1 is an essential signal transducer in intestinal epithelial cells infected with bacteria that avoid recognition by toll-like receptors. Infect Immun 2004; 72(3):1487-1495.
57. Rao P, Hsu KC, Chao MV. Upregulation of NF-κB-dependent gene expression mediated by the p75 tumor necrosis factor receptor. J Interferon Cytokine Res 1995; 15(2):171-177.
58. Kuno K, Matsushima K. The IL-1 receptor signaling pathway. J Leukoc Biol 1994; 56(5):542-547.
59. Rothe M, Sarma V, Dixit VM et al. TRAF2-mediated activation of NF-κB by TNF receptor 2 and CD40. Science 1995; 269(5229):1424-1427.
60. Schutze S, Potthoff K, Machleidt T et al. TNF activates NF-κB by phosphatidylcholine-specific phospholipase C-induced "acidic" sphingomyelin breakdown. Cell 1992; 71(5):765-776.
61. Lee SY, Kaufman DR, Mora AL et al. Stimulus-dependent synergism of the antiapoptotic tumor necrosis factor receptor-associated factor 2 (TRAF2) and nuclear factor kappaB pathways. J Exp Med 1998; 188(7):1381-1384.
62. Grumont RJ, Rourke IJ, Gerondakis S. Rel-dependent induction of A1 transcription is required to protect B cells from antigen receptor ligation-induced apoptosis. Genes Dev 1999; 13(4):400-411.
63. Wang CY, Mayo MW, Korneluk RG et al. NF-κB antiapoptosis: Induction of TRAF1 and TRAF2 and c-IAP1 and c-IAP2 to suppress caspase-8 activation. Science 1998; 281(5383):1680-1683.
64. Wu MX, Ao Z, Prasad KV et al. IEX-1L, an apoptosis inhibitor involved in NF-κB-mediated cell survival. Science 1998; 281(5379):998-1001.
65. Zong WX, Edelstein LC, Chen C et al. The prosurvival Bcl-2 homolog Bfl-1/A1 is a direct transcriptional target of NF-κB that blocks TNFα-induced apoptosis. Genes Dev 1999; 13(4):382-387.
66. Mora AL, Corn RA, Stanic AK et al. Antiapoptotic function of NF-κB in T lymphocytes is influenced by their differentiation status: Roles of Fas, c-FLIP, and Bcl-xL. Cell Death Differ 2003; 10(9):1032-1044.
67. Hellin AC, Bentires-Alj M, Verlaet M et al. Roles of nuclear factor-kappaB, p53, and p21/WAF1 in daunomycin-induced cell cycle arrest and apoptosis. J Pharmacol Exp Ther 2000; 295(3):870-878.
68. Henry DO, Moskalenko SA, Kaur KJ et al. Ral GTPases contribute to regulation of cyclin D1 through activation of NF-κB. Mol Cell Biol 2000; 20(21):8084-8092.
69. Guttridge DC, Albanese C, Reuther JY et al. NF-κB controls cell growth and differentiation through transcriptional regulation of cyclin D1. Mol Cell Biol 1999; 19(8):5785-5799.
70. Hildeman DA, Zhu Y, Mitchell TC et al. Molecular mechanisms of activated T cell death in vivo. Curr Opin Immunol 2002; 14(3):354-359.
71. Mitchell TC, Hildeman D, Kedl RM et al. Immunological adjuvants promote activated T cell survival via induction of Bcl-3. Nat Immunol 2001; 2(5):397-402.
72. Mitchell TC, Teague TK, Hildeman DA et al. Stronger correlation of bcl-3 than bcl-2, bcl-xL, costimulation, or antioxidants with adjuvant-induced T cell survival. Ann NY Acad Sci 2002; 975:114-131.
73. Lemaitre B, Nicolas E, Michaut L et al. The dorsoventral regulatory gene cassette spatzle/Toll/cactus controls the potent antifungal response in Drosophila adults. Cell 1996; 86(6):973-983.
74. Rosetto M, Engstrom Y, Baldari CT et al. Signals from the IL-1 receptor homolog, Toll, can activate an immune response in a Drosophila hemocyte cell line. Biochem Biophys Res Commun 1995; 209(1):111-116.
75. Engstrom Y. Induction and regulation of antimicrobial peptides in Drosophila. Dev Comp Immunol 1999; 23(4-5):345-358.
76. Hultmark D. Immune reactions in Drosophila and other insects: A model for innate immunity. Trends Genet 1993; 9(5):178-183.
77. Medzhitov R, Janeway Jr CA. An ancient system of host defense. Curr Opin Immunol 1998; 10(1):12-15.
78. Belvin MP, Anderson KV. A conserved signaling pathway: The Drosophila toll-dorsal pathway. Annu Rev Cell Dev Biol 1996; 12:393-416.

79. Meng X, Khanuja BS, Ip YT. Toll receptor-mediated Drosophila immune response requires Dif, an NF-κB factor. Genes Dev 1999; 13(7):792-797.
80. Rutschmann S, Jung AC, Hetru C et al. The Rel protein DIF mediates the antifungal but not the antibacterial host defense in Drosophila. Immunity 2000; 12(5):569-580.
81. Imler JL, Hoffmann JA. Signaling mechanisms in the antimicrobial host defense of Drosophila. Curr Opin Microbiol 2000; 3(1):16-22.
82. Lemaitre B, Kromer-Metzger E, Michaut L et al. A recessive mutation, immune deficiency (imd), defines two distinct control pathways in the Drosophila host defense. Proc Natl Acad Sci USA 1995; 92(21):9465-9469.
83. Georgel P, Naitza S, Kappler C et al. Drosophila immune deficiency (IMD) is a death domain protein that activates antibacterial defense and can promote apoptosis. Dev Cell 2001; 1(4):503-514.
84. Dushay MS, Asling B, Hultmark D. Origins of immunity: Relish, a compound Rel-like gene in the antibacterial defense of Drosophila. Proc Natl Acad Sci USA 1996; 93(19):10343-10347.
85. Lu Y, Wu LP, Anderson KV. The antibacterial arm of the drosophila innate immune response requires an IκB kinase. Genes Dev 2001; 15(1):104-110.
86. Rutschmann S, Jung AC, Zhou R et al. Role of Drosophila IKKγ in a toll-independent antibacterial immune response. Nat Immunol 2000; 1(4):342-347.
87. Silverman N, Zhou R, Stoven S et al. A Drosophila IκB kinase complex required for Relish cleavage and antibacterial immunity. Genes Dev 2000; 14(19):2461-2471.
88. Hedengren M, Asling B, Dushay MS et al. Relish, a central factor in the control of humoral but not cellular immunity in Drosophila. Mol Cell 1999; 4(5):827-837.
89. Janeway Jr CA. The immune system evolved to discriminate infectious nonself from noninfectious self. Immunol Today 1992; 13(1):11-16.
90. Medzhitov R, Preston-Hurlburt P, Janeway Jr CA. A human homologue of the Drosophila Toll protein signals activation of adaptive immunity. Nature 1997; 388(6640):394-397.
91. Wu LP, Anderson KV. Regulated nuclear import of Rel proteins in the Drosophila immune response. Nature 1998; 392(6671):93-97.
92. Hirschfeld M, Kirschning CJ, Schwandner R et al. Cutting edge: Inflammatory signaling by Borrelia burgdorferi lipoproteins is mediated by toll-like receptor 2. J Immunol 1999; 163(5):2382-2386.
93. Hoshino K, Takeuchi O, Kawai T et al. Cutting edge: Toll-like receptor 4 (TLR4)-deficient mice are hyporesponsive to lipopolysaccharide: Evidence for TLR4 as the Lps gene product. J Immunol 1999; 162(7):3749-3752.
94. Takeuchi O, Hoshino K, Kawai T et al. Differential roles of TLR2 and TLR4 in recognition of gram-negative and gram-positive bacterial cell wall components. Immunity 1999; 11(4):443-451.
95. Medzhitov R. Toll-like receptors and innate immunity. Nat Rev Immunol 2001; 1(2):135-145.
96. Kirschning CJ, Wesche H, Merrill Ayres T et al. Human toll-like receptor 2 confers responsiveness to bacterial lipopolysaccharide. J Exp Med 1998; 188(11):2091-2097.
97. Muenzner P, Naumann M, Meyer TF et al. Pathogenic Neisseria trigger expression of their carcinoembryonic antigen-related cellular adhesion molecule 1 (CEACAM1; previously CD66a) receptor on primary endothelial cells by activating the immediate early response transcription factor, nuclear factor-kappaB. J Biol Chem 2001; 276(26):24331-24340.
98. Yang RB, Mark MR, Gray A et al. Toll-like receptor-2 mediates lipopolysaccharide-induced cellular signalling. Nature 1998; 395(6699):284-288.
99. Gewirtz AT, Navas TA, Lyons S et al. Cutting edge: bacterial flagellin activates basolaterally expressed TLR5 to induce epithelial proinflammatory gene expression. J Immunol 2001; 167(4):1882-1885.
100. Hayashi F, Smith KD, Ozinsky A et al. The innate immune response to bacterial flagellin is mediated by Toll-like receptor 5. Nature 2001; 410(6832):1099-1103.
101. Zhang D, Zhang G, Hayden MS et al. A toll-like receptor that prevents infection by uropathogenic bacteria. Science 2004; 303(5663):1522-1526.
102. Ohashi K, Burkart V, Flohe S et al. Cutting edge: Heat shock protein 60 is a putative endogenous ligand of the toll-like receptor-4 complex. J Immunol 2000; 164(2):558-561.
103. Kurt-Jones EA, Popova L, Kwinn L et al. Pattern recognition receptors TLR4 and CD14 mediate response to respiratory syncytial virus. Nat Immunol 2000; 1(5):398-401.
104. Roeder A, Kirschning CJ, Rupec RA et al. Toll-like receptors and innate antifungal responses. Trends Microbiol 2004; 12(1):44-49.
105. Diebold SS, Kaisho T, Hemmi H et al. Innate antiviral responses by means of TLR7-mediated recognition of single-stranded RNA. Science 2004; 303(5663):1529-1531.

106. Du X, Poltorak A, Wei Y et al. Three novel mammalian toll-like receptors: Gene structure, expression, and evolution. Eur Cytokine Netw 2000; 11(3):362-371.
107. Heil F, Hemmi H, Hochrein H et al. Species-specific recognition of single-stranded RNA via toll-like receptor 7 and 8. Science 2004; 303(5663):1526-1529.
108. Lund JM, Alexopoulou L, Sato A et al. Recognition of single-stranded RNA viruses by Toll-like receptor 7. Proc Natl Acad Sci USA 2004; 101(15):5598-5603.
109. Alexopoulou L, Holt AC, Medzhitov R et al. Recognition of double-stranded RNA and activation of NF-κB by Toll-like receptor 3. Nature 2001; 413(6857):732-738.
110. Hemmi H, Takeuchi O, Kawai T et al. A Toll-like receptor recognizes bacterial DNA. Nature 2000; 408(6813):740-745.
111. Chuang T, Ulevitch RJ. Identification of hTLR10: A novel human Toll-like receptor preferentially expressed in immune cells. Biochim Biophys Acta 2001; 1518(1-2):157-161.
112. Lazarus R, Raby BA, Lange C et al. Toll-like Receptor 10 (TLR10) Genetic Variation is Associated with Asthma in Two Independent Samples. Am J Respir Crit Care Med 2004.
113. Muzio M, Bosisio D, Polentarutti N et al. Differential expression and regulation of toll-like receptors (TLR) in human leukocytes: Selective expression of TLR3 in dendritic cells. J Immunol 2000; 164(11):5998-6004.
114. Huang Q, Liu D, Majewski P et al. The plasticity of dendritic cell responses to pathogens and their components. Science 2001; 294(5543):870-875.
115. Poltorak A, He X, Smirnova I et al. Defective LPS signaling in C3H/HeJ and C57BL/10ScCr mice: Mutations in Tlr4 gene. Science 1998; 282(5396):2085-2088.
116. Janssens S, Beyaert R. Functional diversity and regulation of different interleukin-1 receptor-associated kinase (IRAK) family members. Mol Cell 2003; 11(2):293-302.
117. Kanakaraj P, Ngo K, Wu Y et al. Defective interleukin (IL)-18-mediated natural killer and T helper cell type 1 responses in IL-1 receptor-associated kinase (IRAK)-deficient mice. J Exp Med 1999; 189(7):1129-1138.
118. Marr KA, Balajee SA, Hawn TR et al. Differential role of MyD88 in macrophage-mediated responses to opportunistic fungal pathogens. Infect Immun 2003; 71(9):5280-5286.
119. Darrah PA, Monaco MC, Jain S et al. Innate immune responses to Rhodococcus equi. J Immunol 2004; 173(3):1914-1924.
120. Takeuchi O, Hoshino K, Akira S. Cutting edge: TLR2-deficient and MyD88-deficient mice are highly susceptible to Staphylococcus aureus infection. J Immunol 2000; 165(10):5392-5396.
121. Seki E, Tsutsui H, Tsuji NM et al. Critical roles of myeloid differentiation factor 88-dependent proinflammatory cytokine release in early phase clearance of Listeria monocytogenes in mice. J Immunol 2002; 169(7):3863-3868.
122. Edelson BT, Unanue ER. MyD88-dependent but Toll-like receptor 2-independent innate immunity to Listeria: No role for either in macrophage listericidal activity. J Immunol 2002; 169(7):3869-3875.
123. Way SS, Kollmann TR, Hajjar AM et al. Cutting edge: Protective cell-mediated immunity to Listeria monocytogenes in the absence of myeloid differentiation factor 88. J Immunol 2003; 171(2):533-537.
124. Feng CG, Scanga CA, Collazo-Custodio CM et al. Mice lacking myeloid differentiation factor 88 display profound defects in host resistance and immune responses to Mycobacterium avium infection not exhibited by Toll-like receptor 2 (TLR2)- and TLR4-deficient animals. J Immunol 2003; 171(9):4758-4764.
125. Scanga CA, Aliberti J, Jankovic D et al. Cutting edge: MyD88 is required for resistance to Toxoplasma gondii infection and regulates parasite-induced IL-12 production by dendritic cells. J Immunol 2002; 168(12):5997-6001.
126. Scanga CA, Bafica A, Feng CG et al. MyD88-deficient mice display a profound loss in resistance to Mycobacterium tuberculosis associated with partially impaired Th1 cytokine and nitric oxide synthase 2 expression. Infect Immun 2004; 72(4):2400-2404.
127. Liu N, Montgomery RR, Barthold SW et al. Myeloid differentiation antigen 88 deficiency impairs pathogen clearance but does not alter inflammation in Borrelia burgdorferi-infected mice. Infect Immun 2004; 72(6):3195-3203.
128. Matsumoto M, Yamada T, Yoshinaga SK et al. Essential role of NF-κB-inducing kinase in T cell activation through the TCR/CD3 pathway. J Immunol 2002; 169(3):1151-1158.
129. Yogi Y, Endoh M, Banba T et al. Susceptibility to Mycobacterium leprae of ALY (alymphoplasia) mice and IFN-γ induction in the culture supernatant of spleen cells. Int J Lepr Other Mycobact Dis 1998; 66(4):464-474.

130. Korenaga M, Akimaru Y, Shamsuzzaman SM et al. Impaired protective immunity and T helper 2 responses in alymphoplasia (aly) mutant mice infected with Trichinella spiralis. Immunology 2001; 102(2):218-224.
131. Karrer U, Althage A, Odermatt B et al. Immunodeficiency of alymphoplasia mice (aly/aly) in vivo: Structural defect of secondary lymphoid organs and functional B cell defect. Eur J Immunol 2000; 30(10):2799-2807.
132. Artis D, Speirs K, Joyce K et al. NF-κB1 is required for optimal CD4+ Th1 cell development and resistance to Leishmania major. J Immunol 2003; 170(4):1995-2003.
133. Sha WC, Liou HC, Tuomanen EI et al. Targeted disruption of the p50 subunit of NF-κB leads to multifocal defects in immune responses. Cell 1995; 80(2):321-330.
134. Artis D, Shapira S, Mason N et al. Differential requirement for NF-κB family members in control of helminth infection and intestinal inflammation. J Immunol 2002; 169(8):4481-4487.
135. Fox JG, Rogers AB, Whary MT et al. Gastroenteritis in NF-κB-deficient mice is produced with wild-type Camplyobacter jejuni but not with C. jejuni lacking cytolethal distending toxin despite persistent colonization with both strains. Infect Immun 2004; 72(2):1116-1125.
136. Erdman S, Fox JG, Dangler CA et al. Typhlocolitis in NF-κB-deficient mice. J Immunol 2001; 166(3):1443-1447.
137. Tomczak MF, Erdman SE, Poutahidis T et al. NF-κB is required within the innate immune system to inhibit microflora-induced colitis and expression of IL-12 p40. J Immunol 2003; 171(3):1484-1492.
138. Caamano J, Tato C, Cai G et al. Identification of a role for NF-κB2 in the regulation of apoptosis and in maintenance of T cell-mediated immunity to Toxoplasma gondii. J Immuno. 2000; 165(10):5720-5728.
139. Speirs K, Caamano J, Goldschmidt MH et al. NF-κB2 is required for optimal CD40-induced IL-12 production but dispensable for Th1 cell Differentiation. J Immunol 2002; 168(9):4406-4413.
140. Schwarz EM, Krimpenfort P, Berns A et al. Immunological defects in mice with a targeted disruption in Bcl-3. Genes Dev 1997; 11(2):187-197.
141. Franzoso G, Carlson L, Scharton-Kersten T et al. Critical roles for the Bcl-3 oncoprotein in T cell-mediated immunity, splenic microarchitecture, and germinal center reactions. Immunity 1997; 6(4):479-490.
142. Sacks D, Noben-Truath N. The immunology of susceptibility and resistance to Leishmania major in mice. Nat Rev Immunol 2002; 2(11):845-858.
143. Weih F, Durham SK, Barton DS et al. Both multiorgan inflammation and myeloid hyperplasia in RelB-deficient mice are T cell dependent. J Immunol 1996; 157(9):3974-3979.
144. Weih F, Warr G, Yang H et al. Multifocal defects in immune responses in RelB-deficient mice. J Immunol 1997; 158(11):5211-5218.
145. Caamano J, Alexander J, Craig L et al. The NF-κB family member RelB is required for innate and adaptive immunity to Toxoplasma gondii. J Immunol 1999; 163(8):4453-4461.
146. Cariappa A, Liou HC, Horwitz BH et al. Nuclear factor kappa B is required for the development of marginal zone B lymphocytes. J Exp Med 2000; 192(8):1175-1182.
147. Sanjabi S, Hoffmann A, Liou HC et al. Selective requirement for c-Rel during IL-12 P40 gene induction in macrophages. Proc Natl Acad Sci USA 2000; 97(23):12705-12710.
148. Mason NJ, Liou HC, Hunter CA. T cell-intrinsic expression of c-Rel regulates Th1 cell responses essential for resistance to Toxoplasma gondii. J Immunol 2004; 172(6):3704-3711.
149. Grumont RJ, Rourke IJ, O'Reilly LA et al. B lymphocytes differentially use the Rel and nuclear factor kappaB1 (NF-κB1) transcription factors to regulate cell cycle progression and apoptosis in quiescent and mitogen-activated cells. J Exp Med 1998; 187(5):663-674.
150. Harling-McNabb L, Deliyannis G, Jackson DC et al. Mice lacking the transcription factor subunit Rel can clear an influenza infection and have functional anti-viral cytotoxic T cells but do not develop an optimal antibody response. Int Immunol 1999; 11(9):1431-1439.
151. Mora AL, Chen D, Boothby M et al. Lineage-specific differences among CD8+ T cells in their dependence of NF-κB/Rel signaling. Eur J Immunol 1999; 29(9):2968-2980.
152. Lavon I, Goldberg I, Amit S et al. High susceptibility to bacterial infection, but no liver dysfunction, in mice compromised for hepatocyte NF-κB activation. Nat Med 2000; 6(5):573-577.
153. Tato CM, Villarino A, Caamano JH et al. Inhibition of NF-κB activity in T and NK cells results in defective effector cell expansion and production of IFN-γ required for resistance to Toxoplasma gondii. J Immunol 2003; 170(6):3139-3146.
154. Puel A, Picard C, Ku CL et al. Inherited disorders of NF-κB-mediated immunity in man. Curr Opin Immunol 2004; 16(1):34-41.

155. Haraguchi S, Day NK, Nelson Jr RP et al. Interleukin 12 deficiency associated with recurrent infections. Proc Natl Acad Sci USA 1998; 95(22):13125-13129.
156. Kuhns DB, Long Priel DA, Gallin JI. Endotoxin and IL-1 hyporesponsiveness in a patient with recurrent bacterial infections. J Immunol 1997; 158(8):3959-3964.
157. Medvedev AE, Lentschat A, Kuhns DB et al. Distinct mutations in IRAK-4 confer hyporesponsiveness to lipopolysaccharide and interleukin-1 in a patient with recurrent bacterial infections. J Exp Med 2003; 198(4):521-531.
158. Picard C, Puel A, Bonnet M et al. Pyogenic bacterial infections in humans with IRAK-4 deficiency. Science 2003; 299(5615):2076-2079.
159. Ouaaz F, Arron J, Zheng Y et al. Dendritic cell development and survival require distinct NF-κB subunits. Immunity 2002; 16(2):257-270.
160. Franzoso G, Carlson L, Xing L et al. Requirement for NF-κB in osteoclast and B-cell development. Genes Dev 1997; 11(24):3482-3496.
161. Carragher D, Johal R, Button A et al. A stroma-derived defect in NF-κB2-/- mice causes impaired lymph node development and lymphocyte recruitment. J Immunol 2004; 173(4):2271-2279.
162. Mikkola ML, Thesleff I. Ectodysplasin signaling in development. Cytokine Growth Factor Rev 2003; 14(3-4):211-224.
163. Iotsova V, Caamano J, Loy J et al. Osteopetrosis in mice lacking NF-κB1 and NF-κB2. Nat Med 1997; 3(11):1285-1289.
164. Day N, Tangsinmankong N, Ochs H et al. Interleukin receptor-associated kinase (IRAK-4) deficiency associated with bacterial infections and failure to sustain antibody responses. J Pediatr 2004; 144(4):524-526.
165. Smahi A, Courtois G, Vabres P et al. Genomic rearrangement in NEMO impairs NF-κB activation and is a cause of incontinentia pigmenti. The International Incontinentia Pigmenti (IP) Consortium. Nature 2000; 405(6785):466-472.
166. Aradhya S, Courtois G, Rajkovic A et al. Atypical forms of incontinentia pigmenti in male individuals result from mutations of a cytosine tract in exon 10 of NEMO (IKK-gamma). Am J Hum Genet 2001; 68(3):765-771.
167. Orange JS, Geha RS. Finding NEMO: Genetic disorders of NF-κB activation. J Clin Invest 2003; 112(7):983-985.
168. Carrol ED, Gennery AR, Flood TJ et al. Anhidrotic ectodermal dysplasia and immunodeficiency: The role of NEMO. Arch Dis Child 2003; 88(4):340-341.
169. Orange JS, Jain A, Ballas ZK et al. The presentation and natural history of immunodeficiency caused by nuclear factor kappaB essential modulator mutation. J Allergy Clin Immunol 2004; 113(4):725-733.
170. Zonana J, Elder ME, Schneider LC et al. A novel X-linked disorder of immune deficiency and hypohidrotic ectodermal dysplasia is allelic to incontinentia pigmenti and due to mutations in IKK-gamma (NEMO). Am J Hum Genet 2000; 67(6):1555-1562.
171. Mansour S, Woffendin H, Mitton S et al. Incontinentia pigmenti in a surviving male is accompanied by hypohidrotic ectodermal dysplasia and recurrent infection. Am J Med Genet 2001; 99(2):172-177.
172. Jain A, Ma CA, Liu S et al. Specific missense mutations in NEMO result in hyper-IgM syndrome with hypohydrotic ectodermal dysplasia. Nat Immunol 2001; 2(3):223-228.
173. Doffinger R, Smahi A, Bessia C et al. X-linked anhidrotic ectodermal dysplasia with immunodeficiency is caused by impaired NF-κB signaling. Nat Genet 2001; 27(3):277-285.
174. Dupuis-Girod S, Corradini N, Hadj-Rabia S et al. Osteopetrosis, lymphedema, anhidrotic ectodermal dysplasia, and immunodeficiency in a boy and incontinentia pigmenti in his mother. Pediatrics 2002; 109(6):e97.
175. Brodeur SR, Angelini F, Bacharier LB et al. C4b-binding protein (C4BP) activates B cells through the CD40 receptor. Immunity 2003; 18(6):837-848.
176. Orange JS, Brodeur SR, Jain A et al. Deficient natural killer cell cytotoxicity in patients with IKK-γ/NEMO mutations. J Clin Invest 2002; 109(11):1501-1509.
177. Frix IIIrd CD, Bronson DM. Acute miliary tuberculosis in a child with anhidrotic ectodermal dysplasia. Pediatr Dermatol 1986; 3(6):464-467.
178. Sitton JE, Reimund EL. Extramedullary hematopoiesis of the cranial dura and anhidrotic ectodermal dysplasia. Neuropediatrics 1992; 23(2):108-110.
179. Abb J. Prevention and therapy of herpesvirus infections. Zentralbl Bakteriol Mikrobiol Hyg [B] 1985; 180(2-3):107-120.
180. Courtois G, Smahi A, Reichenbach J et al. A hypermorphic IκBα mutation is associated with autosomal dominant anhidrotic ectodermal dysplasia and T cell immunodeficiency. J Clin Invest 2003; 112(7):1108-1115.

181. Karban AS, Okazaki T, Panhuysen CI et al. Functional annotation of a novel NFKB1 promoter polymorphism that increases risk for ulcerative colitis. Hum Mol Genet. 2004; 13(1):35-45.
182. Girardin SE, Hugot JP, Sansonetti PJ. Lessons from Nod2 studies: Towards a link between Crohn's disease and bacterial sensing. Trends Immunol 2003; 24(12):652-658.
183. Bonen DK, Ogura Y, Nicolae DL et al. Crohn's disease-associated NOD2 variants share a signaling defect in response to lipopolysaccharide and peptidoglycan. Gastroenterology 2003; 124(1):140-146.
184. Ogura Y, Bonen DK, Inohara N et al. A frameshift mutation in NOD2 associated with susceptibility to Crohn's disease. Nature 2001; 411(6837):603-606.
185. Popkin DL, Virgin HWt. Murine cytomegalovirus infection inhibits tumor necrosis factor α responses in primary macrophages. J Virol 2003; 77(18):10125-10130.
186. Moutaftsi M, Brennan P, Spector SA et al. Impaired lymphoid chemokine-mediated migration due to a block on the chemokine receptor switch in human cytomegalovirus-infected dendritic cells. J Virol 2004; 78(6):3046-3054.
187. Lin M, Rikihisa Y. Ehrlichia chaffeensis downregulates surface Toll-like receptors 2/4, CD14 and transcription factors PU.1 and inhibits lipopolysaccharide activation of NF-κB, ERK 1/2 and p38 MAPK in host monocytes. Cell Microbiol 2004; 6(2):175-186.
188. Bowie A, Kiss-Toth E, Symons JA et al. A46R and A52R from vaccinia virus are antagonists of host IL-1 and toll- like receptor signaling. Proc Natl Acad Sci USA 2000; 97(18):10162-10167.
189. Orth K, Palmer LE, Bao ZQ et al. Inhibition of the mitogen-activated protein kinase kinase superfamily by a Yersinia effector. Science 1999; 285(5435):1920-1923.
190. Orth K, Xu Z, Mudgett MB et al. Disruption of signaling by Yersinia effector YopJ, a ubiquitin-like protein protease. Science 2000; 290(5496):1594-1597.
191. Ruckdeschel K, Harb S, Roggenkamp A et al. Yersinia enterocolitica impairs activation of transcription factor NF- κB: Involvement in the induction of programmed cell death and in the suppression of the macrophage tumor necrosis factor α production. J Exp Med 1998; 187(7):1069-1079.
192. Ruckdeschel K, Mannel O, Richter K et al. Yersinia outer protein P of Yersinia enterocolitica simultaneously blocks the nuclear factor-kappa B pathway and exploits lipopolysaccharide signaling to trigger apoptosis in macrophages. J Immunol 2001; 166(3):1823-1831.
193. Ruckdeschel K, Richter K, Mannel O et al. Arginine-143 of Yersinia enterocolitica YopP crucially determines isotype-related NF-κB suppression and apoptosis induction in macrophages. Infect Immun 2001; 69(12):7652-7662.
194. Yao T, Mecsas J, Healy JI et al. Suppression of T and B lymphocyte activation by a Yersinia pseudotuberculosis virulence factor, yopH. J Exp Med 1999; 190(9):1343-1350.
195. Klumpp DJ, Weiser AC, Sengupta S et al. Uropathogenic Escherichia coli potentiates type 1 pilus-induced apoptosis by suppressing NF-κB. Infect Immun 2001; 69(11):6689-6695.
196. Dhib-Jalbut S, Xia J, Rangaviggula H et al. Failure of measles virus to activate nuclear factor-kappa B in neuronal cells: Implications on the immune response to viral infections in the central nervous system. J Immunol 1999; 162(7):4024-4029.
197. Oie KL, Pickup DJ. Cowpox virus and other members of the orthopoxvirus genus interfere with the regulation of NF-κB activation. Virology 2001; 288(1):175-187.
198. Akari H, Bour S, Kao S et al. The human immunodeficiency virus type 1 accessory protein Vpu induces apoptosis by suppressing the nuclear factor kappaB-dependent expression of antiapoptotic factors. J Exp Med 2001; 194(9):1299-1312.
199. Bour S, Perrin C, Akari H et al. The human immunodeficiency virus type 1 Vpu protein inhibits NF-κB activation by interfering with βTrCP-mediated degradation of IκB. J Biol Chem 2001; 276(19):15920-15928.
200. Telepnev M, Golovliov I, Grundstrom T et al. Francisella tularensis inhibits Toll-like receptor-mediated activation of intracellular signalling and secretion of TNF-α and IL-1 from murine macrophages. Cell Microbiol 2003; 5(1):41-51.
201. Neish AS, Gewirtz AT, Zeng H et al. Prokaryotic regulation of epithelial responses by inhibition of IκBα ubiquitination. Science 2000; 289(5484):1560-1563.
202. Revilla Y, Callejo M, Rodriguez JM et al. Inhibition of nuclear factor kappaB activation by a virus-encoded IκB-like protein. J Biol Chem 1998; 273(9):5405-5411.
203. Pahlevan AA, Wright DJ, Andrews C et al. The inhibitory action of Mycobacterium ulcerans soluble factor on monocyte/T cell cytokine production and NF-κB function. J Immunol 1999; 163(7):3928-3935.
204. Dreyfus DH, Nagasawa M, Pratt JC et al. Inactivation of NF-κB by EBV BZLF-1-encoded ZEBRA protein in human T cells. J Immunol 1999; 163(11):6261-6268.

205. Butcher BA, Kim L, Johnson PF et al. Toxoplasma gondii tachyzoites inhibit proinflammatory cytokine induction in infected macrophages by preventing nuclear translocation of the transcription factor NF-κB. J Immunol 2001; 167(4):2193-2201.
206. Shapira S, Speirs K, Gerstein A et al. Suppression of NF-κB activation by infection with Toxoplasma gondii. J Infect Dis 2002; 185(4 Suppl 1):S66-72.
207. Dobbin CA, Smith NC, Johnson AM. Heat shock protein 70 is a potential virulence factor in murine Toxoplasma infection via immunomodulation of host NF-κB and nitric oxide. J Immunol 2002; 169(2):958-965.
208. Trottein F, Nutten S, Angeli V et al. Schistosoma mansoni schistosomula reduce E-selectin and VCAM-1 expression in TNF-α-stimulated lung microvascular endothelial cells by interfering with the NF-κB pathway. Eur J Immunol 1999; 29(11):3691-3701.
209. Ghosh S, Bhattacharyya S, Sirkar M et al. Leishmania donovani suppresses activated protein 1 and NF-κB activation in host macrophages via ceramide generation: Involvement of extracellular signal-regulated kinase. Infect Immun 2002; 70(12):6828-6838.
210. Ruckdeschel K. Yersinia species disrupt immune responses to subdue the host. ASM News 2000; 66(Number 8):470-477.
211. Cahir-McFarland ED, Davidson DM, Schauer SL et al. NF-κB inhibition causes spontaneous apoptosis in Epstein-Barr virus-transformed lymphoblastoid cells. Proc Natl Acad Sci USA 2000; 97(11):6055-6060.
212. Luftig M, Yasui T, Soni V et al. Epstein-Barr virus latent infection membrane protein 1 TRAF-binding site induces NIK/IKK α-dependent noncanonical NF-κB activation. Proc Natl Acad Sci USA 2004; 101(1):141-146.
213. Xiao G, Cvijic ME, Fong A et al. Retroviral oncoprotein Tax induces processing of NF-κB2/p100 in T cells: Evidence for the involvement of IKKα. EMBO J 2001; 20(23):6805-6815.
214. Heussler VT, Machado Jr J, Fernandez PC et al. The intracellular parasite Theileria parva protects infected T cells from apoptosis. Proc Natl Acad Sci USA 1999; 96(13):7312-7317.
215. Palmer GH, Machado Jr J, Fernandez P et al. Parasite-mediated nuclear factor kappaB regulation in lymphoproliferation caused by Theileria parva infection. Proc Natl Acad Sci USA 1997; 94(23):12527-12532.
216. Greten FR, Eckmann L, Greten TF et al. IKKβ links inflammation and tumorigenesis in a mouse model of colitis-associated cancer. Cell 2004; 118(3):285-296.
217. Sil AK, Maeda S, Sano Y et al. IκB kinase-α acts in the epidermis to control skeletal and craniofacial morphogenesis. Nature 2004; 428(6983):660-664.
218. Maeda S, Chang L, Li ZW et al. IKKβ is required for prevention of apoptosis mediated by cell-bound but not by circulating TNFα. Immunity 2003; 19(5):725-737.
219. Chen LW, Egan L, Li ZW et al. The two faces of IKK and NF-κB inhibition: Prevention of systemic inflammation but increased local injury following intestinal ischemia-reperfusion. Nat Med 2003; 9(5):575-581.

CHAPTER 9

Molecular Basis of Oncogenesis by NF-κB: From a Bird's Eye View to a RELevant Role in Cancer

Yongjun Fan, Jui Dutta, Nupur Gupta and Céline Gélinas*

Abstract

The Rel/NF-κB transcription factors are renowned for their fundamental contribution to normal immune, inflammatory and acute phase responses. A growing body of evidence also underscores their important role in the control of cellular gene expression, cell proliferation and apoptosis. Thus, it comes as no surprise that sustained Rel/NF-κB activity has emerged as a hallmark of many human cancers. Experimental evidence indicates a strong correlation between the transcriptional activity of Rel/NF-κB and its role in malignant cell transformation. The important role of NF-κB in the control of the apoptotic response also supports its participation in the resistance of tumor cells to therapeutic treatment. This review focuses on the mechanisms that underlie the contribution of Rel/NF-κB to cancer and highlights how appreciation of its role in this context has evolved from a bird's eye view to a true recognition of its RELevant function in oncogenesis.

Introduction

The Rel/NF-κB transcription factors have been the focus of numerous studies aimed at elucidating their role in the development and function of the immune system and at unveiling the signaling pathways that control their activity (see accompanying chapters by M. Karin, S.C. Sun, R. Sen, U. Siebenlist, H.C. Liou, Y. Chen, and C. Hunter). In recent years, there has been considerable progress in appreciating their contribution to oncogenesis and in understanding the mechanisms involved. Inappropriate Rel/NF-κB activity is observed in many different types of human cancers. Hyperactivation of the NF-κB signaling cascade, mutations that inactivate the inhibitory IκB subunits or chromosomal aberrations involving various *rel/nf-κb* genes have been noted in many human tumors.[1,2] Consistent with the transforming activity of the viral Rel/NF-κB oncoprotein v-Rel and its cellular homologue c-Rel in primary cells and in animal models, NF-κB is also critically involved in malignant cell transformation by viruses such as the human T-cell leukemia virus type I (HTLV-1) and Epstein-Barr virus (EBV).[3] Collectively, these findings justify the vast body of literature exploring the molecular basis for the role of Rel/NF-κB in cancer. Important findings center on its ability to regulate cellular gene expression, to affect cell proliferation and survival, and on important regulatory mechanisms that control its activity — all of which have important consequences for effective anti-cancer therapy.[4-7]

*Corresponding Author: Céline Gélinas—Center for Advanced Biotechnology and Medicine, Department of Biochemistry, Robert Wood Johnson Medical School, University of Medicine and Dentistry of New Jersey, 679 Hoes Lane, Piscataway, New Jersey 08854-5638, U.S.A. Email: gelinas@cabm.rutgers.edu

NF-κB/Rel Transcription Factor Family, edited by Hsiou-Chi Liou.
©2006 Landes Bioscience and Springer Science+Business Media.

Constitutive Rel/NF-κB Activity Is a Hallmark of Many Human Cancers

Sustained activation of NF-κB is a feature of many human leukemia, lymphoma and solid tumors.[1] Immunohistochemistry, gel mobility shift assays and gene expression profiling of primary tumor specimens and tumor-derived cell lines have highlighted the persistent nuclear localization of NF-κB subunits compared to normal controls (for example see refs. 8-11). The dimer comprised of the p50/p65 subunits is the most frequently reported NF-κB complex to be activated in human cancer, although there is evidence that clearly implicates c-Rel-containing complexes in certain tumor types, like breast cancer.[8,9] The important implication of sustained NF-κB activity for the survival and proliferation of tumor cells is underscored by the growth arrest and rapid onset of apoptosis observed in many tumor-derived cell lines upon introduction of a degradation-resistant form of IκB (IκB super-repressor) to inhibit endogenous NF-κB activity (for example see refs. 12,13).

Activation of the NF-κB Signaling Cascade

Persistent activation of the NF-κB pathway is observed in many different human cancers. By virtue of its ability to trigger the N-terminal phosphorylation of the NF-κB inhibitory subunit IκBα on serines 32 and 36, the IKK kinase complex promotes degradation of IκB via the ubiquitin/proteasome pathway. This enables NF-κB dimers to accumulate in the nucleus where they promote transcription of specific gene programs.[14] Although the detailed mechanisms responsible for sustained IKK activation in many human tumors remain unknown, there are several potential mechanisms (Table 1).[15,16]

IKK Complex Activation

Since no mutation has yet been identified to affect IKK subunits in human tumors, unrelenting activation of NF-κB is likely to result from alterations in upstream signaling components. In many types of cancer, sustained IKK activation is achieved via autocrine loops involving cytokines and growth factors that activate the NF-κB pathway and are themselves transcriptional targets of NF-κB (Tables 1, 2).[17] For instance, IL-1 activates NF-κB in pancreatic carcinoma cell lines and is in turn induced by NF-κB.[18] Likewise CD40, the receptor for CD40 ligand, constitutively activates NF-κB in malignant Reed-Sternberg (H/RS) cells of Hodgkin's disease (HD) and is upregulated in these cells.[19] Another mechanism for constitutive activation of the IKK complex involves deregulation of TRAF adaptor proteins in human tumors. TRAF2 is a critical component of receptor-triggered signaling pathways involving NF-κB, JNK and p38. Recent work showed that loss of the TRAF2- and IKKγ/NEMO-interacting tumor suppressor protein CYLD, a de-ubiquitinating enzyme for TRAF2, leads to constitutive activation of IKK coincident with increased cell resistance to apoptosis.[20-23] Loss of CYLD causes cylindromatosis, an autosomal dominant syndrome that predisposes patients to benign tumors of hair follicles and sweat and scent glands.

Interestingly, recent work unveiled a new NF-κB-independent role for IKK in cancer. IKKβ expression in primary breast cancer specimens is correlated with poor survival and studies in primary breast cancer cell lines showed that IKK negatively regulates the forkhead transcription factor FOXO3a, independent of NF-κB activation.[24] Indeed, IKK-mediated phosphorylation of FOXO3a promoted its nuclear export and proteolysis via the ubiquitin proteasome pathway to promote cell growth and tumorigenesis. It will be interesting to see if the newly reported abilities of IKKα and IKKγ/NEMO to localize to the nucleus and respectively modify histones and interact with CBP to regulate NF-κB gene expression imply that these subunits can also act on other nuclear targets to affect oncogenesis.[25-27]

Activation by Other Kinases, Oncogenes and Viruses

Other means to constitutively activate NF-κB signaling in human tumors entail various kinases other than IKK, as well as oncogenes and viruses (Table 1). One example involves the

Table 1. Mechanisms for constitutive NF-κB activation in human cancer

Mechanism	Type of Cancer
IKK complex activation	
Unknown mechanism	Hodgkin's lymphoma
	Childhood acute lymphoblastic leukemia
	Breast carcinoma
	Colon carcinoma
	Ovarian carcinoma
	Pancreatic carcinoma
	Thyroid carcinoma
	Bladder carcinoma
	Prostate carcinoma
	Melanoma
	Squamous cell carcinoma
Interleukin-1 autocrine loop	Pancreatic carcinoma cell line
Interleukin-13 autocrine loop	Hodgkin's lymphoma
Tumor necrosis factor-α autocrine loop	T-cell lymphoma
Loss of CYLD	Turban tumor syndrome
Activation by other kinases	
Bcr-Abl	Acute lymphoblastic leukemia
	Chronic myelogenous leukemia
Activation by oncogenes	
Ras	Acute lymphoblastic leukemia
	Chronic myelogenous leukemia
API2/MALT1	Mucosa-associated lymphoid tissue lymphoma
Her2/Neu	Breast carcinoma
Activation by viruses	
Human T-cell leukemia virus-1 (HTLV-1)	Adult T-cell leukemia
Epstein Barr virus (EBV)	Burkitt's lymphoma
	Nasopharyngeal carcinoma
	Hodgkin's lymphoma
	Immunoblastic lymphoma
	Gastric carcinoma
Hepatitis B virus (HBV)	Hepatocellular carcinoma
Human herpes virus-8 (HHV-8)	Kaposi's sarcoma
***iκb* gene mutations**	
iκbα Mutation	Hodgkin's lymphoma
iκbε Mutation	Hodgkin's lymphoma
bcl-3 Rearrangement / Overexpression	B-cell non-Hodgkin's lymphoma
	B-cell chronic lymphocytic leukemia
***nf-κb* gene alterations**	
c-rel Amplification	Hodgkin's lymphoma
	Follicular B-cell lymphoma
	Diffuse large cell lymphoma
	Primary mediastinal B-cell lymphoma
Rearrangement / Overexpression	Follicular lymphoma
	Diffuse large cell lymphoma
	Non-small cell lung carcinoma

Table continued on next page

Table 1. Continued

Mechanism		Type of Cancer
nf-κb gene alterations		
relA	Rearrangement / Overexpression	B-cell non-Hodgkin's lymphoma
		Multiple myeloma
		Non-small cell lung carcinoma
		Thyroid carcinoma cell lines
	Amplification	Diffuse large cell lymphoma
		Squamous head and neck carcinoma
		Breast adenocarcinoma
		Stomach adenocarcinoma
nf-κb1	Rearrangement/Overexpression	Acute lymphoblastic leukemia
		Non-small cell lung carcinoma
		Colon cancer cell lines
		Prostate cancer cell lines
		Breast cancer cell lines
		Bone cancer cell lines
		Brain cancer cell lines
nf-κb2	Rearrangement/Overexpression	Cutaneous T-cell lymphoma
		B-cell non-Hodgkin's lymphoma
		B-cell chronic lymphocytic leukemia
		Multiple myeloma
		Breast carcinoma
		Colon carcinoma

PI3-kinase to Akt kinase signaling pathway in response to overexpression of the epidermal growth factor (EGF) receptor family member c-erbB2/Her-2/Neu in breast cancer.[28] IκBα degradation in this context is mediated by the protease calpain. Another example is casein kinase II (CKII) that phosphorylates serines in IκBα distinct from those targeted by IKK and triggers calpain-mediated cleavage of IκBα.[29] Upregulation of CKII activity was suggested as a possible contributing factor to hepatocarcinoma induced by TGF-β1.[30]

A number of oncogenes mediate their transforming function by virtue of NF-κB activation. These include the chimeric oncoprotein tyrosine kinase Bcr-Abl implicated in acute lymphocytic leukemia (ALL) and chronic myelogenous leukemia (CML), and the Ras oncogene. Bcr-Abl enhances nuclear translocation of NF-κB and the transactivation function of NF-κB subunit p65/RelA via MEKK1 and p38 MAPK and also partially requires Ras function.[31-33] Ras is another well-known oncogene mutated in human tumors that utilizes NF-κB to achieve oncogenesis.[34,35] The anti-oncogenic effect of lysyl oxidase on Ras-transformed cells was recently demonstrated to involve suppression of NF-κB activation.[36] An interesting new report showed that the API2/MALT1, a chimeric protein between inhibitor of apoptosis c-IAP2 and the MALT1 paracaspase, participates in the transformation process of mucosa-associated lymphoid tissue (MALT) lymphoma by activating NF-κB dimers comprised of RelB/p50.[37] It will be interesting to see whether other oncogenes act in a similar manner.

Lastly, many viruses achieve their oncogenic effects via the NF-κB signaling cascade (Table 1). A notable example relevant to human cancer is the human T-cell leukemia virus-1 (HTLV-1) implicated in acute T-cell leukemia (ATL). Persistent activation of NF-κB by HTLV-1 Tax causes nuclear accumulation of NF-κB dimers, helps to overcome their inhibition by the p105/ NF-κB1 subunit, and is an essential step in the transformation of T cells.[38-41] Additionally, Tax stimulates phosphorylation-dependent processing of NF-κB2/p100, and hence activates both

Table 2. A sample of NF-κB-regulated gene products implicated in human cancer

Function	Protein	Role in v-Rel-Mediated Transformation
Regulators of apoptosis	Bcl-xL	complements weakly transforming mutants
	Bcl-2	complements weakly transforming mutants
	Bfl-1/A1	
	c-IAP2	c-IAP1 - essential
	c-FLIP	
	GADD45β	
	A20	
	CD95	
	TRAF1	
	IEX1	
Cell cycle regulators	Cyclin D1	
	Cyclin D2	
Transcription factors	JunB	c-Jun - essential
	IRF1	
	IRF4	IRF4 - essential
	c-Myc	
	Stat5a	
Tumor suppressors	p53	
Chemokines/cytokines/ growth factors	Interleukin 1	
	Interleukin 6	
	Interleukin 8	
	Interleukin 13	
	MIP-1α	
	GM-CSF	
	TNFα	
	VEGF	
Cell surface receptors	CD40	
	CD44	
	CD86	
	CCR7	
	CXCR4	
Cell adhesion molecules	ICAM-1	
	VCAM-1	
Metalloproteinases	MMP-9	

the canonical and noncanonical NF-κB pathways.[42] Another virus that contributes to human cancer via NF-κB is the Epstein-Barr virus (EBV) implicated in Burkitt's and Hodgkin's lymphomas. The EBV nuclear antigen (EBNA)-2 and latent membrane protein (LMP)-1 enhance NF-κB activity thereby preventing apoptosis in EBV-transformed B cells.[41,43] This is consistent with the ability of LMP-1 to induce expression of NF-κB-dependent anti-apoptotic proteins such as Bfl-1/A1.[44,45] Akin to Tax, LMP-1 induces proteolytic processing of p100/NF-κB2 to its p52 form, consistent with the high levels of p52 found in Hodgkin's lymphoma and nasopharyngeal carcinoma from EBV-infected patients.[46,47]

iκb Gene Mutations

Although much less frequent than upstream activation of the NF-κB pathway, there have been a few reports of *iκb* gene mutations implicated in constitutively activating NF-κB in human tumors. Mutations that suppress the inhibitory activity of IκBα or IκBε were observed in some Hodgkin's lymphomas and a large B-cell lymphoma cell line (Table 1).[48-53] The fact that bi-allelic mutation was needed for IκBα loss-of-function in Hodgkin's lymphoma raised the suggestion that it may act as a tumor suppressor.

nf-κb Gene Rearrangement, Amplification and/or Overexpression in Human Cancer

While sustained activation of NF-κB signaling is the most common mode of NF-κB activation in human tumors, there are a number of cases in which *rel/nf-κb* gene amplification, rearrangement and/or overexpression was documented (Table 1).[1] The majority of human *rel* and *nf-κb* genes (i.e., *c-rel, relA, nf-κb1* and *nf-κb2*) have been targeted in this fashion, although *nf-κb2* and *c-rel* are the most commonly affected.

Chromosomal rearrangements disrupting the 3' coding region of the *nf-κb2* gene are frequently observed in cutaneous T-cell lymphoma and also in a small number of B-cell non-Hodgkin lymphoma, chronic lymphocytic leukemia and multiple myeloma.[54-58] The resulting C-terminally truncated p100/NF-κB2 proteins primarily localize to nuclei and bind to NF κB DNA motifs. However, how tumor-derived truncated p100 proteins contribute to oncogenesis remains to be clarified. Loss of the C-terminal ankyrin motifs in tumor-derived p100 mutants was proposed to abolish the IκB-like function of p100, resulting in abnormal NF-κB activity. A more recent study suggested another mechanism for oncogenic activation, i.e., that loss of a putative C-terminal death domain in tumor-derived p100 mutants might abrogate a proapoptotic effect of p100,[59] although this has been a subject of debate.[60] It is interesting to note that homozygous deletion of the C-terminal ankyrin repeats of p100 leads to gastric and lymph node hyperplasia in mice, suggesting that overexpression of p52/NF-κB2 contributes to oncogenesis.[61-63] In support of this hypothesis, tumor-derived rearranged p100 proteins undergo constitutive processing to produce functional p52, due to deletion of a C-terminal processing-inhibitory domain (PID).[64] Moreover, overexpression of p52 was detected in several malignancies including T-cell leukemia and breast and colon carcinoma.[9,65,66] The ability of p52 homodimers to function as transcriptional activators in combination with RelB or the IκB-related Bcl-3 transcription factor to promote expression of antiapoptotic and proproliferative genes such as *bcl-2* and *cyclin D1* is consistent with this model.[9,67]

The human *c-rel* locus is amplified in a significant proportion of diffuse lymphoma with a large cell component (DLLC; 23%) and also in primary mediastinal (thymic) B-cell lymphoma, classical Hodgkin's lymphoma and certain follicular large cell lymphoma.[1] However, the extent to which *c-rel* gene amplification causes elevated nuclear c-Rel protein levels is unclear. While some found a correlation between amplification of the *c-rel* locus and nuclear c-Rel protein accumulation in Hodgkin's lymphoma and mediastinal large B-cell lymphoma (MLBCL),[68,69] others found no close association between the two or with NF-κB target gene expression profiles in diffuse large B cell lymphoma (DLBCL).[70,71] These findings suggest that if *c-rel* plays a role, its function may be heterogeneous in different lymphomas or that it might play a role early in the history of some of these tumors that is no longer required later on. Although *c-rel*'s contribution to some of these tumors remains a point of contention,[72] future studies will undoubtedly provide important information on the subject.

Molecular Basis for Oncogenesis by Rel/NF-κB

Studies with the retroviral NF-κB oncoprotein v-Rel and its cellular Rel/NF-κB homologues have provided important insights into the oncogenic properties of Rel/NF-κB factors and the functional mechanisms involved. These are reviewed in this section.

The Viral NF-κB Oncoprotein v-Rel: A Potent Transforming Factor

Evidence pointing to a role for Rel/NF-κB in cancer came about long before the discovery of the *rel/nf-κb* gene family, with the isolation in 1958 of the Rev-T retrovirus from the liver of a diseased turkey.[73,74] The culprit Rev-T-encoded oncogene was identified many years later as v-*rel*, the first member of the Rel/NF-κB family.[75] v-*rel* immortalizes and transforms immature and mature B and T lymphoid, myeloid and dendritic cells from chicken spleen and bone marrow and induces aggressive and fatal leukemia/lymphoma in infected young birds.[76-79] v-Rel can also transform chicken embryo fibroblasts that induce tumors in immunocompetent young chicks.[80]

The oncogenic activity of v-*rel* was believed for some time to be restricted to avian species, as efforts to stably express it in rodent fibroblasts or lymphoid B cells resulted in apparent cytotoxicity.[81-83] Although the molecular basis for this effect remains to be clarified, stable expression was recently achieved in mouse fibroblasts using a mouse stem cell virus (MSCV),[84] but it remains to be seen if MSCV-driven v-*rel* will be transforming in mouse lymphoid cells. Yet, the discovery that transgenic mice expressing v-*rel* under the control of the *lck* promoter developed aggressive T-cell leukemia/lymphoma provided unambiguous proof of its oncogenic potential in mammals.[85] It is noteworthy however that the onset of tumor development in transgenic mice is remarkably slower than in infected chickens, with mice succumbing between 6 to 10 months of age compared to 7 to 10 days in young chicks. Another distinction between the avian and mammalian systems is the fact that tumors arising in v-*rel* transgenic mice are oligoclonal, rather than polyclonal in nature, and that they fail to transplant in syngeneic animals.[85] This suggests that additional cytogenetic alterations are necessary for manifestation of v-*rel*'s tumorigenic potential in mammals. In this regard, chickens lack p16^{INK4a} and express a truncated but functional ARF protein.[86] This raises the possibility that tumor suppressors such as those encoded by the *ink4b-arf-ink4a* locus perhaps contribute to the increased susceptibility of chickens to v-*rel*-induced transformation. Nevertheless, in light of the rapidly increasing number of studies implicating Rel/NF-κB activity in human tumors, v-*rel* is a highly prized tool to unravel the molecular basis for the oncogenic activity of cellular Rel/NF-κB factors.

A Role for c-Rel in Cell Transformation and Tumorigenesis: Lessons from Birds and Mice

Since v-*rel* arose by recombination of the non-transforming Rev-A retrovirus with the turkey c-*rel* proto-oncogene, it is not surprising to find that overexpression of the chicken, mouse or human c-*rel* genes transforms primary chicken cells in culture that induced tumor development in animal models, albeit at a lower efficiency than v-*rel*.[87-91] However when tested under similar conditions, other mammalian Rel/NF-κB subunits namely RelA, RelB, p50/NF-κB1 or p52/NF-κB2 failed to transform lymphoid cells.[87] Together, these findings suggest that overexpression of the c-Rel protein in some tumors showing c-*rel* gene amplification and/or constitutive activation of c-Rel-containing NF-κB complexes might contribute to certain human leukemia/lymphoma.

Importantly, c-*rel* was recently shown to also exhibit an oncogenic capacity in a mammalian system. Indeed, 31% of transgenic mice expressing the mouse c-*rel* gene under the control of the mouse mammary tumor virus (MMTV) promoter developed mammary tumors at an average age of 19.9 months.[92] Tumor development coincided with nuclear localization of NF-κB subunits and upregulation of many NF-κB-target genes including *cyclin D1*, *c-myc*, and *bcl-xl* (Table 2; see below). The significance of these findings is highlighted by the fact that many human breast cancer specimens show elevated NF-κB activity.[8,9,93-95]

Rel/NF-κB Functions Necessary for Cell Transformation

As a result of its acquisition and evolution in the context of the Rev-T retrovirus, v-*rel* encodes a truncated and mutated version of the turkey c-Rel protein fused to remnants of the Rev-A retroviral *env* gene. v-Rel carries a number of deletions and point mutations compared to c-Rel, including the loss of 118 C-terminal amino acids that correspond to a strong transactivation domain (TAD) in c-Rel.[74] Many of these differences contribute to the increased oncogenicity of v-Rel compared to c-Rel. For example selection for C-terminal truncation of c-Rel, reminiscent of that seen in v-Rel, was observed in tumors that arose following retroviral-mediated delivery of c-rel into young chickens.[96] Recent work from our group indicates that c-rel gene deletion or mutation is not necessary for lymphoid cell transformation "per se", but that it may rather be selected for during tumor progression to confer enhanced tumorigenicity, enable escape from immune surveillance and/or facilitate cell adaptation to growth in culture.[87]

A model for v-Rel-mediated oncogenesis has emerged that invokes its ability to transactivate κB site-dependent gene transcription as being critical for cell transformation. Mutations that decrease its DNA-binding or transactivation functions are detrimental to cell transformation, whereas those that increase these activities enhance its transforming potential.[97-106] Consistent with this model, v-Rel shuttles between the nucleus and the cytoplasm, and a threshold of nuclear v-Rel is necessary to transform cells.[107] Other factors also contribute to the enhanced oncogenicity of v-Rel compared to c-Rel.[74] These include the fact that: (1) v-Rel is less susceptible than c-Rel to inhibition by IκBα.[108,109] This agrees with the partial nuclear distribution of v-Rel/IκBα complexes compared to predominantly cytoplasmic NF-κB/IκBα complexes in unstimulated cells.[110,111] Despite its reduced susceptibility to inhibition by IκBα, v-Rel is nevertheless subject to IκBα control, as overexpression of IκBα in v-rel transgenic mice attenuated its tumorigenic phenotype.[112] (2) v-Rel binds to a broader range of NF-κB DNA sites compared to c-Rel and other NF-κB subunits. Nehyba et al identified mutation clusters in v-Rel responsible for this difference, and Phelps and Ghosh recently pinpointed amino acid differences between the Rel-homology domains (RHDs) of v-Rel and c-Rel in this effect.[109,113] (3) The particular dimers in which v-Rel participates also dictate its oncogenic potential. Mutational analysis revealed a critical role for v-Rel homodimers in cell transformation.[114] Although v-Rel/p50 heterodimers and v-Rel homodimers are the major DNA-binding complexes in v-rel transgenic mice, transgenic expression of v-rel in a p50 knockout background led to a more aggressive tumor phenotype.[85] (4) Recent work from our group indicated that critical determinants for the different oncogenic potentials of individual Rel/NF-κB subunits reside within their divergent TADs.[87] While RelA fails to transform primary chicken spleen cells, substitution of its TAD by that of the transforming v-Rel or c-Rel proteins conferred a strong transforming phenotype both in vitro and in vivo. Intrinsic differences between individual Rel/NF-κB TADs might confer distinct oncogenic potentials owing to differences in the repertoire of genes that they activate, as suggested by preliminary microarray analyses (Gupta, Fan, Delrow and Gélinas, unpublished data). Furthermore, the strength of individual Rel/NF-κB TADs is inversely correlated with their transforming potential, indicating that the magnitude of gene activation must be within a suitable range. For example, deletion of either of the two human c-Rel TADs reduced its transcriptional activity and increased its transforming efficiency.[115] Since strong TADs such as that of RelA perhaps activate gene expression to a level that is incompatible with cell transformation, it is tempting to speculate that RelA mutants with decreased transactivation potency might be capable of transformation. Preliminary data from our group suggest that this may indeed be the case (Fan and Gélinas, unpublished data). Overall, these findings underscore a fundamental role for gene transactivation in the transforming ability of Rel/NF-κB.

Functional Consequences of Rel/NF-κB-Mediated Gene Activation in Oncogenesis

The Rel/NF-κB transcription factors activate a wide variety of target genes that influence its oncogenicity. These include cell death inhibitors, cell cycle regulators, transcription factors and oncoproteins, cytokines and receptors, and cell surface and adhesion molecules. This section reviews how these contribute to the transformation process by affecting the regulation of apoptosis, cell proliferation, angiogenesis and metastasis.

Suppression of Apoptosis

Escape from apoptosis is a major factor in oncogenesis and in the resistance of tumor cells to therapy. It is therefore not surprising that NF-κB's anti-apoptotic activity has been linked to many different cancers and that it impedes effective treatment.[116,117] Consistent with the fact that Rel/NF-κB inhibits cell death by activating expression of antiapoptotic genes that can at least partially substitute for NF-κB to suppress cell death, many tumor-derived cell lines display elevated expression of NF-κB-dependent antiapoptotic factors (Table 2).[4] For instance, therapy-resistant DLBCL tumors and malignant H/RS cells have elevated levels of *bfl-1/a1* transcripts compared to controls, and *FLIP* is upregulated in DLBCL while *c-iap2* is induced in H/RS cells.[11,19,71,118] The important role for NF-κB in these cancers is highlighted by the fact that many tumor-derived cell lines, including those derived from HD, DLBCL and breast cancer undergo spontaneous apoptosis, or are sensitized to death-inducing stimuli, following NF-κB inhibition (for example see refs. 6,8,12,13,119-121). Although ectopic expression of Bcl-xL could rescue apoptosis of H/RS cells in which NF-κB activity was suppressed,[19] it should be noted that in other cases multiple apoptosis inhibitors appear to act in concert to promote survival in NF-κB-associated tumors.[12] These findings agree with the observation that sustained expression of v-*rel* is necessary to maintain the viability of transformed lymphoid cells and that v-*rel*-mediated transformation requires expression of specific apoptosis inhibitors (Table 2).[100,106,122-127] Though a majority of studies emphasize a fundamental role for the cytoprotective activity of NF-κB in oncogenesis, there are exceptions. For instance, survival in Bcr-Abl-induced leukemia was reported to be independent of NF-κB's antiapoptotic activity.[31]

Alternative mechanisms have emerged in which NF-κB promotes cell viability by interacting with other factors. For example, NF-κB interferes with the transcriptional function of the pro-apoptotic tumor suppressor p53 by competing for coactivators.[128-130] Similarly, p65/RelA sequesters coactivator p300 to inhibit expression of tumor suppressor PTEN and allow cell survival in lung and thyroid cancer cells.[131] The recently discovered capacity of IKKβ to increase expression of Mdm2 and decrease p53 stability to suppress chemotherapy-induced cell death is another example.[132] Lastly, NF-κB-mediated suppression of the p53-related p73 factor antagonizes apoptosis in antigen-stimulated naïve T cells.[133] This raises the possibility that a similar mechanism might operate in an oncogenic setting, although this remains to be established.

Cell Proliferation

Independent studies highlight a link between Rel/NF-κB's effects on cell proliferation and oncogenesis. Consistent with the critical role of the c-Rel subunit in B cell proliferation,[134,135] lymphoid cells transformed by a temperature-sensitive mutant of v-Rel fail to proliferate at the restrictive temperature under conditions where apoptosis is rescued by cell death inhibitor Bcl-2.[127] Aside from generating autocrine loops to constitutively activate the NF-κB pathway in tumor cells (see above), NF-κB activates expression of factors that influence cell cycle entry such as cyclins D1, D2 and D3 (Table 2).[19,135-137] These findings concur with the elevated levels of cyclin D2 in malignant H/RS cells and cyclin D1 in mantle cell lymphoma and breast cancers that display sustained NF-κB activity, as well as in MMTV-c-*rel*-induced mouse mammary carcinoma.[19,92,138]

NF-κB can also enhance proliferation by activating other transcription factors. Some of them directly contribute to v-Rel-mediated transformation of lymphoid cells (Table 2). For example transcription factor AP-1(c-Jun) is essential for v-Rel's transforming activity in primary lymphoid cells and fibroblasts.[139] Both c-Jun and JunB are aberrantly expressed in malignant H/RS cells of HD, where upregulation of JunB is NF-κB-dependent.[140] These factors act together with NF-κB to stimulate H/RS cell proliferation and expression of cyclin D2, Bcl-xL, c-*met* and chemokine receptor CCR7. Other examples are interferon regulatory factor 4 (IRF 4) that decreases induction of the anti-proliferative IFN pathway and facilitates v-Rel-mediated transformation[141] and c-Myc, a target of p50/c-Rel dimers that is induced in MMTV-c-*rel* mouse mammary tumors.[92,142] Stat5a recently joined this group as an NF-κB target that is activated constitutively in HD and is linked to cell growth regulation.[10]

Angiogenesis and Metastasis

The ability of tumor cells to acquire sustained angiogenesis, invade surrounding tissues and metastasize to remote sites is one of the most significant factors contributing to cancer patient mortality. Here too NF-κB makes an important contribution by inducing expression of factors that promote angiogenesis (Table 2). Elevated NF-κB activity in cancer cells enables deregulated production of chemokines and chemokine receptors, like IL-8 and CXCR4, which increase migratory activity and promote angiogenesis.[143,144] NF-κB-mediated induction of vascular endothelial growth factor (VEGF) is another important contributing factor.[144,145] NF-κB also promotes invasion of surrounding tissues by inducing various cell adhesion molecules and matrix metalloproteinases.[7,145] Together, these factors contribute to the pathogenesis of NF-κB in cancer.

Other Means for NF-κB to Participate in Oncogenesis

Several protein interactions and post-translational modifications modulate the transcriptional activity of NF-κB, and in some cases its contribution to oncogenesis.

Interaction with Transcription Factors and Coactivators

v-Rel and its cellular homologues c-Rel and RelA functionally interact with basal transcription factors and transcriptional coactivators to synergistically enhance gene transcription (for example see refs. 146-148). In some instances, interactions were confirmed in transformed lymphoid cells.[146] Although various coactivators like CBP/p300, TAFII105 or TAFII250 mediate NF-κB dependent transcription of anti-apoptotic genes, their implication in a tumor context is awaiting.[149-153] However PARP (Poly-ADP ribose polymerase-1) that behaves as a coactivator for NF-κB was recently implicated in NF-κB-mediated susceptibility to skin cancer induced by DMBA and TPA in mice but their coordinate action in human carcinogenesis remains to be verified.[154,155]

Post-Translational Modifications

Phosphorylation, acetylation and ubiquitination of NF-κB subunits influence their activity, although in many cases evidence of their role in NF-κB-associated tumors has yet to be obtained.

Various kinases including IKK, casein kinase II and AKT enhance the transcriptional activities of p65/RelA and c-Rel by phosphorylating their TAD.[156-162] Moreover, mutation of certain serines in the v-Rel TAD decreases its transcriptional activity and impairs transformation of lymphoid cells.[106] In addition, the catalytic subunit of PKA (PKAc), MSK1 and PKCζ phosphorylate the RHD of p65/RelA and modification by PKAc is necessary for p65 interaction with coactivator p300.[163-166] Of interest, serine phosphorylation in the C-terminal domain of Bcl-3 by GSK3 affects its interaction with HDACs and is correlated with attenuation of its transforming potential in a mouse model.[167] Acetylation bestows another level of regulation, as exemplified by reversible p300/CBP- and P/CAF-mediated acetylation of p65/RelA that blocks association with IκBα and promotes p65 nuclear localization, DNA binding, and transactivation.[168-171]

A role for the ubiquitin-proteasome pathway in directly regulating the stability of NF-κB subunits came to light in work showing that C-terminally truncated c-Rel mutants and v-Rel display reduced proteasome-mediated turnover coincident with oncogenic transformation.[172] Since then, poly-ubiquitination of p65/RelA was implicated in terminating the NF-κB response.[173] In this regard, the peptidyl-prolyl isomerase Pin1 was recently described to enhance NF-κB activity by associating with nuclear p65 to prevent its SOCS-1-mediated ubiquitination and degradation.[174] The fact that Pin1 is highly overexpressed in human breast cancer suggests a possible role for Pin1 in enhancing the oncogenic activity of NF-κB in certain tumors.

A Tumor Suppressor Role for NF-κB

Despite a large body of evidence supporting a positive role for Rel/NF-κB in oncogenesis, a growing number of studies indicate that NF-κB can behave as a tumor suppressor in some circumstances.[175,176] Indeed RelA opposes the action of TNFR1 and JNK to curb epidermal cell growth, and suppression of NF-κB in skin cooperates with oncogenic lesions such as oncogenic Ras to favor development of squamous cell carcinoma.[177-181] Consistent with this, immortalized *relA-/-* fibroblasts induce tumors in SCID mice.[182] Moreover, RelA actively represses transcription of anti-apoptotic genes in response to certain stimuli such as UV-C and chemotherapeutic agents doxorubicin or daunorubicin,[183] although others found NF-κB to be protective in this context.[184,185]

The interaction of NF-κB with tumor suppressors to downregulate proproliferative or antiapoptotic genes, or to induce expression of pro-death factors further supports the notion that NF-κB can inhibit tumor growth in certain settings (Table 2). ARF, best known for its role in activating p53 via inhibition of Mdm2, inhibits p65/RelA-mediated transcription by inducing p65 association with histone deacetylase HDAC1.[186] This effect is promoter specific, as it leads to downregulation of anti-apoptotic NF-κB target Bcl-xL but not IκBα. Similarly, tumor suppressor p53 converts transcriptionally active Bcl-3/p52 complexes into transcriptionally inactive p52/HDAC1 complexes that inhibit cyclin D1 expression to induce cell cycle arrest.[187] The significance of these findings is highlighted by the observation that c-Rel, p52 and Bcl-3 are activated in human breast cancer.[9,95] Interestingly, tumor suppressor BRCA1 physically interacts with p65/RelA to enhance NF-κB-mediated transcription of proapoptotic gene Fas, while ING4 controls brain tumor angiogenesis by negatively regulating RelA's transcriptional activity.[188,189] Lastly, NF-κB was reported to be required for p53-dependent apoptosis,[190] and recent work implicated the serine/threonine kinase ribosomal S6 kinase 1 (RSK1) in its p53-mediated activation.[191] While one study found that NF-κB activation by doxorubicin decreases p53 stability,[132] another showed that NF-κB stabilizes p53 to provoke apoptosis in response to genotoxic stress.[192] Regardless of the mechanisms involved, the capacity of NF-κB to sometimes behave as a tumor suppressor has obvious implications for its role in oncogenesis and for therapeutic approaches aimed at inhibiting its activity (see below).

Conclusions and Perspectives for Therapy

While additional work is needed to pinpoint the precise role of Rel/NF-κB in human cancer and the mechanisms involved, a parallel has emerged between activities needed for lymphoid cell transformation and lymphomagenesis induced by the v-Rel and c-Rel proteins and those observed in many human tumors displaying constitutive NF-κB activity. These include inappropriate activation of cellular gene expression and aberrant expression of proproliferative and antiapoptotic genes. While the Rel proteins are potently oncogenic in avian species, the delayed onset of tumors in transgenic mice expressing v-*rel* or an MMTV-c-*rel* transgene suggests that additional cytogenetic alterations are necessary for NF-κB to manifest its oncogenic phenotype in mammals.[85,92] Recent studies uncovered a crucial role for interaction between inflammatory cells and precancerous cells in tumor promotion and showed that NF-κB is critical in this process.[193-195] Thus, identification of the genes and pathways that act cooperatively with NF-κB in human cancer is an important goal in the field.

The NF-κB pathway constitutes an important therapeutic target.[145] NF-κB is implicated in the intrinsic resistance of cancer cells to apoptosis and the induced chemoresistance of many tumors to anti-cancer drugs, and its inhibition often enhances the effectiveness of treatment (for example see refs. 119,121,184). However recent advances uncovered a more complex scenario, by revealing that NF-κB can sometimes repress transcription of anti-apoptotic genes in response to "atypical" activators including several chemotherapeutic agents.[175,183] Its interaction with tumor suppressors such as p53, ARF and BRCA1 is another potentially important factor in the outcome of cancer therapy. Careful attention should thus be given to the tumor cell type, the death-inducing agent and perhaps other cytogenetic alterations in these tumors. Ongoing studies in the field promise to further advance our understanding of the transcriptional, anti-/pro-proliferative and anti/pro-apoptotic functions of NF-κB, the modifications and factors that modulate its activity and their impact on the oncogenic process. Together these will help to develop and improve appropriate strategies for cancer therapy tailored to particular contexts.

Acknowledgements

We are very grateful to N. Perkins for sharing a review article before publication. We apologize to many investigators whose work could not be cited due to space limitations. Research in this laboratory on the roles of Rel/NF-κB and its anti-apoptotic target Bfl-1/A1 in apoptosis and oncogenesis is supported by grants from the National Institutes of Health – National Cancer Institute CA54999 and CA83937 to CG. The first three authors contributed equally to this review.

References

1. Rayet B, Gélinas C. Aberrant rel/nf-κb genes and activity in human cancer. Oncogene 1999; 18:6938-6947.
2. Gilmore TD, Kalaitzidis D, Liang MC et al. The c-Rel transcription factor and B-cell proliferation: A deal with the devil. Oncogene 2004; 23:2275-2286.
3. Gilmore TD, Mosialos G. Viruses as intruders in the Rel/NF-kappaB signaling pathway. In: Beyaert R, ed. The Nuclear Factor κB: Regulation and Role in Defense. The Netherlands: Kluwer Academic Publishers, 2003:91-115.
4. Kucharczak J, Simmons MJ, Fan Y et al. To be, or not to be: NF-kappaB is the answer-role of Rel/NF-kappaB in the regulation of apoptosis. Oncogene 2003; 22:8961-8982.
5. Karin M, Lin A. NF-kappaB at the crossroads of life and death. Nat Immunol 2002; 3:221-227.
6. Lin A, Karin M. NF-kappaB in cancer: A marked target. Semin Cancer Biol 2003; 13:107-114.
7. Baldwin AS. Control of oncogenesis and cancer therapy resistance by the transcription factor NF-kappaB. J Clin Invest 2001; 107:241-246.
8. Sovak MA, Bellas RE, Kim DW et al. Aberrant nuclear factor-kappaB/Rel expression and the pathogenesis of breast cancer. J Clin Invest 1997; 100:2952-2960.
9. Cogswell PC, Guttridge DC, Funkhouser WK et al. Selective activation of NF-κB subunits in human breast cancer: Potential roles for NF-κB2/p52 and for Bcl-3. Oncogene 2000; 19:1123-1131.
10. Hinz M, Lemke P, Anagnostopoulos I et al. Nuclear factor κB-dependent gene expression profiling of Hodgkin's disease tumor cells, pathogenic significance, and link to constitutive signal transducer and activator of transcription 5a activity. J Exp Med 2002; 196:605-617.
11. Alizadeh A, Eisen M, Davis R et al. Distinct types of diffuse large B-cell lymphoma identified by gene expression profiling. Nature 2000; 403:503-511.
12. Davis RE, Brown KD, Siebenlist U et al. Constitutive nuclear factor kappaB activity is required for survival of activated B cell-like diffuse large B cell lymphoma cells. J Exp Med 2001; 194:1861-1874.
13. Bargou R, Emmerich F, Krappmann D et al. Constitutive nuclear factor-kappaB-RelA activation is required for proliferation and survival of Hodgkin's disease tumor cells. J Clin Invest 1997; 100:2961-2969.
14. Karin M, Ben-Neriah Y. Phosphorylation meets ubiquitination: The control of NF-kappaB activity. Annu Rev Immunol 2000; 18:621-663.
15. Sun S-C, Xiao G. Deregulation of NF-kappaB and its upstream kinases in cancer. Cancer Metastasis Rev 2003; 22:405-422.
16. Ravi R, Bedi A. NF-kappaB in cancer—a friend turned foe. Drug Resist Updat 2004; 7:53-67.
17. Lu T, Sathe SS, Swiatkowski SM et al. Secretion of cytokines and growth factors as a general cause of constitutive NFkappaB activation in cancer. Oncogene 2004; 23:2138-2145.
18. Vale T, Ngo TT, White MA et al. Raf-induced transformation requires an interleukin 1 autocrine loop. Cancer Res 2001; 61:602-607.
19. Hinz M, Loser P, Mathas S et al. Constitutive NF-kappaB maintains high expression of a characteristic gene network, including CD40, CD86, and a set of antiapoptotic genes in Hodgkin/Reed-Sternberg cells. Blood 2001; 97:2798-2807.
20. Brummelkamp TR, Nijman SM, Dirac AM et al. Loss of the cylindromatosis tumour suppressor inhibits apoptosis by activating NF-kappaB. Nature 2003; 424:797-801.

21. Kovalenko A, Chable-Bessia C, Cantarella G et al. The tumour suppressor CYLD negatively regulates NF-kappaB signalling by deubiquitination. Nature 2003; 424:801-805.
22. Trompouki E, Hatzivassiliou E, Tsichritzis T et al. CYLD is a deubiquitinating enzyme that negatively regulates NF-kappaB activation by TNFR family members. Nature 2003; 424:793-796.
23. Wilkinson KD. Signal transduction: Aspirin, ubiquitin and cancer. Nature 2003; 424:738-739.
24. Hu MC, Lee DF, Xia W et al. IkappaB kinase promotes tumorigenesis through inhibition of forkhead FOXO3a. Cell 2004; 117:225-237.
25. Yamamoto Y, Verma UN, Prajapati S et al. Histone H3 phosphorylation by IKK-alpha is critical for cytokine-induced gene expression. Nature 2003; 423:655-659.
26. Anest V, Hanson JL, Cogswell PC et al. A nucleosomal function for IkappaB kinase-alpha in NF-kappaB-dependent gene expression. Nature 2003; 423:659-663.
27. Verma UN, Yamamoto Y, Prajapati S et al. Nuclear role of I kappa B Kinase-gamma/NF-kappa B essential modulator (IKK gamma/NEMO) in NF-kappa B-dependent gene expression. J Biol Chem 2004; 279:3509-3515.
28. Pianetti S, Arsura M, Romieu-Mourez R et al. Her-2/neu overexpression induces NF-κB via a PI3-kinase/Akt pathway involving calpain-mediated degradation of IκBα that can be inhibited by the tumor supressor PTEN. Oncogene 2001; 20:1287-1299.
29. Romieu-Mourez R, Landesman-Bollag E, Seldin DC et al. Protein kinase CK2 promotes aberrant activation of nuclear factor-kappaB, transformed phenotype, and survival of breast cancer cells. Cancer Res 2002; 62:6770-6778.
30. Cavin LG, Romieu-Mourez R, Panta GR et al. Inhibition of CK2 activity by TGF-beta1 promotes IkappaB-alpha protein stabilization and apoptosis of immortalized hepatocytes. Hepatology 2003; 38:1540-1551.
31. Reuther JY, Reuther GW, Cortez G et al. A requirement for NF-κB activation in Bcr-Abl-mediated transformation. Genes Dev 1998; 12:968-981.
32. Nawata R, Yujiri T, Nakamura Y et al. MEK kinase 1 mediates the antiapoptotic effect of the Bcr-Abl oncogene through NF-kappaB activation. Oncogene 2003; 22:7774-7780.
33. Korus M, Mahon GM, Cheng L et al. p38 MAPK-mediated activation of NF-kappaB by the RhoGEF domain of Bcr. Oncogene 2002; 21:4601-4612.
34. Finco TS, Westwick JK, Norris JL et al. Oncogenic Ha-Ras-induced signaling activates NF-kappaB transcriptional activity, which is required for cellular transformation. J Biol Chem 1997; 272:24113-24116.
35. Mayo MW, Wang CY, Cogswell PC et al. Requirement of NF-kappa B activation to suppress p53-independent apoptosis induced by oncogenic Ras. Science 1997; 278:1812-1815.
36. Jeay S, Pianetti S, Kagan HM et al. Lysyl oxidase inhibits ras-mediated transformation by preventing activation of NF-kappa B. Mol Cell Biol 2003; 23:2251-2263.
37. Stoffel A, Chaurushiya M, Singh B et al. Activation of NF-kappaB and inhibition of p53-mediated apoptosis by API2/mucosa-associated lymphoid tissue 1 fusions promote oncogenesis. Proc Natl Acad Sci USA 2004; 101:9079-9084.
38. Yamaoka S, Inoue H, Sakurai M et al. Constitutive activation of NF-kappa B is essential for transformation of rat fibroblasts by the human T-cell leukemia virus type I Tax protein. EMBO J 1996; 15:873-887.
39. Rousset R, Desbois C, Bantignies F et al. Effects on NF-κB1/p105 processing of the interaction between the HTLV-1 transactivator Tax and the proteasome. Nature 1996; 381:328-331.
40. Li XH, Gaynor RB. Mechanisms of NF-kappaB activation by the HTLV type 1 tax protein. AIDS Res Hum Retroviruses 2000; 16:1583-1590.
41. Mosialos G. The role of Rel/NF-kappa B proteins in viral oncogenesis and the regulation of viral transcription. Seminars in Cancer Biology 1997; 8:121-129.
42. Xiao G, Cvijic ME, Fong A et al. Retroviral oncoprotein tax induces processing of NF-kappaB2/p100 in T cells: Evidence for the involvement of IKKalpha. EMBO J 2001; 20:6805-6815.
43. Cahir-McFarland ED, Davidson DM, Schauer SL et al. NF-kappa B inhibition causes spontaneous apoptosis in Epstein-Barr virus-transformed lymphoblastoid cells. Proc Natl Acad Sci USA 2000; 97:6055-6060.
44. Cahir-McFarland E, Kieff E. NF-kappaB inhibition in EBV-transformed lymphoblastoid cell lines. Recent Results Cancer Res 2002; 159:44-48.
45. D'Souza BN, Edelstein LC, Pegman PM et al. Nuclear factor kappa B-dependent activation of the antiapoptotic bfl-1 gene by the Epstein-Barr virus latent membrane protein 1 and activated CD40 receptor. J Virol 2004; 78:1800-1816.
46. Eliopoulos AG, Caamano JH, Flavell J et al. Epstein-Barr virus-encoded latent infection membrane protein 1 regulates the processing of p100 NF-kappaB2 to p52 via an IKKgamma/NEMO-independent signalling pathway. Oncogene 2003; 22:7557-7569.

47. Perkins ND. Oncogenes, tumor suppressors and p52 NF-kappaB. Oncogene 2003; 22:7553-7556.
48. Cabannes E, Khan G, Aillet F et al. Mutations in the IκBα gene in Hodgkin's disease suggest a tumour suppressor role for IkappaBalpha. Oncogene 1999; 18:3063-3070.
49. Krappmann D, Emmerich F, Kordes U et al. Molecular mechanisms of constitutive NF-kappaB/Rel activation in Hodgkin/Reed-Sternberg cells. Oncogene 1999; 18:943-953.
50. Wood KM, Roff M, Hay RT. Defective IkappaBalpha in Hodgkin cell lines with constitutively active NF-kappaB. Oncogene 1998; 16:2131-2139.
51. Jungnickel B, Staratschek-Jox A, Brauninger A et al. Clonal deleterious mutations in the IkappaBalpha gene in the malignant cells in Hodgkin's lymphoma. J Exp Med 2000; 191:395-402.
52. Emmerich F, Theurich S, Hummel M et al. Inactivating I kappa B epsilon mutations in Hodgkin/Reed-Sternberg cells. J Pathol 2003; 201:413-420.
53. Kalaitzidis D, Davis RE, Rosenwald A et al. The human B-cell lymphoma cell line RC-K8 has multiple genetic alterations that dysregulate the Rel/NF-κB signal transduction pathway. Oncogene 2002; 21:8759-8768.
54. Chang C-C, Zhang J, Lombardi L et al. Mechanism of expression and role in transcriptional control of the proto-oncogene NFκB-2/LYT-10. Oncogene 1994; 9:923-933.
55. Fracchiolla NS, Lombardi L, Salina M et al. Structural alterations of the NF-kappa B transcription factor lyt-10 in lymphoid malignancies. Oncogene 1993; 8:2839-2845.
56. Migliazza A, Lombardi L, Rocchi M et al. Heterogeneous chromosomal aberrations generate 3' truncations of the NFκB2/lyt-10 gene in lymphoid malignancies. Blood 1994; 84:3850-3860.
57. Neri A, Fracchiolla NS, Roscetti E et al. Molecular analysis of cutaneous B- and T-cell lymphomas. Blood 1995; 86:3160-3172.
58. Thakur S, Lin HC, Tseng WT et al. Rearrangement and altered expression of the NFκB2 gene in human cutaneous T-lymphoma cells. Oncogene 1994; 9:2335-2344.
59. Wang Y, Cui H, Schroering A et al. NF-kappa B2 p100 is a pro-apoptotic protein with anti-oncogenic function. Nat Cell Biol 2002; 4:888-893.
60. Hacker H, Karin M. Is NF-kappaB2/p100 a direct activator of programmed cell death? Cancer Cell 2002; 2:431-433.
61. Caamano JH, Rizzo CA, Durham SK et al. Nuclear factor (NF)-kappa B2 (p100/p52) is required for normal splenic microarchitecture and B cell-mediated immune responses. J Exp Med 1998; 187:185-196.
62. Franzoso G, Carlson L, Poljak L et al. Mice deficient in nuclear factor (NF)-kappa B/p52 present with defects in humoral responses, germinal center reactions, and splenic microarchitecture. J Exp Med 1998; 187:147-159.
63. Ishikawa H, Carrasco D, Claudio E et al. Gastric hyperplasia and increased proliferative responses of lymphocytes in mice lacking the COOH-terminal ankyrin domain of NF-kappaB2. J Exp Med 1997; 186:999-1014.
64. Xiao G, Harhaj E, Sun SC. NF-κB-inducing kinase regulates the processing of NF-κB2 p100. Mol Cell 2001; 7:401-409.
65. Dejardin E, Bonizzi G, Bellahcene A et al. Highly expressed p100/p52 (NFκB2) sequesters other NFκB-related proteins in the cytoplasm of human breast cancer cells. Oncogene 1995; 11:1835-1841.
66. Bours V, Dejardin E, Goujon-Letawe F et al. The NF-κB transcription factor and cancer: High expression of NF-κB- and IκB-related proteins in tumor cell lines. Biochem Pharmacol 1994; 47:145-149.
67. Viatour P, Bentires-Alj M, Chariot A et al. NF- kappa B2/p100 induces Bcl-2 expression. Leukemia 2003; 17:1349-1356.
68. Barth TF, Martin-Subero JI, Joos S et al. Gains of 2p involving the REL locus correlate with nuclear c-Rel protein accumulation in neoplastic cells of classical Hodgkin lymphoma. Blood 2003; 101:3681-3686.
69. Savage KJ, Monti S, Kutok JL et al. The molecular signature of mediastinal large B-cell lymphoma differs from that of other diffuse large B-cell lymphomas and shares features with classical Hodgkin lymphoma. Blood 2003; 102:3871-3879.
70. Houldsworth J, Olshen AB, Cattoretti G et al. Relationship between REL amplification, REL function, and clinical and biologic features in diffuse large B-cell lymphomas. Blood 2004; 103:1862-1868.
71. Rosenwald A, Staudt LM. Gene expression profiling of diffuse large B-cell lymphoma. Leuk Lymphoma 2003; 44(Suppl 3):S41-47.
72. Gilmore TD, Starczynowski DT, Kalaitzidis D et al. RELevant gene amplification in B-cell lymphomas? Blood 2004; 103:3243-3245.
73. Theilen G, Zeigel R, Tweihaus M. Biological studies with RE virus (strain T) that induces reticuloendotheliosis in turkeys, chickens, and Japanese quails. J Natl Cancer Inst 1966; 37:747-749.

74. Gilmore TD. Multiple mutations contribute to the oncogenicity of the retroviral oncoprotein v-Rel. Oncogene 1999; 18:6925-6937.
75. Rice NR, Copeland TD, Simek S et al. Detection and characterization of the protein encoded by the v-rel oncogene. Virology 1986; 149:217-229.
76. Zhang JY, Olson W, Ewert D et al. The v-rel oncogene of avian reticuloendotheliosis virus transforms immature and mature lymphoid cells of the B cell lineage in vitro. Virology 1991; 183:457-466.
77. Barth CF, Ewert DL, Olson WC et al. Reticuloendotheliosis virus REV-T(REV-A)-induced neoplasia: Development of tumors within the T-lymphoid and myeloid lineages. J Virol 1990; 64:6054-6062.
78. Boehmelt G, Madruga J, Dorfler P et al. Dendritic cell progenitor is transformed by a conditional v-rel estrogen receptor fusion protein v-relER. Cell 1995; 80:341-352.
79. Beug H, Muller H, Doederlein G et al. Hematopoietic cells transformed in vitro by REV-T avian reticuloendotheliosis virus express characteristics of very immature lymphoid cells. Virology 1981; 115:295-309.
80. Moore BE, Bose HR. Expression of the v-rel oncogene in reticuloendotheliosis virus-transformed fibroblasts. Virology 1988; 162:377-387.
81. Gélinas C, Temin HM. The v-rel oncogene encodes a cell-specific transcriptional activator of certain promoters. Oncogene 1988; 3:349-355.
82. Hannink M, Temin HM. Transactivation of gene expression by nuclear and cytoplasmic rel proteins. Mol Cell Biol 1989; 9:4323-4336.
83. Schwartz RC, Witte ON. A recombinant murine retrovirus expressing v-rel is cytopathic. Virology 1988; 165:182-190.
84. Gilmore TD, Jean-Jacques J, Richards R et al. Stable expression of the avian retroviral oncoprotein v-Rel in avian, mouse, and dog cell lines. Virology 2003; 316:9-16.
85. Carrasco D, Rizzo CA, Dorfman K et al. The v-rel oncogene promotes malignant T-cell leukemia/ lymphoma in transgenic mice. EMBO J 1996; 15:3640-3650.
86. Kim SH, Mitchell M, Fujii H et al. Absence of p16INK4a and truncation of ARF tumor suppressors in chickens. Proc Natl Acad Sci USA 2003; 100:211-216.
87. Fan Y, Rayet B, Gélinas C. Divergent C-terminal transactivation domains of Rel/NF-κB proteins are critical determinants of their oncogenic potential in lymphocytes. Oncogene 2004; 23:1030-1042.
88. Hrdlickova R, Nehyba J, Humphries EH. v-rel induces expression of three avian immunoregulatory surface receptors more efficiently than c-rel. J Virol 1994; 68:308-319.
89. Nehyba J, Hrdlickova R, Humphries EH. Evolution of the oncogenic potential of v-rel: Rel-induced expression of immunoregulatory receptors correlates with tumor development and in vitro transformation. J Virol 1994; 68:2039-2050.
90. Kralova J, Schatzle JD, Bargmann W et al. Transformation of avian fibroblasts overexpressing the c-rel proto-oncogene and a variant of c-rel lacking 40 C-terminal amino acids. J Virol 1994; 68:2073-2083.
91. Gilmore TD, Cormier C, Jean-Jacques J et al. Malignant transformation of primary chicken spleen cells by human transcription factor c-Rel. Oncogene 2001; 20:7098-7103.
92. Romieu-Mourez R, Kim DW, Shin SM et al. Mouse mammary tumor virus c-rel transgenic mice develop mammary tumors. Mol Cell Biol 2003; 23:5738-5754.
93. Nakshatri H, Bhat-Nakshatri P, Martin DA et al. Constitutive activation of NF-kappaB during progression of breast cancer to hormone-independent growth. Mol Cell Biol 1997; 17:3629-3639.
94. Kim DW, Sovak MA, Zanieski G et al. Activation of NF-kappaB/Rel occurs early during neoplastic transformation of mammary cells. Carcinogenesis 2000; 21:871-879.
95. Romieu-Mourez R, Landesman-Bollag E, Seldin DC et al. Roles of IKK kinases and protein kinase CK2 in activation of nuclear factor-kappaB in breast cancer. Cancer Res 2001; 61:3810-3818.
96. Hrdlickova R, Nehyba J, Humphries EH. In vivo evolution of c-rel oncogenic potential. J Virol 1994; 68:2371-2382.
97. Romero P, Humphries EH. A mutant v-rel with increased ability to transform B lymphocytes. J Virol 1995; 69:301-307.
98. Kumar S, Rabson AB, Gélinas C. The RxxRxRxxC motif conserved in all Rel/κB proteins is essential for the DNA-binding activity and redox regulation of the v-Rel oncoprotein. Mol Cell Biol 1992; 12:3094-3106.
99. Morrison LE, Boehmelt G, Enrietto PJ. Mutations in the rel-homology domain alter the biochemical properties of v-rel and render it transformation defective in chicken fibroblasts. Oncogene 1992; 7:1137-1147.

100. White DW, Gilmore TD. Temperature-sensitive transforming mutants of the v-rel oncogene. J Virol 1993; 67:6876-6881.
101. Walker WH, Stein B, Ganchi PA et al. The v-rel oncogene: Insights into the mechanism of transcriptional activation, repression and transformation. J Virol 1992; 66:5018-5029.
102. Kamens J, Richardson P, Mosialos G et al. Oncogenic transformation by vRel requires an amino-terminal activation domain. Mol Cell Biol 1990; 10:2840-2847.
103. Sarkar S, Gilmore TD. Transformation by the v-Rel oncoprotein requires sequences carboxy-terminal to the Rel homology domain. Oncogene 1993; 8:2245-2252.
104. Chen C, Agnès F, Gélinas C. Mapping of a serine-rich domain essential for the transcriptional, antiapoptotic, and transforming activities of the v-Rel oncoprotein. Mol Cell Biol 1999; 19:307-316.
105. Chen C, Agnès F, Gélinas C. Mapping of a serine-rich domain essential for the transcriptional, antiapoptotic, and transforming activities of the v-Rel oncoprotein- Author's Correction. Mol Cell Biol 2001; 21:7115.
106. Rayet B, Fan Y, Gélinas C. Mutations in the v-Rel transactivation domain indicate altered phosphorylation and identify a subset of NF-κB-regulated cell death inhibitors important for v-Rel transforming activity. Mol Cell Biol 2003; 23:1520-1533.
107. Sachdev S, Diehl JA, McKinsey TA et al. A threshold nuclear level of the v-Rel oncoprotein is required for transformation of avian lymphocytes. Oncogene 1997; 14:2585-2594.
108. Sachdev S, Hannink M. Loss of IκBα-mediated control over nuclear import and DNA binding enables oncogenic activation of c-Rel. Mol Cell Biol 1998; 18:5445-5456.
109. Phelps CB, Ghosh G. Discreet mutations from c-Rel to v-Rel alter kappaB DNA recognition, IkappaBalpha binding, and dimerization: implications for v-Rel oncogenicity. Oncogene 2004; 23:1229-1238.
110. Malek S, Chen Y, Huxford T et al. IkappaBbeta, but not IkappaBalpha, functions as a classical cytoplasmic inhibitor of NF-kappaB dimers by masking both NF-kappaB nuclear localization sequences in resting cells. J Biol Chem 2001; 276:45225-45235.
111. Tam WF, Wang W, Sen R. Cell-specific association and shuttling of IkappaBalpha provides a mechanism for nuclear NF-kappaB in B lymphocytes. Mol Cell Biol 2001; 21:4837-4846.
112. Carrasco D, Perez P, Lewin A et al. IκBα overexpression delays tumor formation in v-rel transgenic mice. J Exp Med 1997; 186:279-288.
113. Nehyba J, Hrdlickova R, Bose Jr HR. Differences in kappaB DNA-binding properties of v-Rel and c-Rel are the result of oncogenic mutations in three distinct functional regions of the Rel protein. Oncogene 1997; 14:2881-2897.
114. Liss AS, Bose Jr HR. Mutational analysis of the v-Rel dimerization interface reveals a critical role for v-Rel homodimers in transformation. J Virol 2002; 76:4928-4939.
115. Starczynowski DT, Reynolds JG, Gilmore TD. Deletion of either C-terminal transactivation subdomain enhances the in vitro transforming activity of human transcription factor REL in chicken spleen cells. Oncogene 2003; 22:6928-6936.
116. Karin M, Cao Y, Greten FR et al. NF-kappaB in cancer: From innocent bystander to major culprit. Nat Rev Cancer 2002; 2:301-310.
117. Wang CY, Guttridge DC, Mayo MW et al. NF-kappaB induces expression of the Bcl-2 homologue A1/Bfl-1 to preferentially suppress chemotherapy-induced apoptosis. Mol Cell Biol 1999; 19:5923-5929.
118. Mathas S, Lietz A, Anagnostopoulos I et al. c-FLIP Mediates resistance of Hodgkin/Reed-Sternberg cells to death receptor-induced apoptosis. J Exp Med 2004; 199:1041-1052.
119. Wang CY, Cusack Jr JC, Liu R et al. Control of inducible chemoresistance: Enhanced anti-tumor therapy through increased apoptosis by inhibition of NF-kappaB. Nat Med 1999; 5:412-417.
120. Cusack Jr JC, Liu R, Baldwin Jr AS. Inducible chemoresistance to 7-ethyl-10-[4-(1-piperidino)-1-piperidino]-carbonyloxycamptothe cin (CPT-11) in colorectal cancer cells and a xenograft model is overcome by inhibition of nuclear factor-kappaB activation. Cancer Res 2000; 60:2323-2330.
121. Baldwin Jr AS. Series introduction: the transcription factor NF-kappaB and human disease. J Clin Invest 2001; 107:3-6.
122. White DW, Roy A, Gilmore TD. The v-Rel oncoprotein blocks apoptosis and proteolysis of IκBα in transformed chicken spleen cells. Oncogene 1995; 10:857-868.
123. Zong WX, Farrell M, Bash J et al. v-Rel prevents apoptosis in transformed lymphoid cells and blocks TNFalpha-induced cell death. Oncogene 1997; 15:971-980.
124. Lee RM, Gillet G, Burnside J et al. Role of Nr13 in regulation of programmed cell death in the bursa of Fabricius. Genes Dev 1999; 13:718-728.
125. You M, Bose Jr HR. Identification of v-Rel oncogene-induced inhibitor of apoptosis by differential display. Methods 1998; 16:373-385.

126. Kralova J, Liss AS, Bargmann W et al. Differential regulation of the inhibitor of apoptosis ch-IAP1 by v-rel and the proto-oncogene c-rel. J Virol 2002; 76:11960-11970.
127. White DW, Gilmore TD. Bcl-2 and CrmA have different effects on transformation, apoptosis and the stability of IκBα in chicken spleen cells transformed by temperature-sensitive v-Rel oncoproteins. Oncogene 1996; 13:891-899.
128. Perkins ND, Felzien LK, Betts JC et al. Regulation of NF-kappaB by cyclin-dependent kinases associated with the p300 coactivator. Science 1997; 275:523-527.
129. Wadgaonkar R, Phelps KM, Haque Z et al. CREB-binding protein is a nuclear integrator of nuclear factor-kappaB and p53 signaling. J Biol Chem 1999; 274:1879-1882.
130. Webster G, Perkins N. Transcriptional cross talk between NF-kappaB and p53. Mol Cell Biol 1999; 19:3485-3495.
131. Vasudevan KM, Gurumurthy S, Rangnekar VM. Suppression of PTEN expression by NF-kappa B prevents apoptosis. Mol Cell Biol 2004; 24:1007-1021.
132. Tergaonkar V, Pando M, Vafa O et al. p53 stabilization is decreased upon NFkappaB activation: A role for NFkappaB in acquisition of resistance to chemotherapy. Cancer Cell 2002; 1:493-503.
133. Wan YY, DeGregori J. The survival of antigen-stimulated T cells requires NFkappaB-mediated inhibition of p73 expression. Immunity 2003; 18:331-342.
134. Grumont RJ, Rourke IJ, O'Reilly LA et al. B lymphocytes differentially use the Rel and nuclear factor κB1 (NF-κB1) transcription factors to regulate cell cycle progression and apoptosis in quiescent and mitogen-activated cells. J Exp Med 1998; 187:663-674.
135. Hsia CY, Cheng S, Owyang AM et al. c-Rel regulation of the cell cycle in primary mouse B lymphocytes. Int Immunol 2002; 14:905-916.
136. Guttridge DC, Albanese C, Reuther JY et al. NF-kappaB controls cell growth and differentiation through transcriptional regulation of cyclin D1. Mol Cell Biol 1999; 19:5785-5799.
137. Hinz M, Krappmann D, Eichten A et al. NF-κB function in growth control: Regulation of cyclin D1 expression and G0/G1-to-S-phase transition. Mol Cell Biol 1999; 19:2690-2698.
138. Pham LV, Tamayo AT, Yoshimura LC et al. Inhibition of constitutive NF-kappa B activation in mantle cell lymphoma B cells leads to induction of cell cycle arrest and apoptosis. J Immunol 2003; 171:88-95.
139. Kralova J, Liss AS, Bargmann W et al. AP-1 factors play an important role in transformation induced by the v-rel oncogene. Mol Cell Biol 1998; 18:2997-3009.
140. Mathas S, Hinz M, Anagnostopoulos I et al. Aberrantly expressed c-Jun and JunB are a hallmark of Hodgkin lymphoma cells, stimulate proliferation and synergize with NF-kappa B. EMBO J 2002; 21:4104-4113.
141. Hrdlickova R, Nehyba J, Bose Jr HR. Interferon regulatory factor 4 contributes to transformation of v-Rel-expressing fibroblasts. Mol Cell Biol 2001; 21:6369-6386.
142. Lee H, Arsura M, Wu M et al. Role of Rel-related factors in control of c-myc gene transcription in receptor-mediated apoptosis of the murine B cell WEHI 231 line. J Exp Med 1995; 181:1169-1177.
143. Helbig G, Christopherson 2nd KW, Bhat-Nakshatri P et al. NF-kappaB promotes breast cancer cell migration and metastasis by inducing the expression of the chemokine receptor CXCR4. J Biol Chem 2003; 278:21631-21638.
144. Huang S, Robinson JB, Deguzman A et al. Blockade of nuclear factor-kappaB signaling inhibits angiogenesis and tumorigenicity of human ovarian cancer cells by suppressing expression of vascular endothelial growth factor and interleukin 8. Cancer Res 2000; 60:5334-5339.
145. Garg A, Aggarwal BB. Nuclear transcription factor-kappaB as a target for cancer drug development. Leukemia 2002; 16:1053-1068.
146. Xu X, Prorock C, Ishikawa H et al. Functional interaction of the v-Rel and c-Rel oncoproteins with the TATA-binding protein and association with transcription factor IIB. Mol Cell Biol 1993; 13:6733-6741.
147. Schmitz ML, Stelzer G, Altmann H et al. Interaction of the COOH-terminal transactivation domain of p65 NF-kappa B with TATA-binding protein, transcription factor IIB, and coactivators. J Biol Chem 1995; 270:7219-7226.
148. Kerr LD, Ransone LJ, Wamsley P et al. Association between proto-oncoprotein Rel and TATA-binding protein mediates transcriptional activation by NF-kappa B. Nature 1993; 365:412-419.
149. Yamit-Hezi A, Dikstein R. TAFII105 mediates activation of anti-apoptotic genes by NF-κB. EMBO J 1998; 17:5161-5169.
150. Yamit-Hezi A, Nir S, Wolstein O et al. Interaction of TAFII105 with selected p65/RelA dimers is associated with activation of subset of NF-kappa B genes. J Biol Chem 2000; 275:18180-18187.
151. Edelstein LC, Lagos L, Simmons M et al. NF-kappa B-dependent assembly of an enhanceosome-like complex on the promoter region of apoptosis inhibitor Bfl-1/A1. Mol Cell Biol 2003; 23:2749-2761.

152. Sheppard KA, Rose DW, Haque ZK et al. Transcriptional activation by NF-kappaB requires multiple coactivators. Mol Cell Biol 1999; 19:6367-6378.
153. Guermah M, Malik S, Roeder RG. Involvement of TFIID and USA components in transcriptional activation of the human immunodeficiency virus promoter by NF-kappaB and Sp1. Mol Cell Biol 1998; 18:3234-3244.
154. Hassa PO, Buerki C, Lombardi C et al. Transcriptional coactivation of nuclear factor-kappaB-dependent gene expression by p300 is regulated by poly(ADP)-ribose polymerase-1. J Biol Chem 2003; 278:45145-45153.
155. Martin-Oliva D, O'Valle F, Munoz-Gamez JA et al. Crosstalk between PARP-1 and NF-kappaB modulates the promotion of skin neoplasia. Oncogene 2004; 23:5275-5283.
156. Wang D, Baldwin Jr AS. Activation of NF-κB-dependent transcription by tumor necrosis factor-alpha is mediated through phosphorylation of RelA/p65 on serine 529. J Biol Chem 1998; 273:29411-29416.
157. Sakurai H, Chiba H, Miyoshi H et al. IkappaB kinases phosphorylate NF-kappaB p65 subunit on serine 536 in the transactivation domain. J Biol Chem 1999; 274:30353-30356.
158. Martin AG, Fresno M. Tumor necrosis factor-alpha activation of NF-kappa B requires the phosphorylation of Ser-471 in the transactivation domain of c-Rel. J Biol Chem 2000; 275:24383-24391.
159. Wang D, Westerheide SD, Hanson JL et al. Tumor necrosis factor alpha-induced phosphorylation of RelA/p65 on Ser529 is controlled by casein kinase II. J Biol Chem 2000; 275:32592-32597.
160. Mayo MW, Madrid LV, Westerheide SD et al. PTEN blocks tumor necrosis factor-induced NF-kappa B-dependent transcription by inhibiting the transactivation potential of the p65 subunit. J Biol Chem 2002; 277:11116-11125.
161. Sizemore N, Lerner N, Dombrowski N et al. Distinct roles of the Ikappa B kinase alpha and beta subunits in liberating nuclear factor kappa B (NF-kappa B) from Ikappa B and in phosphorylating the p65 subunit of NF-kappa B. J Biol Chem 2002; 277:3863-3869.
162. Yang F, Tang E, Guan K et al. IKKbeta plays an essential role in the phosphorylation of RelA/p65 on serine 536 induced by lipopolysaccharide. J Immunol 2003; 170:5630-5635.
163. Zhong H, Voll RE, Ghosh S. Phosphorylation of NF-kappa B p65 by PKA stimulates transcriptional activity by promoting a novel bivalent interaction with the coactivator CBP/p300. Mol Cell 1998; 1:661-671.
164. Zhong H, SuYang H, Erdjument-Bromage H et al. The transcriptional activity of NF-kappaB is regulated by the IkappaB-associated PKAc subunit through a cyclic AMP-independent mechanism. Cell 1997; 89:413-424.
165. Duran A, Diaz-Meco MT, Moscat J. Essential role of RelA Ser311 phosphorylation by zetaPKC in NF-kappaB transcriptional activation. EMBO J 2003; 22:3910-3918.
166. Vermeulen L, De Wilde G, Van Damme P et al. Transcriptional activation of the NF-kappaB p65 subunit by mitogen- and stress-activated protein kinase-1 (MSK1). EMBO J 2003; 22:1313-1324.
167. Viatour P, Dejardin E, Warnier M et al. GSK3-mediated BCL-3 phosphorylation modulates its degradation and its oncogenicity. Mol Cell 2004; 16:35-45.
168. Ashburner BP, Westerheide SD, Baldwin Jr AS. The p65 (RelA) subunit of NF-kappaB interacts with the histone deacetylase (HDAC) corepressors HDAC1 and HDAC2 to negatively regulate gene expression. Mol Cell Biol 2001; 21:7065-7077.
169. Chen LF, Mu Y, Greene WC. Acetylation of RelA at discrete sites regulates distinct nuclear functions of NF-kappaB. EMBO J 2002; 21:6539-6548.
170. Chen LF, Greene WC. Regulation of distinct biological activities of the NF-kappaB transcription factor complex by acetylation. J Mol Med 2003; 81:549-557.
171. Kiernan R, Bres V, Ng RW et al. Post-activation turn-off of NF-kappa B-dependent transcription is regulated by acetylation of p65. J Biol Chem 2003; 278:2758-2766.
172. Chen E, Hrdlickova R, Nehyba J et al. Degradation of proto-oncoprotein c-Rel by the ubiquitin-proteasome pathway. J Biol Chem 1998; 273:35201-35207.
173. Saccani S, Marazzi I, Beg AA et al. Degradation of promoter-bound p65/RelA is essential for the prompt termination of the nuclear factor kappaB response. J Exp Med 2004; 200:107-113.
174. Ryo A, Suizu F, Yoshida Y et al. Regulation of NF-kappaB signaling by Pin1-dependent prolyl isomerization and ubiquitin-mediated proteolysis of p65/RelA. Mol Cell 2003; 12:1413-1426.
175. Campbell KJ, Perkins ND. Reprogramming RelA. Cell Cycle 2004; 3:869-872.
176. Perkins ND. NF-kappaB: Tumor promoter or suppressor? Trends Cell Biol 2004; 14:64-69.
177. Zhang JY, Green CL, Tao S et al. NF-kappaB RelA opposes epidermal proliferation driven by TNFR1 and JNK. Genes Dev 2004; 18:17-22.
178. Lind MH, Rozell B, Wallin RP et al. Tumor necrosis factor receptor 1-mediated signaling is required for skin cancer development induced by NF-kappaB inhibition. Proc Natl Acad Sci USA 2004; 101:4972-4977.

179. Dajee M, Lazarov M, Zhang JY et al. NF-κB blockade and oncogenic Ras trigger invasive human epidermal neoplasia. Nature 2003; 421:639-643.
180. van Hogerlinden M, Rozell BL, Ahrlund-Richter L et al. Squamous cell carcinomas and increased apoptosis in skin with inhibited Rel/nuclear factor-kappaB signaling. Cancer Res 1999; 59:3299-3303.
181. van Hogerlinden M, Auer G, Toftgard R. Inhibition of Rel/Nuclear Factor-kappaB signaling in skin results in defective DNA damage-induced cell cycle arrest and Ha-ras- and p53-independent tumor development. Oncogene 2002; 21:4969-4977.
182. Gapuzan ME, Yufit PV, Gilmore TD. Immortalized embryonic mouse fibroblasts lacking the RelA subunit of transcription factor NF-κB have a malignantly transformed phenotype. Oncogene 2002; 21:2484-2492.
183. Campbell KJ, Rocha S, Perkins ND. Active repression of antiapoptotic gene expression by RelA(p65) NF-kappa B. Mol Cell 2004; 13:853-865.
184. Wang CY, Mayo MW, Baldwin Jr AS. TNF- and cancer therapy-induced apoptosis: Potentiation by inhibition of NF-kappaB. Science 1996; 274:784-787.
185. Tergaonkar V, Bottero V, Ikawa M et al. IkappaB kinase-independent IkappaBalpha degradation pathway: Functional NF-kappaB activity and implications for cancer therapy. Mol Cell Biol 2003; 23:8070-8083.
186. Rocha S, Campbell KJ, Perkins ND. p53- and Mdm2-independent repression of NF-kappa B transactivation by the ARF tumor suppressor. Mol Cell 2003; 12:15-25.
187. Rocha S, Martin AM, Meek DW et al. p53 represses cyclin D1 transcription through down regulation of Bcl-3 and inducing increased association of the p52 NF-kappaB subunit with histone deacetylase 1. Mol Cell Biol 2003; 23:4713-4727.
188. Benezra M, Chevallier N, Morrison D et al. BRCA1 augments transcription by the NF-kappaB transcription factor by binding to the Rel domain of the p65/RelA subunit. J Biol Chem 2003; 278:26333-26341.
189. Garkavtsev I, Kozin SV, Chernova O et al. The candidate tumour suppressor protein ING4 regulates brain tumour growth and angiogenesis. Nature 2004; 428:328-332.
190. Ryan KM, Ernst MK, Rice NR et al. Role of NF-κB in p53-mediated programmed cell death. Nature 2000; 404:892-897.
191. Bohuslav J, Chen LF, Kwon H et al. p53 induces NF-kappaB activation by an IkappaB kinase-independent mechanism involving phosphorylation of p65 by ribosomal S6 kinase 1. J Biol Chem 2004; 279:26115-26125.
192. Fujioka S, Schmidt C, Sclabas GM et al. Stabilization of p53 is a novel mechanism for proapoptotic function of NF-kappaB. J Biol Chem 2004; 279:27549-27559.
193. Greten FR, Eckmann L, Greten TF et al. IKKβ links inflammation and tumorigenesis in a mouse model of colitis-associated cancer. Cell 2004; 118:285-296.
194. Pikarsky E, Porat RM, Stein I et al. NF-κB functions as a tumour promoter in inflammation-associated cancer. Nature 2004; 431:461-466.
195. Balkwill F, Coussens LM. An inflammatory link. Nature 2004; 431:405-406.

CHAPTER 10

NF-κB in Human Cancers

Elaine J. Schattner,* Richard R. Furman and Alejandro Bernal

Introduction

Clinical oncology is increasingly based in science, such that physicians are routinely administering drugs designed to inhibit specific growth and survival pathways in cancer cells. Perhaps the best-established targeted therapy is Imatinib Mesylate (Gleevec, STI571), a tyrosine kinase inhibitor that impedes signaling through the hybrid enzyme generated by genetic translocation of the *c-abl* protooncogene in chronic myelogenous leukemia (CML). This drug was promoted based on its capacity to inhibit the abnormal tyrosine kinase activity resulting from the translocation, and is now a standard treatment for patients with CML.[1-3] Of interest to the reader, it was investigators in David Baltimore's group who originally described *c-abl* and recognized its likely role in CML oncogenesis, ultimately leading to targeted therapy for patients with this disease.[4,5]

Nuclear Factor kappa B (NF-κB) proteins were initially reported in the mid-1980s as regulators of immunoglobulin (Ig) kappa gene transcription in B lymphocytes, also based on work in Baltimore's laboratory.[6,7] As a group, NF-κB and related proteins exert a profound influence on the transcriptional response to immune stimuli, thereby controlling both healthy and pathologic inflammatory reactions.[8-10] Over the past two decades, five mature mammalian NF-κB family members and their structural precursors have been elucidated. These are p50, p52, p65 (RelA), c-rel and RelB. NF-κB proteins are distinguished by a conserved region at the N-terminus of approximately 300 amino acids, termed the Rel Homology Domain (RHD). Like other transcriptional regulators, NF-κB proteins control gene expression by binding DNA at consensus sequences for which they have structural affinity.[11,12] It is through the RHDs that NF-κB molecules form dimers, interact with inhibitor of κB (IκB) proteins, enter the nucleus upon activation and, ultimately, bind DNA (Fig. 1).

Whereas in most cell types NF-κB enters the nucleus and induces transcription of pro-inflammatory genes upon activation of an IκB kinase (IKK), in mature B cells, macrophages and neurons some NF-κB proteins reside constitutively in the nucleus.[13-15] The human IκB molecules, including IκBα, IκBβ, IκBγ and IκBε contain characteristic ankyrin-repeat domains which afford their capacity to interact with other cytoplasmic proteins. As they bind and render NF-κB proteins inactive, IκB proteins are also key regulators of gene transcription, cell survival, and in some circumstances, tumor growth and potential therapy. Bcl-3, an atypical and inducible member of the IκB family, can directly regulate gene transcription by its interaction in the nucleus with p52.[13,16] It is noteworthy that the *bcl-3* locus was identified and cloned based on its translocation and over-expression in some cases of chronic B cell leukemia (CLL),[17,18] although the precise relevance of bcl-3 to CLL pathogenesis remains elusive.

The IKK complex includes IKKα and -β, as well as a protein that regulates activation of the complex by binding its other elements, termed NF-kappa B essential modulator (NEMO,

---

*Corresponding Author: Elaine J. Schattner—Department of Medicine, Weill Medical College of Cornell University, New York, New York, U.S.A. Email: ejsch@med.cornell.edu

NF-κB/Rel Transcription Factor Family, edited by Hsiou-Chi Liou.
©2006 Landes Bioscience and Springer Science+Business Media.

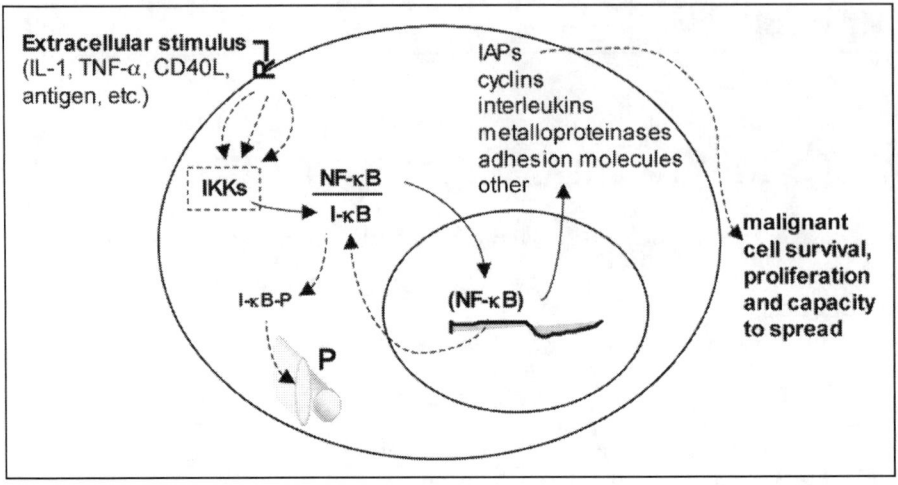

Figure 1. Extracellular stimuli and intracellular signals can lead to NF-κB activation, consequent gene transcription, cellular proliferation and survival. Inactive NF-κB is generated and cleaved in the cytosol, where it is lies bound to inhibitor, IκB. Upon triggering of cell surface receptors such as that for interleukin-1 (IL-1), CD40, the T cell receptor (TcR) or the B cell receptor (BCR) for antigen, cytoplasmic kinases activate the IκB kinase (IKK) complex. Upon specific phosphorylation of IκB, NF-κB is released from its inhibitor to enter the nucleus, and IκB is degraded along the ubiquitin-proteosome pathway (P). In the nucleus, NF-κB complexes regulate transcription of genes that promote cellular proliferation, adhesion and survival.

IKKγ.[19,20] NEMO in itself can regulate transcription by its capacity to enter the nucleus and interact there with p65 and cAMP-responsive element-binding protein-binding protein (CBP).[21] Because NEMO regulates NF-κB activity directly and indirectly through its "scaffolding" relationship with IKK-α and -β, NEMO-IKK interactions are important potential targets for pharmacology. For instance, the inflammatory responses engendered by IKKβ activation can be prevented by application of a small molecule designed precisely to block the IKK-NEMO binding site, termed NEMO-binding domain (NBD) peptide.[20] NEMO is subject to ubiquitination by bcl-10,[22] a molecule implicated in lymphoma pathogenesis,[23] and thereby may be vulnerable to some proteosome inhibitors as are now used in some B cell tumors including multiple myeloma (see below).

The link between NF-κB and cancer was established along several lines, including the early recognition of p50 and p65 as mammalian homologs of v-rel, the transforming gene of the avian reticuloendotheliosis virus (REV-T) and its cellular counterpart, c-rel.[24-28] In the 1960s, the c-rel locus was postulated as a potential factor in oncogenesis based on observed genetic abnormalities at this locus in chickens, turkeys and quail with lymphoid malignancies.[29,30] Over time it was determined that c-rel can transform hematopoietic cells in vitro and induce tumors in chickens.[31,32] In humans, c-rel sequences were localized to chromosome 2p1, the site of some chromosomal translocations in cancer.[33]

NF-κB Controls Cellular Proliferation, Adhesion and Survival

The significance of NF-κB activity in tumor cell growth and survival is now widely appreciated.[12,34-38] This is largely a function of the capacity of NF-κB to confer a proliferative and survival advantage through induction of cell-cycle and apoptosis regulatory genes. For oncologists, knowing how and when distinct NF-κB components are generated, activated such that they can enter the nucleus and bind DNA, and degraded, affords insight regarding current and future therapies based on NF-κB inhibition.

A number of cell cycle regulatory elements are NF-κB target genes (Fig. 1). One of the best-studied of these is cyclin D1.[16,39,40] Over-expression of cyclin D1 is implicated in mantle cell lymphoma, some forms of breast cancer, and other tumors.[41-47] Of interest, NF-κB-dependent cyclin D1 expression and consequent proliferation of mammary epithelial cells appears to be a physiologic response to NF-κB activation in pregnancy.[48,49] In other cell types, interleukins such as IL-6 and IL-8 are induced upon NF-κB activation, resulting in increased cell turnover and inflammation.[19,50-52] In B cells, c-rel is essential to cell cycle progression upon stimulation through surface IgM (sIgM).[53] Upon sIgM crosslinking, NF-κB triggers c-*myc* expression, fostering the proliferative response to antigen for which the B cell has affinity.[54] Induction of c-*myc* via NF-κB pathway activation is implicated in diverse cell types, including fibroblasts.[55] In theory, NF-κB-dependent induction of c-*myc* would be most relevant to pathogenesis and potential therapy of B-cell tumors in which malignant cell growth and survival is enhanced by engagement of the B cell receptor (BCR) by antigen for which the tumor-specific Ig has affinity, such as Burkitt's lymphoma and CLL.[56-59]

The capacity of NF-κB activity to inhibit apoptosis is central to its role in tumor development and resistance of cells to cancer treatment (Fig. 1). In this regard, one of the earliest observations about NF-κB function was that Rel A (p65) protects cells from apoptosis upon exposure to TNF-α, ionizing radiation, and the chemotherapeutic reagent daunorubicin.[60,61] Several anti-apoptotic bcl-2 family members, including bcl-x_L and bfl-1 (A1), are well-established NF-κB target genes;[62-66] through induction of these potent survival factors, stimulation of the NF-κB system enhances the propensity of cells to survive under stress. The gene encoding Bcl-2, the prototypic member of this key family of apoptosis regulatory molecules, contains several NF-κB binding motifs in its promotor. Some evidence suggests that bcl-2 transcription is turned on in the nucleus by p50 or p52 homodimers upon their association with bcl-3.[67] Similarly, the cellular inhibitor of apoptosis-1 (c-IAP-1) and -2, and X-linked inhibitor of apoptosis (XIAP) are induced by NF-κB.[68-70] In addition, both viral and cellular inhibitors of Fas-mediated apoptosis, v-*FLIP* (FLICE, Fas-associated death domain (FADD)-like interleukin-1 β-converting enzyme, caspase 8) Inhibitory Protein, and the cellular homolog, c-*FLIP*, are NF-κB target genes which promote survival of tumor cells by inhibition of caspase 8.[19] Overall, these observations reflect that NF-κB induces a powerful cell survival program, conferring apoptosis resistance in cells through induction of bcl-2 family members, IAPs, and other molecules that inhibit death of malignant cells.

NF-κB stimulates expression of adhesion molecules, metalloproteinases and other inflammatory mediators that play a role in tumor cell growth and dissemination. Adhesion molecules that are thought to facilitate metastatic spread and are NF-κB-dependent include intercellular adhesion molecule 1 (ICAM-1) and endothelial leukocyte adhesion molecule 1 (ELAM-1)[71-74] Among the inflammatory molecules, such powerful cytokines as Granulocyte Macrophage Colony Stimulating Factor (GM-CSF), IL-1, IL-8, and beta interferon (β-IFN) are induced by NF-κB.[75-77] Cyclooxygenase 2 (COX-2), implicated in pathologic inflammatory states and, in some cancer models, tumorogenesis, is also an NF-κB target gene.[78] The metalloproteinase-9 (MMP-9), which is involved in tumor metastatic potential and invasiveness, is also NF-κB -responsive.[79]

Taken together, these responses support that NF-κB activity has a role in formation of numerous tumor types by promoting cell survival, proliferation in some circumstances, and by generation of adhesion molecules rendering cells hardier and more aggressive in phenotype.

Tumors in Which NF-κB Is Implicated in Pathogenesis

Aberrant and increased NF-κB activity are implicated in a variety of human malignancies, including tumor types derived from cells of immune origin such as lymphocytes, monocytes, and myelocytes, as well as tumors arising from cells of nonimmune origin (Table 1). Among the solid tumors, abundant evidence links NF-κB to pathogenesis in breast cancer, hepatocellular carcinoma, renal cell carcinoma and other conditions. Several oncogenic viruses exert their effects, at least in part, by induction of NF-κB in malignant cells.

Table 1. Human tumors in which aberrant or increased NF-κB activity is implicated in pathogenesis

Leukemias and Lymphomas	
Adult T-cell leukemia lymphoma (ATLL)	80-82
Acute lymphoblastic leukemia (ALL)	83,84
Acute myeloid leukemia (AML)	85
Chronic lymphocytic leukemia (CLL)	86,87,123
Diffuse large B cell Lymphoma (B-DLCL)	88-90
Hodgkin's lymphoma	91-94
Mantle cell Lymphoma	95
Mucosa-associated lymphoid tissue (MALT) lymphoma	23,96-98
Multiple myeloma	99-101
Solid Tumors	
Brain cancer (glioma)	102
Breast cancer	48,103-105
Head and neck cancer	106
Gastric carcinoma	107
Hepatocellular carcinoma	108
Melanoma	109-111
Nasopharyngeal Ca	112
Pancreatic cancer	113-115
Prostate cancer	
Renal cell carcinoma	116
Thyroid cancer	117

B lymphocytes constitute the cell of origin in the majority of adult lymphomas, including both Hodgkin's and nonHodgkin's lymphomas, and in related tumors such as multiple myeloma and some forms of lymphocytic leukemia. Constitutive NF-κB activity is a distinctive feature of B lymphocytes and is due, at least in part, to constant degradation of IκB.[118] One physiologic stimulus of inducible NF-κB in B cells is CD40 ligand (CD154), a TNF-like molecule that is expressed in activated CD4+ T cells and, is itself, an NF-κB target gene.[119,120] Virtually all mature B cells, normal and malignant, express CD40, a cell surface receptor for CD40 ligand which promoted differentiation and survival of B cells.[121-123] Here, the pro-survival effects of CD40 ligation are mediated largely, but not completely, by NF-κB.[124-127]

Diffuse large cell lymphoma (DLCL) of B cell origin (B-DLCL) comprise approximately 30% of adult lymphomas. In B-DLCL tumor cells, autocrine expression and binding of the CD40 receptor and its ligand, CD154, can lead to continuous NF-κB activation.[89] Other lines of evidence point to activation of NF-κB in B-DLCL via alternative TNF receptors, such as the receptors for BAFF (B cell Activating Factor of the TNF Family) and APRIL (A Proliferation Induced Ligand). Ligation of BAFF and APRIL result in NF-κB induction and expression of NF-κB-dependent proliferation and survival factors such as c-myc and bcl-x_L, respectively.[90] In Hodgkin's lymphoma, the role of NF-κB is supported by the finding of p50 and p65 in the nuclei of Reed Sternberg cells in fresh tumor specimens.[128] Additional evidence stems from observations that constitutive NF-κB is necessary for growth of Reed Sternberg cell lines.[91] Aberrant, constitutive NF-κB activity in Hodgkin's lymphoma is associated with abnormal and mutated IκB-α.[92] In this tumor, signaling via CD30, another TNFR family member, is implicated.[93,94,129,130]

An interesting connection of NF-κB activity in inflammation and cancer occurs in MALT (Mucosal Associated Lymphoid Tumors) lymphomas, particularly in the stomach, where infection by the bacterium *helicobactor pylori* and subsequent inflammation play a role in tumor pathogenesis. Here, NF-κB is implicated at several levels, including the initial inflammatory response to the bacterium.[97,131] In the subset of MALT tumors containing the t(1;14) (p22;q32) involving the *bcl*-10 locus, a truncated form of bcl-10 is expressed that activates NF-κB and has transforming properties.[132] As considered above, bcl-10 functions as an adaptor protein targeting NEMO (IKK-γ) for ubiquitination;[22] this may result in high NF-κB activity and contribute to survival of these B lymphoma cells. Among MALT lymphomas, nuclear NF-κB activity appears to correlate with the t(1;14) translocation detection of, bcl-10 in the cytoplasm, and tumor aggressiveness.[133]

In chronic lymphocytic leukemia (CLL), another tumor of B lymphocytes, our group has demonstrated that the malignant B cells have high levels of active, nuclear NF-κB immediately upon isolation from patients' blood.[86] The most evident activity comprises p50, p65 and c-rel, consistent with the postulated responsiveness of CLL cells to extracellular, inducible stimuli in vivo.[58] Consistent with this hypothesis, NF-κB activity dissipates over a period of hours upon isolation of the cells ex vivo, but this activity and survival can be rescued by CD40 engagement of CLL cells in vitro. NF-κB in CLL cells is enhanced upon simultaneous stimulation of the BCR, which may in itself, be a modest inducer of NF-κB in CLL.[57] Based on these observations, we put forth a model for CLL pathogenesis based on chronic stimulation of the tumor cells in vivo by such factors as CD40 ligand and antigen for which the tumor cells have avidity.[123] In vivo, CLL cells maintain high levels of inducible NF-κB due to constant costimulatory and BCR-derived signals, leading to an intensely anti-apoptotic phenotype and drug resistance.[58] Among the NF-κB-target genes expressed in CLL are the anti-apoptotic bcl-2 family members bcl-x_L, bfl-1 (A1) and bcl-2,[57,67] cIAP-1 and -2, and XIAP (A. Bernal and E. Schattner, unpublished data). Other investigators have shown that in the microenvironment of CLL proliferation centers, CLL express survivin. This effect is located in the bone marrow and appears to be mediated by tumor infiltrating CD4+ T cells expressing CD40L.[134]

One of the first tumors for which NF-κB was considered as a therapeutic target is multiple myeloma, a cancer of malignant plasma cells.[99] In multiple myeloma, NF-κB dependent IL-6 is a potent growth factor, and additional NF-κB-mediated transcripts confer resistance to apoptotic stimuli such as TNF-α.[100] Bortezomib (PS341) was the first FDA-approved drug targeting the proteosome, a large, cytosolic multi-component protein complex that degrades proteins marked for degradation by a ubiquitin ligase. Among the proteins degraded by the proteosome is IκB. Based on a vast amount of in vitro data supporting the capacity of PS341 to promote cell death of myeloma cells in vitro, Bortezomib was tested in humans and determined to be effective in patients with refractory cases of multiple myeloma.[101,135] Subsequent studies of Bortezomib have demonstrated activity in numerous malignancies including solid tumors.[136] However, one problem that applies to the interpretation of these studies is that the proteosome degrades many cytoplasmic proteins, including some with pro-apoptotic functions. It remains to be formally demonstrated that altered NF-κB is responsible for the efficacy of Bortezomib in patients with myeloma.

Virus-Associated Tumors

Viruses can turn on NF-κB activity in tumor cells by coopting normal cellular activation pathways. One of the best-delineated NF-κB-activation pathways occurs in adult human T cell leukemia lymphoma (ATLL). In this rare neoplasm, the malignant CD4+ T cells are infected by human T-cell lymphotrophic virus-1 (HTLV-1), of which the *tax* gene product induces NF-κB by activating IKK.[80-82,137] The viral tax protein effects T-cell survival and proliferation through a variety of pathways, including NF-κB-dependent induction of *c-myc*.[138]

Viral-mediated NF-κB activation contributes to pathogenesis in B cell tumors infected with Epstein Barr Virus (EBV). Here, the EBV Latent Membrane Protein (LMP)-1 interacts directly with the cytosolic domain of CD40, so as to activate IKKβ, induce expression of bfl-1,

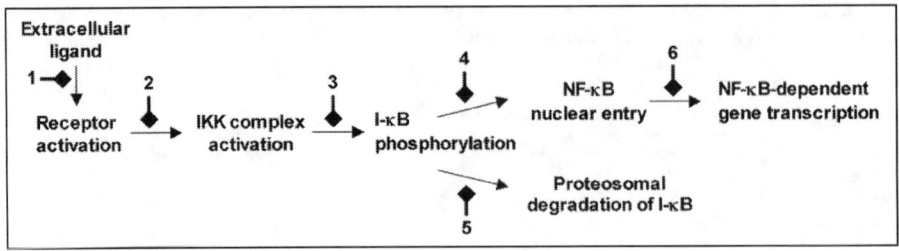

Figure 2. Blocking NF-κB in cancer. Potential sites for disruption of NF-κB activation and downstream effects include 1) impeding ligand-receptor interaction at the cell surface, such as by application of antibody to the ligand or to the receptor; 2) dampening activation of the IKK, such as by use of a small peptide that binds NEMO (NF-κB essential modifier, IKKγ) where it associates with IKKα and IKKβ; 3) preventing phosphorylation of IκB, such as by introduction a "super-repressor" IκB molecule that lacks key phosphorylation sites; 4) preventing NF-κB nuclear entry, possibly through the identification and targeting of a putative NF-κB chaperone protein; 5) reducing IκB degradation in the proteosome; 6) targeting NF-κB activity in the nucleus, such as by interference with RHD domains or implementation of antisense to particular NF-κB target genes.

and enhance B-cell survival.[139] NF-κB activation along this route is thought to be an oncogenic factor in EBV-containing tumors including post-transplant lymphoproliferative disorders (PTLDs), some cases of Burkitt's lymphoma, Hodgkin's lymphoma, and primary effusion lymphomas (PELs). In human immunodeficiency virus (HIV)-associated PELs, infected with HHV-8 and in some cases also with EBV, NF-κB activity is high.[140] Exposure of PEL cells to Bay-11, an NF-κB inhibitor, results in PEL cell death in vitro and in vivo, further supporting the role of NF-κB in proliferation and survival of virally-infected malignant cells.[140,141]

Virus-mediated induction of NF-κB is relevant also in pathogenesis of some solid tumors. For example, in patients with hepatocellular carcinoma and hepatitis B, expression of the hepatitis X protein (HBx) is related to increased NF-κB activity, IL-8 production, and tumor aggressiveness.[108,142] In nasopharyngeal cancer associated with EBV, p50 homodimers are present in the nucleus bound to the transcriptional coactivator Bcl-3. Chromatin immunoprecipitation experiments suggest that the p50 homodimers are bound to NF-κB consensus motifs within the promoter for the endothelial growth factor receptor (EGFR).[112]

The role of NF-κB in breast cancer is evident from several lines of investigation. As considered above, in mammary epithelial cells cyclin D1 expression and cellular proliferation depend on NF-κB.[48] In breast cancer cells stimulated to proliferate in vitro, application of the IKK inhibitory peptide NBD blocked NF-κB induction and the cells died.[104] Some studies have established that high levels of nuclear p65 in biopsy specimens are most evident in tumors that express ErbB2 (Her2/neu), the molecule to which the commonly-used therapeutic antibody Trastuzumab (Herceptin) binds. In these cases, exposure of tumor cells to Trastuzumab resulted in loss of NF-κB activity and cell death. Studies of NF-κB in breast cancer by another group point to the p50 and p52 proteins, as well as bcl-3, rather than p65, in breast cancer pathogenesis.[143] Transfection of mammary tumor cell line with a dominant negative IκB, or with an NF-κB decoy, render cells vulnerable to apoptosis.

Taken together these data support that NF-κB activity contributes to chemotherapy resistance in a variety of tumor types.[34,103]

Inhibition of NF-κB in Cancer Treatment

In principle, interruption of NF-κB activity in tumor cells can be achieved at multiple levels (Fig. 2). Treatment might be aimed at the cell surface, where NF-κB can be activated by particular receptors that could be blocked using appropriate antibodies. For example, in a tumor in which CD40 engagement promotes survival largely via NF-κB-mediated effects, blocking that receptor might be sufficient to allow apoptosis of otherwise resistant tumor cells.

Table 2. NF-κB inhibitors in humans

A. Current, FDA-approved pharmacologic agents that inhibit NF-κB activity:

Substance	Mechanism	Indication	Refs.
arsenic trioxide	inhibition of IKK	leukemia	146-148
Bortezomib	proteosome inhibition	multiple myeloma	149,150
butyrate	proteosome inhibition	investigational	151
cox-2 inhibitors	inhibition of IKK inhibition of DNA binding	anti-inflammatory	152
glucocorticoids	increase IκB synthesis interference with DNA binding	anti-inflammatory	153-158
N-acetylcysteine	oxygen radical scavenging	acetaminophen overdose	159
rapamycin	inhibition of IKK	immune suppressant	160,161
salicylates: including sodium salicylate and acetylsalicylic acid	inhibition of IκB degradation inhibition of TNF signaling inhibition of IKK	anti-inflammatory, inflammatory bowel disease	
sulfasalazine	inhibition of IκB phosphorylation inhibition of IKK		
mesalamine	inhibition of IκB phosphorylation inhibition of RelA phosphorylation		
sulindac	inhibition of IKK		162-170
thalidomide	inhibition of IKK	leprosy	171
ursodeoxycholic acid (URSA)	interferes with p65-glucocorticoid receptor interaction	primary biliary cirrhosis	172

Table continued on next page

Alternatively, therapy may be based on the capacity of some compounds to prevent degradation of IκB, such as is now accomplished in a relatively nonspecific manner by means of proteosomal inhibition (Table 2). Other treatments might target enzymes of the IKK complex,[144] such as the NBD peptide and its effects on NEMO. Other therapies might block NF-κB nuclear entry, which may be feasible once more is understood about the requirements and possible chaperoning of specific NF-κB proteins when they enter the nucleus. Finally, treatments can be designed to impede NF-κB binding to consensus DNA sequences in the nucleus and subsequent gene transcription. For example, downregulation of p65 expression by RNA interference may enhance sensitivity of some tumor cells to chemotherapeutic drugs such as irinotecan.[145]

Conclusions

Twenty years since their discovery, our knowledge regarding the specific role of NF-κB proteins in particular tumor types, the interaction of these proteins with other transcriptional regulators in the nucleus, and our ability to tract and abrogate NF-κB activity, effectively and specifically, remain the subject of intense research efforts in cancer therapy. While NF-κB activity is the principle subject of this chapter and the target of several early clinical trials conceptually aimed at transcriptional regulation, other transcription factors also are potential targets for therapy. Understanding the relationship between NF-κB activity with other, parallel and interacting signaling pathways, is essential to development of effective and safe targeted therapies in cancer treatment.

Table 2. Continued

B. Natural compounds that inhibit NF-κB activity

Substance	Mechanism	Source	Refs.
anethole	inhibition of IκB degradation	fennel, anise	173,174
caffeic acid phenethyl (CAPE)	inhibition of NF-κB translocation	tree resin	175
capsaicin	Inhibition of IκB degradation	red chili peppers	176
curcumin	inhibition of IKK	curry	177,178
dicoumarol	inhibition of NF-κB activation	sweet clover	179
E330	inhibition of IκB degradation, suppression of oxygen radical production, interferes with DNA binding	quinone derivative	180,181
EGCG	inhibition of IKK	teas	182
epoxomicin	proteosome inhibitor	Actinomycetes	183
eugenol	inhibition of IκB degradation	cloves	174
flavonoids	inhibition of NF-κB	pomegranates	184
genistein	inhibition of IkB degradation	legumes	185,186

Substance	Mechanism	Source	Refs.
oleandrin	inhibition of IκB phosphorylation	plant leaves	187
resveratrol	inhibition of IKK	grapes	188,189
sesquiterpene lactones	inhibition of IκB degradation	plant remedies	190
sanguinarine	inhibition of IκB phosphorylation	root	191
sulforaphane	inhibition of DNA binding	cruciferous vegetables	192
silymarin	inhibition of IκB phosphorylation		193
s-allylcysteine	inhibition of NF-κB activation	garlic	194
theaflavins	inhibition of IKK	teas	182
ursolic acid	inhibition of IKK	berries, basil, rosemary, fruits	195

References

1. Topaly J, Zeller WJ, Fruehauf S. Synergistic activity of the new ABL-specific tyrosine kinase inhibitor STI571 and chemotherapeutic drugs on BCR-ABL-positive chronic myelogenous leukemia cells. Leukemia 2001; 15(3):342-347.
2. Druker BJ, Talpaz M, Resta DJ et al. Efficacy and safety of a specific inhibitor of the BCR-ABL tyrosine kinase in chronic myeloid leukemia synergistic activity of the new ABL-specific tyrosine kinase inhibitor STI571 and chemotherapeutic drugs on BCR-ABL-positive chronic myelogenous leukemia cells. N Engl J Med 2001; 344(14):1031-1037.
3. Kantarjian H, Sawyers C, Hochhaus A et al. Hematologic and cytogenetic responses to imatinib mesylate in chronic myelogenous leukemia efficacy and safety of a specific inhibitor of the BCR-ABL tyrosine kinase in chronic myeloid leukemia synergistic activity of the new ABL-specific tyrosine kinase inhibitor STI571 and chemotherapeutic drugs on BCR-ABL-positive chronic myelogenous leukemia cells. N Engl J Med 2002; 346(9):645-652.
4. Wang JY, Baltimore D. Cellular RNA homologous to the Abelson murine leukemia virus transforming gene: Expression and relationship to the viral sequence. Mol Cell Biol 1983; 3(5):773-779.
5. Bernards A, Rubin CM, Westbrook CA et al. The first intron in the human c-abl gene is at least 200 kilobases long and is a target for translocations in chronic myelogenous leukemia. Mol Cell Biol 1987; 7(9):3231-3236.
6. Sen R, Baltimore D. Multiple nuclear factors interact with the immunoglobulin enhancer sequences. Cell 1986; 46(5):705-716.

7. Lenardo M, Pierce JW, Baltimore D. Protein-binding sites in Ig gene enhancers determine transcriptional activity and inducibility. Science 1987; 236(4808):1573-1577.
8. Li Q, Verma IM. NF-kappaB regulation in the immune system. Nat Rev Immunol 2002; 2(10):725-734.
9. Burke JR. Targeting I kappa B kinase for the treatment of inflammatory and other disorders. Curr Opin Drug Discov Devel 2003; 6(5):720-728.
10. Tak PP, Firestein GS. NF-kappaB: A key role in inflammatory diseases. J Clin Invest 2001; 107(1):7-11.
11. Ghosh S, May MJ, Kopp EB. NF-kappa B and rel proteins: Evolutionarily conserved mediators of immune responses. Annu Rev Immunol 1998; 16:225-260.
12. Darnell Jr JE. Transcription factors as targets for cancer therapy. Nat Rev Cancer 2002; 2(10):740-749.
13. Franzoso G, Carlson L, Poljak L et al. Mice deficient in nuclear factor (NF)-kappa B/p52 present with defects in humoral responses, germinal center reactions, and splenic microarchitecture. J Exp Med 1998; 187(2):147-159.
14. Liou HC, Sha WC, Scott ML et al. Sequential induction of NF-kappa B/Rel family proteins during B-cell terminal differentiation. Mol Cell Biol 1994; 14(8):5349-5359.
15. Kaltschmidt C, Kaltschmidt B, Neumann H et al. Constitutive NF-kappa B activity in neurons. Mol Cell Biol 1994; 14(6):3981-3992.
16. Westerheide SD, Mayo MW, Anest V et al. The putative oncoprotein Bcl-3 induces cyclin D1 to stimulate G(1) transition. Mol Cell Biol 2001; 21(24):8428-8436.
17. Kerr LD, Duckett CS, Wamsley P et al. The proto-oncogene bcl-3 encodes an I kappa B protein. Genes Dev 1992; 6(12A):2352-2363.
18. Wulczyn FG, Naumann M, Scheidereit C. Candidate proto-oncogene bcl-3 encodes a subunit-specific inhibitor of transcription factor NF-kappa B. Nature 1992; 358(6387):597-599.
19. Li X, Massa PE, Hanidu A et al. IKKalpha, IKKbeta, and NEMO/IKKgamma are each required for the NF-kappa B-mediated inflammatory response program. J Biol Chem 2002; 277(47):45129-45140, (Epub 42002 Sep 45126).
20. May MJ, D'Acquisto F, Madge LA et al. Selective inhibition of NF-kappaB activation by a peptide that blocks the interaction of NEMO with the IkappaB kinase complex. Science 2000; 289(5484):1550-1554.
21. Verma UN, Yamamoto Y, Prajapati S et al. Nuclear role of I kappa B Kinase-gamma/NF-kappa B essential modulator (IKK gamma/NEMO) in NF-kappa B-dependent gene expression. J Biol Chem 2004; 279(5):3509-3515, (Epub 2003 Nov 3503).
22. Zhou H, Wertz I, O'Rourke K et al. Bcl10 activates the NF-kappaB pathway through ubiquitination of NEMO. Nature 2004; 427(6970):167-171, (Epub 2003 Dec 2024).
23. Ohshima K, Muta H, Kawasaki C et al. Bcl10 expression, rearrangement and mutation in MALT lymphoma: Correlation with expression of nuclear factor-kappaB. Int J Oncol 2001; 19(2):283-289.
24. Ghosh S, Gifford AM, Riviere LR et al. Cloning of the p50 DNA binding subunit of NF-kappa B: Homology to rel and dorsal. Cell 1990; 62(5):1019-1029.
25. Kieran M, Blank V, Logeat F et al. The DNA binding subunit of NF-kappa B is identical to factor KBF1 and homologous to the rel oncogene product. Cell 1990; 62(5):1007-1018.
26. Ruben SM, Dillon PJ, Schreck R et al. Isolation of a rel-related human cDNA that potentially encodes the 65-kD subunit of NF-kappa B. Science 1991; 251(5000):1490-1493.
27. Nolan GP, Ghosh S, Liou HC et al. DNA binding and I kappa B inhibition of the cloned p65 subunit of NF-kappa B, a rel-related polypeptide. Cell 1991; 64(5):961-969.
28. Steward R. Dorsal, an embryonic polarity gene in Drosophila, is homologous to the vertebrate proto-oncogene, c-rel. Science 1987; 238(4827):692-694.
29. Theilen GH, Zeigel RF, Twiehaus MJ. Biological studies with RE virus (strain T) that induces reticuloendotheliosis in turkeys, chickens, and Japanese quail. J Natl Cancer Inst 1966; 37(6):731-743.
30. Gilmore TD. Multiple mutations contribute to the oncogenicity of the retroviral oncoprotein v-Rel. Oncogene 1999; 18(49):6925-6937.
31. Sylla BS, Temin HM. Activation of oncogenicity of the c-rel proto-oncogene. Mol Cell Biol 1986; 6(12):4709-4716.
32. Gilmore TD, Cormier C, Jean-Jacques J et al. Malignant transformation of primary chicken spleen cells by human transcription factor c-Rel. Oncogene 2001; 20(48):7098-7103.
33. Brownell E, O'Brien SJ, Nash WG et al. Genetic characterization of human c-rel sequences. Mol Cell Biol 1985; 5(10):2826-2831.
34. Lin A, Karin M. NF-kappaB in cancer: A marked target. Semin Cancer Biol 2003; 13(2):107-114.

35. Karin M, Cao Y, Greten FR et al. NF-kappaB in cancer: From innocent bystander to major culprit. Nat Rev Cancer 2002; 2(4):301-310.
36. Aggarwal BB. Nuclear factor-kappaB: The enemy within. Cancer Cell 2004; 6(3):203-208.
37. Panwalkar A, Verstovsek S, Giles F. Nuclear factor-kappaB modulation as a therapeutic approach in hematologic malignancies. Cancer 2004; 100(8):1578-1589.
38. Debatin KM. Apoptosis pathways in cancer and cancer therapy. Cancer Immunol Immunother 2004; 53(3):153-159, (Epub 2004 Jan 2029).
39. Guttridge DC, Albanese C, Reuther JY et al. NF-kappaB controls cell growth and differentiation through transcriptional regulation of cyclin D1. Mol Cell Biol 1999; 19(8):5785-5799.
40. Hinz M, Krappmann D, Eichten A et al. NF-kappaB function in growth control: Regulation of cyclin D1 expression and G0/G1-to-S-phase transition. Mol Cell Biol 1999; 19(4):2690-2698.
41. Bosch F, Jares P, Campo E et al. PRAD-1/cyclin D1 gene overexpression in chronic lymphoproliferative disorders: A highly specific marker of mantle cell lymphoma. Blood 1994; 84(8):2726-2732.
42. de Boer CJ, van Krieken JH, Kluin-Nelemans HC et al. Cyclin D1 messenger RNA overexpression as a marker for mantle cell lymphoma. Oncogene 1995; 10(9):1833-1840.
43. Lebwohl DE, Muise-Helmericks R, Sepp-Lorenzino L et al. A truncated cyclin D1 gene encodes a stable mRNA in a human breast cancer cell line. Oncogene 1994; 9(7):1925-1929.
44. Musgrove EA, Lee CS, Buckley MF et al. Cyclin D1 induction in breast cancer cells shortens G1 and is sufficient for cells arrested in G1 to complete the cell cycle. Proc Natl Acad Sci USA 1994; 91(17):8022-8026.
45. Zukerberg LR, Yang WI, Gadd M et al. Cyclin D1 (PRAD1) protein expression in breast cancer: Approximately one-third of infiltrating mammary carcinomas show overexpression of the cyclin D1 oncogene. Mod Pathol 1995; 8(5):560-567.
46. Lovec H, Sewing A, Lucibello FC et al. Oncogenic activity of cyclin D1 revealed through cooperation with Ha-ras: Link between cell cycle control and malignant transformation. Oncogene 1994; 9(1):323-326.
47. Ewen ME, Lamb J. The activities of cyclin D1 that drive tumorigenesis. Trends Mol Med 2004; 10(4):158-162.
48. Cao Y, Bonizzi G, Seagroves TN et al. IKKalpha provides an essential link between RANK signaling and cyclin D1 expression during mammary gland development. Cell 2001; 107(6):763-775.
49. Cao Y, Karin M. NF-kappaB in mammary gland development and breast cancer. J Mammary Gland Biol Neoplasia 2003; 8(2):215-223.
50. Zhang YH, Lin JX, Vilcek J. Interleukin-6 induction by tumor necrosis factor and interleukin-1 in human fibroblasts involves activation of a nuclear factor binding to a kappa B-like sequence. Mol Cell Biol 1990; 10(7):3818-3823.
51. Libermann TA, Baltimore D. Activation of interleukin-6 gene expression through the NF-kappa B transcription factor. Mol Cell Biol 1990; 10(5):2327-2334.
52. Shimizu H, Mitomo K, Watanabe T et al. Involvement of a NF-kappa B-like transcription factor in the activation of the interleukin-6 gene by inflammatory lymphokines. Mol Cell Biol 1990; 10(2):561-568.
53. Cheng S, Hsia CY, Leone G et al. Cyclin E and Bcl-xL cooperatively induce cell cycle progression in c-Rel-/- B cells. Oncogene 2003; 22(52):8472-8486.
54. La Rosa FA, Pierce JW, Sonenshein GE. Differential regulation of the c-myc oncogene promoter by the NF-kappa B rel family of transcription factors. Mol Cell Biol 1994; 14(2):1039-1044.
55. Kessler DJ, Duyao MP, Spicer DJ et al. NF-kappa B-like factors mediate interleukin 1 induction of c-myc gene transcription in fibroblasts. J Exp Med 1992; 176(3):787-792.
56. Schattner EJ, Elkon KB, Yoo DH et al. CD40 ligation induces Apo-1/Fas expression on human B lymphocytes and facilitates apoptosis through the Apo-1/Fas pathway. J Exp Med 1995; 182(5):1557-1565.
57. Bernal A, Pastore RD, Asgary Z et al. Survival of leukemic B cells promoted by engagement of the antigen receptor. Blood 2001; 98(10):3050-3057.
58. Schattner EJ. Apoptosis in lymphocytic leukemias and lymphomas. Cancer Invest 2002; 20(5-6):737-748.
59. Schattner EJ, Friedman SM, Casali P. Inhibition of Fas-mediated apoptosis by antigen: Role in lymphoma pathogenesis. Autoimmunity 2002; 35(4):283-289.
60. Wang CY, Mayo MW, Baldwin Jr AS. TNF- and cancer therapy-induced apoptosis: Potentiation by inhibition of NF-kappaB. Science 1996; 274(5288):784-787.
61. Beg AA, Baltimore D. An essential role for NF-kappaB in preventing TNF-alpha-induced cell death. Science 1996; 274(5288):782-784.

62. Edelstein LC, Lagos L, Simmons M et al. NF-kappa B-dependent assembly of an enhanceosome-like complex on the promoter region of apoptosis inhibitor Bfl-1/A1. Mol Cell Biol 2003; 23(8):2749-2761.
63. Chen C, Edelstein LC, Gelinas C. The Rel/NF-kappaB family directly activates expression of the apoptosis inhibitor Bcl-x(L). Mol Cell Biol 2000; 20(8):2687-2695.
64. Zong WX, Edelstein LC, Chen C et al. The prosurvival Bcl-2 homolog Bfl-1/A1 is a direct transcriptional target of NF-kappaB that blocks TNFalpha-induced apoptosis. Genes Dev 1999; 13(4):382-387.
65. Wang CY, Guttridge DC, Mayo MW et al. NF-kappaB induces expression of the Bcl-2 homologue A1/Bfl-1 to preferentially suppress chemotherapy-induced apoptosis. Molecular & Cellular Biology 1999; 19(9):5923-5929.
66. Bui NT, Livolsi A, Peyron JF et al. Activation of nuclear factor kappaB and Bcl-x survival gene expression by nerve growth factor requires tyrosine phosphorylation of IkappaBalpha. J Cell Biol 2001; 152(4):753-764.
67. Viatour P, Bentires-Alj M, Chariot A et al. NF- kappa B2/p100 induces Bcl-2 expression. Leukemia 2003; 17(7):1349-1356.
68. Chu ZL, McKinsey TA, Liu L et al. Suppression of tumor necrosis factor-induced cell death by inhibitor of apoptosis c-IAP2 is under NF-kappaB control. Proc Natl Acad Sci USA 1997; 94(19):10057-10062.
69. Wang CY, Mayo MW, Korneluk RG et al. NF-kappaB antiapoptosis: Induction of TRAF1 and TRAF2 and c-IAP1 and c- IAP2 to suppress caspase-8 activation. Science 1998; 281(5383):1680-1683.
70. Stehlik C, de Martin R, Kumabashiri I et al. Nuclear factor (NF)-kappaB-regulated X-chromosome-linked iap gene expression protects endothelial cells from tumor necrosis factor alpha-induced apoptosis. J Exp Med 1998; 188(1):211-216.
71. Voraberger G, Schafer R, Stratowa C. Cloning of the human gene for intercellular adhesion molecule 1 and analysis of its 5'-regulatory region. Induction by cytokines and phorbol ester. J Immunol 1991; 147(8):2777-2786.
72. van de Stolpe A, Caldenhoven E, Stade BG et al. 12-O-tetradecanoylphorbol-13-acetate- and tumor necrosis factor alpha-mediated induction of intercellular adhesion molecule-1 is inhibited by dexamethasone. Functional analysis of the human intercellular adhesion molecular-1 promoter. J Biol Chem 1994; 269(8):6185-6192.
73. Whelan J, Ghersa P, Hooft van Huijsduijnen R et al. An NF kappa B-like factor is essential but not sufficient for cytokine induction of endothelial leukocyte adhesion molecule 1 (ELAM-1) gene transcription. Nucleic Acids Res 1991; 19(10):2645-2653.
74. Schindler U, Baichwal VR. Three NF-kappa B binding sites in the human E-selectin gene required for maximal tumor necrosis factor alpha-induced expression. Mol Cell Biol 1994; 14(9):5820-5831.
75. Schreck R, Baeuerle PA. NF-kappa B as inducible transcriptional activator of the granulocyte-macrophage colony-stimulating factor gene. Mol Cell Biol 1990; 10(3):1281-1286.
76. Hiscott J, Alper D, Cohen L et al. Induction of human interferon gene expression is associated with a nuclear factor that interacts with the NF-kappa B site of the human immunodeficiency virus enhancer. J Virol 1989; 63(6):2557-2566.
77. Lenardo MJ, Fan CM, Maniatis T et al. The involvement of NF-kappa B in beta-interferon gene regulation reveals its role as widely inducible mediator of signal transduction. Cell 1989; 57(2):287-294.
78. Yamamoto K, Arakawa T, Ueda N et al. Transcriptional roles of nuclear factor kappa B and nuclear factor-interleukin-6 in the tumor necrosis factor alpha-dependent induction of cyclooxygenase-2 in MC3T3-E1 cells. J Biol Chem 1995; 270(52):31315-31320.
79. Bond M, Fabunmi RP, Baker AH et al. Synergistic upregulation of metalloproteinase-9 by growth factors and inflammatory cytokines: An absolute requirement for transcription factor NF-kappa B. FEBS Lett 1998; 435(1):29-34.
80. Geleziunas R, Ferrell S, Lin X et al. Human T-cell leukemia virus type 1 Tax induction of NF-kappaB involves activation of the IkappaB kinase alpha (IKKalpha) and IKKbeta cellular kinases. Mol Cell Biol 1998; 18(9):5157-5165.
81. Yin MJ, Christerson LB, Yamamoto Y et al. HTLV-I Tax protein binds to MEKK1 to stimulate IkappaB kinase activity and NF-kappaB activation. Cell 1998; 93(5):875-884.
82. O'Mahony AM, Montano M, Van Beneden K et al. Human T-cell lymphotropic virus type 1 tax induction of biologically Active NF-kappaB requires IkappaB kinase-1-mediated phosphorylation of RelA/p65. J Biol Chem 2004; 279(18):18137-18145, (Epub 12004 Feb 18112).

83. Weston VJ, Austen B, Wei W et al. Apoptotic resistance to ionizing radiation in pediatric B-precursor acute lymphoblastic leukemia frequently involves increased NF-kappaB survival pathway signaling. Blood 2004; 104(5):1465-1473, (Epub 2004 May 1413).
84. Munzert G, Kirchner D, Ottmann O et al. Constitutive NF-kappab/Rel activation in philadelphia chromosome positive (Ph+) acute lymphoblastic leukemia (ALL). Leuk Lymphoma 2004; 45(6):1181-1184.
85. Bueso-Ramos CE, Rocha FC, Shishodia S et al. Expression of constitutively active nuclear-kappa B RelA transcription factor in blasts of acute myeloid leukemia. Hum Pathol 2004; 35(2):246-253.
86. Furman RR, Asgary Z, Mascarenhas JO et al. Modulation of NF-kappa B activity and apoptosis in chronic lymphocytic leukemia B cells. J Immunol 2000; 164(4):2200-2206.
87. Cuni S, Perez-Aciego P, Perez-Chacon G et al. A sustained activation of PI3K/NF-kappaB pathway is critical for the survival of chronic lymphocytic leukemia B cells. Leukemia 2004; 18(8):1391-1400.
88. Davis RE, Brown KD, Siebenlist U et al. Constitutive nuclear factor kB activity is required for survival of activated B cell-like diffuse large B cell lymphoma cells. J Exp Med 2001; 194:1861-1874.
89. Pham LV, Tamayo AT, Yoshimura LC et al. A CD40 Signalosome anchored in lipid rafts leads to constitutive activation of NF-kappaB and autonomous cell growth in B cell lymphomas. Immunity 2002; 16(1):37-50.
90. He B, Chadburn A, Jou E et al. Lymphoma B cells evade apoptosis through the TNF family members BAFF/BLyS and APRIL. J Immunol 2004; 172(5):3268-3279.
91. Bargou RC, Emmerich F, Krappmann D et al. Constitutive nuclear factor-kappaB-RelA activation is required for proliferation and survival of Hodgkin's disease tumor cells. J Clin Invest 1997; 100(12):2961-2969.
92. Emmerich F, Meiser M, Hummel M et al. Overexpression of I kappa B alpha without inhibition of NF-kappaB activity and mutations in the I kappa B alpha gene in Reed-Sternberg cells. Blood 1999; 94(9):3129-3134.
93. Horie R, Watanabe T, Morishita Y et al. Ligand-independent signaling by overexpressed CD30 drives NF-kappaB activation in Hodgkin-Reed-Sternberg cells. Oncogene 2002; 21(16):2493-2503.
94. Hinz M, Loser P, Mathas S et al. Constitutive NF-kappaB maintains high expression of a characteristic gene network, including CD40, CD86, and a set of antiapoptotic genes in Hodgkin/Reed-Sternberg cells. Blood 2001; 97(9):2798-2807.
95. Pham LV, Tamayo AT, Yoshimura LC et al. Inhibition of constitutive NF-kappa B activation in mantle cell lymphoma B cells leads to induction of cell cycle arrest and apoptosis. J Immunol 2003; 171(1):88-95.
96. Lucas PC, Yonezumi M, Inohara N et al. Bcl10 and MALT1, independent targets of chromosomal translocation in malt lymphoma, cooperate in a novel NF-kappa B signaling pathway. J Biol Chem 2001; 276(22):19012-19019, (Epub 12001 Mar 19021).
97. Gascoyne RD. Molecular pathogenesis of mucosal-associated lymphoid tissue (MALT) lymphoma. Leuk Lymphoma 2003; 44(Supp l3):S13-20.
98. Stoffel A, Chaurushiya M, Singh B et al. Activation of NF-kappaB and inhibition of p53-mediated apoptosis by API2/mucosa-associated lymphoid tissue 1 fusions promote oncogenesis. Proc Natl Acad Sci USA 2004; 101(24):9079-9084, (Epub 2004 Jun 9077).
99. Chauhan D, Uchiyama H, Akbarali Y et al. Multiple myeloma cell adhesion-induced interleukin-6 expression in bone marrow stromal cells involves activation of NF-kappa B. Blood 1996; 87(3):1104-1112.
100. Hideshima T, Chauhan D, Hayashi T et al. Proteasome inhibitor PS-341 abrogates IL-6 triggered signaling cascades via caspase-dependent downregulation of gp130 in multiple myeloma. Oncogene 2003; 22(52):8386-8393.
101. Mitsiades N, Mitsiades CS, Poulaki V et al. Molecular sequelae of proteasome inhibition in human multiple myeloma cells. Proc Natl Acad Sci USA 2002; 99(22):14374-14379, (Epub 12002 Oct 14321).
102. Tran NL, McDonough WS, Savitch BA et al. The tumor necrosis factor-like weak inducer of apoptosis (TWEAK)-fibroblast growth factor-inducible 14 (Fn14) signaling system regulates glioma cell survival via NFkappaB pathway activation and BCL-XL/BCL-W expression. J Biol Chem 2005; 280:3483-92.
103. Zahir N, Lakins JN et al. Autocrine laminin-5 ligates alpha6beta4 integrin and activates RAC and NFkappaB to mediate anchorage-independent survival of mammary tumors. Cell Biol 2003; 163(6):1397-1407.
104. Biswas DK, Shi Q, Baily S et al. NF-kappa B activation in human breast cancer specimens and its role in cell proliferation and apoptosis. Proc Natl Acad Sci USA 2004; 101(27):10137-10142, (Epub 12004 Jun 10125).

105. Lin MT, Chang CC, Chen ST et al. Cyr61 expression confers resistance to apoptosis in breast cancer MCF-7 cells by a mechanism of NF-kappaB-dependent XIAP up-regulation. J Biol Chem 2004; 279(23):24015-24023, (Epub 22004 Mar 24024).
106. Chang AA, Van Waes C. Nuclear factor-KappaB as a common target and activator of oncogenes in head and neck squamous cell carcinoma. Adv Otorhinolaryngol 2005; 62:92-102.
107. Yamanaka N, Morisaki T, Nakashima H et al. Interleukin 1beta enhances invasive ability of gastric carcinoma through nuclear factor-kappaB activation. Clin Cancer Res 2004; 10(5):1853-1859.
108. Chan CF, Yau TO, Jin DY et al. Evaluation of nuclear factor-kappaB, urokinase-type plasminogen activator, and HBx and their clinicopathological significance in hepatocellular carcinoma. Clin Cancer Res 2004; 10(12 Pt 1):4140-4149.
109. Aggarwal S, Takada Y, Mhashilkar AM et al. Melanoma differentiation-associated gene-7/IL-24 gene enhances NF-kappa B activation and suppresses apoptosis induced by TNF. J Immunol 2004; 173(7):4368-4376.
110. Munshi A, Kurland JF, Nishikawa T et al. Inhibition of constitutively activated nuclear factor-kappaB radiosensitizes human melanoma cells. Mol Cancer Ther 2004; 3(8):985-992.
111. Kashani-Sabet M, Shaikh L, Miller 3rd JR et al. NF-kappa B in the vascular progression of melanoma. J Clin Oncol 2004; 22(4):617-623.
112. Thornburg NJ, Pathmanathan R, Raab-Traub N. Activation of nuclear factor-kappaB p50 homodimer/Bcl-3 complexes in nasopharyngeal carcinoma. Cancer Res 2003; 63(23):8293-8301.
113. Li L, Aggarwal BB, Shishodia S et al. Nuclear factor-kappaB and IkappaB kinase are constitutively active in human pancreatic cells, and their down-regulation by curcumin (diferuloylmethane) is associated with the suppression of proliferation and the induction of apoptosis. Cancer 2004; 101(10):2351-2362.
114. Niu J, Li Z, Peng B et al. Identification of an autoregulatory feedback pathway involving interleukin-1alpha in induction of constitutive NF-kappaB activation in pancreatic cancer cells. J Biol Chem 2004; 279(16):16452-16462, (Epub 12003 Dec 16416).
115. Muerkoster S, Arlt A, Sipos B et al. Increased expression of the E3-ubiquitin ligase receptor subunit betaTRCP1 relates to constitutive nuclear factor-kappaB activation and chemoresistance in pancreatic carcinoma cells. Cancer Res 2005; 65(4):1316-1324.
116. Qi H, M O. The von hippel-lindau tumor suppressor protein sensitizes renal cell carcinoma cells to tumor necrosis factor-induced cytotoxicity by suppressing the nuclear factor-kB-dependent antiapoptotic pathway. Cancer Research 2003; 63:7076-7080.
117. Pacifico F, Mauro C, Barone C et al. Oncogenic and anti-apoptotic activity of NF-kappa B in human thyroid carcinomas. J Biol Chem 2004; 279(52):54610-54619, (Epub 52004 Oct 54618).
118. Miyamoto S, Chiao PJ, Verma IM. Enhanced I kappa B alpha degradation is responsible for constitutive NF-kappa B activity in mature murine B-cell lines. Mol Cell Biol 1994; 14(5):3276-3282.
119. Lederman S, Yellin MJ, Krichevsky A et al. Identification of a novel surface protein on activated CD4+ T cells that induces contact-dependent B cell differentiation (help). J Exp Med 1992; 175(4):1091-1101.
120. Schubert LA, Cron RQ, Cleary AM et al. A T cell-specific enhancer of the human CD40 ligand gene. J Biol Chem 2002; 277(9):7386-7395, (Epub 2001 Dec 7319).
121. Liu YJ, Joshua DE, Williams GT et al. Mechanism of antigen-driven selection in germinal centres. Nature 1989; 342(6252):929-931.
122. Uckun FM, Gajl-Peczalska K, Myers DE et al. Temporal association of CD40 antigen expression with discrete stages of human B-cell ontogeny and the efficacy of anti-CD40 immunotoxins against clonogenic B-lineage acute lymphoblastic leukemia as well as B-lineage nonHodgkin's lymphoma cells. Blood 1990; 76(12):2449-2456.
123. Schattner EJ. CD40 ligand in CLL pathogenesis and therapy. Leuk Lymphoma 2000; 37(5-6):461-472.
124. Lalmanach-Girard AC, Chiles TC, Parker DC et al. T cell-dependent induction of NF-kappa B in B cells. J Exp Med 1993; 177(4):1215-1219.
125. Berberich I, Shu GL, Clark EA. Cross-linking CD40 on B cells rapidly activates nuclear factor-kappa B. J Immunol 1994; 153(10):4357-4366.
126. Zarnegar B, He JQ, Oganesyan G et al. Unique CD40-mediated biological program in B cell activation requires both type 1 and type 2 NF-kappaB activation pathways the IkappaB-NF-kappaB signaling module: Temporal control and selective gene activation two pathways to NF-kappaB. Proc Natl Acad Sci USA 2004; 101(21):8108-8113, (Epub 2004 May 8117).
127. Andjelic S, Hsia C, Suzuki H et al. Phosphatidylinositol 3-kinase and NF-kappa B/Rel are at the divergence of CD40-mediated proliferation and survival pathways. J Immunol 2000; 165(7):3860-3867.
128. Bargou RC, Leng C, Krappmann D et al. High-level nuclear NF-kappa B and Oct-2 is a common feature of cultured Hodgkin/Reed-Sternberg cells. Blood 1996; 87(10):4340-4347.

129. Lee SY, Kandala G, Liou ML et al. CD30/TNF receptor-associated factor interaction: NF-kappa B activation and binding specificity. Proc Natl Acad Sci USA 1996; 93(18):9699-9703.
130. Duckett CS, Gedrich RW, Gilfillan MC et al. Induction of nuclear factor kappaB by the CD30 receptor is mediated by TRAF1 and TRAF2. Mol Cell Biol 1997; 17(3):1535-1542.
131. Varro A, Noble PJ, Pritchard DM et al. Helicobacter pylori induces plasminogen activator inhibitor 2 in gastric epithelial cells through nuclear factor-kappaB and RhoA: Implications for invasion and apoptosis. Cancer Res 2004; 64(5):1695-1702.
132. Willis TG, Jadayel DM, Du MQ et al. Bcl10 is involved in t(1;14)(p22;q32) of MALT B cell lymphoma and mutated in multiple tumor types. Cell 1999; 96(1):35-45.
133. Kuo SH, Chen LT, Yeh KH et al. Nuclear expression of BCL10 or nuclear factor kappa B predicts Helicobacter pylori-independent status of early-stage, high-grade gastric mucosa-associated lymphoid tissue lymphomas. J Clin Oncol 2004; 22(17):3491-3497.
134. Granziero L, Ghia P, Circosta P et al. Survivin is expressed on CD40 stimulation and interfaces proliferation and apoptosis in B-cell chronic lymphocytic leukemia. Blood 2001; 97(9):2777-2783.
135. Richardson PG, Barlogie B, Berenson J et al. A phase 2 study of bortezomib in relapsed, refractory myeloma. N Engl J Med 2003; 348(26):2609-2617.
136. Aghajanian C, Soignet S, Dizon DS et al. A phase I trial of the novel proteasome inhibitor PS341 in advanced solid tumor malignancies. Clin Cancer Res 2002; 8(8):2505-2511.
137. Chu ZL, DiDonato JA, Hawiger J et al. The tax oncoprotein of human T-cell leukemia virus type 1 associates with and persistently activates IkappaB kinases containing IKKalpha and IKKbeta. J Biol Chem 1998; 273(26):15891-15894.
138. Duyao MP, Kessler DJ, Spicer DB et al. Transactivation of the murine c-myc gene by HTLV-1 tax is mediated by NFkB. AIDS Res Hum Retroviruses 1992; 8(5):752-754.
139. D'Souza BN, Edelstein LC, Pegman PM et al. Nuclear factor kappa B-dependent activation of the antiapoptotic bfl-1 gene by the Epstein-Barr virus latent membrane protein 1 and activated CD40 receptor. J Virol 2004; 78(4):1800-1816.
140. Keller SA, Schattner EJ, Cesarman E. Inhibition of NF-kappaB induces apoptosis of KSHV-infected primary effusion lymphoma cells. Blood 2000; 96(7):2537-2542.
141. Keller SA, Hernandez-Hopkins D, Hyjek E et al. NF-kB is essential for progression of KSHV- and EBV-infected lymphomas in vivo. Blood 2006; in press.
142. Mahe Y, Mukaida N, Kuno K et al. Hepatitis B virus X protein transactivates human interleukin-8 gene through acting on nuclear factor kB and CCAAT/enhancer-binding protein-like cis-elements. J Biol Chem 1991; 266(21):13759-13763.
143. Cogswell PC, Guttridge DC, Funkhouser WK et al. Selective activation of NF-kappa B subunits in human breast cancer: Potential roles for NF-kappa B2/p52 and for Bcl-3. Oncogene 2000; 19(9):1123-1131.
144. Yamamoto Y, Gaynor RB. IkappaB kinases: Key regulators of the NF-kappaB pathway. Biochem Sci 2004; 29(2):72-79.
145. Guo J, Verma UN, Gaynor RB et al. Enhanced chemosensitivity to irinotecan by RNA interference-mediated down-regulation of the nuclear factor-kappaB p65 subunit. Clin Cancer Res 2004; 10(10):3333-3341.
146. Wei LH, Lai KP, Chen CA et al. Arsenic trioxide prevents radiation-enhanced tumor invasiveness and inhibits matrix metalloproteinase-9 through downregulation of nuclear factor kappaB. Oncogene 2005; 24(3):390-398.
147. Kapahi P, Takahashi T, Natoli G et al. Inhibition of NF-kappa B activation by arsenite through reaction with a critical cysteine in the activation loop of Ikappa B kinase. J Biol Chem 2000; 275(46):36062-36066.
148. Roussel RR, Barchowsky A. Arsenic inhibits NF-kappaB-mediated gene transcription by blocking IkappaB kinase activity and IkappaBalpha phosphorylation and degradation. Arch Biochem Biophys 2000; 377(1):204-212.
149. Palombella VJ, Conner EM, Fuseler JW et al. Role of the proteasome and NF-kappaB in streptococcal cell wall-induced polyarthritis. Proc Natl Acad Sci USA 1998; 95(26):15671-15676.
150. Hideshima T, Richardson P, Chauhan D et al. The proteasome inhibitor PS-341 inhibits growth, induces apoptosis, and overcomes drug resistance in human multiple myeloma cells. Cancer Res 2001; 61(7):3071-3076.
151. Yin L, Laevsky G, Giardina C. Butyrate suppression of colonocyte NF-{kappa}B activation and cellular proteasome activity. J Biol Chem 2001; 25:25.
152. Niederberger E, Tegeder I, Schafer C et al. Opposite effects of rofecoxib on nuclear factor-kappaB and activating protein-1 activation. J Pharmacol Exp Ther 2003; 304(3):1153-1160.
153. Auphan N, Di Donato JA, Rosette C et al. Immunosuppression by glucocorticoids: Inhibition of NF-kappa B activity through induction of I kappa B synthesis. Science 1995; 270(5234):286-290.

154. Scheinman RI, Cogswell PC, Lofquist AK et al. Role of transcriptional activation of I kappa B alpha in mediation of immunosuppression by glucocorticoids. Science 1995; 270(5234):283-286.
155. Mukaida N, Morita M, Ishikawa Y et al. Novel mechanism of glucocorticoid-mediated gene repression. Nuclear factor-kappa B is target for glucocorticoid-mediated interleukin 8 gene repression. J Biol Chem 1994; 269(18):13289-13295.
156. Heck S, Bender K, Kullmann M et al. I kappaB alpha-independent downregulation of NF-kappaB activity by glucocorticoid receptor. EMBO J 1997; 16(15):4698-4707.
157. De Bosscher K, Schmitz ML, Vanden Berghe W et al. Glucocorticoid-mediated repression of nuclear factor-kappaB-dependent transcription involves direct interference with transactivation. Proc Natl Acad Sci USA 1997; 94(25):13504-13509.
158. Brostjan C, Anrather J, Csizmadia V et al. Glucocorticoid-mediated repression of NFkappaB activity in endothelial cells does not involve induction of IkappaBalpha synthesis. J Biol Chem 1996; 271(32):19612-19616.
159. Staal FJ, Roederer M, Herzenberg LA. Intracellular thiols regulate activation of nuclear factor kappa B and transcription of human immunodeficiency virus. Proc Natl Acad Sci USA 1990; 87(24):9943-9947.
160. Romano MF, Avellino R, Petrella A et al. Rapamycin inhibits doxorubicin-induced NF-kappaB/Rel nuclear activity and enhances the apoptosis of melanoma cells. Eur J Cancer 2004; 40(18):2829-2836.
161. Tunon MJ, Sanchez-Campos S, Gutierrez B et al. Effects of FK506 and rapamycin on generation of reactive oxygen species, nitric oxide production and nuclear factor kappa B activation in rat hepatocytes. Biochem Pharmacol 2003; 66(3):439-445.
162. Kopp E, Ghosh S. Inhibition of NF-kappa B by sodium salicylate and aspirin. Science 1994; 265(5174):956-959.
163. Alpert D, Vilcek J. Inhibition of IkappaB kinase activity by sodium salicylate in vitro does not reflect its inhibitory mechanism in intact cells. J Biol Chem 2000; 275(15):10925-10929.
164. Yin MJ, Yamamoto Y, Gaynor RB. The anti-inflammatory agents aspirin and salicylate inhibit the activity of I(kappa)B kinase-beta [see comments]. Nature 1998; 396(6706):77-80.
165. Grilli M, Pizzi M, Memo M et al. Neuroprotection by aspirin and sodium salicylate through blockade of NF-kappaB activation. Science 1996; 274(5291):1383-1385.
166. Wahl C, Liptay S, Adler G et al. Sulfasalazine: A potent and specific inhibitor of nuclear factor kappa B. J Clin Invest 1998; 101(5):1163-1174.
167. Weber CK, Liptay S, Wirth T et al. Suppression of NF-kappaB activity by sulfasalazine is mediated by direct inhibition of IkappaB kinases alpha and beta. Gastroenterology 2000; 119(5):1209-1218.
168. Yan F, Polk DB. Aminosalicylic acid inhibits ikappab kinase alpha phosphorylation of ikappabalpha in mouse intestinal epithelial cells. J Biol Chem 1999; 274(51):36631-36636.
169. Egan LJ, Mays DC, Huntoon CJ et al. Inhibition of interleukin-1-stimulated NF-kappaB RelA/p65 phosphorylation by mesalamine is accompanied by decreased transcriptional activity. J Biol Chem 1999; 274(37):26448-26453.
170. Yamamoto Y, Yin MJ, Lin KM et al. Sulindac inhibits activation of the NF-kappaB pathway. J Biol Chem 1999; 274(38):27307-27314.
171. Keifer JA, Guttridge DC, Ashburner BP et al. Inhibition of NF-kappa B activity by thalidomide through suppression of IkappaB kinase activity. J Biol Chem 2001; 276(25):22382-22387.
172. Miura T, Ouchida R, Yoshikawa N et al. Functional modulation of the glucocorticoid receptor and suppression of NF-kappaB-dependent transcription by ursodeoxycholic acid. J Biol Chem 2001; 276(50):47371-47378.
173. Sen CK, Traber KE, Packer L. Inhibition of NF-kappa B activation in human T-cell lines by anetholdithiolthione. Biochem Biophys Res Commun 1996; 218(1):148-153.
174. Chainy GB, Manna SK, Chaturvedi MM et al. Anethole blocks both early and late cellular responses transduced by tumor necrosis factor: Effect on NF-kappaB, AP-1, JNK, MAPKK and apoptosis. Oncogene 2000; 19(25):2943-2950.
175. Natarajan K, Singh S, Burke Jr TR et al. Caffeic acid phenethyl ester is a potent and specific inhibitor of activation of nuclear transcription factor NF-kappa B. Proc Natl Acad Sci USA 1996; 93(17):9090-9095.
176. Singh S, Natarajan K, Aggarwal BB. Capsaicin (8-methyl-N-vanillyl-6-nonenamide) is a potent inhibitor of nuclear transcription factor-kappa B activation by diverse agents. J Immunol 1996; 157(10):4412-4420.
177. Singh S, Aggarwal BB. Activation of transcription factor NF-kappa B is suppressed by curcumin (diferuloylmethane) [corrected] [published erratum appears in J Biol Chem 1995; 270(50):30235]. J Biol Chem 1995; 270(42):24995-25000.

178. Bharti AC, Donato N, Singh S et al. Curcumin (diferuloylmethane) down-regulates the constitutive activation of nuclear factor-kappa B and IkappaBalpha kinase in human multiple myeloma cells, leading to suppression of proliferation and induction of apoptosis. Blood 2003; 101(3):1053-1062.
179. Cross JV, Deak JC, Rich EA et al. Quinone reductase inhibitors block SAPK/JNK and NFkappaB pathways and potentiate apoptosis. J Biol Chem 1999; 274(44):31150-31154.
180. Goto M, Yamada K, Katayama K et al. Inhibitory effect of E3330, a novel quinone derivative able to suppress tumor necrosis factor-alpha generation, on activation of nuclear factor-kappa B. Mol Pharmacol 1996; 49(5):860-873.
181. Hiramoto M, Shimizu N, Sugimoto K et al. Nuclear targeted suppression of NF-kappa B activity by the novel quinone derivative E3330. J Immunol 1998; 160(2):810-819.
182. Nomura M, Ma W, Chen N et al. Inhibition of 12-O-tetradecanoylphorbol-13-acetate-induced NF-kappaB activation by tea polyphenols, (-)-epigallocatechin gallate and theaflavins. Carcinogenesis 2000; 21(10):1885-1890.
183. Meng L, Mohan R, Kwok BH et al. Epoxomicin, a potent and selective proteasome inhibitor, exhibits in vivo antiinflammatory activity. Proc Natl Acad Sci USA 1999; 96(18):10403-10408.
184. Schubert SY, Neeman I, Resnick N. A novel mechanism for the inhibition of NF-kappaB activation in vascular endothelial cells by natural antioxidants. FASEB J 2002; 16(14):1931-1933.
185. Tabary O, Escotte S, Couetil JP et al. Genistein inhibits constitutive and inducible NFkappaB activation and decreases IL-8 production by human cystic fibrosis bronchial gland cells. Am J Pathol 1999; 155(2):473-481.
186. Li Y, Ellis KL, Ali S et al. Apoptosis-inducing effect of chemotherapeutic agents is potentiated by soy isoflavone genistein, a natural inhibitor of NF-kappaB in BxPC-3 pancreatic cancer cell line. Pancreas 2004; 28(4):e90-95.
187. Manna SK, Sah NK, Newman RA et al. Oleandrin suppresses activation of nuclear transcription factor-kappaB, activator protein-1, and c-Jun NH2-terminal kinase [In Process Citation]. Cancer Res 2000; 60(14):3838-3847.
188. Manna SK, Mukhopadhyay A, Aggarwal BB. Resveratrol suppresses TNF-induced activation of nuclear transcription factors NF-kappa B, activator protein-1, and apoptosis: Potential role of reactive oxygen intermediates and lipid peroxidation. J Immunol 2000; 164(12):6509-6519.
189. Holmes-McNary M, Baldwin Jr AS. Chemopreventive properties of trans-resveratrol are associated with inhibition of activation of the IkappaB kinase. Cancer Res 2000; 60(13):3477-3483.
190. Hehner SP, Heinrich M, Bork PM et al. Sesquiterpene lactones specifically inhibit activation of NF-kappa B by preventing the degradation of I kappa B-alpha and I kappa B-beta. J Biol Chem 1998; 273(3):1288-1297.
191. Chaturvedi MM, Kumar A, Darnay BG et al. Sanguinarine (pseudochelerythrine) is a potent inhibitor of NF-kappaB activation, IkappaBalpha phosphorylation, and degradation. J Biol Chem 1997; 272(48):30129-30134.
192. Heiss E, Herhaus C, Klimo K et al. Nuclear factor-{kappa}B is a molecular target for sulforaphane-mediated anti-inflammatory mechanisms. J Biol Chem 2001; 15:15.
193. Manna SK, Mukhopadhyay A, Van NT et al. Silymarin suppresses TNF-induced activation of NF-kappa B, c-Jun N-terminal kinase, and apoptosis. J Immunol 1999; 163(12):6800-6809.
194. Geng Z, Rong Y, Lau BH. S-allyl cysteine inhibits activation of nuclear factor kappa B in human T cells. Free Radic Biol Med 1997; 23(2):345-350.
195. Shishodia S, Majumdar S, Banerjee S et al. Ursolic acid inhibits nuclear factor-kappaB activation induced by carcinogenic agents through suppression of IkappaBalpha kinase and p65 phosphorylation: Correlation with down-regulation of cyclooxygenase 2, matrix metalloproteinase 9, and cyclin D1. Cancer Res 2003; 63(15):4375-4383.

CHAPTER 11

NF-κB in Neurons:
Behavioral and Physiologic Roles in Nervous System Function

Jonathan M. Levenson, Marina Pizzi and J. David Sweatt

Cells in general and neurons in particular display an amazing ability to respond to several different types of environmental stimuli and integrate this response physiologically. What is even more surprising is that some of these responses can outlive the original stimulus by days, weeks or even longer. Long-term changes in physiology that occur in response to an external stimulus are almost always mediated at least in part by changes in gene expression. To effect these changes, cells have developed an impressive repertoire of signaling systems designed to modulate the activity of numerous transcription factors.

Transduction of cytoplasmic signaling events from the plasma membrane to the nucleus in most cells is straightforward; the nucleus is surrounded by a relatively spherical cytoplasm (Fig. 1A). Thus, most cells consist of one extra-nuclear signaling compartment- the cytoplasm. The unique morphology of neurons however, presents an intriguing wrinkle to the problem of nuclear signal transduction. Neurons consist of three very different cytoplasmic compartments: soma, dendrite and axon (Fig. 1B). The soma is the central portion of the neuron, and contains the nucleus. The dendrites extend from the soma and consist of long, thin projections that form tree-like arborizations. A single axon projects from the soma and can connect to several hundred other neurons. Information flow in a neuron typically starts at the dendrite, travels to the soma and can pass down the axon if the signal is strong enough. When one considers that a majority of the neuronal cytoplasm resides in the dendrites, the question of how a neuron can translate distal signaling events into changes in nuclear transcription becomes very complex.

A Synaptic Messenger

A great deal of work has centered on characterizing a signaling system that translates synaptic activity to gene expression. This signaling system has been postulated to consist of a "retrograde" messenger that would travel from an activated synapse to the nucleus to regulate transcription.[1] A molecule must conform to five basic principles to be considered a synaptically activated "retrograde" messenger that signals the nucleus. (1) The putative messenger must be present in synapses in the inactive form. (2) Enzymes that activate the messenger must be present in synapses. (3) The enzymes that activate the messenger must be regulated by physiologically relevant stimuli. (4) Once activated, the messenger must be retrogradely transported to the nucleus. (5) Once inside the nucleus, the messenger must bind to DNA and regulate gene transcription. Researchers have identified several molecules that fulfill some of the criteria required of a synaptic retrograde messenger. In the following section, we will illustrate how the NF-κB signaling system meets all of the criteria required for a neuronal retrograde messenger whose function is to couple synaptic activity to gene expression.

*Corresponding Author: Jonathan M. Levenson—Department of Neuroscience, Baylor College of Medicine, Houston, Texas 77030, U.S.A. Email: jlevenson@wisc.edu

NF-κB/Rel Transcription Factor Family, edited by Hsiou-Chi Liou.
©2006 Landes Bioscience and Springer Science+Business Media.

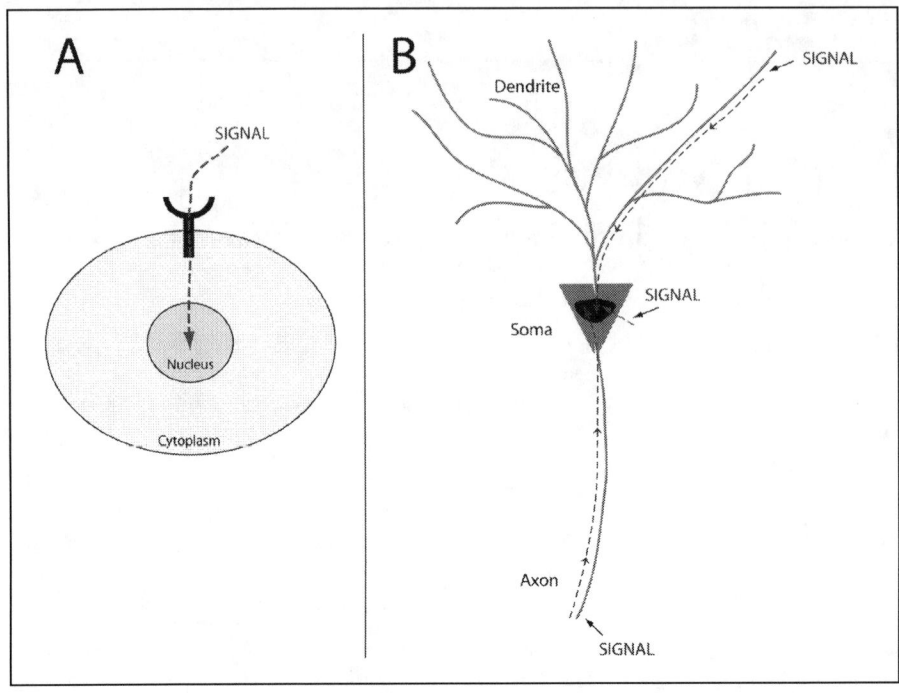

Figure 1. Neuronal morphology provides a unique challenge to retrograde nuclear signaling. A) Most cells have a simple morphology. There are only two signaling compartments: cytoplasm and nucleus. Therefore, movement of signals from the plasma membrane to the nucleus is straightforward. B) Neurons are highly specialized cells containing unique morphologies. This creates compartmentalization, complicating transport of signals from the plasma membrane to the nucleus.

The first evidence for the presence of NF-κB in the brain came in 1989, when NF-κB binding activity was observed in grey matter.[2] After this initial discovery, several labs have since localized NF-κB protein and binding activity in synapses located in several different regions of the brain.[3-5] Moreover, the synaptic NF-κB activity identified in these studies was only present if induced by exposure to a detergent, indicating that the native NF-κB proteins were present in an inhibited complex.[3-5] Therefore, *NF-κB exists in synaptic regions in an inactive form.*

As described in more detail elsewhere in this book, the NF-κB (also called NF-κB/Rel) family of dimeric transcription factors include p50, p52, p65 or RelA, RelB and c-Rel proteins, that use the Rel homology domain for dimerization and DNA binding.[6] In the cytoplasm NF-κB factors are bound to inhibitory proteins known as IκBs. Activation of signaling pathways that engage NF-κB signaling causes IκB to be phosphorylated by a protein complex known as IκB kinase (IKK). After it is phosphorylated, IκB is polyubiquitinated and degraded. The phosphorylation event releases NF-κB from IκB, allowing NF-κB to translocate to the nucleus and regulate transcription. The IKK complex consists of two catalytic (IKKα, IKKβ)[7] and one regulatory subunit (IKKγ).[8,9] To date, expression of only the α and β subunits of the IKK complex have been localized to synapses in the brain.[10,11] Despite the lack of information regarding the localization of IKKγ, the great deal of evidence regarding activation of NF-κB proteins by synaptic stimuli (see next section) coupled with the synaptic localization of the α and β subunits to the brain indicate that *the enzymes involved in activation of quiescent NF-κB protein are present in the brain, at synapses.*

Several lines of evidence indicate that activation of NF-κB can occur via physiologically relevant stimuli. Perhaps the most relevant physiologic stimulus in the brain is synaptic activity. NF-κB is regulated by both low- and high-frequency synaptic stimuli.[3] This observation likely explains why several researchers have observed constitutively active NF-κB DNA binding activity in the brain.[5,12-14] Glutamate, the primary excitatory neurotransmitter of the central nervous system, activates NF-κB by acting through a variety of glutamate receptors.[14-19] In addition, different cytokines, growth factors and lipopolysaccharide have been shown to modulate NF-κB activity in the brain. Therefore, *activation of NF-κB in the brain occurs in response to physiologically relevant stimuli*.

Recently, an additional level of regulation of NF-κB activity, that may involve post-translational modification of NF-κB subunits in the nuclear compartment and protein-protein interaction with other promoter-bound factors, has emerged. These new forms of regulation have been primarily characterized in nonneuronal cells, and include phosphorylation and acetylation of NF-κB factors.[20,21] One study has provided evidence that acetylation of NF-κB occurs in the brain during memory formation.[22] These results indicate that NF-κB activity is highly regulated, and that at least some of the pathways involved in fine-tuning NF-κB activity exist in the brain.

Activated NF-κB must translocate from the synapse to the nucleus to affect transcription. Therefore, stimuli that activate NF-κB must also promote its movement into the nucleus. Several methods have been employed to show that activation of NF-κB leads to increases in nuclear NF-κB amount and activity. Increases in the amount of active NF-κB have been measured immunocytochemically after activation with glutamate or glutamate receptor agonists.[15-17] NF-κB DNA binding activity in nuclear fractions has been shown to increase in response to several different types of synaptic stimuli.[3] Moreover, recent studies have utilized a chimeric protein containing both p65 and a protein isolated from jellyfish that fluoresces without excitation, known as green fluorescent protein, to visualize p65 subcellular localization in living neurons.[19] Upon stimulation of neurons with either glutamate, kainate, or high K^+ solutions, p65 was observed to rapidly redistribute from the synaptic regions of neurons to the nuclei.[19] These results indicate that *upon activation, synaptic NF-κB is retrogradely transported to the nucleus*.

The final step in demonstrating that a synaptic protein is a transcriptional retrograde messenger is to *show that transcription of genes regulated by the transcription factor is affected by the same stimuli that activate the messenger*. In this vein, the expression of NF-κB genes, including p50, p65 and IkBα, have been shown to be increased after stimulation of the NF-κB pathway.[23-25] Upon stimulation of neurons, expression of p50, p65 and IkBα have all been shown to increase.[3] Moreover, using reporter gene constructs several laboratories have shown that NF-κB-mediated gene transcription occurs in neurons and can be regulated by neuronal activity and levels of Ca^{++}.[5,12,13]

Signaling Pathways That Regulate Neuronal NF-κB

A great deal of work over the last decade has focused on characterizing the expression, regulation and function of NF-κB in the nervous system. NF-κB was the first transcription factor discovered in the brain that was present in synaptic regions, hinting that NF-κB might serve a crucial role in transducing synaptic events directly to the nucleus.[3,4] In the following sections we will discuss the regulation of NF-κB in the brain and its role in memory formation. While we will not discuss it here, it should be noted that a great deal of evidence exists indicating that NF-κB is also important in neuronal development and in neurodegeneration.[6]

To fully understand the role a transcription factor plays in any cell type, the signaling pathways that activate it must be identified. Neurons, arguably, have the most sophisticated complement of signaling systems known. Thus, the task of identifying the signaling pathways that play a role in regulation of NF-κB is not trivial. Ultimately, activation of these upstream signaling pathways must impinge on the IKKs to affect NF-κB function. Full characterization of the upstream kinases and signaling molecules is still lacking in neurons, despite efforts by numerous laboratories. We will review what is currently known about the signaling pathways that modulate NF-κB activity in neurons.

Glutamate

Signaling events usually begin at the plasma membrane of a cell, and neurons are highly adapted to respond to various neurotransmitters present present at their cell surface. The primary excitatory neurotransmitter in the central nervous system is glutamate; most neurons in the brain either release glutamate as a neurotransmitter and/or possess receptors that are responsive to glutamate. Therefore, as an initial hypothesis one would expect neuronal NF-κB to be responsive to glutamate. In fact, several studies have shown that NF-κB is regulated by glutamate.[14,15,17,18,26] Further investigation has revealed that different subtypes of glutamate receptors contribute to the regulation of NF-κB by glutamate. Activation of NF-κB occurs upon stimulation of kainate receptors[16,17,19,27] or NMDA receptors,[5,14,15,18] and may be modulated by stimulation of metabotropic receptors.[28]

Growth Factors

In the brain, growth factors play important roles in regulation of synaptic plasticity, neuronal survival and differentiation. Therefore, growth factor signaling is extremely relevant to and vital for normal neuronal function. A great deal of evidence indicates that nerve growth factor (NGF) activates NF-κB through the p75 neurotrophin receptor.[29-35] Subsequent studies have shown that NF-κB is activated by a variety of growth factors including epidermal growth factor,[36] pigment epithelium-derived factor,[37] and brain-derived neurotrophic factor.[38] Interestingly, transforming growth factor β2 appears to inhibit NF-κB activity.[39] All of these observations suggest that NF-κB plays a critical role in neuronal growth factor signaling, and suggests that NF-κB coordinates a variety of transcriptional responses during synaptic plasticity and neuronal survival.

Calcium

Synaptic activity and stimulation of glutamate receptors almost always results in changes in intracellular Ca^{++} concentrations. Given that dormant NF-κB is concentrated at synaptic terminals, one would expect NF-κB to be regulated by Ca^{++}. The NMDA receptor mediates many of the Ca^{++}-dependent signaling processes that occur in neurons, including activation of NF-κB.[5,14,15,18] In addition to the NMDA receptor, voltage-gated Ca^{++} channels and opening of In3P receptors associated with the intracellular stores of Ca^{++} also contribute to activation of NF-κB.[5,16,40]

Reactive Oxygen Species

Reactive oxygen species (ROS) have been implicated in both pathology and normal signaling in neurons.[41] ROS refers to any molecule derived from molecular oxygen. There is growing evidence indicating that generation of ROS is necessary for the formation of long-term forms of synaptic plasticity and memory.[42-45] One signal for the generation of ROS in neurons is the Ca^{++} influx caused by the activation of NMDA receptors.[46] Some evidence exists that suggests activation of NF-κB in neurons also requires ROS.[17]

Protein Kinases

Signaling events that begin at the membrane result in the generation of small second messenger signaling molecules, which ultimately activate downstream kinase pathways. Several different kinase cascades have been implicated in regulation of neuronal NF-κB. Influx of Ca^{++} activates the calcium-calmodulin dependent protein kinases (CaMK) via binding calmodulin and interacting with regulatory subunits on these kinases, and in an important recent study the

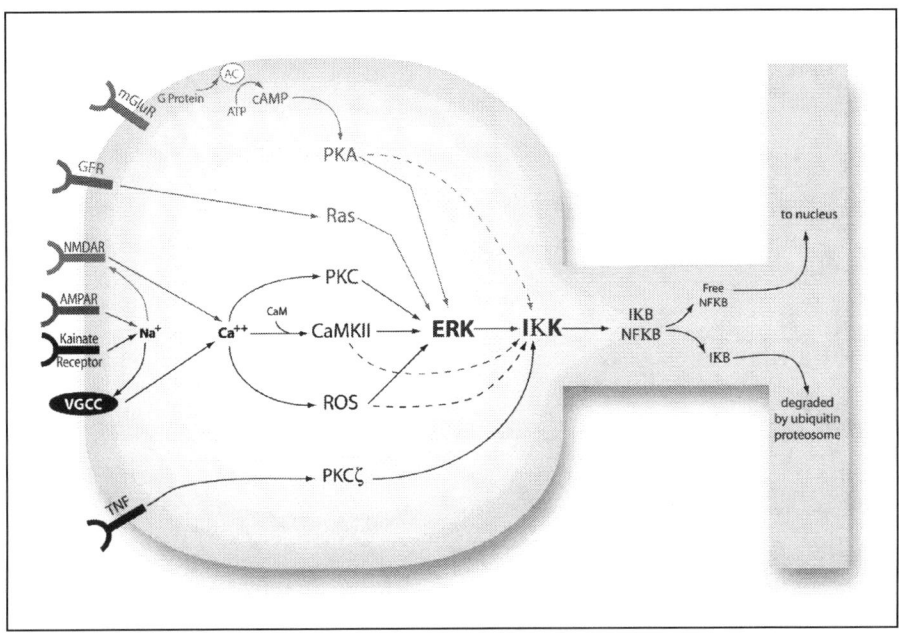

Figure 2. Regulation of NF-κB within neurons. Regulation of NF-κB in neurons can occur through multiple signaling pathways. Activation of various receptors at the plasma membrane engages many cytoplasmic signaling pathways that converge on the ras-MEK-ERK/MAP kinase system. ERK then activates IKK, either directly or indirectly, which leads to phosphorylation of IκB and release of NF-κB.

specific isoform CaMKII has been implicated in neuronal NF-κB signaling.[5] Many isoforms of protein kinase C (PKC) are expressed in the brain, however only the atypical PKCζ has been directly implicated in regulation of NF-κB in the brain.[16,47-50] The extracellular signal regulated kinase (ERK), which is a member of the MAP kinase superfamily, has also been shown to be involved in regulation of NF-κB activity.[16] Other kinases, such as the cAMP-dependent protein kinase and casein kinase, have been shown to phosphorylate IKK in vitro, however there is no convincing in vivo evidence for their importance in regulation of neuronal NF-κB.[51]

A Model for Activation of NF-κB in Neurons

From the previous sections, one can appreciate the variety and scope of cellular stimuli that can modulate NF-κB function. Drawing from what is known about basic signaling pathways in the brain, we can begin to assemble the puzzle of neuronal NF-κB regulation (Fig. 2). No comprehensive studies of the activation of NF-κB in neurons have been performed to date, so much of what we will discuss here is speculation. Activation of NF-κB begins at the plasma membrane, where activation of any number of receptors initiates second messenger signaling (Fig. 2). Key second messengers include Ca^{++} and cAMP. In addition to second messenger-mediated signaling, activation of growth factor receptors engages both ras and PKCζ[49] (Fig. 2). TNF activates NF-κB signaling via PKC.[52] In the immune system, TNF receptor-mediated arachidonic acid signaling activates NF-κB,[53] but data in neurons suggests that arachidonic acid signaling through TNFRs actually inhibits NF-κB.[49] Nevertheless, activation of all of these signaling cascades leads to phosphorylation of ERK MAP kinase[54] (Fig. 2). ERK activates IKK either directly or indirectly, which promotes dissociation of IκB from NF-κB, allowing NF-κB to move retrogradely to the nucleus to regulate transcription (Fig. 2).

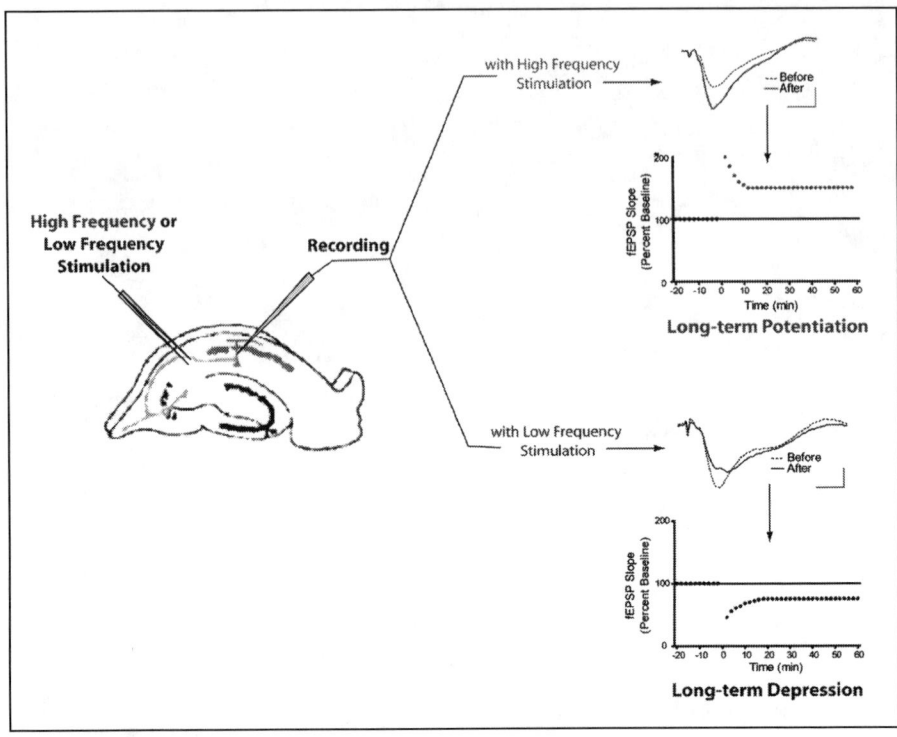

Figure 3. Synaptic plasticity. Synaptic plasticity refers to the ability of a synapse to change strength in response to synaptic stimuli. One preparation used to study the phenomenon of synaptic plasticity is the acute hippocampal slice. A stimulating electrode is placed along the presynaptic axon fibers and a recording electrode is placed in the dendritic field. Extracellular potentials provide an index of postsynaptic depolarization generated upon stimulation of the slice. High frequency stimulation leads to a lasting increase in the synaptic response, referred to as long-term potentiation. Conversely, low-frequency stimulation stably decreases the synaptic response, referred to as long-term depression.

Synaptic Plasticity

Neuron-to-neuron connections are referred to as synapses. There are two types of synapses: chemical and electrical. Chemical synapses are ones in which information from one neuron to another is transmitted by release of a specific chemical, referred to as a neurotransmitter. An electrical synapse is one where two neurons are directly connected, and changes in the electrical potential of one neuron are directly transmitted to the other neuron. Perhaps the biggest functional difference between these two types of synapses is the total inability of an electrical synapse to exhibit changes in strength in response to previous synaptic activity. These activity-dependent changes in synaptic strength, commonly referred to as synaptic plasticity, are thought mediate higher cognitive functions such as memory formation. Therefore, by studying synaptic plasticity, researchers can learn a great deal about the basic mechanisms that underlie very complex cognitive processes, such as memory formation, that occur in the brain.

Synaptic plasticity can take two basic forms: potentiation and depression (Fig. 3). Synaptic potentiation occurs when previous synaptic stimuli result in a larger, or enhanced synaptic response (Fig. 3). Conversely, depression occurs when previous synaptic stimuli result in a diminished synaptic response (Fig. 3). In general, high-frequency stimulation (≥ 0.2 Hz) usually leads to potentiation while low-frequency stimulation (~1 Hz) usually induces depression.

Changes in synaptic strength can exist for either very brief periods of time, lasting seconds to minutes, or persist for extended periods of time lasting several hours or longer. In the following paragraphs, we will focus specifically on two forms of synaptic plasticity referred to as long-term potentiation (LTP) and long-term depression (LTD). There are early and late phases of both LTP and LTD, which are distinguished by their reliance on protein synthesis. For our purposes, early-phases of plasticity refer to forms of plasticity that require only posttranslational modification, such as phosphorylation, but not new protein synthesis.[55] These early phases of plasticity generally do not persist for longer than 1-2 hours. Late-phase plasticity refers to forms of plasticity that require protein synthesis for their expression, and generally persist for at least several hours if not longer.[55]

NF-κB is a particularly attractive candidate transcription factor for the induction of synaptic plasticity because it is regulated by many of the same signaling pathways that are involved in induction of LTP and LTD (Fig. 2, see also ref. 54). Induction of LTP and some forms of LTD require activation of the NMDA class of glutamate receptors.[56-58] The small messenger molecules Ca^{++} and ROS are required for induction of LTP.[42,59,60] Moreover, the kinases PKA, PKC, CaMKII and ERK are all necessary for induction of LTP.[61-65] Induction of various forms of LTD requires PKC, CaMKII and ERK.[66-69] Therefore, induction of synaptic plasticity employs the same complement of signaling pathways that activate NF-κB, suggesting that long-term forms of synaptic plasticity may involve NF-κB-mediated gene transcription.

Initial studies into the regulation of NF-κB by synaptic activity focused on the expression of p50 and p65. Application of high-frequency stimulation to the perforant pathway input into the dentate gyrus induced LTP and increased expression of genes encoding p50, p65 and IkBα in dentate granule neurons.[3] The increases in all of these genes were time-dependent, with maximal increases occurring 1 h after induction of LTP.[3] Interestingly, low-frequency stimulation of the perforant pathway did not induce any form of synaptic plasticity, yet the expression of p50, p65 and IkBα were all increased to levels seen in after induction of LTP, suggesting that activation of the NF-κB pathway is sensitive to synaptic activity in general.[3] In concert with these findings, recent studies in cultures of hippocampal neurons demonstrated that NF-κB DNA binding activity was decreased by treatments that inhibited synaptic activity.[5] Moreover, both NF-κB DNA binding activity and NF-κB-mediated transcription were increased by treatments that increased synaptic activity.[5] All of these results demonstrate that the NF-κB pathway is activated not only by induction of synaptic plasticity, but also by basal levels of activity.

The activation of the NF-κB signaling pathway may not be *sufficient* to trigger lasting changes in synaptic strength, however several lines of evidence indicate that proper NF-κB function is *necessary* for induction of synaptic plasticity. One technique used to assess the involvement of NF-κB in induction of synaptic plasticity utilized a response element decoy technique. Acute brain slices were exposed to DNA oligonucleotides containing the NF-κB responsive element for several hours prior to induction of LTP.[70,71] The decoy DNA acts to competitively inhibit binding of NF-κB transcription factors to the nuclear DNA, effectively blocking the ability of NF-κB to regulate gene expression. Exposure of slices to the NF-κB decoy DNA attenuated induction of LTP in the hippocampus, and blocked induction of late-phase LTP in the amygdala.[70,71] These results suggest that NF-κB-mediated gene expression is necessary for induction of transcription-dependent synaptic plasticity in the amygdala and hippocampus.[70,71]

A second class of experiments investigating the role of NF-κB in synaptic plasticity focused on the cytokine tumor necrosis factor (TNF). TNF is expressed in the brain, and its production appears to be regulated by synaptic stimulation.[72,73] TNF normally binds to either the p55 or p75 TNF-receptors (TNFR), which results in activation of the NF-κB signaling pathway.[74-78] Early studies of the possible effects of TNF on synaptic physiology revealed that very brief treatments of hippocampal slices with exogenous TNFα increased synaptic strength, and inhibiting the normal action endogenous TNFα with exogenous TNF receptor fragments decreased synaptic strength.[79] The effects of TNFα were attributed to regulation of glutamate receptor expression.[79] Subsequent studies of the long-term effects of TNFα on synaptic physiology revealed that TNFα actually downregulated AMPA and NMDA receptor-mediated current, and increased voltage-gated Ca^{++} currents at the plasma membrane.[27] These results indicate that TNF can exert a variety of effects on synaptic physiology depending on the temporal kinetics of TNF signaling.

Treatment of neurons with TNF has profound effects on the ability to induce synaptic plasticity. Exposure of acute hippocampal slices to TNFα inhibits induction of LTP in area CA1 and the dentate gyrus.[80,81] Moreover, genetic knockout of both TNF receptor (TNFR) isoforms has no effect on induction of LTP, however, induction of LTD is severely impaired.[70] To fully understand the effects that TNF might have on synaptic plasticity, recall that induction of both LTP and LTD are dependent upon influx of Ca^{++} and the frequency of synaptic activity,[82] and exposure of neurons to TNF enhances their ability to maintain Ca^{++} homeostasis via NF-κB-dependent transcription.[27,83,84] Therefore, TNF may inhibit the induction of LTP by upregulating the expression of Ca^{++}-buffering enzymes while causing the loss of TNFRs and thus, may lead to erroneous induction of LTP by diminished expression of Ca^{++}-buffering enzymes. Alternatively, exposure of slices to TNF might result in regulation of glutamate receptor trafficking that prevents or occludes expression of LTP. All of these effects are presumably mediated by TNF-induced activation of NF-κB. Therefore, all of the studies above indicate that interference with the normal function of NF-κB inhibits proper induction of synaptic plasticity. Thus, NF-κB-mediated transcription appears to be necessary for proper induction or expression of synaptic plasticity.

Memory Formation

Synaptic plasticity is a candidate mechanism that may contribute to the formation of memory in vivo. As outlined above, several lines of evidence indicate that NF-κB plays a role in induction of various forms of synaptic plasticity. This suggests the intriguing possibility that NF-κB might be involved in the formation of memory in the behaving animal. Studies into the molecular mechanisms of memory formation have been greatly facilitated in the past decade with the advent of many genetic mouse model systems that mimic human disorders of memory formation.[85] Several experimental paradigms have been developed to probe several different kinds of memory. Researchers have just begun to develop mouse models suitable for the study of NF-κB in memory formation; therefore the number of experiments performed to date is relatively low. In the following sections we will review the behavioral paradigms used to investigate the role of NF-κB in memory formation, and the results.

Fear Motivated Learning: An Explanation

At its most basic level, learning involves exposure to an external stimuli that results in a lasting change in behavior. The study of learning and memory in animal model systems is difficult in that the animals cannot directly communicate whether they remember a certain event. Therefore, researchers have designed a number of simple behavioral tasks to test different forms of learning. A relatively popular paradigm currently in use is fear-motivated learning. Emotionally motivated learning paradigms evoke strong, easily quantifiable responses. Fear-motivated learning relies on the amygdala, an almond-shaped region of the brain that is part of the limbic system. Fear learning involves exposure of animals to a noxious stimulus, usually a mild electric shock to the paws, in conjunction with a previously innocuous stimulus such as a noise or light. When these 2 unrelated stimuli are presented together, the animal can then form an association between the noxious and the innocuous stimulus. Thus, an animal learns that an *unconditioned stimulus* (the footshock) that always evokes fear can be predicted by a *conditioned stimulus* (the light or noise) that never evoked a fearful response prior to training.

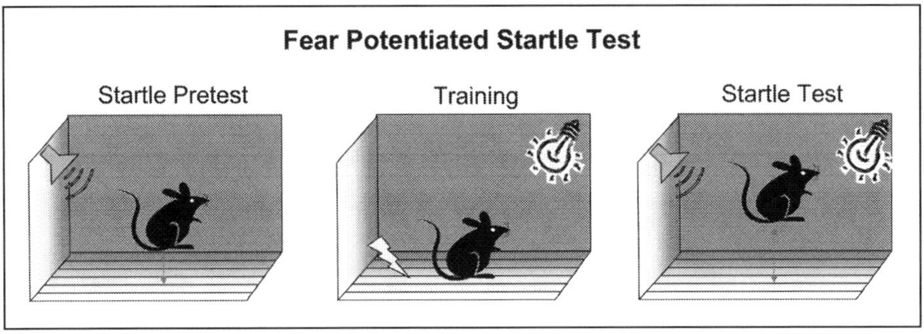

Fear Potentiated Startle Test

Figure 4. Fear potentiated startle. Fear potentiated startle is an associative learning paradigm that couples a noxious stimulus (electric shock) with an innocuous stimulus (light). A) Baseline startle responses induced by a brief, loud tone are measured. B) The subject is exposed to a series of training sessions that pair the presentation of a light cue with an electric shock. During the training period, the subject learns to associate the light with the shock. C) Startle responses are tested in the presence and absence of the light cue. Startle responses elicited in the presence of the light are usually 2-3 times greater than in the absence. This phenomenon is called fear potentiated startle.

Fear Potentiated Startle

Most animals exhibit defensive startle reflexes in response to sudden, unexpected environmental stimuli. These reflexes evolved to deal with the sudden presence of predators or other threats. Due to their nature, defensive reflexes are sensitive to the emotional state of an animal. For example, if an animal is frightened, then expression of defensive reflexes is enhanced. Researchers have utilized this phenomenon to measure fear-motivated learning in a learning paradigm known as fear potentiated startle (Fig. 4). In this learning paradigm, an animal's startle reflex is first assessed by presentation of a series of brief, but very loud audible tones (Fig. 4A). These unpredictable auditory stimuli cause the animal to flinch, which is measured with an accelerometer. Then, the animal is exposed to a series of training sessions that involve exposure to a light and an electric shock (Fig. 4B). During this training, the animal learns to associate the light with the electric shock. After training, the startle response is tested again, however in the presence and absence of the light cue (Fig. 4C). When the animal sees the light, the startle response is generally 2-3 times greater than pretraining values.

Some evidence exists for the involvement of NF-κB mediated gene transcription in fear potentiated startle memories. Fear potentiated startle training increases levels of NF-κB protein, acetylation of NF-κB and its DNA binding activity selectively in the amygdala.[22,71] These increases in DNA binding activity are accompanied by decreases in IκB and increases in the activity of IKK, indicating that the NF-κB signaling system has been activated.[71] Moreover, acquisition of fear potentiated startle memory is blocked by administration of NF-κB inhibitors or κB response element decoys directly to the amygdala.[71] Interestingly, some evidence exists that suggests fear potentiated startle memory can be enhanced by agents that increase acetylation of NF-κB.[22] Together, these data demonstrate that NF-κB is upregulated by fear potentiated startle, and that NF-κB function in the amygdala is necessary for the formation of fear potentiated startle memory.

Fear Conditioning

Fear conditioning is another fear-motivated learning paradigm currently in use. Fear conditioning also involves pairing a noxious unconditioned stimulus with a previously innocuous conditioned stimulus. When rodents become frightened they decrease their spontaneous activity, which is referred to as "freezing". Researchers quantify acquisition of fear conditioning by measuring freezing behavior (Fig. 5). There are two different forms of fear conditioning. Cued

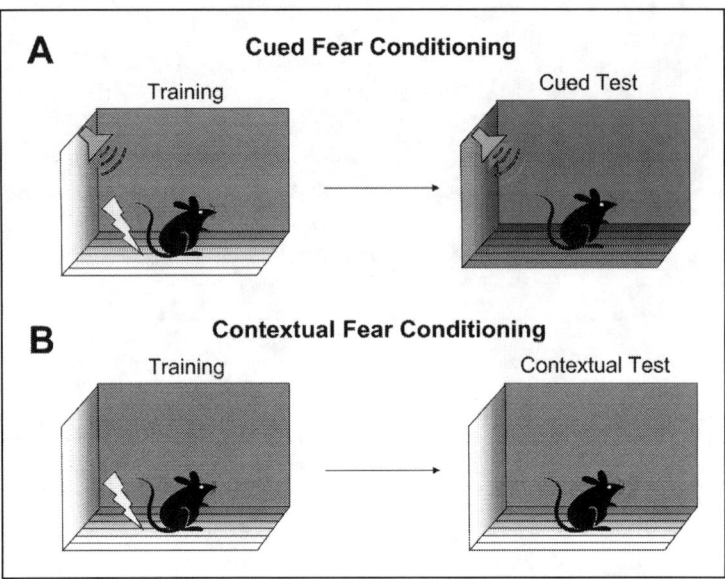

Figure 5. Associative fear conditioning. In the fear conditioning paradigm, subjects associate a previously innocuous stimulus with an aversive electric shock. A) In the cued fear conditioning paradigm, the subject is trained to associate a tone with an electric shock. Testing the subject in the presence of the tone elicits freezing behavior, indicating successful acquisition of the associative fear memory. B) In contextual fear conditioning, the animal is placed into a novel environment and receives a series of shocks. No other cues are provided. Therefore, the animal associates the aversive stimuli with the environment. Placement of the subject back into the environment where the shocks were administered elicits freezing behavior, indicating successful acquisition of the associative fear memory.

fear conditioning involves paired exposure of animals to a tone and electric shock (Fig. 5A). In the cued paradigm, the animals learn to associate the tone with the shock (Fig. 5A). Contextual fear conditioning involves placing the animal in a novel environment and administering a series of mild shocks (Fig. 5B). In the contextual paradigm, the animal learns to associate the environment with the shock (Fig. 5B). Formation of cued fear memories requires the amygdala.[86] Formation of contextual fear memory requires both the hippocampus and the amygdala.[86,87]

In a recent study the role of *c-rel* was ascertained in the formation of long-term associative fear memory.[88] Through a series of mRNA expression profiling and bioinformatics studies, it was discovered that c-Rel-mediated gene transcription is important for the formation of long-term contextual fear memory in the hippocampus.[88] To confirm the results, mice that lacked *c-rel* (*c-rel$^{-/-}$*) through genetic knockout[89] were tested in both cued and contextual fear conditioning paradigms. There were no differences between *c-rel$^{-/-}$* and normal (*c-rel$^{+/+}$*) mice in the cued fear conditioning paradigm, indicating that *c-rel* was not necessary for the formation of long-term memory subserved exclusively by the amygdala. This is in agreement with other studies in the amygdala which have shown that c-Rel DNA binding activity is not upregulated in the amygdala after induction of LTP.[71] However, *c-rel$^{-/-}$* animals exhibited significant deficits in contextual fear conditioning.[88] In addition to testing behavior, these experiments also provided some of the first characterization of nonmemory behaviors in any mouse model containing genetic alterations in the NF-κB system. *c-rel$^{-/-}$* animals were less active overall than *c-rel$^{+/+}$* animals.[88] However, *c-rel$^{-/-}$* animals had similar levels of anxiety and nociception, indicating that genetic disruption of *c-rel* did not lead to gross abnormalities in sensory perception or emotional state.[88] Taken together, the above results indicate that *c-rel* is necessary for the formation of hippocampus-dependent long-term contextual fear memory.[88]

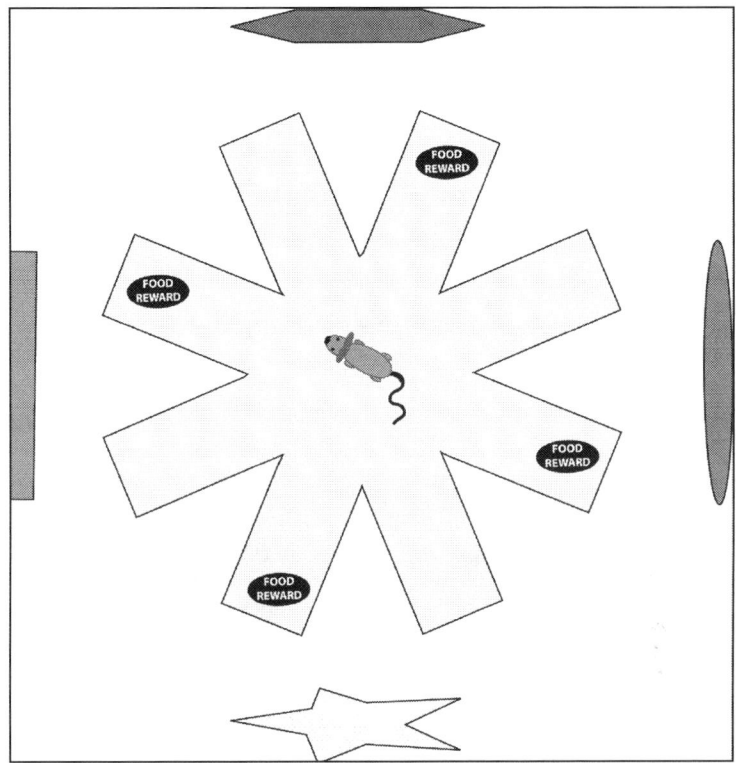

Figure 6. Radial arm maze. The radial arm maze tests the accuracy of an animal to navigate in a simple maze. Food rewards are placed in half of the arms. Animals begin in the center of the maze. Accuracy is scored based on either how many times an animal reenters arms of the maze, or how many errors the animal makes in the process of finding all of the rewards. Prominent spatial cues serve as visual landmarks, allowing the animal to form a mental map of the maze.

Radial Arm Maze

The radial arm maze test is a learning paradigm that can be used to investigate spatial and working memory. Remembering where your favorite restaurant is located is an example of spatial memory. Remembering at any given time where you have left your fork while you are eating there is an example of a working memory. The radial arm maze is used to model working memory and consists of at least 8 symmetric arms that protrude from a central platform (Fig. 6). Food is used to motivate the animals to learn the layout of the maze, so feeding is restricted in animals 1 week prior to the start of training and testing. In the spatial version of the radial arm maze, several distinct visual cues are placed such that the animal can use them as landmarks to navigate to specific areas of the maze (Fig. 6). In the cued version, the visual landmarks are removed and baited arms are indicated by either lights or flags (Fig. 6). The spatial version of the radial arm maze requires the hippocampus, while the cued version requires dorsal striatum.[90]

Trials are typically performed for over 2 weeks. Two different kinds of trials are performed each day; in one trial, several arms of the maze are blocked off. The arms that are not blocked are baited with food pellets. The animal is placed into the center of the maze and allowed 5 min to explore. For the second trial, all arms of the maze are open, and food pellets are placed only into the arms that were previously blocked. Animals are scored based on how many "mistakes" they make as assessed by how many times animals enter an arm that has been previously visited. The number of correct arm entries out of the first few entries can also be scored as an index of whether the animal has successfully learned the task.

The radial arm maze task was recently utilized to assess the role of p65 in hippocampus and striatum-dependent memory formation.[5] Mice that lacked both p65 and TNFR1 (p65$^{-/-}$) through genetic knockout were tested in the spatial and cued versions of the radial arm maze. Initially, p65$^{-/-}$ mice performed worse than p65$^{+/+}$ in the spatial version of the radial arm maze.[5] Over time however, the p65$^{-/-}$ mice were able to learn the task and perform equally as well as p65$^{+/+}$ mice. In the cued version of the task, p65$^{-/-}$ mice performed equally as well as p65$^{+/+}$ mice. These results suggest that p65 is involved in facilitating the formation of hippocampus-dependent memory, but not striatal memory.[5] It should be noted that in these studies, all of the mice tested lacked TNFR1.[5] Recall that disruption of TNFRs leads to derangement of synaptic plasticity in in vitro preparations.[70] Therefore, the memory impairments observed in p65$^{-/-}$ mice may not be entirely due to loss of p65 function. Nevertheless, these experiments provide additional evidence supporting a role for the NF-κB signaling pathway in the formation of hippocampus-dependent memory.

Summary

The neuronal NF-κB signaling system represents a novel and potentially vital component involved in linking activity at the synapse to activity within the nucleus. NF-κB was the first transcription factor to be localized to the synapse. Since its initial characterization in the nervous system, NF-κB has been shown to be regulated by a variety of stimuli, suggesting that it may play a role in integration of numerous different types of information within the nervous system. Ample evidence exists demonstrating that NF-κB factors are engaged in and are necessary for formation of synaptic plasticity and long-term memory. Even though we are only beginning to understand the contribution of distinct NF-κB family members to the regulation of gene transcription in the brain, all of the evidence collected thus far indicates that NF-κB may represent a vital part of the molecular machinery involved in mammalian cognition.

References

1. Schmitt RO, Dev P, Smith BH. Electrotonic processing of information by brain cells. Science 1976; 193(4248):114-120.
2. Korner M, Rattner A, Mauxion F et al. A brain-specific transcription activator. Neuron 1989; 3(5):563-572.
3. Meberg PJ, Kinney WR, Valcourt EG et al. Gene expression of the transcription factor NF-kappa B in hippocampus: Regulation by synaptic activity. Brain Res Mol Brain Res 1996; 38(2):179-190.
4. Kaltschmidt C, Kaltschmidt B, Baeuerle PA. Brain synapses contain inducible forms of the transcription factor NF-kappa B. Mech Dev 1993; 43(2-3):135-147.
5. Meffert MK, Chang JM, Wiltgen BJ et al. NF-kappa B functions in synaptic signaling and behavior. Nat Neurosci 2003; 6(10):1072-1078.
6. O'Neill LA, Kaltschmidt C. NF-kappa B: A crucial transcription factor for glial and neuronal cell function. Trends Neurosci 1997; 20(6):252-258.
7. Zandi E, Rothwarf DM, Delhase M et al. The IkappaB kinase complex (IKK) contains two kinase subunits, IKKalpha and IKKbeta, necessary for IkappaB phosphorylation and NF-kappaB activation. Cell 1997; 91(2):243-252.
8. Yamaoka S, Courtois G, Bessia C et al. Complementation cloning of NEMO, a component of the IkappaB kinase complex essential for NF-kappaB activation. Cell 1998; 93(7):1231-1240.
9. Rothwarf DM, Zandi E, Natoli G et al. IKK-gamma is an essential regulatory subunit of the IkappaB kinase complex. Nature 1998; 395(6699):297-300.
10. Hu MC, Wang Y. IkappaB kinase-alpha and -beta genes are coexpressed in adult and embryonic tissues but localized to different human chromosomes. Gene 1998; 222(1):31-40.
11. Hu MC, Wang Y, Qiu WR et al. Hematopoietic progenitor kinase-1 (HPK1) stress response signaling pathway activates IkappaB kinases (IKK-alpha/beta) and IKK-beta is a developmentally regulated protein kinase. Oncogene 1999; 18(40):5514-5524.
12. Kaltschmidt C, Kaltschmidt B, Neumann H et al. Constitutive NF-kappa B activity in neurons. Mol Cell Biol 1994; 14(6):3981-3992.

13. Rattner A, Korner M, Walker MD et al. NF-kappa B activates the HIV promoter in neurons. EMBO J 1993; 12(11):4261-4267.
14. Guerrini L, Blasi F, Denis-Donini S. Synaptic activation of NF-kappa B by glutamate in cerebellar granule neurons in vitro. Proc Natl Acad Sci USA 1995; 92(20):9077-9081.
15. Scholzke MN, Potrovita I, Subramaniam S et al. Glutamate activates NF-kappaB through calpain in neurons. Eur J Neurosci 2003; 18(12):3305-3310.
16. Cruise L, Ho LK, Veitch K et al. Kainate receptors activate NF-kappaB via MAP kinase in striatal neurones. Neuroreport 2000; 11(2):395-398.
17. Kaltschmidt C, Kaltschmidt B, Baeuerle PA. Stimulation of ionotropic glutamate receptors activates transcription factor NF-kappa B in primary neurons. Proc Natl Acad Sci USA 1995; 92(21):9618-9622.
18. Pizzi M, Goffi F, Boroni F et al. Opposing roles for NF-kappa B/Rel factors p65 and c-Rel in the modulation of neuron survival elicited by glutamate and interleukin-1beta. J Biol Chem 2002; 277(23):20717-20723.
19. Wellmann H, Kaltschmidt B, Kaltschmidt C. Retrograde transport of transcription factor NF-kappa B in living neurons. J Biol Chem 2001; 276(15):11821-11829.
20. Schmitz ML, Bacher S, Kracht M. I kappa B-independent control of NF-kappa B activity by modulatory phosphorylations. Trends Biochem Sci 2001; 26(3):186-190.
21. Chen LF, Greene WC. Regulation of distinct biological activities of the NF-kappaB transcription factor complex by acetylation. J Mol Med 2003; 81(9):549-557.
22. Yeh SH, Lin CH, Gean PW. Acetylation of nuclear factor-kappaB in rat amygdala improves long-term but not short-term retention of fear memory. Mol Pharmacol 2004; 65(5):1286-1292.
23. Cogswell PC, Scheinman RI, Baldwin Jr AS. Promoter of the human NF-kappa B p50/p105 gene. Regulation by NF-kappa B subunits and by c-REL. J Immunol 1993; 150(7):2794-2804.
24. Sun SC, Ganchi PA, Ballard DW et al. NF-kappa B controls expression of inhibitor I kappa B alpha: Evidence for an inducible autoregulatory pathway. Science 1993; 259(5103):1912-1915.
25. Sun SC, Ganchi PA, Beraud C et al. Autoregulation of the NF-kappa B transactivator RelA (p65) by multiple cytoplasmic inhibitors containing ankyrin motifs. Proc Natl Acad Sci USA 1994; 91(4):1346-1350.
26. Grilli M, Goffi F, Memo M et al. Interleukin-1beta and glutamate activate the NF-kappaB/Rel binding site from the regulatory region of the amyloid precursor protein gene in primary neuronal cultures. J Biol Chem 1996; 271(25):15002-15007.
27. Furukawa K, Mattson MP. The transcription factor NF-kappaB mediates increases in calcium currents and decreases in NMDA- and AMPA/kainate-induced currents induced by tumor necrosis factor-alpha in hippocampal neurons. J Neurochem 1998; 70(5):1876-1886.
28. Wang Y, Qin ZH, Nakai M et al. Costimulation of cyclic-AMP-linked metabotropic glutamate receptors in rat striatum attenuates excitotoxin-induced nuclear factor-kappaB activation and apoptosis. Neuroscience 1999; 94(4):1153-1162.
29. Wood JN. Regulation of NF-kappa B activity in rat dorsal root ganglia and PC12 cells by tumour necrosis factor and nerve growth factor. Neurosci Lett 1995; 192(1):41-44.
30. Carter BD, Kaltschmidt C, Kaltschmidt B et al. Selective activation of NF-kappa B by nerve growth factor through the neurotrophin receptor p75. Science 1996; 272(5261):542-545.
31. Hamanoue M, Middleton G, Wyatt S et al. p75-mediated NF-kappaB activation enhances the survival response of developing sensory neurons to nerve growth factor. Mol Cell Neurosci 1999; 14(1):28-40.
32. Gu Z, Toliver-Kinsky T, Glasgow J et al. NGF-mediated alteration of NF-kappaB binding activity after partial immunolesions to rat cholinergic basal forebrain neurons. Int J Dev Neurosci 2000; 18(4-5):455-468.
33. Bui NT, Livolsi A, Peyron JF et al. Activation of nuclear factor kappaB and Bcl-x survival gene expression by nerve growth factor requires tyrosine phosphorylation of IkappaBalpha. J Cell Biol 2001; 152(4):753-764.
34. Bui NT, Konig HG, Culmsee C et al. p75 neurotrophin receptor is required for constitutive and NGF-induced survival signalling in PC12 cells and rat hippocampal neurones. J Neurochem 2002; 81(3):594-605.
35. Culmsee C, Gerling N, Lehmann M et al. Nerve growth factor survival signaling in cultured hippocampal neurons is mediated through TrkA and requires the common neurotrophin receptor P75. Neuroscience 2002; 115(4):1089-1108.
36. Zelenaia O, Schlag BD, Gochenauer GE et al. Epidermal growth factor receptor agonists increase expression of glutamate transporter GLT-1 in astrocytes through pathways dependent on phosphatidylinositol 3-kinase and transcription factor NF-kappaB. Mol Pharmacol 2000; 57(4):667-678.

37. Yabe T, Wilson D, Schwartz JP. NFkappaB activation is required for the neuroprotective effects of pigment epithelium-derived factor (PEDF) on cerebellar granule neurons. J Biol Chem 2001; 276(46):43313-43319.
38. Rodriguez-Kern A, Gegelashvili M, Schousboe A et al. Beta-amyloid and brain-derived neurotrophic factor, BDNF, up-regulate the expression of glutamate transporter GLT-1/EAAT2 via different signaling pathways utilizing transcription factor NF-kappaB. Neurochem Int 2003; 43(4-5):363-370.
39. Kaltschmidt B, Kaltschmidt C. DNA array analysis of the developing rat cerebellum: Transforming growth factor-beta2 inhibits constitutively activated NF-kappaB in granule neurons. Mech Dev 2001; 101(1-2):11-19.
40. Lilienbaum A, Israel A. From calcium to NF-kappa B signaling pathways in neurons. Mol Cell Biol 2003; 23(8):2680-2698.
41. Knapp LT, Klann E. Role of reactive oxygen species in hippocampal long-term potentiation: Contributory or inhibitory? J Neurosci Res 2002; 70(1):1-7.
42. Klann E, Roberson ED, Knapp LT et al. A role for superoxide in protein kinase C activation and induction of long-term potentiation. J Biol Chem 1998; 273(8):4516-4522.
43. Thiels E, Urban NN, Gonzalez-Burgos GR et al. Impairment of long-term potentiation and associative memory in mice that overexpress extracellular superoxide dismutase. J Neurosci 2000; 20(20):7631-7639.
44. Gahtan E, Auerbach JM, Groner Y et al. Reversible impairment of long-term potentiation in transgenic Cu/Zn-SOD mice. Eur J Neurosci 1998; 10(2):538-544.
45. Levin ED, Brady TC, Hochrein EC et al. Molecular manipulations of extracellular superoxide dismutase: Functional importance for learning. Behav Genet 1998; 28(5):381-390.
46. Bindokas VP, Jordan J, Lee CC et al. Superoxide production in rat hippocampal neurons: Selective imaging with hydroethidine. J Neurosci 1996; 16(4):1324-1336.
47. Sanz L, Diaz-Meco MT, Nakano H et al. The atypical PKC-interacting protein p62 channels NF-kappaB activation by the IL-1-TRAF6 pathway. EMBO J 2000; 19(7):1576-1586.
48. Sanz L, Sanchez P, Lallena MJ et al. The interaction of p62 with RIP links the atypical PKCs to NF-kappaB activation. EMBO J 1999; 18(11):3044-3053.
49. Macdonald NJ, Perez-Polo JR, Bennett AD et al. NGF-resistant PC12 cell death induced by arachidonic acid is accompanied by a decrease of active PKC zeta and nuclear factor kappa B. J Neurosci Res 1999; 57(2):219-226.
50. Esteve PO, Chicoine E, Robledo O et al. Protein kinase C-zeta regulates transcription of the matrix metalloproteinase-9 gene induced by IL-1 and TNF-alpha in glioma cells via NF-kappa B. J Biol Chem 2002; 277(38):35150-35155.
51. Baeuerle PA, Henkel T. Function and activation of NF-kappa B in the immune system. Annu Rev Immunol 1994; 12:141-179.
52. Sippy BD, Hofman FM, Wright AD et al. Induction of intercellular adhesion molecule-1 by tumor necrosis factor-alpha through the 55-kDa receptor is dependent on protein kinase C in human retinal pigment epithelial cells. Invest Ophthalmol Vis Sci 1996; 37(4):597-606.
53. Moghaddami N, Costabile M, Grover PK et al. Unique effect of arachidonic acid on human neutrophil TNF receptor expression: Up-regulation involving protein kinase C, extracellular signal-regulated kinase, and phospholipase A2. J Immunol 2003; 171(5):2616-2624.
54. Adams JP, Sweatt JD. Molecular psychology: Roles for the ERK MAP kinase cascade in memory. Annu Rev Pharmacol Toxicol 2002; 42:135-163.
55. Roberson ED, English JD, Sweatt JD. A biochemist's view of long-term potentiation. Learn Mem 1996; 3(1):1-24.
56. Harris EW, Ganong AH, Cotman CW. Long-term potentiation in the hippocampus involves activation of N-methyl-D-aspartate receptors. Brain Res 1984; 323(1):132-137.
57. Morris RG, Anderson E, Lynch GS et al. Selective impairment of learning and blockade of long-term potentiation by an N-methyl-D-aspartate receptor antagonist, AP5. Nature 1986; 319(6056):774-776.
58. Desmond NL, Colbert CM, Zhang DX et al. NMDA receptor antagonists block the induction of long-term depression in the hippocampal dentate gyrus of the anesthetized rat. Brain Res 1991; 552(1):93-98.
59. Lynch G, Browning M, Bennett WF. Biochemical and physiological studies of long-term synaptic plasticity. Fed Proc 1979; 38(7):2117-2122.
60. Dunwiddie TV, Lynch G. The relationship between extracellular calcium concentrations and the induction of hippocampal long-term potentiation. Brain Res 1979; 169(1):103-110.
61. Matthies H, Reymann KG. Protein kinase A inhibitors prevent the maintenance of hippocampal long-term potentiation. Neuroreport 1993; 4(6):712-714.
62. Lovinger DM, Wong KL, Murakami K et al. Protein kinase C inhibitors eliminate hippocampal long-term potentiation. Brain Res 1987; 436(1):177-183.
63. Silva AJ, Stevens CF, Tonegawa S et al. Deficient hippocampal long-term potentiation in alpha-calcium-calmodulin kinase II mutant mice. Science 1992; 257(5067):201-206.

64. English JD, Sweatt JD. Activation of p42 mitogen-activated protein kinase in hippocampal long term potentiation. J Biol Chem 1996; 271(40):24329-24332.
65. English JD, Sweatt JD. A requirement for the mitogen-activated protein kinase cascade in hippocampal long term potentiation. J Biol Chem 1997; 272(31):19103-19106.
66. Funauchi M, Tsumoto T, Nishigori A et al. Long-term depression is induced in Ca2+/calmodulin kinase-inhibited visual cortex neurons. Neuroreport 1992; 3(2):173-176.
67. Crepel F, Krupa M. Activation of protein kinase C induces a long-term depression of glutamate sensitivity of cerebellar Purkinje cells. An in vitro study. Brain Res 1988; 458(2):397-401.
68. Norman ED, Thiels E, Barrionuevo G et al. Long-term depression in the hippocampus in vivo is associated with protein phosphatase-dependent alterations in extracellular signal-regulated kinase. J Neurochem 2000; 74(1):192-198.
69. Thiels E, Kanterewicz BI, Norman ED et al. Long-term depression in the adult hippocampus in vivo involves activation of extracellular signal-regulated kinase and phosphorylation of Elk-1. J Neurosci 2002; 22(6):2054-2062.
70. Albensi BC, Mattson MP. Evidence for the involvement of TNF and NF-kappaB in hippocampal synaptic plasticity. Synapse 2000; 35(2):151-159.
71. Yeh SH, Lin CH, Lee CF et al. A requirement of nuclear factor-kappaB activation in fear-potentiated startle. J Biol Chem 2002; 277(48):46720-46729.
72. Minami M, Kuraishi Y, Satoh M. Effects of kainic acid on messenger RNA levels of IL-1 beta, IL-6, TNF alpha and LIF in the rat brain. Biochem Biophys Res Commun 1991; 176(2):593-598.
73. de Bock F, Dornand J, Rondouin G. Release of TNF alpha in the rat hippocampus following epileptic seizures and excitotoxic neuronal damage. Neuroreport 1996; 7(6):1125-1129.
74. Schall TJ, Lewis M, Koller KJ et al. Molecular cloning and expression of a receptor for human tumor necrosis factor. Cell 1990; 61(2):361-370.
75. Loetscher H, Pan YC, Lahm HW et al. Molecular cloning and expression of the human 55 kd tumor necrosis factor receptor. Cell 1990; 61(2):351-359.
76. Smith CA, Davis T, Anderson D et al. A receptor for tumor necrosis factor defines an unusual family of cellular and viral proteins. Science 1990; 248(4958):1019-1023.
77. Nophar Y, Kemper O, Brakebusch C et al. Soluble forms of tumor necrosis factor receptors (TNF-Rs). The cDNA for the type I TNF-R, cloned using amino acid sequence data of its soluble form, encodes both the cell surface and a soluble form of the receptor. EMBO J 1990; 9(10):3269-3278.
78. Kohno T, Brewer MT, Baker SL et al. A second tumor necrosis factor receptor gene product can shed a naturally occurring tumor necrosis factor inhibitor. Proc Natl Acad Sci USA 1990; 87(21):8331-8335.
79. Beattie EC, Stellwagen D, Morishita W et al. Control of synaptic strength by glial TNFalpha. Science 2002; 295(5563):2282-2285.
80. Tancredi V, D'Arcangelo G, Grassi F et al. Tumor necrosis factor alters synaptic transmission in rat hippocampal slices. Neurosci Lett 1992; 146(2):176-178.
81. Cunningham AJ, Murray CA, O'Neill LA et al. Interleukin-1 beta (IL-1 beta) and tumour necrosis factor (TNF) inhibit long-term potentiation in the rat dentate gyrus in vitro. Neurosci Lett 1996; 203(1):17-20.
82. Bear MF. Mechanism for a sliding synaptic modification threshold. Neuron 1995; 15(1):1-4.
83. Cheng B, Christakos S, Mattson MP. Tumor necrosis factors protect neurons against metabolic-excitotoxic insults and promote maintenance of calcium homeostasis. Neuron 1994; 12(1):139-153.
84. Mattson MP, Goodman Y, Luo H et al. Activation of NF-kappaB protects hippocampal neurons against oxidative stress-induced apoptosis: Evidence for induction of manganese superoxide dismutase and suppression of peroxynitrite production and protein tyrosine nitration. J Neurosci Res 1997; 49(6):681-697.
85. Weeber EJ, Levenson JM, Sweatt JD. Molecular genetics of human cognition. Mol Intervent 2002; 2(6):376-391.
86. Phillips RG, LeDoux JE. Differential contribution of amygdala and hippocampus to cued and contextual fear conditioning. Behav Neurosci 1992; 106(2):274-285.
87. Kim JJ, Rison RA, Fanselow MS. Effects of amygdala, hippocampus, and periaqueductal gray lesions on short- and long-term contextual fear. Behav Neurosci 1993; 107(6):1093-1098.
88. Levenson JM, Choi S, Lee S-Y et al. A Bioinformatics analysis of memory consolidation reveals involvement of the transcription factor c-Rel. J Neurosci 2004; 24(16):3933-43.
89. Tumang JR, Owyang A, Andjelic S et al. c-Rel is essential for B lymphocyte survival and cell cycle progression. Eur J Immunol 1998; 28(12):4299-4312.
90. McDonald RJ, White NM. A triple dissociation of memory systems: Hippocampus, amygdala, and dorsal striatum. Behav Neurosci 1993; 107(1):3-22.

CHAPTER 12

Inhibitors of NF-κB Activity:
Tools for Treatment of Human Ailments

Vinay Tergaonkar, Qiutang Li and Inder M. Verma*

Summary

Apart from being a paradigm for understanding cellular signaling, the NF-κB pathway has been thoroughly investigated over the last two decades due to its involvement in a number of human diseases. In the post genomic era, improved knowledge and novel technologies have contributed immensely to the discovery of several hitherto unknown cellular processes that regulate NF-κB. Identification of covalent modifications of many NF-κB pathway components, both in the cytoplasm and the nucleus has shed light on novel mechanisms that regulate NF-κB activity. Similarly, study of a number of cellular and viral proteins that regulate this pathway has added to our understanding of the molecular mechanisms and molecular targets in the NF-κB pathway for drug development.

Introduction

NF-κB signaling plays a pivotal role in several cellular and developmental processes in metazoans.[1] Deregulation of this pathway is suspected in and also causally linked to the initiation and progression of many human pathologies.[2] Since its discovery in the B cells and the relatively greater understanding of its role in the cells of the immune system, uncontrolled NF-κB activity has most often been associated with inflammatory disorders. In recent years, hyperactivity of NF-κB has also been linked to insulin resistance,[3] cachexia,[4] Alzheimer's disease,[5] transplant tolerance/graft rejection,[6] organ ischemia/reperfusion injury,[7] incontinentia pigmenti[8-10] and several neoplastic malignancies.[11,12] These observations warrant the development of strategies for modulation and inhibition of NF-κB activity. A wealth of information regarding the efforts in this direction has been a topic of exhaustive reviews in the literature. The aim of this chapter is to update the progress made in the very recent years and to direct the reader towards much of the established facts for more detailed perusal.

NF-κB is a family of transcription factors and the term NF-κB usually refers to a dimer of two similar or heterologous subunits of the family. In resting cells, these proteins are found associated with IκB family of inhibitory proteins and are predominantly localized in the cytoplasm.[13] The common denominator for NF-κB activation is the removal of IκB proteins from the DNA binding subunits of NF-κB.[14] Except in response to select stimuli, this removal is mediated by proteasomal degradation and requires that the IκB molecules are phosphorylated at specific serine residues by the IκB kinases (IKK).[15] Ample biochemical and genetic evidence exists that βTrCP1 and βTrCP2 (vertebrate homologues of *Drosophila* Slimb; HOS)[16] form the receptor component

---

*Corresponding Author: Inder M. Verma—Laboratory of Genetics, The Salk Institute for Biological Studies, 10010 North Torrey Pines Road, La Jolla, California 92037, U.S.A. Email: verma@salk.edu.

NF-κB/Rel Transcription Factor Family, edited by Hsiou-Chi Liou.
©2006 Landes Bioscience and Springer Science+Business Media.

of the SCF$^{\beta TrCP/HOS}$ E3 ubiquitin ligase along with Skp1, cullin1 and Roc1/Rbx1 components[17] and target IκB[18-21] and p100[22] molecules for ubiquitin-dependent proteolysis only when their conserved DSGXXS motif, is phosphorylated in a stimulus dependent manner.

Although, the rate-limiting step in the activation of NF-κB is the stimulus dependent degradation of IκB proteins, it is now apparent that mere release from IκB molecules is not enough for activation of NF-κB. Several modifications such as those of upstream molecules that activate the IKKs and NF-κB itself have been reported to be important for functional NF-κB activity.[23] Sub-cellularly, the targets of covalent modifications that regulate NF-κB activity range from adaptors such as TRAFs on the cell membrane, the IKKs in the cytoplasm to the p65 and p50 subunit in the nucleus.[23] Molecularly, the known post translational modifications include s-nitrosylation, addition/removal of phosphate, acetyl, ubiquitin, and sumo moieties and might include cis/trans isomerization by enzymes such as pin1.[24] While some of these modifications are crucial for NF-κB activity, others might play a subtler role in defining the duration and extent of NF-κB activity. The fact that NF-κB is activated by over 200 stimuli and it in turns activates an equally large subset of target genes in different cell types poses a major challenge in designing pathway specific inhibitors of NF-κB.[25] Knowledge of all relevant modifications of NF-κB pathway members is imperative to comprehend how specificity is generated in such a promiscuous pathway and this would be critical in designing inhibitors that do not have pleiotropic effects.

Inhibition of NF-κB Can Be Achieved at Multiple Points in the Pathway

Inhibition of NF-κB in various cellular compartments can be achieved by (a) inhibiting the receptors and the adaptors on the membrane, (b) the upstream kinases and IKKs, in the cytoplasm and (c) NF-κB DNA binding and transactivation in the nucleus. In this review, we describe the chemical and biological inhibitors of NF-κB known in the literature and then will classify them based on the specific molecular reaction they are known to inhibit. The biological inhibitors will be further classified based on their origin as cellular, viral, bacterial or natural products. Based on our knowledge of how NF-κB is activated in response to various stimuli, all cellular proteins essential for activation are targets for drug development. A common method to inhibit any signaling pathway is to make dominant negative versions of activators such that they now act as inhibitors. We will also discuss such biological molecules that do not exist in nature but which have been derived from proteins known to modulate the pathway. The chemical inhibitors will be classified as synthetic and natural. Given the relatively greater nonspecificity of chemicals inhibitors, we will describe the mode of action of only a few well-characterized class of compounds.

Biological Inhibitors

Cellular Inhibitors

IκB and Related Cellular Proteins Involved in Limiting NF-κB Activity

The most potent mechanism that inactivates NF-κB function is its association with IκB proteins. This process primarily prevents NF-κB DNA binding and also sequesters the complex in the cytoplasm.[1] Members of the IκB family have been proposed to play different roles in activation of NF-κB in response to different stimulation.[23] The knock-in of IκBβ in IκBα locus has clearly demonstrated that in the absence of IκBα, its biochemical function can be substituted by IκBβ and that the functional differences between these proteins might just be due to their differential expression patterns and/or association with other proteins.[26] Several viral proteins like HBX[27] and cellular proteins like G3BP2,[28] can associate with and regulate the subcellular localization and degradation of IκB proteins. It is conceivable that IκBα, IκBβ and IκBϵ proteins interact with different sets of cytoplasmic proteins and that this interaction, in a cell type specific manner regulates the kinetics and the degree of their degradation. The κB-ras proteins are a case in point, since they specifically regulate the activation of IκBβ bound NF-κB complexes.[29] Recently β-arrestin2 has been shown to regulate the extent of IκBα phosphorylation and degradation.[30]

Although much of the attention in the field has focused on mechanisms that activate NF-κB, the events that regulate the degree of NF-κB activity and lead to efficient termination of the activity are only now beginning to be unraveled. At the level of IKK activation, inhibitory auto-phosphorylation of IKKs[31] and deubiquitination of adaptor molecules by enzymes such as CYLD[32,33] have been documented to prevent prolonged NF-κB activity. An abundant cellular protein, hnRNPU has been shown to regulate the SCF$^{βTrCp/HOS}$ complex and thereby impinge on NF-κB activation by limiting IκB degradation in the nucleus.[34] These cellular signaling molecules and the partners of IκB:NF-κB complex could be efficiently used to limit NF-κB activity.

Regulation of NF-κB by Transformation/Cell Cycle Related Proteins

Aberrant regulation of NF-κB is observed in several human diseases. Although the role of NF-κB in regulating molecules involved in inflammation might be the primary unifying mechanism underlying most of the diseases, other activities of NF-κB that aid in the development of diseases have now been uncovered. One prominent activity attributed to NF-κB that might contribute towards its role in some human disorders is its ability to protect cells from apoptosis[1] in response to a diverse set of physiological stimuli. Indeed the relevance of NF-κB in mitigating apoptosis is now becoming evident in the evolution of several human malignancies.[2] These malignancies include those of haematological origin, helicobacter pylori-associated carcinogenesis, and cancer of the breast, colon, liver and cervix.[12,35] Apart from the NF-κB mediated cytokines and adhesion molecules that play an important role in tumorigenesis,[36] NF-κB has also been described to interact with molecules involved in growth, differentiation and transformation. Inhibitors of NF-κB activity have been documented to augment existing chemotherapy regimens.[37,38] NF-κB mediated expression of Cyclin D1 has been shown to be important for growth and proliferation[39,40] and mammary gland development[41] and the p65 subunit of NF-κB is known to interact with cyclin E-CDK2 complex.[42] IKK2 mediated NF-κB activation can destabilize the levels of p53 in fibroblasts[43] and intestinal epithelial cells.[44] Similarly, NF-κB has been documented to play an important role in subverting p53 and p73 induced cell death in lymphocytes[45] and in MALT.[46] As evidence mounts that NF-κB mediated regulation of p53 is important in regulating apoptosis resistance and transformation, repression of NF-κB by tumor suppressor genes including p53 is also emerging as a potential mechanism of overcoming apoptosis resistance.

At least four tumors suppressors have now been documented to inhibit NF-κB function. The tumor suppressor ARF can inhibit NF-κB function by recruitment of the histone deacetylase, HDAC1 to the transcriptional activation domain of p65.[47] The candidate tumor suppressor gene ING4 involved in regulating brain tumor growth and angiogenesis, has also been demonstrated to physically interact with p65.[48] Infact, repression of p65 transactivation by ING is postulated as a potential mechanism of tumor suppressor function of ING.[48] Similarly, mutations of CYLD in familial cylindromatosis,[49] and consequent loss of its negative regulatory effects on NF-κB signaling might be linked to progression of this malignancy. Finally, transcriptional cross talk between p53 and NF-κB[50] has been documented and it is evident that p53 represses NF-κB activity,[51,52] probably as an autoregulatory mechanism. Similarly, the oncogenic protein *twist*, a known target gene of NF-κB, has now been documented to autoregulate its expression by negatively regulating NF-κB activity through repressing p65 function.[53]

Cellular Proteins Involved in Inhibiting Inflammation

Failure to downregulate NF-κB transcriptional activity leads to chronic inflammation and cell death. Hence, the factors that are responsible for preventing sustained NF-κB activity in inflammatory diseases are important to understand. Mice deficient for the cellular zinc finger protein A20 display cachexia due to severe inflammation and die prematurely because they fail to terminate NF-κB activity.[54] Recently, a unique mechanism of downregulating NF-κB by A20 has been highlighted.[55] A20 mediated downregulation of NF-κB involves two sequential processes including deubiquitination and then polyubiquitination and proteasomal degradation of RIP, an essential component of TNF receptor 1 signaling complex.[55]

Repression of NF-κB by transforming growth factor-β1 (TGF-β1) is believed to be one of the mechanisms that is responsible for termination of production of pro-inflammatory cytokines in the gut.[56] Macrophages pretreated with the anti-inflammatory cytokine IL10 show reduced expression of TNFα but not IL-6 in response to LPS, due to reduced binding of p50:p65 complexes on the TNFα promoter.[57] Deciphering the mechanism of selective repression of certain NF-κB promoters would be crucial in designing strategies tailored towards inhibiting specific subsets of genes.

Mutations in cold-induced autoinflammatory syndrome 1 (CIAS1) genes are associated with chronic inflammatory disorders.[58] Multiple isoforms of CIAS1 protein inhibit p65 nuclear translocation,[58] suggesting a plausible mechanism for sustained inflammatory activity in patients with these mutations. Heparin-binding growth factor-like growth factor (HB-EGF), a member of the EGF family, is known to significantly decrease cytokine-induced NO production by preventing IκB degradation and thus NF-κB activation.[59] The anti-inflammatory effects of nitric oxide (NO), a free radical, involve repression of IKK2 kinase activity by S-nitrosylation of cysteine 179[60] and cysteine 62 of p50.[61] Cysteine residues of IKKs have been targets of many NF-κB inhibitors including arsenite,[62] H_2O_2,[63] 4-hydroxy-2-nonenal[64] and cyclopentenone prostaglandins.[65]

Cellular Proteins Involved in Inhibiting Viral Replication

Two cellular factors have recently been identified to regulate HIV1 replication by inhibiting NF-κB. While RelA-associated inhibitor (RAI),[66] identified by a two hybrid screen, is localized in the nucleus and prevents p65 DNA binding upon overexpression, Murr1, a gene previously known to be involved in copper metabolism was found to be essential to inhibit NF-κB activation and HIV replication in CD4+T cells.[67]

Miscellaneous Cellular Proteins Known to Repress NF-κB

Notch-1 proteins and their receptors are important for several developmental decisions including cell fate specifications.[68] The N terminal portion of human notch-1 (Notch^{IC}) was shown to inhibit NF-κB activity in the nucleus. Similarly, Fas-associated factor 1 (FAF1) has been documented to inhibit nuclear accumulation of NF-κB.[69] The inhibitor of Cdk4 (INK4), which contains ankyrin repeats, like IκB proteins has been reported to bind and modulate the activity of p65 in certain cell types.[70] IFNα and not IFNγ has been shown to inhibit HBx mediated NF-κB activation.[71] Expression of c-Myc has been shown to repress NF-κB activity by blocking the transactivation potential of p65.[72] Overexpression of manganese superoxide dismutase (MnSOD) can inhibit IκB degradation and NF-κB activation.[73] NRF, nuclear protein was found to inhibit NF-κB by steric hindrance, specifically on the IFNβ promoter.[74] Also, androgen receptor signaling represses NF-κB activity in the androgen sensitive LNCaP cells.[75] Understanding the molecular mechanisms underlying these repressions needs further experimentation.

Variants of Cellular Proteins That Inhibit NF-κB

Dominant Negative Molecules

One potential avenue of inhibiting NF-κB signaling is to make decoy receptors of ligands that activate this pathway. Indeed such approaches have been attempted and several TNF variants that could inhibit the endogenous TNF signaling identified by a structure-based rationally designed screen.[76] Similar efforts with other NF-κB pathway components could yield more specific and potent inhibitors to specific stimuli. The most common and tested method of inhibiting NF-κB is to exogenously deliver IκB molecules that are not degraded following stimulation. The IκBαM[77] molecule is a prototype IκBα used in many studies, wherein the serine 32 and 36 which are sites of IKK phosphorylation are mutated to alanines. In addition, all the serines and threonines in the C terminal PEST domain of IκBα are mutated to alanines in IκBαM.[77] The advantage of using such molecules is that they bind and inhibit almost all NF-κB homo and heterodimers and thus achieve almost complete blockade of the pathway. Although all IκBαM overexpression results are interpreted to be because of inhibition of NF-κB activity, a cautionary note should be added since IκBαM has also been shown to bind and inhibit the function of cyclin dependent kinase-4.[78] In another variation of a similar strategy the lysine residues 21/22 (which are the site of ubiquitination) of IκBα are converted to alanines thus blocking its degradation.[79] Phosphorylation of tyrosine 42 of IκBα is another mechanism known to activate NF-κB which operates via removal of IκB from NF-κB.[80] The nonphosphorylatable IκBαY42F protein has also been used to block NF-κB activity and to prevent blockage of bone erosion associated with inflammatory arthritis.[81]

Cell Permeable Peptides

Recent advances in designing cell permeable peptides has led to the development of small peptide molecules that could interfere with protein-protein interactions essential for NF-κB signaling. These peptides can work at various subcellular locations. A peptide from interacting domain of TIRAP has been known to block its association to TLR4 and inhibit LPS-induced NF-κB activation.[82] Since NEMO or IKKγ is essential for IKK and NF-κB activation, another strategy used to inhibit NF-κB is to make a peptide derived from the extreme carboxy terminus of IKK2 and IKK1 that binds and squelches free NEMO in the cell.[83] In vivo studies with this peptide (NBD), are indeed showing promising results.[83,84] NBD peptide administered into mice prior to induction of inflammatory arthritis efficiently blocks in vivo osteoclast recruitment, inhibits focal bone erosion, and ameliorates inflammatory responses in the joints of arthritic mice.[85] Ben-Neriah and colleagues have used cell permeable phosphopeptides derived from IκBα as a means of competitively inhibiting the $SCF^{\beta TrCP}$ ligase.[86] Interestingly a range of cellular and viral proteins have recently been documented to inhibit the NF-κB pathway by competing for the $SCF^{\beta TrCP}$. It has been documented that a mutant of βTrCP1 that has been deleted of the F box (ΔF-βTrCP1) can function as a dominant negative in the process of IκBα degradation, since it can bind the substrate but cannot recruit it for efficient ubiquitination.[16] This method is also able to block processing of other IκB members like p100 and also prevent degradation of small amounts of phospho-IκBα generated by weak NF-κB activating stimuli such as DNA damage.[87] A twelve-residue-peptide from p65 covering serine residues 276, whose phosphorylation is required for NF-κB activation is fused to membrane permeable domain PTD was shown to selectively inhibit NF-κB activation induced by various inflammatory stimuli.[88] A forty-one-residue-peptide consisting the nuclear localization sequences of p50 can also effectively inhibit NF-κB.[89,90] However, since such an approach nonspecifically clogs the nuclear import machinery, it offers a very nonspecific method of NF-κB inhibition.[89,90] Theoretically peptides or phosphopeptides derived from all essential components of the IKK complex, such as ELKS[91] can also be used to make dominant negative inhibitors of IKK and NF-κB.

Bacterial Inhibitors

Many bacterial strains of the *Yersinia* species, known to colonize both plant and animal hosts, encode a set of effector proteins called the Yops (Yersinia outer proteins).[92] YopJ, a protein encoded by both plant and animal pathogens has been documented to block the NF-κB pathway.[93] It is now known that YopJ has a ubiquitin like protease (ULP) activity and may target several cellular proteins with ubiquitin domains,[93] thus explaining its inhibitory effect on many signaling pathways.[92] With the documented role of ubiquitination in activation of IKK[94,95] and thus NF-κB pathways, it is most likely that the YopJ mediated inhibition of

NF-κB activity operates by inhibition of IKK activation by upstream MAP kinase signaling. Peptides derived from YopJ might be useful tools to block IKK activity but this hasn't been tested yet. The periodontal pathogen *Porphyromonas gingivalis* has been documented to inhibit p65 DNA binding as a mode of inducing tolerance.[96]

Viral Inhibitors

Viruses have been known to evade the host immune system by a variety of mechanisms.[97] Since NF-κB is required for activation of many cytokines that are crucial to mount an effective immune response, inhibition of NF-κB activity seems a worthwhile strategy for viruses to utilize to escape immune surveillance.[98] Pichinde virus is an arenavirus which displays tropism towards macrophages and causes Lassa fever in humans.[99] Virulent, but not attenuated strains of this virus have been documented to cause repression of NF-κB activation which leads to decreased macrophage activation, possibly as a means to evade the host immune system.[99] In a similar vein, the African swine fever virus (ASFV), which replicates in macrophages, encodes a truncated version of IκB protein, which lacks the N terminal signal responsive regions but binds and inhibits NF-κB activity in response to all forms of activation.

The use of adenoviruses, which are large double strand DNA viruses in gene delivery is largely impeded by their immunogenicity. However these viruses are postulated to have evolved mechanisms to evade the immune system. Two adenoviral protein complexes, the E1A[100] and the E3-10.4/14.5[101] have been shown to inhibit IKK and thus NF-κB activation, possibly as a means to limit the action of interferon mediated immune action. Human cytomegalovirus (HCMV), a significant factor in infections of immune compromised patients, encodes the pp65 protein which inhibits NF-κB DNA binding.[102] Similarly, the NS1 protein of influenza A virus inhibits NF-κB activation in response to the double strand RNA generated during the course of infection as a means to inhibit IFN production.[103] The Epstein-Barr virus (EBV) ZEBRA protein can also inhibit NF-κB activation in T cells during acute EBV infections.[104] Vaccinia virus K1L gene product blocks IκB degradation as a mechanism of inhibiting NF-κB activity.[105]

Like the cellular protein hnRNPU,[34] two viral proteins, have also recently been documented to interact with SCF$^{βTrCp/HOS}$ complex and control the extent of NF-κB activation. The HIV-1 membrane protein, Vpu has been proposed to interact with βTrCP1,[106] while the latent membrane protein 1 (LMP1) encoded by EBV have been documented to interact with βTrCP2.[107] Since βTrCP1, unlike βTrCP2 is nuclear,[34,108] Vpu has been proposed to sequester βTrCP1 in the cytoplasm and thereby inhibit its function during viral infection.[106] Inhibition of NF-κB by Vpu and LMP1 could certainly represent a mechanism for HIV1 and EBV to counterbalance innate immunity pathways in response to viral infections.

Natural Inhibitors

Dietary and natural products have long been used to reduce the risk for inflammatory disorders and cancers as complement regimens.[109] Molecular understanding of these natural NF-κB modulators may help us to design inhibitory small molecules compounds. Along with the rapid advances in our understanding of NF-κB signaling, large number of natural compounds are now being tested and reassessed for their actions in modulating NF-κB function. Phenolic compounds are one such group which contains a number of NF-κB inhibitory compounds such as EGCG, wogonin, oroxylin A, resveratrol, hepericin, curcumin, and silybin. The tea polyphenol EGCG is a potent inhibitor of TNFα-mediated NF-κB activity and it possibly functions by inhibiting IKK activity.[110] Curcumin found in the spice turmeric exhibits anti-inflammatory, anti-oxidant, and chemo preventive activities. It is suggested that curcumin most likely inhibits cell proliferation, cell-mediated cytotoxicity, and cytokine production through suppression of IKK-mediated NF-κB activation.[111] Some of other famous natural compound groups include isoprenoid compounds (such as kaurene diterpenes, kaurane diterpenes and cyclic diterpene) and class of sesquiterpene lactones (such as parthenolide and helenalin), triterpenoids (such as avicin, pristimerin, and oleandrin).[109] These compounds inhibit different steps in the NF-κB activation cascade. Compounds like kaurene diterpeniods, parthenolide, oleandrin directly inhibit the IKK complex, while others such as avicin and helenalin suppress p65 function.[109]

Finally, cellular bioproducts of inflammatory reactions such as cyclopentenone prostaglandins inhibit IKK2[65] and NF-κB[112] by covalent modification as a mechanism of limiting prolonged NF-κB activity and this is important for the resolution of inflammation.

Synthetic Inhibitors

A number of conventionally used-inflammatory drugs and natural products have demonstrated NF-κB inhibitory activity by targeting various network components including IKK complex activity, IκB degradation, NF-κB nuclear localization, and NF-κB DNA binding and transcriptional capacity.[113] These agents include proteasome inhibitors glucocorticoids, nonsteroidal anti-inflammatory drugs, anti-inflammatory cytokines, and natural compounds. The finding of inhibition of NF-κB by multiple anti-inflammatory and anti-cancer pharmacologic agents reassures us of the involvement of NF-κB in pathogenesis of these diseases and its potential as drug target. Hence, development of pharmacologic inhibitors specific for NF-κB might provide safer and more efficacious drugs for the treatment of diseases ranging from inflammation to cancer.

Proteasome Inhibitors

Given the importance of the ubiquitination system in NF-κB activation, this system is a hot target for development of new anti-inflammatory therapies.[114,115] Proteasome inhibition is an effective way to block NF-κB activity generated by the canonical, noncanonical[114] and DNA damage pathways.[43] The proteasomal inhibitor Bortezomib (PS-341), is a potent inhibitor of NF-κB[116] activity, however like other proteasomal inhibitors, it is very toxic over a long period of time and each treatment lasts for a few days only.[117] Recently, PS-341 was approved by US food and Drug administration for the treatment of relapsed and refractory multiple myeloma.[118,119] It has demonstrated impressive antitumor activity in combination with other drugs. Although, evaluation of the mechanisms that underlie the antitumor effects of proteasome inhibitors reveals a considerable contribution of NF-κB blockade, it is evident that modulation of other cell cycle proteins and other pro- and antiapoptotic pathways also plays critical roles. In part, the nonspecific effects of proteasomal inhibition also stem from the fact the proteasome controls the physiological activity of several key cellular proteins, like p53[120] and β-catenin[18] to name a few.

Glucocorticoids

Glucocorticoids (GCs) such as synthetic steroid dexamethasome and hydrocortisone have been widely prescribed as anti-inflammatory and immunosuppressive drugs.[121] They inhibit expression of many genes including cytokines, chemokines, cell-surface receptors, adhesion molecules, tissue factor, degradative proteinases, and enzymes such as cyclooxygenase 2 (COX-2) and induced nitric oxide synthase (iNOS), which produce mediators of inflammation. Two mechanisms are proposed for the anti-inflammatory action of glucocorticoids. The first depends on its receptor binding activity directly to target transcription. And the second is dependent on its direct interaction and interference with other inflammatory transcription factors. Interfering with NF-κB signaling is thought as a major underlying mechanism for the anti-inflammatory capacity of GCs. Although the exact mechanism of GC mediated repression of NF-κB is not completely understood, several mechanisms have been proposed. GCs have been shown to increase the expression of IκBα, which retains NF-κB in cytoplasm and block its transcription activity.[122,123] GC-binding receptor also directly interacts with p65 in nucleus and inhibits activation of pro-inflammatory genes by blocking its transcriptional activity through modulating functions of RNA polymerase II and recruiting histone deacetylase 2.[124-126]

Nonsteroidal Anti-Inflammatory Drugs (NSAIDs)

NSAIDs are believed to operate by inhibiting cyclooxygenase activity to prevent pro-inflammatory prostaglandin (e.g., PGE2) synthesis. Interestingly, COX-2 is a NF-κB regulated gene and the inhibition of NF-κB further suppress the COX-2 expression. However interfering with NF-κB system is believed to contribute considerably to the anti-inflammatory effects of NSAIDs.[113] Several NSAIDs including aspirin and sodium salicylate, Ibuprofen, acetaminophen, sulindac, and tepoxalin are capable of inhibiting NF-κB activity. At a suprapharmacological dosage (mM range), both sodium salicylate and its semi-synthetic derivative, aspirin, bind and block the ATP binding site of IKK2.[127] Likewise, Sulindac and its metabolites, have been shown to inhibit the NF-κB-mediated signals through inhibition of IKK2 by direct interaction.[128,129] Celecoxib, a selective COX-2 inhibitor,[130] has recently been approved for the treatment of colon carcinogenesis, rheumatoid arthritis, and other inflammatory diseases and it can suppress NF-κB activation induced by various agents through inhibition of IKK and Akt.[35]

A recent estimate by Coussens and Werb[131] suggests that upwards of 15% of all cancers in humans are attributable to inflammation. The molecular mechanisms that sensitize sites of chronic inflammation to be malignant are virtually unknown. NF-κB is a key determinant of inflammatory responses and is known to be hyperactivated in many tumors. Use of nonsteroidal anti-inflammatory agents such as asprin and others that also block IKK2, reduce the risk of gastric, colon and lung cancers.[131] Taken together, these observations present a case that activation NF-κB could be the missing link between inflammation and cancer and that apart from their relevance in treating inflammatory disorders, NF-κB inhibitors could be useful means of preventing human malignancies.

Antioxidants

A number of NF-κB activating stimuli such LPS, H_2O_2, TNF, and IL-1 UV light and ionizing radiation increase cellular levels of reactive oxygen species ROS. Although the mechanistic details are still not completely clear, considerable evidence implicates ROS as common second messages in NF-κB activation. Introduction of various antioxidants such as ascorbic acid (vitamin C), vitamin E, NADPH, glutathione peroxidase, or MnSOD, along with these stimuli abolished NF-κB activation induced by these agents.[132] Interestingly, a direct linkage between the redox state of vitamin C and NF-κB signaling has been identified recently. It was shown that ascorbic acid quenches ROS intermediate involved in the activation of NF-κB and is oxidized to dehydroascorbic acid, which directly inhibit IKK2 and IKK1 kinase activity.[133] The list of anti-oxidant and inhibitors that prevent phosphorylation and degradation of IκBα in response to general and specific stimuli has been complied in a pervious review.[134]

IKK Inhibitors

Although a range of natural compounds and synthetic drugs are able to inhibit NF-κB activation pathway, they are not selective inhibitors of NF-κB activation pathway. Since undesirable side effects are often associated with such therapeutics, designing IKK selective inhibitors is a promising method to generate more NF-κB specific and safer therapeutics for the treatment of inflammatory diseases and cancer. Many drug companies are currently focusing on generating small molecules to specifically block IKK1 or IKK2 activities.

Several small molecule compounds have been identified which are highly selective, and orally bioavailable IKK2 inhibitors such as quinazoline analogues (SPC839), β-carboline derivatives (PS-1145), imidazoquinoxaline derivative (BMS-345541), amino-thiophenecar derivative (SC-514), ureido-thiophenecarboxamide derivative, diarylpyridine derivative.[135] The development of IKK1 specific inhibitor is also gaining momentum since IKK1 has a key role in the development of tooth,[136] B cells,[137] osteoclasts and breast epithelium.[41] Karin et al[135] and Bruke[138] summarize the efforts that are underway towards developing IKK inhibitors in recent reviews. Most of those inhibitors are still in preclinical stages of development and their safety and efficacy in the treatment of inflammatory disorder still remains to be determined.

Transcription Factor Decoy

Another approach towards inhibiting NF-κB signaling is to introduce double-stranded oligodeoxynucleotides (ODNs) into cells as decoy *cis*-elements that bind NF-κB and thus alter target gene expression. Decoy oligodeoxynucleotides to NF-κB have been designed and used in several studies[139-141] but they suffer from a number of limitations, including their solubility across membranes, their sensitivity to polymerases, lack of sequence specificity, and their tendency to activate cytokines production.[142] To overcome such problems, circular dumbbell decoys have also been designed which exhibit high resistance to nucleases, are easily taken up by cells, and have a nontoxic unmodified backbone that resembles natural DNA.[142]

Antisense Oligodeoxynucleotide

Antisense ODNs are short synthetic nucleotide sequences formulated to be complementary to a specific RNA message. Through binding of these ODNs to a target mRNA sequence, translation of the desired target genes can be selectively blocked, and the disease process generated by those genes can be halted.[143]

SiRNA As Inhibitors

Several recent reports have demonstrated the use of short hairpin RNAs (shRNAs) (for example against p65[144] and the upstream TAK1[145]) as an efficient means of inhibiting NF-κB pathway. Since mammalian cells are not known to make shRNA and although shRNAs could be delivered by a virus mediated route,[146] (or by direct transfection[147]) we have categorized them as synthetic inhibitors. However given that shRNAs have been reported to have several off-target effects,[148] the use of shRNA mediated inhibition should only be contingent upon rigorous evaluation of nonspecific effects. Also the use of regulated shRNAs[144] could be one way of limiting the off target effects of shRNAs.

Indirect Inhibitors

Several kinases including MEKK1, MEKK3, NIK, AKT, CKII, RSK, TAK1/T2K/NAK, GSK3β, MSK1 and MSK2 have been shown to be required for NF-κB activity.[1,149] Inhibitors that principally target these enzymes are also likely inhibitors of NF-κB, albeit with other side effects. Indeed wortmannin and other PI3K kinase inhibitors are potent inhibitors of NF-κB activity.[150] The involvement of AKT pathway in cancers has led to the development of several inhibitors of this pathway,[151] which could be used to inhibit NF-κB under certain situations for example in tumors with over active AKT. Similarly inhibitors of GSK3β have been shown to inhibit NF-κB activity.[152] Also, mitogen-activated protein kinase inhibitors as well as a mitogen and stress activated protein kinase (such as MSK1) inhibitors have been effective in blocking NF-κB activity in vitro.[153] Transforming growth factor beta 1 (TGFβ1) can stabilize the protein levels of IκBα by inhibiting CKII activity in hepatocytes.[154] Targeting of Hsp90, a component of the IKK complex,[1] by geldanamycin, a benzoquinone ansamycin isolated many years ago from Streptomyces, is known to inhibit NF-κB activity. However, these compounds also inhibit several other kinases due to the destabilizing affects they have on the general chaperone hsp90.[155]

Future Directions

The rapid advance in our understanding NF-κB signaling pathway has led us to a number of nodes where NF-κB activity could be regulated. Figure 1 summarizes the major inhibitors and the nodes of the pathway that they have been documented to operate upon.

Research on basic mechanisms of current pharmacological agents may provide potential for developing more specific and potent drugs with less toxicological side effects. Since blocking IκB degradation and its control by IKK complex are seminal steps in NF-κB activation, targeting this node for NF-κB specific blockade without a global inhibition is worth exploring further. This is particularly challenging since IKK1 and IKK2 that are highly similar have now

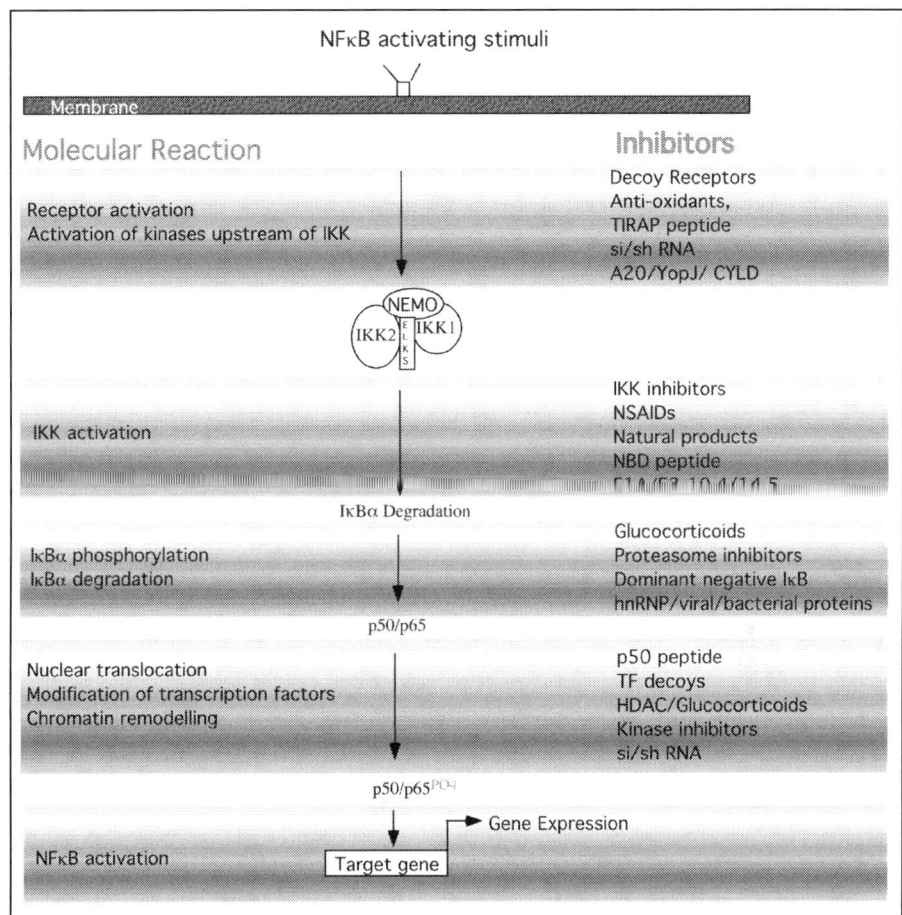

Figure 1. Major nodes in the NFκB pathway that have been inhibited experimentally. The molecular steps in the activation of NF-κB are depicted on the left and the inhibitors that have been documented to work have been shown on the right of the pathway. Only certain representative proteins or class of compounds have been shown. Please refer to text for details.

been shown to control distinct biological processes that may or may not work via NF-κB. Screening of bioactive substances from natural sources, against defined targets such as NF-κB and developing derivatization leads toward drugs should be a promising strategy for drug discovery. A large-scale genetic screen would be ideal towards rapidly understanding the distinct roles of IKK1 and IKK2 and in designing pathway specific inhibitors which do not have other systemic effects. Further, since it is becoming more evident that NF-κB is required for resolution of inflammation (since it is also critical for the activation of anti-inflammatory cytokines[156,157]), inhibitors of NF-κB will have to be designed to work efficaciously in a short period of time. Such inhibitors of NF-κB would also be beneficial to limit or block bystander effects on nontarget cells when systemic inhibition of NF-κB is called for.

Screening in Fish and Fly

The immense power of forward genetic screens have been very successful in identifying important signal transduction pathways involved in many biological processes in Drosophila.[158]

With the characterization of the NF-κB pathway in drosophila,[159] and now in zebrafish[160] these model systems represent an attractive tool for chemical and genetic screens to identify modulators and inhibitors of the NF-κB pathway. The rapid developmental stages and permeability of zebrafish embryos to small molecule inhibitors, and the strong homology between kinases from zebrafish and humans,[161,162] has made this system highly suited for high-throughput inhibitor screens. Zebrafish is now also used as a vertebrate model organism in immunologic research.[163] While present screening approaches target select candidate genes, recent organism-based screens have led to the discovery of chemicals that mitigate complex dysmorphic syndromes even without targeting the affected gene directly.[164,165]

Conclusion

In the last two decades we have learnt a lot about the basic mechanisms that activate NF-κB. With the advent of high throughput technologies, rational drug design techniques and novel model systems like drospophila and zebrafish, screening for specific modulators and inhibitors of NF-κB is sure to accelerate in the future. The challenge of the future would be to wade through a wealth of drug target hits found by such screens and to find the most effective and selective inhibitor of a given physiological process. The knowledge of how specific modifications regulate NF-κB activity in response to distinct stimuli would undoubtedly propel our efforts towards designing more specific inhibitors of NF-κB.

Acknowledgements

VT is supported by a career development fellowship from the Leukemia and Lymphoma Society. QL is supported by Wyeth Pharmaceuticals. I.M.V. is an American Cancer Society Professor of Molecular Biology and supported by grants from the NIH, the March of Dimes, the Wayne and Gladys Valley Foundation, and the Larry L. Hillblom Foundation, Inc., the Lebensfeld Foundation, the H.N. and Frances C. Berger Foundation.

References

1. Li Q, Verma IM. NF-kappaB regulation in the immune system. Nat Rev Immunol 2002; 2(10):725-734.
2. Rayet B, Gelinas C. Aberrant rel/nfkb genes and activity in human cancer. Oncogene 1999; 18(49):6938-6947.
3. Shoelson SE, Lee J, Yuan M. Inflammation and the IKK beta/I kappa B/NF-kappa B axis in obesity- and diet-induced insulin resistance. Int J Obes Relat Metab Disord 2003; 27(Suppl 3):S49-52.
4. von Haehling S, Genth-Zotz S, Anker SD et al. Cachexia: A therapeutic approach beyond cytokine antagonism. Int J Cardiol 2002; 85(1):173-183.
5. Mattson MP, Camandola S. NF-kappaB in neuronal plasticity and neurodegenerative disorders. J Clin Invest 2001; 107(3):247-254.
6. Tsoulfas G, Geller DA. NF-kappaB in transplantation: Friend or foe? Transpl Infect Dis 2001; 3(4):212-219.
7. Valen G, Yan ZQ, Hansson GK. Nuclear factor kappa-B and the heart. J Am Coll Cardiol 2001; 38(2):307-314.
8. Nichols TC, Fischer TH, Deliargyris EN et al. Role of nuclear factor-kappa B (NF-kappa B) in inflammation, periodontitis, and atherogenesis. Ann Periodontol 2001; 6(1):20-29.
9. Bell S, Degitz K, Quirling M et al. Involvement of NF-kappaB signalling in skin physiology and disease. Cell Signal 2003; 15(1):1-7.
10. Smahi A, Courtois G, Rabia SH et al. The NF-kappaB signalling pathway in human diseases: From incontinentia pigmenti to ectodermal dysplasias and immune-deficiency syndromes. Hum Mol Genet 2002; 11(20):2371-2375.
11. Garg A, Aggarwal BB. Nuclear transcription factor-kappaB as a target for cancer drug development. Leukemia 2002; 16(6):1053-1068.
12. Karin M, Cao Y, Greten FR et al. NF-kappaB in cancer: From innocent bystander to major culprit. Nat Rev Cancer 2002; 2(4):301-310.
13. Verma IM, Stevenson JK, Schwarz EM et al. Rel/NF-kappa B/I kappa B family: Intimate tales of association and dissociation. Genes Dev 1995; 9(22):2723-2735.

14. Karin M, Ben-Neriah Y. Phosphorylation meets ubiquitination: The control of NF-[kappa]B activity. Annu Rev Immunol 2000; 18:621-663.
15. Brown K, Gerstberger S, Carlson L et al. Control of I kappa B-alpha proteolysis by site-specific, signal-induced phosphorylation. Science 1995; 267(5203):1485-1488.
16. Spencer E, Jiang J, Chen ZJ. Signal-induced ubiquitination of IkappaBalpha by the F-box protein Slimb/beta-TrCP. Genes Dev 1999; 13(3):284-294.
17. Deshaies RJ. SCF and Cullin/Ring H2-based ubiquitin ligases. Annu Rev Cell Dev Biol 1999; 15:435-467.
18. Winston JT, Strack P, Beer-Romero P et al. The SCFbeta-TRCP-ubiquitin ligase complex associates specifically with phosphorylated destruction motifs in IkappaBalpha and beta-catenin and stimulates IkappaBalpha ubiquitination in vitro. Genes Dev 1999; 13(3):270-283.
19. Yaron A, Hatzubai A, Davis M et al. Identification of the receptor component of the IkappaBalpha-ubiquitin ligase. Nature 1998; 396(6711):590-594.
20. Suzuki H, Chiba T, Kobayashi M et al. IkappaBalpha ubiquitination is catalyzed by an SCF-like complex containing Skp1, cullin-1, and two F-box/WD40-repeat proteins, betaTrCP1 and betaTrCP2. Biochem Biophys Res Commun 1999; 256(1):127-132.
21. Wu C, Ghosh S. beta-TrCP mediates the signal-induced ubiquitination of IkappaBbeta. J Biol Chem 1999; 274(42):29591-29594.
22. Fong A, Sun SC. Genetic evidence for the essential role of beta-transducin repeat-containing protein in the inducible processing of NF-kappa B2/p100. J Biol Chem 2002; 277(25):22111-22114.
23. Ghosh S, Karin M. Missing pieces in the NF-kappaB puzzle. Cell 2002; 109(Suppl):S81-96.
24. Ryo A, Suizu F, Yoshida Y et al. Regulation of NF-kappaB signaling by Pin1-dependent prolyl isomerization and ubiquitin-mediated proteolysis of p65/RelA. Mol Cell 2003; 12(6):1413-1426.
25. Pahl HL. Activators and target genes of Rel/NF-kappaB transcription factors. Oncogene 1999; 18(49):6853-6866.
26. Cheng JD, Ryseck RP, Attar RM et al. Functional redundancy of the nuclear factor kappa B inhibitors I kappa B alpha and I kappa B beta. J Exp Med 1998; 188(6):1055-1062.
27. Weil R, Sirma H, Giannini C et al. Direct association and nuclear import of the hepatitis B virus X protein with the NF-kappaB inhibitor IkappaBalpha. Mol Cell Biol 1999; 19(9):6345-6354.
28. Prigent M, Barlat I, Langen H et al. IkappaBalpha and IkappaBalpha /NF-kappa B complexes are retained in the cytoplasm through interaction with a novel partner, RasGAP SH3-binding protein 2. J Biol Chem 2000; 275(46):36441-36449.
29. Fenwick C, Na SY, Voll RE et al. A subclass of Ras proteins that regulate the degradation of IkappaB. Science 2000; 287(5454):869-873.
30. Gao H, Sun Y, Wu Y et al. Identification of beta-arrestin2 as a G protein-coupled receptor-stimulated regulator of NF-kappaB pathways. Mol Cell 2004; 14(3):303-317.
31. Delhase M, Hayakawa M, Chen Y et al. Positive and negative regulation of IkappaB kinase activity through IKKbeta subunit phosphorylation. Science 1999; 284(5412):309-313.
32. Brummelkamp TR, Nijman SM, Dirac AM et al. Loss of the cylindromatosis tumour suppressor inhibits apoptosis by activating NF-kappaB. Nature 2003; 424(6950):797-801.
33. Kovalenko A, Chable-Bessia C, Cantarella G et al. The tumour suppressor CYLD negatively regulates NF-kappaB signalling by deubiquitination. Nature 2003; 424(6950):801-805.
34. Davis M, Hatzubai A, Andersen JS et al. Pseudosubstrate regulation of the SCF(beta-TrCP) ubiquitin ligase by hnRNP-U. Genes Dev 2002; 16(4):439-451.
35. Shishodia S, Aggarwal BB. Nuclear factor-kappaB activation mediates cellular transformation, proliferation, invasion angiogenesis and metastasis of cancer. Cancer Treat Res 2004; 119:139-173.
36. Richmond A. Nf-kappa B, chemokine gene transcription and tumour growth. Nat Rev Immunol 2002; 2(9):664-674.
37. Orlowski RZ, Baldwin Jr AS. NF-kappaB as a therapeutic target in cancer. Trends Mol Med 2002; 8(8):385-389.
38. Haefner B. NF-kappa B: Arresting a major culprit in cancer. Drug Discov Today 2002; 7(12):653-663.
39. Hinz M, Krappmann D, Eichten A et al. NF-kappaB function in growth control: Regulation of cyclin D1 expression and G0/G1-to-S-phase transition. Mol Cell Biol 1999; 19(4):2690-2698.
40. Guttridge DC, Albanese C, Reuther JY et al. NF-kappaB controls cell growth and differentiation through transcriptional regulation of cyclin D1. Mol Cell Biol 1999; 19(8):5785-5799.
41. Cao Y, Bonizzi G, Seagroves TN et al. IKKalpha provides an essential link between RANK signaling and cyclin D1 expression during mammary gland development. Cell 2001; 107(6):763-775.
42. Perkins ND, Felzien LK, Betts JC et al. Regulation of NF-kappaB by cyclin-dependent kinases associated with the p300 coactivator. Science 1997; 275(5299):523-527.

43. Tergaonkar V, Bottero V, Ikawa M et al. IkappaB Kinase-independent IkappaBalpha degradation pathway: Functional NF-kappaB activity and implications for cancer therapy. Mol Cell Biol 2003; 23(22):8070-8083.
44. Egan LJ, Eckmann L, Greten FR et al. IkappaB-kinasebeta-dependent NF-kappaB activation provides radioprotection to the intestinal epithelium. Proc Natl Acad Sci USA 2004; 101(8):2452-2457.
45. Wan YY, DeGregori J. The survival of antigen-stimulated T cells requires NFkappaB-mediated inhibition of p73 expression. Immunity 2003; 18(3):331-342.
46. Stoffel A, Chaurushiya M, Singh B et al. Activation of NF-kappaB and inhibition of p53-mediated apoptosis by API2/mucosa-associated lymphoid tissue 1 fusions promote oncogenesis. Proc Natl Acad Sci USA 2004; 101(24):9079-9084.
47. Rocha S, Campbell KJ, Perkins ND. p53- and Mdm2-independent repression of NF-kappa B transactivation by the ARF tumor suppressor. Mol Cell 2003; 12(1):15-25.
48. Garkavtsev I, Kozin SV, Chernova O et al. The candidate tumour suppressor protein ING4 regulates brain tumour growth and angiogenesis. Nature 2004; 428(6980):328-332.
49. Bignell GR, Warren W, Seal S et al. Identification of the familial cylindromatosis tumour-suppressor gene. Nat Genet 2000; 25(2):160-165.
50. Webster GA, Perkins ND. Transcriptional cross talk between NF-kappaB and p53. Mol Cell Biol 1999; 19(5):3485-3495.
51. Wadgaonkar R, Phelps KM, Haque Z et al. CREB-binding protein is a nuclear integrator of nuclear factor-kappaB and p53 signaling. J Biol Chem 1999; 274(4):1879-1882.
52. Ravi R, Mookerjee B, van Hensbergen Y et al. p53-mediated repression of nuclear factor-kappaB RelA via the transcriptional integrator p300. Cancer Res 1998; 58(20):4531-4536.
53. Sosic D, Richardson JA, Yu K et al. Twist regulates cytokine gene expression through a negative feedback loop that represses NF-kappaB activity. Cell 2003; 112(2):169-180.
54. Lee EG, Boone DL, Chai S et al. Failure to regulate TNF-induced NF-kappaB and cell death responses in A20-deficient mice. Science 2000; 289(5488):2350-2354.
55. Wertz IE, O'Rourke KM, Zhou H et al. De-ubiquitination and ubiquitin ligase domains of A20 downregulate NF-kappaB signalling. Nature 2004.
56. Monteleone G, Mann J, Monteleone I et al. A failure of transforming growth factor-beta1 negative regulation maintains sustained NF-kappaB activation in gut inflammation. J Biol Chem 2004; 279(6):3925-3932.
57. Kuwata H, Watanabe Y, Miyoshi H et al. IL-10-inducible Bcl-3 negatively regulates LPS-induced TNF-alpha production in macrophages. Blood 2003; 102(12):4123-4129.
58. O'Connor W, Harton JA, Zhu X et al. Cutting edge: CIAS1/cryopyrin/PYPAF1/NALP3/ CATERPILLER 1.1 is an inducible inflammatory mediator with NF-kappa B suppressive properties. J Immunol 2003; 171(12):6329-6333.
59. Mehta VB, Besner GE. Inhibition of NF-kappa B activation and its target genes by heparin-binding epidermal growth factor-like growth factor. J Immunol 2003; 171(11):6014-6022.
60. Reynaert NL, Ckless K, Korn SH et al. Nitric oxide represses inhibitory kappaB kinase through S-nitrosylation. Proc Natl Acad Sci USA 2004; 101(24):8945-8950.
61. Marshall HE, Stamler JS. Inhibition of NF-kappa B by S-nitrosylation. Biochemistry 2001; 40(6):1688-1693.
62. Kapahi P, Takahashi T, Natoli G et al. Inhibition of NF-kappa B activation by arsenite through reaction with a critical cysteine in the activation loop of Ikappa B kinase. J Biol Chem 2000; 275(46):36062-36066.
63. Korn SH, Wouters EF, Vos N et al. Cytokine-induced activation of nuclear factor-kappa B is inhibited by hydrogen peroxide through oxidative inactivation of IkappaB kinase. J Biol Chem 2001; 276(38):35693-35700.
64. Ji C, Kozak KR, Marnett LJ. IkappaB kinase, a molecular target for inhibition by 4-hydroxy-2-nonenal. J Biol Chem 2001; 276(21):18223-18228.
65. Rossi A, Kapahi P, Natoli G et al. Anti-inflammatory cyclopentenone prostaglandins are direct inhibitors of IkappaB kinase. Nature 2000; 403(6765):103-108.
66. Yang JP, Hori M, Sanda T et al. Identification of a novel inhibitor of nuclear factor-kappaB, RelA-associated inhibitor. J Biol Chem 1999; 274(22):15662-15670.
67. Ganesh L, Burstein E, Guha-Niyogi A et al. The gene product Murr1 restricts HIV-1 replication in resting CD4+ lymphocytes. Nature 2003; 426(6968):853-857.
68. Artavanis-Tsakonas S, Rand MD, Lake RJ. Notch signaling: Cell fate control and signal integration in development. Science 1999; 284(5415):770-776.
69. Park MY, Jang HD, Lee SY et al. Fas-associated factor-1 inhibits nuclear factor-kappaB (NF-kappaB) activity by interfering with nuclear translocation of the RelA (p65) subunit of NF-kappaB. J Biol Chem 2004; 279(4):2544-2549.

70. Wolff B, Naumann M. INK4 cell cycle inhibitors direct transcriptional inactivation of NF-kappaB. Oncogene 1999; 18(16):2663-2666.
71. Ohata K, Ichikawa T, Nakao K et al. Interferon alpha inhibits the nuclear factor kappa B activation triggered by X gene product of hepatitis B virus in human hepatoma cells. FEBS Lett 2003; 553(3):304-308.
72. You Z, Madrid LV, Saims D et al. c-Myc sensitizes cells to tumor necrosis factor-mediated apoptosis by inhibiting nuclear factor kappa B transactivation. J Biol Chem 2002; 277(39):36671-36677.
73. Manna SK, Zhang HJ, Yan T et al. Overexpression of manganese superoxide dismutase suppresses tumor necrosis factor-induced apoptosis and activation of nuclear transcription factor-kappaB and activated protein-1. J Biol Chem 1998; 273(21):13245-13254.
74. Nourbakhsh M, Oumard A, Schwarzer M et al. NRF, a nuclear inhibitor of NF-kappaB proteins silencing interferon-beta promoter. Eur Cytokine Netw 2000; 11(3):500-501.
75. Altuwaijri S, Lin HK, Chuang KH et al. Interruption of nuclear factor kappaB signaling by the androgen receptor facilitates 12-O-tetradecanoylphorbolacetate-induced apoptosis in androgen-sensitive prostate cancer LNCaP cells. Cancer Res 2003; 63(21):7106-7112.
76. Steed PM, Tansey MG, Zalevsky J et al. Inactivation of TNF signaling by rationally designed dominant-negative TNF variants. Science 2003; 301(5641):1895-1898.
77. Van Antwerp DJ, Martin SJ, Kafri T et al. Suppression of TNF-alpha-induced apoptosis by NF-kappaB [see comments]. Science 1996; 274(5288):787-789.
78. Li J, Joo SH, Tsai MD. An NF-kappaB-specific inhibitor, IkappaBalpha, binds to and inhibits cyclin-dependent kinase 4. Biochemistry 2003; 42(46):13476-13483.
79. Baldi L, Brown K, Franzoso G et al. Critical role for lysines 21 and 22 in signal-induced, ubiquitin-mediated proteolysis of I kappa B-alpha. J Biol Chem 1996; 271(1):376-379.
80. Imbert V, Rupec RA, Livolsi A et al. Tyrosine phosphorylation of I kappa B-alpha activates NF-kappa B without proteolytic degradation of I kappa B-alpha. Cell 1996; 86(5):787-798.
81. Clohisy JC, Roy BC, Biondo C et al. Direct inhibition of NF-kappa B blocks bone erosion associated with inflammatory arthritis. J Immunol 2003; 171(10):5547-5553.
82. Horng T, Barton GM, Medzhitov R. TIRAP: An adapter molecule in the Toll signaling pathway. Nat Immunol 2001; 2(9):835-841.
83. May MJ, D'Acquisto F, Madge LA et al. Selective inhibition of NF-kappaB activation by a peptide that blocks the interaction of NEMO with the IkappaB kinase complex. Science 2000; 289(5484):1550-1554.
84. May MJ, Marienfeld RB, Ghosh S. Characterization of the Ikappa B-kinase NEMO binding domain. J Biol Chem 2002; 277(48):45992-46000.
85. Dai S, Hirayama T, Abbas S et al. The IKK inhibitor, NEMO-binding domain peptide, blocks osteoclastogenesis and bone erosion in inflammatory arthritis. J Biol Chem 2004.
86. Yaron A, Gonen H, Alkalay I et al. Inhibition of NF-kappa-B cellular function via specific targeting of the I-kappa-B-ubiquitin ligase. EMBO J 1997; 16(21):6486-6494.
87. Huang TT, Feinberg SL, Suryanarayanan S et al. The zinc finger domain of NEMO is selectively required for NF-kappa B activation by UV radiation and topoisomerase inhibitors. Mol Cell Biol 2002; 22(16):5813-5825.
88. Takada Y, Singh S, Aggarwal BB. Identification of a p65 peptide that selectively inhibits NF-kappa B activation induced by various inflammatory stimuli and its role in down-regulation of NF-kappaB-mediated gene expression and up-regulation of apoptosis. J Biol Chem 2004; 279(15):15096-15104.
89. Kolenko V, Bloom T, Rayman P et al. Inhibition of NF-kappa B activity in human T lymphocytes induces caspase-dependent apoptosis without detectable activation of caspase-1 and -3. J Immunol 1999; 163(2):590-598.
90. Torgerson TR, Colosia AD, Donahue JP et al. Regulation of NF-kappa B, AP-1, NFAT, and STAT1 nuclear import in T lymphocytes by noninvasive delivery of peptide carrying the nuclear localization sequence of NF-kappa B p50. J Immunol 1998; 161(11):6084-6092.
91. Ducut Sigala JL, Bottero V, Young DB et al. Activation of transcription factor NF-kappaB requires ELKS, an IkappaB kinase regulatory subunit. Science 2004; 304(5679):1963-1967.
92. Orth K. Function of the Yersinia effector YopJ. Curr Opin Microbiol 2002; 5(1):38-43.
93. Orth K, Xu Z, Mudgett MB et al. Disruption of signaling by Yersinia effector YopJ, a ubiquitin-like protein protease. Science 2000; 290(5496):1594-1597.
94. Deng L, Wang C, Spencer E et al. Activation of the IkappaB kinase complex by TRAF6 requires a dimeric ubiquitin-conjugating enzyme complex and a unique polyubiquitin chain. Cell 2000; 103(2):351-361.
95. Wang C, Deng L, Hong M et al. TAK1 is a ubiquitin-dependent kinase of MKK and IKK. Nature 2001; 412(6844):346-351.

96. Hajishengallis G, Genco RJ. Downregulation of the DNA-binding activity of nuclear factor-kappaB p65 subunit in Porphyromonas gingivalis fimbria-induced tolerance. Infect Immun 2004; 72(2):1188-1191.
97. Alcami A, Koszinowski UH. Viral mechanisms of immune evasion. Immunol Today 2000; 21(9):447-455.
98. Hiscott J, Kwon H, Genin P. Hostile takeovers: Viral appropriation of the NF-kappaB pathway. J Clin Invest 2001; 107(2):143-151.
99. Fennewald SM, Aronson JF, Zhang L et al. Alterations in NF-kappaB and RBP-Jkappa by arenavirus infection of macrophages in vitro and in vivo. J Virol 2002; 76(3):1154-1162.
100. Shao R, Karunagaran D, Zhou BP et al. Inhibition of nuclear factor-kappaB activity is involved in E1A-mediated sensitization of radiation-induced apoptosis. J Biol Chem 1997; 272(52):32739-32742.
101. Friedman JM, Horwitz MS. Inhibition of tumor necrosis factor alpha-induced NF-kappa B activation by the adenovirus E3-10.4/14.5K complex. J Virol 2002; 76(11):5515-5521.
102. Browne EP, Shenk T. Human cytomegalovirus UL83-coded pp65 virion protein inhibits antiviral gene expression in infected cells. Proc Natl Acad Sci USA 2003; 100(20):11439-11444.
103. Wang X, Li M, Zheng H et al. Influenza A virus NS1 protein prevents activation of NF-kappaB and induction of alpha/beta interferon. J Virol 2000; 74(24):11566-11573.
104. Dreyfus DH, Nagasawa M, Pratt JC et al. Inactivation of NF-kappaB by EBV BZLF-1-encoded ZEBRA protein in human T cells. J Immunol 1999; 163(11):6261-6268.
105. Shisler JL, Jin XL. The vaccinia virus K1L gene product inhibits host NF-kappaB activation by preventing IkappaBalpha degradation. J Virol 2004; 78(7):3553-3560.
106. Besnard-Guerin C, Belaidouni N, Lassot I et al. HIV-1 Vpu sequesters beta-transducin repeat-containing protein (betaTrCP) in the cytoplasm and provokes the accumulation of beta-catenin and other SCFbetaTrCP substrates. J Biol Chem 2004; 279(1):788-795.
107. Tang W, Pavlish OA, Spiegelman VS et al. Interaction of Epstein-Barr virus latent membrane protein 1 with SCFHOS/beta-TrCP E3 ubiquitin ligase regulates extent of NF-kappaB activation. J Biol Chem 2003; 278(49):48942-48949.
108. Lassot I, Segeral E, Berlioz-Torrent C et al. ATF4 degradation relies on a phosphorylation-dependent interaction with the SCF(betaTrCP) ubiquitin ligase. Mol Cell Biol 2001; 21(6):2192-2202.
109. Bremner P, Heinrich M. Natural products as targeted modulators of the nuclear factor-kappaB pathway. J Pharm Pharmacol 2002; 54(4):453-472.
110. Pan MH, Lin-Shiau SY, Ho CT et al. Suppression of lipopolysaccharide-induced nuclear factor-kappaB activity by theaflavin-3,3'-digallate from black tea and other polyphenols through down-regulation of IkappaB kinase activity in macrophages. Biochem Pharmacol 2000; 59(4):357-367.
111. Jobin C, Bradham CA, Russo MP et al. Curcumin blocks cytokine-mediated NF-kappa B activation and proinflammatory gene expression by inhibiting inhibitory factor I-kappa B kinase activity. J Immunol 1999; 163(6):3474-3483.
112. Straus DS, Pascual G, Li M et al. 15-deoxy-delta 12,14-prostaglandin J2 inhibits multiple steps in the NF- kappa B signaling pathway. Proc Natl Acad Sci USA 2000; 97(9):4844-4849.
113. D'Acquisto F, May MJ, Ghosh S. Inhibition of Nuclear Factor Kappa B (NF-B): An emerging theme in anti-inflammatory therapies. Mol Intervent 2002; 2(1):22-35.
114. Amit S, Ben-Neriah Y. NF-kappaB activation in cancer: A challenge for ubiquitination- and proteasome-based therapeutic approach. Semin Cancer Biol 2003; 13(1):15-28.
115. Magnani M, Crinelli R, Bianchi M et al. The ubiquitin-dependent proteolytic system and other potential targets for the modulation of nuclear factor-kB (NF-kB). Curr Drug Targets 2000; 1(4):387-399.
116. Sunwoo JB, Chen Z, Dong G et al. Novel proteasome inhibitor PS-341 inhibits activation of nuclear factor-kappa B, cell survival, tumor growth, and angiogenesis in squamous cell carcinoma. Clin Cancer Res 2001; 7(5):1419-1428.
117. Adams J. Preclinical and clinical evaluation of proteasome inhibitor PS-341 for the treatment of cancer. Curr Opin Chem Biol 2002; 6(4):493-500.
118. Richardson PG, Barlogie B, Berenson J et al. A phase 2 study of bortezomib in relapsed, refractory myeloma. N Engl J Med 2003; 348(26):2609-2617.
119. Adams J, Kauffman M. Development of the proteasome inhibitor Velcade (Bortezomib). Cancer Invest 2004; 22(2):304-311.
120. Tergaonkar V, Pando M, Vafa O et al. p53 stabilization is decreased upon NFkappaB activation: A role for NFkappaB in acquisition of resistance to chemotherapy. Cancer Cell 2002; 1(5):493-503.
121. Saklatvala J. Glucocorticoids: Do we know how they work? Arthritis Res 2002; 4(3):146-150.
122. Heck S, Bender K, Kullmann M et al. I kappaB alpha-independent downregulation of NF-kappaB activity by glucocorticoid receptor. EMBO J 1997; 16(15):4698-4707.

123. Auphan N, Didonato JA, Helmberg A et al. Immunoregulatory genes and immunosuppression by glucocorticoids. Arch Toxicol Suppl 1997; 19:87-95.
124. Doucas V, Shi Y, Miyamoto S et al. Cytoplasmic catalytic subunit of protein kinase A mediates cross- repression by NF-kappa B and the glucocorticoid receptor [In Process Citation]. Proc Natl Acad Sci USA 2000; 97(22):11893-11898.
125. Ito K, Jazrawi E, Cosio B et al. p65-activated histone acetyltransferase activity is repressed by glucocorticoids: Mifepristone fails to recruit HDAC2 to the p65-HAT complex. J Biol Chem 2001; 276(32):30208-30215.
126. Nissen RM, Yamamoto KR. The glucocorticoid receptor inhibits NFkappaB by interfering with serine-2 phosphorylation of the RNA polymerase II carboxy-terminal domain. Genes Dev 2000; 14(18):2314-2329.
127. Kopp E, Ghosh S. Inhibition of NF-kappa B by sodium salicylate and aspirin [see comments]. Science 1994; 265(5174):956-959.
128. Fuchs SY, Adler V, Buschmann T et al. JNK targets p53 ubiquitination and degradation in nonstressed cells. Genes Dev 1998; 12(17):2658-2663.
129. Yamamoto Y, Yin MJ, Lin KM et al. Sulindac inhibits activation of the NF-kappaB pathway. J Biol Chem 1999; 274(38):27307-27314.
130. Warner TD, Mitchell JA. Cyclooxygenases: New forms, new inhibitors, and lessons from the clinic. FASEB J 2004; 18(7):790-804.
131. Coussens LM, Werb Z. Inflammation and cancer. Nature 2002; 420(6917):860-867.
132. Kretz-Remy C, Mehlen P, Mirault ME et al. Inhibition of I kappa B-alpha phosphorylation and degradation and subsequent NF-kappa B activation by glutathione peroxidase overexpression. J Cell Biol 1996; 133(5):1083-1093.
133. Carcamo JM, Pedraza A, Borquez-Ojeda O et al. Vitamin C is a kinase inhibitor: Dehydroascorbic acid inhibits I{kappa}B{alpha} kinase {beta}. Mol Cell Biol 2004; 24(15):6645-6652.
134. Epinat JC, Gilmore TD. Diverse agents act at multiple levels to inhibit the Rel/NF-kappaB signal transduction pathway. Oncogene 1999; 18(49):6896-6909.
135. Karin M, Yamamoto Y, Wang QM. The IKK NF-kappa B system: A treasure trove for drug development. Nat Rev Drug Discov 2004; 3(1):17-26.
136. Ohazama A, Hu Y, Schmidt-Ullrich R et al. A dual role for Ikk alpha in tooth development. Dev Cell 2004; 6(2):219-227.
137. Senftleben U, Cao Y, Xiao G et al. Activation by IKKalpha of a second, evolutionary conserved, NF-kappa B signaling pathway. Science 2001; 293(5534):1495-1499.
138. Burke JR. Targeting I kappa B kinase for the treatment of inflammatory and other disorders. Curr Opin Drug Discov Devel 2003; 6(5):720-728.
139. Penolazzi L, Lambertini E, Borgatti M et al. Decoy oligodeoxynucleotides targeting NF-kappaB transcription factors: Induction of apoptosis in human primary osteoclasts. Biochem Pharmacol 2003; 66(7):1189-1198.
140. Giannoukakis N, Bonham CA, Qian S et al. Prolongation of cardiac allograft survival using dendritic cells treated with NF-kB decoy oligodeoxyribonucleotides. Mol Ther 2000; 1(5 Pt 1):430-437.
141. Hess DC, Howard E, Cheng C et al. Hypertonic mannitol loading of NF-kappaB transcription factor decoys in human brain microvascular endothelial cells blocks upregulation of ICAM-1. Stroke 2000; 31(5):1179-1186.
142. Lee IK, Ahn JD, Kim HS et al. Advantages of the circular dumbbell decoy in gene therapy and studies of gene regulation. Curr Drug Targets 2003; 4(8):619-623.
143. Crooke ST. Antisense strategies. Curr Mol Med 2004; 4(5):465-487.
144. Tiscornia G, Singer O, Ikawa M et al. A general method for gene knockdown in mice by using lentiviral vectors expressing small interfering RNA. Proc Natl Acad Sci USA 2003; 100(4):1844-1848.
145. Takaesu G, Surabhi RM, Park KJ et al. TAK1 is critical for IkappaB kinase-mediated activation of the NF-kappaB pathway. J Mol Biol 2003; 326(1):105-115.
146. Pinkenburg O, Platz J, Beisswenger C et al. Inhibition of NF-kappaB mediated inflammation by siRNA expressed by recombinant adeno-associated virus. J Virol Methods 2004; 120(1):119-122.
147. Guo J, Verma UN, Gaynor RB et al. Enhanced chemosensitivity to irinotecan by RNA interference-mediated down-regulation of the nuclear factor-kappaB p65 subunit. Clin Cancer Res 2004; 10(10):3333-3341.
148. Sledz CA, Holko M, de Veer MJ et al. Activation of the interferon system by short-interfering RNAs. Nat Cell Biol 2003; 5(9):834-839.
149. Yamamoto Y, Gaynor RB. IkappaB kinases: Key regulators of the NF-kappaB pathway. Trends Biochem Sci 2004; 29(2):72-79.

150. Sizemore N, Leung S, Stark GR. Activation of phosphatidylinositol 3-kinase in response to interleukin-1 leads to phosphorylation and activation of the NF-kappaB p65/RelA subunit. Mol Cell Biol 1999; 19(7):4798-4805.
151. Mitsiades CS, Mitsiades N, Koutsilieris M. The Akt pathway: Molecular targets for anti-cancer drug development. Curr Cancer Drug Targets 2004; 4(3):235-256.
152. Hoeflich KP, Luo J, Rubie EA et al. Requirement for glycogen synthase kinase-3beta in cell survival and NF-kappaB activation. Nature 2000; 406(6791):86-90.
153. Vermeulen L, De Wilde G, Van Damme P et al. Transcriptional activation of the NF-kappaB p65 subunit by mitogen- and stress-activated protein kinase-1 (MSK1). EMBO J 2003; 22(6):1313-1324.
154. Cavin LG, Romieu-Mourez R, Panta GR et al. Inhibition of CK2 activity by TGF-beta1 promotes IkappaB-alpha protein stabilization and apoptosis of immortalized hepatocytes. Hepatology 2003; 38(6):1540-1551.
155. Maloney A, Workman P. HSP90 as a new therapeutic target for cancer therapy: The story unfolds. Expert Opin Biol Ther 2002; 2(1):3-24.
156. Yamamoto Y, Gaynor RB. Therapeutic potential of inhibition of the NF-kappaB pathway in the treatment of inflammation and cancer. J Clin Invest 2001; 107(2):135-142.
157. Tak PP, Firestein GS. NF-kappaB: A key role in inflammatory diseases. J Clin Invest 2001; 107(1):7-11.
158. St Johnston D. The art and design of genetic screens: Drosophila melanogaster. Nat Rev Genet 2002; 3(3):176-188.
159. Silverman N, Maniatis T. NF-kappaB signaling pathways in mammalian and insect innate immunity. Genes Dev 2001; 15(18):2321-2342.
160. Correa RG, Tergaonkar V, Ng JK et al. Characterization of NF-kappa B/I kappa B proteins in zebra fish and their involvement in notochord development. Mol Cell Biol 2004; 24(12):5257-5268.
161. Chan J, Bayliss PE, Wood JM et al. Dissection of angiogenic signaling in zebrafish using a chemical genetic approach. Cancer Cell 2002; 1(3):257-267.
162. Pichler FB, Laurenson S, Williams LC et al. Chemical discovery and global gene expression analysis in zebrafish. Nat Biotechnol 2003; 21(8):879-883.
163. Trede NS, Langenau DM, Traver D et al. The use of zebrafish to understand immunity. Immunity 2004; 20(4):367-379.
164. Peterson RT, Shaw SY, Peterson TA et al. Chemical suppression of a genetic mutation in a zebrafish model of aortic coarctation. Nat Biotechnol 2004; 22(5):595-599.
165. MacRae CA, Peterson RT. Zebrafish-based small molecule discovery. Chem Biol 2003; 10(10):901-908.

Index

Symbols

4-1BB 27, 36

A

A20 19, 33, 86, 116, 165
Acetaminophen 137, 169
Acute lymphoblastic leukemia (ALL) 114, 115, 134
Acute myeloid leukemia (AML) 134
Adaptive immunity 70, 71, 75, 79, 91, 92, 94, 96, 100
Adaptor 19, 26, 28, 30-33, 61, 64, 71, 77, 93-96, 101, 113, 135, 163, 164
Adenovirus 167
Adult T-cell leukemia lymphoma (ATLL) 134, 135
Alzheimer's disease 162
Amino-thiophenecar derivative 169
Amygdala 153-156
Anethole 138
Angiogenesis 27, 36, 119, 121, 122, 164
Ankyrin repeat 3, 5, 7, 8, 12, 13, 42, 44, 117, 165
Ankyrin repeat domain (ARD) 2, 3, 5, 7, 9, 42, 46
Antigen presenting cell (APC) 26, 72, 87, 88, 95
Antigen receptor 13, 14, 16, 19, 20, 26-29, 32, 72, 76
Antioxidants 169
Antisense ODNs 170
Apoptosis 1, 17, 20, 21, 27, 32, 33, 51-57, 59-62, 65, 74, 77, 78, 84-88, 95, 101, 102, 112, 113, 115, 116, 119, 120, 122, 132, 133, 136, 164
Apoptotic survival 74
APRIL 35, 63, 134
Arsenic trioxide 137
Ascorbic acid 169
Aspirin 169
Autoimmune disease 16, 21, 71, 75, 79, 84, 85, 87, 88
Autoimmunity 57, 74, 78, 84-88
Autoinflammatory syndrome 1 165
Autoreactive effector T cells 57
Avicin 167, 168

B

B cell receptor 26, 54, 58, 64, 72, 132, 133
B lymphocyte 1, 15, 21, 51, 53, 70, 72, 77-79, 131, 134, 135
B lymphocyte stimulator 15
β-carboline derivatives 169
β-catenin 168
β-TrCP1 42, 45, 162, 166, 167
B-cell chronic lymphocytic leukemia 114, 115
B-cell non-Hodgkin's lymphoma 114, 115
B7.1 73, 79
B7.2 73, 79
Bacteria 70-72, 93-95, 98, 100-102
Bacterial infection 71, 93, 99, 100
BAFF receptor 35, 63, 65
Bcl-2 36, 53, 56, 62-64, 74, 116, 117, 120, 133, 135
Bcl-10 19-21, 28-30, 32, 57, 58, 61, 64, 77, 132, 135
Bcl-X 36, 59, 61, 74, 77-79
Bcl-X transgene 59
Bcl-xL 53, 64, 93, 116, 118, 120, 122, 133-135
BCR 19, 20, 26-29, 53, 54, 58-62, 64, 65, 72, 77, 78, 86, 132, 133, 135
Bfl-1 77, 79, 116, 120, 133, 135
Bfl-1/A1 116, 120, 133, 135
Bioinformatics 156
BLR1 61
Blys 15, 16, 21, 92, 96
Blys receptor 15
BMS-345541 169
Borrelia burgdorferi 95
Bortezomib 135, 137, 168
BR3 15, 16
Breast adenocarcinoma 115
Breast carcinoma 114, 115
Butyrate 137

C

C-erbB2/Her-2/Neu 115
C-FLIP 116, 133
C-IAP1 19, 116
C-myc 55, 60, 86, 87, 116, 118, 120, 133-135, 165
C-type lectin 78
Ca^{++}-dependent signaling 150
Cachexia 162, 165
Caffeic acid phenethyl 138
Calcium-calmodulin dependent protein kinase (CaMK) 150, 151, 153
Camplybacter jejuni 95, 96
Cancer 21, 37, 71, 91, 102, 112-118, 120-123, 131-137, 164, 167-170, 172
Canonical pathway 12, 15-17, 35, 96
Capsaicin 138
CARMA1 20, 29, 57, 61, 77
Casein kinase II (CKII) 9, 17, 34, 115, 121, 170
CCR7 61, 101, 116, 120
$CD4^+$ single positive (SP) thymocyte 54, 74
$CD8^+$ single positive (SP) thymocyte 54, 55, 74
$CD4^-CD8^-$ double negative (DN) thymocyte 54, 74
$CD4^+CD8^+$ double positive 52, 54, 74, 84
CD27 27, 36
CD28 28, 73, 75, 76, 78, 87
CD30 27, 36, 78, 134
CD40 16, 17, 27, 34, 35, 36, 61, 73, 78, 86, 92, 97, 99, 100, 113, 116, 132, 134-136
CD40 ligand (CD40L) 16, 17, 35, 60, 73, 74, 78, 96, 97, 99, 113, 134, 135
CD95 116
CD154 35, 134
CDK 164
CDK2 164
Cell cycle 61, 70, 74, 75, 77, 78, 88, 93, 116, 119, 120, 122, 133, 164, 168
Central tolerance 85
Cold-induced autoinflammatory syndrome 1 (CIAS1) 165
Colon carcinoma 114, 115, 117
COX-2 inhibitor 137, 169
CRM1 43, 44
Curcumin 138, 167
Cutaneous T-cell lymphoma 115, 117
CXCR4 116, 121
CXCR5 61

Cyclin 36, 60, 74, 77-79, 93, 116-118, 120, 122, 133, 136, 164, 166
Cyclin-dependent kinase inhibitor (CKI) 60
Cyclooxygenase 2 (COX-2) 133, 137, 168, 169
Cyclopentenone prostaglandins 165, 168
Cylindromatosis tumor suppressor gene (CYLD) 19, 113, 114, 164
Cytokine 1, 2, 20, 27, 30, 32, 35, 46, 51, 55, 70-72, 74-79, 84, 86-88, 91, 92, 94, 95, 98, 100, 101, 113, 116, 119, 133, 149, 153, 164, 165, 167, 168, 170, 171

D

Decoy oligodeoxynucleotides (ODNs) 170
Dendritic cell 17, 35, 36, 70, 72, 74, 78, 79, 87, 88, 92, 94, 96, 118
Dendritic cell maturation 92
Diabetes 87, 88
Diarylpyridine derivative 169
Dicoumarol 138
Dif 93
Diffuse large cell lymphoma (DLCL) 114, 115, 134
Dimerization domain 4-9
DNA binding 1, 3, 4, 6, 9, 12, 14, 15, 17-19, 41, 42, 45, 100, 101, 121, 137, 138, 148, 149, 153, 155, 156, 162, 163, 165, 167, 168
Double negative (DN) thymocyte 52, 54-57, 74, 75, 79
Double-stranded oligodeoxynucleotides 170
DR4 27
DR5 27

E

E1A 167
E2F 77-79
EGCG 138, 167
EGF receptor 27
Epoxomicin 138
Epstein-Barr virus (EBV) 44, 101, 102, 112, 114, 116, 135, 136, 167
Escherichia coli 96, 101
Eugenol 138
Experimental autoimmune encephalomyelitis (EAE) 86, 87

F

Fas 20, 27, 87, 122, 133, 165
Fas-associated factor 1 (FAF1) 165
Fear conditioning 155, 156
Fear potentiated startle 155
Fear-motivated learning 154, 155
Flavonoids 138
Follicular B cell 16, 61-64, 114
FOXO3a 113

G

G3BP2 163
GADD45β 33, 116
Gastric carcinoma 114, 134
GATA3 76
Genistein 138
Germinal center 35, 72, 77-79
Germinal center immune response 78
Glioma 134
Glucocorticoid (GC) 137, 168
Glutamate receptor 149, 150, 153, 154
Glutathione peroxidase 169
Graft rejection 71, 162
Graft tolerance 71
Granulocyte macrophage colony stimulating factor (GM-CSF) 27, 76, 88, 95, 116, 133

H

Haemophilus influenzae 96
Head and neck cancer 134
Helenalin 167, 168
Helicobacter hepaticus 95, 96
Helminth 92, 96, 97
Heparin-binding growth factor-like growth factor (HB-EGF) 165
Hepatitis X (HBx) 136, 163, 165
Hepatocellular carcinoma 114, 133, 134, 136
Hepericin 167
Herceptin 136
Hippocampus 153, 156-158
Histone deacetylase 3 (HDAC3) 18
hnRNP-U 45
Hodgkin's disease (HD) 113, 120
Human T-cell leukemia virus type I (HTLV-1) 1, 112, 114, 115, 135

I

Ibuprofen 169
IFN-γ 74, 76, 79, 87, 92, 95-100, 165
Ig class switching 78, 79
IκB 1-4, 6-9, 12-15, 17, 18, 26, 34, 41, 42, 44-47, 51, 55-57, 59, 74, 76, 84, 86, 91, 93, 101, 112-114, 116, 117, 131, 132, 134-138, 148, 151, 155, 162-168, 170
IKK 1, 3, 12-21, 26, 28-37, 44-47, 51-57, 61, 63-65, 77, 85, 91, 93, 94, 99, 101, 113-115, 120, 121, 131, 132, 135-138, 148, 149, 151, 155, 162-171
IKK-activating signalsome 29, 30
IL-1 1, 13, 19-21, 30, 46, 71, 72, 79, 86, 93, 95, 113, 114, 116, 132, 133, 169
IL-1 receptor (IL-1R) 13, 27, 30, 31, 36, 71, 93, 94
IL-1 receptor associated kinase (IRAK) 30-32, 71, 93-95, 98, 99
IL-4 76, 78, 87, 88, 92, 96
IL-6 3, 13, 30, 71, 77-79, 94, 95, 99, 100, 116, 133, 135, 165
IL-7 59, 60, 65
IL-8 116, 121, 133, 136
IL-12 30, 73, 74, 76, 79, 86, 87, 91, 92, 95, 97
IL-12p35 74, 87
IL-12p40 74, 86, 87
IL-13 76, 78, 92, 114, 116
Imidazoquinoxaline derivative 169
Immature B cell 53, 54, 58-61, 77, 78
Immediate early response gene X-1 (IEX-1) 87, 116
Immune deficiency gene (imd) 93
Immune receptor 4, 36
Immunoreceptor tyrosine-based activation motifs (ITAM) 28, 29
Induced nitric oxide synthase (iNOS) 71, 79, 168
Infection 21, 30, 70-72, 75, 76, 78, 79, 91, 93-100, 102, 135, 167
Inflammation 1, 21, 27, 32, 33, 75, 85, 86, 91, 94-97, 133, 135, 164, 165, 168, 169, 171
ING4 122, 164
Inhibitor of Cdk4 (INK4) 165
Innate immunity 71, 72, 91, 93, 167
Insulin resistance 162
Integrin 27, 36
Intercellular adhesion molecule 1 (ICAM-1) 2, 73, 116, 133

Ionizing radiation 133, 169
IRF1 116
IRF4 116
Isoprenoid compounds 167

J

JunB 116, 120

K

κB-Ras 7, 14, 41, 44, 45, 164
Kinase 1, 4, 9, 14, 15, 17, 19-21, 26, 28-35, 42, 44-47, 55, 56, 60, 63, 64, 71, 75, 77, 84, 85, 91, 93, 94, 96, 98, 101, 113, 114, 115, 121, 122, 131, 132, 148, 150, 151, 153, 162, 163, 165-167, 169, 170, 172
Kip1 60
KSHV-GPCR 27

L

Latent membrane protein 1 (LMP-1) 44, 116, 135, 167
Leishmania major 95-97
Leucine-rich repeats (LRR) 3, 30
LFA-1 73
Lipopolysaccharide (LPS) 1, 20, 21, 30, 32, 58, 59, 71, 87, 88, 91, 94, 99, 100, 149, 165, 166, 169
Listeria monocytogenes 71, 95-98
Long-term contextual fear memory 156
Long-term depression (LTD) 152-154
Long-term memory 156, 158
Long-term potentiation (LTP) 153, 154, 156
Lymphocyte development 20, 51, 54
Lymphocyte proliferation 84, 88
Lymphocyte survival 77, 86
Lymphocytic choriomeningitis virus (LCMV) 76, 96, 97
Lymphopoiesis 51, 52, 54, 77
Lymphotoxin beta receptor (LTβR) 15-17, 27, 34, 35

M

MALT1 19, 20, 28-30, 57, 64, 77, 114, 115
Manganese superoxide dismutase (MnSOD) 165, 169
Mantle cell lymphoma 120, 133, 134
Marginal zone (MZ) B cells 53, 61-65, 97
Mcl-1 77, 79

MEKK 15, 18, 21, 33, 115, 170
Melanoma 114, 134
Memory formation 149, 152, 154, 158
Memory T cell 57, 79
Mesalamine 137
Metalloproteinase-9 (MMP-9) 116, 133
Metastasis 119, 121
MIP-1α 116
MRL and NZB mouse strains 86
MRL/+ mice 86
Mucosa-associated lymphoid tissue (MALT) 19, 20, 28-30, 57, 64, 77, 114, 115, 134, 135, 164
Multiple myeloma 115, 117, 132, 134, 135, 137, 168
Multiple sclerosis 85
Myasthenia gravis 85
Mycobacteria leprae 95, 96
Myeloid differentiation protein 88 (MyD88) 30-32, 36, 95, 96, 101

N

N-acetylcysteine 137
NADPH 169
NAK 17, 170
Nasopharyngeal carcinoma 114, 116
Negative selection 52-56, 60, 65, 71, 74, 75, 79, 85
NEMO-binding domain (NBD) 132, 136, 137, 166
Neurodegeneration 149
NF-AT 6, 47, 74-77
NF-κB 1-9, 12-21, 26-37, 41-48, 51-66, 70-79, 84-88, 91-102, 112-123, 131-138, 147-151, 153-156, 158, 162-172
NF-κB inducing kinase (NIK) 15-17, 21, 33, 35, 36, 46, 63, 75, 85, 88, 95, 96, 170
NF-κB signaling 4, 12, 18-20, 26, 28-34, 36, 60, 71, 77, 79, 91, 93, 96, 98-102, 112, 113, 115, 117, 147, 148, 151, 153, 155, 158, 162, 164-170
NF-κB/DNA complex 6, 41
NKT cell 52, 57, 58
NMDA receptor 150, 154
NOD 27, 37, 86, 91
NOD mice 86
Non-small cell lung carcinoma 114, 115
Noncanonical pathway 15, 17, 34-36, 56, 96, 97
Nonsteroidal anti-inflammatory drugs (NSAIDs) 168, 169

Notch 165
Nuclear export sequences/signal (NES) 14, 42-45, 47, 48
Nuclear localization sequence/signal (NLS) 1, 4-6, 8, 9, 13, 14, 41, 43, 44, 47, 48, 166
Nucleocytoplasmic shuttling 41, 43, 46
NZB/W mice 86

O

Oleandrin 138, 167, 168
Oncogene 12, 58, 113-115, 117, 118
Oroxylin A 167
OX40 36

P

26S proteasome 15, 44
p27 60
p35 74, 87
p40 74, 86, 87
p53 87, 116, 120, 122, 164, 168
p300/CBP 18, 121
Pancreatic cancer 134
Parasite 70, 91, 92, 94, 95, 97, 101, 102
Pathogen-associated molecular patterns (PAMPs) 30, 91, 94
Pattern recognition receptors (PRRs) 27, 30, 70, 71, 91, 94
PDGF-R 27
PEST 2, 3, 7-9, 13, 166
Phosphoinositide 3-kinase (PI3K) 17, 35, 78, 170
Phosphorylation 3, 4, 8, 13-17, 26, 28, 29, 31, 32, 34, 44-46, 55, 60, 61, 85, 86, 91, 101, 113, 115, 121, 132, 136-138, 148, 149, 151, 153, 164, 166, 169
PKA 17, 121, 153
PKCβ 19-21, 29, 61
PKCθ 20, 21, 28, 29, 57, 61, 77
PKCζ 17, 34, 121, 151
Positive selection 52, 54, 55, 65, 74, 75, 79, 85
Pre-B cell 53, 54, 58-60
Proliferation 20, 27, 29, 32, 35, 36, 51, 56, 60, 62, 65, 74, 76, 78, 79, 84-88, 92, 95, 96, 99, 102, 112, 113, 119, 120, 132-136, 164, 167
Prostate cancer 115, 134
Proteasomal inhibitor 132, 138, 168
Protein-protein interaction 3, 8, 21, 149, 166
PS-341 168
PS-1145 169

Q

Quinazoline analogues 169

R

Radial arm maze 157, 158
RANK 27, 34, 36
Rapamycin 137
Reactive oxygen species (ROS) 150, 153, 169
Recirculating B cell 61, 65
Regulatory T cell (Tr) 52, 57, 74, 75, 85
Rel 2-4, 12, 13, 28, 41-44, 46-48, 51-53, 56, 57, 61-63, 65, 70, 71, 73-79, 84-88, 95, 97, 112-114, 116-122, 131-133, 135, 148, 156
Rel homology region (RHR) 2-7, 9
Renal cell carcinoma 133, 134
Resveratrol 138, 167
Retrograde messenger 147, 149
Rheumatoid arthritis 85, 169
RNA aptamer 1, 5, 9

S

S-allylcysteine 138
Salicylate 137, 169
Sanguinarine 138
SC-514 169
SCF 44, 45, 163, 164, 166, 167
Sesquiterpene lactones 138, 167
Signal transduction 1, 12, 28, 44, 147, 171
Silybin 167
Silymarin 138
siRNA 33, 44
SLP-76 28, 29
Small molecule inhibitors 172
SOCS-1 121
Sodium salicylate 137, 169
SP thymocyte 52, 55-57, 74, 75, 79
SPC839 169
Staphylococcus aureus 95, 98, 99
STAT 6, 9
Stat5a 116, 120
Stomach adenocarcinoma 115
Structural biology 3
Sulfasalazine 137
Sulforaphane 138
Sulindac 137, 169
Syk 28, 60
Synaptic activity 147, 149, 150, 152-154
Synaptic messenger 147

Synaptic plasticity 150, 152-154, 158
Systemic lupus erythematosus 85

T

T cell receptor (TCR) β chain rearrangement 54
T lymphocyte 52, 54
T2K 14, 17, 170
TAK1 18, 19, 21, 31-33, 170
TBK 17
TCR 13, 19-21, 26-30, 52, 54-58, 60, 64, 72-77, 85, 87, 101
TCRα 54, 74
Tepoxalin 169
Th1 97
T_H1 cytokine 76
T_H2 cytokine 74, 76, 78
Thalidomide 137
Theaflavins 138
Therapy 112, 119, 120, 122, 123, 131, 133, 137
Thymocyte 20, 21, 27, 45, 52, 54-57, 74, 75, 79, 84, 85
Thyroid cancer 120, 134
Thyroid carcinoma 114, 115
TICAM-1 30
TIR 27, 30, 31, 36, 94, 98, 100
TIR domain-containing adaptor protein (TIRAP) 30, 166
TNF receptor (TNFR) 17, 20, 26, 27, 30, 32-36, 134, 151, 153, 154, 165
Toll-like receptor (TLR) 13, 18, 20, 21, 26, 27, 30-32, 36, 71-74, 91, 93-96, 98-101, 166
Toxoplasma gondii 95-98, 101
TRAF 16-21, 30-33, 35, 36, 94, 113, 116, 163
TRAM 30, 31
Transcription 1-4, 6, 9, 12-15, 17, 26, 30, 32, 33, 41, 44, 46-48, 53-56, 58, 59, 65, 66, 70, 73-79, 84-86, 88, 91-94, 96-98, 100, 102, 112, 113, 116, 117, 119-122, 131-133, 137, 147-149, 151, 153-156, 158, 162, 168, 170
Transcription activation domain (TAD) 2, 3, 12, 13, 47, 48, 118, 119, 121, 164
Transcription factor 1-3, 6, 9, 12, 26, 30, 32, 33, 46-48, 54-56, 58, 59, 65, 66, 70, 73, 75-79, 84-86, 91-93, 96-98, 100, 112, 113, 116, 117, 119-121, 137, 147-149, 153, 158, 162, 168, 170

Transforming growth factor-β1 (TGF-β1) 115, 165
Transitional 1 B cell 53, 61, 63, 64
Transitional 2 B cell 61, 63, 64
Transitional B cell 36, 53, 60-64
Transplant tolerance 162
Trastuzumab 136
Trichinella spiralis 95-97
Trichuris muris 95-97
TRIF 30-32
Triterpenoids 167
Tumor 15, 19, 26, 32, 71, 102, 112-123, 131-137, 153, 164, 168-170
Tumor necrosis factor (TNF) 1, 15, 17, 20, 21, 26, 30, 32-34, 51, 62, 71, 76, 85, 86, 88, 91, 93, 95, 100, 114, 133-135, 137, 151, 153, 154, 165, 169

U

UbcH5 44
Ubiquitin 18, 19, 30-33, 44, 46, 77, 101, 113, 121, 132, 135, 163, 166
Ubiquitin-conjugating enzyme 30, 32, 44
Ubiquitination 14, 15, 18, 19, 29-33, 42, 44, 45, 55, 99, 101, 121, 132, 135, 164-166, 168
Ureido-thiophenecarboxamide derivative 169
Ursodeoxycholic acid 137
Ursolic acid 138
UV 13, 122, 169

V

v-rel 12, 112, 116-122, 132
Vascular endothelial growth factor (VEGF) 116, 121
VCAM-1 35, 116
Vesicular stomatitis virus (VSV) 55, 95, 96
Viral infection 99, 167
Virus 1, 21, 44, 58, 70, 72, 95-97, 99-102, 112-116, 118, 132, 133, 135, 136, 167, 170
Vitamin C 169
Vitamin E 169

W

Wogonin 167

X

X-linked inhibitor of apoptosis (XIAP) 33, 133, 135
X-ray crystallographic 4, 7, 8, 43, 47

Y

Y. enterocolitica 101
Yersinia outer protein (Yop) 101, 166, 167

Bode · Durst (Eds.)
High Performance Computing in Science and Engineering, Garching 2004

Arndt Bode · Franz Durst
Editors

High Performance Computing in Science and Engineering, Garching 2004

Transactions of the KONWIHR Result Workshop,
October 14–15, 2004, Technical University of Munich,
Garching, Germany

With 177 Figures, 13 in Color, and 16 Tables

 Springer

Editors

Arndt Bode

Technische Universität München
Lehrstuhl für Rechnertechnik/Rechnerorganisation
Institut für Informatik
Boltzmannstr. 3
85748 Garching, Germany
e-mail: bode@in.tum.de

Franz Durst

Universität Erlangen-Nürnberg
Lehrstuhl für Strömungsmechanik
Cauerstr. 4
91058 Erlangen, Germany
e-mail: durst@lstm.uni-erlangen.de

Library of Congress Control Number: 2005927484

Mathematics Subject Classification (2000): 65Cxx, 65K05, 68M20, 68U20, 70-08, 74F10, 74F15, 74L99, 76-04, 76G25, 76Txx, 81-04, 81-08, 81V05, 81V10, 85-08, 86-04, 86A15, 92-04, 92Exx

ISBN-10 3-540-26145-1 Springer Berlin Heidelberg New York
ISBN-13 978-3-540-26145-2 Springer Berlin Heidelberg New York

This work is subject to copyright. All rights are reserved, whether the whole or part of the material is concerned, specifically the rights of translation, reprinting, reuse of illustrations, recitation, broadcasting, reproduction on microfilm or in any other way, and storage in data banks. Duplication of this publication or parts thereof is permitted only under the provisions of the German Copyright Law of September 9, 1965, in its current version, and permission for use must always be obtained from Springer. Violations are liable for prosecution under the German Copyright Law.

Springer is a part of Springer Science+Business Media

springeronline.com

© Springer-Verlag Berlin Heidelberg 2005

The use of general descriptive names, registered names, trademarks, etc. in this publication does not imply, even in the absence of a specific statement, that such names are exempt from the relevant protective laws and regulations and therefore free for general use.

Typesetting: by the authors using a Springer TEX macro package
Production: LE-TEX Jelonek, Schmidt & Vöckler GbR, Leipzig
Cover design: Erich Kircher, Heidelberg, Germany

Printed on acid-free paper   46/3142/YL - 5 4 3 2 1 0

Preface

This volume of High Performance Computing in Science and Engineering is fully dedicated to the final report of KONWIHR, the Bavarian Competence Network for Technical and Scientific High-Performance Computing. It includes the transactions of the final KONWIHR workshop, that was held at Technische Universität München, October 14-15, 2004, as well as additional reports of KONWIHR research groups.

KONWIHR was established by the Bavarian State Government from 2000 to 2004 with 4.6 million Euro in order to support the broad application of high performance computing in science and technology throughout the country. KONWIHR is a supporting action to the installation of the German supercomputer Hitachi SR 8000 in the Leibniz Computing Center of the Bavarian Academy of Sciences. KONWIHR projects not only support the development, implementation and application of scientific codes for high performance computing, but also general measures such as conferences, workshops, tutorials, the invitation of international research partners and the adaptation of existing application codes.

KONWIHR projects have been reviewed by a panel of researchers: Prof. Dr. Friedel Hossfeld, Jülich (spokesman), Prof. Dr. Albert Gilg, Munich, Prof. Dr. Ernst Heinrich Hirschel, Munich, Prof. Dr. Rolf Rannacher, Heidelberg, Prof. Dr. Hanns Ruder, Tübingen, Prof. Dr. Dieter Seitzer, Munich, Prof. Dr. Dr. h.c. Eberhard Witte, Munich, and Norbert Willisch, as the representative of the Bavarian States Ministry for Science, Research and Art. We are most grateful to the board, which has supported strongly KONWIHR over its existence.

High performance computing requires the interdisciplinary cooperation between one or more application areas, mathematicians for the algorithmic structure and computer scientists for the implementation and optimization of the code with various tools. Development, implementation and evaluation of codes for high performance computing needs large investment and long term support. KONWIHR projects have, therefore, primarily been supported, if they included interdisciplinary work, application partners in the industry and additional funding from universities and other third parties.

According to this schema, around 40 individual actions were funded by KONWIHR. Table 1 gives an overview over the projects and the subjects covered by them.

For the larger projects, this report includes scientific documentations ordered in groups of the main application areas. Part I covers 8 projects from fluid dynamics, part II 7 projects in the area of computer science and mathematics and part III 8 projects from natural science.

We gratefully acknowledge the continued support by the State of Bavaria in promoting high performance computing. We finally thank Springer for publishing this volume and making this contribution public to the international scientific community.

Munich,   *Arndt Bode*
November 2004   *Franz Durst*

	Multidisziplinary	Biology	Chemistry	Elec. Engineering	Geology	Computer Science	Mathematics	Physics	Comp. Fluid Dynamics	Material Science	Advisory	Service	Guests	Industry	Conference	Page number
KONWIHR-Project																
AG-SMP															X	
DiSiVGT	X							X						X		19
EuroPar															X	
FPGA						X										103
FPGA-Exhibition						X									X	
GAMM															X	
Kurzlehrgang															X	
LIA												X				
LRZ											X	X				129
MethWerk	X					X			X					X		141
NBW					X											251
PADDA-Workshop															X	
PARAGAUSS	X		X			X										285
ParBaum	X	X				X										275
Peridot						X										193
SkvG	X					X			X							71
Vajtersic												X				
VISimLab	X							X						X		83
Bestwihr	X							X						X		3
Biswas												X				
cxHPC										X						97
CUHE								X								205
Enzymech			X													213
FlowNoise	X			X		X			X							31
Flusib	X								X					X		43
FreeWIHR	X					X				X						225
gridlib	X					X			X							117
HQS@HPC	X					X	X							X		237
LBA-Workshop															X	
OOPCV								X								263
OPTILAS	X					X								X		153
PAR-CHEM			X													51
Par-EXPDE	X					X	X									167
ParRICHY	X					X	X							X		181
Popa													X			
RexSim									X							63
Conference															X	
Administration												X				

Contents

Part I Fluid Dynamics

BESTWIHR: Testing of a Closure Assumption for Fully Developed Turbulent Channel Flow with the Aid of a Lattice Boltzmann Simulation
Peter Lammers, Kamen N. Beronov, Thomas Zeiser, Franz Durst 3

DiSiVGT: Validation of a novel turbulence model using direct numerical simulation
J. Kreuzinger, J. Jovanović, R. Friedrich 19

FlowNoise: Flow Induced Noise Computation on Hitachi SR8000-F1
M. Escobar, I. Ali, M. Kaltenbacher, S. Becker, F. Hülsemann 31

FLUSIB: Fully Three-Dimensional Coupling of Fluid and Thin-Walled Structures
Dominik Scholz, Ernst Rank, Markus Glück, Michael Breuer, Franz Durst 43

ParChem: Efficient Numerical Methods for Chemical Problems related to MOVPE
E. Mesic, M. Mukinovic, L. Kadinski and G. Brenner 51

RexSim: Monte Carlo Simulations of Radiative Heat Transfer in Parallel Computer Architectures
G. Brenner, L. Kadinski, J.G. Marakis, ... 63

SkvG: Cache-Optimal Parallel Solution of PDEs on High Performance Computers Using Space-Trees and Space-Filling Curves
Markus Langlotz, Miriam Mehl, Tobias Weinzierl, Christoph Zenger 71

VISimLab: Optimizing an Interactive CFD Simulation on a Supercomputer for Computational Steering in a Virtual Reality Environment
Petra Wenisch, Oliver Wenisch, Ernst Rank 83

Part II Computer Science and Mathematics

cxHPC: Setting up ByGRID — First Steps Towards an e-Science Infrastructure in Bavaria
Georg Hager, Thomas Zeiser, Helmut Heller 97

FPGA: Exploration of the possibilities for the direct synthesis of concurrent C programs on high-performance computers in FPGAs
Peter Urbanek and Stefan May 103

gridlib: A Parallel, Object-oriented Framework for Hierarchical-hybrid Grid Structures in Technical Simulation and Scientific Visualization
Frank Hülsemann, Stefan Meinlschmidt, Ben Bergen, Günther Greiner, Ulrich Rüde 117

LRZ: The Suitability of Contemporary Processors for Quantum Chemical Computations
Ludger Palm 129

MethWerk: Scalable Mesh-based Simulation on Clusters of SMPs
Amitava Gupta, Peter Luksch, Andreas C. Schmidt 141

OPTILAS: Numerical Optimization as a Key Tool for the Improvement of Advanced Multi-Beam Laser Welding Techniques
Verena Petzet, Christof Büskens, Hans Josef Pesch, Victor Karkhin, Maksym Makhutin, Andrey Prikhodovsky, Vasily Ploshikhin 153

ParEXPDE: Expression Templates and Advanced PDE Software Design on the Hitachi SR8000
Christoph Freundl, Ben Bergen, Frank Hülsemann, Ulrich Rüde 167

ParRichy: Parallel Simulation of Bioreactive Multicomponent Transport Processes in Porous Media
S. Kräutle, M. Bause, A. Prechtel, F. Radu, P. Knabner 181

Peridot: Towards Automated Runtime Detection of Performance Bottlenecks
Karl Fürlinger, Michael Gerndt 193

Part III Natural Sciences

CUHE: Electron-Spin Interaction in High-T_c Superconductors
Zhongbing Huang, Werner Hanke, and Enrico Arrigoni 205

ENZYMECH: Computer Simulations of Enzyme Reaction Mechanisms: Application of a Hybrid Genetic Algorithm for the Superimposition of Three-Dimensional Chemical Structures
Alexander von Homeyer, Johann Gasteiger 213

FreeWIHR: Lattice Boltzmann Methods with Free Surfaces and their Application in Material Technology
Carolin Körner, Thomas Pohl, Ulrich Rüde, Nils Thürey, Torsten Hofmann .. 225

HQS@HPC: Comparative numerical study of Anderson localisation in disordered electron systems
Gerald Schubert, Alexander Weiße, Gerhard Wellein, Holger Fehske 237

NBW: Computational Seismology: Narrowing the Gap Between Theory and Observations
Bernhard Schuberth, Michael Ewald, Heiner Igel, Markus Treml, Haijiang Wang, Gilbert Brietzke 251

OOPCV: Phasediagram and Scaling Properties of the Projected SO(5) Model in Three Dimensions
Martin Jöstingmeier, Ansgar Dorneich, Enrico Arrigoni, Werner Hanke, S.C. Zhang 263

ParBaum: A Fast Program for Phylogenetic Tree Inference with Maximum Likelihood
Alexandros P. Stamatakis, Thomas Ludwig and Harald Meier 275

ParaGauss: The Density Functional Program ParaGauss for Complex Systems in Chemistry
Notker Rösch, Sven Krüger, Vladimir A. Nasluzov, Alexei V. Matveev 285

Appendix Color figures

Color figures 297

Part I

Fluid Dynamics

Franz Durst

Institute of Fluid Mechanics
Friedrich-Alexander-Universität Erlangen-Nürnberg
Cauerstr. 4
91058 Erlangen, Germany

Computational Fluid Mechanics is one of the subjects that benefited considerably from support received through KONWIHR to apply high performance computer codes on large computers to study fluid flows and related problems. The success of seven of the projects was impressively demonstrated by the presentations of results obtained. In this section, the written versions of the project presentations are given to complete the information already made available at the closing colloquium of KONWIHR in Munich on the 14th and 15th October 2004.

BESTWIHR was a project within KONWIHR that aimed at the application of the developed computer code BEST to numerically predict turbulent channel flows. Most DNS-predictions were carried out in the past using spectral methods or finite-volume techniques; the aim within the BESTWIHR-project was to utilize the Lattice-Boltzmann techniques for direct numerical simulations of turbulent channel flows. Results obtained are summarised in the first paper in this section. A good agreement is shown with existing data for the mean velocity profile, the turbulence intensity distribution and also some higher-order moments. Only the higher-order moments of the v-component (perpendicular to the wall) show the same defect as predicted with other methods.

The project DiSiVGT was set up within KONWIHR in order to utilise data from direct numerical simulations obtained within the project and also within BEST-WIHR for verifying the assumptions introduced into statistical turbulence models. The work took into account that the development of turbulence closure for flow predictions was heavily limited by the lack of adequate data. For nearly all important quantities which are required in the application of the closure assumptions, experimental measurements are difficult, or sometimes even impossible. With the appearance of direct numerical simulations, significant progress has been made in this respect by exploring numerical databases. These databases contain the complete budgets of all important quantities which are necessary for the direct verification of the closure assumptions currently employed. New directions in turbulence research closely follow computer technology, which is in an advanced stage of development.

Some of the new fundamental contributions to understanding turbulence phenomena have been obtained by exploring these databases. Furthermore, new modelling concepts based on the anisotropy invariants were evaluated with these DNS data.

FlowNoise is the third paper in this section. It describes how the computer code FASTEST-3D was used for large-eddy simulations in order to obtain the source terms for the noise production in fluid flows. FASTEST-3D was coupled with a program called CFS++ to predict flow induced in noise. The data were analyzed to understand physically how noise is produced. In KONWIHR all the work could be done to run both programs efficiently on supercomputers. FlowNoise will provide the basis for efficient studies of sources of noise in technical equipment.

FLUSIB was a project within KONWIHR that studied the coupling of fluid flows within thin-walled structures. Structural simulations were carried out using three-dimensional higher-order elements and the flow computations were performed with finite-volume methods using the large eddy simulation approach. Special attention was given to effective coupling of the structural and fluid information in order to maintain the efficiency of both programs in the combined fluid flow and structural computations. Finally, several technical applications were considered, e.g. the wind-excitation of a thin-walled tower.

Within RexSim simulations were carried out to incorporate efficiently the computation of combined radiative and conductive heat transfer into numerical computer codes. Monte-Carlo simulations were carried out and a specially formulated solution method was employed to avoid the time consuming rate raising part of the Monte-Carlo method. Efficiency tests were performed and the method was found to work satisfactorily on the Hitachi supercomputer.

Within VISimLab a first but big step was made to couple a supercomputer with the virtual reality environment. This coupling permitted interactions with the computations by adding and removing objects in rooms. The flow computations aimed at efficient ways for computational steering of fluid flow computations. Impressive results have been obtained although some of the results are still demonstrative in nature.

Flows with chemical reactions become important when treating in detail the performance of reactors. In KONWIHR, the metal organic vapour phase epitaxy (MOVPE) was examined in order to support development work on optimising epitaxy methods at various institutes at Bavarian Universities and also in industry. It is shown in the enclosed paper that the AlGaN-growth process could be reliably treated with numerical codes that solve the basic equations of fluid flows and take, in addition, chemical reactions into account.

The project SkvG deals with parallel solution procedures for partial differential equations in order to ensure good performance of computer codes even if large problems are solved. The paper outlines the approach and stresses that the numerical methods used should be based on hierarchical multi-level data and adaptively refined data structure. Space-filling curves are introduced to order the store data in such a way that highly efficient computations can be carried out. This is an approach that justifies new efforts in constructing solvers for partial differential equations for future computations.

As a whole, the projects aiming for fluid flow predictions on supercomputers were successfully completed within KONWIHR and the set of papers presented in this section shows this.

BESTWIHR: Testing of a Closure Assumption for Fully Developed Turbulent Channel Flow with the Aid of a Lattice Boltzmann Simulation

Peter Lammers, Kamen N. Beronov, Thomas Zeiser, and Franz Durst

Institute of Fluid Mechanics
University of Erlangen–Nuremberg
Cauerstraße 4
91058 Erlangen, Germany
{$plammers,kberonov,thzeiser,durst$} @lstm.uni-erlangen.de

1 Introduction

The objective in classical turbulence modeling is to construct models for the unknown Reynolds stress tensor in the Reynolds–averaged Navier–Stokes (RANS) equations. One way to do it is to consider the evolution equations for the Reynolds stresses, which can be derived from the original Navier-Stokes equations. In these additional evolution equations, further unknown correlations appear, whose evolution depends in turn on still more unknown correlations. The art of modeling is to a great extent the choice of closure for this infinite system of equations. Here, a closure at the level of the evolution equations for the Reynolds stresses R_{ij} and for their dissipation rate tensor ϵ_{ij} (defined respectively in Eq. (4) and Eq. (8) below) is considered. For all unknown correlations appearing in these evolution equations are modeled, using rational closure assumptions.

To justify a particular closure of this kind, it is necessary to check the approximate validity of these assumptions, at least for some basic flows, such as homogeneous anisotropic incompressible turbulence or developed (time-independent) inhomogeneous turbulence. Reliable databases for all unknown correlations are needed for the selected kinds of turbulent test flows. Since many components of velocity, pressure, and their derivatives are required, such databases are best generated by direct numerical simulation (DNS) of the test flows, because the complete three-dimensional instantaneous flow fields, from which any required statistics can be obtained. This is in many cases not possible with the available measurement techniques like hot-wire anemometry or laser Doppler anemometry.

In spite of the steady increase of computer performance, however, DNS of turbulent flows remains very expensive. Its high memory and CPU time requirements can be met only by supercomputers and only when simulations are restricted to moderate Reynolds numbers. This continues to motivate the development of less expensive numerical schemes for fluid flow simulation, especially for turbulence. Such schemes must exploit the specific features and advantages of existing and future supercomputers efficiently. The lattice Boltzmann method is a candidate for such a numerical scheme.

This was clearly shown in the work [15] and [16], in which one–point statistics up to higher order were examined. Building on that work, the present paper aims at two goals: First, the lattice Boltzmann results in fully developed plane channel flow are checked, see Sect. 2.1, for the terms, arising in the balance equations of the Reynolds stresses and the dissipation rate, against standard pseudospectral DNS databases. Second, to use lattice Boltzmann simulation to check closure models for the three unknown correlations in the Reynolds stress equations. The esults are presented in Sect. 3.2. The considered closure models are explained in Sect. 2.2.

The computations are done on the Hitachi SR8000–F1 at the Leibniz Rechenzentrum (LRZ) computing center in Munich. The architecture of this computer allows highly flexible programming strategies. Parallelization can be done either by assigning one MPI process to each processor (MPP mode) or in a hybrid way by shared memory parallelization inside one node (SMP mode) and one MPI process on each node. The differences between these two modes are examined in Sect. 3.1, using the computer code $\mathcal{BEST}$ (**B**oltzmann **E**quation **S**olver **T**ool). This is a 3D lattice Boltzmann solver developed at the authors' institute.

2 Formulation

2.1 The physical problem: fully developed, turbulent, plane channel flow

A test problem for the simulation of turbulence in wall bounded shear flows is the DNS of the so–called minimal channel, a restricted specification [12] based on the well known plane channel flow. The geometry and the used coordinate system are shown in Fig. 1. It is natural to define the Reynolds number by the wall–shear

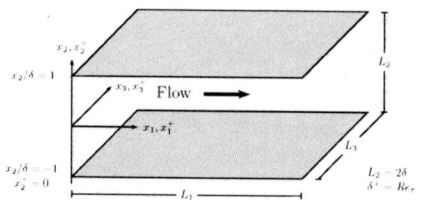

Fig. 1. Computational domain for plane channel flow and coordinate system.

velocity u_τ. From the momentum equation the dependence on the pressure gradient in streamwise direction $\partial p/\partial x_1$ is derived according to

$$u_\tau = \sqrt{\frac{\tau_w}{\rho}} = \sqrt{\frac{\delta}{\rho}\frac{\partial p}{\partial x_1}}, \qquad (1)$$

with δ being the channel half-width, ρ the fluid density, and τ_w the wall shear stress. All lengths can then be measured in the "wall units" of velocity length, $,u_\tau$ and ν/u_τ, and are indicated by a $^+$ superscript, as usual. A corresponding Reynolds number can be defined,

$$Re_\tau = \delta u_\tau/\nu = \delta^+. \qquad (2)$$

In order to compare with the databases described in [18] and [23], both of which have

$$Re_\tau = 180,$$

the same Re_τ is chosen which for the present DNS, as well. The computations are performed on a grid of 4096:256^2 points, ensuring a uniform spatial resolution of $\Delta x_i^+ \approx 1.4$.

In the simulated flow, the velocity field can be considered periodic in streamwise direction x_1 and in the spanwise direction x_3, provided that the computational grid is large enough to accommodate the correlations lengths in these two direction. The x_1 length must then be especially large. Along the remaining coordinate, the non-slip condition is imposed at the walls, located at $x_2 = \pm \delta$.

It is documented in the literature on DNS of turbulence that a step size of $\Delta = 1.5\eta - 2\eta$, where η is the dissipative (Kolmogorov) length scale, is an upper limit, above which the fine structure of turbulence is not resolved, while $\Delta = \eta$ guarantees full resolution. For the present 2D–channel turbulence, it is estimated [19, exercise 7.8] that $\eta^+ \approx 1.5$ at the wall and η increases with increasing distance from the wall. A uniform grid step size $\Delta^+ \approx 1.5$ in wall units would therefore guarantee a fully resolved DNS, which is the case for the present simulation.

As initial condition, a superposition of an approximation to the expected turbulent mean velocity profile and of streamwise and spanwise vortices is taken. The simulation is continued until $100\,\tau$, before the averaging is started. The time unit $\tau = \delta/u_\tau$ is a physical measure of convective transport by the mean flow. According to experience, initial transients last typically $30\,\tau$, so that the turbulence used for computing the statistics reported below can be considered fully developed indeed. All statistics are obtained by averaging over the $x_1 - x_3$ plane and additionally in time.

2.2 The turbulence model: a Reynolds–stress closure

By introducing the Reynolds decomposition for the velocity and pressure fields,

$$u_i = \overline{u}_i + u'_i, \quad p = \overline{p} + p',$$

into the governing incompressible Navier–Stokes equations, one can obtain equations (RANS) for the mean fields $\overline{u}_i, \overline{p}$ and for the disturbances u'_i, p'. In the RANS-equations, the so-called Reynolds stress tensor, which is the average of the second-order, one-point moments of fluctuating velocity u'_i, is unknown. Often, further equations are taken into account to relegate the closure problem from the level of the RANS to the higher level of equations for the Reynolds stresses themselves and other related second–order moments. This approach was first put forward in [13].

By manipulating the equation for the disturbances, a transport equation for the Reynolds stresses is obtained (see [9] for example):

$$DR_{ij}/Dt = P_{ij} + T_{ij} - 2\epsilon_{ij} + \Pi_{ij} + D_{ij} \qquad (3)$$

The following physical meanings are assigned to the tensors in this equation [5]:

$$R_{ij} = \overline{u'_i u'_j} \qquad \text{Reynolds stress component} \qquad (4)$$

$$D\overline{u'_i u'_j}/Dt = \partial_t \overline{u'_i u'_j} + \overline{u}_k \partial_k \overline{u'_i u'_j} : \qquad \text{Total change in Reynolds stress} \qquad (5)$$

$$P_{ij} = -(\overline{u'_j u'_k} \partial_k \overline{u}_i + \overline{u'_i u'_k} \partial_k \overline{u}_j) : \qquad \text{Production} \qquad (6)$$

$$T_{ij} = -\partial_k \overline{u'_i u'_j u'_k} : \qquad \text{Third order velocity correlation, turbulent transport} \qquad (7)$$

$$\epsilon_{ij} = \nu \overline{\partial_k u'_i \cdot \partial_k u'_j} : \qquad \text{Dissipation} \qquad (8)$$

$$\Pi_{ij} = -\frac{1}{\rho}\left(\overline{u'_i \partial_j p'} + \overline{u'_j \partial_i p'}\right) : \qquad \text{Velocity/pressure gradient correlation} \qquad (9)$$

$$D_{ij} = \nu \partial_k^2 \overline{u'_i u'_j} : \qquad \text{Viscous diffusion.} \qquad (10)$$

Obviously, the three correlations T_{ij}, Π_{ij}, and ϵ_{ij} are unknown even if the mean flow velocity and pressure were known. By contracting Eq. (5), a similar equation can be obtained for the evolution of the turbulent kinetic energy

$$k = \frac{q^2}{2} = \frac{\overline{u'_i u'_i}}{2}$$

More complicated is the situation in the case of the transport equation for the full tensor of the turbulent dissipation rate. Even if only its trace is considered, as done in the very popular k-ϵ models, see for example [7], the resulting equation contains even more terms that need to be modeled and are difficult to compute or measure. Most of them are modeled in a simple–fashioned way in the two–equation (k-ϵ, k-ω, etc.) models. Here, this equation is written out in its full form:

$$D\epsilon/Dt = P_\epsilon^1 + P_\epsilon^2 + P_\epsilon^3 + P_\epsilon^4 + T_\epsilon + \Pi_\epsilon - \gamma + D_\epsilon. \qquad (11)$$

with following definitions of the individual terms (see [21]):

$$D\epsilon/Dt = \partial_t \epsilon + \overline{u}_k \partial_k \epsilon : \qquad \text{Total change in dissipation} \qquad (12)$$

$$\begin{aligned} P_\epsilon^1 &= -2\epsilon_{ik}\partial_k \overline{u}_i \\ &= -2\nu \overline{\partial_l u'_i \cdot \partial_l u'_k}\partial_k \overline{u}_i \end{aligned} \qquad \text{Production due to mean velocity gradient} \qquad (13)$$

$$\begin{aligned} P_\epsilon^2 &= -2\tilde{\epsilon}_{lk}\partial_l \overline{u}_k \\ &= -2\nu \overline{\partial_l u'_i \cdot \partial_k u'_i}\partial_l \overline{u}_k \end{aligned} \qquad \text{Production due to mean velocity gradient} \qquad (14)$$

$P_\epsilon^3 = -2\nu \overline{u'_k \partial_l u'_i \cdot \partial_l \partial_k \overline{u}_i}$: Production due to mixed effects of the gradients of mean and fluctuating velocities (mixed production) (15)

$P_\epsilon^4 = -2\nu \overline{\partial_l u'_i \cdot \partial_k u'_i \cdot \partial_l u'_k}$: Production due to deformation of the vortices (vortex stretching) (16)

$T_\epsilon = -\nu \partial_k \overline{(u'_k \partial_l u'_i \cdot \partial_l u'_i)}$: Diffusive transport due to turbulent fluctuations (17)

$\Pi_\epsilon = -\frac{2\nu}{\rho} \overline{\partial_l u'_i \partial_i \partial_l p'}$: Diffusive transport due to turbulent pressure fluctuations (18)

$-\gamma = -2\nu^2 \overline{(\partial_k \partial_l u'_i)^2}$: Viscous destruction (19)

$D_\epsilon = \nu \partial_k^2 \epsilon$: Viscous diffusion (20)

In Eq. (11), seven terms are unknown: P_ϵ^1, P_ϵ^2, P_ϵ^3, P_ϵ^4, T_ϵ, Π_ϵ, and γ.

In order to close the equations for the Reynolds stresses, the kinetic turbulent energy and the dissipation rate, the unknown correlations appearing on the right-hand sides of the corresponding evolution equations must be expressed in terms of $\overline{u}_i$ and $\overline{u'_i u'_j}$. In this paper, the lattice Boltzmann DNS results will be used to check closure models for the three correlations of Eq. (5), following the model described in detail in [10]. In particular, reliable modeling of the dissipation rate is needed. As shown by [14], the dissipation can be decomposed according to

$$\epsilon_{ij} = \epsilon_{ij}^h + \epsilon_{ij}^{inh} = -\nu \Delta_\xi (\overline{u'_i u''_j})_0 + \frac{1}{4} D_{ij}. \qquad (21)$$

Here $\overline{u'_i u''_j}$ is a twopoint correlation in the limit of zero separation in space, $\boldsymbol{\xi} = 0$. For the model itself it is therefore reasonable to solve an equation for the homogeneous part ϵ_h of the ϵ instead of Eq. (11). To elaborate Eq. (21) further, use is made of the observation that for axisymmetric disturbances all involved tensors of rank two are linearly ligned [11]. For example,

$$e_{ij} = \frac{\epsilon_{ij}^h}{\epsilon^h} - \frac{1}{3}\delta_{ij} = A a_{ij} = A \left(\frac{\overline{u'_i u'_j}}{q^2} - \frac{1}{3}\delta_{ij} \right). \qquad (22)$$

The tensor a_{ij} is the anisotropy tensor, introduced first by [17]. Combining the above relations, one may write for the case of axisymmetric turbulence

$$\epsilon_{ij} \approx A\epsilon_h a_{ij} + \frac{1}{3}\epsilon_h \delta_{ij}. + \frac{1}{4}D_{ij}. \qquad (23)$$

The function A can depend only on the invariants II_a and II_e of the respective tensors,

$$A = A(II_a, II_e) = \sqrt{II_a/II_e}$$

or alternatively,

$$A = A(II_a, III_a, Re_\lambda) = 1 - J(W - 1), \qquad (24)$$

with
$$W(Re_\lambda) = 0.626 \left(-0.049 Re_\lambda + \tfrac{1}{2}\sqrt{0.009604 Re_\lambda^2 + 10.208}\right) \quad (25)$$
and
$$J(II_a, III_a) = 1 - 9\left(\tfrac{1}{2}II_a - III_a\right), \quad (26)$$

where $Re_\lambda = \lambda q/\nu$ is the Taylor microscale Reynolds number. The Taylor microscale λ itself is defined as $\lambda = \sqrt{5\nu q^2/\epsilon_h}$.

The velocity/pressure–gradient correlation can be split into two parts traditionally called "slow" and "fast" part. Coming from an analytical expression by Chou [5] for homogeneous turbulence, and exact solution for initially-isotropic is given in [6] for turbulence exposed to rapid distortion:

$$\Pi_{ij} = \tfrac{2}{5} q^2 S_{ij}, \quad II_a \to 0. \quad (27)$$

¿From the dynamic equation of the anisotropy tensor, conclusions are drawn for the behavior of Π_{ij} in two–component turbulence and in decaying, homogeneous, axisymmetric turbulence. Taking the concept of realizability introduced in [22] into account, the following model can be derived,

$$\Pi_{ij} \approx \underbrace{a_{ij} P_{ss} + F\left(\tfrac{1}{3} P_{ss} \delta_{ij} - P_{ij}\right)}_{\Pi_{ij}^{fast}} + \underbrace{C \epsilon_h a_{ij}}_{\Pi_{ij}^{slow}}, \quad (28)$$

with
$$F(II_a, III_a) = \tfrac{3}{5} + \tfrac{18}{5}\left(\tfrac{1}{2}II_a - III_a\right), \quad (29)$$
and
$$C(II_a, III_a, Re_\lambda) = 4.78\,(W - 1)\,J. \quad (30)$$

The invariant functions F and C interpolate between the different turbulent states.

The analytical treatment of the turbulent transport correlation T_{ij} is rather difficult because one has to deal with correlation higher than third order in the transport equation for T_{ij}. Supposing once again axisymmetry, the suggested model becomes

$$T_{ij} = c_q \partial_l \frac{\overline{u'_i u'_j}}{q^2} \frac{k^2}{\epsilon_h} J \partial_l k^2 \quad \text{with} \quad c_q = 0.5. \quad (31)$$

This is supposed to be an acceptable approximation only if the transport term is small.

2.3 The numerical method: 3D lattice Boltzmann

The numerical method for the present simulation utilizes the fact that the (macroscopic) velocity field **u** and the pressure field p of an viscous fluid can be obtained by solving an kinetic equation for a one–particle, one–point probability density distribution function f, instead of solving the Navier-Stokes equation directly. The function $f = f(\boldsymbol{\xi}, \mathbf{x}, t)$ depends on a microscopic velocity $\boldsymbol{\xi}$, the spatial coordinate **x**, and the time t. The hydrodynamic fields are given by the moments of the distribution function f. The oldest and best known kinetic equation is the Boltzmann equation. A radical simplification of that equation and of many other kinetic equations of similar form has been proposed by [1]. This now very popular BGK ansatz for the collision operator is

$$(\partial_t + \boldsymbol{\xi} \cdot \boldsymbol{\nabla}_r) f(\boldsymbol{\xi}, \mathbf{r}, t) = -\frac{f(\boldsymbol{\xi}, \mathbf{r}, t) - f^{eq}(\boldsymbol{\xi}, \mathbf{r}, t)}{\lambda}. \tag{32}$$

The function f^{eq} is the equilibrium distribution (of Maxwell–Boltzmann kind) and λ is the single parameter of the model, called relaxation time.

The first step to specify a lattice Boltzmann model is the choice of a lattice to discretize the dependence on $\boldsymbol{\xi}$. A finite set of velocities $\mathbf{c}_i$ is chosen. For every $\mathbf{c}_i$ a discrete distribution function f_i is defined. After discretization in space and time, Eq. (32) finally reduces to [4]

$$f_i(\mathbf{x} + \mathbf{c}_i, t+1) - f_i(\mathbf{x}, t) = -\omega \left(f_i(\mathbf{x}, t) - f_i^{eq}(\rho, \mathbf{u}, t) \right). \tag{33}$$

Eq. (33) appears as an explicit first-order scheme but is in fact second-order in time. In order to proof the equivalence of Eq. (33) and the Navier–Stokes equations the Chapman–Enskog procedure [3] is applied to Eq. (33). By means of this procedure,

$$f_i^{eq} = t_p \rho \left\{ 1 + \frac{c_{i\alpha} u_\alpha}{c_s^2} + \frac{u_\alpha u_\beta}{2 c_s^2} \left(\frac{c_{i\alpha} c_{i\beta}}{c_s^2} - \delta_{\alpha\beta} \right) \right\} \tag{34}$$

can be shown to be an appropriate procedure to recover approximately the incompressible Navier–Stokes dynamics under the condition that the Mach number $|u/c_s| \ll 1$, where c_s is the speed of sound of the model. The parameters t_p depend on the set of discrete velocities $\mathbf{c}_i$ and can be found by a rational algebraic procedure [20]. For turbulent flows the authors have made good experience with the D3Q19 lattice model which is used for the present simulations, as well. As mentioned above, the hydrodynamic quantities are given by moments of f_i, namely

$$\rho = \sum_{i=0}^{n} f_i, \qquad \rho \mathbf{u} = \sum_{i=0}^{n} \mathbf{c}_i f_i, \qquad p = c_s^2 \rho, \tag{35}$$

with $n = 18$ and $c_s^2 = 1/3$ for the chosen model. ¿From the Chapman–Enskog procedure also the expression for the viscosity is deduced which comes out to depend on the relaxation parameter ω as follows,

$$\nu = \frac{1}{6} \left(\frac{2}{\omega} - 1 \right).$$

In the present simulation a slight modification of Eq. (33) is used, namely the socalled incompressible D3Q19I model [8, 15]. The procedure to add the pressure gradient into Eq. (33) is described in [2]. To ensure no–slip boundary conditions at the solid walls, at $x_2 = \pm \delta$, the so-called bounce-back rule is applied: The density f_i leaving a node along the direction of $\boldsymbol{\xi}_i$ and bound to cross into the solid wall, is returned to the same node of departure at the subsequent time step as the density associated with the opposite discrete velocity $-\boldsymbol{\xi}_i$.

3 Results

3.1 Hitachi SR8000-F1: MPP versus SMP mode

With the SR8000-series the Japanese vendor of high performance computers Hitachi intended to bring together its MPP architecture SR2201 and its vector architecture S-3000. The final installation at the Leibniz Rechenzentrum consists of 1344 superscalar RISC CPU's with 1.5 Gflops (Giga *floating point operations per second*) peak performance. Each CPU possesses 128 KByte L1–Cache und 160 floating point register . The instruction set allows two ways loading data from main memory to the floating point register . Either by direct transfer of a data element from the main memory to the floating point register (preload) or via prefetch of complete cache lines to the cache followed by a load to the floating point register (prefetch). This second mechanism is used by the compiler in $\mathcal{BEST}$. The procedure of load (from floating point register) + operation + store to memory in combination with preceding prefetch instructions is called PVP (*Pseudo Vector Processing*) by Hitachi. The CPU's are integrated into shared memory node of 6.5 GByte. Each of these SMP (*Shared Memory Processing*) nodes consists of 8 CPU's. The nodes are connected by a crossbar network with a bandwidth of 770 MByte/s in the case of $\mathcal{BEST}$[1]. In Fig. 2(a,b) the performance of $\mathcal{BEST}$ for the MPP (*Massively Parallel Processing*) mode is measured for a plane channel geometry with a cross-section consisting of 128^2 grid points. The domain decomposition is done in streamwise direction. In all cases the definition of the speedup is

$$\text{Speedup} = \frac{nT(N,1)}{T(nN,n)}, \qquad (36)$$

where T is the compute time, N the problem size on one node or one processor respectively and n the number of nodes (processors). For parallel communication, MPI (*Message Passing Interface*) is employed. In Fig. 2(a) two nodes are benchmarked by three different grid sizes, with 256, 512 and 1024 points in streamwise direction. As expected, performance increases with increasing compute/communication time ratio. The maximum is 20 MLUP/s (mega lattice side updates per second). This is measured for a processor topology in which neighboring domains are located on the same node as far as possible. In the default setting (round robin) the first domain is on node one, the second on node two, the third domain again on node one and so on. Therefore, communication takes place over the network instead of making use of shared memory inside one node. A noticeable loss in performance is the consequence. In Fig. 2(b) the basic grid size on one node is 512×128^2. The speedup is measured up to 64 processors. A significant performance loss can be observed only for the round-robin setting.

Mode parallelization can be done not only using MPP, but also within each node by running shared memory parallelization with eight threads. Hitachi has named this strategy COMPAS (*Co-operative Micro Processors in single Address Space*). In this mode, the memory bandwidth is 32 GBytes/s which matches the aggregated single processor bandwidth. For the largest grid the SMP mode is slightly slower than the MPP mode. But it does not show significant sensibility with respect to

---

[1] The SR8000-F1 offers a special feature which extent the bandwidth to up to 950 MByte/s.

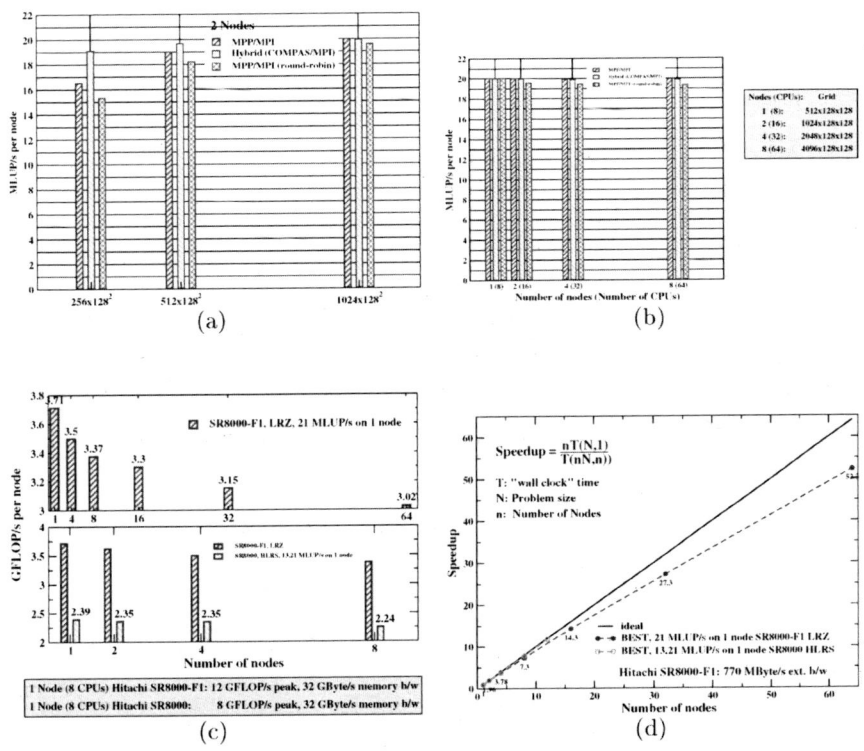

Fig. 2. Performance and speedup of $\mathcal{BEST}$ for Hitachi SR8000-series

grid size. The performance loss between SMP and MPP mode (no round-robin) on the smallest grid is about 14%, as can be seen from Fig. 2(a).

Between the nodes MPI is used again. The communication between the nodes is done by one thread only (hybrid/masteronly). In Fig. 2(c) the machines in Munich with a clock frequency of 350 MHz and in Stuttgart with 250 MHz are compared. Both machines have the same memory bandwidth. The compiler is able to crudely estimate the performance of loop constructs. For the most time consuming loop this information reveals that four floating point operations per load operation should be possible. Therefore $\mathcal{BEST}$ is not limited by the memory bandwidth excessively and should benefit from a higher frequency which is indeed the case[2]. Finally, Fig. 2(d) shows the speedup measurement up to 64 nodes. The SMP mode is used inside the nodes.

---

[2] The measurements on the SR8000-F1 are done with an older version of $\mathcal{BEST}$ than that used for measurements on the SR8000 in Stuttgart. In the meantime, some additional subroutine calls which do not pertain to the algorithm itself cause a slight performance loss.

3.2 Tensor balance equations for plane channel flow

DNS databases can be used to check turbulent model assumptions as it is shown in the following with the database of the present lattice Boltzmann simulation for the plane channel flow and the model explained in Sect. 2.2. First the results for Eq. (3) and Eq. (11) are validated against the pseudospectral simulations of [18] and [23]. The last simulation uses 256^3 points for the same extensions of the computational domain as in [18], resulting in a twice finer spatial resolution.

In Fig. 3 all quantities from Eq. (3) for the $\overline{u'_1 u'_1}$-component of the Reynolds stress are plotted. This component is the most important one. All terms vanish in the

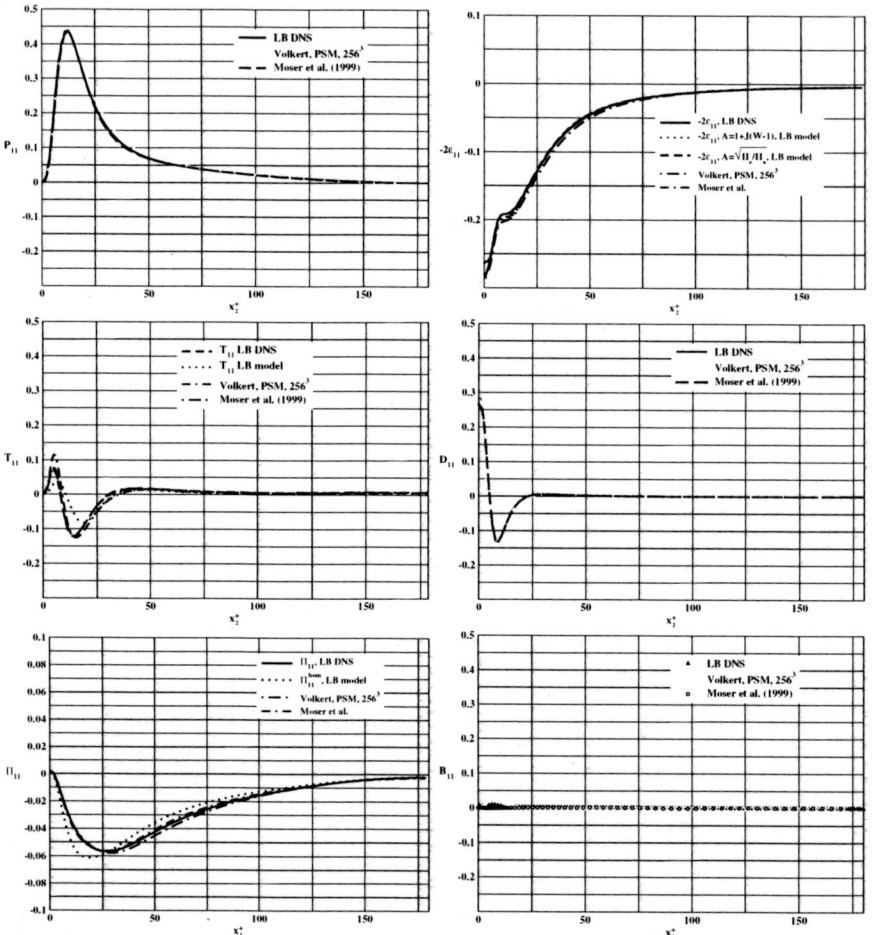

Fig. 3. Comparison of the terms in the balance for $\overline{u'_1 u'_1}$, Eq. (3), with the databases [18] und [23]. All terms are scaled with u_τ^4/ν.

center of the channel. Globally, the balance equation for the $\overline{u'_1 u'_1}$ component of the

Reynolds stress tensor is clearly dominated by the production and the dissipation term. The largest contribution of the turbulent production by the mean velocity gradient takes place in the buffer layer. Near the wall production becomes zero and the dissipation -2ϵ reaches its minimum. Here the equation is balanced by large viscous diffusion of $\overline{u'_1 u'_1}$ from the buffer layer into the viscous sublayer. Inside the buffer layer, the diffusion D_{11} becomes negative. The behavior of the turbulent transport term is similar, except that T_{11} is forced by the boundary conditions to be zero at the wall. The velocity/pressure–gradient correlation influences the balance by a negative contribution reaching the minimum also in the buffer layer. Globally the prediction of the terms is identical for all three simulations. The most significant difference arises for T_{11}. Its extrema in the buffer layer are more distinct in the case of the pseudospectral simulation with finer resolution. But the cross-check of the results done by calculating the balance B_{11} reveals that the balance is not fulfilled so well for this simulation, as it is for the other two simulations.

Along with the DNS results for the individual terms in the balance, the model expressions for the dissipation, the turbulent transport and the velocity/pressure–gradient correlation are plotted in Fig. 3. All modeled correlations vanish in the center of the channel. The model for the dissipation rate ϵ_{11} covers the physics exactly in the whole channel, regardless in which way the function A is obtained. The inhomogeneous part of the dissipation is calculated from the dissipation itself and the viscous diffusion. For T_{11} and Π_{11} the physical behavior is correctly reproduced. Only the extrema are overestimated.

There are three further nontrivial Reynolds stress components, $\overline{u'_2 u'_2}$, $\overline{u'_3 u'_3}$, and $\overline{u'_1 u'_2}$. ¿From these we chose $\overline{u'_1 u'_2}$ as one further example. The component ϵ_{12} of the dissipation tensor is not so important as the other ones. The equation is dominated by a negative production and a positive velocity/pressure–gradient correlation, both reaching their extrema in the viscous and the buffer layer. Turbulent transport makes also a significant contribution to the balance. The results for Π_{12} from the three independent DNS match each other globally, but not in detail. The same is the case for T_{12}. Again in the viscous buffer layer the balance is not fulfilled for the pseudospectral simulation with the finer resolution. The reasons are the predictions for the turbulent transport and the velocity/pressure–gradient correlation. Here the extrema are significantly higher (smaller) than in the two other simulations.

The predictive quality of the considered model is very good for the $\overline{u'_1 u'_2}$ component, the transport T_{12}, and the velocity/pressure–gradient correlation Π_{12}, whereas the dissipation is predicted poorly. In [24] a model is given, which overcomes this deficiency by taking the stress rate into account.

Finally, the individual terms in Eq. (11) are plotted in Fig. 5. Only the results for the pseudospectral simulation with the higher spatial resolution are used for comparison. At the wall, the balance appears to be dominated by viscous diffusion D_ϵ. But the only term showing a tendency to balance D_ϵ is the destruction term and it is not small enough at the wall. So, the balance is not fulfilled. In comparison with D_ϵ, all others terms are small. Their prediction by both DNS methods is nearly equal, except for the destruction, diffusion and transport terms in a region very close to the wall. For the destruction and diffusion term, the reversal points can not be reproduced with the chosen resolution of the lattice Boltzmann simulation. For T_ϵ, the pseudospectral DNS gives a higher peak near the wall than the lattice Boltzmann DNS.

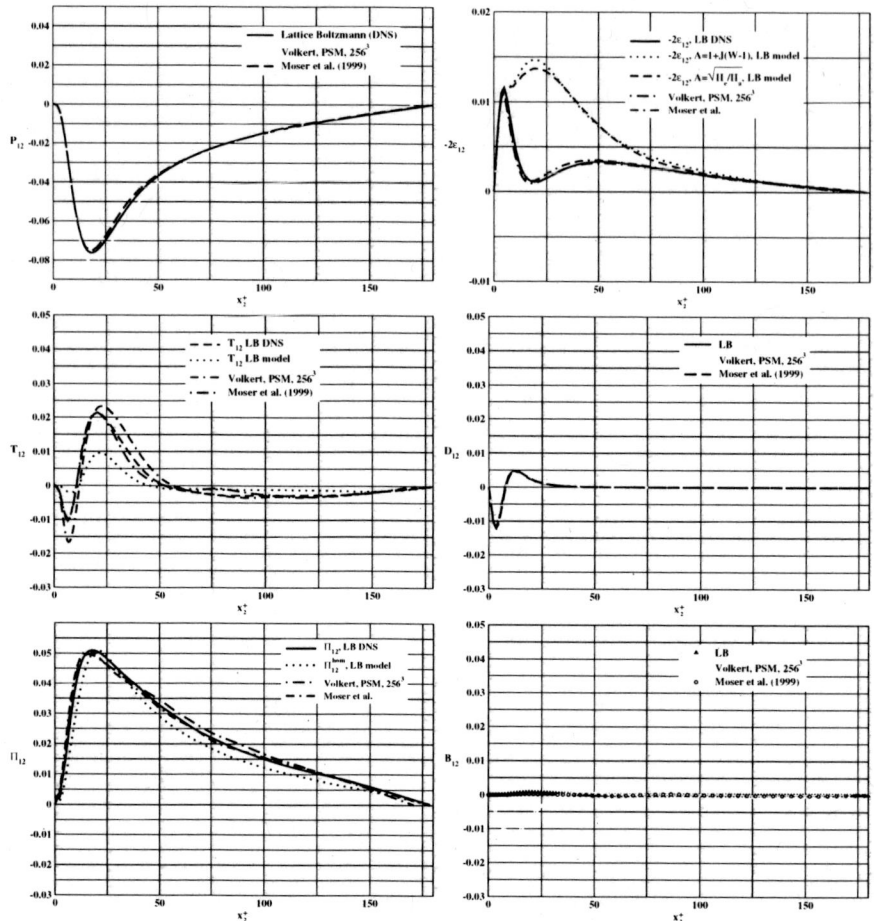

Fig. 4. Comparison of the terms in the balance for $\overline{u'_1 u'_2}$, Eq. (3), with the databases [18] and [23]. All terms are scaled with u_τ^4/ν.

4 Summary

A direct numerical simulation (DNS) of developed, turbulent, plane channel flow at a moderate Reynolds number has been performed, using a lattice Boltzmann method. The turbulent flow field is analyzed in terms of detailed contributions to the balance equations for the Reynolds stress tensor and the turbulent dissipation rate tensor components. Generating databases with such statistics is important for turbulence modeling purposes. An example is given in this paper, taking up a model developed at the authors' institute and designed to match the limits of two–component and axisymmetric turbulence, which are important in describing wall–bounded turbulent flows.

The lattice Boltzmann simulation results are obtained in encouraging agreement with earlier DNS using a standard pseudospectral method. The agreement is partic-

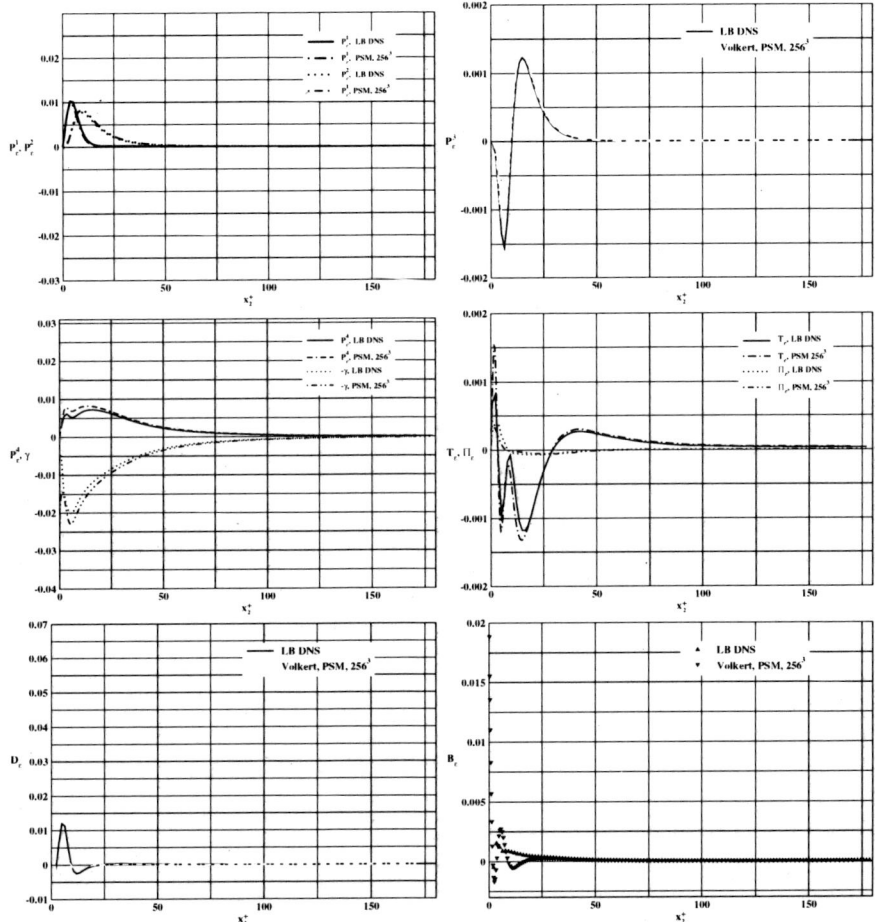

Fig. 5. Comparison of the terms in the balance for ϵ, Eq. (11), with the database [23]. All terms are scaled with u_τ^6/ν^2.

ularly good for the Reynolds stress equations. For the dissipation rate equation, the lattice Boltzmann DNS appears not to have sufficient resolution in the buffer layer in orther to catch some effects present in the most complicated terms of the balance, such as destruction and viscous diffusion of the dissipation rate. Adding a local grid refinement feature to the lattice Boltzmann solver should overcome this problem in general. It may be noted that even the pseudospectral method, which has an inherent strong resolution refinement near the wall and a very high formal precision, experiences problems near the wall, as well. Those with the viscous diffusion term are particularly severe.

With the exception of some benchmarking, all computations have been carried out on the "Bundeshöchstleistungsrechner" computer Hitachi SR8000-F1 at Leibniz Rechenzentrum Munich. This machine can be viewed either as a massively parallel system or as offering a hybrid approach of shared memory nodes connected by a high capacity crossbar network. Computations based on the former view point use the MPP mode; those based on the latter view point use SMP. With one process in SMP mode, a performance comparable to a vector processor is achieved. For the lattice Boltzmann solver $\mathcal{BEST}$, the same performance can obtained for both modes, globally over a large number of nodes.

In detail, the performance depends on the time ratio of computation to communication. Based on the reported performance measurement results, the MPP mode can be recommended when this ratio is high, and the SMP mode when it is low. This is in accordance with general expectations and shows that the code $\mathcal{BEST}$ allows indeed to use the specifics of the particular parallel architecture.

Acknowledgement. The presented work has been funded by KONWIHR, through the BESTWIHR project and through a grant by the Deutsche Forschungsgemeinschaft. The large–scale computations were carried out at the Leibniz Rechenzentrum. Also gratefully acknowledged is the support by the regional computing center at the University Erlangen–Nuremberg (RRZE), that by the John von Neumann–Institut for Computing (NIC) in Jülich, and that by the computing center at the University Bayreuth. R. Volkert at LSTM has kindly made available some of his data from pseudospectral simulations of channel turbulence at the same Re_τ as for the reported $\mathcal{BEST}$ runs.

References

1. P. Bhatnagar, E. P. Gross, and M. K. Krook. A model for collision processes in gases. I. small amplitude processes in charged and neutral one-component systems. *Phys. Rev.*, 94(3):511–525, 1954.
2. J. Buick and C. Greated. Gravity in lattice Boltzmann model. *Phys. Rev. E*, 61(6):5307–5320, 2000.
3. S. Chapman and T. G. Cowling. *The Mathematical Theory of Non-Uniform Gases*. University Press, Cambridge, 1999.
4. S. Chen and G. D. Doolen. Lattice Boltzmann method for fluid flows. *Annu. Rev. Fluid Mech.*, 30:329–364, 1998.
5. P. Y. Chou. On the velocity correlation and the solution of the equation of turbulent fluctuation. *Q. Appl. Maths.*, 3:38–54, 1945.
6. S. C. Crow. Viscoelastic properties of fine-grained incompressible turbulence. *J. Fluid Mech.*, 33:1–12, 1968.
7. D.C.Wilcox. *Turbulence modelling for CFD*. DCW Industries, Inc., La Cañada, California, 1998.
8. X. He and L.-S. Luo. Lattice Boltzmann model for the incompressible Navier-Stokes equation. *J. Stat. Phys.*, 88(3/4):927–944, 1997.
9. J. O. Hinze. *Turbulence*. McGraw-Hill, New York, 2. edition, 1975.
10. J. Jonanovic. *Konwihr-Vorlesung: Turbulenz und Turbulenzmodellierung II.* Vorlesungsmitschrift, Lehrstuhl für Strömungsmechanik, Universität Erlangen-Nürnberg, 2002.
11. J. Jovanović and I. Otić. On the constitutive relation for the reynolds stresses and the prandtl-kolmogorov hypothesis of effective viscosity in axisymmetric strained turbulence. *Transactions of ASME Journal of Fluids Engineering*, 122:48–50, 2000.

12. J. Kim, P. Moin, and R. Moser. Turbulence statistics in fully developed channel flow at low Reynolds number. *J. Fluid Mech.*, 177, 1987.
13. A. N. Kolmogorov. Equations of motion of an incommpressible turbulent fluid. *Izvestiya Akad Nauk SSSR, Ser. Phys*, 6:56–58, 1942.
14. B. A. Kolovandin and I. A. Vatutin. Statistical transfer theory in non-homogeneous turbulence. *Int. J. Heat Mass Transfer*, 15:2371–2383, 1970.
15. P. Lammers, K. Beronov, G. Brenner, and F. Durst. Direct simulation with the lattice Boltzmann code BEST of developed turbulence in channel flows. In S. Wagner, W. Hanke, A. Bode, and F. Durst, editors, *High Performance Computing in Science and Engineering, Munich 2002.* Springer, 2003.
16. P. Lammers, K. Beronov, R. Volkert, G. Brenner, and F. Durst. Lattice Boltzmann Direct Numerical Simulation of Fully Developed 2d–Channel Turbulence. *Computers & Fluids*, submitted.
17. J. L. Lumley and G. Newman. The return to isotropy of homogeneous turbulence. *J. Fluid Mech.*, 82:161–178, 1977.
18. R. Moser, J. Kim, and N. Mansour. Direct numerical simulation of turbulent channel flow up to $Re_\tau = 560$. *Phys. Fluids*, 11, 1999.
19. S. B. Pope. *Turbulent Flows*. Cambridge Univ. Press., 2000.
20. Y. H. Qian, D. d'Humières, and P. Lallemand. Lattice BGK models for Navier-Stokes equation. *Europhys. Lett.*, 17(6):479–484, 1992.
21. T. C. Schenk. *Messung der turbulenten Dissipationsrate in ebenen und achsensymmetrischen Nachlaufströmungen.* PhD thesis, Lehrstuhl für Strömungsmechanik, Universität Erlangen-Nürnberg, 1999.
22. U. Schumann. Realizability of Reynolds stress turbulence models. *Phys. Fluids*, 20:721–725, 1977.
23. R. Volkert. *Bestimmung von Turbulenzgrößen zur verbesserten Turbulenzmodellierung auf der Basis von direkten numerischen Simulationen der ebenen Kanalströmung.* PhD thesis, Lehrstuhl für Strömungsmechanik, Universität Erlangen-Nürnberg, 2004. In Vorbereitung.
24. Q.-Y. Ye. *Die turbulente Dissipation mechanischer Energie in Scherschichten.* PhD thesis, Lehrstuhl für Strömungsmechanik, Universität Erlangen-Nürnberg, 1996.

DiSiVGT: Validation of a novel turbulence model using direct numerical simulation

J. Kgpuztcrpg[1], J. Jdvlcdvtń[2], lco R. Fgtpogtns[1]

[1] FG Sigdb ucrhb pnslctk, Tpnscthns p Uctvpghtil'i Büchspc
Mdaizb lcchig. 15, 85748 Glgnstcr, Gpgblcy
johannes@flm.mw.tu-muenchen.de

[2] Lpsghiusaqig Sigdb ucrhb pnslctk, Uctvpghtil'i Egalcrpc
Clupghig. 4, 91058 Egalcrpc, Gpgblcy
jovan@lstm.uni-erlangen.de

Abstract. B dgp gpad map iugmuapcnp b dopah lgp ugrpciay cppopo qdg isp ltgnglq, luidb dmtap lco nspb tnl a tcouhigtph. Tsp opvpadeb pci dqhuns b dopah cppoh cpw ndcnpeih lco isp vlatolitdc my ndb eapip lco lnnuglip olil hpih. Ic isth wdgk l cdvpaRpyc daoh higphh b dopaslh mppc opvpadepo nl hpo dc tcvlgtlci ispdgy lco iwd- edtci ndggpalitdc ipnsctfup. Tsth b dopah th pxipchtvpay iphipo ndis vtl *a priori* lco *a posteriori* iphitcr nl hpo dc lvltalmap olil hpih. Fdg isp othhtelitdc glip ppfulitdc, isp pxthitcr olil huffpgpo qgdb alnk dqhelitla gphdauitdc. Sd strsay gphdavpo otgpni cub pgtnl ahtb ualitdch slvp mppc nl ggtpo dui id hspo b dgp atrsi dcid isp egdnphhph ophngtmpo my isli pfulitdc. Ic lootitdc, isp cdvpa b dopawl h tb eapb pcipo tc pcrt- cppgtcr hdqwlgp, id b lkpisp cpw opvpadeb pcihlvltalmap qdg ipnsctnl aleeatnlitdch.

1 Introduction

Tsp opvpadeb pci dql cdvpaiugmuapcnp b dopa uhulaay cppoh ispdgy lco esyhtml a tciutitdc tc dgopgi id lggtvp li tb egdvpo nadhugp lhhub eitdch, hpp hpnitdc 2. Tsphp lhhub eitdch lgp ispc iphipo my lipgb my ipgb ndb elgthdc tc hd nlaepo *a priori* iphih (hpnitdc 4). Tdisth pco, gpadmap olil qgdb pxepgtb pcihdg htb ualitdch th cppopo. Ic nl hpo dqis p othhtelitdc glip pfulitdc isp ipgb hnl ccdi mp b plhugpo, spcnp cub pgtnl a htb ualitdch slvp id mp nl ggtpo dui (hpnitdc 3). Evpc tqisp cdvpaiugmuapcnp b dopa egdvtoph rddo gphuaih tc isphp iphih, ti slhid mp hsdwc tc *a posteriori* iphih, isli ti thlmap id ndb euip fldwhl nnuglipay (hpnitdc 5). Ftclaay isp iphipo b dopa nl c mp uhpo tc pcrtcppgtcr leeatnlitdch (hpnitdc 6).

2 Model development

Tsp cdvpaRpyc daoh higphh b dopaslh mppc opvpadepo uhtcr Lub apy'h tcvlgtlci ble [1]. Tsth b le nslglni pgtzphlcy iugmuapcnp hilip my edtcihtc l helnp helccpo my isp

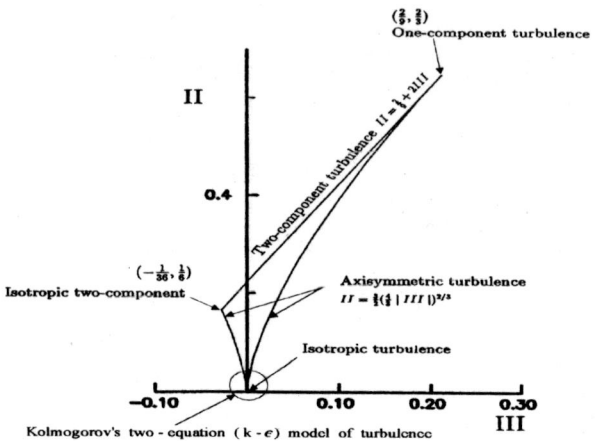

Fig. 1. Lumley's invariant map of turbulence

second and third invariant $II_a = a_{ij}a_{ji}$ and $III_a = a_{ij}a_{jk}a_{ki}$ of the Reynolds stress anisotropy tensor $a_{ij} = \overline{u_i u_j}/\overline{u_k u_k} - \frac{1}{3}\delta_{ij}$ (Fig. 1). Using invariant theory and two-point correlation technique introduced by Chou [2], expressions are derived for the unclosed terms in the transport equations for Reynolds stresses and dissipation rate. At the edges of the invariant triangle, most of the remaining scalar functions can exactly be determined. Otherwise e.g. numerical data available for homogeneous flows is used. For turbulent states in the inner region of the triangle the scalar functions are interpolated. This modeling work is extensively described by Jovanović [3].

The resulting transport equations for the turbulent stresses can be written in the following closed form:

$$\frac{\partial \overline{u_i u_j}}{\partial t} + \overline{U}_k \frac{\overline{u_i u_j}}{\partial x_k} \simeq P_{ij} + a_{ij}P_{ss} + \mathcal{F}(\frac{1}{3}P_{ss}\delta_{ij} - P_{ij}) + (\mathcal{C} - 2\mathcal{A}\epsilon_h)a_{ij}$$

$$-\frac{2}{3}\epsilon_h \delta_{ij} + c_q \frac{\partial}{\partial x_k}\left(\frac{\overline{u_i u_j}}{q^2}\frac{k^2}{\epsilon_h}J\frac{\partial k}{\partial x_k}\right) + \frac{1}{2}\nu \frac{\partial^2 \overline{u_i u_j}}{\partial x_k \partial x_k}, \quad (1)$$

where P_{ij} is the production tensor. Validation of the analytical concepts for modeling the dissipation rate correlations (ϵ_{ij}) utilizing numerical databases support the proposed model for the homogeneous part of the turbulent dissipation rate $\epsilon_h = \overline{\frac{\partial u_i}{\partial x_j}\frac{\partial u_i}{\partial x_j}} - \frac{1}{4}\frac{\partial^2 \overline{u_i u_i}}{\partial x_j^2}$ which takes the following form:

$$\frac{\partial \epsilon_h}{\partial t} + \overline{U}_k \frac{\partial \epsilon_h}{\partial x_k} = -2\mathcal{A}\frac{\epsilon_h \overline{u_i u_k}}{k}\frac{\partial \overline{U}_i}{\partial x_k} - \psi\frac{\epsilon_h^2}{k} + c_\epsilon \frac{\partial}{\partial x_k}\left(\frac{k^2}{\epsilon_h}J\frac{\partial \epsilon_h}{\partial x_k}\right) + \frac{1}{2}\nu \frac{\partial^2 \epsilon_h}{\partial x_k \partial x_k}, \quad (2)$$

where scalar functions J, $\mathcal{A}$, $\mathcal{F}$, $\mathcal{C}$ and ψ depend on the anisotropy invariants II_a, III_a and the turbulent Reynolds number $R_\lambda = q\lambda/\nu = \sqrt{5}\,\overline{u_i u_i}/\sqrt{\nu \epsilon_h}$. These functions are given in [3].

The first term on the RHS of eq.(2) is the production of dissipation due to mean gradients, the second one, the decay term, represents the difference of the produc-

Validation of a novel turbulence model using DNS 21

tion due to vortex stretching and viscous destruction, the third term is turbulent transport of dissipation rate and the last term is its viscous diffusion.

3 Direct numerical simulation

Three different flows have been simulated: homogeneous shear flow, turbulent channel flow and turbulent mixing layer. In the case of homogeneous shear flow and turbulent mixing layer, no DNS data for terms appearing in the dissipation rate equation was available so far in the literature. In the case of channel flow, the available data was restricted to low Reynolds numbers and some terms, needed to test the special form of the dissipation rate equation in the new model, have not been computed up to now.

3.1 Numerical scheme and boundary conditions

The DNS code integrates the compressible Navier-Stokes equations for the primitive variables pressure p, velocity u_i and entropy s using a characteristic type formulation (Sesterhenn [4]). Two different discretizations of these equations are used: Fifth order compact upwind schemes for the convective terms along with sixth order compact central schemes for the viscous terms (cases S12, S13, K09, K13, see below) and sixth order compact central schemes for all terms together with a low-pass filter to avoid numerical instability (case M10). The time integration is performed by a third order low storage Runge Kutta scheme. The code has been shown to produce results comparable to those of a spectral code (Lechner [5]).

In streamwise (x) and spanwise (y) directions the computational domain is periodic in all cases. So, no boundary condition is required. For homogeneous shear flow a shear periodic boundary condition is applied in z-direction. To avoid inhomogeneities at the boundaries this requires an explicit derivation scheme in z-direction. A fifth order upwind scheme and the usual sixth order explicit scheme with a seven point stencil are used. The channel flows are simulated using isothermal walls and a forcing term to drive the flow [5]. For the calculation of the mixing layer non-reflective characteristic boundary conditions are used in the normal direction and the viscous terms are set to zero at the boundaries.

3.2 Simulated flow cases

In tables 1 to 3 the physical and numerical parameters of the performed simulations are presented. The homogeneous shear flow simulations differ by the initial ratio between turbulent time scale and the time scale of the mean flow $S_0^* = \frac{Sk}{\epsilon_h}$ where S is the gradient of the mean streamwise velocity. In the case of plane channel flow two different Reynolds numbers are simulated. The data for the time dependent mixing layer is averaged over a time interval where all quantities, even the terms in the dissipation rate equation, have achieved a state of self-similarity. The Reynolds number given in table 3 is the one in the middle of that interval. Re_θ is based on the momentum thickness δ_θ, that is 1/5 of the vorticity thickness in this flow.

Fig. 2 shows the behavior of these different flows simulated in the invariant map. The homogeneous shear flows start with isotropic turbulence and after a transient

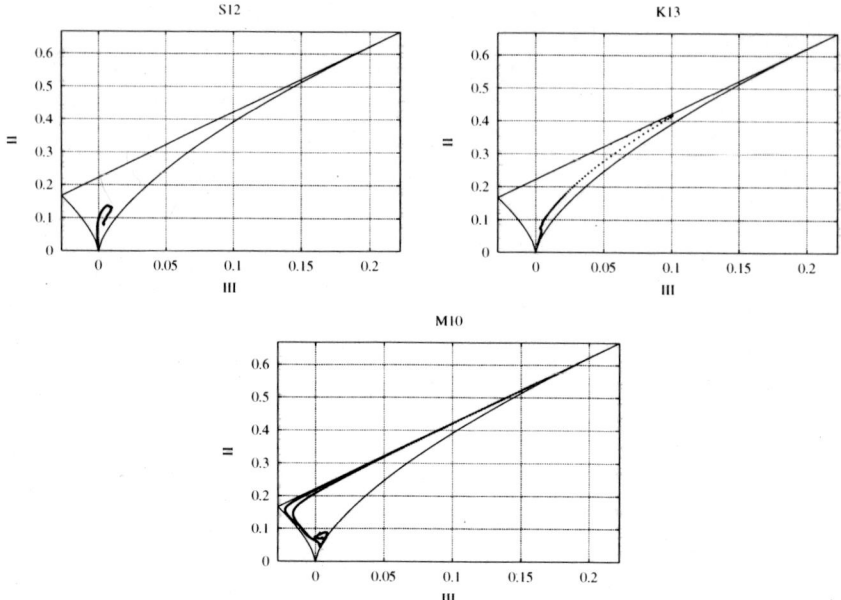

Fig. 2. The simulated flow cases mapped into the invariant triangle

settle down to a point which is close to the experimental values $II_a = 0.1$ and $III_a = 0.009$ found by Tavoularis [6]. S13 does not reach that point, the simulation has to be stopped before, because of lack of resolution of the small scales. The channel flows start with two-component turbulence at the wall, then cross the triangle in the buffer layer and follow the curve for axisymmetric flow down to nearly isotropic turbulence

Table 1. Parameters of the simulated homogeneous turbulent shear flows

	$Re_{\lambda 0}$	S_0^*	box size/$\frac{2\pi}{k_0}$	$\Delta x, \Delta y, \Delta z / \frac{2\pi}{k_0}$	number of grid points
S12	48.56	6.84	$(2 \cdot 0.9 \cdot 1) \cdot 12$	0.0625, 0.0422, 0.025	$47.2 \cdot 10^6$
S13	48.53	16.6	$(3 \cdot 1 \cdot 1) \cdot 12$	0.09, 0.05, 0.025	$46.1 \cdot 10^6$

(k_0 is the peak wave number of the initial spectrum.)

Table 2. Parameters of the simulated turbulent channel flows

	Re_τ	box size/H	$\Delta x^+, \Delta y^+, \Delta z^+$	number of grid points
K09	180	$9.6 \cdot 6.0 \cdot 2$	4.5, 3.75, 0.65..2.31	$25.6 \cdot 10^6$
K13	550	$6.4 \cdot 2.0 \cdot 2$	6.9, 5.8, 0.68..4.06	$45.4 \cdot 10^6$

Table 3. Parameters of the simulated turbulent mixing layer

	Re_θ	box size/δ_θ	$\Delta x_i/\delta_\theta$	number of grid points
M10	1050	$52 \cdot 13 \cdot 26$	0.05	$134.2 \cdot 10^6$

in the core region. The center of the mixing layer behaves like homogeneous shear turbulence. Towards the edges of the mixing layer, the turbulence is nearly two-component turbulence caused by the vanishing spanwise contribution. Outside the mixing layer the one-component limit is approached, only the normal component remains.

3.3 Validation of the data

Fig. 3. From left to right: Instantaneous fields of $\sqrt{\omega_i \omega_i}$, the sum of production terms in the ϵ-equation and the destruction term Υ. Additionally selected vortex lines in a structure are shown. Low values are white, high values black.

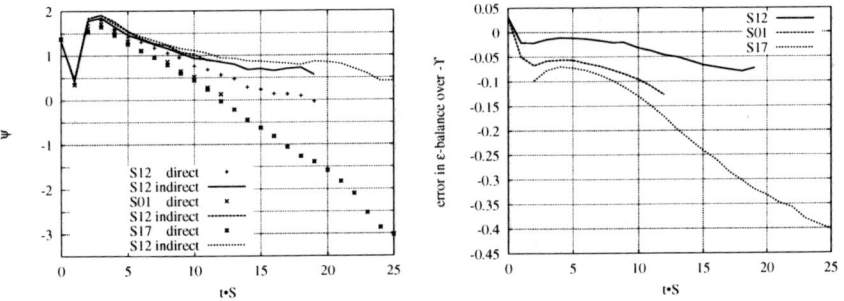

Fig. 4. left: ψ calculated direct and indirect for different resolutions and box sizes; right: error in the balance of ϵ normalized by $-\Upsilon$

The most important point in validating the reliability of the DNS data concerns the resolution of both the large and the small scales. The resolution of the small scales is demanding, because the terms in the dissipation rate equation get their large contributions from very fine scales (especially the destruction term Υ, see Mansour [7], that is part of the decay term in eq.(2)). A measure for the proper resolution of the small scales is the error in the dissipation rate balance. It has to be specified, how big the maximally acceptable error can be. Fig. 3 shows instantaneous pictures of the square root of twice the enstrophy (which is similar to dissipation rate), the sum of the production terms in the dissipation rate equation and the destruction term Υ. The enstrophy and the production terms have similar structure. The destruction term, however, has twice as fine structures, so its resolution is more demanding than the resolution of the dissipation range.

As an alternative to using the flow-field data, the value of Υ can be calculated indirectly using the dissipation rate balance given in Mansour [7] where all other terms are known. Υ is needed for the computation of ψ in *a priori* tests. Fig. 4 (left) shows the values of ψ using the directly (points) and indirectly (lines) calculated values of Υ from simulations with the same physical parameters, but a grid-spacing refined by a factor of 1.5 from case S17 over S01 to S12. Fig. 4 (right) shows the error in the balance of the dissipation rate normalized by Υ for the same simulations. The indirect values are close together, while the direct values differ much. For finer grids and small errors the direct values approach the indirect ones, see S12 for non-dimensional time $t \cdot S < 10$. So this value is assumed to be computed correctly. The indirect value of ψ for simulation S01 collapses with the one of S12. Simulation S01 has a maximal error in the balance of the dissipation rate of 10% of Υ in the range $t \cdot S < 10$. So an error of 10% is regarded as acceptable in a good calculation of ψ. The error is below this value for all the flow data presented here, except for case S13 when $t \cdot S > 12$ and the wall point in the channel flow simulations.

In order to test the resolution of the large scales simulations of homogeneous shear turbulence have been performed using 4/3 and 5/3 times bigger boxes. No change in the large scales has been observed. In the other flow cases the box size has been chosen according to experience and former studies.

The flows are simulated at low Mach numbers to obtain incompressible data. In the case of homogeneous shear flow a simulation at one third of the initial Mach number ($M_t = 0.033$ instead of 0.1) showed no variation in the statistical quantities. The channel flow data obtained at $M = 0.3$ has been compared with data from incompressible spectral simulations. The data sets collapse even for quantities like pressure strain terms. For mixing layers the Langley experimental curve shows compressibility effects for $M_c > 0.3$. Here a value of 0.15 is chosen. Additionally, it was tested that explicit compressibility terms in the balances for turbulent kinetic energy and dissipation rate are always negligible.

4 Results of *a priori* tests of models in the ϵ_h-equation

The *a priori* tests show good performance of the model for all the terms appearing in the Reynolds stress equations. There are only minor deviations in the turbulent transport of the wall normal and spanwise Reynolds stress and the off-diagonal value of ϵ_{ij} in the region of channel flow where the flow deviates from the edges of the invariant triangle to go from the two-component to the axisymmetric limit. This is not shown here, since this work is mainly focused on the dissipation rate equation.

Figures 5 to 7 show *a priori* tests for the terms modeled in the dissipation rate equation (2): the production by mean gradients, the turbulent transport and the coefficient ψ in the model for the difference between production by vortex stretching and destruction.

The production by mean gradients is well estimated by the model. We especially emphasize its perfect prediction in the mixing layer M10. Only in situations where the turbulence is not close to the limits of the invariant triangle this term is sometimes overestimated (buffer layer of channel flows K09 (not shown) and K13, high shear case S13). The turbulent transport is also well predicted. Once again there are some deviations, if the flow is not close to an edge of the invariant triangle. All these deviations are acceptable.

Concerning the coefficient ψ let us first consider channel flow: The peak of ψ close to the wall is overestimated by $\approx 20\%$ (Fig. 6, bottom). This is not a very severe deviation. Also there is a possible explanation: The assumptions used in modeling the dissipation rate equation are not valid in the rapid distortion limit $S \cdot k/\epsilon_h \gg \frac{1}{10} Re_\lambda$ (Jovanović et al. [8]). In the buffer layer $5 < z^+ < 30$ the inequality $S \cdot k/\epsilon_h > \frac{1}{10} Re_\lambda$ holds, therefore this can explain the deviation between model and simulation. In the core region the *a priori* test shows good behavior, the error oscillates between 5 and 20%. So, in channel flow the model works well. The homogenous shear flows showed strong deviations between model and simulation, e.g. $\psi \approx 1.8$ resp. 0.7 in S12. In the case of high shear S13 (not shown), this deviation once again can be explained by the inability of the model to predict the rapid distortion limit properly. But, in case S12 the mismatch cannot be explained up to now. This mismatch is confirmed by the data for the turbulent mixing layer M10. In the core region the flow is very similar to the homogeneous shear flow with moderate shear rate S12. The model predicts a value of ψ around 2.0, while the simulation shows $\psi \approx 0.75$ in the center of the shear layer.

5 Results of *a posteriori* tests

5.1 Homogeneous flows

Using the present turbulence model we have calculated 17 homogeneous flows reviewed by Ferziger [9], using the initial values of the Reynolds stress components $\overline{u_i u_j}$ and of the dissipation rate ϵ_0 of these experiments. The model was able to predict all these flows within an accuracy of $\pm 10\%$. This is especially remarkable in

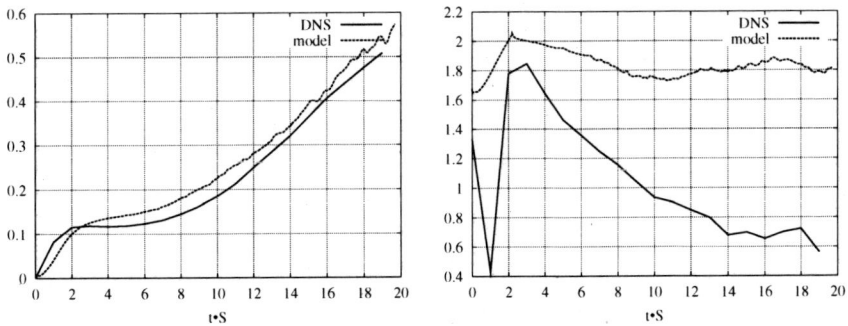

Fig. 5. *A priori* tests of models in the dissipation rate equation, flow case S12. left: Production of ϵ_h by mean gradients; right: ψ. Abscissa: time non-dimensionalized by mean velocity gradient S.

the case of plain strain, that is shown here as an example (Fig. 8 (left)), because this flow is not restricted to follow the edges of the invariant triangle.

The good agreement in the case of homogeneous shear turbulence (Fig. 8 (right)) does not contradict the *a priori* tests in section 4. In the experiment the flow is investigated only at low values of non-dimensional time up to 3.33. This is only the transient period.

5.2 Inhomogeneous flows

Much of our effort in modeling the turbulence dynamics was devoted to the treatment of the equation for the homogeneous part of the turbulent dissipation rate (2). The adequacy of this equation was checked by predicting the dissipation using the numerical results for the mean flow and turbulent stresses as an input.

Figure 9 (left) shows the comparison of the model prediction and the numerical data for plane channel flow. This figure demonstrates the good performance of the model and the solid agreement with the numerical data across the entire flow field.

Turbulent channel flow has also been computed using a wall function approach. This is a standard way to treat flows involving solid boundaries since it allows to avoid extremely fine grid spacing close to the wall. The predictions displayed in Fig. 9 (right) were obtained by integrating the model equations starting from the

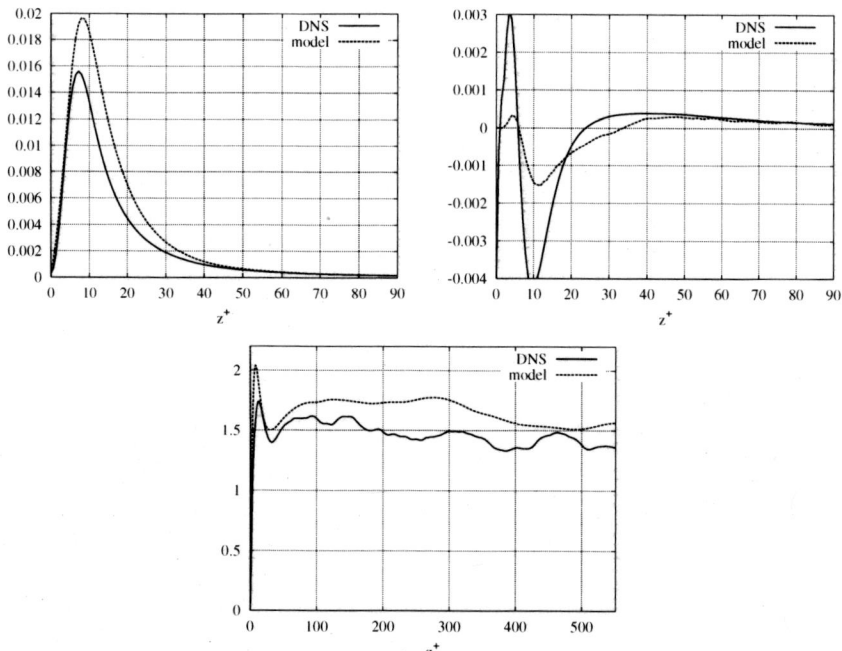

Fig. 6. *A priori* tests of models in the dissipation rate equation, flow case K13. left top: Production of ϵ_h by mean gradients; right top: turbulent transport of ϵ_h; bottom: ψ. Abscissa: wall distance in plus units.

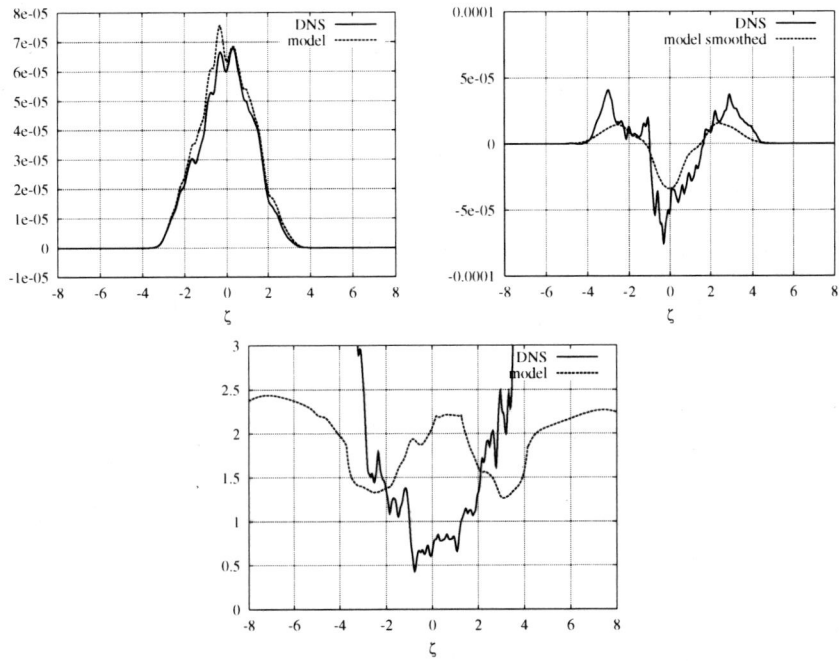

Fig. 7. *A priori* tests of models in the dissipation rate equation, flow case M10. left top: Production of ϵ_h by mean gradients; right top: turbulent transport of ϵ_h (The curve of the modeled value is very wiggly, so it has been filtered.); bottom: ψ. Abscissa: transverse coordinate normalized by momentum thickness

outer edge of the buffer region and by utilizing the numerical data for the turbulent stresses and the dissipation rate at this location as the boundary condition. The high degree of agreement achieved between the model predictions and numerical results again demonstrates the good performance of the model.

6 Implementation of the model into engineering software

Implementation of the novel turbulence model into a commercial software code is a logical step towards improving engineering flow predictions.

We decided to implement the model into the commercial program Comet which solves the mass, momentum and energy balance equations in integral form employing a finite volume method. The code Comet offers to the user a choice of several spatial and temporal discretization schemes, nonlinear equation solvers and other parameters. These parameters affect either the accuracy or computational efficiency. The size of the time step, the values of the under-relaxation parameters and convergence criteria are examples of control parameters. An appropriate choice saves computational time and prevents possible problems like slow convergence or even divergence, numerical diffusion, etc. In most cases there is no theory of how to select

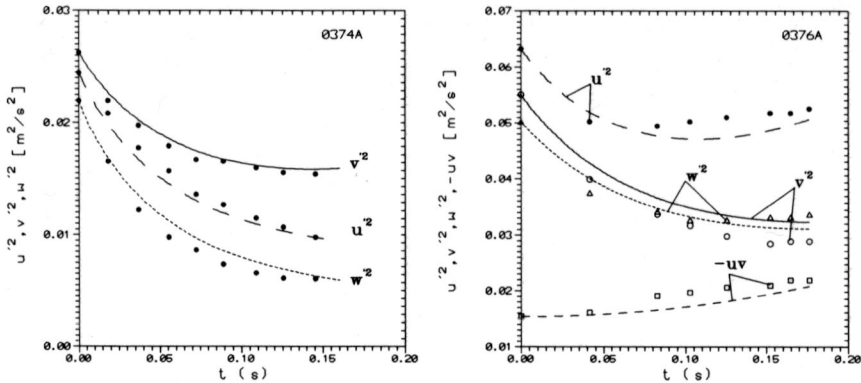

Fig. 8. Histories of Reynolds stresses. Left: homogeneous turbulence undergoing plain strain: •, Townsend (1956); — — — $\overline{u^2}$, —— $\overline{v^2}$, ----- $\overline{w^2}$, Present computation; right: sheared homogeneous turbulence: •, ○, △, □, Champagne, Harris & Corrsin (1970); — — — $\overline{u^2}$, —— $\overline{v^2}$, ----- $\overline{w^2}$, ---- $\overline{uv}$, Present computation.

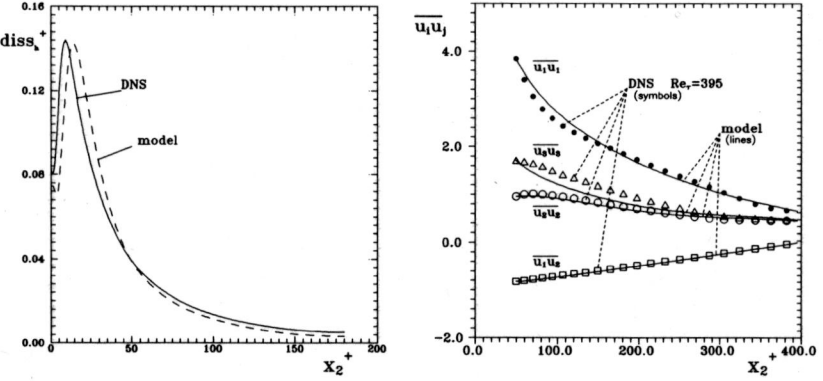

Fig. 9. Left: Predicted profile of the homogeneous part of the turbulent dissipation rate in a plane channel flow; right: Predictions of the turbulent stresses in a plane channel flow.

the right parameters for a particular problem so that one has to rely upon ones own experience.

In close cooperation with the staff of the Computational Dynamics LTD office located in Nürnberg we implemented the novel turbulence model as described in [3] into the Comet program. Initially this was a very complicated and demanding task owing to the complexity of the scalar functions which involve non-linear dependence on the anisotropy invariants. In the present state the model is fully implemented in the Comet program and various control parameters which influence the numerical performance are tested in different flow configurations. As with all second-moment closure schemes the numerical performance turned out to be critical especially with respect to the convergence problem.

The rate of convergence of several flow quantities in a sample calculation of a plane mixing layer using the Comet software is shown in Fig. 10. Available further results cannot be shown due to lack of space.

7 Conclusions

A new data basis for the development of turbulence models, especially for the modeling of the dissipation rate equation has been produced by direct numerical simulations of homogeneous sheared turbulence, turbulent channel flow and turbulent mixing layer. This data set was used for *a priori* tests of a novel Reynolds stress model. During the course of this investigation it was found that the proposed model approximations closely fit most of the results obtained from numerical simulations except for the decay term ψ of the dissipation rate equation. This term represents the difference of the production due to the vortex stretching and the destruction of the dissipation rate. For the term mentioned above, some of the numerical results were found to deviate significantly from the model, indicating the necessity for further refinements which are currently on the way to be implemented into the model and tested for a wide class of homogeneous and inhomogeneous turbulent shear flows.

It is worth noting that ψ is the equivalent to C_ϵ^2 in usual k-ϵ modeling. Usually $C_\epsilon^2 = 1.92$ is assumed, that is obtained from isotropic decaying turbulence. Here, the simulations for homogeneous shear flow and mixing layer show considerably lower values. Additionally, the value of $2\mathcal{A}$ is in these situations also lower than the corresponding $C_\epsilon^1 = 1.44$. Perhaps the effects of too high values of C_ϵ^1 and C_ϵ^2 cancel in the case of homogeneous shear: A too high production term is compensated by a too high decay term.

Parallel to the testing and further development of the model, it was implemented into a commercial software.

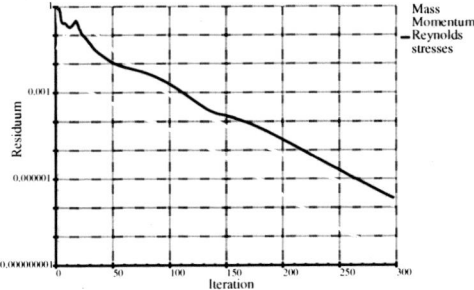

Fig. 10. Examples of the convergence rate for the predictions of the plane mixing layer using Comet software.

References

1. J.L. Lumley. Computational modeling of turbulent flows. *Advances in Applied Mechanics*, 18:123–176, 1978.
2. P.Y. Chou. On the velocity correlation and the solution of the equation of turbulent fluctuation. *Quart. Appl. Math.*, 3:38–54, 1945.
3. J. Jovanović. *The Statistical Dynamics of Turbulence*. Springer, Berlin Heidelberg New York, 2004.
4. J. Sesterhenn. A characteristic-type formulation of the equations for high order upwind schemes. *Computers & Fluids*, 30:37–67, 2001.
5. R. Lechner, J. Sesterhenn, and R. Friedrich. Turbulent supersonic channel flow. *J. Turb.*, 2, 2001.
6. S. Tavoularis and U. Karnik. Further experiments on the evolution of turbulent stresses and scales in uniformly sheared turbulence. *JFM*, 204:457–478, 1989.
7. N.N. Mansour, J. Kim, and P. Moin. Reynolds-stress and dissipation-rate budgets in a turbulent channel flow. *J. Fluid Mech.*, 194:15–44, 1988.
8. J. Jovanović, I. Otić, and P. Bradshaw. On the anisotropy of axisymmetric strained turbulence in the dissipation range. *J. Fluids Eng.*, 125:1–13, 2003.
9. J. Ferziger. Homogeneous turbulent flows. In S.J. Kline, B.J. Cantwell, and G.M. Lilley, editors, *Proceedings 1980-81 AFOSR-HTTM Stanford Conference on Complex Turbulent Flows*, pages 405–433. Stanford University, 1980.

FlowNoise: Flow Induced Noise Computation on Hitachi SR8000-F1

M. Escobar[1], I. Ali[2], M. Kaltenbacher[1], S. Becker[2] and F. Hülsemann[3]

[1] Dept. of Sensor Technology (LSE), Paul-Gordan-St. 3/5, 91052 University of Erlangen-Nuremberg, Erlangen, Germany.
{*max.escobar,manfred.kaltenbacher*}*@lse.eei.uni-erlangen.de*

[2] Institute of Fluid Mechanics (LSTM), Cauerstrasse 4, 91058 University of Erlangen-Nuremberg, Erlangen.
{*irfan.ali, stefan.becker*}*@lstm.uni-erlangen.de*

[3] Dept. of System Simulation (LSS), Cauerstrasse 6, 91058 University of Erlangen-Nuremberg, Erlangen.
frank.huelsemann@informatik.uni-erlangen.de

Abstract. We have developed a simulation environment for the efficient numerical computation of flow induced sound. Thereby, the fluid flow program FASTEST-3D has been coupled via MpCCI to CFS++ (Coupled Field Simulation), which performs the sound field computation. Thereby, different computational domains as well as grids for the fluid field and acoustic field can be chosen. As an practical example, we discuss the computation of the emitted noise from a square cylinder within a turbulent flow.

1 Introduction

A large amount of the total noise in our daily life is generated by turbulent flows (e.g. airplanes, cars, air conditioning systems, etc.). The physics behind the generation process is quite complicated and still not fully understood. The use of numerical simulation tools is one important way to analyze the generation of flow induced sound. Currently the most often used physical model is Lighthill's analogy [8], in which the acoustic sound field is computed by solving an inhomogeneous wave equation, where the inhomogeneous part is calculated from fluid flow data. In practice, the two main problems arise:

- The acoustic domain is in most cases large compared to the fluid flow domain.
- The spatial discretization for the flow computation is much finer as compared to the discretization of the acoustic domain.

Therefore, we have setup our computational environment in such a way, that both the computational domain as well as its discretization can be different for fluid dynamics and acoustics. The realization is based on the CFD-code FASTEST-3D

(LSTM), CFS++ (LSE) and MpCCI, which performs the data exchange between the two computational grids. The whole simulation environment runs on Hitachi SR8000-F1 at LRZ Munich.

2 Flow Induced Noise Computation

2.1 Fluid Flow Computation

Numerical fluid simulation is carried out with FASTEST-3D [4], a CFD tool developed at LSTM. The underlying numerical scheme is based on a procedure described by Perić [9], consisting of a fully conservative second-order finite volume space discretization with a collocated arrangement of variables on non-orthogonal grids, a pressure correction method of the SIMPLE type for the iterative coupling of velocity and pressure and an iterative ILU decomposition method for the solution of the sparse linear systems for velocity components, pressure correction and temperature. For time discretisation an implicit second-order scheme is employed, while a nonlinear multigrid scheme, in which the pressure correction method acts as a smoother on the different grid levels, is used for convergence acceleration. For the LES computations, the Smagorinsky model is used as implemented by Glück [6]. The concept of block structured grid is employed to handle complex geometries and for the ease of parallelization. The parallel implementation is based on grid partitioning with automatic load balancing and follows the message-passing concept, ensuring a high degree of portability.

2.2 Sound Field Computation

A common approach for solving aeroacoustic problems is the acoustic analogy. It is based on the assumption that one can perform the fluid flow simulation and acoustic field computation separately. This assumption means that the noise induced by the turbulent flow does not affect the flow characteristics of the fluid itself.

We applied Lighthill's analogy [8], which results in the following inhomogeneous wave equation

$$\frac{\partial^2 \rho'}{\partial t^2} - c^2 \frac{\partial^2 \rho'}{\partial x_i^2} = \frac{\partial^2 \bar{T}_{ij}}{\partial x_i \partial x_j} . \tag{1}$$

In (1), ρ' denotes the density fluctuation, c the speed of sound and T_{ij} the components of Lighthill's stress tensor computed by

$$T_{ij} \approx \rho \partial v_i \, v_j \tag{2}$$

with v_i the velocity components obtained from the flow simulation. Multiplying (1) by an arbitrary test function $w \in H_0^1$, we arrive at the weak form given by

$$\int_\Omega w \frac{\partial^2 \rho'}{\partial t^2} \, d\Omega + c^2 \int_\Omega \nabla w \cdot \nabla \rho' \, d\Omega = -\int_\Omega (\nabla \cdot \bar{T}_{ij}) \nabla w \, d\Omega + c^2 \int_\Gamma w \frac{\partial \rho'}{\partial \mathbf{n}} \, d\Gamma$$

$$+ \int_\Gamma (w \, (\nabla \cdot \bar{T}_{ij}) \cdot \mathbf{n}) \, d\Gamma . \tag{3}$$

By solving (3), we obtain the fluctuation of the density, which, for linear acoustics, has the following relationship to the acoustic pressure p_a

$$p_a = p - p_0 = c^2(\rho - \rho_0) = c^2 \rho'. \qquad (4)$$

By applying the finite element method to (3), we solve for the acoustic field directly in the time domain. Time discretization is performed by a predictor-corrector method of the Newmark family [7]. At the boundary of our acoustic domain, we apply a set of first order absorbing boundary conditions derived from Padé approximations to account for free field radiation.

3 Coupling of FASTEST-3D and CFS++ via MpCCI

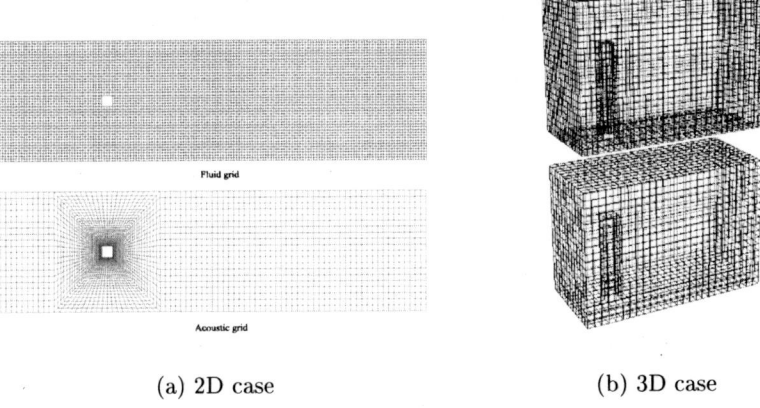

(a) 2D case   (b) 3D case

Fig. 1. Coupling of the fluid mesh and the corresponding region of the acoustic mesh via MpCCI.

In order to compute the acoustic field, we need to interpolate the fluid data from the fluid mesh to the corresponding part of the acoustic mesh. For this coupling of the two codes we employ the *Mesh-based parallel Code Coupling Interface* (MpCCI) [1].

The data exchange process carried out with MpCCI is based on MPI communication [2], which has established itself as standard for the parallelization on distributed-memory computers.

On both sides, a grid definition is carried out in the initialization phase. For the 2D case, both programs use elements of type cci_elem_quad with four nodes per element, whereas for the 3D case elements of the type cci_elem_hexahedron are employed (see Fig. 1).

The sequence of events of the coupled simulation is presented in Fig. 2. During the actual simulations (shaded in gray) the velocity field computed by FASTEST-3D is transfered to CFS++. The values are exchanged at the nodal positions of the grids by means of interpolation algorithms provided by MpCCI. Subsequently, with

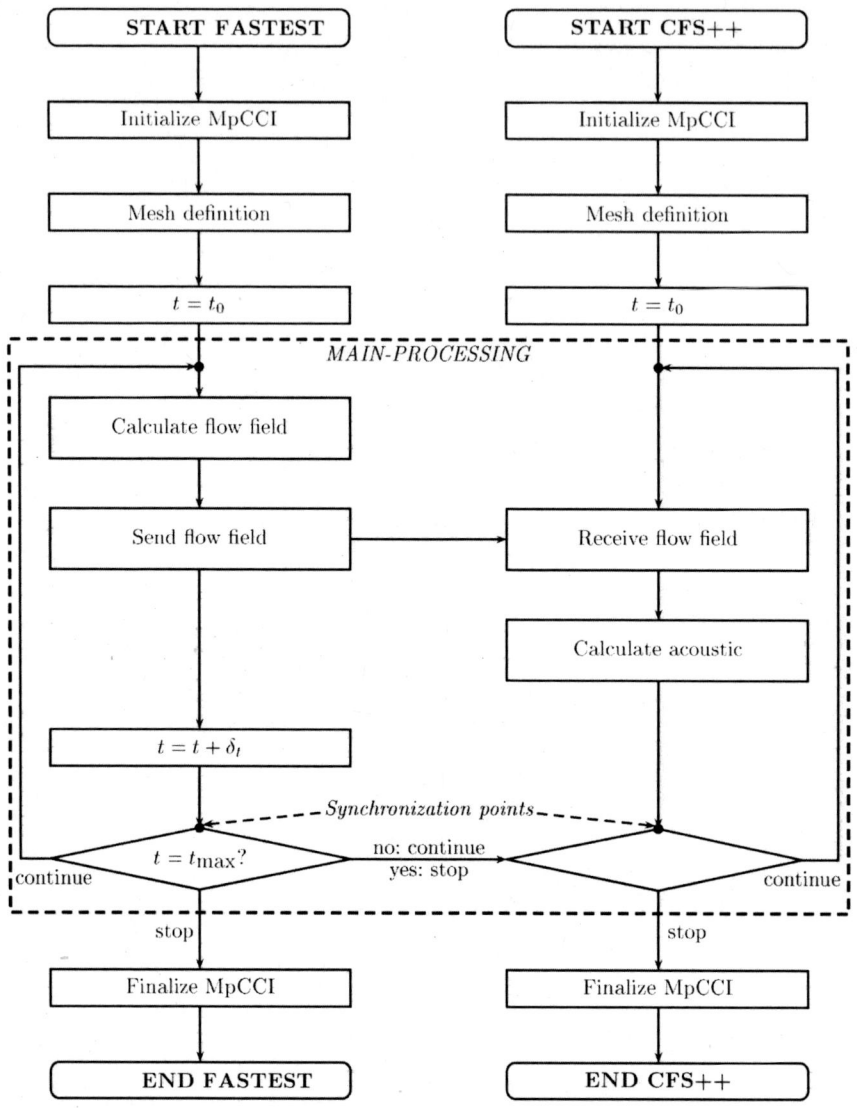

Fig. 2. Schema of a coupled simulation between FASTEST-3D and CFS++ with MpCCI.

these values it is possible to carry out the acoustic computation. The exchange of the data between FASTEST-3D and CFS++ is synchronized at each time step until the end of the coupled simulation.

3.1 Numerical Case Study: Flow around a Cylinder

Our numerical case study concentrates on a square cylinder immersed in a turbulent flow resulting in a Reynolds number of about 15.000 [3]. Figure 3 represents the configuration chosen for the simulations. The inner region surrounding the square cylinder represents the computational domain for the fluid simulation. The square cylinder has a length a of 20 mm in a fluid flow domain of length $L_x = 800$ mm and width $L_y = 100$ mm, respectively.

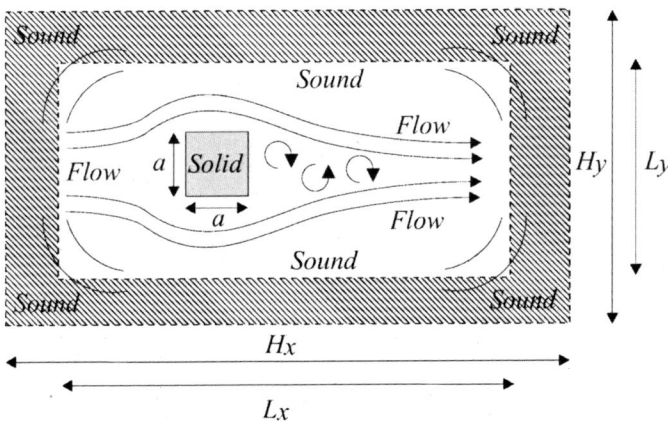

Fig. 3. Principle setup.

For the fluid computation, the simulation domain with boundary conditions is displayed in Fig. 4. We performed the LES (Large Eddy Simulation) computations using the Smagorinsky model and a time step value Δt of 1 ms. The flow visualization using stream lines is shown in Fig. 5, whereas the velocity component plots are displayed in Fig. 6. The vortex shedding shows a dominant frequency at 66 Hz (see Fig. 7), which results in a Strouhal number of 0.13.

For the acoustic computation, we have extended the domain of the flow simulation to a sphere with a radius of 50 m (corresponds to about five times the acoustic wavelength) in order to reduce reflection of non-orthogonal waves impinging on the boundary (see Fig. 8).

Figures 9 and 10 represent the acoustic pressure at point (x, y)=(0.0 m, 5.0 m) in time and frequency domain respectively. The acoustic pressure at points at the top and bottom of the cylinder (x, y)=(0.0 m, ±0.002 m) is shown in Fig. 11. All three acoustic pressure signals (see Fig. 9 and 11) show a main frequency component at 63 Hz. The acoustic sound field for a fixed time is displayed in Fig. 12.

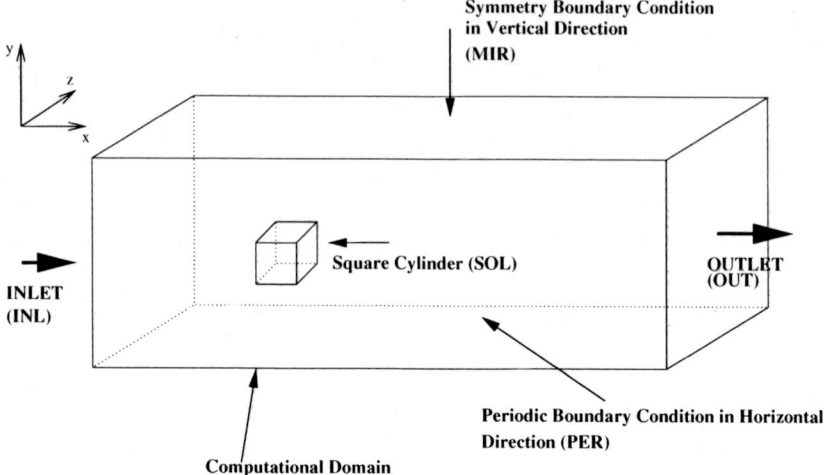

Fig. 4. Block diagram of the fluid flow computation

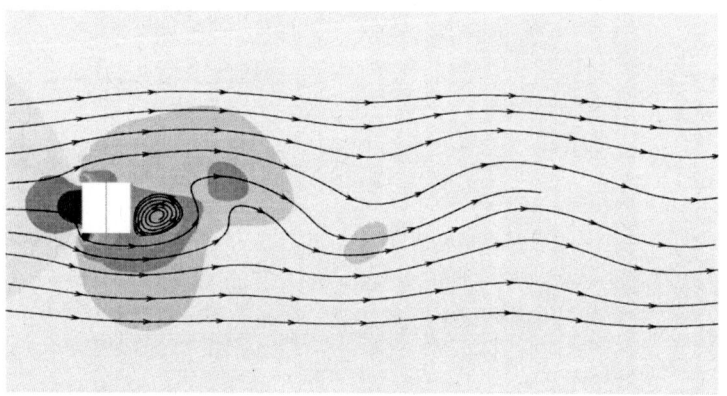

Fig. 5. Streamlines around the cylinder

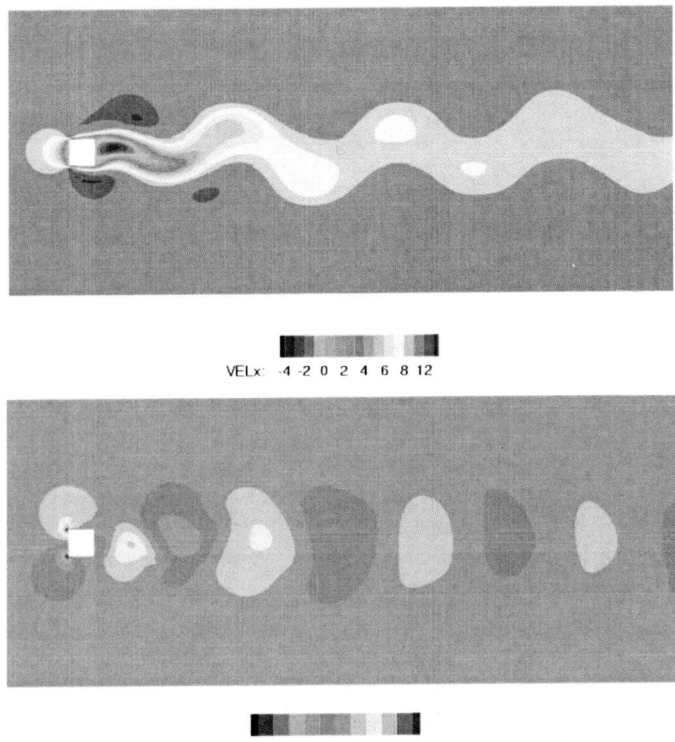

Fig. 6. Velocity profiles

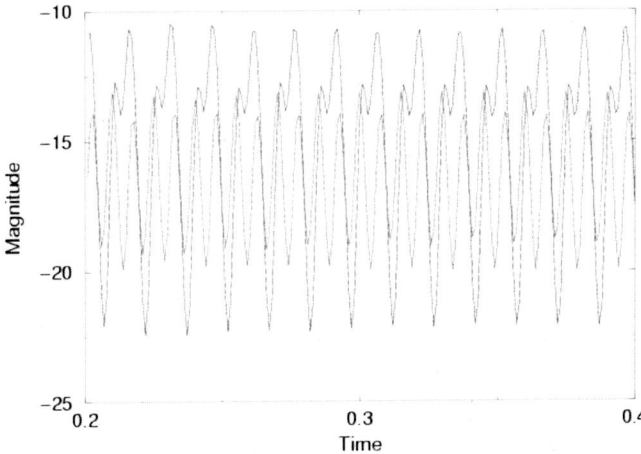

Fig. 7. Pressure signal: red corresponds to a point in the middle after the cylinder and blue to a point near the edge of the cylinder

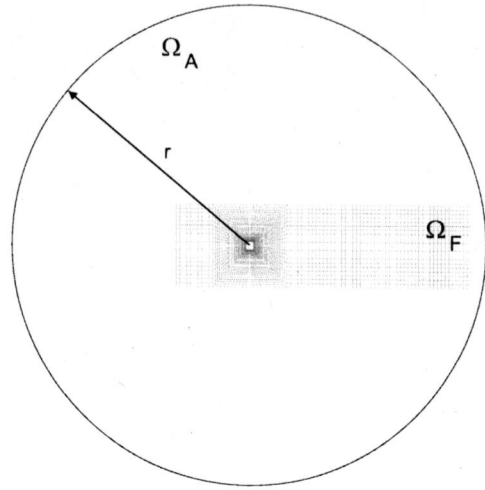

Fig. 8. Total computational domain for acoustic computation (Ω_F: fluid domain)

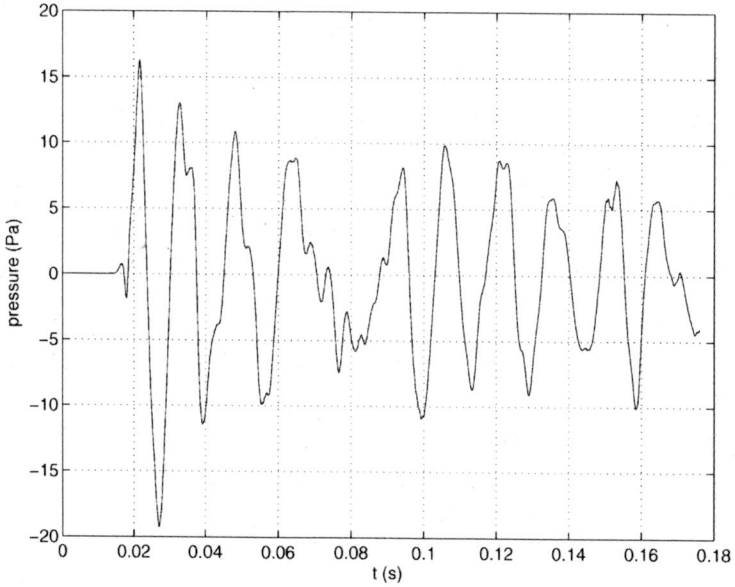

Fig. 9. Pressure solution at point $(x,y)=(0.0\,\mathrm{m},\,5.0\,\mathrm{m})$

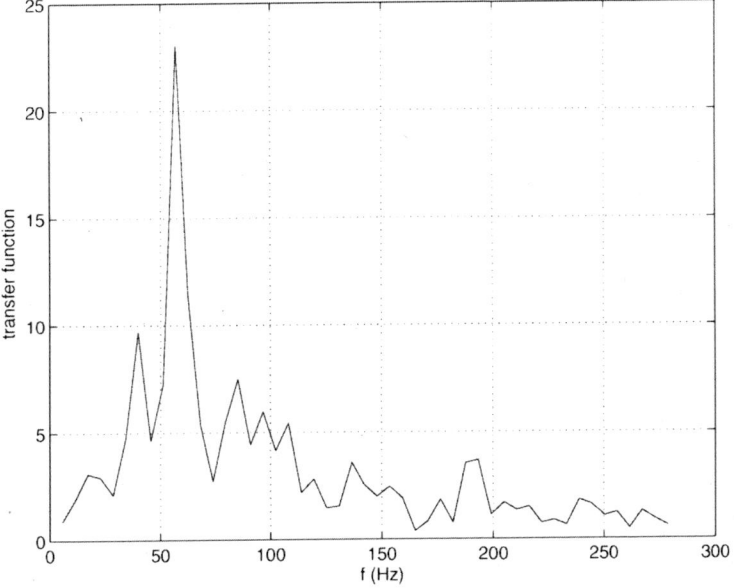

Fig. 10. Pressure spectra at point $(x,y)=(0.0\,\text{m},\ 5.0\,\text{m})$

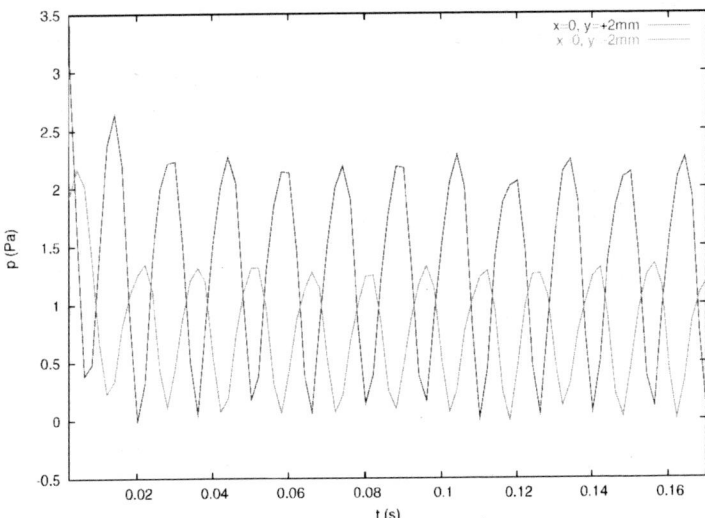

Fig. 11. Acoustic pressure solution for two points next to obstacle at $(x,y)=(0.0\,\text{m},\ \pm 0.002\,\text{m})$

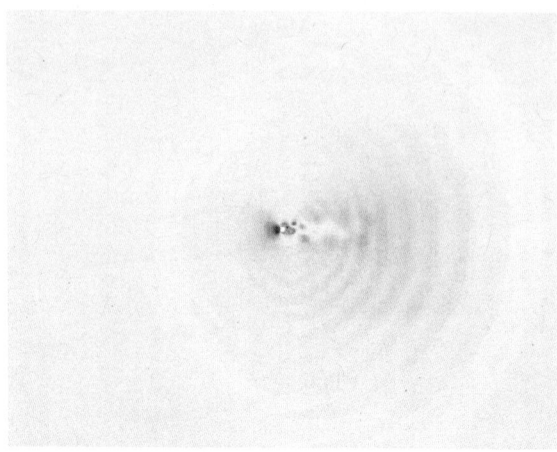

Fig. 12. Acoustic sound field around the square cylinder

3.2 Computational Aspects

The fluid flow domain has been discretized by 2.621.440 control volumes (smaller grid) and 5.242.880 control volumes (larger grid). For the the parallel computation on 8 processors, we achieved a load balance efficiency of 100%. Communication between the individual processes is carried out with MPI. Therewith, FASTED-3D used about 92% from the overall CPU-time compared to about 8% used by CFS++. The acoustic mesh itself consisted of 230.000 finite elements. The coupling via MpCCI resulted in a noticeable increase in MPI communication time compared to uncoupled (single physics) simulations.

CFS++, coded in C++, was ported to the Hitachi SR8000-F1 . To speed up 3D simulations, different parallelization options have been considered [5]. For the computations reported here, we have preconditioned the conjugate gradient solver with an overlapping additive Schwarz method. The domain decomposition itself is done on the discrete level (matrix graph partitioning).

References

1. http://www.mpcci.org/, 2003.
2. http://www.mpi-forum.org/, 2003.
3. I. Ali, M. Escobar, C. Hahn, M. Kaltenbacher, and S. Becker, *Numerical and Experimental Investigation on Flow Induced Noise from a Square Cylinder*, AIAA, London, 2004.
4. F. Durst and M. Schäfer, *A Parallel Block-Structured Multigrid Method for the Prediction of Incompressible Flows*, Int. J. Num. Methods Fluids **22** (1996), 549–565.

5. M. Escobar, M. Kaltenbacher, J. Thies, and F. Hülsemann, *Large Scale Computations in Acoustics*, GAMM, 2004.
6. M. Glück, *Ein Beitrag zur numerischen Simulation von Fluid-Structure-Interaction-Grundlagenuntersuchungen und Anwendung auf Membrantragwerke*, Ph.D. thesis, University of Erlangen, Institute of Fluid Mechanics, Erlangen, 2002.
7. T. J. R. Hughes, *The finite elemente method*, 1 ed., Prentice-Hall, New Jersey, 1987.
8. M.J. Lighthill, *On sound generated aerodynamically i. general theory*, Proc. Roy. Soc. Lond. (1952), no. A 211, 564–587.
9. M. Perić, *A Finite Volume Method for the Prediction of Three-Dimesional Fluid Flow in Complex Ducts*, Ph.D. thesis, University of London, 1985.

FLUSIB: Fully Three-Dimensional Coupling of Fluid and Thin-Walled Structures

Dominik Scholz[1], Ernst Rank[1], Markus Glück[2], Michael Breuer[2] and Franz Durst[2]

[1] Institute of Computer Science in Civil Engineering
Technical University of Munich
Arcisstraße 21, 80290 München, Germany
{d.scholz,rank}@bv.tum.de

[2] Institute of Fluid Mechanics
University of Erlangen-Nürnberg
Cauerstraße 4, 91058 Erlangen, Germany
{glueck,breuer,durst}@lstm.uni-erlangen.de

Abstract. In this contribution, fully three-dimensional models are used for the numerical simulation of both the structure and the fluid in fluid-structure interaction computations. A partitioned, but fully implicit coupling algorithm is employed. As an example, the wind-excitation of a thin-walled tower is investigated.

1 Introduction

Interaction phenomena between fluids and structures can be found in many engineering and also medical disciplines such as civil, mechanical and medical engineering, shipbuilding and biotechnology. Although the simulation tool presented in this paper was designed for *civil engineering* applications, it could also be applied to other fields. The *partitioned coupling approach* for time-dependent fluid-structure interactions, which is described in more detail in *Glück et al.* [8] can be applied to thin-walled structures with large displacements. The frame algorithm connects a three-dimensional, finite-volume based multi-block flow solver for incompressible fluids [5] with a three-dimensional, high-order finite-element code for geometrically non-linear, dynamic structural problems using a commercial coupling interface (MpCCI [1]). The advantages of the fully three-dimensional modeling of both the structure and the fluid are outlined in this article.

2 Structural Simulation Using Three-Dimensional, High-Order Elements

When modeling plate-like or thin-walled structures, which can be sensitive to fluid-structure interaction, it would be advantageous to use three-dimensional solid elements. Three-dimensional effects could be described, e.g. at supports, and the transition from thin-walled to massive structures would not require transition elements.

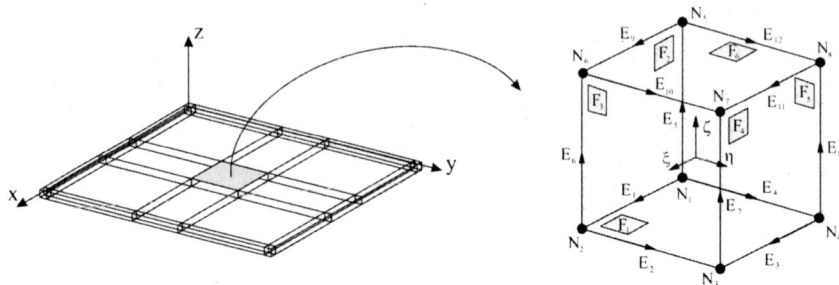

Fig. 1. Discretization of plate-like structure with high-order hexahedral elements

The problem usually arising here, is that standard low-order elements are very sensitive to large aspect ratios, and locking can occur. Therefore, one would have to use a very high number of small elements having a width and length in the same range as the thickness. In contrast, high-order elements can cope with high aspect ratios, provided, that the polynomial degree is large enough. They can be realized using a mapping between the standard element and the thin-walled structure (see Fig. 1).

A very important feature in the context of high-order elements are anisotropic Ansatz spaces. Using the approach of [6, 11] one can define different polynomial degrees in the different local directions of the hexahedral element. For the plate-like structure in Fig. 1, it makes sense to choose high polynomial degrees for the in-plane directions, whereas in thickness direction a lower polynomial degree can be used in order to reduce the computational effort. In our implementation, it is not only possible to define different polynomial degrees for the different local directions, but also for the different displacement fields.

With this approach the "model error" of 2D-plate and shell theories, using fixed kinematic assumptions over the thickness, is replaced by a 3D-discretization error. The big advantage is, that this error can be controlled by varying the polynomial degree over the thickness in a sequence of computations. This error could not be controlled when using fixed kinematic assumptions on which plate or shell theories are usually based.

For fluid-structure interaction problems, where the fluid and the structural fields are coupled at the surface of the structure, the use of hexahedral elements for thin-walled structures has another, very important advantage: Since we use 3D-elements, the numerical and the geometrical model are always consistent. The interface geometry is given explicitly by the numerical model. In contrast, when using dimensionally reduced models like shells, one has to reconstruct the interface from the middle-surface and an offset. This is not necessary with a fully three-dimensional structural model.

Even, when the geometry of surfaces is arbitrarily complex, we can still use the advantages of high-order elements, which are usually large. In order to set up a mapping between the standard element and the complex geometry, the blending function method is used, which describes the geometry exactly [9].

For the spatial discretization of the structural problem, high-order elements are used, whereas the time domain is discretized using the generalized-α-method, which is second-order accurate and has favorable numerical damping properties [4].

3 Simulation of Fluid in Three Dimensions

3.1 Large-Eddy Simulation

Large-eddy simulations have been performed in order to predict the flow around bluff bodies more accurately. For this purpose, the *Smagorinsky model* has been implemented into the CFD code *FASTEST-3D*. The anisotropic part of the subgrid-scale stress tensor is modeled as follows:

$$\tau_{ij} = -2\,\mu_\mathrm{t}\,\widetilde{S}_{ij}\,, \qquad (1)$$

whereas $\widetilde{S}_{ij}$ is the tensor of the shear rate:

$$\widetilde{S}_{ij} = \frac{1}{2}\left(\frac{\partial \widetilde{U}_i}{\partial x_j} + \frac{\partial \widetilde{U}_j}{\partial x_i}\right). \qquad (2)$$

According to the eddy-viscosity models of the RANS concept, the turbulent viscosity μ_t is a function of the grid-scale variables $\widetilde{U}_i$. First, it is assumed that μ_t is the product of a characteristic length scale L_c and a characteristic velocity U_c

$$L_c = C_\mathrm{s}\,\Delta\,, \qquad U_c = L_c\,|\widetilde{S}_{ij}| \qquad (3)$$

with the filter width Δ and the Smagorinsky constant C_s. This yields

$$\mu_\mathrm{t} = \rho\,L_c\,U_c = \rho\,C_\mathrm{s}^2\,\Delta^2\,|\widetilde{S}_{ij}| \quad \text{with} \quad |\widetilde{S}_{ij}| = \sqrt{2\,\widetilde{S}_{ij}\,\widetilde{S}_{ij}}\,. \qquad (4)$$

The filter width Δ is calculated from the grid sizes Δx, Δy, and Δz as follows:

$$\Delta = (\Delta x\,\Delta y\,\Delta z)^{1/3} = \Delta V^{1/3}\,. \qquad (5)$$

Breuer [3] found out, that the optimal Smagorinsky constant is dependent on the Reynolds number and on the flow configuration. For practically relevant shear flows with inhomogeneous and anisotropic turbulence C_s should be within the range of 0.065 and 0.1.

For high Reynolds numbers the boundary layers become very thin. This requires a very fine spatial resolution. In order to avoid a too high total number of grid points, the wall model of *Werner and Wengle* [12] has been implemented. The turbulent boundary layer is separated in only two parts – the *viscous sublayer* with a linear relation between $\widetilde{U}^+$ and y^+ and the *fully turbulent outer region*, which is approximated by a power law:

$$\widetilde{U}^+ = y^+ \qquad \text{for} \qquad 0 \leq y^+ < 11.81\,, \qquad (6)$$

$$\widetilde{U}^+ = A\,y^{+B} \qquad \text{for} \qquad 11.81 \leq y^+ < 1000\,. \qquad (7)$$

Werner and Wengle [12] suggested the parameters $A = 8.3$ and $B = 1/7$. The wall model is based on a phase coincidence assumption between the *transient* (non averaged) wall shear stress and the *transient* tangential velocity component in the first grid cell.

3.2 Increase of Efficiency of Fluid Code on Hitachi SR8000-F1

A very fruitful cooperation with the Local Computing Center in Erlangen (Regionales Rechenzentrum, Dr. Wellein, Dipl.-Ing. Deserno) enabled a code optimization, which led to an increase of the performance of *FASTEST-3D* from 1342 MFlops to 1870 MFlops for a single SMP node of the *HITACHI SR8000-F1*. This was reached by means of the following steps:

- parallelization of *outer* loops,
- segmentation of *inner* loops in several single loops (leads to a more efficient use of cache),
- rearrangement of IF requests,
- vectorization of special parts of the code, including the solver *SIPSOL* (especially of the LU decomposition and of the backward substitution).

In addition to the reference version on *HITACHI SR8000-F1*, portings and performance tests were carried out on the following platforms (see Fig. 2):

- Fujitsu VPP 300,
- MIPS R14000 (SGI),
- Intel Itanium 2,
- Intel Pentium IV,
- IBM Power4.

In Fig. 2 the measured CPU times are plotted. These times are normalized in such a way that the speed-up factor based on the CPU time on an SGI workstation can be read off. The highest speed-up (more than one order of magnitude) was reached on *HITACHI SR8000-F1* using the auto-parallelization on eight processors of one node and the above listed optimizations of *FASTEST-3D* ("-DHSROPT" in the legend of the diagram).

4 Coupling in 3D

4.1 Partitioned Solution Approach and Data Transfer

For the coupling between fluid and structure, a standard partitioned solution approach is used with a predictor/corrector algorithm, since both CFD and CSD are implicit time-stepping schemes. First, the fluid code is predicting the wind loads using the displacements of the previous time step or any extrapolation from there to the new time step. Then the structure code computes the corresponding displacements, and these are used by the fluid code to correct the wind loads. This iteration is repeated until convergence is reached, e.g., until the changes in displacements of the structure are below a certain tolerance.

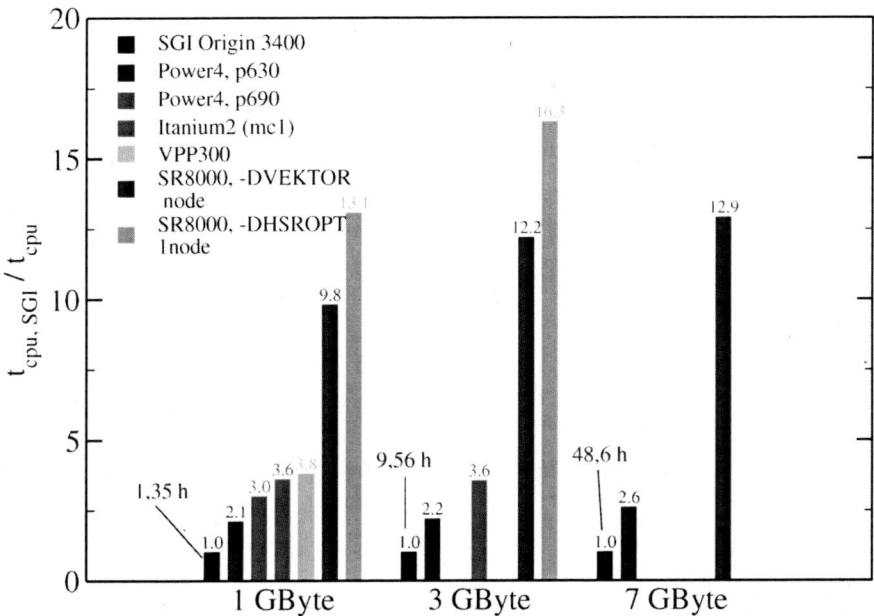

Fig. 2. Measured CPU times for several computer architectures (each related to the SGI workstation)

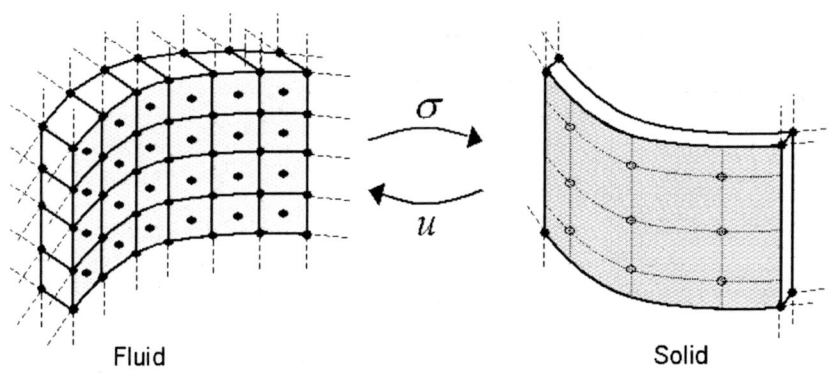

Fig. 3. Data exchange between fluid and solid interface

The data transfer between both simulations is shown in Fig. 3.

In the left part, one can see a cut-out of the finite volumes of the fluid simulation and the interface to the structure, where the pressure and stresses are given at the center points of the interface cells. These load data have to be transfered to the structural interface.

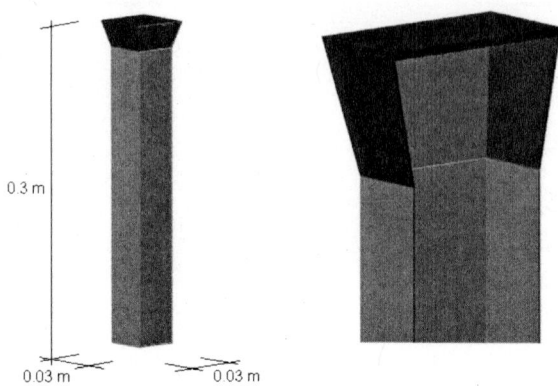

Fig. 4. Entire system and discretization of the tower

On the right part, one hexahedral element is shown with its interface to the fluid. In order to compute the load vector, stress data are needed at the Gauss points of the structural element being interpolated bilinearly from the fluid elements.

The structural computation yields a displacement field being defined by high-order polynomials over each element. These data can be evaluated at the grid points of the fluid interface.

For the coupling, the commercial interface (MpCCI [1]) is used.

4.2 Example: Thin-Walled Tower

The following example is a simple wind tunnel model of a thin-walled tower with low stiffness in a turbulent flow. The tower has a height of 0.3 m. In the section on the right hand side of Fig. 4 the three-dimensional discretization with hexahedral elements can be seen. The inflow velocity is 30 m/s which corresponding to a Reynolds number of 60,000. The fluid flow is computed using a large-eddy simulation. The fluid-structure interaction starts at $t=0$, when the tower is at rest and undeflected.

In order to show the behavior of the tower due to the fluid forces, the motion of a point on the top of the tower is observed. Figure 5 shows large displacements in flow direction in the beginning of the computation, which are then damped by the surrounding fluid. On the other hand, the cross-flow displacements in Fig. 5 are low in the beginning, but amplify very strongly due to vortex-shedding. In order to simulate such effects, the fluid code must necessarily compute the three-dimensional behavior of the fluid.

5 Conclusions

A partitioned, but fully implicit algorithm for the simulation of fluid-structure interaction problems was presented, where both disciplines employ three-dimensional models for the numerical simulation. The advantages of the fully three-dimensional

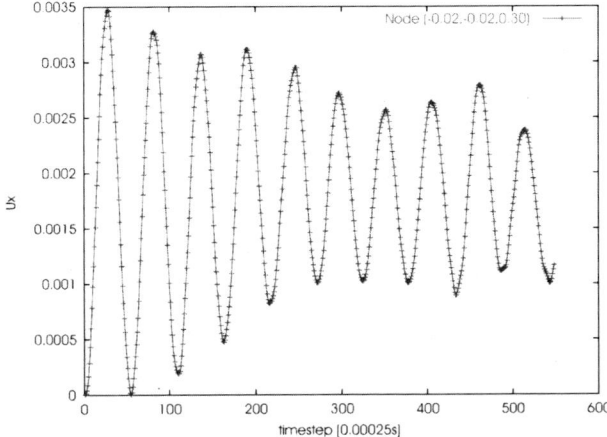

Fig. 5. Point on top of tower: displacements in flow direction (u_x) over time

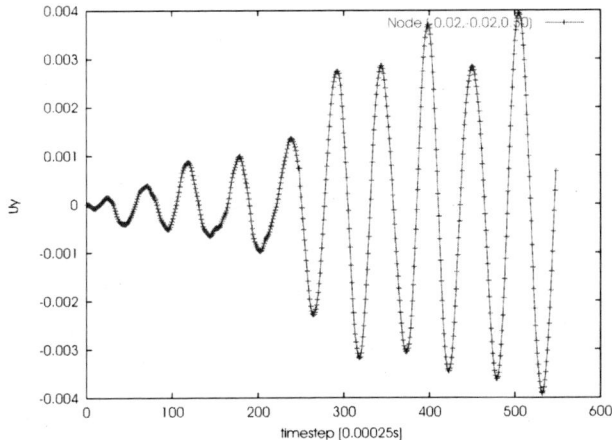

Fig. 6. Point on top of tower: cross-flow displacements (u_y) over time

modeling were discussed. As an example a thin-walled tower was investigated showing a non-trivial spatial interaction of fluid and structure.

Acknowledgement. The financial support of FLUSIB by the Bavarian State Ministry for Science, Research and the Arts in the Competence Network KONWIHR is gratefully acknowledged. The authors want to thank the HLRB Munich (Germany) providing the SMP cluster Hitachi SR8000-F1 used to perform the numerical simulations. The authors also want to thank Dr. rer. nat. G. Wellein and Dipl.-Ing. F. Deserno from the Computing Center in Erlangen (Germany) for the optimization of the fluid code for the above mentioned SMP cluster.

References

1. Ahrem, R., Hackenberg, M.G., Post, P., Redler, R., Roggenbuck, J. (2000): MpCCI – Mesh Based Parallel Code Coupling Interface. Institute for Algorithms and Scientific Computing (SCAI), GMD, http://www.mpcci.org/.
2. Brehm, M., Bader, R., Ebner, R. (2001): Höchstleistungsrechner in Bayern (HLRB): The Hitachi SR8000-F1. http://www.lrz-muenchen.de/services/compute/hlrb/.
3. Breuer, M. (2002): *Direkte Numerische Simulation und Large-Eddy-Simulation turbulenter Strömungen auf Hochleistungsrechnern*. Habilitationsschrift, Technische Fakultät, Universität Erlangen-Nürnberg, Berichte aus der Strömungstechnik, ISBN: 3-8265-9958-6, Shaker Verlag, Aachen.
4. Chung, J., Hulbert, G. (1993): A Time Integration Algorithm for Structural Dynamics with Improved Numerical Dissipation: The Generalized-α-Method. J. of Applied Mechanics, vol. 60, pp. 1562–1566.
5. Durst, F., Schäfer, M. (1996): A Parallel Block-Structured Multigrid Method for the Prediction of Incompressible Flows. Int. J. Num. Methods Fluids, vol. 22, pp. 549–565.
6. Duester, A. (2002): High-Order Finite Elements for Three-Dimensional, Thin-Walled Nonlinear Continua. Dissertation, Technische Universiät München, Shaker-Verlag, Aachen.
7. Glück, M., Breuer, M., Durst, F., Halfmann, A., Rank, E. (2001): Computation of Fluid-Structure Interaction of Lightweight Structures. J. Wind Eng. Ind. Aerodyn., vol. 89/14-15, pp. 1351–1368.
8. Glück, M., Breuer, M., Durst, F., Halfmann, A., Rank, E. (2003): Computation of Wind-Induced Vibrations of Flexible Shells and Membranous Structures. J. of Fluids and Structures, vol. 17, pp.739–765.
9. Gordon, W.J., Hall, C.A. (1973): Construction of Curvilinear Co-ordinate Systems and Applications to Mesh Generation. Int. J. Num. Meth. Eng., vol. 7, pp. 461–477.
10. Halfmann, A. (2002): Ein geometrisches Modell zur numerischen Simulation der Fluid-Struktur-Interaktion windbelasteter, leichter Flächentragwerke. Dissertation, Lehrstuhl für Bauinformatik, Technische Universität München.
11. Szabo, B.A., Babuska, I. (1991) Finite Element Analysis. John Wiley & Sons.
12. Werner, H. & Wengle, H. (1991): *Large-Eddy Simulation of Turbulent Flow Over and Around a Cube in a Plate Channel*. 8th Symposium on turbulent shear flow, Technical University of Munich, Germany, Sept. 9-11, 1991.

ParChem: Efficient Numerical Methods for Chemical Problems related to MOVPE

E. Mesic[1,2], M. Mukinovic[1], L. Kadinski[2] and G. Brenner[1,2]

[1] present address: Institute of Applied Mechanics, TU-Clausthal,
Adolph-Roemer-Strasse 2A, D-38678 Clausthal-Zellerfeld,
Gunther.Brenner@tu-clausthal.de
[2] formerly: Institute of Fluid Mechanics, University of Erlangen-Nürnberg,
Cauerstraße 4, D-91058 Erlangen

Abstract. The prediction of MOVPE processes requires the modeling of numerous coupled transport phenomena for momentum, mass and heat including temperature dependent physical properties and chemical reactions. In the present paper, the numerical simulation is used as a tool to identify and distinguish systematically the chemical parameters in the AlGaN growth process in order to obtain a reliable and efficient prediction of the process.

1 Introduction

The fabrication of many semi conductor devices is based on the so-called III nitride compounds such as gallium, aluminium and indium nitride. The favorable properties of these materials are the wide band gap range and high temperature stability. Additionally, the band gap can be increased through the use of ternary alloys of these nitrides, e.g. $Al_xGa_{1-x}N$ [ZKS95]. In the last decade, the Metal–organic vapor phase epitaxy (MOVPE) has become the most widely used technique for growing thin films of III–V compound semiconductors [BCC99]. However, the details of the process with regard to the chemistry are far from being well understood. Consequently, the numerical simulation of the MOVPE process is usually based on simplified chemical kinetics besides several idealizations for the thermal and chemical boundary conditions. As a remedy, a lot of effort has been spend through experiments and numerical work to provide a better insight into the growth process of III–V compounds. The development of detailed kinetic models for the growth of III nitride compounds is complicated by the large number of possible gas–phase and surface reactions. Adduct formation and decomposition, oligomerization and dissociation are found to be the essential growth pathways in the formation of III nitride compounds [SRK99]. The kinetic data for these pathways have to be obtained either by measurements or by theoretical methods. In that context, theoretical and numerical studies are extremely desirable to investigate MOVPE processes systematically and to improve the data used in chemistry models. The focus of the present work is to investigate the sensitivity of the chemical kinetics in the gas phase with respect to the predicted growth rates. However, there are many other physical parameters entering into a simulation model which are not presented here. Instead, we refer to a paper [MMB04] were we systematically investigated the influence of radiation, transport phenomena as well as thermal and chemical boundary conditions among others on the prediction of growth rates. The computations are performed using the computer code FASTEST 3D. The numerical solution method implemented in this code is described in [DSc96]. The module for the calculation of radiative heat transport is introduced in [KKL02, IDW96].

2 Problem Description and Experimental Setup

The growth process is performed in an AIX200 RF industrial horizontal reactor (AIXTRON AG) with standard precursors TMGa, TMAl and NH_3 using hydrogen as a carrier gas. Two sets of experimental data, denoted as A and B, are provided by AIXTRON AG [KBS00]. A sketch of the reactor is shown in Fig. 1. The inlet of metal–organic precursors (metal–organic line: TMGa + TMAl + H_2) is separated from the inlet of hydride precursors (hydride line: NH_3 + H_2). A graphite susceptor is inductively heated, reaching a temperature of 1170 °C. The 2 inch diameter wafer rotates very slowly at 10 rpm. The pressure of 50 mbar in the reactor chamber is the same as that in the cooler part. Regarding to the inflow rates two sets of experimental data are considered. Thus, hydrogen and ammonia are supplied in a ratio of approximately 4:1 and 8:1, whereas the ratio of the inflow rates of TMGa and TMAl is approximately 1:10 and 1:15 for the cases A and B, respectively. The metal–organic components are highly diluted. Hydrogen is also used as a coolant with an inflow rate of 7 slm. The coolant (water) and the inlet gases are assumed to be at room temperature, i.e. 27 °C.

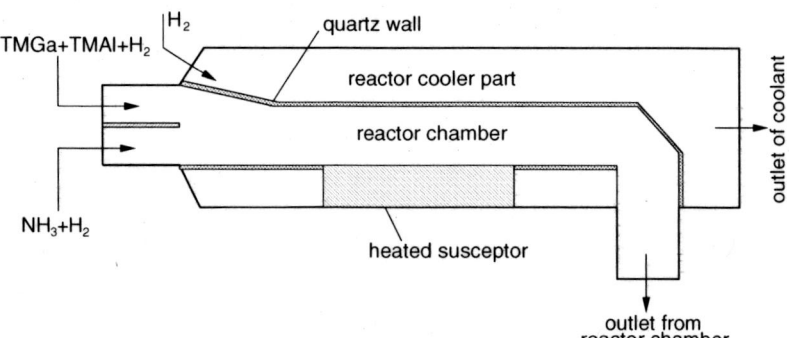

Fig. 1. Sketch of the AIX200 RF reactor with cooling part

3 Physical Models

The motion of the fluid in the present process is governed by conservation equations describing the rate of change of mass, momentum and energy due to convective and diffusive transport processes. The resulting system of nonlinear partial differential equations is completed by suitable models for the radiative heat transfer and for the conversion of species due to chemical reactions. In modeling the diffusion of mass, ordinary as well as thermal diffusion is taken into account. The transport coefficient, i.e. the viscosity, the thermal conductivity and the solute diffusivity, have to be specified for mixtures of gases. A detailed description of the model equations and of the numerical scheme may be found in [MMB04]. MOVPE processes for III nitride compounds are characterized by a strong temperature gradient between the susceptor and the upper reactor wall [Str99]. Due to this temperature gradient a realistic modeling of the heat transfer phenomena plays a crucial role for reliable prediction of the chemical vapor deposition process. In the present paper, the gas is considered to be transparent and surface to surface radiation is considered by a view-factor method. The transport coefficients for pure gases are defined in terms of intermolecular potential energy functions, i.e. the collision diameter (σ) and the attraction energy (ϵ/k). The calculation of the gas-mixture transport coefficients is derived on the basis of empirical expressions. These expressions are defined in terms of transport coefficients of the components of the mixture and their mole fractions. Though based on the same arguments, significant differences in these mixture rules defined e.g. by Hirschfelder et al. [HCB64], Reid et al. [RPP87], Kleijn [Kle91] or Mihopoulos [Mih99] are observed. See [MMB04] for a critical evaluation of the influence of the different transport models. In the present study, a binary mixture of ammonia and hydrogen is considered. This is justified since the multi-component species system consists of approximately 99% ammonia and hydrogen and 1% of other species. The transport properties of the mixture are computed according to [HCB64], [RPP87], [Kle91] and [Mih99]. Besides the physical model, the formulation of boundary conditions, in particular for the thermal fluxes, exhibits a significant influence on the predicted growth rates. The thermal boundary conditions are fixed using a heat transfer coefficient (HTC) obtained for the natural convection around a horizontal, heated cylinder. Since experimental results for surface temperatures are not available, this is the only way to model thermal boundary conditions. Nn the present study, a $HTC \approx 7$ is used corresponding to a Nusselt number of $Nu_D \approx 18$ at a Grashof number of the problem of about $Gr_D \approx 2.5 \cdot 10^6$ [IDW96]. Undesirable parasitic deposition takes place at the inner reactor walls. Due to this effect, species are deposited outside the wafer area causing a reduced growth rate at the wafer itself. Depending on the reactor wall temperature, this deposition is limited either by surface kinetics if the wall temperature is low or by diffusive mass transport through the gas phase for high wall temperatures. In the present process, the temperatures of the susceptor are quite high. Thus, here the effect of surface kinetics may not be taken into account. However, parasitic deposition takes place at the reactor walls, where the temperatures are significantly lower. Here, a critical temperature is defined for which an onset of the parasitic deposition is obtained [MMB04]. The definition of this critical temperature is motivated by physical arguments. However, an exact value cannot be derived from theory. In the numerical scheme, Dirichlet boundary conditions are defined for fixed species concentrations at the wall if the wall temperature is larger than the critical value and Neumann conditions are used vice versa.

4 Gas–Phase Chemistry Models

The chemistry models used in this work may be classified as follows: Simple models (*i*) have only one gas–phase reaction and consider only the pyrolise of metal–organic precursors to mono methyl derivatives of gallium [JPr63] and aluminium [Mih99]. Literature based models (*ii*) usually include some more reactions. In the present work, the models of Theodoropoulos, Mihopoulos and Safvi for the GaN growth [Mih99, TMM00, PTS00, SRT97, TKu96, WDK00, TBS01, MGJ98, TBS01, TBS97] and from Mihopoulos [MGJ98] for the AlN growth process are used. Finally, optimized models (*iii*) are used, that are described in [MMB04].

5 Results and Discussion

The results presented in this section show the sensitivity of the the chemistry models with respect to the GaN and AlN growth rates. Only the final results are presented for both set of operating conditions, A and B, whereas all other results are presented only for case A.

5.1 Investigation of the chemistry models for GaN and AlN growth process

The central part in this study is the analysis of the chemistry models. All computations presented in this subsection are performed using the transport data of Hirschfelder et al. [HCB64] (σ^*, $(\epsilon/k)^*$), assuming parasitic deposition (T_{crit}=1050 K for GaN and 1100 K for AlN) and for a heat transfer coefficient of $\alpha = 50$ W/m^2 K. Thus, only the reaction chamber is taken into account to reduce the computational effort. The investigation of the chemistry model is done basically in two steps. First, an appropriate chemistry mechanism is chosen based on available information in the literature. Thereafter the kinetic data for the corresponding mechanism are investigated. This is done only for the pathways which are suspected to be most important. In the GaN growth process, this is the oligomer decomposition while for the AlN growth process, the adduct formation has to be considered.

GaN chemistry models

In the first step of the evaluation of the GaN chemistry model, the results based on the four detailed models (ii) are compared against the results obtained by the simple model (Fig. 2). The figure presents the minimum and maximal value of the growth rate found on the susceptor surface for each set of parameters. These four models are from Theodoropoulos et al. [TMM00, PTS00], Mihopoulos [Mih99], Safvi et al. [SRT97] and the fourth model is obtained combining the models of Mihopoulos and Safvi. For set A of the operating conditions, the Theodoropoulos model extremely under-predicts the growth rate compared to experimental results. The small difference between the other three models shows that the pathways for the adduct formation and for the oligomerization are less important with regard to the GaN growth rate than the pathway for the oligomer decomposition. This is visible from the comparison of the species distributions obtained by the different models. Fig. 4 shows the species distributions of the $(DMGa:NH_2)_3$ (trimer of $DMGa:NH_2$) and of the oligomer decomposition product (monomer of $MMGa:NH$). The results presented are obtained applying the Theodoropoulos and Safvi models. In the Theodoropoulos model, the low concentration of $(DMGa:NH_2)_3$ in the reactor region near the wafer indicates a stronger convective mass transport in comparison with the diffusive mass transport. This comes from the fact that the molecules of oligomeric species are very large and heavy. Thus, the solute diffusivity of these species through the gas mixture of hydrogen and ammonia is low. Therefore, the species are transported behind the wafer before they reach the wafer surface, inhibiting a deposition at the wafer itself. In Safvi's model, the oligomers decompose to the monomers before they are transported behind the wafer. The monomers have a smaller molecular size and higher solute diffusivity through the carrier gas than the oligomers. Consequently, the monomers are transported faster by diffusion to the wafer surface than the oligomers, providing a higher growth rate at the wafer. Next, the kinetic data for the most important pathways – the oligomers decomposition pathways are investigated. In addition, the kinetic data for reactions G4 and G5 are also studied. The results of this study are not presented in details, but rather the optimized values are used in the further computations presented here. The kinetic data for reactions G6 and G7 used in the investigation are given in Table 1. The results obtained are shown in Fig. 3. Almost no deviation in the growth rate

Table 1. Variation of the pre–exponential factors and activation energies in reactions G6 and G7

E_A [kcal/mol] \ A_0 [1/s]	4.3×10^{13}	1.0×10^{15}	1.0×10^{17}	1.0×10^{20}	1.0×10^{22}
43.4	m1	m2	m3	m4	m5
48.7	m6	m7	m8	m9	m10
54.0	m11	m12	m13	m14	m15
60.0	m16	m17	m18	m19	m20

is observed. Due to this, the results for models m1–m5, m7–m10, m13–m15, m19 and m20 in Fig. 3 are grouped together and denoted as "others". A slight deviation is obtained for lower values of the reaction rates (models m16, m11, m17, m6, m12 and m18). These rates are computed for higher values of the activation energies and lower values of the pre–exponential factors. In these cases, a higher growth rate is obtained. This might be explained by the fact that the lower reaction rates cause a slower conversion of the oligomers to the monomers. The higher value of the activation energy increases the residence time of oligomers in the reactor. For these reasons, the region with the maximum concentration of the monomer species is moved downstream and placed just in the region above the wafer. Consequently, a larger mass flux at the wafer surface is provided, yielding a higher growth rate. However, comparing the growth rate deviations due to the different chemistry mechanisms and due to the different kinetic data for the corresponding mechanism, one can conclude that the choice of the correct chemistry mechanism is the major task in the chemistry model investigation. The pre–exponential factors and activation energies for reactions G4–G7 in the final model, which is used in the final computations, are as follows:

G4: $A_0 = 1.0e^{21}$ $1/s$ and $E_A = 49.0$ $kcal/mol$
G5: $A_0 = 2.5e^{17}$ $1/s$ and $E_A = 0.0$ $kcal/mol$
G6: $A_0 = 4.3e^{13}$ $1/s$ and $E_A = 48.7$ $kcal/mol$
G7: $A_0 = 4.3e^{13}$ $1/s$ and $E_A = 48.7$ $kcal/mol$.

It is found that these parameters fit the experimental data well as shown below.

AlN chemistry models

Basically, only one chemistry mechanism introduced by Mihopoulos et al. [MGJ98] is available. As the reference model for comparison, the so-called simple model is used. This model is actually a sub–model of the Mihopoulos model and consists of only one reaction, i.e. the first reaction in Mihopoulos model. In Fig. 5 the results for the AlN growth rate obtained by these two models are compared. It is obvious, that the solid particle formation driven by the reversible reactions is the most influential pathway in the AlN growth process. By the reversible reactions, also called the parasitic reactions, the first adducts are formed and decomposed. The material depletion rate depends directly on the kinetics of these parasitic reactions. The growth rates based on the Mihopoulos model are nearly equal to zero. This indicates, that the material depletion rate is very high. With respect to this growth rate, one can say that the kinetic data for the parasitic reactions are overestimated in the Mihopoulos model. Right these kinetic data are investigated here and the resulting optimized model based on these data is called the final model. Following the study by Timoshkin et al. [TBS01], the kinetic data investigated are found to be much higher due to the more stable reactants at low temperature. This is clearly visible from the comparison of the AlN growth rates obtained using the Mihopoulos and the final models (see Fig. 5). Unlike in the GaN growth process, the kinetic data for a certain mechanism in the AlN growth process are much more influential and cause very high deviations of the growth rate. The activation energies for reversible reactions used in the final model are found to be 43.69 and 65.69 kcal/mol for the forward and reverse reactions, respectively. These parameters provide the best agreement with experiment (see Section 5.3).

5.2 Final computations

In this section results are presented that were obtained using the optimized parameters that have been worked out in the above study. Thus, the transport properties are computed using σ^* and $(\epsilon/k)^*$, the approach for the computation of the gas–mixture properties introduced by Hirschfelder [HCB64], a heat transfer coefficient of $\alpha = 120$ W/m^2 K and the final chemistry models are imposed in the computations, simulating the transport processes in the full reactor. In addition, the parasitic deposition at the reactor walls is assumed to take place if the wall temperature is above 1050 K for GaN and 1100 K for AlN. The comparisons between the numerical and experimental results for both, the GaN and AlN growth rates are shown in Fig. 6 and for the both set of operating conditions, A and B. A better agreement between the experimental and numerical results is observed in the case A then in the case B for both, the GaN and AlN growth rate. The growth rate deviations obtained in the middle of the wafer and at the wafer periphery do not have the same trend switching from A to B. While the GaN growth rate at the periphery is under-predicted in case A, it is over-predicted in case B. Also the deviation for GaN in A is lower at the wafer center than at the wafer periphery, whereas in B is vice versa. The similar disagreement can be seen for AlN growth rate. The diversity deviation trends clearly indicate that the results obtained strongly depends on the numerical parameters tuned. Thus, the growth process for operating conditions B must be systematically studied too, in order to provide more accurate and generalised numerical parameters. However, the deviations might be explained by the non–homogeneous temperature distribution at the susceptor, gas gap between the static and rotating parts of the susceptor, unknown surface chemistry at the wafer and the possible formation of an adduct between Ga– and Al–containing species. The efficiency for both the GaN and AlN growth processes is illustrated in Fig. 7 again for cases A and B. A very high GaN growth efficiency is observed for both cases, A and B. However, the higher growth efficiency for case B shows that depending on the overall operating conditions, a more efficient process can be obtained even if lower inflow rates of the Ga precursor are supplied. The low AlN growth efficiency confirms the fact that the parasitic reactions, which cause the depletion of material, play a much more important role in the AlN than in the GaN growth process. Although the AlN growth is strongly influenced by parasitic reactions, the growth efficiency of around 2000 μm/mol is fairly high. Probably such a relatively high growth efficiency is caused by the low pressure in the reactor.

6 Conclusions

A numerical analysis of the AlGaN growth process in an industrial MOVPE reactor (Aixtron–AIX200 RF) is presented. Due to estimates and uncertainties in defining chemical model parameters used in the simulations, the prediction of the growth rate process in some cases provides a variation of up to 100%. The development of the chemistry model is divided into two basic steps, i.e. the definition of the chemistry mechanism and the determination of the kinetic data. An outcome of the evaluation of the chemistry models in this work is that the main pathways in the GaN process are the oligomerization and the oligomers decomposition. In the AlN process the main pathway is the formation of AlN solid particles. This pathway consists of three parts, adduct formation, oligomerization and oligomer decomposition to solid fragments. A good agreement between the experimental and numerical results is achieved. In spite of that, a more rigorous verification must be made in order to provide the relevant models for the chemistry modeling. Hence, it is necessary to verify the chemistry model in other reactor configurations and under different operating conditions. This will be done in the forthcoming work.

Acknowledgement

The authors are grateful to KONWIHR for partial financial support in the frame of the project ParChem as well as to the German Research Foundation (Deutsche Forschungsgemeinschaft) for financial support provided in the frame of the project "Anorganische Materialien durch Gasphasensynthese: Interdisziplinäre Ansätze zu Entwicklung, Verständnis und Kontrolle von CVD–Verfahren" (DFG SPP–1119) and to AIXTRON AG for providing the experimental data.

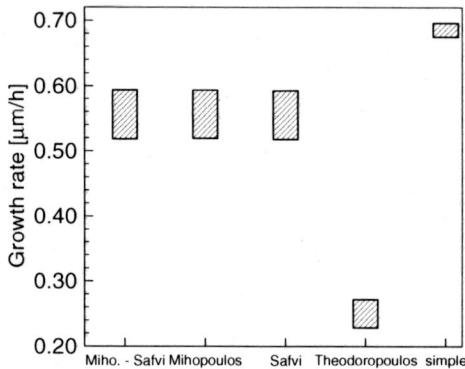

Fig. 2. GaN growth rate applying five different models: combination of the Mihopoulos and Safvi models, Theodoropoulos, Safvi, Mihopoulos and simple model

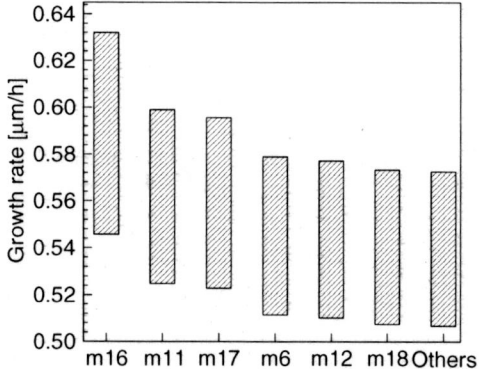

Fig. 3. GaN growth rate by using Safvi mechanism with the optimized kinetic data for reactions G4 and G5 and the kinetic data from Table 1 for reactions G6 and G7

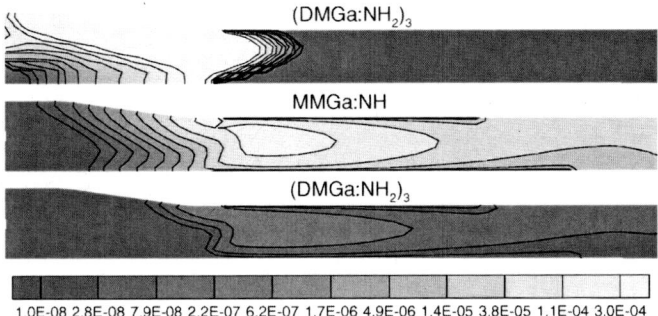

Fig. 4. Species distribution for $(DMGa:NH_2)_3$ (top) and MMGa:NH (middle) obtained by the Safvi model and for $(DMGa:NH_2)_3$ (bottom) obtained by the Theodoropoulos model

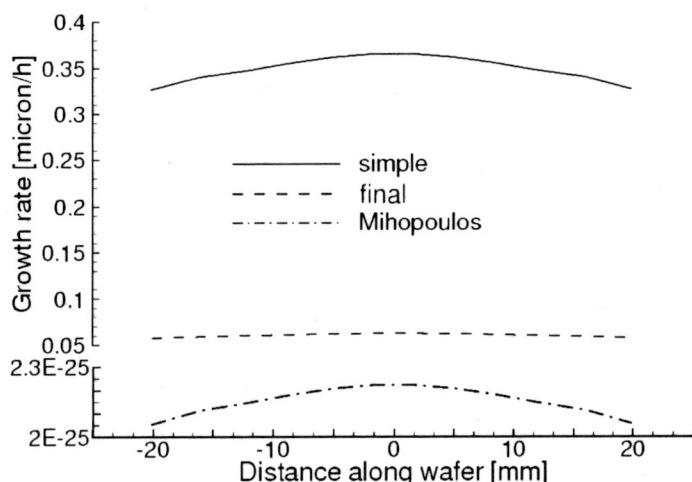

Fig. 5. AlN growth rate applying three different models: simple, Mihopoulos and final models

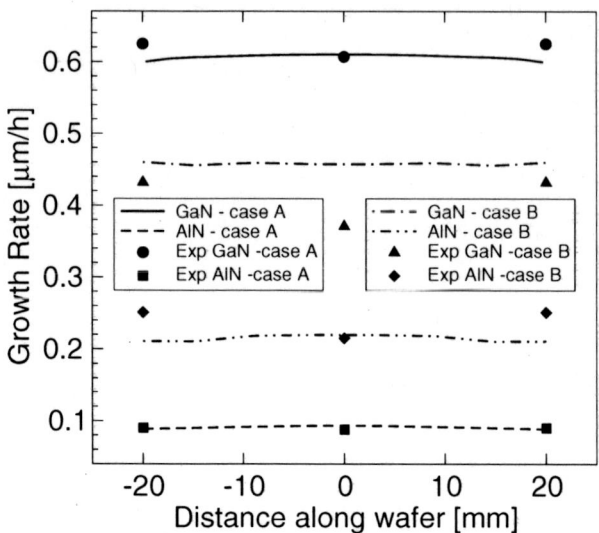

Fig. 6. Comparison of the experimental and numerical growth rates for GaN and AlN

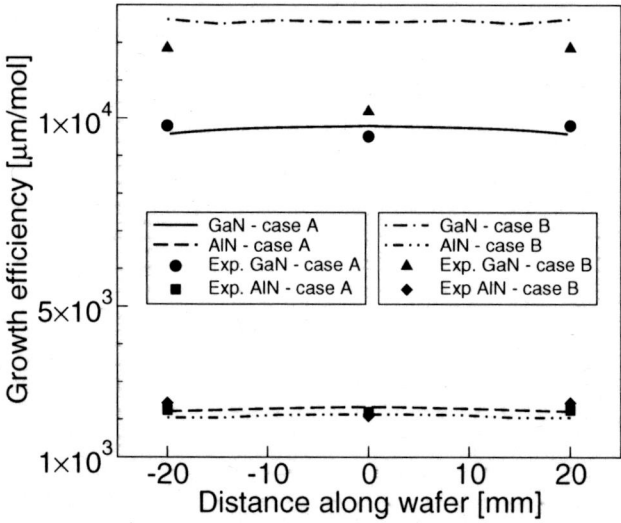

Fig. 7. Comparison of the experimental and numerical growth efficiency for GaN and AlN

References

[ZKS95] X. Zhang, P. Kung, A. Saxler, D. Walker, T.C. Wang, M. Razeghi, *Appl. Phys. Lett.* **67** (1995) p. 1745.
[BCC99] W. Breiland, M. Coltrin, R. Creighton, H. Hou, H. Moffat, J. Tsao, *Mater. Sci. Eng.* **R24** (1999) p. 241.
[SRK99] J. Sun, J. Redwing, T. Kuech, *Phys. Status Solidi A* **176** (1999) p. 693.
[HCB64] J. Hirschfelder, C. Curtiss, B. Bird, *Molecular Theory of Gases and Liquids*, John Wiley & Sons, New York, 1964.
[Hol98] W.L. Holstein, *J. Electrochem. Soc.* **13** (1998) p. 1788.
[Sve62] R.A. Svehla, *NASA Tech. Rep. R-132*, Lewis Research Center, Cleveland, Ohio, 1962.
[RPP87] R. Reid, J. Prausnitz, B. Poling, *The Properties of Gases and Liquids*, 4th edn, McGraw-Hill, New York, 1987.
[Kle95] C. Kleijn, *Computational Modelling in Semiconductor Processing*, M. Mey-yappan Editor, Artech House, Boston, 1995.
[Kad96] L. Kadinski, *Mathematische Modellirung und numerische Simulation von CVD–Prozessen in der Halbleitertechnik*, PhD thesis, Universität Erlangen–Nürnberg, 1996.
[Kle91] C. Kleijn, *Transport Phenomena in Chemical Vapor Deposition Reactors*, PhD thesis, Technische Universiteem Delft, 1991.
[Mih99] T. Mihopoulos, *Reaction and Transport Processes in OMCVD: Selective and Group III-Nitride Growth*, PhD thesis, Massachusetts Institute of Technology, 1999.
[CLS96] C. Chen, H. Lui, D. Steigerwald, W. Imler, C. Kuo, M. Ludowise, S. Lester, J. Amano, *J. Electron. Mater.* **25** (1996) p. 1004.
[JPr63] M.G. Jacko, S.J.W. Price, *Can. J. Chem.* **41** (1963) p. 1560.
[TMM00] C. Theodoropoulos, T. Mountziaris, H. Moffat, J. Han, *J. Cryst. Growth* **217** (2000) p. 65.
[PTS00] R. Pawlowski, C. Theodoropoulos, A. Salinger, T. Mountziaris, H. Moffat, J. Shadid, E. Thrush, *J. Cryst. Growth* **221** (2000) p. 622.
[SRT97] S. Safvi, J. Redwing, M. Tischler, T. Kuech, *J. Electrochem. Soc.* **144** (1997) p. 1789.
[TKu96] A. Thon, T. Kuech, *Appl. Phys. Lett.* **69(1)** (1996) p. 55.
[WDK00] R. Watwe, J. Dumesic, T. Kuech, *J. Cryst. Growth* **221** (2000) p. 751.
[TBS01] A. Timoshkin, H. Bettinger, H. Schaefer, *J. Phys. Chem. A* **105** (2001) p. 3240.
[MGJ98] T. Mihopoulos, V. Gupta, K. Jensen, *J. Cryst. Growth* **195** (1998) p. 733.
[TBS01] A. Timoshkin, H. Bettinger, H. Schaefer, *J. Cryst. Growth* **222** (2001) p. 170.
[TBS97] A. Timoshkin, H. Bettinger, H. Schaefer, *J. Am. Chem. Soc.* **119** (1997) p. 5668.
[DSc96] F. Durst, F. Schaefer, *Int. J. Num. Methods Fluids* **22** (1996) p. 549.
[MMB04] E. Mešić, M. Mukinović, G. Brenner, *Numerical Study of Algan Growth by Movpe in an Aix200 RF Horizontal Reactor*, I. Journal Comp. Mat. Science, 2004.
[KKL02] L. Kadinski, P. Kaufamnn, C. Lindner, F. Durst, *Proc. 3rd Int. FORTWIHR Conf. on HPSEC*, Erlangen, Germany, March 12-14, 2001; published in *Lecture Notes in Computational Science and Engineering Computing*, M. Breuer, F. Durst, C. Zenger, eds. **21**, Springer Verlag, 2002.

[IDW96] Frank P. Incropera, David P. DeWitt, *Fundamentals of Heat and Mass Transfer*, John Wiley & sons, 1996.
[KBS00] A. Krost, J. Bläsing, F. Schulze, O. Schön, A. Alam, M. Heuken, *J. Cryst. Growth* **221** (2000) p. 251.
[Str99] G.B. Stringfellow, *Organometallic Vapour–Phase Epitaxy: Theory and Practice*, 2nd edn, Academic Press, San Diego, 1999.

RexSim: Monte Carlo Simulations of Radiative Heat Transfer in Parallel Computer Architectures

G. Brenner[1,2], L. Kadinski[2] and J.G. Marakis,[3]

[1] present address: Institute of Applied Mechanics, TU-Clausthal, Adolph-Roemer-Strasse 2A, D-38678 Clausthal-Zellerfeld, Gunther.Brenner@tu-clausthal.de
[2] formerly: Institute of Fluid Mechanics, University of Erlangen-Nürnberg, Cauerstraße 4, D-91058 Erlangen,
[3] formerly: National Technical University Athens

Abstract. This work presents a parallel Monte Carlo algorithm for the calculation of combined radiative and conductive heat transfer. The proposed formulation effectively separates the time-consuming ray-tracing part of the Monte Carlo method from the energy computations required in the iterative solution of the energy equation. The method is applied for a simple combined radiative and conductive heat transfer problem and excellent agreement with the benchmark results is found. The ray-tracing part of the algorithm is parallelised and applied in two configurations, which represent the opposite ends of the currently available parallel computer architectures; a PC cluster and the Hitachi SR8000-F1 supercomputer. For sufficiently large sampling sets, the measurements show an almost ideal speed-up.

1 Introduction

The solution of the energy equation in combined radiative and conductive / convective heat transfer problems is a time consuming task. The usual practice in solving this type of problems is to employ an approximate and fast model for radiation and thus to avoid the significant overloading of the iterative solution of the energy equation. The alternative option, the use of an accurate model, appears rarely in the literature. The reasons for that are generally associated with the lack of sufficient computational resources. The present projects exploits the straightforward method to circumvent this deficiency; namely it investigates the parallelization of an accurate radiative heat transfer model, such as the Monte Carlo method. Descriptions of the serial version of this method can be found in textbooks [SiHo92] and [Mod93], or monographs [HaSh88] and [YTK95]. Recently, in [Ho98] numerous applications in the field of radiative heat transfer have been reviewed. The parallelization of this method has been addressed in [FaHo98] and [MCBD01].

In the Monte Carlo method, radiative heat exchange is simulated by stochastically tracing an amount of discrete packets emitted by each surface or volume element. All the possible radiative events (emission, transmission, absorption, scattering, reflection) are described by probability density functions. Macroscopic radiative quantities, such as radiative fluxes or source terms, are formed as averages over a sufficiently large statistical sample. The problem is that each time the temperature field changes, as it happens during the iterative solution of the energy equation, a new sample of energy packets has to be traced. Since even the tracing of a single set of energy packets is time consuming, this repetitive application makes the method unpractical for combined heat transfer problems of engineering interest. An alternative method is to reuse all, or part, of the previous sample. This approach has been proposed in [Kob89] and further elaborated in [SOCS99]. A method to consistently reuse previous samplings is to introduce, as an intermediate step in the algorithm, the calculation of the direct exchange factors. This approach originates from the zonal method presented in [HS67] and it will be described in the next section.

The specific problem investigated in this work is the determination of the temperature field between two infinite parallel plates. The medium between the plates is absorbing, emitting and conducting. This simple configuration, apart from the interest it has in many applications, such as modelling of melting/solidification and heat transfer in semitransparent crystals, it also has the advantage of an accurate semi-analytic solution. It may therefore serve as a benchmark solution for the validation of the combined Monte Carlo and finite volume code used in this study. Another focal point is the evaluation of the efficiency of the parallel Monte Carlo implementation. This issue is investigated by running the ray-tracing part of the algorithm in the two platforms which represent the lowest and the highest ends of the current parallel computing architectures, namely a 10-node PC cluster and the Hitachi SR-8000 F1 supercomputer of the Leibniz Supercomputing Center in Munich.

2 Mathematical Formulation

The steady state energy equation for a gray radiating and conducting medium reads

$$\frac{\partial}{\partial x_i}\left(k\frac{\partial T}{\partial x_i}\right) = \frac{\partial q_{r,i}}{\partial x_i} = \kappa\left(4\sigma T^4 - \int_{4\pi} I d\Omega\right) \qquad (1)$$

where k is the thermal conductivity, q_r is the net raditive flux, κ is the absorption coefficient, σ the Stefan-Boltzmann constant, I the radiative intensity and Ω a solid angle. A semi-analytic solution of Eq. (1) based on the exponential integral functions can be found in standard textbooks, e.g. [SiHo92] and [Mod93]. The method adopted in this study is the usual iterative solution of Eq. (1); based on an assumed temperature distribution, the radiative source term, $\frac{\partial q_{r,i}}{\partial x_i}$, is first calculated and then Eq. (1) is solved to obtain an improved estimation of the temperature field. In this iterative procedure, the Monte Carlo method is used to calculate the radiative source term. Details for the implementation of this method, the ray-tracing algorithm on which it is based, as well as a description of a parallelisation strategy

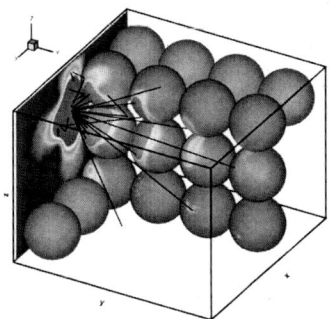

Fig. 1. An illustration of the direct exchange factor isolines on the surface of spheres on a regular packed bed.

can be found in [MCBD01] and therefore will not be repeated here. A schematic picture is shown only for illustration purposes in Fig. 1.

The previous description implies that the Monte Carlo algorithm has to be repeated once the new temperature field has changed. A very significant reduction in the computational time is achieved if the problem is expressed in terms of the direct exchange factors, as shown in [HS67] and [YTK95]. This formulation allows splitting the problem in two parts. In the first part, the ray-tracing part of the Monte Carlo method is used to compute the direct exchange factors. These factors are defined as

$$R_{ij} = \frac{n_j}{n_i} \qquad (2)$$

where n_i is the number of packets emitted by the i-th element and n_j the number of them that is absorbed in the j-th element ($\sum_j n_j = n_i$). This part of the algorithm is "energy-free", i.e. it is considered that the packets do not carry energy. In the second part, the direct exchange factors are used to quantify the energy transfer. This is done by computing the total amount of energy absorbed by the j-th element as

$$Q_{abs,j} = \sum_i R_{ij} Q_{emit,i} \qquad (3)$$

and $Q_{emit,i}$ is either

$$Q_{emit,i} = \varepsilon_i \sigma T_i^4 \triangle A_i \qquad (4)$$
$$Q_{emit,i} = 4\kappa_i \sigma T_i^4 \triangle V_i, \qquad (5)$$

depending on whether i is a surface or volume element. In Eqs. (4) and (5), ε is the emissivity, κ is the absorption coefficient and $\triangle A_i$, or $\triangle V_i$, is the size of the i-th element.

The advantage of this formulation is that the time consuming ray-tracing algorithm is performed only once. Therefore it has the potential to reduce the computational time by a factor in the order of the number of iterations necessary for the solution of Eq. (1). Due to the very strong non-linearity of this equation, a large number of iterations are needed and therefore this saving is expected to be significant. The restrictions for the application of the direct exchange formulation are, first, that the spatial distribution of the radiative properties of the medium should remain unchanged with iterations, and second, that these properties are also temperature independent. Both restrictions are met in a wide range of applications, especially those involving a solid matrix embedded in a gas phase, as it happens for example in porous burners. In the cases where the aforementioned restrictions are not met, the repetition of the ray-tracing algorithm is necessary to recalculate the new set of direct exchange factors. In these cases, the direct exchange formulation may turn to represent a slight computational overhead, instead of saving, when compared with typical Monte Carlo implementations.

Another improvement reported in this work is the introduction of a set of cutoff criteria. This improvement is based on the observation that only a small group of the direct exchange factors represent almost completely the radiative heat exchange. It is therefore desirable to identify which factors belong in this group and subsequently to perform the energy calculations using only this small subset. The benefit is significant because, first the database for storing the direct exchange factors scales with n^2, the number of the discretisation elements, and second, the size of the loop implied by Eq. (3) also scales with the number of the direct exchange factors. The reduction of the database is achieved by sorting the factors in descending order and storing them until any of the cutoff criteria is reached. Two criteria have been implemented, the first referring to the value of a single factor and the second on their cumulative value which is calculated as they are sorted. The advantage of these threshold parameters is that they offer a consistent method to manage the accuracy vs. the computational requirements of the Monte Carlo algorithm.

3 Results

The problem considered in this section is to determine the temperature distribution in an absorbing, emitting and conducting medium confined between two infinite parallel, isothermal, gray, emitting and diffusely reflecting plates. The aim is twofold; first to validate the combination of the Monte Carlo algorithm and a finite volume code used to solve the energy equation, and second to demonstrate the feasibility and the benefits of the cutoff criteria.

The temperature distribution between the plates calculated by the semi-analytic method presented in [Mod93] is shown in Fig. 2 for three values of the conduction-to radiation parameter N. The latter is defined as

$$N = \frac{k\kappa}{4\sigma T_{\max}^3} \qquad (6)$$

and it is a measure of the relative importance of conductive and radiative fluxes for optically thick media. The curves shown in Fig. 2 correspond to a slab with optical thickness $\tau_L = 1$, black walls ($\varepsilon_1 = \varepsilon_2 = 1.0$), a temperature ratio of the cold and hot plates equal to 0.5 and temperature independent thermal conductivity. In this figure, $\theta = \frac{T}{T_{\max}}$ is the non-dimensional temperature and $\xi = \frac{x}{L}$ the non-dimensional distance across the slab. For decreasing values of N, the temperature distributions are shown to depart progressively from the linear profile corresponding to pure conduction towards the pure radiation profile. Monte Carlo results are also shown

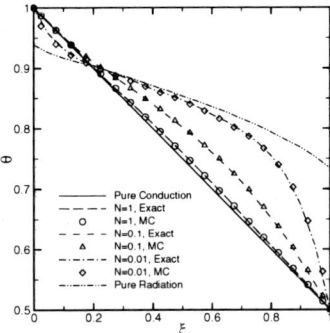

Fig. 2. Comparison between Monte Carlo results and benchmark data [Mod93] for three values of the conduction-to-radiation parameter.

in Fig. 2. They have been obtained by discretising the geometry into $21 \times 11 \times 11$ elements. The elements are equidistant in each direction and they have an aspect ratio of 10, in reference to their thickness, to represent the infinite width of the geometry. The direct exchange factors have been calculated after tracing 150,000 packets from each element. The data mentioned in this paragraph are common for all of the examined test cases. The agreement between the Monte Carlo results and the benchmark data shown in Fig. 2 is very satisfactory.

In order to evaluate the feasibility of the cutoff criteria, a new set of cases has been examined. In these cases, $N = 0.01$, while the cutoff criterion has been decreased from the value of 1.0, which corresponds to a complete database of the direct exchange factors, up to 0.85 by taking the intermediate values shown in Fig. 3. It is shown in this figure that the accuracy progressively diminishes with decreasing cutoff. The solutions are sufficiently accurate for a range between 0.99 and 0.95, while lower cutoff values produce less accurate results. The reduction in size of the database that is achieved for various values of the cutoff parameter is shown in Fig. 3. The results in this figure correspond to three values of the absorption coefficient and $N = 0.01$. The very rapid reduction of the database with decreasing cutoff can be easily observed in Fig. 3. For a moderately low criterion of 0.95 for example, the database is shrunk by a factor ranging from 3, for the optically thin medium, and up to 5 for the optically thick case.

The ray-tracing part of the algorithm has been parallelized following the Single-Program-Multiple-Data (SPMD) paradigm and using the Message Passing Interface (MPI) library. Since the packets are propagating independently of each other the parallelisation is made by tracing concurrently as many of them as possible in each of the available computing nodes. If p homogeneous processing nodes are available, then decomposing each sample of N energy packets into $N_p = \frac{N}{p}$ equal subsets accomplishes this task. This simple parallelisation strategy was tested in two very different homogeneous parallel computing architectures, namely a dedicated PC cluster made of 10 nodes and the Hitachi SR-8000 F1 supercomputer.

For the PC cluster, Fig. 4 shows the relation between the number of processors and the size of the sample subset. The reduction of this sample is in principle desirable because it helps to avoid unnecessary calculations after a certain level

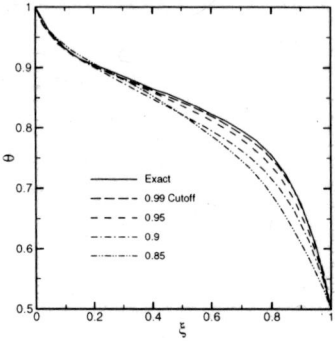
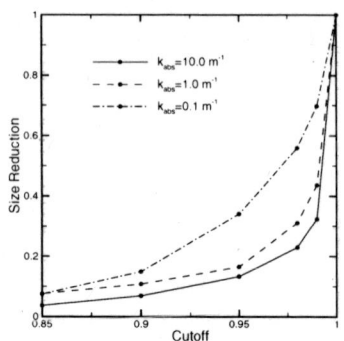

Fig. 3. The feasibility of the cutoff criterion; In the left: influence of the cutoff on the accuracy of the results; In the right: reduction of the size of the database achieved by decreasing the cutoff in reference with the optical thickness of the medium.

of confidence has been reached. However, it is evident from Fig. 4 that it is also associated with a degradation of the parallel efficiency. The latter is attributed to the increased communication overhead that results from the more frequent gathering of data from all the processors. The same tendency has been observed in the Hitachi SR8000 supercomputer. Measurements taken in this machine are shown in Fig. 5. An almost ideal speed-up was achieved for very large subset sample size (0.5 million packets) for up to 512 processors. However, such a large sample is rarely used in practical calculations. More interesting is the range of 50,000 packets, where the total sample size remains in reasonable size, while the parallel efficiency, as shown in Fig. 5, is still satisfactorily high.

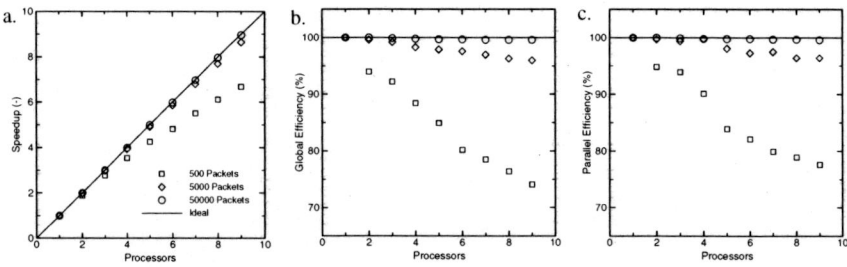

Fig. 4. Influence of the sample size on: Speed Up (left), $S_p = \frac{t_1}{t_p}$, Global Efficiency (center), $E_g = \frac{t_1}{pt_p}$ and Parallel Efficiency (right), E_p, where t_1 and t_p are the execution times using 1 and p nodes, respectively, and $E_g = E_{load} E_{num} E_p$. The machine was a dedicated cluster of 10 PCs.

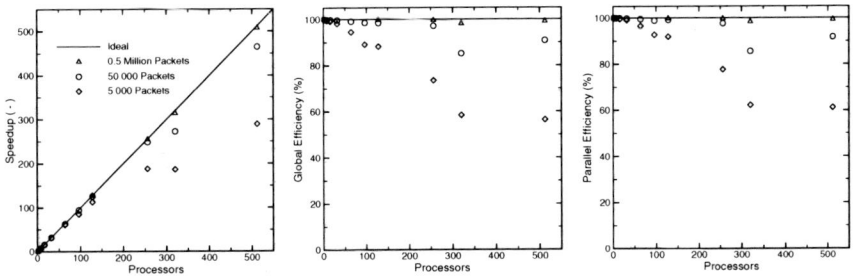

Fig. 5. Influence of the sample size on: Speedup, Global Efficiency and Parallel Efficiency for the Hitachi SR8000-F1 with 96 nodes (512 processors).

4 Conclusions

A Monte Carlo method for combined radiative and conductive heat transfer has been applied in a simple configuration for which a semi-analytic solution is available. A very good agreement between the Monte Carlo results and the benchmark data was obtained. Then a series of test cases were examined aiming to determine whether the size of the direct exchange factor database can be reduced by introducing a cutoff parameter. It was shown that a very significant reduction is achievable for a usable range of the cutoff between 0.95 and 0.99. Further reduction of this parameter, though it continuously reduces the database size, is not recommended because it also reduces the accuracy of the solution.

The implementation of the ray-tracing algorithm in a dedicated network of workstations showed that very high efficiencies could be obtained from the computing resources readily available in most laboratories. This is because the histories of the energy packets are by definition independent of each other and therefore only the statistics of a sampling set need to be transferred among the processors. This minimal communication requirement between the nodes helps to avoid any potential network bottleneck caused by limited network bandwidth. The application of the ray-tracing algorithm in a supercomputer infrastructure showed that the very high parallel efficiency is maintained for as many as 512 processors. Therefore, large and computational demanding problems can now be solved using these high performance architectures.

5 Acknowledgements

This study has been partially financially supported through KONWIHR within the RexSIM project as well as the Commission of the European Communities through the TMR-RADIARE network "Fundamental improvements in radiative heat transfer". The authors would also like to acknowledge the Leibniz Supercomputing Center in Munich for making available the Hitachi SR-8000 F1 computer as well as the team of the KONWIHR project BAUWIHR to support the optimization of the code.

References

[SiHo92] Siegel, R., Howell, J.R., Thermal Radiation Heat Transfer. 3rd ed., Taylor and Francis, Washington DC (1992)

[Mod93] Modest, M.F., Radiative Heat Transfer. McGraw-Hill, New York (1993)

[HaSh88] Haji-Sheikh, A., Monte Carlo Methods. In Minkowycz, W.J., Sparrow, E.M., Schneider, G. E. and Pletcher, R. H. (Eds.), Handbook of Numerical Heat Transfer, Wiley, New York (1988)

[YTK95] Yang, W.J., Tanigushi, H., Kudo, K., Radiative heat transfer by the Monte Carlo method. in Hartnett, J.P. and Irvine, T. (Eds.), Advances in Heat Transfer **27**, 1–215, Academic Press, San Diego (1995)

[Ho98] Howell, J.R., The Monte Carlo method in radiative heat transfer. Journal of Heat Transfer **120**, 547–560 (1998)

[FaHo98] Farmer, J.T. and Howell, J.R., Comparison of Monte Carlo strategies for radiative transfer in participating media, In Hartnett, J.P. and Irvine, T. (Eds.), Advances in Heat Transfer **131**, 1–97, Academic Press, San Diego (1998)

[MCBD01] Marakis, J. G., Chamico, J., Brenner, H. and Durst, F., Parallel ray tracing for radiative heat transfer: Application in a distributed computing environment. Int. J. Numer. Methods Heat and Fluid Flow, **11**(7), 663–681 (2001)

[Kob89] Kobiyama, M., Reduction of computing time and improvement of convergence stability of the Monte Carlo method applied to radiative heat transfer with variable properties. Journal of Heat Transfer **111**, 135–140 (1989)

[SOCS99] Schweiger, H., Oliva, A., Costa, M. and Pérez Segarra, C.D., A Monte Carlo method for the simulation of transient radiation heat transfer: Application to compound honeycomb transparent insulation. Numerical Heat Transfer, Part B **35**, 113–136 (1999)

[HS67] Hottel, H.C. and Sarofim, A.F., Radiative Transfer. McGraw-Hill, New York (1967)

SkvG: Cache-Optimal Parallel Solution of PDEs on High Performance Computers Using Space-Trees and Space-Filling Curves

Markus Langlotz[1], Miriam Mehl[1], Tobias Weinzierl[1], Christoph Zenger[1]

Institut für Informatik
TU München
Boltzmannstraße 3
85748 Garching, Germany
{guenthef,mehl,poegl,zenger}@in.tum.de

Abstract. Facing the problem of implementing an efficient solver for partial differential equations, we are, in general, confronted with a certain quandary between numerical efficiency and efficiency in the usage of hardware resources: Modern numerical methods require the handling of hierarchical multilevel data on adaptively refined data structures, which are mostly represented by trees. On the other hand, as data access is one of the most important bottlenecks in high performance computing, we would wish to process data linearly with a high locality in time and space to be able to exploit the capability of cache hierarchies. In this paper, we show an approach based on space-filling curves as an odering mechanism for the cells of space-tree grids, with the help of which we can transform our (inherently highly non-local) data respresentation by trees to a few linearly processed data sets. As a consequence, we reach extremely high cache hit-rates above $99,9\%$. In addition, the used methods make both parallelization and multigrid algorithms on adaptive grids with hierarchical data very straightforward and efficient.

1 Introduction

Developing numerical algorithms for PDEs on modern high performance computer architectures, it is not sufficient to concentrate solely on the numerical efficiency of the used methods. In fact, we can even loose overall efficiency (measured in terms of runtimes, for example) if we try to implement the most efficient numerical methods like adaptive grid refinement and multigrid methods without taking care of some essential hardware features. In particular, we have to take into account two important trends: First, high performance computers achieve their power by massive parallelism. Thus, our programs should be massivly parallelizable with a good load balance and a moderate communication overhead. Second, the computing speed of processors grows much faster than the access rate to memory. As a consequence, the most important bottleneck is memory access and not computing power. From the hardware side, this problem is attenuated by cache hierarchies to some extend. From the software side we have to ensure that our algorithms run efficiently on cache hierarchies, which means that they should process data locally in time and in space to prevent data from being eliminated from the cache before they are used for the last time. Unfortunately, these requirements are not easy to achieve at least in the case of modern PDE-solvers.

Modern PDE-solvers usually work on grids – so called adaptive grids – where the mesh width should be to some extent (automatically) adaptable to the properties of the solution. Thus, the amount of computational work per part of the domain depends on the local position in the domain and, in addition, changes during the computation. This makes both load balancing and a small communication overhead difficult to achieve. Moreover, implementations of such a concept usually are based on pointer structures and, thus, the algorithms access the memory non-locally.

A second complication for numerical PDE-solvers is the interaction between local operators and global effects of local changes, which has to be reflected by the solver if we want to get an optimal $(O(N))$ efficiency. Since classical iteration schemes only work with local operators and, thereby, transport changes in far away regions very slowly, we have to use multigrid schemes, which lead to hierarchical data structures represented by trees. This causes problems for an efficient parallelization since the coarse level grid has too little grid points to be easily distributed on a high number of processes. In addition, hierarchical data structures with dependencies over all levels again lead to non-local memory access.

In many applications, a third problem we have to face is to represent complicated and potentially changing geometries in an appropriate way minimizing administration overhead. In this context, structured approaches like space-trees are much cheaper than unstructured grids [BBFM02] and usually general enough to allow an adequate grid in the interior of the computational domain as long as the problem does not need an unisotropic refinement of the grid not aligned with the coordinate system. The problem of representing boundaries with sufficient accuracy can be overcome by modifying the discretization at boundary cells (see for example [Fra00]). Due to these reasons, we will concentrate on adaptive grids represented by space-trees in the following.

For such grids, space-filling curves have turned out to be very useful to construct domain decompositions with good load balance and quasi-minimal communication overhead (see for example [Zum00]). In addition, it was observed that also cache-performance is improved by ordering computational cells according to a space-filling curve [ABA00]. Even for hierarchical multilevel, these concepts can be used in a straightforward way [GZ99, GZ02].

In this paper, we show in a specialized context that the use of Peano space-filling curves[1] lead to an extremely high cache performance. For this, we enhanced the concepts mentioned above by a more consequent usage of the space-filling curve for the construction of data structures which are, in combination with the processing order of cells along the Peano curve, "cache-oblivious" in the sense of [Pro99, FLPR99, Dem02] and allow fast parallel multilevel schemes on adaptive grids without destroying data locality.

In section 2, we will shortly describe the basic numerical algorithms used in our codes. Section 3 displays the connection between space-trees and space-filling curves

---

[1] see [Sag94] for an overview on space-filling curves

and shows properties of the Peano-curve wich are essential for the mapping of grid data to our memory model (section 4). Finally, we present two application examples, the parallel solution of the Poisson equation on a complicated three-dimensional domain and the solution of the two-dimensional Navier-Stokes eqations, in section 5.

2 Numerical Algorithms

In this section, we will not discuss details of concrete discretizations (see section 5 for the discretization methods for two examples). Instead, we concentrate on basic algorithmic details which are the prerequisites of our method. In fact, the memory model desribed in section 4 can be applied to any discretization that can be described in a cell-oriented way (see 2.1). As a solver for the resulting systems of linear equations, we use an additive multigrid method working on a generating system (see 2.2).

2.1 Cell-Oriented Operator Evaluation

In terms of cache-efficiency, the common point-wise operator evaluation is not an optimal choice as we have to access data of several neighbouring points to evaluate a discrete operator at a certain point in our grid. Thus, we switch to a cell-oriented way to process the grid and to evaluate the occuring operators, that is, in each cell, we access only data owned to this cell. We assume that the unknowns of the PDEs are associated to the cells vertices[2]. In each cell, we evaluate a cell-part of the operators associated to all four vertices[3]. The complete operator value is accumulated from the parts of all four (in $2d$) or eight (in $3d$) cells involved.

For adaptively refined grids, this method can be carried over very naturally. We always evaluate the cell-parts of the operators the same way for all cells (up to and eventual scaling with the mesh width). An explicit computation of special operators at local refinement boundaries is needless. We only have to ensure that vertices of cells which do not correspond to degrees of freedom like for example hanging nodes are correctly interpolated and that, corresponding to this interpolation, fine grid partial operators are transported to the coarser vertices by a suitable restriction at the boundary to a locally coarser grid.

To make this cell-oriented algorithm work reasonably, we of course have to find suitable data structures which can be handled efficiently in terms of cache-hierarchies and avoid the separate storage of a data point in each cell involved. The solution to this problem is based on the usage of the Peano-curve as an odering mechanism for our grid cells (see section 3) and will be proposed in section 4.

For more complicated PDEs, in particular systems of PDEs like for example the Stokes or Navier-Stokes equations we get additional unknowns associated to the cells midpint. In this case, to maintain the cell-oriented operator evaluation, some

---

[2] see the remark below for an extension to data at the cells midpoint
[3] This method is well-known for finite element methods (see [Bra01]) but can be generalized also to finite difference or finite volume discretizations. In a finite element context, the cell-part of the operator corresponds to the integrals over the actual cell

operators might need an evaluation in two stages, whereas in the first stage the values are transported to the vertices of the cell (with suitable factors depending on the desired operator) and in the second stage these values are collected together to compute the finite value of the operator at the cells midpoint again. See 5 for an example.

2.2 Additive Multigrid

For our multigrid method, we work on a hierarchical generating system. One multigrid cycle corresponds to one top-down depth-first sweep over the space-tree associated to our adaptive grid.

Before and after each multigrid cycle, data are distributed over the generating system in an arbitrary manner. Thus, we first have to assemble the nodal values of the unknown variables on the way 'down' in the tree[4]. On the finest level, we evaluate the respective operator(s) cell-wise as described above and smoothen the solution on the finest grid. In the bottom-up part of the grid-traversal, we restrict the fine-grid residuals to the vertices of the coarser level and, again, apply the resulting residual to the unknowns situated at this level. We end up with a standard additive multigrid v-cycle with a Jacobi smoother.

3 Space-Trees and Space-Filling Curves

3.1 Space-Filling Curves as a Tool for Ordering Grid Cells

In the context of numerical simulations on space-tree grids, space-filling curves are nowaday widely used as a tool for a cheap and – with respect to communication costs – quasi-optimal tool for a load balanced partitioning of the grid [Zum00,GZ99,GZ02]. In addition, the ordering of grid cells along a space-filling curve leads – due to the good locality properties of the curve – to an increase in cache-efficiency [ABA00]. All this works for any connected space-filling curve [Zum00].

For reasons we will see at the end of this section, we will concentrate on the Peano-curve. The Peano-curve is a recursively defined, self-similar space-filling curve with a rectangular decomposition of the domain[5] and, thus, defined as the limit of a refinement process which starts with a generating template and proceeds by repeated decomposition of the domain and (mirrored and/or rotated) application of the template in each subdomain. As we look for an ordering mechanism of the cells of a grid with finite resolution, we are – in fact – not interested in the Peano-curve itself but in its iterates, the so-called discrete curves which are associated to the actual grid. Figure 1 shows the template, two subsequent iterates and the discrete curve associated to adaptive grids in two dimensions.

According to the recursive top-down depth-first definition of the Peano-curve, we get an ordering mechanism not only for the cells of the (locally) finest grid level but also for the cells on all levels, which will be needed for our multilevel algorithm. Figure 2 shows the numbering of cells for a two-dimensional example.

---

[4] in contrast to the common notation for v-cycles, 'down' denotes the way from the coarsest (root of the space-tree) to the finest (leaves of the space-tree) grid cells

[5] Another prominent representative of this curve is the Hilbert-curve. For more informations on space-filling curves and their definition see [Sag94].

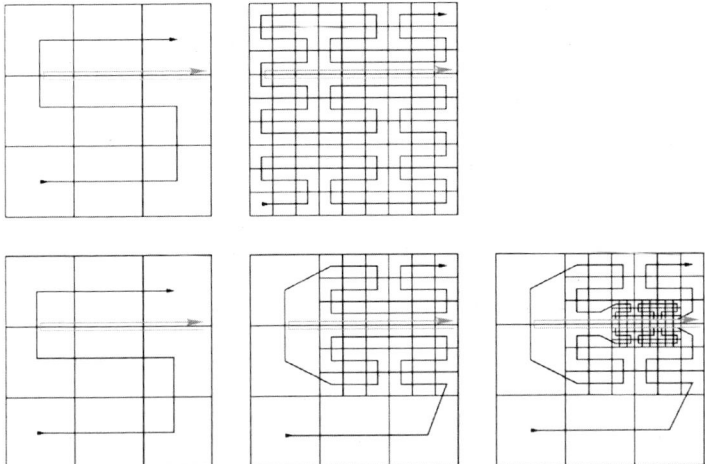

Fig. 1. Template and first iterates of the two-dimensional Peano-curve for a regular (upper row) and an adaptive (lower row) refinement. The grey arrows mark a line of grid points which is processed linearly forward and, afterwards, backward.

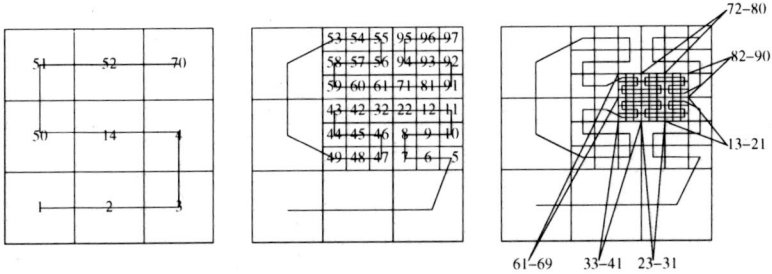

Fig. 2. Example for the top-down, depth-first numbering of cells in the adaptive two-dimensional grid from figure 1

3.2 Particular Properties of the Peano-Curve

In contrast to other curves like for example the Hilbert-curve, the definition of the Peano-curve is recursive not only with respect to the refinement depth but also with respect to the dimension. The following algorithm is a schematic demonstration of the numbering of cells on different grid levels according to the Peano-curve. In particular, the parameter lists are incomplete.

```
peano(dimension) {
if(dimension>0) {
  peano(dimension-1);
  x_dimension += direction(dimension); invert_direction(dimension-1);
```

```
    peano(dimension-1);
    x_dimension += direction(dimension); invert_direction(dimension-1);
    peano(dimension-1);
  }
  else {
    give_actual_cell_a_number();
    if(further_refinement) {
      peano(dimension_max,direction);
      invert_all_directions_above_1();
    }
  }
}
```

This 'double' recursivity of the Peano-curve in dimension and in refinement depth leads to two conclusions:

First, a grid line (plane or hyperplane), which is not intersected by the curve when created on a certain refinemement level, will not be intersected by further refinements of the curve as well. This is a simple consequence of the fact that refinements of the curve are strictly local, that is, each (local) refinement of the curve creates a new curve *within* existing cells of the actual refinement level. This holds for every recursively defined space-filling curve, thus also for the Hilbert-curve.

Second, if we look at the order in which grid points along such a line (plane, hyperplane) are visited, we get an inversion of order once we pass this line (plane, hyperplane) on our way along the Peano-curve. This nice behaviour can easily be concluded from the inversion of directions in the dimension-recursive definition of the Peano-curve. In the three-dimensional and higher-dimensional case, we could not find any Hilbert-curve for which such an inversion of processing order along lower dimensional elements within the domain holds. There are good reasons to assume that there is even no such three-dimensional Hilbert-curve.

4 Memory Model

The property of the Peano-curve that grid points on certain hyperplanes, planes, and lines are visited in alternating directions if we process the cells along the curve is now used to construct cache-efficient data.structures to store data asociated to the vertices of grid cells. In [GKLMPZ04, GMPZ04, GMPZ, Gue04, Poe04] we could show that we can group togehter some of the hyperplanes/planes/lines mentioned above and achieve eight (in $2D$) or 26 (in $3D$) data sets wich are processed linearly forward and backward. We can map these data sets and their access during the run of one iteration over all cells to the following memory model:

The memory can be considered as the band of a Turing machine, but, instead of one read/write head we use a fixed small number of r/w heads. For each of the data sets mentioned above, we need one such r/w head. As we know that grid data on these geometrical objects are visited in strict linear order, it is sufficient to allow only read and write operations during our 'stay' in one cell and movements of the positions of the r/w heads by one position to the right or to the left to execute the switch to the next cell. Each entity consisting of the 'stay' in a cell and the switch to the next cell is interpreted as on cycle of the Turing machine.[6] As an alternative model we can use a fixed number of stacks where in every step from one cell to another at most one push- or pop-operation per stack is executed.

The fact that the access to memory can change in one step only by one memory unit (spatial locality) already leads to a good cache-performance. This performance is further improved by the fact that most of the r/w heads change directions frequently due to the locality of the Peano-curve. This so called locality in time further reduces the probability that the memory units are overwritten before being used the last time. Our strategy does not need any information about the parameters of the cache hierarchy[7] and it works also in the case that data have to be stored on disk [Poe04]. Adress calculations for the access of grid points are totally avoided. Such, the organizational overhead for managing data structures is very small.

5 Numerical Examples

The work described in this paper is still in progress and the code is optimized only with respect to cache-performance. But with respect to this aspect, it is nearly optimal as can be seen from the comparison of a lower bound for the number of cache misses with the number of misses measured in a simulation run. The lower bound can be computed very easily: Each data unit needed during the run of our computation has to be loaded from the main memory to the cache at least once. For all computed examples, the number of cache misses in our program is only about 12% bigger than this lower bound. So a further improvement can be only marginal.

5.1 Parallel Solution of the Poisson Equation on a Star-Shaped Domain

The parallelization of the algorithm follows the approach described in [Zum01] with the additional feature that the multilevel structure is also reflected in the structure of the subprocesses communicating with one another.

For our particular algorithm, the difficulty is to preserve all properties of the data structures discribed in the previous section und, thus, to carry over the high cache-efficiency of the sequential code to the parallel version. A cut of the sequence of cells at a certain position leads to vertex data belonging to several processes at boundaries between several partitions. Thus, the evaluation of operators at these vertices is distributed over several process and the assembling of the complete operator values has to be done by inter-process communication.

To be able to start the traversion of the grid along the Peano-curve at any point in the domain using analogue data structures as in the sequential case, we provide a grid covering the whole domain for each process but staying as coarse as possible within these parts of the domain, which do not belong to the process. In some of the

---

[6] In some cases we have to read two (or more) data points with one r/w head if we proceed to the next cell. Thus, we need two (or more) cycles of the Turing machine for the switch to the next cell.

[7] Algorithms wich are cache-aware by concept without detailed knowledge of the cache parameters are also called cache-oblivious [Pro99, FLPR99, Dem02]

according coarse grid nodes, we store extra information about the exact location of the inter-process boundaries at the next finer level. As the amount of such coarse grid data is very small in comparison to fine grid data, the overhead is very small.

With this technique, each process computes its parts of coarse grid operators/residual automatically. Thus, also the implementation of a parallel multigrid algorithm becomes very natural. The coarse grid residuals are assembled by inter-process communication completely analogue to the fine grid residuals at inter-process boundaries. We do not need any conceptual extensions like for example a master process administrating coarse grid data.

As a first example, we solved the three-dimensional Poisson

$$\Delta u((\mathbf{x})) = \sin(\pi x_1)\sin(\pi x_2)\sin(\pi x_3) \qquad (1)$$

on a star-shaped domain (see figure 3 for a picture of the geometry). The cache-hit-rates carry over from the sequential codes (see [Gue04,Poe04,GKLMPZ04,GMPZ04, GMPZ]) and stay above 99.5%. Figure 3 shows iteration and communication times as well as the parallel efficiency for an a priori adaptively refined grid with 6.7 million grid cells. The computations were performed on a dual Intel XEON 2.4 GHz with 4 GB of RAM. The implementation of the communication routines is still in progress and by far not optimized yet. Thus, these results will be improved essentially in the near future. The computing times per iteration scale very well with growing number of processes. Thus, the overhead due to the storage of extra coarse grid cells could be proven to be really almost neglectable.

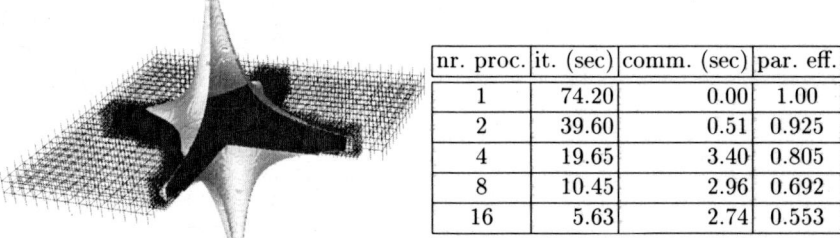

nr. proc.	it. (sec)	comm. (sec)	par. eff.
1	74.20	0.00	1.00
2	39.60	0.51	0.925
4	19.65	3.40	0.805
8	10.45	2.96	0.692
16	5.63	2.74	0.553

Fig. 3. Three-dimensional star-shaped domain with a cutout of an adaptive grid and parallel runtimes and efficiency for the Poisson equation on a star-shaped domain solved on dual Intel XEON 2.4 GHz with 4 GB of RAM

5.2 Two-Dimensional Navier-Stokes Equations

To show that our approach is not restricted to simple scalar problems we studied the solution of the time dependant incompressible Navier-Stokes equations. The equations are given by

$$\frac{\partial}{\partial t}\mathbf{u} + (\mathbf{u}\cdot\nabla)\mathbf{u} - \frac{1}{Re}\Delta\mathbf{u} + \nabla p = 0, \qquad (2)$$

$$\mathbf{div}\,\mathbf{u} = 0. \qquad (3)$$

For time discretization we used a combination of the standard Chorin projection method and explicit Forward-Euler time-steps [GDN95, FP99]:

$$\mathbf{u}^{(t+\tau)} = \mathbf{u}^{(t)} + \tau \left(\frac{1}{Re} \Delta \mathbf{u}^{(t)} - (\mathbf{u}^{(t)} \cdot \nabla)\mathbf{u}^{(t)} - \nabla p \right)$$

In this approach, a Poisson equation for the pressure follows from the side condition $\mathbf{div}\,\mathbf{u}^{(t+\tau)} = 0$ and, thus, has to be solved for every Cauchy-step:

$$\mathbf{div}\,\nabla p = \mathbf{div}\frac{1}{\tau}\mathbf{u}^{(t)} + \mathbf{div}\left(-(\mathbf{u}^{(t)}\nabla)\mathbf{u}^{(t)} + \frac{1}{Re}\Delta\mathbf{u}^{(t)}\right) \tag{4}$$

For the discretization in space we use a partially staggered grid with velocities and pressure gradients associated to the vertices of the grid cells and the pressure values assigned to the cells midpoints. To apply our memory model to such a staggered grid, we only have to enhance the size of one memory unit of the Turing machine to be able to store additional values like the second velocity component and the pressure gradients. As the pressure values are visited in a strict linear order along the Peano curve, it is sufficient to add one additional r/w head to store and read the corresponding degrees of freedom.

Our discretization leads to an automatic pointwise fulfillment of the continuity equation if only the discrete continuity equation is fulfilled. To achieve this property, we use specialized Finite Elements preserving continuity pointwise [Bla04]. By this approach standard FEM theory can be applied instead of the mixed element theory. Moreover, discrete energy and momentum conservation is guaranteed automatically [VV01, Bla04]. The so-called checkerboard instability can be avoided easily by appropriately chosen boundary conditions. To solve the Poisson equation for the pressure, we use an additive multigrid scheme.

As a first example, we implemented the driven-cavity-scenario with Reynolds numbers from $Re \leq 1$ up to $Re = 2800$ showing that the approach works efficiently and, in particular, cache-efficiency is preserved. Detailed results shall be reported in a separate paper.

6 Conclusion

As we have seen in the previous sections, we could show that our approach to construct cache-optimal data structures for solving PDEs based on space-tree grids and the Peano-curve leads to a very high cache-efficiency as well as an easy, balanced, and cheap parallelization. Our work on more sophisticated applications like flows in complicated and/or changing geometries, n-dimensional PDEs for financial pricing methods, PDEs with non-constant coefficients, and many others is by far not finished yet, but as far as we know now, there's no general obstacle preventing us from carrying over the high efficiency of our method to these applications, too. In addition, we will enhance some numerical aspects like the accuracy of our discretization at the boundary in the near future. Detailed results achieved by enhanced versions of our program will be published soon.

References

[ABA00] Aftosmis, M.J., Berger, M.J., Adomavivius, G.: A Parallel Multilevel Method for adaptively Refined Cartesian Grids with Embedded Boundaries. AIAA Paper, 2000.

[BBFM02] Bader, M., Bungartz, H.-J., Frank, A., Mundani, R.-P.: Space tree structures for PDE software. In: P.M.A. Sloot, et. al. (editors). Computational Science - ICCS 2002, International Conference, Amsterdam. *Proceedings, Part III*, Lecture Notes in Computer Science 2331, pages 662-671, Springer, 2002.

[Bla04] Blanke, C.: Kontinuitätserhaltende Finite-Elemente-Diskretisierung der Navier-Stokes-Gleichungen. Diplomarbeit, Institut für Informatik, TU München, 2004.

[Bra01] Braess: Finite Elements. Theory, Fast Solvers and Applications in Solid Mechanics. Cambridge University Press, 2001.

[CS00] Chatterjee, S., Sen, S.: Chache-Efficient Matrix Transposition. In: Proceedings of HPCA-6, pages 195-205, Toulouse, France, January 2000.

[CLPT99] Chatterjee, S., Lebeck, A.R., Patnala, P.K., Thottethodi, M.: Recursive array layouts and fast parallel matrix multiplication. In: Proceedings of Eleventh Annual ACM Symposium on Parallel Algorithms and Architectures, pages 222-231, Saint-Malo, France, 1999.

[Cla96] W. Clarke. Key-based parallel adaptive refinement for FEM. Bachelor thesis, Australian National Univ., Dept. of Engineering, 1996.

[Dem02] Demaine, E.D.: Cache-Oblivious Algorithms and Data Structures. In: Lecture Notes from the EEF Summer School on Massive Data Sets, Lecture Notes in Computer Science, BRICS, University of Aarhus, Denmark, June 27-July 1, 2002, to appear.

[FP99] Ferziger, J.H., Peri"c, M.: Computational Methods for Fluid Dynamics. Springer-Verlag, Berlin Heidelberg, 1999.

[Fra00] Frank, A.: Organisationsprinzipien zur Integration von geometrischer Modellierung, numerischer Simulation und Visualisierung. Doctoral thesis, Institut für Informatik, TU München, 2000.

[FLPR99] Frigo, M., Leierson, C.E., Prokop, H., Ramchandran, S.: Cach-oblivious algorithms. In: Proceedings of the 40th Annual Sympoisium on Foundations of Computer Science, pages 285-297, New York, October 1999.

[GR86] Girault, V., Raviart, P.: Finite Element Methods for Navier-Stokes Equations. Springer-Verlag, Berlin Heidelberg, 1986.

[GSE98] Gresho, P.M., Sani, R.L., Engelman, M.S.: Incompressible Flow and the Finite Element Method - Advection-Diffusion and Isothermal Laminar Flow. John Wiley & Sons Ltd, Chichester, 1998.

[GDN95] Griebel, M., Dornseifer, T., Neunhoeffer, T.: Numerische Simulation in der Str"omungsmechanik. Vieweg, 1995.

[Gri93] Griebel, M.: Multilevelverfahren als Iterationsmethoden über Erzeugendensystemen. Habilitationsschrift, TU München, 1993.

[GKZC04] Griebel, M., Knapek, S., Zumbusch, G., Caglar, A.: Numerische Simulation in der Moleküldynamik. Numerik, Algorithmen, Parallelisierung, Anwendungen, Springer, Berlin, Heidelberg, 2004.

[GZ99] Griebel, M., Zumbusch, G.W.: Parallel multigrid in an adaptive PDE solver based on hashing and space-filling curves. Parallel Computing, **25**, 827-843, 1999.

[GZ02] Griebel, M., Zumbusch, G.W.: Hash based adaptive parallel multilevel methods with space-filling curves. In: H. Rollnik and D. Wolf, editors, NIC Series, **9**, 479-492, Germany, 2002. Forschungszentrum Jülich.

[Gue04] Günther, F.: Eine cache-optimale Implementierung der Finite-Elemente-Methode. Doctoral thesis, Institut für Informatik, TU München, 2004.

[GKLMPZ04] Günther, F., Krahnke, A., Langlotz, M., Mehl, M., Pögl, M., Zenger, Ch.: On the Parallelization of a Cache-Optimal Iterative Solver for PDEs Based on Hierarchical Data Structures and Space-Filling Curves. Conference Proceedings EuroPVMMPI 2004, Budapest, September 2004, LNCS, Springer, to appear.

[GMPZ04] Günther, F., Mehl, M., Pögl, M., Zenger, Ch.: A cache-aware algorithm for PDEs on hierarchical data structures. Conference Proceedings PARA '04, Kopenhagen, June 2004, LNCS, Springer, submitted.

[GMPZ] Günther, F., Mehl, M., Pögl, M., Zenger, Ch.: A cache-aware algorithm for PDEs on hierarchical data structures based on space-filling curves. SIAM Journal on Scientific Computing, in review.

[OPF94] Oden, J.T., Para, A., Feng, Y.: Domain decomposition for adaptive hp finite element methods. In: D.E. Keyes and J. Xu, editors, Domain decomposition methods in scientific and engineering computing, proceedings of the 7th int. conf. on domain decomposition, vol. 180 of Contemp. Math., pages 203-214, 1994, Pennsylvania State Universitiy.

[PLL99] Patra, A.K., Long, J., Laszloff, A.: Efficient Parallel Adaptive Finite Element Methods Using Self-Scheduling Data and Computations. HiPC, pages 359-363, 1999.

[Poe04] Pögl, M.: Entwicklung eines cache-optimalen 3D Finite-Element-Verfahrens für große Probleme. Doctoral thesis, Institut für Informatik, TU München, 2004.

[Pro99] Prokop, H.: Cache-Oblivious Algorithms. Master Thesis, Massachusetts Institute of Technology, 1999.

[RKCC98] S. Roberts, S. Klyanasundaram, M. Cardew-Hall, and W. Clarke. A key based parallel adaptive refinement technique for finite element methods. In: B.J. Noye, M.D. Teubner, and A.W. Gill, editors, Proc. Computational Techniques and Apüplications: CTAC '97, pages 577-584, World Scientific, Singapore, 1998.

[Sag94] Sagan, H.: Space-Filling Curves. Springer-Verlag, New York, 1994.

[SLP83] Stevens, R.J., Lehar, A.F., Preston, F.H.: Manipulation and Presentation of Multidimensional Image Data Using the Peano Scan. IEEE Trans. Pattern An. and Machine Intelligence, Vol PAMI-5, pages 520-526, 1983.

[VMG91] Velho, L., de Miranda Gomes, J.: Digital Halftoning with Space-Filling Curves. Computer Graphics, **25**, 81-90, 1991.

[VV01] Verstappen, R.W.C.P., Veldman, A.E.P.: Symmetry-Preserving Discretization of Turbulent Channel Flow. In: M. Breuer, F.Durst, C.Zenger (editors): High Performance Scientific and Engineering Computing, Springer-Verlag, Berlin Heidelberg, 2001.

[Zum01] Zumbusch, G.W.: Adaptive Parallel Multilevel Methods for Partial Differential Equations. Habilitationsschrift, Universität Bonn, 2001.

[Zum00] Zumbusch, G.W.: On the quality of space-filling curve induced partitions. Z. Angew. Math. Mech., **81**, 25-28, 2001. Suppl. 1, also as report SFB 256, University Bonn, no. 674, 2000.

VISimLab: Optimizing an Interactive CFD Simulation on a Supercomputer for Computational Steering in a Virtual Reality Environment

Petra Wenisch[1], Oliver Wenisch[2], and Ernst Rank[1]

[1] Lehrstuhl für Bauinformatik, Arcisstraße 21, 80290 München
{*wenisch,rank*}*@bv.tum.de*
[2] Leibniz-Rechenzentrum, Barer Straße 21, 80333 München
wenisch@lrz.de

Abstract

This article presents a computational steering research project coupling a supercomputer with a virtual reality (VR) environment to allow for interaction during a real-time CFD simulation using an immersive high-end visualization interface. Interaction comprises not only the changing of parameters, but also the modification of geometry, e.g. removing, adding and transforming objects in a virtual CAD-generated room during runtime.
The underlying CFD computation and grid generation is processed on a high-performance computer (HPC) to enable a real-time simulation which instantly reacts to user manipulations. Based on the Lattice-Boltzmann method, the simulation kernel shows good parallel efficiency on the Hitachi SR8000 pseudo-vector supercomputer at the Leibniz Computing Center (LRZ) in Munich. For post-processing and steering in virtual reality a stereoscopic projection screen and a tracked wand input device is used. To achieve optimal immerson a head-tracked view is supported and complemented by a context-sensitive 3D VR menu.

1 Introduction

Numerical CFD simulations are gaining increasing importance as a valuable supplement of classical wind tunnel experiments. However, the accurate computation of a fluid flow scenario is still a time-consuming process and requires resources of powerful clusters or supercomputers to keep the computation time reasonably short. Since the planning phase of a building is characterized by an iterative process, it is of considerable interest to be able to perform several case studies within an acceptable period of time. The aim is therefore to devise an interactive application to achieve a preliminary qualitative estimation of the flow field quickly and conveniently and also to conserve HPC resources.

With regard to post-processing issues, a VR environment represents an efficient tool for analyzing the huge amount of time-variant data that is generated while running a CFD application. 3D visualization provides a combined graphical representation of the CAD-based flow geometry including textured surfaces together with mapped simulation results. In addition, the immersive steering and post-processing facilitates the handling of large, multi-dimensional datasets and the control of simulation parameters. This may have a beneficial effect especially during the planning phase of a building, when engineers, architects, property owners and a large number or further participants of different disciplines have to collaborate.

The first part of this article will introduce the computational steering concept of the interactive CFD simulation and the underlying communication scheme. Next, the topics of grid generation and optimization on the simulation side are dealt with. The means of optimization and speed-up to provide interactivity are discussed in detail. Finally, the newly developed visualization framework and its integration into a virtual reality environment are presented.

2 Computational Steering

Typically, large and compute-intensive CFD simulations are run non-interactively as batch processes on queuing systems of high-performance computers. Following the pre-processing step, i.e. after mapping the CAD data to a computational grid and defining the boundary conditions, a file or a database describing this setup information is generated and submitted to a batch queue. As soon as the required resources (a sufficient number of free processors and amount of memory) are available, computation starts and saves its output to disk. The user evaluates the simulation results during the subsequent post-processing step. These steps are usually carried out on different hardware architectures, e.g. desktop PC, supercomputer and graphics workstation.

This kind of workflow is expedient for precise problems when one is interested in obtaining detailed information in a fluid flow investigation. For the wide variety and the explorative environment needed in case studies, the user would yet want to interact with a running simulation and to visualize the current results immediately. This is the basic idea behind computational steering [BCHJPPP03, MWL99]. It requires the integration of the above-mentioned steps, viz. pre-processing, computation and post-processing, into a single environment.

When a user is interacting with a steering terminal, this event is forwarded to the computation kernel (cf. Figure 1). Of course, the user expects to see the corresponding simulation output, preferably without delay. Thus, a computational steering application has to provide short-latency responses from the underlying simulation process. To fulfill this requirement, a fast CFD solver running on a supercomputer or cluster has to be coupled to a steering and visualization workstation by an efficient communication concept. Furthermore, the computation kernel has to allow for a fast grid modification during the running simulation, and both steering and visualization should be available to the user within a single environment. These aspects will be described below.

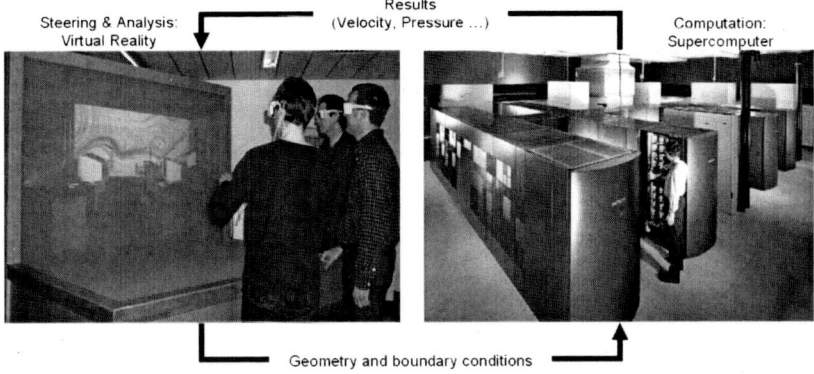

Fig. 1. During an interactive simulation run modifications of geometry or boundary conditions are sent to the high-performance computer. The simulation kernel immediately incorporates the new configuration, computes the resulting fluid flow updates and transmits the flow data to the VR system for visualization.

3 Coupling a Supercomputer with a Virtual Reality Environment

In order to enable low-latency responses, it is important to connect the components of a computational steering application as efficiently as possible [WTR04]. Figure 2 shows the different components of the application and the corresponding data flows. Since the computation has to run on a supercomputer which is normally inadequate for visualization, the steering and post-processing interface (VIS) is run as a separate application on a suitably equipped graphics workstation. A communication node (SIM-M) on the supercomputer exchanges data between simulation and steering application. This node collects the results of all the computation processes and forwards them in a single message to the visualization workstation to eliminate multiple-message latencies. Driven by user interaction, the steering application has to send a corresponding message (e.g. modified boundary conditions, new geometry descriptions or flow parameters) to the simulation master, which controls distribution and assignment of this information to the slaves.

As the network between the supercomputer and the graphics workstation is not necessarily a high speed connection (and may be in concurrent use by several supercomputer users), the main advantage of this node is to avoid deficits due to network latency. It also permits a continual computation of the slaves with minimized interruption by communication. The compute slaves (SIM-S) calculate the fluid flow and need to communicate with their simulation master (SIM-M) only.

With this communication concept, three different types of process have to exchange data within a heterogeneous hardware architecture setup (e.g. Hitachi SR8000 and an SGI Onyx2 at the LRZ). Thus, an MPI implementation supporting cross-architectural communication between the different platforms is required. In the application presented here, the GLOBUS-enabled MPICH-G2 library is used for the communication between simulation and visualization processes, hence the cutting-

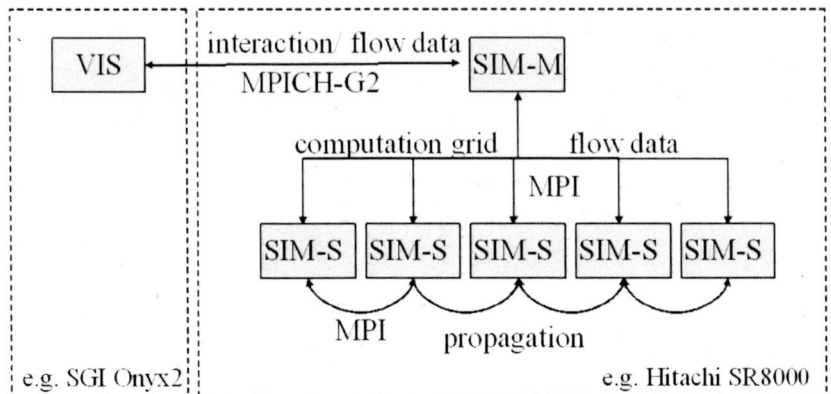

Fig. 2. Application scheme showing the visualization (VIS) and the simulation, with its communication node (SIM-M) and the compute slaves (SIM-S). In the setup at the LRZ the VIS process runs on a separate graphics workstation (SGI Onyx2). The simulation SIM uses several nodes on the Hitachi SR8000, one of which is dedicated as the simulation master SIM-M. Within the SR8000 all processes communicate via a vendor-optimized version of MPI. Cross-machine communication between VIS and SIM-M is implemented with a Grid-enabled flavor of MPICH.

edge capabilities of Grid-Computing are readily available. Furthermore, by relying on this standardized MPI interface, it is possible to use various steering and visualization front ends. For communication within the supercomputer a vendor-optimized version of MPI is used to achieve the most efficient data exchange between the compute slaves and their master node.

Due to the different structure of the processes, a non-clocked communication is more efficient; only between slaves does data have to be exchanged after each individual time step of flow computation. Without user interaction the slaves send their results to the communication node from where they are forwarded to the visualization client in regular, user-adjustable intervals. All the remaining communication activity is completely event-driven. As soon as the user interacts, e.g. by modifying geometry or changing boundary conditions, this information is sent to the supercomputer. The master node, which is polling for new input, notices any in-coming message and evaluates it. Based on the new information, the parameters and/or the computational grid are updated (see also 4) and, according to the fluid domain of each slave, sent to the computation itself. The slaves also recognize any new data that arrives and update their computation accordingly. To give the user a direct response to his interaction, the slaves send their new results to the visualization station/location after just a few time steps and subsequently continue their normal communication cycle with the user-defined, regular forwarding intervals. This communication concept supports different hardware architectures, does not interrupt the computation process more than necessary and sends messages only as required.

4 Grid Generation

To avoid unnecessary feedback delay in a computational steering project, the application must be able to quickly transfer the user's modifications to the computation kernel. With regard to fluid-flow simulation in CAD-generated virtual rooms, the most time-consuming/labor-intensive tasks are the geometric modification of obstacles or boundary conditions of the fluid domain. Therefore, special focus has been put on the development of a fast grid generator [WW04]. Lattice-Boltzmann solvers generally use finite-difference discretization schemes on Cartesian grids. The mapping of CAD-generated geometry onto a discrete voxel model can be done automatically. This voxel representation is created using a space-partitioning (i.e. octree-based) algorithm. To begin with, the algorithm starts with a facetted approximation of the CAD object imported from an STL file (stereo lithography file).

The object's bounding cube is recursively divided into eight smaller cubes (shown in Figure 4) resulting in a hierarchical octree structure. At each level of refinement every cube is tested for intersection with the object's surface facets using a fast box-triangle intersection routine implemented by Tomas Akenine-Möller [Ake02]. To reduce the complexity of the algorithm with regard to the number of triangles (cf. Figure 5), only facets which already have been lying within the superordinate octant of the preceding subdivision level are tested for intersection again. The subdivision process is continued until the voxels describing the surface are identified and properties according to the boundary condition of the facet are attached. Therefore, the level of refinement depends on the resolution, i.e. the voxel size, of the Cartesian grid.

The voxelization is processed on the communicator node of the supercomputer. Thus, only small messages describing object transformation have to be sent to the supercomputer instead of sending the whole new grid following each interaction. New obstacle descriptions have to be sent only once and afterwards a single transforma-

Fig. 3. Volume discretization of a CAD-generated facet model needed by the CFD simulation: On the left an object approximated by 30014 triangles is shown. The discrete representation on the right results from an octree-based voxelization algorithm which also sets the boundary condition at the same time. (computation time on an AMD 2.0 GHz Opteron: 0.67 sec)

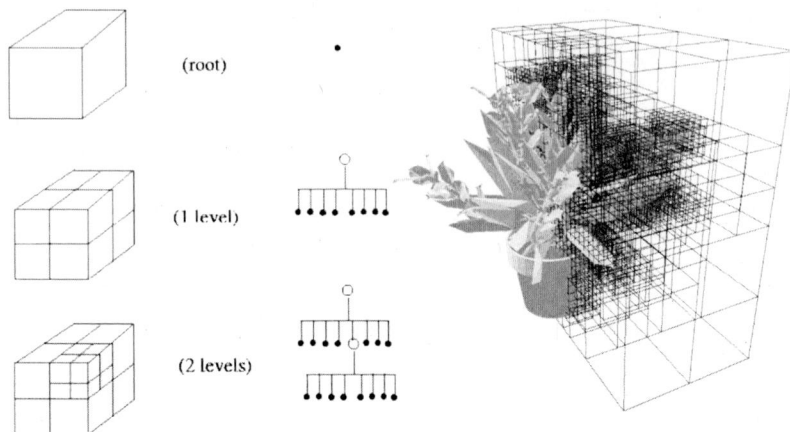

Fig. 4. Octree-based voxelization of an object can be computed by a recursive algorithm which starts from a 'root' octant, which is the bounding cube of a given object. In each recursion the current octant is split into eight new octants as long as intersections with object triangles can be found. This procedure is continued until the octant size corresponds to the voxel size of the computational grid. In the right part of the figure a illustration shows the path of voxel identification for a potted plant.

tion matrix is sufficient to describe the modifications of each object. This significant decrease in the amount of transferred data is especially advantageous with regard to the often less powerful network connection to the outside world of a supercomputer. Another aspect is to avoid the waste of powerful supercomputer resources, since the node would not be fully utilized by communication only (as its primary occupation is waiting for user input and forwarding/collecting data fragments from the slaves). Grid generation can be parallelized in a particularly straightforward manner on the Hitachi SR8000 with its eight processors per node, by letting each of the eight processors refine one of the eight octants after the first subdivision.

5 Optimizing the Lattice-Boltzmann Kernel

The Lattice-Boltzmann method (LBM) has emerged as a complementary technique for the computation of fluid flow phenomena [Kra01, Toe01]. Common numerical methods for the simulation of fluid dynamics are based on a discretization of the Navier-Stokes nonlinear, partial differential equations. LBM represents an alternative approach by starting with a discrete microscopic model combined with statistical physics, i.e. one utilizes dynamic particle densities for a discrete number of velocities and directions at each grid point. By construction, quantities such as mass and momentum are conserved to fulfill the hydrodynamic laws.

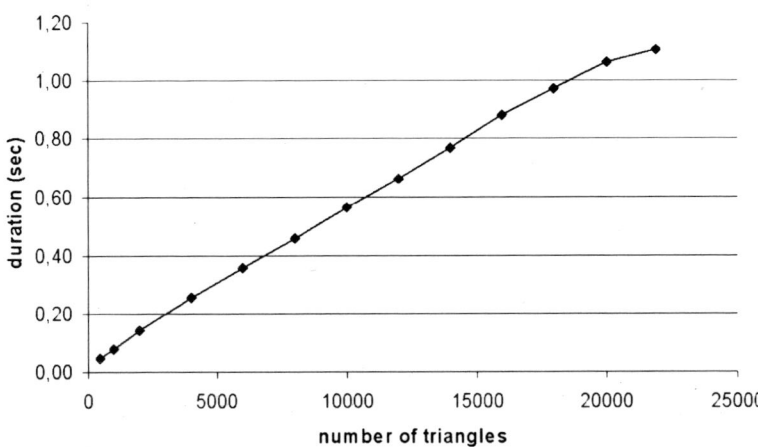

Fig. 5. For this graph, the shape of an object was approximated by means of a varying number of triangles while the corresponding number of voxels remained constant. The octree-based voxelization algorithm shows a time complexity of $O(n)$.

Basically, within each simulation time step the Lattice-Boltzmann algorithm computes the 'collision' of microscopic, 'virtual particles' and updates the velocity distribution functions at each grid point. Collision is represented by the evaluation of the new velocity distribution functions and luckily does not require data exchange with adjacent grid nodes. 'Propagation' is the migration of these distribution functions to the neighbors of a given cell. This approach to solving the Navier-Stokes equation has been shown to be particularly well-suited for taking advantage of the parallelization capabilities of the (2 TFlop/s) high-performance supercomputer Hitachi SR8000 at the LRZ in Munich [PDTRLWZ04].

Special care has to be taken to fully profit from the compute capabilities of the specialized hardware. It often turns out, that straightforward programs which would perform well on a standard workstation are not adequate for high performance computing. Special code optimization may yield dramatic improvements. Figure 6 illustrates the performance and scaling of the simulation kernel before and after the optimizations for the SR8000. The benchmarks were performed for a grid size of 50x50x50 nodes per slave. As the diagram shows, we succeeded in significantly improving the performance of the computation and were almost able to preserve the very good scaling behavior of the kernel.

The improvements were accomplished by specifically rewriting the main computation loops to enable the optimizing Hitachi compiler to use its vectorizing capabilities much more efficiently. As a first step, the array of particle densities was rearranged to benefit from data locality acceleration within the 'collision' step and the 'propagation' was adjusted accordingly. This had the positive side effect that the loops were able to run over continuous data fields, making it possible to fuse loops over the

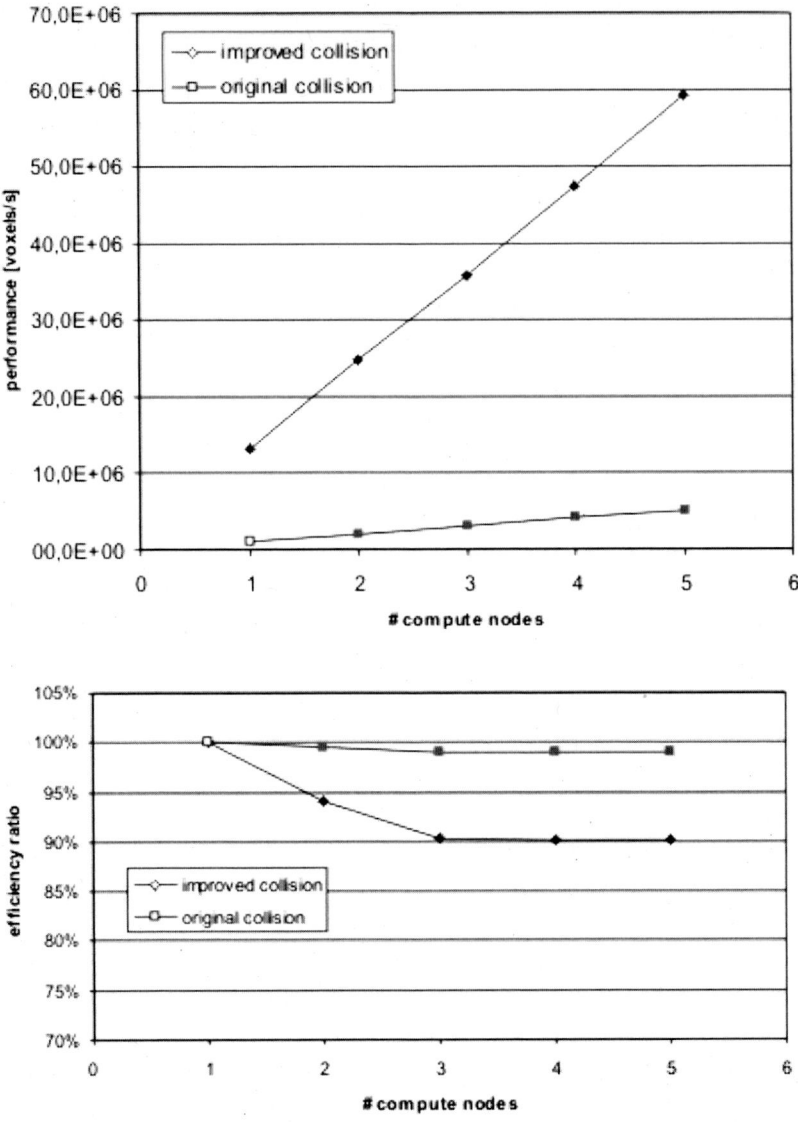

Fig. 6. Performance and scaling benchmark for the simulation before and after optimizing the kernel.

x-, y-, and z-dimension into a single one. Furthermore, conditional statements have been substituted by equivalent floating point operations, i.e. boolean expressions were mapped onto real-valued coefficient arrays for multiplication. Finally, pointer arithmetic has been eliminated. Hence, the new loop structure is much simpler and therefore easier to analyze for the compiler, which is now able to deploy fast vector instructions much more often.

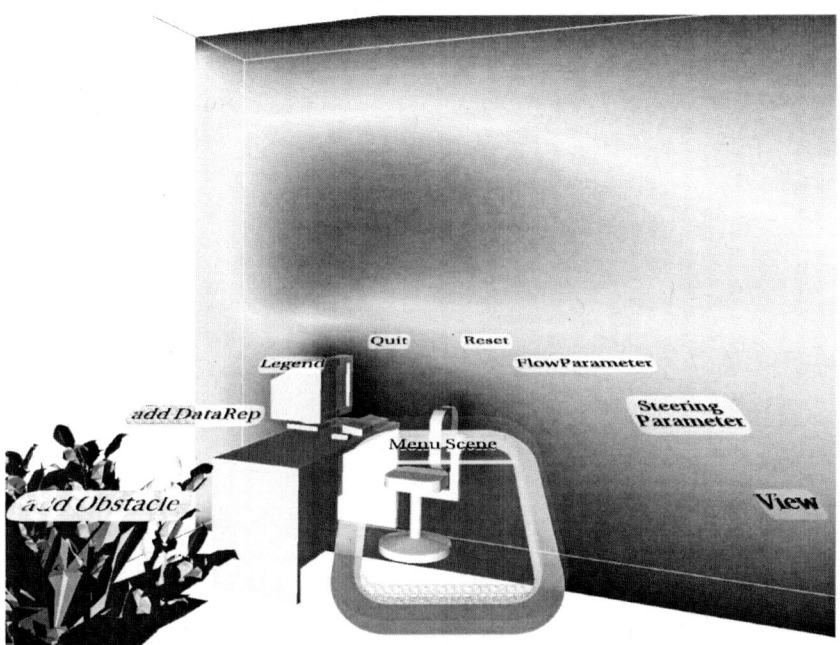

Fig. 7. If required, the immersive 3D user interface shown here can be displayed within the virtual scene. Controlling the application is simplified by a context sensitive menu structure. The menu is rendered transparent to preserve the view onto the simulated scene.

6 Visualization and Interaction in Virtual Reality

VR environments represent an efficient tool for analyzing the huge amount of time-varying data produced by a CFD simulation [BL92]. Investigating the three-dimensional structures in fluid flows is much more intuitive and illustrative.
TGS Open Inventor (www.tgs.com) is used as a scene-graph toolkit for generating the three-dimensional visualization. It is supplemented by VR-Juggler (www.vrjuggler.org) to support VR devices like head-tracked shutter glasses and the tracked wand input device. User interactions are accepted for manipulating object geometry and position, navigation and menu handling. Thanks to the support of different modes of navigation within the virtual world, i.e. a walk and fly mode, the user can truly immerse into the virtual room augmented by the fluid flow visualization. This emphasizes the need for a capable 3D menu to control and steer the

application. The menu itself is a dynamic 3D object and part of the scene graph (see Figure 7). It is designed as a view-fixed control panel for the user's convenience while he is exploring the current scene [Mar04]. To avoid constraining the view by the menu, it is only displayed on request. In addition, its layout and functionality adapt to the context from which it was invoked. For example, double-clicking an object causes an object manipulation menu to pop up with transformation options (rotation, scaling, cloning, removal, etc.). If, however, the menu is invoked by clicking on the background, it offers object import, post-processing features and global parameter controls. For post-processing, common tools such as iso-surfaces, stream lines, cutting planes and vector fields are provided.

7 Conclusion and Outlook

We have presented a computational steering environment (cf. Figure 8) for the simulation of fluid flows on the basis of a Lattice-Boltzmann simulation kernel. While the CFD kernel is processing continuously in the background, it sends simulation results to the visualization station/location at regular intervals. At the same time, the simulated scene can be modified interactively and the kernel will instantly include the parameter changes. Due to the efficient grid generation, the modification of complex geometries is well supported. Using the grid-enabled GLOBUS MPI provides simplified usage and improved portability due to standardized communication interfaces. The post-processing is done within the steering interface and allows an immersive exploration of the fluid data in a virtual world.

Further improvements will cover the support of turbulent convective flows and the evaluation of comfort parameters.

8 Acknowledgements

The authors would like to express their gratitude to Dr. Matthias Brehm (LRZ) for assisting during the optimization process, Irene Geiseler (also LRZ) for help with the complex scheduling of the interactive application. This project has received financial support from KONWIHR, which is gratefully acknowledged.

References

[Ake02] Akenine-Möller, T.: Fast 3D Triangle-Box Overlap Testing, Journal of Graphics Tools, Volume 6 Issue 1, Pages: 29–33 (2002)

[BCHJPPP03] Brooke, J. M., Coveney, P. V., Harting, J., Jha, S., Pickles, s. M., Pinning, R. L., Porter, A. R.: Computational Steering in RealityGrid, Proceedings of the UK e-Science All Hands Meeting 2003, published online http://www.nesc.ac.uk/events/ahm2003/AHMCD/pdf/179.pdf

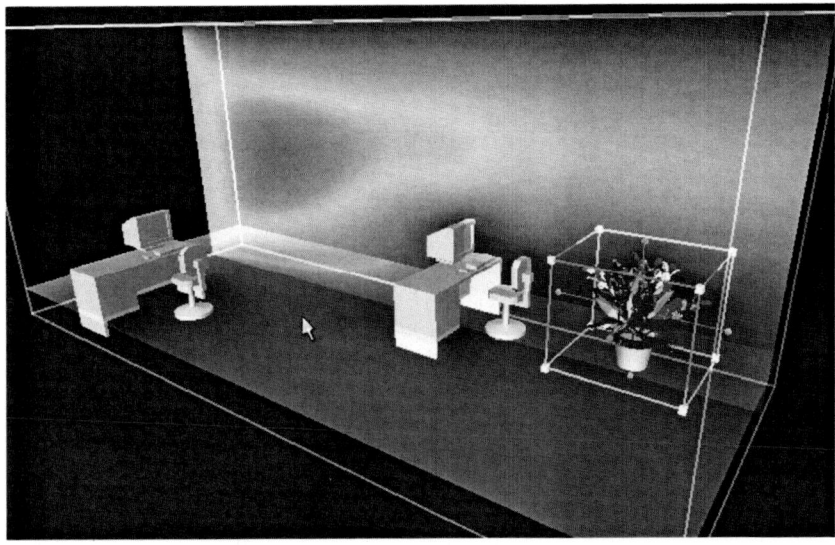

Fig. 8. This picture shows an interactive session of an air-flow analysis in an office room. The colored cutting planes depict the distribution of flow velocities (component alng x axis) in the fluid domain.

[BL92] Bryson, S., Levit, C.: The Virtual Windtunnel: An environment for the exploration of three-dimensional unsteady fluid flow, IEEE Computer graphics and Applications, **12(4)**,25–34 (1992)

[Kra01] Krafczyk, M.: Gitter-Boltzmann Methoden: Von der Theorie zur Anwendung, Professoral dissertation, LS Bauinformatik, TU München (2001)

[Mar04] Marcheix, L.: A 3D User Imterface for a Virtual Reality Environment, diploma thesis, LS Bauinformatik, TU München (2004)

[MWL99] Mulder, J. D., Wijk, J. van, Liere, R. van: A Survey of Computational Steering Environments, Future generation computer systems, 15(2), (1999)

[PDTRLWZ04] Pohl, T., Deserno, F., Thürey, N., Rüde, U., Lammers, P., Wellein, G., Zeiser, T.: Performance Evaluation of Parallel Large-Scale Lattice Boltzmann Applications on Three Supercomputer Architectures, submitted, Supercomputing 2004, Pittsburgh, USA (2004)

[Toe01] Tölke, J.: Gitter-Boltzmann-Verfahrenzur Simulation von Zweiphasenströmungen, Phd thesis, LS Bauinformatik, TU München (2001)

[WTR04] Wenisch, P., van Treeck, C., Rank, E.: Interactive Indoor Air Flow Analysis using High Performane Computing and Virtual Reality Techniques, Proceedings of RoomVent 2004, Coimbra, Portugal (2004)

[WW04] Wenisch, P., Wenisch, O.: Fast octree-based Voxelisation of 3D Brepobjects, techncal report, LS Bauinformatik, TU München (2004)

Part II

Computer Science and Mathematics

Christoph Zenger

Technische Universität München
Fakultät für Informatik
Boltzmannstr. 3
85748 Garching

Computer Science and Mathematics are disciplines which are basic for virtually every project in high performance computing. In this section papers are collected where this disciplines play a predominant role or where the topic of the project is part of theses disciplines. In this introduction we shortly characterize the papers in this section.

The cxHPC project (Hager, Zeiser, Heller) provides the toolkit ByGrid based on the Open Source software Globus. It offers a simplified user interface for the users of the computing centres in Munich (LRZ) and Erlangen (RRZE) and can be considered as a prototype of grid technology in a special environment. The tools are successfully used by researchers using both computer centres and switching frequently for specific tasks from one to the other.

The FPGA-project (Urbanek, May) can be considered as a case study for the transformation of algorithms written in a concurrent high level programming langage (like C in combination wth OpenMP) to a hardware implementation on an FPGA described in a hardware description language like VHDL. It is very important to support this process by software running on a high performance computer to make the process cheaper, faster and more reliable. Some tools like the programming language Handel-C are already available. The project investigated this process for a prototype system "Intelligent Camera" and reports the problems and difficulties and also the progress achieved so far.

Reflecting the fact that numerical simulation always needs complex visualization tools for the representation of the results the gridlib project aims at a closer integration of simulation and visualization. It provides its own grid structure which is well suited for multi level algorithms. It offers both online and offline visualization functionality and is based on object oriented programming methodology without neglecting performance issues.

As already reported in the previous volume of this series the efficient implementation of Computational chemistry codes on the SR8000 of the LRZ was considered as a difficult task and it took some time to settle this problem. The paper "The suitability of Contemporary Processors for Quantum Chemical Computations" (Palm) investigates in detail the suitability of some of the most important modern processor architectures for Computational Chemistry.

Domain decomposition is still the most important strategy for the parallel solution of PDEs. Mesh partitioning has to be done in such a way that a good load balance is achieved in connection with acceptable communication overhead. The project MethWerk (Gupta, Luksch, Schmidt) presents a two-level hybrid approach applied to the industrial CFD code trace.

Process optimization of processes is usually much more time consuming than the mere simulation of the underlying process because the process has to be simulated for many different choices of the parameters to be optimized. Optimization is therefore in many cases a good candidate for high performance computing. The project OPTILAS (Petzet et al.) investigates the optimization of Multi-Beam Laser Welding Techniques where the prevention of hot cracking is an important goal to ensure the quality of the product. A combination of analytical studies and numerical simulation yields practically useful solutions with reasonable effort.

Software design respecting modern object oriented software engineering concepts sometimes results in bad performance on modern computer architectures. The paper "Advanced PDE Software Design on the Hitachi SR 8000" (Freundl, Bergen, Hülsemann, Rüde) shows that the concept of so called "expression templates" can overcome this problem on modern architectures including the SR8000.

The project ParRichy (Kränke, Bause, Prechtel, Radu, Knabner) is a case study for the interplay of complicated chemical processes combined with material transport in porous media as it is typical for many contamination problems in environmental engineering. Progress in modelling and simulation of these phenomena could help to show the usefulness of high performance computing to a broader audience.

The Peridot project (Fürlinger, Gerndt) addresses an ongoing important issue for every user of high performance computers: The detection of performance bottlenecks. The emphasis in the paper is on the reduction of performance information gained from the performance monitors in such a way that the user gets hints for improving his program without being lost by the presentation of too many details where a useful interpretation is almost impossible.

cxHPC: Setting up ByGRID — First Steps Towards an e-Science Infrastructure in Bavaria

Georg Hager[1], Thomas Zeiser[1], and Helmut Heller[2]

[1] Regionales Rechenzentrum Erlangen (RRZE), Martensstr. 1, D-91058 Erlangen, Germany
{Georg.Hager,Thomas.Zeiser}@rrze.uni-erlangen.de
[2] Leibniz-Rechenzentrum der Bayerischen Akademie der Wissenschaften (LRZ), Barer Str. 21, D-80333 München, Germany
heller@lrz.de

Abstract. ByGRID is a Grid project based on the GLOBUS toolkit. Its purpose is to (i) demonstrate the functionality of Grid middleware, delivering to end users the benefits, but not the complexity of Grid-based workflow and to (ii) share, for the benefit of the users, unique resources between the two computing centers involved, RRZE and LRZ. We examine ByGRID from the computing center's view, reporting on setup, encountered problems, and administration, and from the user's point of view, explaining a practical usage scenario.

1 Introduction

The GLOBUS toolkit [1], being developed by the Globus Alliance, is an Open Source software that can be used for building grids. It contains libraries and utilities for resource allocation, security, monitoring and data movement within and across institutions. At a higher level, it is — like the related UNICORE software [5] — a tool for enhancing scientific workflow with numerous virtualization facilities that serve to decouple users from unnecessary details.

ByGRID builds on the GLOBUS toolkit. Users from RRZE and LRZ can mutually use compute resources in both centers through the use of Grid technology. Having started as a mere proof of concept, ByGRID is now applied by some users as part of their daily work. The following are the main goals which have been defined within the project framework:

1. Collect experience in running GLOBUS middleware in a production environment.
2. Lower the entry threshold to GLOBUS and supply users with appropriate documentation and guidance.
3. Build a testbed for resource balancing between LRZ and RRZE.
4. Establish a pilot scheme with a long-term vision of incorporating other HPC systems and sites in Bavaria.

2 Technical setup

One long-term goal of the project is to establish a testbed for balancing computer resource (i.e. cycle) usage between LRZ and RRZE. Because of the more or less standardized system environment (Linux) and the attractiveness of clusters for the majority of scientific users, the IA32 Linux clusters of both centers have been chosen for an initial setup. At the time of writing, LRZ has also other systems at its disposal, for instance the federal supercomputer (HLRB) SR8000, a sophisticated graphics workstation (SGI Holobench), and an IBM p690 on all of which GLOBUS is available as well.

A technical impediment was posed by the fact that the RRZE's IA32 compute cluster uses private IP addresses and can thus not be accessed directly from outside the university's campus network. In order to overcome this limitation, a special Linux server machine has been set up that has a valid, routable IP address and from which batch job requests into the cluster can be submitted (Fig. 1). This 'grid gateway' is not only used for GLOBUS and ByGRID but carries all the required, non-machine-specific software and services for grid applications in general. On the LRZ cluster, GLOBUS is installed directly on the frontend.

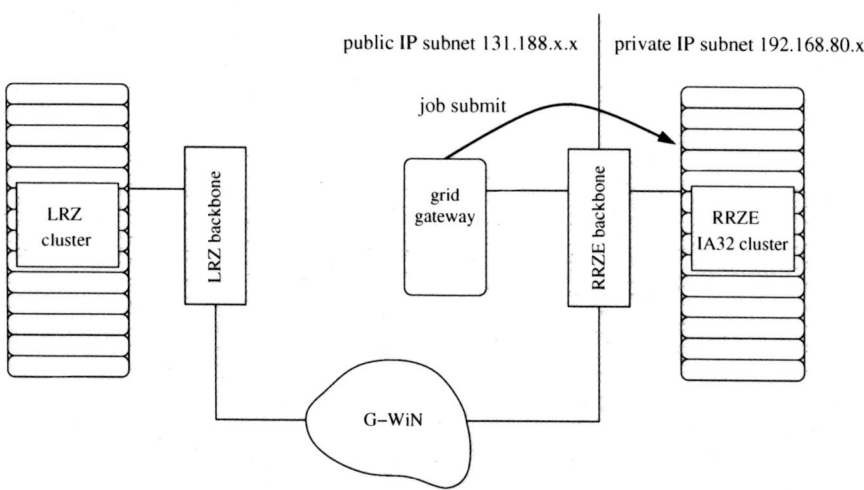

Fig. 1. Illustration of the ByGRID environment with LRZ and RRZE cluster installations.

A further problem was introduced by the use of AFS (Andrew File System) at LRZ for user's home directories. Instead of the well-known UNIX permissions on files and directories (read, write, execute for user, group, and others), AFS introduces a much more fine-grained scheme which can give file access rights to individual users. To validate a user for AFS, a so-called *token* is required. Native GLOBUS is not aware of these tokens and thus cannot access files on AFS, including important information in the user's home folder that must be available for authentication. There is, however, a GLOBUS add-on called `gsiklog(d)`, written by Doug Engert of Argonne National Lab [3]. gsiklog(d) makes GLOBUS AFS-aware and provides AFS tokens based on GLOBUS proxy certificates. This has the added advantage that users from non-AFS sites like RRZE can now access LRZ's AFS system without ever having to worry about tokens — the GLOBUS middleware layer hides the complications introduced by AFS from the users.

The GLOBUS installations on both ends allow for interactive execution of programs as well as submission to the respective queueing systems (OpenPBS/Maui at RRZE and Sun Grid Engine at LRZ). Serial and parallel jobs are possible. In order to ensure availability for testing purposes as well as an enhanced service level for users, special resources have been allocated on the RRZE cluster. There, some compute nodes (usually eight CPUs; more as demand requires) are reserved for exclusive use by ByGRID participants. On the LRZ cluster, ByGRID jobs are merely highly prioritized.

At the time of writing, ByGRID uses GLOBUS in version 2.4.x.

3 User's perspective

From a user's perspective, the GLOBUS infrastructure is a means to provide the first two two of the three traditional AAA services (authentication, authorization, accounting), together with a variety of more advanced functionality like job submission, file transfer and the like. However, there are several obstacles potential GLOBUS users have to face before they can effectively go into production mode:

- The user must apply for a *GLOBUS certificate* that can be used later for authentication. This certificate must be mapped to the local user account on each machine on which access is required. In the past, certificate requests were often a tedious undertaking, requiring a user to travel in person to the certification authority (CA).
- The GLOBUS toolkit contains a rather low-level collection of command-line tools that form a rudimentary interface. Taking into account the convenient environment users have at their disposal when connecting directly to a machine via ssh, it is sometimes hard to justify why GLOBUS should be preferred over direct login.
- A machine with GLOBUS installed must be equipped with a valid host certificate for machine authentication. Due to this restriction, installation of the toolkit on the user's own workstation or even laptop requires even more commitment from their side, further prolonging time to production.

It is one of the goals of ByGRID to establish an environment where those obstacles pose the least possible hassle for the potential user, minimizing overhead and time-to-result. Unfortunately, the third obstacle cannot be removed within the GLOBUS framework as machine authentication is a key security issue.

Traditional "early adopters" are usually ready to invest a considerable amount of work into emerging technologies. ByGRID, however, is eventually targeting the average scientific user who rather likes to generate results than mess around with obscure tools. The following sections describe what measures have been taken to lower the entry barrier.

3.1 Certificate requests

In order to ensure the authenticity of a user applying for a GLOBUS certificate, it is usually required for them to appear in person at the CA, showing some proof of identity. This procedure is unacceptably tedious, especially because there is only a very limited number of CAs available in Germany, e. g., at DFN-PCA in Hamburg. The obvious solution, viz, to establish a CA at every single computing center, would be a waste of resources. In ByGRID, RRZE's HPC services group (cxHPC) serves as a Registration Authority (RA), channeling certificate requests to the CA at LRZ. In this way, a user in Erlangen has to prove their identity to their local computing center only — no traveling is required. The actual request is submitted to the CA by the RA and is cryptographically secured and authenticated. The mapping between a GLOBUS certificate and the corresponding local user account can then be done at the user's request.

Given that some sensible standard for local user identification can be agreed upon, this model could easily be extended to all computing centers where RA services are sufficient. Simplifying the process even more by the use of standardized web services is being considered.

3.2 Support of user projects

The RRZE and LRZ compute clusters were, among other goals, acquired for the purpose of providing a major resource for serial throughput computing, a field where Grid middleware of any kind excels by definition. Identification of suitable user projects that benefit from 'gridization' is not difficult if the requirements of large-scale throughput applications are well-known by local computing centers. The main task here is to communicate the benefits of GLOBUS-based workflow as opposed to the usual login-submit-wait-check cycle. Unfortunately, GLOBUS in the version used does not fully support complete job data staging so that there is still the need for separate, manual submission and staging, albeit with reduced complexity due to GLOBUS's authentication and virtualization functionality. Supplementary software that uses GLOBUS tools in order to provide full staging and even direct remote file access is under development. For some time into the future, such software will always be to some extent application-specific.

A considerable amount of GLOBUS documentation has been provided by LRZ [2], with some supplementary, ByGRID-specific material contributed by RRZE [4]. This kind of information is, however, only the entry point for a new GLOBUS user. New adopters are continuously accompanied by both HPC teams in order to pinpoint problems right from the start.

3.3 Accounting

Up to now, accounting is strictly based on the local user accounts on the machines the user has access to. No GLOBUS information (e. g., certificates) is used. Although there do exist projects that aim for a certificate- i. e., identity-based accounting scheme [6], such a setup is beyond the scope of ByGRID at the moment.

3.4 Open problems

While GLOBUS is able to virtualize a lot of details about authentication, queueing systems and file access, there is still much room for improvement. Some problems tend to arise when using GLOBUS as a production tool:

- The location of fast and/or large local filesystems must be conveyed to a job. GLOBUS has no facilities to support automatic configuration of temporary storage.
- It is not possible for MPI-parallel jobs (jobtype="mpi") to use a more sophisticated setup than just starting a program with, e. g., mpirun, at least not without some unportable workarounds.
- GLOBUS commands often have a very long execution latency (seconds or even tens of seconds).

These deficiencies could, in our opinion, be remedied by optimizations to the GLOBUS software, which is of course still work in progress. On the other hand, some problems emerge from an insufficient integration of GLOBUS into standard procedures of computing centers:

- Use of GLOBUS commands inside job scripts is usually unsupported. This is a problem especially on cluster machines. Even if the commands themselves are available, an existing proxy certificate is invisible because standard queueing systems do not use GLOBUS for starting jobs on compute nodes. If the GLOBUS proxies were put into a directory which can be accessed by every system at a well-defined mount point, it would be possible to do, e.g., data staging from inside a GLOBUS batch job.
- The default temporary storage for stdout and stderr data is the .globus folder in the user's home directory. This leads to problems if there are restrictive quota limits or small file systems. Similar to the previous problem, this could be circumvented by substituting the .globus directory with a link to an institution-wide temporary folder that must of course be accessible everywhere using the same path. Automatic high watermark deletion or other mechanisms could then prevent this folder from being filled up with old data and at the same time relieve the users from having to clean up manually.
- The process of providing the GLOBUS certificate for a new user and setting up a Grid-friendly working environment is not integrated into the standard way of offering user accounts in computing centers. If the previous problems were solved, this should not pose any significant difficulties.

4 Current status and outlook

ByGRID has established a functioning, GLOBUS-based Grid testbed between the two major Bavarian computing centers. After an intense testing phase, real user projects are now using the special resources with good success. Adoption by other centers or institutions and inclusion of more computing resources at LRZ or RRZE would pose no significant problems. Current work includes improved, GLOBUS-based facilities for data transfer (staging, distributed storage schemes). Evaluation of UNICORE, especially with respect to the (GLOBUS-)compatibility of certificates, is also underway.

It is our opinion that if the use of Grid middleware is to be permeating scientific workflow, it is not only vital that users are provided with more than the ubiquitous web pages, but that Grid topics find their way into standard HPC workshops and tutorials. Computing centers must engage in a proactive search for Grid-suitable applications. Apart from the well-known showcase projects, e.g., in the particle physics community, for which LRZ provides a NorduGrid [7] installation on the Linux cluster, tendencies along those lines are still in their infancy right now.

This work is partly supported by the Competence Network for Scientific High Performance Computing in Bavaria (KONWIHR). Due to an uncertain funding situation, RRZE does currently not see itself in a position to continue support for ByGRID in the mid-term future.

References

1. http://www.unix-globus.org/toolkit/
2. http://www.lrz-muenchen.de/services/compute/globus/
3. deengert@anl.gov
4. http://www.rrze.uni-erlangen.de/dienste/arbeiten-rechnen/hpc/grid.shtml
5. http://unicore.sourceforge.net/
6. http://www.gridlab.org/
7. http://www.nordugrid.org/

FPGA: Exploration of the possibilities for the direct synthesis of concurrent C programs on high-performance computers in FPGAs

Peter Urbanek and Stefan May

Fachhochschule Nürnberg
Fachbereich efi
Wassertorstr. 10
90489 Nürnberg
peter.urbanek@fh-nuernberg.de

Abstract. This report describes the exploration of the possibilities for the direct synthesis of concurrent C programs on high-performance computers in FPGAs

Key words: FPGA, VHDL, Handel-C, Nios, OpenMP, "intelligent camera"

1 Introduction

The way of creating suitable prototypes and functional models from application oriented research results often leads from the design, the simulation and the optimisation on a mainframe to a subsequent conversion in a system consisting of hardware und software components. In such a case, single chip solutions are prefered, which integrate the results found on the mainframe in a specific chip.

For this purpose, the software solution running on the mainframe is rewritten from scratch in a hardware description language such as VHDL. This method is both time and money consuming. Moreover, during the new implementation, new errors are admitted into the design, which already had been eliminated in the prior software solution.

It would be an enormous improvement, if the hardware could also be described in the high-level language C and if concurrent processes could be converted directly. There are already first approaches on the basis of PCs. One of these is the programming language Handel-C [1, 8, 9], developed at Oxford University. It offers the possibility to synthesize systems described in C into ASICs, especially FPGAs [5].

The ideal conversion procedure would be, if the scientist could solve his problem, e.g. an algorithm, on the mainframe using its high performance as well as the excellent development environment and consequently synthesize a chip directly from these results.

2 Objective

This specific form of downright synthesizing concurrent software, running on a mainframe in silicon, is to be analysed in this paper. We shall start from today's possibilities which contain plain constrictions. Instead of using mainframe software, we must fall back on C with the extension OpenMP [10], because there are only synthesis tools for C on the hardware side at the moment.

The conclusions drawn from this project could certainly be converted to Fortran, which will only make sense once there are synthesis tools on the hardware side for that language.

Due to its good availability, C in connection with OpenMP was used on the mainframe side and Handel-C was used on the FPGAs of the chip side. The studies are held in general and carry no evaluation of the tools. Only the method of converting mainframe programs is of interest and the effort and quality required for that. Is there an advantage in this new approach over the known redevelopment in VHDL? If this question can be answered positively, then research results which need to be proven by a realisation can be converted faster and safer in the future.

The purpose of this research is to convert concurrent mainframe software, running on several processors, into prototypical hardware designs with FPGAs.

Procedure:

At first the conversion of simple algorithms was examined. Since there are no standards to evaluate the C synthesis on the hardware side, the Handel-C realisations were compared to known processor realisations.

After that an image processing system was built to compare complex algorithms. This image processing system, later appearing under the name "Intelligent Camera", served as a development platform on the one hand and as a demonstration tool on various fairs and events on the other. It contains all the algorithms which took the intended path from the mainframe to the FPGA. An evaluation rounds off the research.

All project results are covered by examples and proved by measurements. The statements are held in general and can be transfered to other mainframes as well as other FPGAs. Many thoughts can be transfered directly to other algorithmic programming languages.

3 Concurrent Computing on the Mainframe

In comparison to PCs, mainframes differ a lot from each other. For this study however, the differences in architecture are of most importance. Out of the various classifications for rough distinctions, Flynn's was the most successful because of its simplicity. Due to the high efficiency of standard processors and their swift development, the MIMD principle [4] is the most widely spread over all mainframes today. This group is distinguished between multi-computer systems (distributed memory machines) and multi-processor systems (shared memory machines) [11, 12]. A third part consists of computers with virtually shared memory. In this group, computers with distributed address space are united and form a logical address space. With regard to the conversion to FPGAs shared memory systems are easier to handle, although not imperative (see chapter Results).

On the software side, the shared memory architecture for the present considerations is supported by the OpenMP library. The processing is carried out sequentially until a parallel-construct occurs. Then the program flow branches into concurrent threads (see fig. 1).

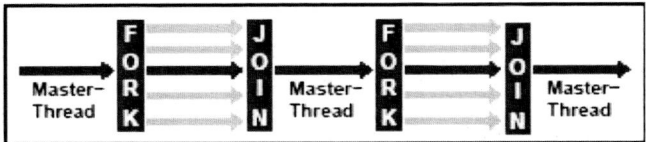

Fig. 1. Concurrent program flow with fork and join

Grammatically, this principle is realized with the instructions fork and join. All threads either carry out the same program on different variables according to the SPMP (Single Program Multiple Data) principle [15] or divide the work between them. Once tasks on the separate program parts are completed, all the threads are synchronized again. All threads except the master thread are finished.

4 Concurrent Computing on the FPGA

Today, FPGAs consist of millions of logical gates. In this project an FPGA from Xilinx was applied. The FPGA in question is a type Virtex-II XC2V1000 with one million system gates. All in all, 11520 logical function units are made available by these gates. On top of this the FPGA disposes of forty 18 bit multipliers. At present Xilinx offers the Virtex-II XC2V8000 with eight million system gates and 104882 function units as the largest platform within this component series. For even higher demands, Xilinx offers the Virtex-II Pro family. Other manufacturers such as Altera, present appropriate chips as well, so that the choice is rather a selection than a limitation.

A design on an FPGA practically allows a hardware independent realisation. The single gates are combined in logic cells and form a function unit. Consequently not only simple mathematical operations can be managed, but also complex applications. The programming can be done in a higher level programming language, such as Handel-C, which is used here. This language is converted into hardware through a synthesis process, i.e. the logic function units are combined intentionally. The knowledge about the intended algorithm is used to link the logical gates together. Due to this fact, the FPGAs achieve their high performance. Mostly whole lines of high language code are handled in only one clock cycle. The parallelism is inherent to the system and its use is self-evident owing to all the signals running concurrently. True parallelism can be converted just as well as pipelining or sequential or feedback structures. Additionally the amount of time for process distribution to the processors and the start of the single processes can be neglected completely and the communication can be replaced ideally by a circuit on the chip.

In contrast to this, the programming of each computer with one or more processors is carried out by operation codes, which are created by assembler instructions, which again are translated from a high level language. The order of events is always sequential. Improvements in performance are achieved mainly through the addition of caches or simply by increasing the clock frequency.

As a result, it is obvious, why these particular mainframe programs with parallel constructs are converted as easily to FPGAs. The basic structure is similar. It is also immediately obvious, why it is of no use to create several micro processors in one FPGA and in vain to attempt to reach the performance of a super computer. Without anticipating the results, it is already clear, why single-CPU PCs must do badly at performance comparisons. The hundredfold speed advantage (clock) of the CPUs is easily compensated by the appropriate concurrence. In addition, no operation codes have to be handled.

In view of of these purely theoretical considerations, it is evident, that the only sensible conversion strategy consists in converting the whole function described in both cases by an algorithmic language. Up to now, the language elements of VHDL and Fortran or C were too varied. However, with the appearance of similar C languages, such as the applied Handel-C, a conversion seems applicable in practice. Handel-C offers constructs to create sequential and concurrent processes, interprocess communication and synchronisation mechanisms. Handel-C was chosen out of the variety of similar C languages, because at the start of this project, the only available synthesis tools which could convert language elements into gates existed for Handel-C. Thus, the choice of the compiler does not limit the universality of the concept, but is a concession to the practical part of the work.

The following chapters first show the practical approaches which also describe the struggle with the different language elements. These problems only need to be considered temporarily. They form no fundamental obstacle. Finally the complete conversion method is assessed.

5 Results: Fundamentals

During the first year some basic, simple examples were converted. A quantitive summary of the most important results can be found in tabular form below.

The following examinations, respectively implementations were executed:

1. Simple basic circuits → Basis for a more complex logic.
2. Creation of an I/O library for the connection of external I/O and memory → Adaption of the hardware used in later examples (Intelligent Camera).
   Result: A synchronous memory interface with 233 MHz was implemented. That proved to be enough for processors with approximately 66 MHz which are synthesisable at present and for all examples planned in this project and more.
3. NIOS 32 bit microprocessor (manufacturer Altera's term for a soft-CPU) on Altera FPGA → Thereby, multi-processor systems could be implemented directly as in the mainframe design. The mechanisms for communication only need to be adapted, not rewritten. A 32 bit processor was synthesized.
   Result: According to expectations, this concept resulted in poor performance and a high demand for resources in the FPGA. Nevertheless the NIOS can prove a solution for all problems, for which a processor realisation appears to be the better way. At present, NIOS kernels are available with 8, 16 and 32 bit clocked by 66 MHz.

In Tab. 1 the image processing algorithm, used in the "Intelligent Camera" was brought in for a comparison between a realisation with a processor (NIOS) and one without a processor (algorithm converted directly into gates).

Table 1. Comparison NIOS to Handel-C

Perf. Parameter of the Conversion	Algorithm on Nios	Algorithm in Handel-C
Space consumption in logic cells	3131 (37%)	677 (20%)
Max. operation frequency in MHz	33 MHz	35 MHz
Images per sec at f_{max}	7	30 (purely sequential)

Since future realisations will be measured in comparison to the hardware description language VHDL, the image processing algorithm mentioned above was once converted into VHDL and once into Handel-C (see Tab. 2). The implementation in Handel-C required two weeks, whereas the conversion to VHDL took approximately six weeks.

Table 2. Comparison Algorithm VHDL to Handel-C

Performance Parameter of the Conversion	VHDL	Handel-C
Space consumption in logic cells	304	562
f_{max} in MHz	99,9	34,5
Required RAM bits	3328	4096

Result: These values show that the C synthesis still has weaknesses, which however, surely will be removed in the course of time, so that it is quite appropriate to continue to consider the C synthesis.

4. Connection of a VGA-LCD → For the graphic display of the subsequent results from the 3D scanner built in the following year, a LC Display was connected directly to the FPGA.
   Result: The hardware was set up and tested.
5. Implementation of a graphics library → At present, no graphics library exists on the market for FPGAs, which is programmed in Handel-C. Therefore, a small proprietary library was set up. It handles 2D and 3D models without fillings, clipping, light and colour effects. No further time was invested on this topic, as such libraries will be available before long.
   Result: A simple graphics library was established which contains only the most necessary functions.
6. Construction of an "Intelligent Camera" a laboratory model. It is the first practical prototype, which gives measurable evidence of the direct success or failure of the C synthesis from a mainframe to FPGAs. The first task was to count the number of spots on a dice (see fig. 2). The algorithm was not very sophisticated — the goal was to judge the conversion from the SGI to the FPGA. → A camera and a display with 7 segments were connected to the FPGA via the I/O interface described above.

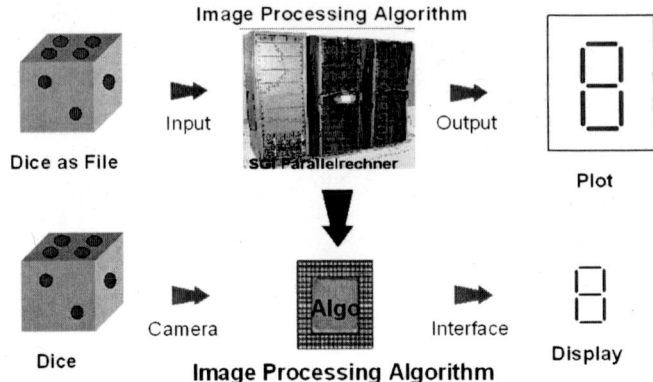

Fig. 2. Recognition of dice spots with the SGI and with the FPGA (same algorithm)

Result: For the first time an algorithm, in this case for image processing, was computed on several CPUs of a SGI in Erlangen and sucessfully converted to FPGAs.

For that purpose scanned images were transfered onto the SGI as data and the algorithm was transfered as a program. Then the results were examined in a text window. In a research project where the aim is to optimize the algorithm, the scientist would now work with the SGI and its excellent development environment until the algorithm is perfect.

After that, the image processing program was converted to Handel-C. Furthermore, a hardware interface for the camera and for the 7 segments display needed to be added. For completion, the communication structure was adjusted. It proved, that the communication mechanism of the SGI was perfectly suited for this kind of conversion.

By means of the applied camera algorithm, fig. 3 shows the increase in performance through concurrent tasks, i.e. the division of the camera's image in equal segments. On the SGI one CPU was used per segment and on the FPGA the algorithm was converted directly into hardware (without the synthesis of a microprocessor).

Parallelism in FPGAs requires the multiple generation of circuit parts. In fig. 4 the influence of parallelism on the chip area is shown. Since a common sequential share is included in the program, the demand for chip area doesn't rise linear.

In the following figure 5 a performance comparison was carried out with a Pentium III with 1 GHz clock frequency. The interesting conclusion related to this comparison is that the same performance is reached with a high degree of parallelism (e.g. 128), but only with 1% of the energy demand!

Summary of results after the first year in research:

For the applied research with the SGI and the involved experimental applications, this result implies that from now on simple, concurrent C programs with any number of processors can be converted to silicon, within the restrictions of the compiler which is available today. This chip is the basis for experiments or demonstrations.

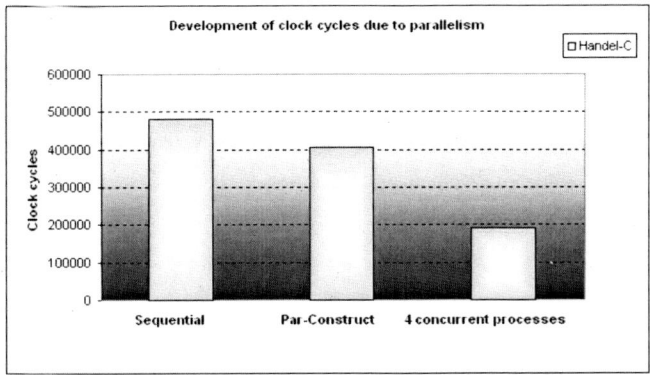

Fig. 3. Increase in performance by parallelism

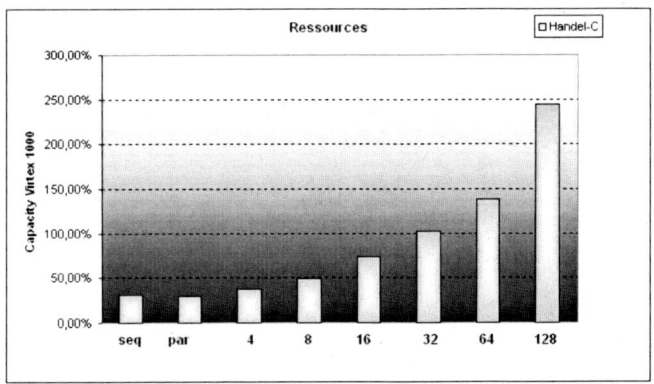

Fig. 4. Effect of parallelism on the chip area

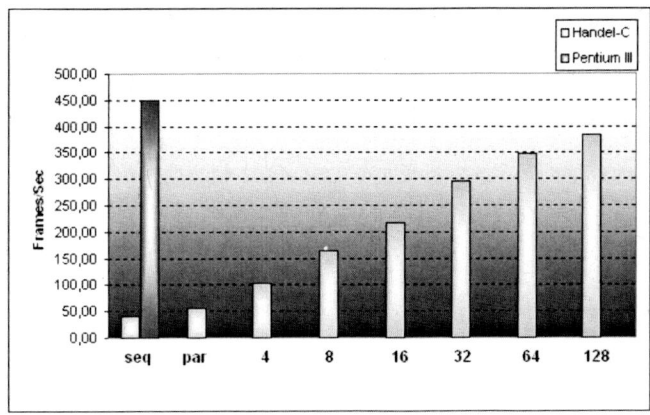

Fig. 5. Effective performance in comparison to a 1 GHz Pentium III-PC

The advantage of these proceedings is, that debugging the realisation on the SGI is much more comfortable and faster to manage as in FPGAs, and the first downloads to the hardware are ideally running faultlessly without the need for further modifications.

The conversion systematic was described in the second year.

6 Results: Converting OpenMP to Handel-C

The concurrent processing with OpenMP and Handel-C was introduced above. In this chapter a systematic approach to conversion is shown. The goal is to reuse concurrent sections, concurrent loops and synchronisation mechanisms in Handel-C, which were developed in OpenMP. The goal is not to find the best algorithm for the mainframe, but to find one for the target platform FPGA. The maximum amount of calculation time and address space will never be a problem for the mainframe. The complete runtime on the FPGA should range in the area of milliseconds. Therefore there was no attempt to minimize the communication, space and synchronisation overhead. These times are not relevant on the FPGA (see below).

The conversion starts with the definition of concurrent regions in the program. These are then converted step by step. Should it be impossible to join the regions due to dependances, then there is a danger of creating unnecessary overhead with multiple Fork-Join mechanisms. At the beginning, each single thread is created by a "`fork`" instruction. These threads can operate in the whole region. At the end of the region all threads will be closed, except the master thread. It is essential to clarify what each regional thread has to do and what it should not do. In Handel-C only the required number of concurrent threads are created. This means that the Fork-Join overhead on the mainframe can easily be taken into the bargain of the conversion, as it will anyhow be eliminated after the conversion.

The following part describes the conversion step by step. For each aspect there are program examples with descriptions and additional informations [16].

6.1 Conversion of Concurrent Sections

The concurrent conversion of sections is done by pragma-directives in OpenMP. An equivalent in Handel-C is found in the simple use of the par-construct. Both constructs show equal behaviour.

Note that the nowait-clause must not be used. In Handel-C there is no possibility to avoid the implicit synchronisation at the end of the par-construct. Generally, this clause is not necessary for the conversion. Should instructions exist after the par-construct which can be handled concurrent to the other threads, then these should be included in the structure immediately.

6.2 Conversion of Concurrent Loops

The concurrent conversion of for-loops was carried out simply by using the parallel for-construct in OpenMP. In Handel-C there is no equivalent to this. However, a comparable construct can be created with a simple trick. It can be generated through a combination of the par-replication with a normal for-loop.

6.3 Conversion of Pipelining Constructs

Pipelining addresses a substantial problem that occurs again and again in different forms. Data must be stored intermediately. Thereby, one must discern between various variables in the size of several bytes and whole memory areas, that are used e.g. for large arrays.

Variables are represented after the conversion by internal RAM cells in the FPGA. Depending on its type, an FPGA has several Kbyte of RAM at its disposal, so it can be estimated which variables can be realized as an internal resource and for which an external memory is necessary. With increasing technological progress, there will be more and more onchip RAM, depite this a moderate consumption is advisable as otherwise the FPGA's operating performance will drop to an inacceptable level.

For larger memory areas, such as the image storage of the "Intelligent Camera" an array is defined as follows:

`#define SELECTMEM(mem) (mem==FALSE ? data1 : data2)`

In it `data1` and `data2` are defined as static. The access to these arrays takes place with the help of the SELECTMEM-instruction (see above). The individual pipeline stages can then be filled with a `do...while`-loop.

Since internal RAM can easily be created with the compiler whereas an external RAM must be especially defined on the circuit board, this decision is of particular importance.

Essential knowledge concerning this problem was derived from the application "Intelligent Camera". This application has two memories, which can be addressed by two separate busses. This makes it possible to switch from one area to another between two clock edges similar to the ability in internal memory. With this knowledge, one can dispense of all the synchronisation mechanisms, data flow control and FIFOs. The gate count drops dramatically in comparison to a memory with preceeding access logic and the access speed to both memories rises to the maximum speed value of the memory interface, which is identical to the system clock rate. This discovery leads to the technique of using several busses and memories. Particularly in this case the advantage of the pre-simulation of the whole functional model on the mainframe is obvious.

6.4 Conversion of Synchronisation Directives

In OpenMP synchronisation directives are described as the `master-`, `critical-` and `barrier`-directives. The language extent of Handel-C offers no direct equivalents. In spite of this, the directives can be converted with simple means.

Conversion of Master and Single Directives

There is no equivalent in Handel-C for the `master-` and `single`-directive of OpenMP. Taking into consideration the sense of these constructs, it becomes evident, that there is in fact no need for a conversion. These directives can be used to prevent another thread from carrying out a certain instruction within one concurrent region. With Handel-C this opportunity does not arise. Within a concurrent region one thread carries out one instruction at a time. If several threads should carry out the same instruction, it must be explicitly declared in all these threads.

Conversion of Critical Directives

The critical-directive of OpenMP can easily be translated with the seq-directive of Handel-C. The difference, which has already been mentioned with the conversion of the master- and single-directives, applies here as well. Single instructions within a concurrent OpenMP region are carried out by each thread, whereas this behaviour must be explicitly declared in Handel-C.

Conversion of the Barrier Directive

The `barrier`-directive in OpenMP serves for the synchronisation of concurrent processes. There is no equivalent in Handel-C. Usually, an explicit synchronisation is redundant because of the implicit synchronisation of the `par`-construct.

Further tips and tricks are explained with code examples in [16].

7 Application: Recognition of Traffic Signs

The conversion of a complete, non-trivial concurrent application demonstrates in practice the conversion rules described above. The "Intelligent Camera" was used to recognize traffic signs. Image analysis makes up a very calculation intensive topic.

What the human brain constantly achieves subconsciously when it processes information gathered from our surroundings, is an enormous challenge for an electronical system. Image analysis in everyday life brings a lot of interference with it. All sorts of disturbances can occur, such as illumination changes, stainings, cover-ups, distortions, etc. In addition these applications must run in realtime.

The procedure that is employed here does not claim to solve these complex problems of image recognition [7, 13]. Several simplifications and restrictions were made so the focus is on parallelism as before and is not obscured by a too intricate process. The result is an algorithm, which can classify a number of traffic signs under the given conditions. This algorithm was checked for all possibilities to realize it concurrently. As a basis for the traffic sign recognition system Ritter's method [14] was used. During this realisation a new, concurrent sorting algorithm was found, which is faster than all known realisations up to now [16].

The complete system (see fig. 6) was displayed at several fairs and exhibitions and was used as an illustrative model in many presentations.

Result: A complete image processing system has been built consisting of a camera, lighting, image processing, communication interface and display of the recognized traffic signs, which can be used for demonstration purposes. The output is simply data, here in form of LEDs, not a picture. This system served as an efficiency proof of the conversion methods described above. All measurements that were taken for the comparisons above were carried out on this system. It can be treated as a development system, because the FPGA kernel with its circuits is of common use. It was proved that the replacement of the image processor with any other sensor or actor creates a new class of "Intelligent Sensors/Actors" with enormous efficiency whose development environment consists of the combination of a high performance mainframe and FPGAs.

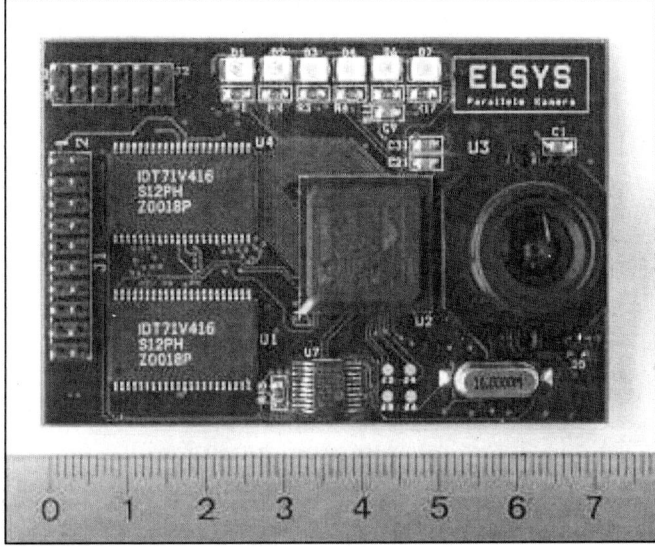

Fig. 6. Intelligent Camera

8 Conclusion

After the introduction the characteristic properties of a mainframe and an FPGA were introduced as far as they were relevant to this project. It is interesting that both platforms have a different evolution history, but on closer inspection show a lot of similarities. Architectural considerations, which have been a topic in the mainframe field for a long time, occur in the same form at the FPGA. An abstracted FPGA can be compared to a multi-computer system, as well as to a multi-processor system as it was done in this paper. From this point of view, various network topologies were developed in the mainframe field between single processors and between processors and memory areas. These topologies have a very high influence on the performance of a system. The solution of extensive problems reveals its strength. The FPGA need not cope with the same size of problems, instead it has to manage "smaller" problems with an extremely high execution speed. The most important considerations under this aspect concern the efficiency of concurrent algorithms. They benefit in a special way from the possibilty of a conversion to an FPGA. In combination with the parallel connection of external peripherals the performance can hardly be beaten by any other system.

The project was handled by a team consisting of one professor, one staff member of the Institute for Electronic Systems, three thesis students, a masters student and several student assistents within a period of two years.

The results were presented at the IENA, the CeBit, the Hannover Fair, the SMT, the Sensor Fair and the Internal Fair of the FH Nürnberg (called "efi-Kolloquium"). The application "Intelligent Camera" won two innovation awards during this time. The findings have been explained to many industrial companies and are being used directly in the lectures. A complete list of publications can be found on the author's homepage.

Results:

A method for the direct conversion of concurrent programs from a mainframe to hardware, in this case FPGAs, was considered and assessed. The core of the examinations is a conversion systematic from mainframe programs to chips with the use of C-based language elements. This conversion was proved by a specially built hardware realizing an "Intelligent Camera". This demonstration system in itself is already a novelty, due to its parallel memories, and shows the team-work between the concurrent hardware and software in a comprehensible way. THrough the strict application of the findings won in this project, it was possible to realize a complete image processing system with communication interface the size of a business card. The result surpasses all PC image processing systems that are presently known to us in efficiency, power consumption and size.

As scientific results a new and very fast concurrent sorting algorithm and the use of parallel memories with parallel bus systems can be emphasized.

It must be kept in mind, that this technology is at the starting point and still holds great potential for optimisation.

Final conclusion:

This paper focuses on the fast conversion of electronic systems simulated on mainframes to realisations in hardware and software.

For the scientific field, this means that theoretical research results, which were simulated on a mainframe, can be converted safely in prototypes for experimental proof with the help of the know-how gained here. Even if the conversion must now be carried out via the programming language C, the theoretical considerations taken here are valid in general and the results are applicable on other algorithmic programming languages.

For industrial applications, the gathered knowledge of how to gain an assessable simulation from the mainframe in a relatively short project term and convert it without the detour over VHDL quickly and savely in a future product is very valuable.

As a "waste product" of this research project the "Intelligent Camera" can be named. It serves as a basic technology for a new class of intelligent sensors or actors in the high performance sector. It is an example for the application of these research results, because in detail, it is neither new to build a small-sized camera nor is it something special to work concurrently on an FPGA. Conventional image processing systems do not exactly look like the one built here, because in comparison to a processor solution it is very costly to develop, test and optimize an image processing algorithm on an FPGA. Timing diagrams are of limited use as a debugging tool. A high performance mainframe on the other hand apears ideal. Only through the team-work between high performance mainframe and FPGA this innovation was made possible.

References

1. Bowen, Matthew, u.a.: Handel-C Language Reference Manual Version 2.1 o.O.: Embedded Solutions Limited, 2001
2. Brünig, Heinz: Vorlesungsskriptum Wissensbasierte Systeme 1 Nürnberg: 2003/2004

3. Haberäcker, Peter: Praxis der Digitalen Bildverarbeitung und Mustererkennung München/Wien: Carl Hanser Verlag, 1995
4. Huber, Walter: Paralleles Rechnen – Eine Einführung München/Wien: Oldenburg Verlag, 1997
5. XILINX: Virtex-II Pro Platform FPGA, Handbook 2002
6. Landrock, Uwe: Mustererkennung mit neuronalen Netzen Diplomarbeit am Fachbereich Nachrichten- und Feinwerktechnik der Georg-Simon-Ohm-Fachhochschule Nürnberg, WS 1997/98
7. Leighton, F. Thomson: Einführung in Parallele Algorithmen und Architekturen – Gitter, Bäume und Hypercubes Himberg: Wiener Verlag, 1997
8. Louis, Dirk: C/C++ – Die praktische Referenz München: Markt+Technik Verlag, 2003
9. o.V.: Handel-C Application Note / Optimisation Celoxica Limited, 2001
10. o.V.: OpenMP C and C++ Application Program Interface Version 2.0 OpenMP Architecture Review Board, 2002
11. o.V.: Net-Lexikon akademie.de asp GmbH, 2004 siehe http://www.net-lexikon.de/MISD.html
12. Rauber, Thomas/ Rünger, Gundull: Parallele und verteilte Programmierung Berlin/Heidelberg: Springer-Verlag, 2000
13. Reß, Harald/ Viebeck, Günter: Datenstrukturen und Algorithmen – Objektorientiertes Programmieren in C++ München/Wien: Carl Hanser Verlag, 2000
14. Ritter, Werner: Automatische Verkehrszeichenerkennung Koblenz: Fölbach Verlag, 1997
15. Stöcklein, Thomas: Einsatz der Programmiersprache Handel-C zur effektiven Entwicklung von synthetisierbaren Systemen und Algorithmen Diplomarbeit am Fachbereich Nachrichten- und Feinwerktechnik der Georg-Simon-Ohm-Fachhochschule Nürnberg, SS 2000
16. May, Stefan: Synthese von parallelen Algorithmen vom Großrechner zum FPGA Masterarbeit im Studiengang M-SE am Fachbereich Elektrotechnik, Feinwerktechnik, Informationstechnik der Georg-Simon-Ohm-Fachhochschule Nürnberg, SS 2004

gridlib: A Parallel, Object-oriented Framework for Hierarchical-hybrid Grid Structures in Technical Simulation and Scientific Visualization

Frank Hülsemann[1], Stefan Meinlschmidt[2], Ben Bergen[1], Günther Greiner[2], and Ulrich Rüde[1]

[1] System Simulation Group, University of Erlangen, Cauerstraße 6, 91058 Erlangen, Germany
{huelsemann,ben.bergen,ulrich.ruede}@informatik.uni-erlangen.de
[2] Computer Graphics Group, University of Erlangen, Am Weichselgarten 9, 91058 Erlangen, Germany
{meinlschmidt,greiner}@informatik.uni-erlangen.de

Abstract. The KONWIHR project *gridlib* has developed a framework for the integration of simulation and visualization for large scale applications. This framework provides its own grid structure, the so called hierarchical hybrid grid, which is well suited for runtime efficient realization of multilevel algorithms. Furthermore, it offers flexible visualization functionality for both local and remote use on number crunchers and workstations. It is based on modern object-oriented software engineering techniques without compromising on performance issues.

1 Introduction

The goal of the *gridlib* project is to develop a modern object-oriented software infrastructure for common grid-based numerical simulation problems on trans-TFLOP/s machines. These supercomputers, like the Hitachi SR8000, and modern scalable algorithms allow numerical simulations to be performed at unprecedented grid resolutions. However, this also tremendously increases the sizes of the data sets, surpassing the capabilities of current pre- and post-processing tools by far. At the same time, pre- and post-processing has become more and more important. Current complex engineering solutions require the automatic generation of problem-specific, time-dependent, adaptive, hybrid 3D grids that can be partitioned for parallel simulation codes. Enormous amounts of data must be presented visually for easy interpretation.

The system hardware of current supercomputers also places non-trivial demands on the software architecture, in particular the gap between the low bandwidth of external communication channels and the available size of local data. This requires the execution of pre- and post-processing steps on the supercomputer, which is a significant problem due to missing generic software support. Other difficulties arise since only special data and software structures can be efficiently handled on the high performance architectures. A naive implementation may lead to unacceptable performance problems.

The *gridlib* project addresses these problems, acting as a middle ware between existing software modules for pre- and post processing and as platform for implementing efficient solvers for complex simulation tasks.

2 System Overview

gridlib is designed to flexibly suit the needs of a given application while honoring the specifics of modern supercomputing.

Therefore *gridlib* implements the concept of hierarchical hybrid grids (HHG). In this approach, an unstructured base mesh is further refined in a structured manner (Sect. 3). This separation into (comparatively) coarse geometry meshes and (extremely) fine compute grids is an efficient trade off between flexibility and maximum runtime efficiency on current supercomputers. The concept of hierarchical hybrid grids has proved successful in delivering high numerical performance on the Hitachi SR8000 at the Leibniz Computing Center (LRZ) in Munich as well as on other systems. A prototype solver has been implemented and integrated with the new flexible interface (Sect. 2.2). The ability of the *gridlib* approach to accommodate and integrate existing solvers, even legacy software that is available as object code only, has been reported in [9].

Interactive flow visualization has been implemented as an example application. It has been integrated into the *GridViewer* (Sect. 4.2) so that it is available along with several other *gridlib* features.

The *gridlib* rendering subsystem features rasterization plug-in code for both hardware accelerated OpenGL based rendering on graphics workstations and parallel software-only rendering on the supercomputer. Additionally, there are integrated viewer applications using standard 3D software (OpenGL, OpenInventor) for interaction with local and remote visualization code.

The remote visualization uses a system independent format to exchange commands as well as binary data, widget descriptions and geometries over a single bidirectional communication channel. This can be stdin/stdout of an external program as well as a socket. The remote viewer presents a widget on behalf of the back end which can be used for user-solver interaction, thus enabling interactive control over batch mode solvers.

2.1 System Architecture

The *gridlib* architecture provides three major abstraction layers (Fig. 1) [11,12]. The lowest one is responsible for encapsulating the actual memory layout of data. Because the next layer entirely relies on this abstraction, the lowest layer can organize the storage freely. In particular, it can format its own memory layout to conform to the memory layout of other third party codes. We exploit this possibility for using a binary-only flow solver.

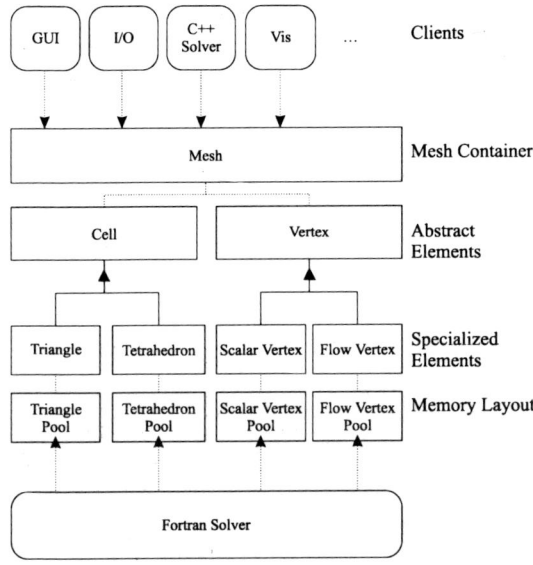

Fig. 1. The *gridlib* core is built around an interface of a mesh that contains abstract elements. The concretizations implement a custom memory layout

The second abstraction layer is the main link to the object-oriented world. It provides interfaces for all primitive elements (triangles, quads, tetrahedra, hexahedra, prisms, pyramids, octahedra), edges, and vertices as regular C++ classes. This sets the *gridlib* apart from other grid management libraries, as most of them do not allow the programmer to actually call methods on the objects.

The topmost layer provides the concept of a mesh container. It does not make any assumption on the mesh topology and implements abstract services like neighborhood setup, subdivision functionality and management, and content iterators.

The mesh container and the element abstraction layer provide powerful object-oriented programming support. For the library user, the *gridlib* further implements several clients that use the three-layer architecture for disk I/O, visualization, and simulation. The performance of the interfaces for the data exchange between the grid management, the solver, and the visualization and rendering subsystems has been evaluated by performing several simulations.

2.2 Flexible Interface

The original container abstraction has been enhanced to provide additional flexibility to the users of the top layer. The new code provides uniform access to any data stored or computed on the mesh, making service routines like visualization and even I/O completely independent of any specifics of the special mesh.

This enables true portability of service code between mesh based applications without any need for intervention by the application programmer.

2.3 Partitioning

The HHG subsystem includes interfaces to the METIS [6] and ParMETIS [7] partitioning suites to allow partitioning of unstructured input grids for the solution of problems in parallel. These interfaces provide an intuitive, easy-to-use means of accessing the partitioning information needed for setting up and solving problems within the HHG framework.

3 Hierarchical Hybrid Grids

The hierarchical hybrid grid subsystem (HHG) is a grid framework for automatically generating block-structured, regularly-refined grid hierarchies from purely unstructured input grids. This approach is useful for obtaining extremely high performance on modern computer architectures when solving systems of equations resulting from discretizations of partial differential equations. In this section we will outline the basic principles behind HHG and present performance results for several different variations of possible memory models that occur on modern super computers. All performance results in this section were obtained by a combination of an overwhelming part of C++ and few routines in Fortran. The reason for the inclusion of some Fortran routines lies in the optimization capabilities of the available compilers and not in the language specifications as such. For more details concerning the implementation and performance results on different platforms, we refer to [2], [5], and [4]. In this section, we concentrate on the performance of this approach on the Hitachi SR8000.

3.1 Basic Principles

The idea behind HHG is as follows: We begin with a purely unstructured input grid. This grid is assumed to be fairly coarse and is only meant to resolve the geometry of the problem being solved. This means, for example, that the input grid will resolve different material parameters but will not attempt to resolve problems such as shocks or singularities. It is also assumed that the desired resolution of the solution is much higher than that of the input grid, so that some type of refinement is necessary, both to ensure the proper resolution and to handle mathematical problems like those already mentioned. We then apply regular refinement to each patch of the input grid. Doing this successively generates a nested grid hierarchy, which is suitable for use with geometric multigrid algorithms. For an example of what such a grid hierarchy looks like consider Fig. 2.

Each grid in the new hierarchy is still logically unstructured. However, by using the right data structures we can now exploit the regularity of the patches. What we would ultimately like is for the neighbors of the interior points to occur at known, regular offsets from those points. This can now be accomplished by allocating separate blocks of memory for each individual patch. Then, patch-wise, we will have memory access patterns similar to a structured grid, thus allowing stencil based implementation of operations such as smoothing, residual calculation, and transfer operations, and thereby avoiding the performance penalties associated with indirect indexing. This is essentially a variant of using block structured grids, with the advantage that the structure and the resolution of the block connections are generated automatically.

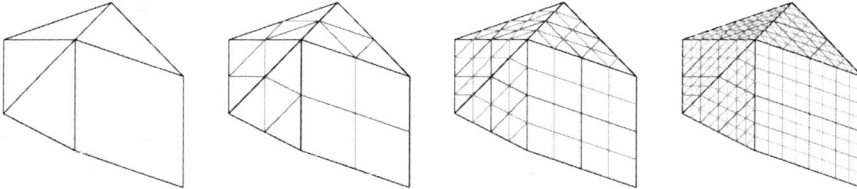

Fig. 2. Beginning with the hybrid input grid on the left, regular refinement is applied once to obtain the next grid, and again to obtain the next, and again to finally obtain the grid on the right. After two levels of refinement(third grid from the left), we can see that each patch has a structured interior. This structure may now be exploited to achieve higher performance

3.2 2D Sequential Results

Figure 3 shows performance results for varying levels of refinement on a single processor on the Hitachi SR8000 at the LRZ at Munich. Here, the left-most bar of the graph shows the MFLOP/s rate for performing colored Gauß-Seidel only on the finest level of refinement using the HHG data structures. This achieves the best overall performance due to the long line lengths in the inner-most loops of the smoothing algorithm. The second and third bars show results using the HHG data structures for smoothing on all levels, and geometric multigrid respectively. In both cases, there is a slight performance loss caused by doing work on the coarse grids. The rightmost bar shows the results for applying Gauß-Seidel to the finest level of refinement when the grid is treated in a purely unstructured manner using a *Compressed Row Storage* [1] storage scheme. Clearly, the HHG algorithms yield a substantial gain in performance over the purely unstructured case. In some cases, even half of the processor's theoretical peak performance is achieved.

3.3 3D SMP Parallel Results

Figure 4 shows performance results for varying levels of refinement on all eight processors of a single node on the LRZ's Hitachi SR8000. These implementations use COMPAS ("Cooperative Microprocessors in a Single Address Space") for producing shared memory parallelization. Here, the left-most bar of the graph shows the results obtained for a highly optimized Gauß-Seidel implementation for a purely unstructured grid using the *Jagged Diagonals Storage* [1] scheme. Great care has been taken in the implementation of this algorithm to exploit the procedural and architectural modifications made respectively to the IBM POWER instruction set, and to the IBM PowerPC processors [3] used in the Hitachi SR8000. In spite of this, the purely unstructured implementation still fails to achieve a high percentage of the node's theoretical peak performance of 12 GFLOP/s. On the other hand, as the second and third bars show, the HHG data structures, when applied to homogeneous tetrahedral and hexahedral meshes respectively, again attain extremely high floating point performance. In fact, for seven levels of refinement, the hexahedral implementation achieves more than half of the theoretical peak.

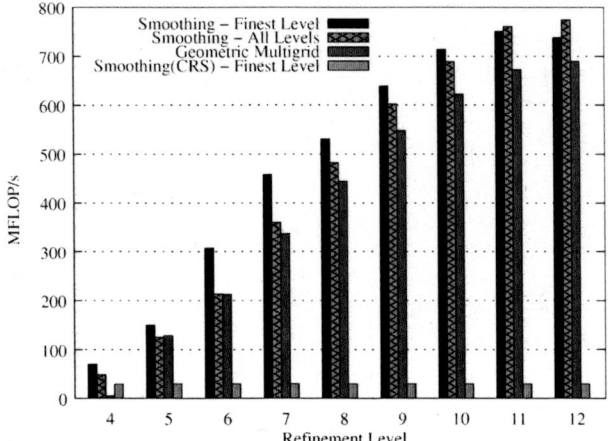

Fig. 3. Results for Hitachi SR8000 using a single CPU with theoretical peak performance of 1500 MFLOP/s

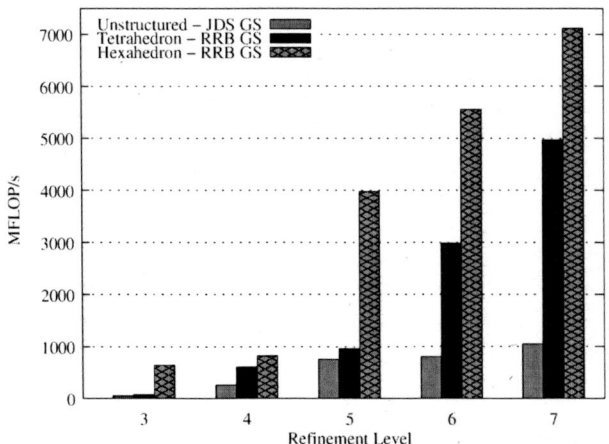

Fig. 4. Results for Hitachi SR8000 using 8 CPUs with combined theoretical peak performance of 12 GFLOP/s. Note that *RRB GS* stands for *Row-wise Red-Black Gauß-Seidel*. This means, that each line is updated using a red-black ordering of points, where all red points are updated, and then all black points. This is done to avoid pipeline stalls

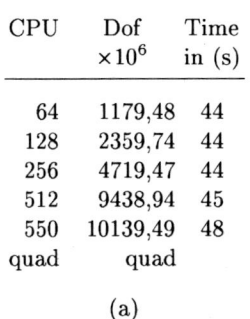 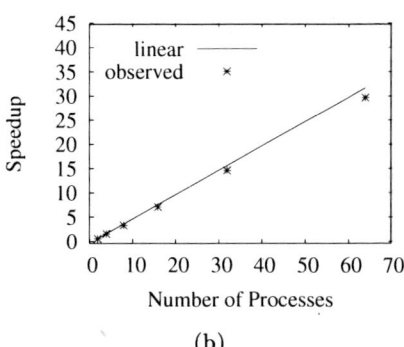

CPU	Dof $\times 10^6$	Time in (s)
64	1179,48	44
128	2359,74	44
256	4719,47	44
512	9438,94	45
550 quad	10139,49 quad	48

(a)   (b)

Fig. 5. Parallel performance results in 3D: (a) Scalability experiments using a Poisson problem with Dirichlet boundary conditions on an L-shaped domain. Each partition in the scalability experiments consist of nine cubes, each of which is regularly subdivided seven times. The timing results given refer to the wall clock time for the solution of the linear system using a full multigrid solver. The program runs with an overall performance, including startup, reading the grid file and all communication, of 220 MFLOP/s per process, which yields an agglomerated node performance of 1.76 GFLOP/s; (b) Speedup results for the same Poisson problem. The L-shaped problem domain consisted of 128 cubes, each of which was subdivided six times. The problem domain was distributed to 2, 4, 8, 16, 32 and 64 processes

3.4 3D Distributed Memory Results

The results for the distributed memory performance results in this section were obtained with a prototype implementation of HHG which is still restricted to hexahedral computational cells and input grids that contain only quadrilateral faces and hexahedra in the interior of the partition. Although this limits the practical applicability, it allows the investigation of the components that dominate the performance, in terms of both computation and communication. Given a sufficient level of refinement, the operations inside the volume elements, in this case hexahedra, account for the largest part of the computational effort, while, at the same time, the data transfer between volume elements and faces dominates the communication.

One aim in the construction of the HHG concept is to exploit the resulting regular structures to manage large problem sizes. The scalability results in the table of Fig. 5 show that a Poisson problem with Dirichlet boundary conditions involving more than 10^{10} unknowns can be solved in less than 50 seconds on 69 nodes of the Hitachi SR8000. This result demonstrates that efficient hierarchical algorithms, in this case full multigrid, in combination with "hardware-aware" (or: *architecture-friendly*) data structures are capable of dealing with large scale problems in an acceptable amount of time. The algorithmic components of the multigrid method are essentially standard: The row-wise red-black Gauss-Seidel smoother of the previous section was combined with full weighting and trilinear interpolation. On each level in the multigrid hierarchy we perform two V(2,2) cycles before prolongating the obtained approximation to the next refinement level. Trilinear finite elements result in a 27 point stencil in the interior of the domain.

The scalability results owe much to the ability of the full multigrid algorithm to arrive at the result with a fixed number of cycles, independent of the problem size. One might think that the speedup experiment represents a harder test, as the amount of communication over the network increases, while the amount of computations per process decreases. However, as shown in Fig. 5, the behavior is close to optimal. In the experiment, an L-shaped domain consisting of 128 cubes is distributed to 2, 4, 8, 16, 32 and 64 processes. Each cube is regularly subdivided six times. The same Poisson problem is solved using the same multigrid algorithm as before.

4 Visualization and Rendering

The visualization and rendering subsystem implements visualization methods for arbitrary planar slices through the unstructured grid [13], direct volume rendering by regular re-sampling, fast isosurface extraction [14, 15], and local exact particle tracing [10] (Fig. 6). All methods use an abstract renderer for geometric primitives

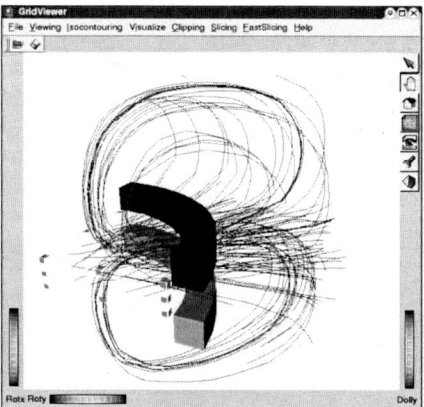

Fig. 6. Example: Magnetic field around a coil visualized by locally exact stream lines. For a reproduction of this figure in colour, see Fig. A.2.

(triangles, quads, ...) for displaying the result (Fig. 7). We have derived several concrete implementations from the abstract renderer that perform the actual image generation. Two types of rasterizer classes are provided. One is a pure software solution, the other one relies on OpenGL for hardware accelerated rendering. This allows generating the visualization image on screen or in an off screen rendering context using hardware acceleration if supported by the computers architecture.

The concept of the visualization and rendering system is very flexible and allows four basic usage scenarios:

- *remote rendering* on the supercomputer, using pure software algorithms for direct visualization on the high resolution simulation grid. Visualization and rendering parameters are passed to the subsystem along with simulation parameters at process startup time.

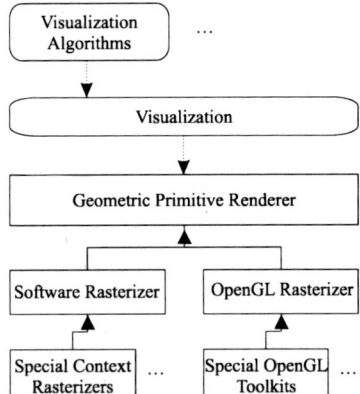

Fig. 7. The visualization system uses an abstract renderer interface that is implemented for both workstations and simulation hosts

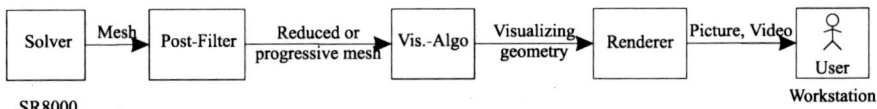

Fig. 8. *gridlib* visualization pipeline: The intermediate stages lie on the simulation or visualization host or the user workstation

- *remote visualization*: The visualization algorithm is run on the supercomputer with access to the full simulation grid. The resulting geometry (e.g. the triangles of an isosurface) is passed to the local desktop computer for (hardware accelerated) rendering.
- *post-processed rendering* on the local desktop computer, using hardware accelerated algorithms to visualize data on a reduced mesh that is the result of a pre-visualization processing step on the supercomputer.
- *hybrid rendering*: By manipulating a reduced geometric representation of the simulation grid on the local desktop computer, the visualization parameters can be tuned interactively. The parameters are sent to the supercomputer for remote rendering. The result is sent back to the desktop where it is integrated into the local model.

The hybrid rendering approach allows especially easy handling of the grid, while maintaining very accurate visualization results. Furthermore, it allows simulation and visualization tasks to be run in parallel, display intermediate results, and control grid management and numerical solvers on-the-fly.

4.1 Rendering

The on-host rendering uses a software-only implementation of the rasterizer subsystem that does not need any special graphics hardware. The frame buffer is implemented in a distributed manner, so that any partitioned geometry can be rendered in parallel [8].

4.2 Interactive Visualization

gridlib comes with the *GridViewer* application that provides access to most of the library's features interactively. This allows for quick tests as well as simple access to the example solvers.

As *GridViewer* includes all the visualization code it can be used as a standalone visualization tool. It implements the full visualization pipeline (see Fig. 8) or allows the use of external solvers via reading mesh files.

Additionally this application can easily be extended by the *gridlib* user and thus serve as a framework for custom applications and as example code for *gridlib* use in general.

4.3 Remote Visualization

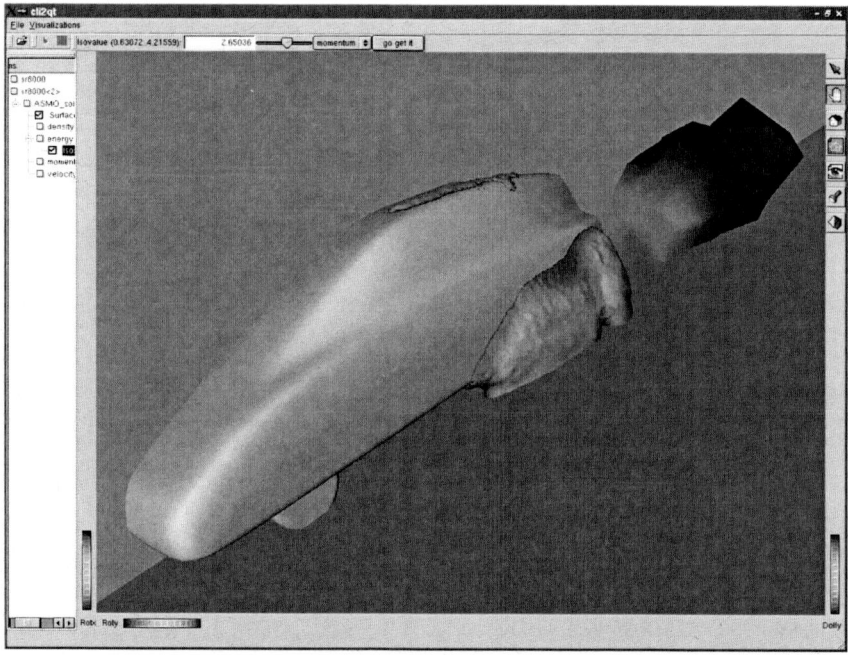

Fig. 9. Flow solution around ASMO – energy isosurface with color mapped momentum. The data set resides and is post processed on the Hitachi SR8000 in Munich while being viewed on a PC in Erlangen. For a reproduction of this figure in colour, see Fig. A.1.

The remote visualization system revolves around a lean viewer application that runs on any graphics capable workstation. Although the post processing code can be linked into this viewer, the main use is to connect the viewer to a second program with integrated post processing features at runtime, either locally or remote.

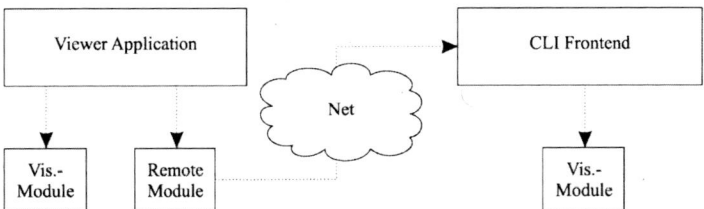

Fig. 10. The visualization viewer expects its data from abstract visualization modules. The modules expect to talk to an abstract application front end. A module and a front end that implement stream communication enable remote visualization

This viewer implements the last two stages of the visualization pipeline (Fig. 8), while its remote post processing counterpart implements the first three and transmits a visualization geometry (see Fig. 9).

The viewer is able to connect to a local socket. Thus, it can attach to an already running program. By this means it is possible to provide interactive visualization even while the simulation is running in batch mode.

4.4 Solver Interaction

The remote visualization protocol includes the ability for the back end code to open a widget in the viewer window. The main use is for interactively tuning visualization parameters (think of an isovalue here). However, close integration of visualization and simulation code gives the solver access to this, too. This way the visualization viewer becomes a steering front end for solvers that support this feature, both locally and remote and for both interactive and batch mode runs.

5 Conclusion

The main conclusion of the *gridlib* project is that modern object-oriented software engineering techniques can successfully be employed even on supercomputers, provided that up to date development tools are available.

Providing for grid adaptivity in the HHG implementation and integrating the existing components into one single framework remain future work.

We have implemented a flexible visualization subsystem and demonstrated that it supports all basic scenarios for the visualization of large data on supercomputers. Thus it adapts to the varying needs of different simulation applications, both interactively and in batch mode.

We have demonstrated that patch wise regular grids can be exploited very successfully for high performance computations on the Hitachi SR8000 while keeping the geometric flexibility of unstructured grids.

References

1. R. Barrett, M. Berry, T. F. Chan, J. Demmel, J. Donato, J. Dongarra, V. Eijkhout, R. Pozo, C. Romine, and H. Van der Vorst. *Templates for the Solution*

of Linear Systems: Building Blocks for Iterative Methods, 2nd Edition. SIAM, Philadelphia, PA, 1994.
2. B. Bergen and F. Hülsemann. Hierarchical hybrid grids: A framework for efficient multigrid on high performance architectures. Technical Report 03-5, Lehrstuhl für Informatik 10, Universität Erlangen-Nürnberg, 2003.
3. C. May et al.(Ed.). The PowerPC architecure: A specification for a new family of RISC processors. Morgan Kaufmann Publishers, 2nd edition, 1994.
4. F. Hülsemann, B. Bergen, and U. Rüde. Hierarchical hybrid grids as basis for parallel numerical solution of PDE. In H. Kosch, L. Böszörményi, and H. Hellwagner, editors, Euro-Par 2003 Parallel Processing, volume 2790 of Lecture Notes in Computer Science, pages 840–843, Berlin, 2003. Springer.
5. F. Hülsemann, P. Kipfer, U. Rüde, and G. Greiner. gridlib: Flexible and efficient grid management for simulation and visualization. In P. Sloot, C. Tan, J. Dongarra, and A. Hoekstra, editors, Computational Science - ICCS 2002, volume 2331 of Lecture Notes in Computer Science, pages 652–661, Berlin, 2002. Springer.
6. G. Karypis and V. Kumar. Metis a software package for partitioning unstructured graphs, partitioning meshes, and computing fill-reducing orderings of sparse matrices. Technical report, Department of Computer Science, University of Minnesota, Minneapolis, 2002.
7. G. Karypis and V. Kumar. Parmetis parallel graph partitioning and sparse matrix ordering library. Technical report, Department of Computer Science, University of Minnesota, Minneapolis, 2002.
8. P. Kipfer and G. Greiner. Parallel rendering within the integrating simulation and visualization framework "gridlib". VMV Conference Proceedings, Stuttgart, 2001.
9. P. Kipfer, F. Hülsemann, S. Meinlschmidt, B. Bergen, G. Greiner, and U. Rüde. gridlib—a parallel, object-oriented framework for hierarchical-hybrid grid structures in technical simulation and scientifiv visualization. In S. Wagner, W. Hanke, A. Bode, and F. Durst, editors, High Performance Computing in Science and Engineering 2000-2002 – Transactions of the First Joint HLRB and KONWIHR Result and Reviewing Workshop, pages 489–501, Berlin, 2003. Springer.
10. P. Kipfer, F. Reck, and G. Greiner. Local exact particle tracing on unstructured grids. Computer Graphics Forum, 2002. submitted.
11. Peter Kipfer. gridlib: System design. Technical Report 4/00, Computer Graphics Group, University of Erlangen-Nürnberg, 2000.
12. Peter Kipfer. gridlib: Numerical methods. Technical Report 2/01, Computer Graphics Group, University of Erlangen-Nürnberg, 2001.
13. U. Labsik, P. Kipfer, and G. Greiner. Visualizing the structure and quality properties of tetrahedral meshes. Technical Report 2/00, Computer Graphics Group, University of Erlangen-Nürnberg, 2000.
14. U. Labsik, P. Kipfer, S. Meinlschmidt, and G. Greiner. Progressive isosurface extraction from tetrahedral meshes. Pacific Graphics Conference Proceedings, Tokio, 2001.
15. M. Schrumpf. Beschleunigte Isoflächenberechnung auf unstrukturierten Gittern. Studienarbeit, 2001. Computer Graphics Group, University of Erlangen-Nürnberg.

LRZ: The Suitability of Contemporary Processors for Quantum Chemical Computations

Ludger Palm

Leibniz–Rechenzentrum der Bayerischen Akademie der Wissenschaften,
Barer Straße 21, D-80333 München
Ludger.Palm@lrz-muenchen.de

Abstract. The suitability of the Intel Pentium 4 Xeon, AMD Opteron, Intel Itanium2, IBM Power4 and Intel Pentium 4 architectures for application of the quantum chemical HF, B3LYP, MP2, CCSD and CCSD(T) methods is investigated, and conclusions on their relative performance are drawn.

1 Introduction

Quantum chemical methods play an ever increasing role in chemical research. This is partially due to the good reliability of today's methods, and partially due to the great compute power of state-of-the-art processors.

The multitude of methods as well as architectures makes it increasingly difficult to find the proper combination for the questions to answer. To assist in selecting the right computer for the method of choice we performed computations with the currently most widely used quantum chemical methods HF, B3LYP, MP2, CCSD and CCSD(T) on five contemporary architectures.

Benchmark computations will also assist the Leibniz Computing Centre (LRZ) in selecting the new installation of the regional high performance computer at the end of 2004.

2 Methods

2.1 The Processors

The computations were run on five different computer systems at the Leibniz Computing Centre[1]:

---

[1] For a detailed description of the various systems please see
http://www.lrz-muenchen.de/services/compute/hlr/.

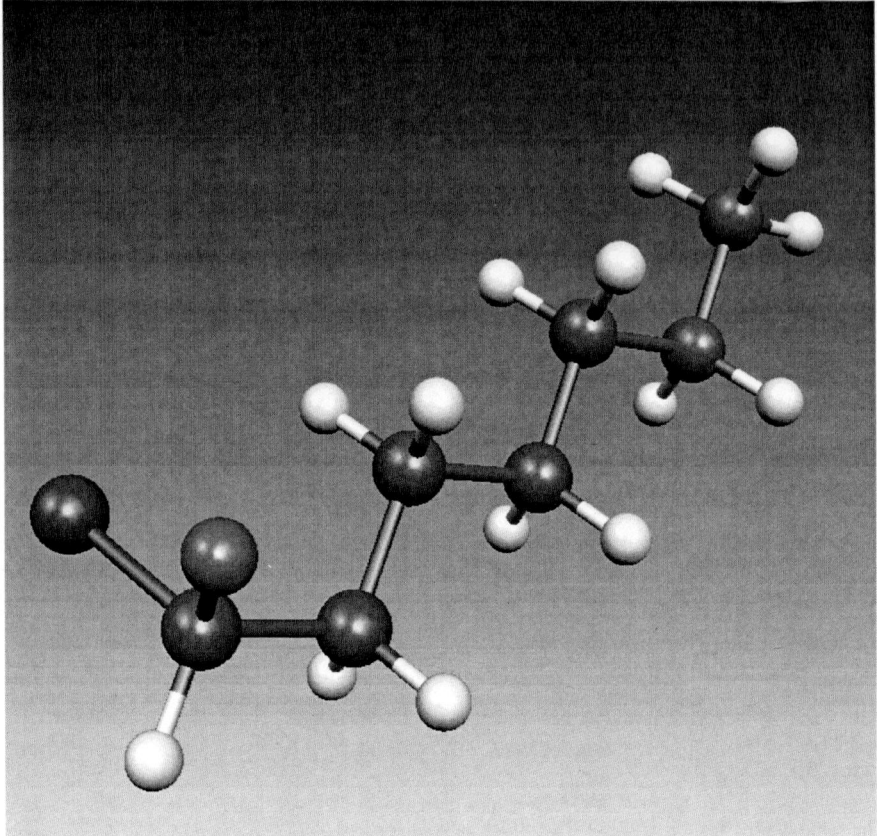

Fig. 1. The molecular structure of 1-chloro-1-fluoroheptane, $C_7H_{14}ClF$, an example of the benchmark set molecules. The image was generated using MOLEKEL [1]

- P4Xeon: Dual Intel Pentium 4 3.06 GHz Xeon, 512 kB L2 cache, 4096 MB memory, 533 MHz front side bus,
- Opteron: Dual 2.0 GHz AMD Opteron, 1 MB L2 cache for each CPU, 4993 MB memory,
- Itanium2: Four-way 1.3 GHz Itanium2, 3 MB L3 cache, 8192 MB memory,
- Power4: Four-way 1.3 GHz IBM Power4 in an eight-way p690 HPC system, 1.44 MB L2 cache per CPU, 128 MB L3 cache for all four processors, 16384 MB memory,
- Pentium4: Single 3.06 GHz Intel Pentium 4, 512 kB L2 cache, 2048 MB memory, 533 MHz front side bus.

All computers are operated under Linux except the Power4, which runs under IBM AIX.

On the computers with two or four processors, two or four, resp., identical processes were run to run the systems at full capacity. This is especially important for computations with high memory demands: while, e.g., in the AMD Opteron shared-memory processor every processor has its own memory interface, in the four-way Intel Itanium2 system the four processors share a common memory interface bus.

The quantity measured in all benchmarks was the wall clock time.

2.2 Program and Quantum Chemical Methods

The Gaussian 98 [2] package was chosen for this investigation due to its great popularity.

The computations were performed using the quantum chemical Hartree–Fock (HF), B3LYP density functional (B3LYP), Møller–Plesset perturbation theory to second order (MP2), coupled cluster with single and double excitations without (CCSD) and with perturbative inclusion of triple excitations (CCSD(T)) methods [3] as they are implemented in Gaussian 98. Program defaults were used except that all electrons were correlated in the MP2, CCSD and CCSD(T) computations.

While HF is the basis of all these methods, B3LYP has been chosen as the most popular density functional method. CCSD(T) is currently often the best one can do from the point of view of achievable precision. MP2 is intermediate in effort and accuracy between HF and CCSD computations, which are the basis for CCSD(T), but often do not deliver the required accuracy for chemical predictions.

As we do not expect that the results change essentially, neither geometry optimization nor frequency computations were performed, which would require much more computation time compared to the plain energy computation. The 6-31G(d)

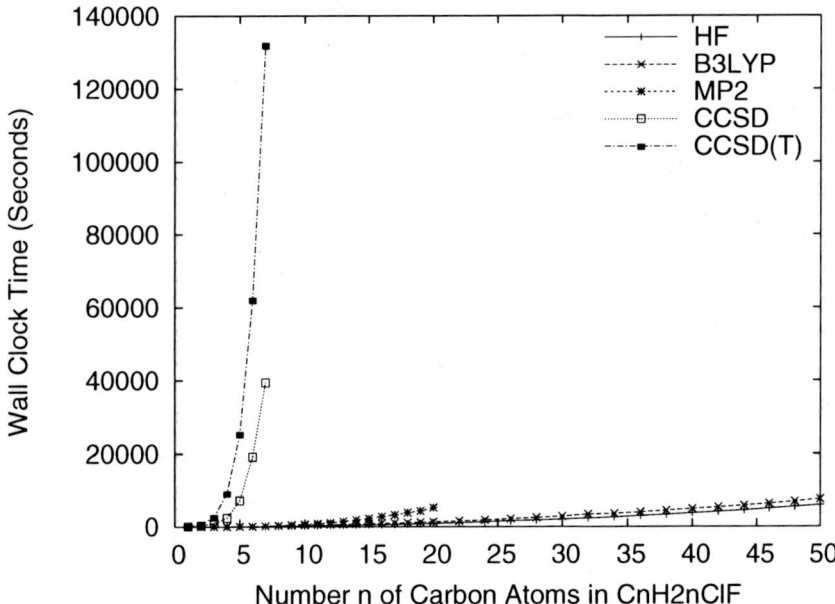

Fig. 2. Wall clock times for the computations on the Pentium 4 with the HF, B3LYP, MP2, CCSD and CCSD(T) methods for the molecules $C_nH_{2n}ClF$ (n = 1–50)

basis set used is a compromise between run time and accuracy. For meaningful chemical results, the basis set should be much larger, especially in the case of MP2, CCSD and CCSD(T) computations, but since we are interested in processor performance, we restricted this investigation to that moderate basis set. Each run was allowed to use up to 1600 MB of memory.

This combination of program, methods and basis set is very typical of what most researchers will choose as a starting point for their own investigations.

A 64-bit version of Gaussian 98 was available for the Itanium2 and the Power4 architecture, but only the 32-bit Pentium version of Gaussian was available for the AMD Opteron. Optimized linear algebra routines were used on all computers.

2.3 The Molecules

As molecular structures we selected a set of unsymmetrical substituted alkanes: 1-chloro-1-fluoro-alkanes $C_nH_{2n}ClF$ with n = 1–50. The structure of 1-chloro-1-fluoroheptane is displayed in Fig. 1 as an example. The bond lengths are: C–H: 110 pm, C–C: 150 pm, C–Cl: 180 pm and C–F: 140 pm; all angles are 109.1°. These molecules are stable, experimental data of many of them are available for comparison, they all have closed shell structures. Due to the 1-chloro-1-fluoro substitution, they have only C_1 symmetry. The input for a series of structures for any molecular size and quantum chemical method can easily be generated using scripting languages.

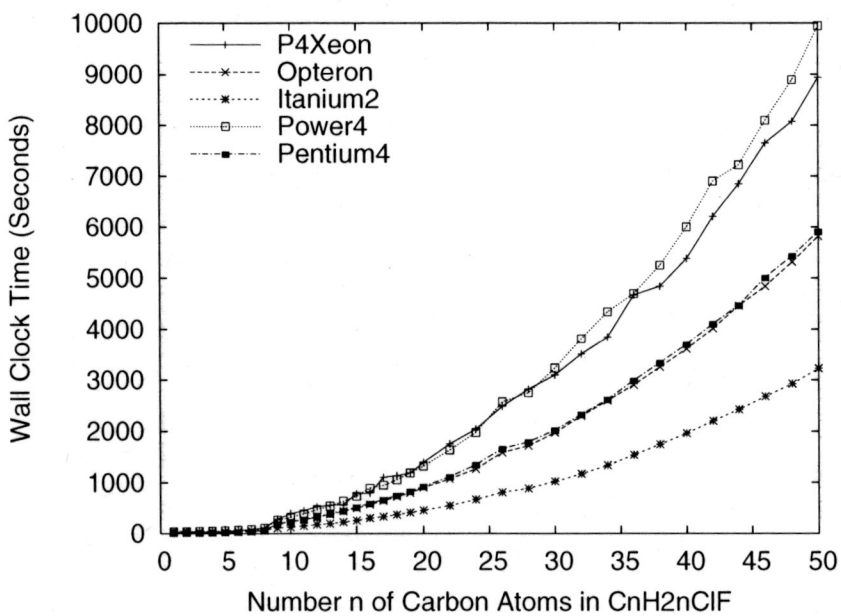

Fig. 3. Wall clock times for the HF computations on the molecules $C_nH_{2n}ClF$ (n = 1–50)

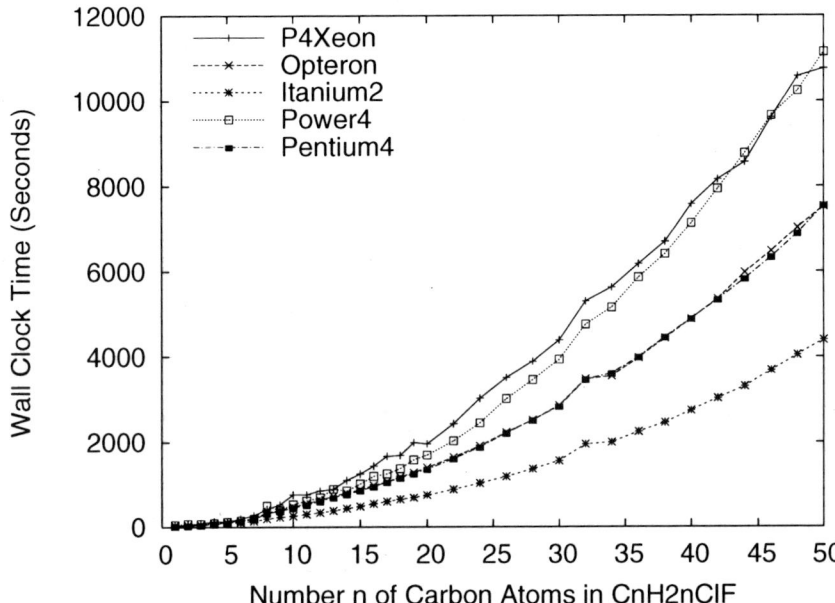

Fig. 4. Wall clock times for the B3LYP computations on the molecules $C_nH_{2n}ClF$ (n = 1–50)

3 Results and Discussion

3.1 Comparison of the Methods

The computational effort for the five methods varies strongly. Theoretically, it rises with $O(N^4)$ for HF, $O(N^5)$ for MP2, $O(N^6)$ for CCSD and $O(N^7)$ for CCSD(T) computations, if N is a measure of the system size. Fig. 2 shows the wall clock times for computations using these methods and B3LYP on the Pentium4 processor. While the HF and B3LYP computations can easily be extended to larger molecules, MP2 and especially CCSD and CCSD(T) computations can only be done for the smallest systems. This is not only due to the long run times but mainly to the high storage (memory and disk file) demands. Thus, the MP2 computations could only be done up to $C_{20}H_{40}ClF$ and CCSD and CCSD(T) could only be done up to $C_7H_{14}ClF$. Coupled cluster computations on larger systems or with larger basis sets thus require much more powerful computing resources, e.g., parallel computers: While the CCSD(T) energy computation with Gaussian 98 on a 3.06 GHz Intel Pentium 4 processor needs 132,029 seconds, the same computation can be run in 16,965 seconds on 64 processors of the Hitachi SR8000 at LRZ with 375 MHz using NWChem 4.5 [11]. However, Gaussian 98 does by default not support distributed memory parallelism at all, and shared memory parallelism only for the HF and density functional methods.

Nevertheless, from Fig. 2 we can see that performing coupled cluster computations on contemporary serial computers can be done for a moderate molecular system size and basis set.

3.2 Computation Times in Detail

On all architectures, Hartree–Fock (HF) and the most popular density functional method B3LYP demand similar compute resources.

Fig. 3 shows the wall clock times for the HF computations and Fig. 4 for the B3LYP computations on the five processors investigated. The increase in computation time is moderate and systematic. Even the SCF energy of the molecule $C_{50}H_{100}ClF$ can be computed in less than three hours.

From Fig. 5 it can be seen that the differences among the processors in the computation times are smaller for the MP2 computations than for the HF and B3LYP computations. MP2 energies of molecules of the size $C_{20}H_{40}ClF$ can be computed using the processors of this investigation in less than three hours.

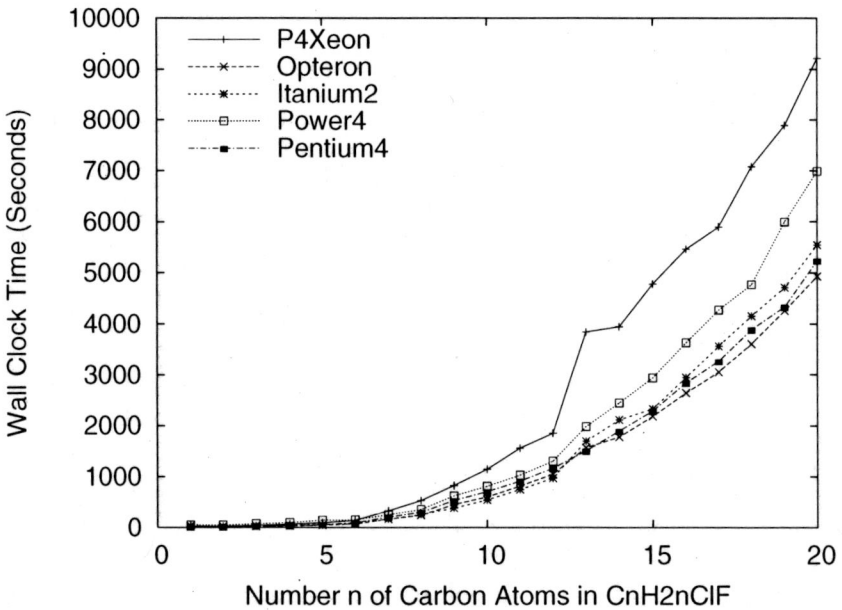

Fig. 5. Wall clock times for the MP2 computations on the molecules $C_nH_{2n}ClF$ (n = 1–20)

The situation changes completely when we look at Fig. 6 and Fig. 7 for the CCSD and CCSD(T) computations. The computation of the energy of the molecular structure $C_7H_{14}ClF$ takes 8–28 hours for the CCSD or 33–61 hours for the CCSD(T) computation and the differences between the five processors are again large.

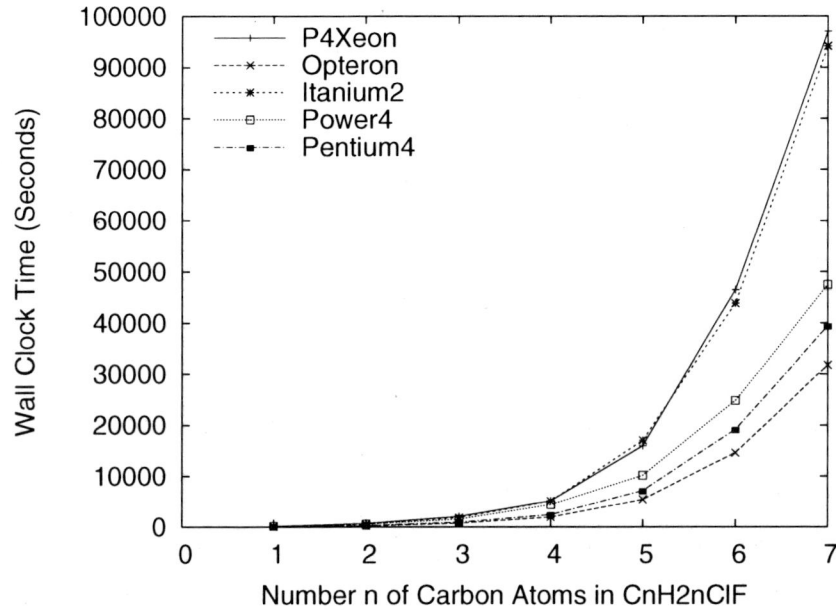

Fig. 6. Wall clock times for the CCSD computations on the molecules $C_nH_{2n}ClF$ (n = 1–7)

3.3 Relative Performance of Contemporary Processors

The relative performance of the Pentium 4 Xeon, Opteron, Itanium2 and Power4 processors with respect to the Pentium 4 processor

$$RelativePerformance(processor) = \frac{T(Pentium4)}{T(processor)}$$

is shown in Fig. 8 for the B3LYP and in Fig. 9 for the CCSD(T) computations, higher numbers representing faster execution.

B3LYP

For B3LYP computations, the Intel Itanium2 shows the best performance. The 1.3 GHz Itanium2 shows double performance in comparison to the 3.06 GHz Pentium 4 processor. We attribute this to the efficient utilisation of the second floating point unit and to the large 3 MB L3 cache for each CPU, running at full system frequency.

The AMD 2.0 GHz Opteron delivers nearly the same performance as the 3.06 GHz Pentium 4. We have to point out that the 32-bit Gaussian 98 had to be used. There may be further speedup when an optimized 64-bit version is available.

The IBM 1.3 GHz Power4 processor performs considerably slower than the Pentium 4.

The Intel Pentium 4 Xeon dual processor system shows only 3/4 of the Pentium 4 single processor performance. As both have the same 533 MHz front side bus, 3.06

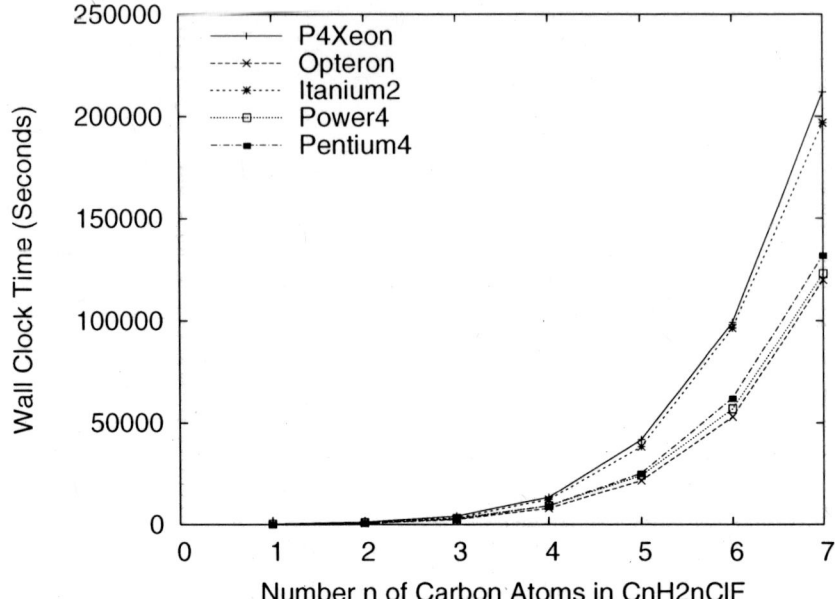

Fig. 7. Wall clock times for the CCSD(T) computations on the molecules $C_nH_{2n}ClF$ (n = 1–7)

GHz processor speed and 512 kB L2 cache for each processor, we attribute the difference to the faster memory access of a single process than of two concurrent processes on the Xeon.

CCSD(T)

The performance of the Pentium 4 Xeon, Opteron, Itanium2 and Power4 processors relative to the Pentium 4 processor for CCSD(T) computations (Fig. 9) differ from those for B3LYP computations (Fig. 8).

The CCSD(T) computations heavily use the memory and thus the speed of the memory interface may be more important than the caches.

The Pentium 4 Xeon dual processor system with two concurrent identical processes performs much worse than the Pentium 4 single processor.

The four-way Intel Itanium2 system performs considerably slower than the Pentium 4 processor. All four Itanium2 processors have to share a common memory interface, slowing down the concurrent identical jobs remarkably.

Each IBM Power4 processor in contrast to the Intel Itanium2 has its own path to the memory. Thus, the performance of the Power4 processor for the larger molecules is much better and nearly identical to the Pentium 4.

In AMD's Opteron dual processor system, again, each processor has its own memory interface. Accordingly, as with the Power4 processor, the Opteron performs nearly identical to the Pentium 4 and much better than the Itanium 2.

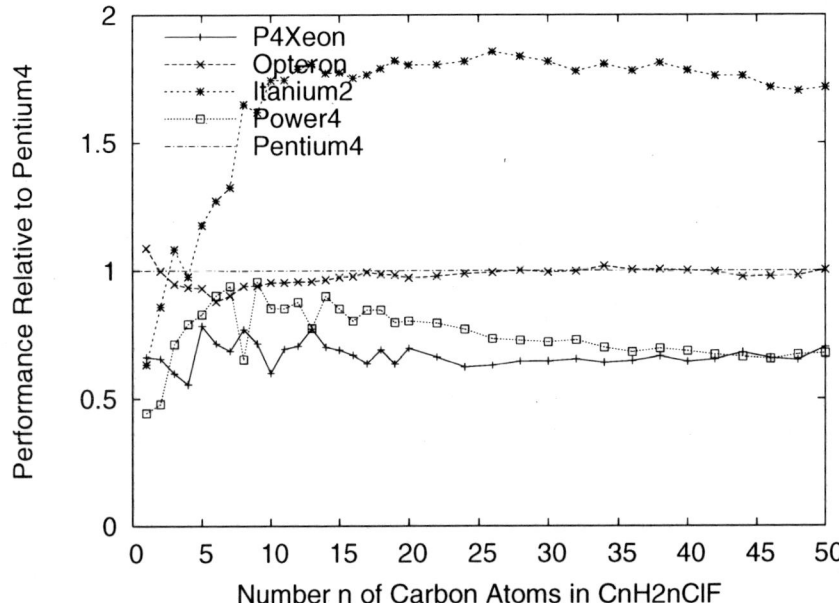

Fig. 8. Relative performance of the different processors compared to the Pentium 4 for the B3LYP computations on the molecules $C_nH_{2n}ClF$ (n = 1–50); see text for details

4 Conclusions

The five contemporary processors Intel 3.06 GHz Pentium 4 Xeon, AMD 2.0 GHz Opteron, Intel 1.3 GHz Itanium2, IBM 1.3 GHz Power4 and Intel 3.06 GHz Pentium 4 can all be used with great success for computations applying the HF, B3LYP, MP2, CCSD or CCSD(T) method.

The Intel Pentium 4 single processor system is a good choice in all cases for which the memory restriction to 2 GigaByte is not a problem. Due to the existing mass market, prices are low and performance is high even for the memory demanding CCSD(T) computations at least if also high quality chip sets, motherboards etc. are used.

The AMD Opteron processor also is a good choice in all cases, especially if one takes into account that its price is comparable to the Pentium 4 processor, furthermore, no 32-bit memory restriction exists if native 64-bit binaries are used.

IBM's Power4 processor will only be advantageous if high memory throughput, 64-bit architecture, excellent compilers and a broad range of available software are required.

Intel's Itanium2 looks excellent as long as processor and cache speed are crucial. It is much faster than any other of the investigated processors when B3LYP computations have to be executed. In the case of the memory demanding CCSD(T) computations, we would prefer the AMD Opteron, IBM Power4, the Pentium 4 sin-

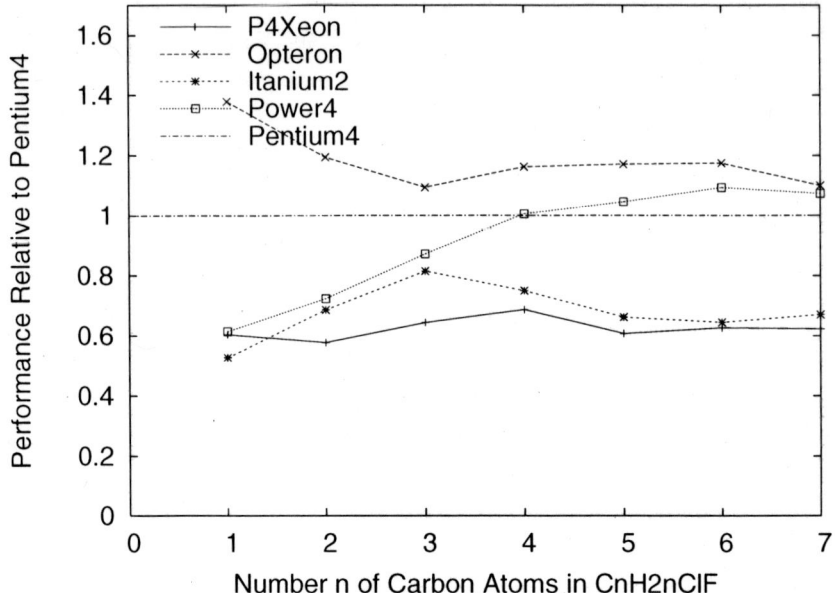

Fig. 9. Relative performance of the different processors compared to the Pentium 4 for the CCSD(T) computations on the molecules $C_nH_{2n}ClF$ (n = 1–7); see text for details

gle processor or a two-way Itanium2 computer with regards to the four-way Itanium2 system.

The Pentium 4 Xeon dual processor system performs much slower than the single processor version in all tests.

With regards to selecting the next generation regional high performance computer at the Leibniz Computing Centre at the end of 2004, we can claim the Itanium2 to be the preferred architecture for this purpose, as most chemical applications will run B3LYP computations on this system.

5 Acknowledgements

Substantial financial support by KONWIHR, the Competence Network for Technical, Scientific High Performance Computing in Bavaria, and helpful discussions with Dr. Reinhold Bader, Dr. Matthias Brehm and Dr. Herbert Huber at the Leibniz Computing Centre are gratefully acknowledged.

References

1. http://www.cscs.ch/molekel/
2. Gaussian 98, Revision A.11, M. J. Frisch, G. W. Trucks, H. B. Schlegel, G. E. Scuseria, M. A. Robb, J. R. Cheeseman, V. G. Zakrzewski, J. A. Montgomery, Jr., R. E. Stratmann, J. C. Burant, S. Dapprich, J. M. Millam, A. D. Daniels, K. N. Kudin, M. C. Strain, O. Farkas, J. Tomasi, V. Barone, M. Cossi, R. Cammi, B. Mennucci, C. Pomelli, C. Adamo, S. Clifford, J. Ochterski, G. A. Petersson, P. Y. Ayala, Q. Cui, K. Morokuma, P. Salvador, J. J. Dannenberg, D. K. Malick, A. D. Rabuck, K. Raghavachari, J. B. Foresman, J. Cioslowski, J. V. Ortiz, A. G. Baboul, B. B. Stefanov, G. Liu, A. Liashenko, P. Piskorz, I. Komaromi, R. Gomperts, R. L. Martin, D. J. Fox, T. Keith, M. A. Al-Laham, C. Y. Peng, A. Nanayakkara, M. Challacombe, P. M. W. Gill, B. Johnson, W. Chen, M. W. Wong, J. L. Andres, C. Gonzalez, M. Head-Gordon, E. S. Replogle, and J. A. Pople, (2001): Gaussian, Inc., Pittsburgh PA
3. For a description of the methods HF, MP2, CCSD and CCSD(T) see any good textbook on quantum chemistry, e.g. [4–6]. B3LYP is described in [7–9]. For applications of the methods with Gaussian see [10]
4. Hehre, W.J., Radom, L., Pople, J.A., Schleyer, P.v.R. (1986): Ab Initio Molecular Orbital Theory. Wiley, New York
5. Szabo, A., Ostlund, N.S. (1989): Modern Quantum Chemistry. McGraw-Hill, Inc., New York
6. Roos, B.O. (Ed.) (1994): Lecture Notes in Quantum Chemistry II. Springer, Berlin
7. Becke, A.D. (1993): Density-functional thermochemistry. III. The role of exact exchange. J. Chem. Phys. **98**, 5648–5652
8. Becke, A.D. (1988): Density-functional exchange-energy approximation with correct asymptotic behavior. Phys. Rev. A **38**, 3098–3100
9. Lee, C., Yang, W., Parr, R.G. (1988): Development of the Colle-Salvetti correlation-energy formula into a functional of the electron density. Phys. Rev. B **37**, 785–789
10. Foresman, J.B., Frisch, Æ. (1996): Exploring Chemistry with Electronic Structure Methods. Gaussian, Inc., Pittsburgh, 2nd ed
11. E. Apra, E. J. Bylaska, W. de Jong, M. T. Hackler, S. Hirata, L. Pollack, D. Smith, T. P. Straatsma, T. L. Windus, R. J. Harrison, J. Nieplocha, V. Tipparaju, M. Kumar, E. Brown, G. Cisneros, M. Dupuis, G. I. Fann, H. Fruchtl, J. Garza, K. Hirao, R. Kendall, J. A. Nichols, K. Tsemekhman, M. Valiev, K. Wolinski, J. Anchell, D. Bernholdt, P. Borowski, T. Clark, D. Clerc, H. Dachsel, M. Deegan, K. Dyall, D. Elwood, E. Glendening, M. Gutowski, A. Hess, J. Jaffe, B. Johnson, J. Ju, R. Kobayashi, R. Kutteh, Z. Lin, R. Littlefield, X. Long, B. Meng, T. Nakajima, S. Niu, M. Rosing, G. Sandrone, M. Stave, H. Taylor, G. Thomas, J. van Lenthe, A. Wong, and Z. Zhang, "NWChem, A Computational Chemistry Package for Parallel Computers, Version 4.5" (2003), Pacific Northwest National Laboratory, Richland, Washington 99352-0999, USA

MethWerk: Scalable Mesh-based Simulation on Clusters of SMPs

Amitava Gupta[1], Peter Luksch[2], and Andreas C. Schmidt[3]

[1] Jadavpur University, Calcutta, India, *amitg@pe.jusl.ac.in*
[2] Universität Rostock, Germany, *Peter.Luksch@inf.uni-rostock.de*
[3] Technische Universität München, Germany, *schmiand@cs.tum.edu*

1 Introduction

Load balancing is essential for efficiency and scalability of parallel computations. Numerical simulations are typically based on a mesh, which is partitioned for SMPD-parallel execution. Load balancing in this case means to find an optimal partitioning of the mesh. This paper presents a two-level, hybrid approach to mesh partitioning for clusters of SMPs. The proposed method has been applied to the industrial CFD code TRACE [1].

A cluster of SMPs essentially consists of a number of compute nodes that are connected by some high-speed network. Each nodes has at least two CPUs that are coupled by shared memory. Graph partitioning methods can produce acceptable task assignment across the nodes. When assigning tasks to processors within a node, one has to take into account the fact that all CPUs on that node have to share a single network interface and hence the node turn-around time is the minimum when the computation and communication activities of the processing elements can be interleaved optimally.

Gao et.al. have attempted to solve the problem in [2],[3] and [4] by using a two step approach involving graph-partitioning and then heuristic graph-matching using the well known A* algorithm. However, in all these cases the heuristic function could not be approximated for this problem and hence the search for optimal mapping of the mesh subdomains was found to be time consuming for practical test cases. In [4], a multi-level refinement based approach has been suggested which makes the method effective for practical test cases with up to 100 mesh subdomains. In this paper, a more effective method is proposed which uses a heuristic function for obtaining the optimal map of the subdomains on the processors of a SMP node within a shorter time, along with the multi-level refinement approach proposed in [4] so that the method becomes effective for larger practical test cases easily. These methods have been tested with TRACE on a variety of platforms viz. the Linux Cluster at MTU Aeroengines GmbH, Munich, Germany, the SMILE cluster at the Institute of Informatics, Technische Universitaet Muenchen and on the SR8000 cluster of the Leibniz Compute Center(LRZ), Munich. In a number of test cases, turnaround time has been reduced by one half compared to the existing parallel implementation.

Simulations of CFD codes involving meshes often use a mesh generation software which actually generates the mesh for the particular problem based on the defined geometry. This software generates subdomains according to geometric necessity. The number of subdomains hence typically is too small to be used as the unit of partitioning for large scale parallel simulation. Therefore we developed a software module that subdivides subdomains in order to obtain a suitable level of granularity for parallel execution.

Long running simulations on large scale clusters are quite likely to encounter failure of some component. Therefore, they require a mechanism for checkpointing, fault detection, and automatic recovery. In a production environment, simulations typically run under control of a job scheduling system, which controls access to clusters that are shared by multiple groups within an organization. Therefore, fault recovery must be integrated with the job scheduler.

At MTU Aero Engines GmbH, Munich, TRACE is used on a Linux cluster under control of the LSF job scheduling system [5]. We have implemented a fault recovery mechanism that is integrated with LSF.

2 Application Integrated Load Balancing: A hybrid approach for SMP clusters

Load distribution for parallel applications on a cluster of Symmetrical Multiprocessors(SMP) poses a challenging problem. A cluster of SMPs consists of a number of compute nodes each comprising of at least two identical, tightly coupled processing elements, the nodes being connected over a network. While approximate methods of load balancing using standard methods like graph-partitioning can produce acceptable task assignment across the nodes, they cannot be applied to obtain optimal task assignment on the processors constituting a node. When assigning tasks to the processors of an SMP cluster, load distribution that is based on computation time alone, does not guarantee minimum turnaround time. This is due to the fact that the processors on one SMP node share a single network interface unit. As an example, consider the following set of task modules:

task module	computation time [ms]	communication time [ms]
T_A	5	15
T_B	18	2
T_C	5	2
T_D	2	5

In an SMP cluster, intra-node communication, i.e. the transfer of data between processors residing on the same SMP node typically is orders of magnitude faster than inter-node communication, i.e., data transfer between processors on different SMP nodes. An SMPD parallel numerical simulation typically performs computation on all the task modules assigned to it first and then starts to communicate results to processors that process neighboring partitions. The processor that finishes computation first will get access to the single network interface first; other processes on the same SMP node will have to wait until the network interface becomes available.

It is clear that if these task modules be mapped on two processors constituting a SMP node, then a task assignment based on equal load distribution is likely to produce a distribution $\{T_A, T_C\}$ on one processor and $\{T_B, T_D\}$ on another, with a turnaround time of 34 ms. On the other hand, the assignment $\{T_A, T_D\}$ on one processor and $\{T_B, T_C\}$ on another will have a node turnaround time of 31 ms. It may be mentioned that a task assignment which aims at equal load distribution might produce either of the two assignments, but a task assignment which can take into account the option of interleaving computation and communication shall produce only the latter assignment. For the case where the task assignment is made based on equal load distribution, the processor with $\{T_A, T_C\}$ shall be busy with computation for 10 ms and thereafter block the communication channel for further 17 ms. Meanwhile, the other processor with $\{T_B, T_D\}$ would have finished computation in (18+2) ms but has to wait for 7 ms more for getting access to the communication channel. It then communicates for another (5+2) ms and so the node's turnaround time is 34 ms.

With interleaved computation and communication amongst the processors constituting the SMP node, the processor with $\{T_A, T_D\}$ computes for 7 ms and then communicates for further 20 ms. Meanwhile, the other processor which takes 23 ms for computation also gets ready to communicate, but has to wait for 4 ms more. Then this processor needs to communicate for (2+2) ms pushing the node turnaround time to 31 ms. With this distribution, the idle time for the second processor is 3 ms less and hence the gain. In this section, a modified version of the methodology presented in [4] is presented. In [4] a multi-level graph-partitioning algorithm is presented which makes the methodology presented in [2] and [3] extensible to practical computational meshes with a large number of subdomains. In this section, this is further modified to harness the strength of a branch and bound heuristic method to find optimal mappings on a cluster of SMPs within acceptable time limits.

The methodology developed is one by which an optimal task assignment on a cluster of SMPs is obtained in two steps comprising of first graph-partitioning and then graph-matching. For this purpose, blocks or subdomains comprising the mesh are viewed as basic indivisible modules of the original mesh and the problem boils down to find a map of the modules on to the processors comprising the SMP cluster. The mesh is viewed as a task graph with the subdomains as the vertices representing task modules.

Metis [7] based graph-partitioning is used to assign a set of modules to each SMP node of the cluster. This produces as many partitions of the original mesh as there are nodes, with the total weight of the edge cut minimized. Thus the computation and communication load is balanced across all nodes and each node would have had the same turnaround time had each been composed of a single processor. Subsequently, graph-matching is used to assign these basic modules to individual processors of each SMP node. The graph-matching phase uses a branch and bound heuristics based methodology which is an adaptation of the methodology proposed by Shen et. al. in [6].

Given the set of task modules assigned to a node of a given SMP cluster, the methodology developed produces an optimal distribution across the processors constituting the node in question in an iterative way, each step involving finding the optimal distribution of a subset of the task modules assigned to the node in question on two processors at a time. This is essentially a refinement methodology because for any step, the initial non-optimal mapping of the subset of task modules for the concerned step on the 2 processors considered must be known.

As in [6], a mapping **S** is defined as a set of tuples, $s_1, s_2, s_3, \ldots, s_n$, each being of the form $\{x,y\}, x \in \mathbf{T}, y = m(x), y \in [0,1,2,\ldots,p-1]$ or the set of processors. **T** is the set of modules assigned to the node in question and $m(x)$ is the function which maps a module x to a processor.

The goal is the mapping in which all elements of **T** are mapped on one of the p processors with the minimum turnaround time for the node in question. The graph matching methodology is a multi-level one. It starts with definition of a default map **S**_default which is obtained by dividing **T** into p equal partitions with weight of edge-cut minimized. It is to be noted that for this, the weight of the edges of the dependency graph corresponding to the sub-task is actually the time taken to transfer the data associated with the edge in question across two adjacent processors in the same node. The partitions are equal with respect to the time spent in computing plus the time spent in communicating to processors in other nodes. To explain the methodology, the following definitions are proposed:

Let **T** be the set of modules assigned to a node in question, also referred to as the subtask assigned to that compute node. **S** is a set of tuples, $s_1, s_2, s_3, \ldots, s_n$, each being of the form $(x,y), x \in \mathbf{T}, y = m(x), y \in [0,1,2,\ldots,p-1]$, $m(x)$ being the function which maps a module x on to a processor. **S** may be viewed as a complete map in the sense all elements of **T** are mapped; **P** is the ordered set of processors, ordered in the ascending order of the time spent by each for computation, corresponding to **S**. $\mathbf{T}_{i,j}$ is the set of modules assigned to processors i and j for any **S**. $\mathbf{S}_{i,j}$ is the optimal mapping for $\mathbf{T}_{i,j}$ on processors i and j;

We define the following operations: *Arrange*(**S**) produces the ordered set **P** for any **S**. *Balance*(i,j) produces an optimal mapping $\mathbf{S}_{i,j}$ of subtask $\mathbf{T}_{i,j}$ on processors i and j using graph-matching, $i,j \in \mathbf{P}$. *Update*(**S**, $\mathbf{S}_{i,j}$) produces a complete mapping **S**' with elements of $\mathbf{T}_{i,j}$ redistributed amongst ith and jth processors of **P**. *Diff*(**P**, **P**') returns k, the index of the first element in which the two ordered sets of processors **P** and **P**' differ. Now, the methodology is explained by the following pseudo code:

Let **S** = **S**_default;   be any initial complete map
Let **P** = *Arrange*(**S**);   be the corresponding ordered set of processors
Complete = *false*; *nos_pass* = 1;
$i = 0;\ j = 1;$
while (not *complete*) **do**
   $\mathbf{S}_{nos\_pass} = Update(\mathbf{S}, Balance(i,j));$
   $\mathbf{P}_{nos\_pass} = Arrange(\mathbf{S}_{nos\_pass});$
   $k = Diff(\mathbf{P}, \mathbf{P}_{nos\_pass});$
   if (k ==p)   The 2 sets are identical for a p processor system
      then $i = i + 1;\ j = j + 1;$
   fi
   if (i==p)
      then *complete* = *true*;   No further balancing is required
      else if $(k == 0)$
            then $i = 0;\ j = 1;$
            else $i = k - 1;\ j = k;$

```
      fi
   fi
   S = S_{nos_pass}; P = P_{nos_pass};
   nos_pass + +;
od
```

The methodology can be explained with the help of Fig. 1. In the first case, we assume that for the successive passes, as the initial complete mapping is updated, the ordered set of processors remains same. A four processor case is assumed.

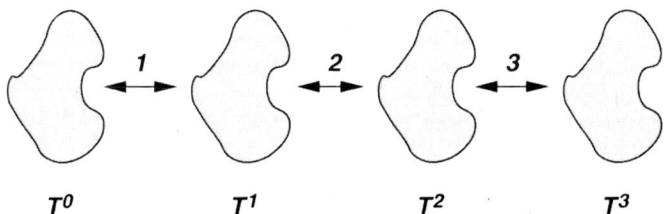

Fig. 1. Multi-level graph partitioning example

In Fig. 1, it is assumed that $\mathbf{T}^0, \mathbf{T}^1, \mathbf{T}^2$ and $\mathbf{T}^3$ represent the partitions of $\mathbf{T}$ on processors corresponding to the elements 0,1,2 and 3 of $\mathbf{P}$ to any $\mathbf{S}$. Since $\mathbf{P}$ is ordered in the ascending order of computation time associated with the processors, the turnaround time for the ith element of $\mathbf{P}$ can be defined as

$$TA_i = \begin{cases} TP_i + TC_i & \text{if } i = 0 \\ \max\{(TP_i + TC_i), (TA_{i-1}, TC_i)\} & \text{if } i \neq 0 \end{cases} \quad (1)$$

TA_k denotes the turnaround time on processor k, TP_k the computation time, and TC_i the communication time. The numbers against the edges represent the pass number at which re-distribution of elements occurs using graph-matching (represented by edges amongst partitions). In this case the set $\mathbf{P}$ does not change with successive passes. To start with, $i = 0$ and $j = 1$, so that $Balance(0, 1)$ distributes the tasks in $\mathbf{T}^0 \cup \mathbf{T}^1$ so that the turnaround time of the processor corresponding to the second element of $\mathbf{P}$ becomes minimum and the complete map $\mathbf{S}$ is updated. $\mathbf{T}^0, \mathbf{T}^1, \mathbf{T}^2$ and $\mathbf{T}^3$ are also suitably updated. In the next pass, $Balance(1, 2)$ distributes the tasks in $\mathbf{T}^1 \cup \mathbf{T}^2$ (updated) so that the turnaround time of the processor corresponding to the third element of $\mathbf{P}$ becomes minimum. Now, it is clear that if $(TP_0 + TC_0) < TP_1$, it might be possible to shift some elements from $\mathbf{T}^2$ after re-distribution by graph-matching such that the turnaround time of the processor corresponding to the second element of $\mathbf{P}$ does not change.[4] This is also achieved by the routine $Balance()$.

The routine $Balance(i, j)$ actually uses the graph matching algorithm put forward in [2]. The processors i and j are two consecutive elements of $\mathbf{P}$ and the task

---

[4] In this case, the first processor in $\mathbf{P}$ has finished computation and communication before the second processor has finished computation and becomes ready to start communication.

graph in question is the current set of tasks mapped on the processors corresponding to processors p_i and p_j, $p_i, p_j \in \mathbf{P}, i = j-1, j \geq 1$. This is denoted by $\mathbf{T}'$ in the following discussion. If $j > 1$, it tries to distribute the sub-task on processors preceding $p_i \in \mathbf{P}, i = j-1$, assuming that elements already mapped to this processor cannot be re-located to either processor p_i or p_j, but elements of the sub-task can be added additionally to this processor by $Balance(i,j)$. Thus elements of the sub-task may be mapped on the processors starting from the 0th to the jth element of $\mathbf{P}$ and this forms the set of permissible processors, $\mathbf{M}$. The routine $Balance(i,j)$ actually uses the graph matching algorithm put forward in [6]. The processors p_i and p_j are two consecutive elements of $\mathbf{P}$ and the task graph in question is the current set of tasks mapped on the processors corresponding to processors p_i and p_j, $p_i, p_j \in \mathbf{P}, i = j-1, j \geq 1$. This is denoted by $\mathbf{T}'$ in the following discussion. If $j > 1$, it tries to distribute the sub-task on processors preceding $p_i \in \mathbf{P}, i = j-1$, assuming that elements already mapped to this processor cannot be re-located to either processor p_i or p_j, but elements of the sub-task can be added additionally to this processor by $Balance(i,j)$. Thus elements of the sub-task may be mapped on the processors starting from the 0th to the jth element of $\mathbf{P}$ and this forms the set of permissible processors $\mathbf{M}$.

The graph-matching phase starts with an initial partial map where the first element of the sub-task in question is assumed to be mapped on the first permissible processor. Entries are then added to the state-space by expanding the partial map, each time mapping a new task on each of the permissible processors. The goal is found using the A* algorithm. Each state-space entry is assumed to be associated with two weights $g(n)$, which is the cost of generating the entry, and $h(n)$, which is the cost of reaching the goal state from the state-space entry. The method guarantees finding an optimal task assignment as long as $h(n)$ is less than a lower bound, the value of which depends upon the problem in hand.

Computation of g(n). For each mapping, let $\mathbf{R} \subseteq \mathbf{T}'$ be the set of tasks mapped; $\forall x \in \mathbf{R}$, let $y = m(x)$ be the processor on which x is assigned by the mapping, let $u(x)$ be the time required by the task for communication with tasks on other nodes, and $v(x)$ be the computation time required. $\forall x, x' \in \mathbf{R}$, let $w(x \to x')$ be the time taken to transfer the data element represented by the edge $x \to x'$, if $m(x) \neq m(x')$. It is assumed that $w(x \to x') = 0$, when $m(x) = m(x')$. Then, if $\mathbf{Q}$ denotes the set of all processors on which at least one of the tasks is mapped by the mapping in question, the computation time associated with a processor $q \in \mathbf{Q}$ is

$$comp(q) = \sum_{x \in \mathbf{R}: m(x) = q} v(x) \qquad (2)$$

Similarly, the communication time associated with the processor $q \in \mathbf{Q}$ can be expressed as

$$\sum_{x, x': m(x) = 1 \wedge m(x') \neq q} w(x \to x') + u(x) \qquad (3)$$

Finally, if the ordered set $\mathbf{P}$ is now formed such that

$$\forall i \geq 1, p_{i-1}, p_i \in \mathbf{Q} : comp(p_i) \geq comp(p_i - 1)$$

then the actual cost of the mapping $g(n)$ can be found out using equation 1.

Computation of h(n). For each mapping, let $\mathbf{L} \subseteq \mathbf{T}'$ be the set of tasks not assigned to any processor by the mapping; let z be the processor which finishes computing last; if one considers the augmented mapping by adding the tuple (x, z), $x \in \mathbf{L}$, $u(x) \neq 0$, let $\Delta_p(x)$ be the increase in the value of $g(n)$ for this mapping; if one considers augmented mappings by adding tuples $(x, y), x \in \mathbf{L}, y \in \mathbf{M}, y \neq z$, and computes the increase in cost $g(n)$ in each case be $\Delta ca(x, y)$, let

$$\Delta_c(x) = \min_{y \in \mathbf{M}: y \neq z} (\Delta_{ca}(x, y)) \qquad (4)$$

be the minimum increase in $g(n)$. Finally, if

$$\Delta_a(x) = \min(\Delta_c(x), \Delta_p(x)) \qquad (5)$$

then

$$h(n) = \min_{x \in \mathbf{L}: u(x) \neq 0} (\Delta_a(x)) \qquad (6)$$

and the total cost of mapping is $g(n) + h(n)$.

The computation of $g(n)$ and $h(n)$ is based on the methodology presented in [6]. However, the general methodology presented in [6] cannot be directly applied to the present graph-matching problem, and therefore, necessary modifications have been made for the present problem.

Using the above formulation, the graph-matching based load balancing algorithm next tries to find an optimal map of $\mathbf{T}^1 \cup \mathbf{T}^2$ on processors corresponding to elements 0,1 and 2 of $\mathbf{P}$. And finally, *Balance*(2, 3) finds the distribution so that the turnaround time for the processor corresponding to the fourth and the final element of $\mathbf{P}$ is minimum. Since $\mathbf{P}$ has been assumed to have remained unaltered during the successive passes, at the end of the third pass, the turnaround time for the processor which finishes the last is minimized and this guarantees minimum node turnaround time. As mentioned before, *Balance*(2, 3) also tries to move some of the elements of $\mathbf{T}^3$ to the first two processors of $\mathbf{P}$ to obtain the minimum turnaround time for the last processor in $\mathbf{P}$ and hence the task, for the present example.

However, as we go on re-distributing task modules amongst partitions, it may so happen that $\mathbf{P}$ gets altered. Thus, after a pass, the partition corresponding to the ith element in $\mathbf{P}$ can have its processing time reduced so that the order of the corresponding process is now $j(j < i)$. It then is necessary to update turnaround times for all processors starting from the jth element of $\mathbf{P}$. When all partitions are refined such that $j = p$, for a p processor system, the complete map for the optimal distribution is achieved.

It is to be noted, that the routine *Diff* updates the value of j to a value less than its original value if and only if some processors succeeding it actually gets a smaller computation time due to re-distribution, than the jth processor or any of its preceding processors. That is to say, suppose after balancing with $j = 2$, it is seen that the processor which was in position 3, has moved to position 1 as an example, the next step would be *Balance*(0, 1) , with $j = 1$, but if the processors in positions 0,1 and 2 change positions amongst themselves, then the next step would be *Balance*(2, 3). This is done to prevent generation of homologous mappings or mappings with identical turnaround times.

It is to be noted that the routine *Diff* updates j to a value less than its original value if and only if re-distribution caused computation time TC of some processor(s) succeeding processor j (in **P**) to decrease. For instance, suppose that after balancing with $j = 2$ the processor that was in position 3 in **P** has moved to position 1. Then the next step would be $Balance(0, 1)$. However, if a change of position occurs within the set of processors originally in positions 0,1, and 2, the next step would be $Balance(2, 3)$. This is done to prevent the generation of mappings with identical turnaround times.

3 Adjusting Granularity by Splitting Meshes

TRACE operates on block-structured grids. The mesh generation software used in the preprocessing phase generates a number of blocks according to the geometry the user has defined by means of a CAD system. The mesh generator typically will generate the minimum number of blocks that are required to express the given geometry. It uses the CFD General Notation System (CGNS) standard as its output format. Using these blocks as the unit of partitioning for parallel execution will result in poor load balancing if the number of processors is large. Often, one large block has more than half of the total mesh nodes, which limits the achievable speedup to 2.

To achieve scalability, we have developed a *splitter* module, that takes as inputs the blocks produced by the mesh generator and automatically subdivides each of them into smaller units, which we call *micro-blocks*. They are used as the basis for partitioning in the subsequent phase.

A block is split by subdividing cells along the i, j, and k dimensions. A user-definable parameter ensures that micro-blocks have a minimum number of layers in each of the three dimensions, thus avoiding degenerate geometries. In addition, the splitting process has to take into account a number of model specific restrictions. The output is again stored in standard CGNS format.

The splitter module adjusts the size of the micro-blocks to the granularity required for the number of partitions to be generated. Subsequently, the mapper module, which implements the multi-level graph-partitioning algorithm described in the previous section, assigns micro-blocks to processors. Micro-blocks that have been assigned to the same processor and that are adjacent in the same block, are joined to form a larger micro-block in order to minimize overhead.

The system thus can dynamically adapt to the available execution environment, if the splitter and mapper module are called after the batch queuing system has selected a pool of resources on which the simulation is to be run.

4 Fault Tolerance – The Auto-Restart Module

In the MTU production environment, a typical TRACE simulation will run for several days on a cluster under control of the control of the LSF batch queuing system. As our load balancing strategy increases scalability, much larger clusters will be used in the future. Hence, component failure will become more likely. Therefore, we developed an automatic fault recovery mechanism that is integrated with LSF.

A monitoring program determines the state of the SMP nodes of the resource pool that has been assigned to the application. It uses LSF API functions and is

Table 1. Runtimes on 24 CPUs of the MTU Linux Cluster (4000 iterations)

		original distribution [sec]			with splitter-mapper [sec]		
Processor No.	Node No.	TP	TC	TA	TP	TC	TA
1	1	4574.20	9389.53	14353.29	4584.69	2724.83	7640.83
2	1	4407.73	9545.18	14349.05	4060.39	3405.67	7640.77
3	2	4063.21	9945.41	14349.06	4874.88	2449.87	7640.85
4	2	4096.10	9920.84	14349.12	4744.50	2350.96	7640.84
5	3	4329.30	9600.94	14349.13	4351.32	2732.32	7640.84
6	3	4241.74	9778.23	14349.10	4760.19	2605.42	7640.81
7	4	4506.06	9557.66	14349.10	4622.49	2555.25	7640.83
8	4	4828.30	9267.79	14349.06	5119.85	2131.49	7640.77
9	5	4640.25	9251.18	14349.17	4494.67	2680.84	7640.79
10	5	4710.0	9339.92	14349.09	4947.44	2141.49	7640.84
11	6	3985.80	10060.04	14349.07	4814.42	2325.04	**7640.88**
12	6	4663.99	9367.00	14349.12	4658.69	2601.71	7640.85
13	7	6714.19	7422.17	14349.12	4852.80	2213.77	7640.83
14	7	4586.14	9405.85	14349.07	4497.80	2552.97	7640.84
15	8	3545.57	10450.86	14349.13	4346.67	3016.68	7640.86
16	8	3576.55	10438.54	14349.06	4953.98	2287.74	7640.79
17	9	12944.75	1273.78	14349.11	4438.38	2955.62	7640.85
18	9	6817.54	7264.70	14349.10	4178.61	3050.82	7640.84
19	10	3572.06	10435.64	14349.13	5062.44	2129.01	7640.81
20	10	3602.37	10401.43	14349.07	5038.58	2306.47	7640.85
21	11	4151.34	9872.06	14349.12	4561.48	2341.59	7640.83
22	11	4157.92	9894.03	14349.05	4877.53	2005.52	7640.77
23	12	4212.97	9840.94	14349.19	4723.46	2451.68	7640.84
24	12	4014.64	9997.63	14349.07	5053.87	2197.62	7640.82

triggered automatically whenever a node failure is detected. We have developed two strategies for recovery. One is to remap subdomains from the faulty nodes to healthy nodes by means of graph matching. This approach performs well as long as the number of subdomains that have to be remapped is small, typically less than 20. If a larger number of subdomains have to be migrated, the second approach performs better: all subdomains are re-assigned.

Table 2. Runtimes on 16 CPUs of the Hitachi SR8000 (100 iterations)

		original distribution [sec]			with splitter-mapper [sec]		
Processor No.	Node No.	TP	TC	TA	TP	TC	TA
1	1	566.84	24.90	604.97	319.33	113.49	440.91
2	1	336.07	257.93	603.45	336.87	94.52	442.95
3	1	331.74	259.87	602.46	327.90	106.50	440.24
4	1	300.90	291.83	602.05	340.37	83.00	441.34
5	2	320.43	270.26	603.73	336.04	94.80	440.90
6	2	319.69	272.51	602.28	324.85	107.07	443.07
7	2	316.36	275.09	602.73	285.73	145.19	439.63
8	2	316.97	273.87	605.13	349.96	81.59	441.62
9	3	324.68	266.48	604.997	355.26	75.53	441.70
10	3	302.59	289.97	603.44	306.96	124.05	441.27
11	3	313.38	276.90	602.98	342.96	87.837	**443.44**
12	3	314.96	276.27	602.09	229.60	200.99	440.77
13	4	306.32	283.77	603.74	401.17	30.89	441.02
14	4	331.84	258.58	**605.15**	363.60	58.19	441.42
15	4	286.47	304.58	602.49	227.96	205.44	439.73
16	4	306.40	284.36	602.98	342.78	83.10	443.00

Currently, automatic recovery is limited by the fact that LSF does not return an error code if a node becomes inaccessible due to network failure. In this case, the job will enter a blocked state and our restart mechanism will not be activated.

5 Experimental Results

The splitter-mapper was used to get runtime results using several practical test cases on two platforms, the Linux cluster at MTU, and the Hitachi SR8000 at the Leibniz Rechenzentrum, Munich. Each node of the MTU Linux cluster is an SMP comprising of 2 Pentium III processors running at 1.1 GHz with a total memory of 2GB per node. Each node of the Hitachi SR8000 is an SMP comprising of 8 RISC processors each with 1.5 GFlop/s peak performance. Several representative test cases were used. Tables 1 and 2 shows the aggregate time each processor spent in computation, TP; the aggregate time each processor spent in communication, TC, and the turnaround time, TA.

It is important to mention that the number of iterations required for convergence in the latter case was found to be marginally more than the number of iterations in the former(original) case. Furthermore, the sets of CPUs used in the 2 cases were not identical, but similar in the sense that they were part of the same cluster and connected in the same way i.e. 2 CPUs per node (same topology and hence homogeneous w.r.t. communication) and were of same strength (hence homogeneous w.r.t. computation). In the former case the task assignment was done using an in-house developed mapping algorithm provided by MTU Aero Engines GmbH. and in the latter case, the integrated splitter-mapper was used. As seen from Table 1, the turnaround time in the former case is 14353.29 secs. and in the latter case, it is 7640.88 secs. implying a speed-up of 1.87 . This test was then repeated on 16 processors of the Hitachi SR8000 at the LRZ, München and the results are presented in Table 2. As seen from Table 2, the speed-up obtained is $605.15/443.44 = 1.36$.

Experiments were then conducted to explore the relationship between speedup and the topology for a particular application and also to determine the gain in speedup due to the optimal intra-node task assignment methodology which takes care of interleaved computation and communication. The results show that the speed-up increased with finer splitting of subdomains and also with increased number of processors. The gain due to optimal intra-node assignment alone, as compared to an assignment using Metis varied from 1.04(4%) on the MTU cluster with 2 processors per node to 1.11(11%) on the SR8000 with 4 processors per node. In general, for an application, the gain in speed-up for a particular application was found to be dependent on the number of processors per node and the ratio of inter-node to intra-node message passing latencies.

6 Conclusion

The integrated load balancing methodology for mesh-based simulations, discussed in this paper, provides an effective means of achieving scalable parallel simulation on large clusters of SMPs. It addresses the limitations of commercial mesh generation software by splitting such meshes using information about the topology of the processors. This guarantees a high scalability and relieves the application programmer from the task of producing finer meshes at the cost of a larger communication and computation. The hybrid load-balancing mechanism presented makes it possible to interleave computation and communication amongst the processors constituting the nodes of the SMP cluster, thus adding to further speed-up. This methodology is, therefore, ideal for explicit parallel processing platforms e.g. the Message Passing Interface (MPI) in the sense it automatically analyses the simulation task and the resources available to derive the explicit task assignment which can be used by the application itself. The restart module provides an option to automatically restart the simulation application without having to abort it in the event of a node failure in a SMP cluster.

References

[1] Eulitz, F., Engel, K. and Gebing, H., Application of a One-equation Eddy-viscosity Model to Unsteady Turbomachinery Flow, Engineering Turbulence Modelling and Experiments 3, Rodi, W. and Bergeles, G (Eds.), Elsevier Science B.V., 1996.
[2] Gao, H. , Schmidt, A. , Gupta, A. and Luksch, P., Load Balancing for Spatial-Grid-Based Parallel Numeric Simulation on Clusters of SMPs - A Case Study from an Industrial CFD Simulation, published as a chapter in High Performance Computing in Science and Engineering, Munich 2002, Wagner, S., Hanke, W., Bode, A. and Durst, F.(Eds.), SpringerVerlag , ISBN 3-540-00474-2, pp. 467–481.
[3] Gao, H. , Schmidt, A. , Gupta, A. and Luksch, P., Load Balancing for Spatial-Grid-Based Parallel Numeric Simulation on Clusters of SMPs, Proceedings of the Euromicro PDP2003 Conference, February 5-7, 2003, Genoa, Italy, IEEE Computer Society Publications, pp. 75–82.
[4] Gao, H. , Schmidt, A. , Gupta, A. and Luksch, P., A Graph-matching based Intra-node Load- Balancing Methodology for Clusters of SMPs, Proceedings of the 7th World Multiconference on Systems, Cubernetics and Informatics (SCI 2003), July 2003.
[5] Platform LSF Family, http://www.platform.com/products/LSFfamily/
[6] Shen, Chien-Chung and Tsai, Wen-Hsiang, A Graph Matching Approach to Optimal Task Assignment in Distributed Computing Systems using a Minimax Criterion, IEEE Transactions on Computers, vol. C-34, No. 3, March 1985.
[7] MeTiS Home Page. http://www-users.cs.umn.edu/~karypis/metis/

OPTILAS: Numerical Optimization as a Key Tool for the Improvement of Advanced Multi-Beam Laser Welding Techniques

Verena Petzet[1], Christof Büskens[1,4], Hans Josef Pesch[1], Victor Karkhin[1,3], Maksym Makhutin[1,2], Andrey Prikhodovsky[2], and Vasily Ploshikhin[2]

[1] Lehrstuhl für Ingenieurmathematik, Universität Bayreuth, D-95440 Bayreuth
{verena.petzet,hans-josef.pesch}@uni-bayreuth.de
[2] Neue Materialien Bayreuth GmbH, Gottlieb-Keim-Str. 60, D-95448 Bayreuth
{andrey.prikhodovsky,vassili.plochikhine}@nmbgmbh.de
[3] now: Department of Welding Engineering, St. Petersburg Polytechnical University, RUS-195251 St. Petersburg
[4] now: Zentrum für Technomathematik, AG Optimierung & Optimale Steuerung, Universität Bremen, D-28334 Bremen

Abstract. Multi-beam laser welding is an advanced welding technique which can successfully prevent hot cracking, cf. [3], [4]. In order to guarantee that this technique prevents the initiation of hot cracks in the solid-liquid region, it is important to choose the positions, sizes, and powers of the additional heat sources suitably, e.g. optimally if an appropriate objective function can be established. In case of inappropriate choices for these parameters, hot cracking can even be enhanced. Until now these quantities are generally chosen by trial and error. This paper aims towards the simulation and optimization of multi-beam laser welding in order to demonstrate the potential of numerical optimization for the further improvement of this welding technique.

For this purpose a constrained nonlinear programming problem is formulated which provides a solution for the hot cracking problem by minimizing the accumulated transverse strain, i.e. the opening displacement, in the solid-liquid region. This approach is based on the so-called strip expansion technique, cf. [6]. For the objective function investigated in this paper it is sufficient to take into account a stationary temperature field in a moving reference frame. It is described by a partial differential equation for which it is possible to find a semi-analytical solution in terms of Bessel functions. Their computation is very time consuming and should be performed in parallel. If an optimization of the process is desired the amount of computation increases even more. This is due to the fact that, in addition to the solution of the partial differential equation, certain sensitivities must be computed in each loop of the optimization iteration, i.e., partial derivatives of the simulation output with respect to the optimization parameters.

Key words: Modeling, simulation, and optimization with partial differential equations, multi-beam laser welding, hot cracking.

1 Introduction

Laser welding is a modern joining technique for new metallic materials and particularly suitable for handling lightweight constructions as they occur, e.g., in automotive, aircraft, rail vehicle and shipbuilding industries because weight is to be reduced and energy is to be minimized. In these industries new materials like aluminum alloys become increasingly important, however their welding can cause severe problems because of the high risk of the appearance of hot cracks, cf. [5].

The idea to use additional heat sources as a mean for the prevention of hot cracking in welds was first suggested in the 70ies (cf. [1], [9]). On the basis of FEM-simulations it was demonstrated that introducing an additional heat source at the right place and at the right time leads to a beneficial compressive stress (or strain) in regions, which are critical with respect to hot cracking, cf. [1], [2]. Despite the apparent simplicity of the suggested idea, additional heating has still not found the expected industrial application. The determination of the optimal positions, sizes, and powers of additional heat sources seems to be the most important reason to retard the application of this technique in practice.

Recently, the multi-beam welding technique has been successfully realized experimentally, cf. [5], [4], and [3].

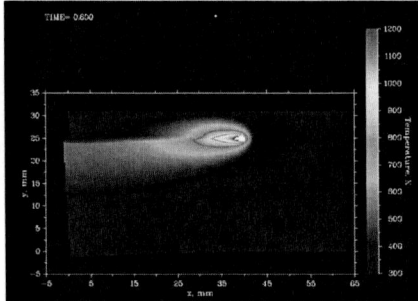 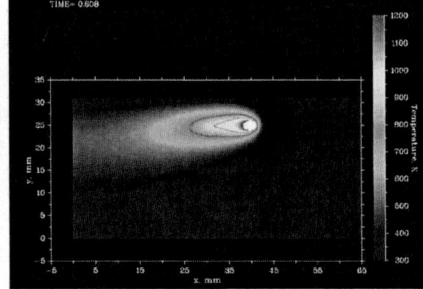

Fig. 1. Simulation of single welding: a crack arises. For a reproduction of this figure in colour, see Fig. A.3.

Fig. 2. Simulation of two-beam welding: no crack arises. For a reproduction of this figure in colour, see Fig. A.4.

Based on a detailed model, which takes into account the temperature field, the thermomechanical deformations, and the metallurgical properties of the weld, a FEM-simulation package has been developed at the Neue Materialien Bayreuth GmbH. Figures 1 and 2 show typical simulation results obtained by this software. Figure 1 shows the arising of a hot crack for a single laser beam performing a dummy seam. Figure 2 shows an equivalent situation where a second laser beam is driven along the lower boundary of the plate. Note that no optimization has been performed so far on the basis of this detailed model. Appropriate parameters for the additional heat sources have been found by a number of trial and error simulations. These numerical results have been validated experimentally, cf. Figs. 3 and 4.

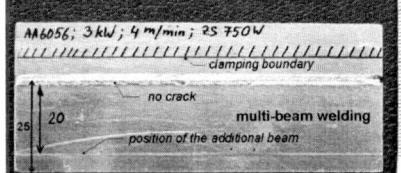

Fig. 3. Performing a dummy seam from left to right. A crack arises since no additional laser beam is employed

Fig. 4. Performing a dummy seam by driving an additional laser beam along the lower boundary of the plate. No crack arises

Hence, there is an obvious need for the application of systematic numerical optimization to improve this technique. This paper presents a first step in this direction.

2 Modeling of Hot Crack Initiation

Hot crack initiation is caused by transverse tensile strains accumulated in the mushy zone, i.e. the solid-liquid region. The boundaries of the mushy zone are defined by the two isotherms, which correspond to the liquidus[5] and to the solidus[6] temperature.

The simplest approach to model hot cracking is based on the so-called strip expansion technique, see [6] and Fig. 5. According to this method the plate to be welded is divided into thin strips. It is assumed that each strip frictionlessly deforms independently of each other.

We assume an ideal plastic behavior of the material. Then the shrinking of each strip during the process of solidification is proportional to the temperature differences in this strip between the beginning and ending of the solidification. Then the accumulated transverse strain (opening displacement) W_{od} can be calculated by the following integral:

$$W_{\text{od}}(x_S) = 2\alpha \int_0^B \left[T(x_L, y) - T(x_S, y)\right] dy . \tag{1}$$

The opening displacement W_{od} will be the objective function to be minimized.

Hereby, T denotes the temperature depending on the spatial coordinates x and y. α is a thermal expansion coefficient, and B is the distance from the centerline of the plate to the restraint.

In order to describe the x-coordinates x_S and x_L, the two closed curves $T = T_S$ and $T = T_L$ in Fig. 5 have to be explained. They are the isotherms of the solidus

---

[5] temperature at which the solidification starts
[6] temperature at which the solidification is completed

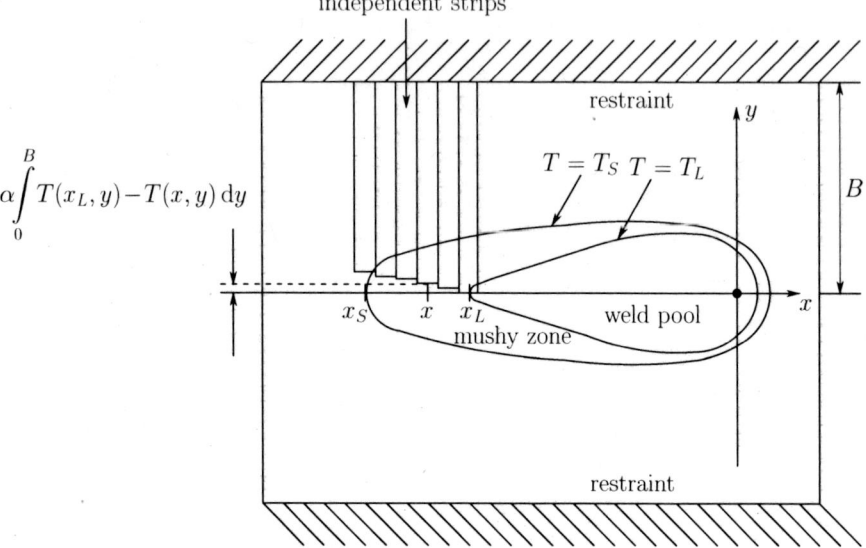

Fig. 5. Illustration of the strip expansion technique and notations

temperature T_S and the liquidus temperature T_L, resp., which enclose the solid-liquid region in which hot crack initiation takes place. The points of intersection of those isotherms with the x-axis are denoted by x_S and x_L. For our problem only those points of intersection are of interest which are behind the main laser beam relative to the direction of its movement.

No hot crack will arise, if the value $W_{\mathrm{od}}(x_S)$ is below a critical value. The value of the critical strain is about 15 μm for the aluminum alloy AA6056 considered here and is determined experimentally, cf. [5].

In order to calculate the opening displacement W_{od} we obviously need to compute the temperature field T, i.e., to solve a partial differential equation.

Mathematical Model for the Temperature Field

By Fourier's law of heat conduction and the conservation law of energy we have the following instationary parabolic differential equation

$$c\rho s \frac{\partial T}{\partial t} = \left[\frac{\partial}{\partial x}\left(\lambda \frac{\partial T}{\partial x}\right) + \frac{\partial}{\partial y}\left(\lambda \frac{\partial T}{\partial y}\right)\right] s + \bar{q}_2\left(x - vt, y\right). \tag{2}$$

If c, ρ, λ, and s are assumed to be constant, this leads to

$$\frac{\partial T}{\partial t} = a\,\Delta T + \frac{\bar{q}_2\left(x - vt, y\right)}{c\rho s}. \tag{3}$$

Here t denotes the time. The quantities c, ρ, and s are the specific heat capacity, the density, and the thickness of the plate, resp. Furthermore λ denotes the thermal conductivity and $\bar{q}_2$ the moving area-specific heat flow density associated with a source moving with constant speed v in the direction of the x-axis. Hereby v is the travel speed of the main laser beam. The quantity $a = \lambda/(c\rho)$ in (3) stands for the thermal diffusivity. All these quantities are put together with their physical units in Table 1 of Appendix A.

The 2D heat equation (3) describes the situation in a fixed coordinate system. Since the laser beam is moving it is obvious to change to a moving reference frame with the position of the main laser beam in the origin. We denote the coordinates of the moving coordinate system traveling with constant speed v by (x, y, t), whereas (x_0, y_0, t_0) are the coordinates of the original system with the origin at the left boundary of the plate. Both x-axis are aligned with the velocity vector of the main laser beam. See Fig. 6.

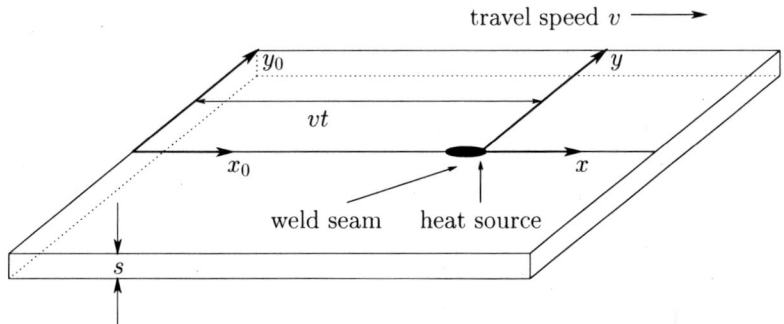

Fig. 6. Original coordinate system and the moving reference frame

From Fig. 6 we immediately obtain the following coordinate transformation:

$$x = x_0 - v\,t_0, \quad y = y_0, \quad \text{and} \quad t = t_0. \tag{4}$$

To see this, take an arbitrary but fixed point $P = (\tilde{x}, \tilde{y})$ of the plate. Then the value of its x-coordinate in the (x, y, t)-system decreases when this systems is moving from left to right with increasing time t.

Now the 2D heat equation in the moving reference frame reads as follows:

$$\frac{\partial T}{\partial t}(x, y, t) = a\,\Delta T(x, y, t) + v\,\frac{\partial T}{\partial x}(x, y, t) + \frac{q_2(x, y)}{c\rho s}. \tag{5}$$

In addition to the diffusion and the source term in (3), one now obtains a convection term, while the source term is now time independent.

Because of the formulation (1) of the objective function it is sufficient to consider only the quasi-stationary state of the temperature field. Hence (5) can be simplified to

$$\Delta T(x, y) + \frac{v}{a}\frac{\partial T}{\partial x}(x, y) + \frac{q_2(x, y)}{\lambda s} = 0. \tag{6}$$

The source term q_2 for the temperature entries of the three laser beams can be modeled as follows:

$$q_2(x,y) = \begin{cases} q_{2_{add}} = \dfrac{q_{add}}{\pi r_{add}^2} & \text{if } (x - x_{add})^2 + (y - y_{add})^2 \leq r_{add}^2 \\ q_{2_{main}} = \dfrac{q}{\pi r^2} & \text{if } x^2 + y^2 \leq r^2 \quad (r \ll 1) \\ 0 & \text{otherwise} \end{cases}$$

All quantities in the definition of q_2 with subscript add are related to the additional laser beams. Thereby x_{add} and y_{add} denote the positions of the additional laser beams in the moving reference frame. The power of both auxiliary laser beams is denoted by q_{add}. It is assumed to be evenly distributed over a domain $D = D_1 \cup D_2$, consisting of two disjoint circles. The radius of both circles is denoted by r_{add}. Power and radius of the main laser beam is denoted by q and r.

In order to have a unique solution of the parabolic Eq. (5) an initial condition and boundary conditions must be imposed. For the sake of simplicity we have chosen

$$T(x,y,0) = T_0 = \text{const} \tag{7}$$
$$T(x,B,t) = T(x,-B,t) = T(\pm\infty, y, t) = T_0 = \text{const}. \tag{8}$$

These conditions imply that the plate is infinitely long in x-direction but with finite length $2B$ in y-direction. T_0 denotes the initial temperature, e.g., the ambient temperature. The Dirichlet type boundary conditions keeping the temperature on a constant level can be interpreted as heat conduction through the restraint devices in an appropriate way.

For the elliptic Eq. (6) only boundary conditions are now allowed to get a well-posed unique solution:

$$T(x,B) = T(x,-B) = T(\pm\infty, y) = T_0 = \text{const}. \tag{9}$$

Solution of the Temperature Field Equation

If the plate is infinite there exists an analytical solution for the temperature field (cf. [7]):

$$T(x,y) = \frac{q}{2\pi\lambda s} \exp\left(-\frac{vx}{2a}\right) K_0\left(\frac{v\sqrt{x^2+y^2}}{2a}\right) + T_{add}(x,y) + T_0 \tag{10}$$

with

$$T_{add}(x,y) = \frac{1}{2\pi\lambda s} \frac{q_{add}}{\pi r_{add}^2} \iint_D \exp\left(-\frac{v(x-\xi)}{2a}\right) K_0\left(\frac{v\sqrt{(x-\xi)^2+(y-\eta)^2}}{2a}\right) d\xi d\eta \tag{11}$$

and $D = D_1 \cup D_2$. Hereby, $K_0(.)$ denotes the modified Bessel function of the second kind of order zero.

Equation (10) shows that the temperature in a point (x,y) consists of essentially two parts. The first term on the right hand side describes the temperature induced by the main laser beam and the second term the temperature due to the additional laser beams.

In order to get a solution which fulfills also the boundary conditions (9), the so-called *method of images* is used. By this method the process of heat propagation in a bounded solid is considered as part of such a process in an infinite solid. Hereby a set of additional sources and sinks are placed in such a way, that the process of heat propagation is influenced in the same way as by the boundary conditions in the finite case (cf. [7]).

The idea is simple: Put additional sources and sinks so that the boundary conditions are preserved. This is obtained by a pattern of infinitely many sources and sinks lying symmetrically to the boundaries of the plate at $y = B$ and $y = -B$. Starting from the positions of the original real laser beams virtual sinks are placed so that they compensate for the entries of the real beams in such a way that the boundary conditions are kept. This process is continued by placing now sources to compensate for the virtual entries of the sinks which have been just placed in the step before, and so on. Applied to the laser welding problem one obtains

$$T(x,y) - T_0 = \sum_{i=-\infty}^{\infty} \frac{q}{2\pi\lambda s} \exp\left(-\frac{vx}{2a}\right)[f(x,y-2i(2B)) - f(x,y-(2i+1)(2B))]$$

$$+ \sum_{i=-\infty}^{\infty} \iint_D \frac{q_{2\text{add}}}{2\pi\lambda s} \exp\left(-\frac{v(x-\xi)}{2a}\right)[f(x-\xi,y-2i(2B)-\eta)$$

$$- f(x-\xi,y-(2i+1)(2B)+\eta)]\,\mathrm{d}\xi\mathrm{d}\eta.$$

(12)

Here, $f(\arg_1, \arg_2) := K_0(v\,\|(\arg_1, \arg_2)\|_2 / (2a))$. This is the unique solution of the elliptic boundary value problem (6), (9).

3 Optimization of the Welding Process

By employing the multi-beam welding technique it is possible to compensate for the tensile strains. Hereby two additional laser beams move with the same travel speed as the main laser beam. This can help to prevent hot cracking since the additional beams may produce compressive strains in the mushy zone of the weld. Crack initiation can even be suppressed if the coinciding positions, sizes, and powers of the two additional heat sources are suitably chosen. However, an inappropriate choice of these parameters may even enhance hot cracking.

Figures 7 and 8 show a comparison between single and multi-beam welding. Now we can combine the minimization of the opening displacement (1) with the semi-analytical formula (12) for the temperature field. The optimization is to be performed with respect to the coinciding optimal positions, sizes, and powers of the additional heat sources. Figure 9 shows the geometry and, except for symmetry, the optimization variables for this problem.

Introducing the vector parameter $p := (p_1, \ldots, p_6)$ the temperature now depends on p, too. Hence, we write $T(.,.;p)$. The parameters p_5 and p_6 are implicitly defined by the equations given in Fig. 9. In the optimization process they are taken into account by penalty terms.

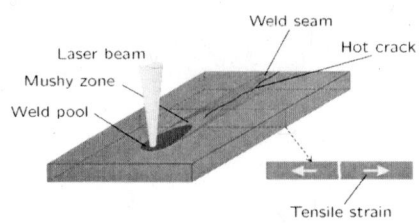

Fig. 7. Conventional laser beam welding with the formation of a hot crack, cf. [6]

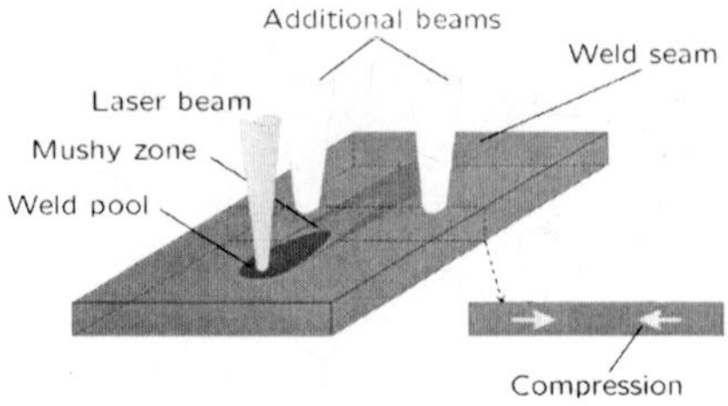

Fig. 8. Multi-beam welding with optimal positions, sizes, and powers of two additional beams to suppress hot crack initiation, cf. [6]

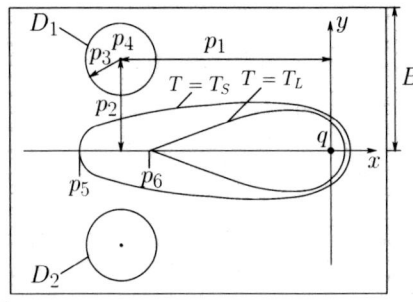

Fig. 9. Top view of the plate

p_1 x_{add} : x-position of the midpoints of the circles

p_2 y_{add} : y-position of the midpoint of the upper circle

p_3 r_{add} : radius of the circles

p_4 q_{add} : power of the additional laser beams

p_5 def. by $T_S = T(p_5, 0; p)$, $p_5 \,\widehat{=}\, x_S$

p_6 def. by $T_L = T(p_6, 0; p)$, $p_6 \,\widehat{=}\, x_L$

The nonlinear constrained programming problem now reads as follows:

$$\min_{p \in \mathbb{R}^6} \left\{ r_1 \left[2\alpha \int_0^B [T(p_6, y; p) - T(p_5, y; p)] \, dy \right] + \right.$$
$$\left. + (1 - r_1) \left[r_2 \left(\frac{(T(p_6, 0; p) - T_L)^2}{T_L^2} + \frac{(T(p_5, 0; p) - T_S)^2}{T_S^2} \right) \right] \right\} \quad (13)$$

subject to the following inequality constraints: $0.5\, p_3 \leq |p_1| \leq 30$, $0.1 \leq p_3 \leq p_2$, $0 \leq p_4 \leq 1200$, $p_2 + p_3 \leq B$, $T_{\max}(x, y; p) \leq T_S$ for $x, y \in D(p) = D_1(p) \cup D_2(p)$. Hereby, the constants r_1 and r_2 are weighting factors with appropriate units to make the objective function dimensionless.

The inequalities describe restrictions which must be obeyed. The x and y distances between the main laser beam and the additional laser beams should be sufficiently separated but not to far from each other which is expressed by the first two constraints. Moreover, some physical restrictions have to be fulfilled. Technically it is not possible to have a smaller radius then 0.1 mm; see the first inequality of the second constraint. The third condition limits the power of the additional laser beams. Note $p_4 = 0$ W means that there are no additional heat sources. If $p_4 = 1200$ W the additional laser beams have the same power as the main laser beam. The next constraint ensures that y_{add} plus the radius of the additional laser beam is not greater than half of the width of the plate. Finally there is a nonlinear condition bounding the maximum temperature at each point $(x, y) \in D$ by the solidus temperature. It makes sure that no melting takes place at the additional heat sources. For the sake of simplicity the maximum temperature is assumed to arise in the point $(p_1 - p_3, p_2)$ since this usually is the hottest point of a circularly radiated single laser beam.

From a physical point of view it is important that the first term of the objective function allows for both positive and negative signs. If this term is greater than the critical value of 15 µm, strain effects dominate. If it is less than that value, the stress effects dominate. Therefore minimizing the integral in (13) will lead to the best possible compensation of the accumulated transverse strain. If the value for W_{od} is below the critical value no hot cracks will arise. For more information see [6].

4 Numerical Results

First some remarks have to be made how to compute the objective function (13). Because of (12) the modified Bessel function of the second kind of order zero must be computed. This is performed by the module K0 of the software package SPECFUN of NETLIB, see [10]. Then the integrals are approximated by the composite midpoint rule. Using this approach the following difficulties must be overcome.

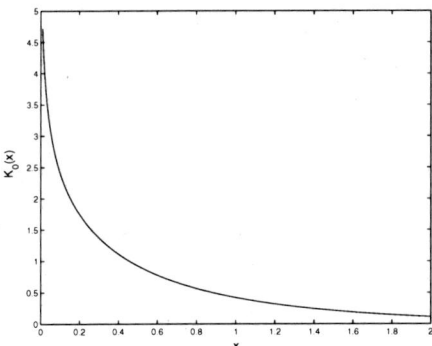

Fig. 10. The modified Bessel function of the second kind of order zero

Since the modified Bessel function K_0 has a singularity at $x = 0$ (cf. Fig. 10), the numerical results for the triple integral (see (12),(13)), as a function of the parameter p, show spurious peaks which vanish if the stepsize of the discretization for the composite midpoint rule tends to zero (cf. Fig. 12).

If the discretization is too coarse the optimizer may get stuck in spurious minima (cf. Fig. 11).

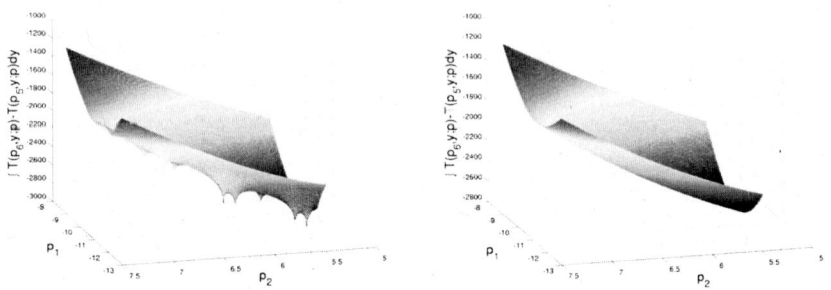

Fig. 11. Spurious minima caused by the singularity of the modified Bessel function (number of discrete points for the approximation of the triple integral is 6)

Fig. 12. Here no spurious minima can be seen (number of discrete points for the approximation of the triple integral is 60)

However, there is a price to be paid for decreasing the stepsize. The amount of computation increases considerably with the number of discretization points, see Fig. 13. More than fifty percent of the computing time is spent evaluating the modified Bessel function.

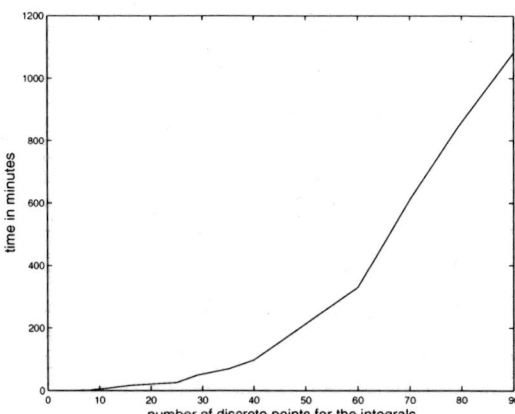

Fig. 13. Computing time for the optimization depending on the number of discrete points for the approximation of the triple integral

Independent of the difficulties caused by the singularity of the Bessel function K_0 the optimization suffers from the existence of many local minima. All local minima which have been found lay on the boundary of the admissible set of the nonlinear programming problem (13). The best local minimum obtained so far is

$p_1 = -11.31\ldots$ [mm] $p_2 = 6.46\ldots$ [mm] $p_3 = 3.53\ldots$ [mm]
$p_4 = 820.65\ldots$ [W] $p_5 = -13.71\ldots$ [mm] $p_6 = -5.31\ldots$ [mm]

Since the objective function is negative (its value is $-0.096\ldots$) hot cracks will, according to the model, not arise as explained before. For all computations the SQP method due to [8] has been used. The result of the optimization is given in Figs. 14 and 15.

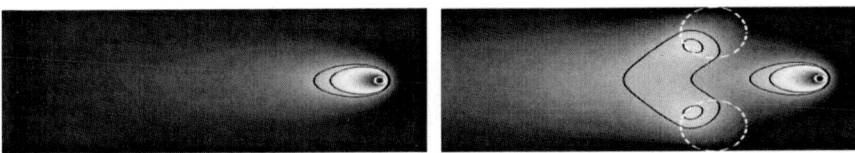

Fig. 14. Temperature field generated by conventional laser beam welding, i.e. without additional laser beams. For a reproduction of this figure in colour, see Fig. A.5.

Fig. 15. Temperature field generated by multi-beam welding, i.e. with additional laser beams. For a reproduction of this figure in colour, see Fig. A.6.

A comparison of the temperature fields show that the multi-beam technique leads to a more evenly distributed temperature and therefore to a much larger solid-liquid region, cf. Fig. 15. Nevertheless the tensile strain in the mushy zone induced by the main laser beam can be compensated by the compression induced by the additional laser beams. Due to this fact no hot crack initiation takes place in the mushy zone assuming the mathematical model is valid. This is due to the fact that the value of the objective function is below the critical value.

As known the optimization of a nonlinear programming problem requires that all functions involved in the problem must be sufficiently smooth. A crucial point is here the approximation of the triple integral as mentioned above.

The use of a large number of discrete points for the approximation of the triple integral leads to a rapid increase of the computing time, see Fig. 13. In order to improve the performance of the calculation, that part of the program calculating the objective function has been parallelized using MPI (Message-Passing Interface).

Calculations with the parallelized program have been made on two computer clusters running under a Linux operation system. The cluster available at the Regionales Rechenzentrum Erlangen (RRZE) of the University of Erlangen-Nuremberg has 82 nodes with Dual Intel Xeon CPUs and is equipped with the Gigabit Ethernet interconnect. Another cluster available at Neue Materialien Bayreuth GmbH has four nodes with Dual AMD processors and the Myrinet2000 interconnect.

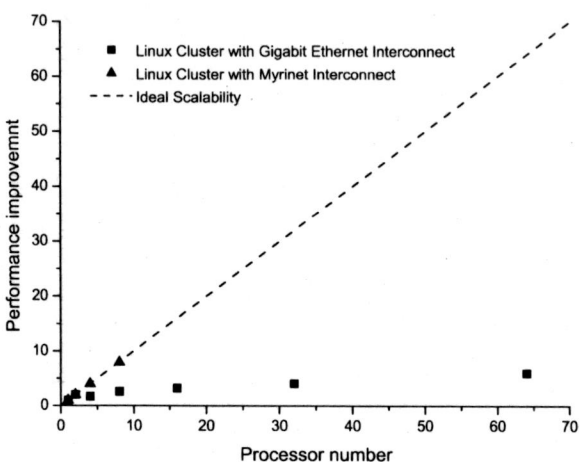

Fig. 16. Performing improvement in computing the triple integral

The relative performance improvement as a function of the number of processors for the optimization is carried out for the approximation of the triple integral using 64 discrete points and is depicted in Fig. 16. This plot shows that the interprocessor communication speed is crucial for the program scalability. On the cluster with the faster interconnect (Myrinet) the performance improvement is much better and the scalability approaches the ideal one. The performance analysis of the other cluster shows that, because of its slower communication, the use of more than 16 CPUs is senseless due to the high communication overhead.

The presented calculation results show that a performance improvement can be achieved using a parallelized version of the program on computer clusters with high-speed low-latency interconnects. The large amount of interprocessor communications prevents getting good performances on clusters equipped with slower cluster interconnects.

5 Conclusions and Outlook

The investigations demonstrate that there is a high potential for the application of numerical optimization for the improvement of advanced laser welding techniques such as the multi-beam welding. Hereby, the amount of computation is crucial so that parallelization is advantageous.

The major task of future work will be the development of a hierarchy of models of increasing complexity. Then we have to investigate how far both simulation and optimization can be performed for this chain of models. Finally the optimal solution of the most detailed model which can be computed within reasonable times has to be implemented into the software package developed by the Neue Materialien Bayreuth GmbH. From a practical point of view this yields an acceptable suboptimal solution.

Mathematically a PDE constrained parameter optimization problem has been obtained. Their solution is a great challenge since PDE constrained optimization is still at the beginning of its theoretical and numerical development. Even more challenging problems of optimal control in the context of laser welding is waiting for their investigations.

The work of the present paper may help to pave the road for future optimization tasks in welding such as for the identification of parameters like the concentrations of welding additives, for the optimal control of the welding processes themselves, and even for shape optimization problems for welded constructions.

Acknowledgement. This work was sponsored by the Bayerische Forschungsstiftung within the Competence Network for Technical Scientific High Performance Computing in Bavaria (KONWIHR) through the project superconducting Optilas (optimization of laser beam welding). The authors would like to thank all those who have supported and financed the project.

A Notation of Physical Quantities

The following table puts together the values and units of all quantities which have been mentioned in the present paper.

Notation	Value	Unit	Notation	Value	Unit
x, y		mm	T_0	293.0	K
t		s	T_{add}		K
s	2.0	mm	T_S	850.0	K
B	10.0	mm	T_L	929.4	K
x_{add}		mm	q	1200	W
y_{add}		mm	q_{add}		W
r_{add}		mm	q_2		$\mathrm{J\,mm^{-2}\,s^{-1}}$
r	0.1	mm	ρ	2.71×10^{-6}	$\mathrm{kg\,mm^{-3}}$
D_1		mm^2	c	894.8	$\mathrm{J\,kg^{-1}\,K^{-1}}$
D_2		mm^2	λ	0.125	$\mathrm{W\,K^{-1}\,mm^{-1}}$
D		mm^2	a	51.55	$\mathrm{mm^2\,s^{-1}}$
v	40.0	mm s^{-1}	α	24.5×10^{-6}	$\mathrm{K^{-1}}$
T		K			

References

1. Akesson, B., Karlsson, L.: Prevention of Hot Cracking of Butt Welds in Steel Panels by Controlled Additional Heating of the Panels. Welding Research International, **6(5)**, 35–52 (1976)
2. Herold, H., Streitenberger, M., Pchennikov, A., Makarov, E.: Modelling of one sided welding to describe hot cracking at the end of longer Butt weld seams. Welding in the World, **43(2)**, 56–64 (1999)
3. Ploshikhin, V., Prikhodovsky, A., Makhutin, M., Zoch, H.-W., Heimerdinger, C., Palm, F.: Multi-beam welding: advanced technique for crack-free laser welding. Proc. 4th Int. Conf., "LANE - Laser Assisted Net Shape Engineering - 2004", Erlangen (Sept. 2004), to be published
4. Ploshikhin, V., Prikhodovsky, A., Zoch, H.-W.: Technologische Maßnahmen zur Vermeidung der Heißrissbildung beim Schweißen von Aluminiumlegierungen, Schweißen und Löten im Luft- und Raumfahrzeugbau. DVS-Verlag, Düsseldorf (2004), p. 46–51
5. Plochikhine, V., Prikhodovsky, A., Zoch, H.-W.: Zum Mechanismus der Heißrissbildung beim Schweißen von Al-Legierungen. HTM, **58(6)**, 357–362 (2003)
6. Plochikhine, V., Zoch, H.-W., Karkhin, V.A., Makhutin, M., Pesch, H.J.: Numerical optimisation of the temperature field for the prevention of solidification cracking during laser beam welding using the multi-beam technique. Proc. of Int. Conf., "Materials Week 2002", Munich (2002)
7. Rykalin, N.N.: Berechnung der Wärmevorgänge beim Schweißen. VEB Verlag Technik, Berlin (1957)
8. Spellucci P.: DONLP2 USERS GUIDE. Technische Universität Darmstadt
9. Shumilin, V.G., Karkhin, V.A., Rakhman, M.I., Gatovsky, K.M.: A Technique of Arc Welding. Patent No. 1109280, USSR (1980)
10. NIST Guide to Available Mathematical Software (GAMS), http://gams.nist.gov/serve.cgi/Module/SPECFUN/K0/9271/, 31.10.2003

ParEXPDE: Expression Templates and Advanced PDE Software Design on the Hitachi SR8000

Christoph Freundl, Ben Bergen, Frank Hülsemann, and Ulrich Rüde

Lehrstuhl für Systemsimulation, Institut für Informatik, Universität Erlangen-Nürnberg, Cauerstraße 6, D-91058 Erlangen, Germany
{Christoph.Freundl,Ben.Bergen,
Frank.Huelsemann,Ulrich.Ruede}@informatik.uni-erlangen.de

Abstract. We demonstrate the use of expression templates for the development of a numerical PDE software library for high performance computers. We discuss the library design and show that expression templates on performance tuned data structures achieve both a user-friendly interface and efficient runtime performance. Our performance results on various architectures including the Hitachi SR8000 illustrate that modern programming techniques like expression templates are well suited for large scale parallel computers.

1 Introduction

The aim of the ParExPDE project is to provide a library for the rapid development of numerical PDE solvers on parallel (super-)computers. To this end, the library features a high level and intuitive user interface that hides the complexities of parallel program development without compromising on efficiency. In this paper we demonstrate how to achieve these goals with the help of expression template programming techniques.

The paper is organized as follows. We start with a description of expression template programming. Then we turn to the types of grids that the library covers and the representation of the discretised differential operators. The latter issue has such an influence on the run time performance that it determines the core storage structures used in this project. Efficient PDE solvers require first and foremost efficient algorithms. Therefore we discuss briefly the merits of multigrid algorithms and we emphasize the fact that our grid structures are well suited for this type of method. Next we illustrate the different possibilities where expression templates can be used in different parts of the library. This is followed by an overview of the library's architecture. After going into some details of the implementation we finally present some performance results on different architectures.

1.1 Basic Expression Templates

The main idea of expression templates as presented in [12] is to enclose arithmetic expressions in a tree-like structure by means of C++ template constructs such that the resulting expression object can be passed to other functions. It also enables fast evaluation of vector and matrix expressions by eliminating temporaries and using inline techniques.

We illustrate this by showing how a simple vector assignment is treated by the compiler without going to deep into the details of expression templates. The assignment is a well-known daxpy operation, in the program context it would look like this:

```
Vector x, y, z;
double a;
z = a * x + y;
```

By appropriately overloading the arithmetic operators, the expression `a * x + y` generates an object of the type `Expr< ExprBinOp< Expr< ExprBinOp< ExprLiteral, ExprVector, OpMult > >, ExprVector, OpAdd > >` as shown in Fig. 1

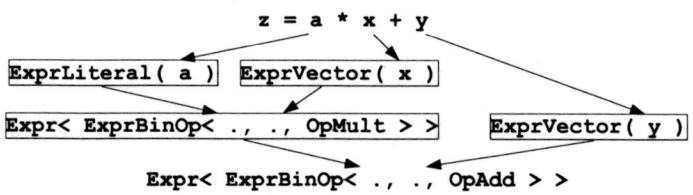

Fig. 1. Generation of the expression object for a daxpy operation

The assignment is performed by the overloaded assignment operator of the `Vector` class.

```
template<class T>
void Vector::operator=( Expr<T>& expr ) {
  for ( int i = 0; i < _size; i++ )
    _values[i] = expr.valueAt( i );
}
```

Each expression class implements the function `valueAt` appropriately, furthermore all `valueAt` functions are declared to be inlined. The resulting assignment in the loop after performing all inlining looks finally like this:

```
_values[i] = a * x._values[i] + y._values[i];
```

An optimizing compiler should have no problems in accelerating the execution of this loop by techniques like loop unrolling or – in case of the Hitachi SR8000 – pseudo-vectorization of the loop (see Sect. 5).

1.2 Structured vs Unstructured Grids

In order to solve a partial differential equation, we first need some type of discretization of the problem. In the current setting, this means that we introduce an unstructured, hexahedral, input grid that provides a discrete representation of the problem domain and optionally apply regular refinement to it until the desired level of resolution is achieved. Such a grid that has had at least one level of regular refinement applied to it will be call a *block–structured grid*. We then apply either the *Finite Difference Method* or the *Finite Element Method*, both of which lead to sparse matrix systems. However, depending on the type of compute grid used to represent the domain, different types of data structures are required to solve these systems on a computer. In the case of the structured grid interiors resulting from regular refinement, we may use stencil based data structures to represent the discretization matrix, as couplings can only occur at regular, known offsets. Unstructured grids sections, like those encountered on the input grid and the shared vertices and edges of the block interfaces, result in generally sparse systems. This implies that, from the point of view of the storage mechanism, the couplings that exist between unknowns cannot be determined beforehand. This makes it necessary to use more complicated data structures to store such systems, and results in added costs during computation. To illustrate this, we will consider the *Jagged Diagonals Storage* scheme for storing a purely unstructured grid, and discuss why it is unable to achieve high performance on modern computer architectures.

The JDS scheme stores the non zero entries of a generally sparse matrix in the following manner: First, the matrix is compressed row-wise by removing all of the zero entries and shifting the remaining terms to the left. The rows are then sorted in descending order by length. Finally, the entries are stored in a linear array by column. These columns are called the jagged diagonals. This scheme requires four arrays to represent the matrix: The *values* array, already mentioned, stores the jagged diagonals. The *columns* array stores the column index of each entry in the *values* array. The *offsets* array stores the extents of each jagged diagonal, and the *permutations* array stores the permutations that were applied during the row sorting. As an example of how the values are accessed, consider the following section from a matrix-vector multiplication,

```
y[i] += values[offsets[j]+i]*x[columns[permutations[offsets[j]+i]]]
```

The important thing to notice in this algorithm, is that indirect indexing is used to access the entries of the matrix. This indirection can cause several problems for the compiler during optimization. Here, we will only consider the most troublesome problem having to do with prefetch optimizations.

The performance of many numerical algorithms is limited by the speed and bandwidth with which main memory can be accessed. Many, if not most, modern architectures have processors that are able to perform many more operations on data than can be supplied by the memory subsystem. This gap in memory speed access versus processor speed is generally referred to as *memory bandwidth limitation*. In order to overcome this limitation, different architectural and compiler strategies and optimizations have been developed. One such strategy that is particularly relevant for our purposes involves both hardware specialization and compiler optimization. This is the so called *prefetch* optimization. Prefetch allows the compiler to insert special scheduling directives to the processor which affect memory access.

This enhanced scheduling cuts down on the *latency* between the time that memory is requested by the CPU and the time that the memory actually begins being streamed by the memory subsystem, and results in a higher percentage of the peak memory bandwidth being available for processing. Our target platform, the Hitachi SR8000, has special hardware modifications that allow it to perform both preload and prefetch operations[1]. These optimizations require that the compiler be able to analyze how data is accessed by a particular algorithm so that it can issue special scheduling instructions to the CPU. This is the root of the problem with the JDS scheme. Because of the additional layer of indirection required to access the matrix entries, the compiler cannot perform this type of optimization, with the result that memory intensive algorithms are unable to achieve a high level of efficiency. For structured grids and stencil based matrix representations, there is no indirection. Therefore, stencil based data structures are more desirable when performance is an issue. To illustrate the difference in performance between structured and unstructured grid representations, consider the following table, which shows the MFLOP/s rates obtained on a single compute node of the Hitachi SR8000 in Munich, Germany.

Table 1. MFLOP/s Rates for Matrix-Vector Multiplication (JDS results courtesy of Dr. Gerhard Wellein, RRZE Erlangen Germany)

# unknowns	729	4913	35937	274625	2146689
JDS	50	260	750	800	1050
Stencils		640	820	3980	5550	7120

Clearly, the stencil based data structures achieve much better performance. The approach presented here draws from experience of the gridlib project [4], [6], [5], in which similar structures for hybrid grids have already achieved high performance results on various platforms [3], [1]. As can be seen from the results in Table 1, the ParExPDE algorithms will have the best performance when several levels of regular refinement have been added to the unstructured input grid.

1.3 Multigrid and Other Iterative Solvers

As stated in the previous section, the types of discretizations considered here lead to a linear system of equations. For solving large systems it is necessary to turn to iterative methods. This is due to the lower algorithmic complexity of iterative solvers. For example, if we were to use standard *Gaußian Elimination* to solve a system of n unknowns, we would need $\mathcal{O}(n^3)$ operations. Suppose that we would like to solve a system with $1,000,000$ unknowns on a CPU that can perform 2 GFLOP/s. We would then need to perform on the order of 10^{18} operations. On our CPU, at peak theoretical performance, this would take $500,000,000$ seconds which is approximately $1,000$ years. In the special case that the linear system is symmetric positive

---

[1] Here, we make the distinction that preload operations load data from main memory directly into a floating point register, in the same manner as a true vector processor. Prefetch, refers to loading data from main memory into one of the caches.

definite, which is true for the types of discretizations we are considering here, iterative schemes such as *Gauß–Seidel* have better complexity $\mathcal{O}(n^2)$. However, these are also unsatisfactory as they lose convergence as the oscillatory error components are eliminated. Fortunately, there is an asymptotically optimal iterative solver for such systems called *full multigrid* with complexity $\mathcal{O}(n)$. Considering the same constraints as in the previous example, we would now only need 5×10^{-4} seconds to solve the problem. It is not a coincidence that regular refinement of an unstructured input grid was discussed in the previous section, as this leads to a suitable, nested, grid hierarchy for performing geometric multigrid, an important component of full multigrid.

2 Expression Templates for Partial Differential Equations

The idea of using expression templates for the treatment of partial differential equations is not new, the development of the ParExPDE library is based on the already existing EXPDE library ([11], [10]). Although there are differences between the two libraries the common basis is that both provide a comfortable interface for the application programmer. An example is given in Fig. 2 which shows how easy and compact the cg method can be formulated in C++ when using the ParExPDE library. Note that we are working on an arbitrary unstructured hexahedral grid where the variables (u, r, g etc.) are defined. The complexity of the evaluation of the right-hand side of the assignments is hidden to the programmer.

```
for(int i = 1; i <= iteration && delta > eps; ++i) {
  g = laplace(d) | interior_points;
  double tau = delta / pxdScalarProduct(d, g, &interior_points);
  u = u + tau * d | interior_points;
  r = r + tau * g | interior_points;
  double delta_prime = pxdScalarProduct(r, r, &interior_points);
  double beta = delta_prime / delta;
  delta = delta_prime;
  d = beta * d - r | interior_points;
  e = u - u_exakt | interior_points;
  double l2_error =
    sqrt(pxdScalarProduct(e, e, &interior_points) / normi);
}
```

Fig. 2. Implementing the cg method using ParExPDE

In the example listing we can see different applications of expression templates:
- Simple arithmetic operations like addition of two variables or multiplication with a scalar.
- Application of a differential operator to a variable (`laplace(d)`).
- Restriction of the evaluation of an expression to only a part of the whole domain. In this case this is done by using the predefined marker `interior_points`.

Another application of expression templates in the ParExPDE library is the definition of differential operators. For the creation of an operator we must simply provide the function to be integrated (in a finite element context) as shown in Fig. 3. The integration routines which are developed by J. Härdtlein in [2] make also use of expression templates for the formulation of the function. Again they are used for simple arithmetic operations but also for the expression of simple differential operators like d_dx for $\frac{d}{dx}$.

```
class pxdLaplaceOperatorFunction {
public:
  static double integrate(_Cuboid_<bgPoint3D>& c,
                          int m, int n, bgPoint3D** cell_point) {
    return c.integrate(d_dx(v(cell_point[m])) *
                       d_dx(v(cell_point[n])) +
                       d_dy(v(cell_point[m])) *
                       d_dy(v(cell_point[n])) +
                       d_dz(v(cell_point[m])) *
                       d_dz(v(cell_point[n])));
  }
};
```

Fig. 3. Definition of the Laplace Operator

3 Software Architecture of ParExPDE

The realization of a library with the desired functionality is clearly not a simple task as many stages of abstraction are involved. A common approach in software development is to structure the library into several *layers*. Each layer has a clearly defined interface and uses only the functionality of lower layers. Ideally, the layers are completely independent of each other if they are strictly based only on the interfaces but not concrete implementations of lower layers. As a result, the implementation of a certain layer could easily be replaced by another implementation without changing the rest of the library. This leads to a clear structure and to a well maintanable library.

The layered architecture of the ParExPDE library is described next. In the current design we can distinguish six different layers as shown in Fig. 4.

Geometry Layer The geometry layer holds and handles all information of the underlying hexahedral grid which includes the following:
- Storing all geometric primitives in the grid: in a hexahedral grid there are vertices, edges, (quadrilateral) faces and cuboids
- Storing neighbour information for each geometric primitive
- The ability to retrieve the coordinates for a given point within one of the primitives

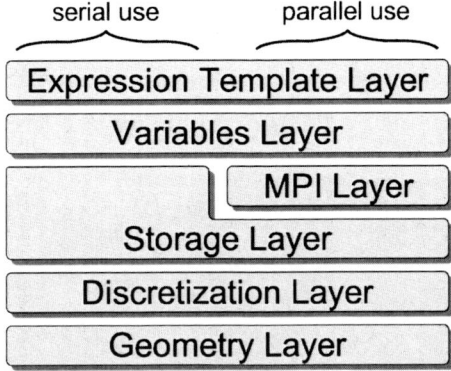

Fig. 4. Software layers of ParExPDE

Discretization Layer This layers stores for each geometric primitive the number of discretization points. Also, periodicity information is handled here: on a periodic boundary there are for each grid element two geometric elements but they are treated identically.

Storage Layer Using the information from the discretization layer, this layer provides the construction of objects containing memory for each element in the grid. Memory is not only needed for the elements themselves but also each element must be able to access certain data from neighbouring elements.

MPI Layer Placed on top of the storage layer, the MPI layer provides mechanisms for the exchange of data between storage objects. MPI is the de facto standard for distributed memory machines, including supercomputers and high performance clusters. This layer manages the communication buffers and the data exchange between elements on different processes.

Variables Layer Up to this layer we only have storage objects that are not related to each other. Putting a storage object for each geometric primitive together in one place we get the notion of a variable defined on the whole grid. By using the MPI layer — if available and wanted — it is easy to keep the variables synchronized among all involved processors.

Expression Template Layer The user interface is realized by the expression template layer. It enables the user to write program statements that are very close to the mathematical language. An efficient evaluation of expressions is possible.

4 Implementation

The first step in implementing the ParExPDE library for the Hitachi SR8000 consisted in using the EXPDE code as a framework to develop the so-called EXPDE-cuboid version which possessed the functionality of EXPDE except having multigrid operators. This version had already used the Hitachi specific features like PVP and COMPAS but it only worked on a single cuboid and lacked MPI parallelization. The next step was to overcome these deficiencies by enabling the library to work on general hexahedral grids and parallelizing the code. It turned out that the insertion of the new geometric primitives and the inclusion of MPI required major program restructuring.

Expression Templates for Heterogeneous Data Structures

The EXPDE-cuboid code worked on just one cuboid. Therefore, the faces, edges and vertices of the cuboid did not require a treatment as separate entities since there were no adjacent neighbours which would have caused them to become regions with arbitrary structure. With the treatment of general hexahedral grids, this is no longer the case as there might be interior edges or vertices which can possibly have an arbitrary number of neighbours.

When we evaluate an expression template applied to a variable defined on a hexahedral grid, we have to iterate over all storage objects of all types. Furthermore, the application of the expression template to the different element types might require different implementations e.g. during the application of a differential operator.

Therefore, the expression templates provide specializations of themselves for a certain grid element. We demonstrate this for the expression type that represents a variable. The general expression class defines types for the element specializations and provides functions that return specializations for certain elements:

```
class pxdExprVariable {
public:
  typedef pxdExprCuboid CuboidReturnType;
  typedef pxdExprFace FaceReturnType;
  typedef pxdExprEdge EdgeReturnType;
  typedef pxdExprVertex VertexReturnType;
private:
  pxdVariable& variable_;
public:
  pxdExprVariable(pxdVariable& variable) : variable_(variable) { }
  pxdExprCuboid getCuboidExpr(int cuboid_index) const {
    return pxdExprCuboid(variable_, cuboid_index); }
  pxdExprFace getFaceExpr(int face_index) const {
    return pxdExprFace(variable_, face_index); }
  pxdExprEdge getEdgeExpr(int edge_index) const {
    return pxdExprEdge(variable_, edge_index); }
  pxdExprVertex getVertexExpr(int vertex_index) const {
    return pxdExprVertex(variable_, vertex_index); }
  ...and other things
};
```

The specialized class for cuboids is presented here, the specializations for the other element types are similar.

```
class pxdExprCuboid {
private:
  pxdVariable& variable_;
  double* memory_;
public:
  pxdExprCuboid(pxdVariable& variable, int cuboid_index) : variable_(variable) {
    memory_ = variable_.getCuboidArray(cuboid_index)->getRawMemory(); }
  double valueAt(int index) const {
    return memory_[index]; }
};
```

Finally, for the application of differential operators, we present the specializations of the expressions for cuboids and edges to illustrate that their evaluation requires different implementations. For the evaluation inside a cuboid we can exploit its regular structure i.e. we know that inside the cuboid we only have 27-point stencils, at least for piecewise linear discretizations.

```cpp
template <class A, class DiffOp>
class pxdExprDiffOpCuboid {
public:
  typedef typename A::Type Type;
private:
  A a_;
  DiffOp op_;
  bgComputeCuboid* compute_cuboid_;
  const bgCuboidOffsets* offsets_;
public:
  pxdExprDiffOpCuboid(const A& a, const DiffOp& op, bgComputeCuboid* compute_cuboid)
    : a_(a), op_(op), compute_cuboid_(compute_cuboid) {
    offsets_ = compute_cuboid_->getOffsets(); }
  Type valueAt(int index) const {
    // apply 27-point stencil
    return op_.getStencilAt(index).apply
      (a_.valueAt(index + offsets_->offset_left_front_bottom),
       a_.valueAt(index + offsets_->offset_front_bottom),
       ...the values for the other neighbours...
       a_.valueAt(index + offsets_->offset_right_back_top)); }
};
```

In contrast to cuboids we do not know the precise number of entries of the stencils on an edge at compile time. This leads to a quite different evaluation of an differential operator there.

```cpp
template <class A, class DiffOp>
class pxdExprDiffOpEdge {
public:
  typedef typename A::Type Type;
private:
  A a_;
  DiffOp op_;
  bgComputeEdge* edge_;
public:
  pxdExprDiffOpEdge(const A& a, const DiffOp& op, bgComputeEdge* edge)
    : a_(a), op_(op), edge_(edge) { }
  Type valueAt(int index) const {
    const bgEdgeOffsets* offsets = edge_->getOffsets();
    // create array for the values of the point and its neighbours
    Type* edge_value = new Type[op_.stencilAt(index).getSize()];
    int vertical_offset =
      edge_->getTotalNumberNeighbours() + 1;
    // values on the edge
    edge_value[0] = a_.valueAt(index-1);
    edge_value[vertical_offset] = a_.valueAt(index);
    edge_value[2*vertical_offset] = a_.valueAt(index+1);

    // values on all neighbouring cuboids
    for (int n = 0; n < edge_->getNumberCuboidNeighbours(); n++) {
      edge_value[2*n+2] =
        a_.valueAt(offsets->offset_neighbour_cuboid[n]+index-1);
      edge_value[vertical_offset+2*n+2] =
        a_.valueAt(offsets->offset_neighbour_cuboid[n]+index);
      edge_value[2*vertical_offset+2*n+2] =
        a_.valueAt(offsets->offset_neighbour_cuboid[n]+index+1); }

    // values on all neighbouring faces
    for (int n = 0; n < edge_->getNumberFaceNeighbours(); n++) {
      similar as above }

    // apply the unstructured stencil to the value array
    typename A::Type result = op_.stencilAt(index).apply(edge_value);
```

```
    delete[] edge_value;
    return result;
  }
};
```

5 Performance Analysis

At first sight it seems that the use of expression templates should cause an overhead which prevents the resulting code to be executed efficiently. But to the opposite, we can demonstrate good performance with our expression template codes.

In order to get good performance on the Hitachi SR8000, the code should be prepared to be optimized using the pseudo-vectorization feature (PVP). In particular, this means that we have to design the data layout carefully. In the storage layer, we have chosen to arrange the data in single large arrays wherever this is possible. PVP should work well for operations on large arrays of known size. This applies not only to the SR8000 architecture but also to other architectures. On cache based machines these data structures are well suited for layout and access modification techniques presented in [9], [7], [8] that take the memory hierarchy into account.

Due to the peculiarities of Hitachi's ANSI-compliant C++ compiler sCC, the application of differential operators could not yet be optimized. Unfortunately, sCC is the only C++ compiler on this machine that makes use of the machine's special features. Therefore, it has not been possible to obtain satifactory performance for programs involving differential operators. However, we can show that the evaluation of expression templates consisting only of simple arithmetic operations shows a reasonably good performance. On other architectures we can demonstrate that we get good performance for both regular arithmetic expressions and expressions with differential operators (see Table 2).

Table 2. MFLOPs obtained on a single processor for the evaluation of expression templates. The "Simple Expression" is of the type "constant + scalar * variable + scalar * variable * variable", the other expression consists just of the application of the Laplace operator to a variable. The evaluation has been performed on a single cuboid containing 128^3 unknowns

	Hitachi SR8000	Pentium 4 2.4 GHz
Simple Expression	513	407
Expr. with Differential Operator	25	918

Performance results that we have obtained with a more procedural-style implementation of the top layer show that the suboptimal treatment of complex expression templates by the Hitachi compiler is responsible for the unsatisfactory performance.

Therefore, the new top layer has been implemented in a mix of C++ and FORTRAN77 and it still uses C++ features like inheritance and templates but in a more conservative way than the original expression template layer.

The components of the geometric multigrid method are well known in the literature. We use a V(2,2) cycle with a Gauß-Seidel smoother on a seven point finite difference discretisation of a Poisson problem with Dirichlet boundary conditions. The problem domains for the scale up experiment were $\Omega_1 = [0,2] \times [0,2] \times [0,2]$, partitioned into 8 hexahedra, $\Omega_2 = [0,4] \times [0,2] \times [0,2]$, partitioned into 16 hexahedra, $\Omega_3 = [0,4] \times [0,4] \times [0,2]$, partitioned into 32 hexahedra. The meshsize was always $h = \frac{1}{256}$. The performance results in Table 3 underline that the ParExPDE layers result in runtime efficient programs. When comparing these performance values with the ones in Table 1, one has to take several differences into account. The most significant difference is that Table 3 states the performance for the whole multigrid program including program start, reading of configuration files and memory allocation in addition to the geometric multigrid part, whereas Table 1 concentrates on one computationally intensive routine. Usually, the computational cost of a geometric multigrid implementation is dominated by the smoothing operation. We utilise a Gauß-Seidel smoother, which is similar to a matrix-vector product in terms of memory access pattern and which, in fact, achieves a similar, if slightly lower, MFLOP performance. However, other components of the parallel multigrid implementation, in particular the data exchanges, do not and cannot run at the same speed, hence it cannot be expected to see a whole program performance on par with the matrix-vector product. Furthermore, the scale up experiments include inter-process communication via MPI for both, inter-node and intra-node data exchange, whereas the matrix-vector routine was run on a single node using Hitachi's shared memory programming model COMPAS. Despite all the differences, the computation of the matrix vector product, or, more precisely, the application of the discrete operator to a problem variable using a stencil based operator representation, does achieve a similar performance as in Table 1.

Table 3. Scale up results for parallel geometric multigrid in 3D: *Dof* are the degrees of freedom, *Time* is the wall clock solution time for the linear system, the value *MFLOPs per node* is computed by multiplying the average performance over all MPI processes by eight, the number of processes per node. The average performance of all MPI processes is taken over the whole program run, including start up, reading of configuration files, initialisation in addition to the numerically intensive part

CPU	Dof $\times 10^6$	Time in (s)	V-cycles	MFLOPs per Node
8	133.4	78	9	2209.6
16	267.1	81	10	2337.6
32	534.7	95	11	2167.2

6 Conclusions

In this paper, we have shown that expression templates are an appropriate programming paradigm for the development of parallel numerical software on supercomputers. With expression templates, ParExPDE provides the application programmer with an intuitive interface to the underlying parallel grid software library. This software infrastructure enables the rapid development of PDE solvers and numerical simulation programs.

Depending on the quality of the compiler, this can be achieved without having to compromise on runtime performance. Unfortunately, the SR8000 is one of the examples where currently only relatively simple expression templates can be executed with full efficiency. With more complex expression templates the optimizer still fails to exploit the full inherent performance potential. The benchmarks of ParExPDE presented above show clearly that, when using a different compiler, the performance degradation does not occur and is due to the limited functionality and optimization capabilities of the current C++ implementation on the Hitachi.

This claim has been cross-checked by presenting results on the Hitachi with a more conservative implementation of a multigrid code (derived from the associated gridlib project) using the same ParExPDE data structures and communication routines. This program variant avoids using more complex expression templates. The performance results show that the basic software architecture has excellent performance, speedup and scalability characteristics. Different from unstructured grid algorithms on the Hitachi, it achieves a fully satisfactory sustained performance and thus justifies the software design of the ParExPDE and gridlib projects. Both projects are based on the assessment that the Hitachi is much better suited for computations on structured grids than on unstructured grids or general sparse matrix implementations.

Summarizing, the ParExPDE project has resulted in a highly efficient software infrastructure for parallel grid based computations. It provides both a very high level expression template interface and a somewhat lower level C++ library interface. The performance degradation that is still observed with complex expression templates on the Hitachi must be attributed to the state of the currently available compiler. This limitation will disappear with increasing compiler maturity so that the use of expression template programming will become even more attractive and feasible.

Acknowledgements

The authors gratefully acknowlegde the support of the Bavarian high performance computing initiative KONWIHR. We are also indebted to Professor C. Pflaum without whom this project would not have been possible.

References

1. B. Bergen and F. Hülsemann. Hierarchical hybrid grids: data structures and core algorithms for multigrid. *Numerical Linear Algebra with Applications*, 11:279–291, 2004.
2. J. Härdtlein, A. Linke, and C. Pflaum. Fast expression templates. To appear.
3. F. Hülsemann, B. Bergen, and U. Rüde. Hierarchical hybrid grids as basis for parallel numerical solution of PDE. In H. Kosch, L. Böszörményi, and H. Hellwagner, editors, *Euro-Par 2003 Parallel Processing*, volume 2790 of *Lecture Notes in Computer Science*, pages 840–843, Berlin, 2003. Springer.

4. F. Hülsemann, P. Kipfer, U. Rüde, and G. Greiner. gridlib: Flexible and efficient grid management for simulation and visualization. In P. Sloot, C. Tan, J. Dongarra, and A. Hoekstra, editors, *Computational Science - ICCS 2002*, volume 2331 of *Lecture Notes in Computer Science*, pages 652–661, Berlin, 2002. Springer.
5. F. Hülsemann, S. Meinlschmidt, B. Bergen, G. Greiner, and U. Rüde. *gridlib* – a parallel, object-oriented framework for hierarchical-hybrid grid structures in technical simulation and scientific visualization. In *High Performance Computing in Science and Engineering, Munich 2004. Transactions of the Second Joint HLRB and KONWIHR Result and Reviewing Workshop*, pages 37–50. Springer-Verlag Berlin Heidelberg New York, 2004.
6. P. Kipfer, F. Hülsemann, S. Meinlschmidt, B. Bergen, G. Greiner, and U. Rüde. gridlib: A parallel, object-oriented framework for hierarchical hybrid grid structures in technical simulation and scientific visualizations. In S. Wagner, W. Hanke, A. Bode, and F. Durst, editors, *High Performance Computing in Science and Engineering 2000-2002*, pages 489–501, Berlin, 2003. Springer.
7. M. Kowarschik, U. Rüde, N. Thürey, and C. Weiß. Performance Optimization of 3D Multigrid on Hierarchical Memory Architectures. In *Proc. of the 6th Int. Conf. on Applied Parallel Computing (PARA 2002)*, volume 2367 of *Lecture Notes in Computer Science (LNCS)*, pages 307–316, Espoo, Finland, 2002. Springer.
8. M. Kowarschik and C. Weiß. An Overview of Cache Optimization Techniques and Cache-Aware Numerical Algorithms. In U. Meyer, P. Sanders, and J. Sibeyn, editors, *Algorithms for Memory Hierarchies — Advanced Lectures*, volume 2625 of *Lecture Notes in Computer Science (LNCS)*, pages 213–232. Springer, 2003.
9. M. Kowarschik, C. Weiß, and U. Rüde. Data Layout Optimizations for Variable Coefficient Multigrid. In P. Sloot, C. Tan, J. Dongarra, and A. Hoekstra, editors, *Proc. of the 2002 Int. Conf. on Computational Science (ICCS 2002), Part III*, volume 2331 of *Lecture Notes in Computer Science (LNCS)*, pages 642–651, Amsterdam, The Netherlands, 2002. Springer.
10. C. Pflaum. *EXPDE — Expression Templates for Partial Differential Equations.* http://www10.informatik.uni-erlangen.de/~pflaum/expde/public_html/.
11. C. Pflaum. Expression templates for partial differential equations. *Computing and Visualization in Science*, 4(1):1–8, November 2001.
12. T. Veldhuizen. Expression templates. *C++ Report*, 7(5):26–31, June 1995.

ParRichy: Parallel Simulation of Bioreactive Multicomponent Transport Processes in Porous Media

S. Kräutle, M. Bause, A. Prechtel, F. Radu, and P. Knabner

Inst. f. Applied Mathematics, University of Erlangen-Nuremberg,
Martensstr. 3, D-91058 Erlangen,
(kraeutle,knabner)@am.uni-erlangen.de

Abstract. Numerical simulations have become an important tool to predict the evolution of groundwater and subsurface contamination by organic compounds. Due to the observation of sharp interfaces on which biochemical degradation of the contaminants takes place, a reliable and accurate prognosis requires high spatial resolution of the considered subsurface domain and/or the application of higher order discretization methods. A successfull realization of these simulations can only be provided by parallel computer systems with high computing power and memory availability. In this article, our concepts and simulation results of "real world" scenarios of organic contamination are presented.

1 Introduction

Groundwater contamination by organic compounds has become a serious and widespread environmental problem in industrialized countries. Major organic contaminants include petroleum fuels (gasoline, diesel), petroleum byproducts (coal tar, creosote), and chlorinated solvents. In many cases, groundwater contains a variety of organic contaminants, either due to the complex mixture in many non-aqueous phase liquids (NAPLs) or due to co-disposal/co-spillage (e.g., landfill leachates). The degradation of these contaminants is controlled to a large extent by the biological and geochemical conditions in the groundwater. Fortunately, biodegradation tends to attenuate at least some organics during goundwater transport. The question of whether costly active remediation is required, or whether natural processes of attenuation will be sufficient is a critical issue in "real world" situations.

Numerical models can be used to help to answer this question and to predict the long-term behaviour of the contaminant plume. However, the decision-making capability of numerical simulations requires that the mathematical model includes the full range of the controlling processes. These processes are often nonlinear, and the large number of (bio-)chemical species and reactions requires large computational power provided by parallel computers. Recently it was observed that biodegradation processes typically occur on small interfaces where the reactants (e.g., the contaminant and oxygen) are mixing ([BaK04] and references therein). Such processes require a high spatial resolution of the computational domain (which usually covers several square kilometers) as well as advanced numerical techniques (e.g., higher order numerical approximation schemes to avoid artificial diffusion).

2 The project

The authors of this article participated in the interdisciplinary network project "Sustainable remediation involving natural attenuation" (Bayerisches Verbundvorhaben "Nachhaltige Altlastenbewältigung unter Einbeziehung des natürlichen Reinigungsvermögens") funded by the *BayStMLU* under coordination of the *Gesellschaft zur Altlastensanierung in Bayern* (GAB). In cooperation with environmental scientists who investigated several contaminated sites in Bavaria, the authors' research group was responsible for the mathematical modelling of the identified processes, the development and implementation of numerical schemes and numerical studies. While the intention of the GAB-project was to develop codes running on PCs and workstations (e.g., in engineers' offices or at local authorities), the support by KONWIHR enabled the implementation and adaption of the codes on parallel computer systems. This gain of cpu power and memory enabled the use of a higher resolution as well as more detailed biogeochemical models, which are crucial factors for prognoses.

In this article we consider two "real world" scenarios: The first one is the study of the long-term behaviour of a BTEX (Benzene Toluene Ethylbenzene Xylene) plume in the earth's subsurface at a site in southern Bavaria (cf. Sections 3,4) investigated in the GAB project. The second one deals with the reactions and biodegradation in the groundwater and in sediments of the metal complexant Ethylene Diamine Tetraacetic Acid (EDTA) which can be found in groundwater systems near military sites (cf. Sect. 5). The main aspect of the first problem was the implementation and comparison of two different numerical approximation schemes (lowest order mixed finite elements vs. second order conforming elements) and different programming platforms: UG/DDD ("unstructured grids", "Dynamic Distributed Data") vs. M++ ("Meshes, Multigrid and More", [Wie04]). The second problem involves 14 (bio)chemical species and 10 reactions and demonstrates the performance of the developed software for larger reaction networks. M++ was designed to have more transparent structures and to enable an easier handling. In contrast to UG/DDD, M++ references all geometric objects by their barycentric coordinates ('distributed point objects') which are stored in hash-tables. Communication and load balancing concepts are similar to those of UG/DDD.

Both programming models are implemented on the Linux cluster of the Institute of Applied Mathematics at the University of Erlangen-Nuremberg. The cluster consists of 18 nodes with 1GB RAM and two 2.4-GHz-Xeon-processors each.

3 Real case study of a BTEX contamination site using UG/DDD and mixed finite elements

3.1 Problem setting

The site investigated by our cooperation partners of the GAB project is the former location of a chemical laundry. Large quantities of mineral oil were spilt into the soil between 1948 and 1989 and significant concentrations of BTEX have been measured between 2001 and 2004. The objective of this numerical study was to predict the long-term behaviour of the xylene contamination plume. The site is approximately 730m wide and 1380m long.

We modelled the biodegradation rate using the Monod model [BaK04, Mer04]. The equations for the concentrations of the mobile electron donator xylene, the mobile electron acceptor oxygen (O_2) and the immobile biomass X read

$$\partial_t(\theta c_D) - \nabla \cdot (D_D \nabla c_D - q c_D) = -R - \rho \partial_t s_D,$$
$$\partial_t(\theta c_A) - \nabla \cdot (D_A \nabla c_A - q c_A) = -\alpha_{A/D} R,$$
$$\partial_t c_X = \frac{Y}{\theta}\left(1 - \frac{c_X}{c_{X,max}}\right) R,$$

where the rate R is given by

$$R = \theta \mu c_X \frac{c_D}{K_D + c_D} \frac{c_A}{K_A + c_A}.$$

The release of xylene is modelled by a linear nonequilibrium desorption from an immobile oil plume,

$$\partial_t s_D = k\left(\phi(c_D) - s_D\right), \quad \phi(c_D) = K_d c_D.$$

The stationary flow field q is computed by solving Richards' equation using a (discontinuous) hydraulic conductivity which is smaller within the oil plume than in the surrounding region (Fig. 1).

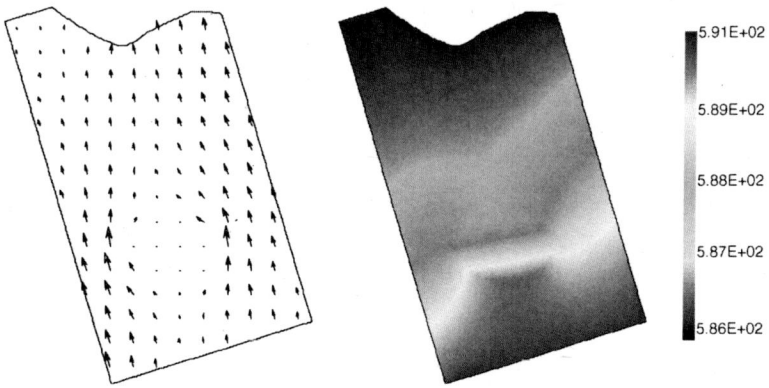

Fig. 1. Groundwater flow field q (left) and the pressure head (right).

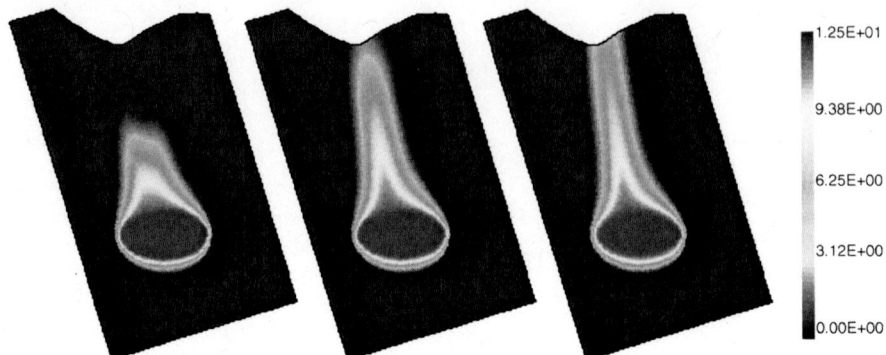

Fig. 2. Concentration of xylene after 3, 5, 10 years.

3.2 Numerical methods

For this simulation we use an implicit Euler time stepping, a damped version of Newton's method and *mixed* finite element methods, i.e., the pressure and the species concentrations are approximated by piecewise constant functions and the flux is approximated by piecewise linear functions. The mixed finite element approach has the advantage of mass conservation, and it leads to more accurate approximations in the case of a discontinuous hydraulic conductivity. However, the numerical effort to solve the resulting linear problems is bigger than for standard conforming elements: Lagrangian multipliers are introduced, and a static condensation leads to small local nonlinear problems on the elements, which are solved, before each global Newton step, by Newton's method again. Since the Lagrangian multipliers converge of higher order than the corresponding element values, the comparison of multipliers and element values can provide a simple a posteriori error indicator and adaptive mesh refinement strategy. More details on the numerical methods and the computational results can be found in [Rad04].

3.3 Numerical results

The numerical simulation (cf. Figs. 2–4) covers a range of 10 years, divided into 73 000 time steps of length 0.05 [d]. A special feature of the simulated biodegradation problem and one main result of the simulation is that the reaction, indicated by the massiv growth of biomass (cf. Fig. 4), takes place in narrow zones where contaminant and oxygen are mixing. The mixing of the reactants is caused mainly by diffusion. It seems obvious that a rather high spatial resolution of the numerical scheme is required to resolve the process. The implementation of the code on a parallel computer system (see Sect. 2) enabled us to use 45 523 triangular elements which corresponds, due to the large size of the computational domain, to an average edge size of the elements of 19 metres.

We measured a computation time of 27 hours on 16 processors (see Table 1 and Fig. 8). A similar computation on a single cpu would take more than 9 days. Since a main purpose of numerical simulations for biodegradation of contaminants lies in

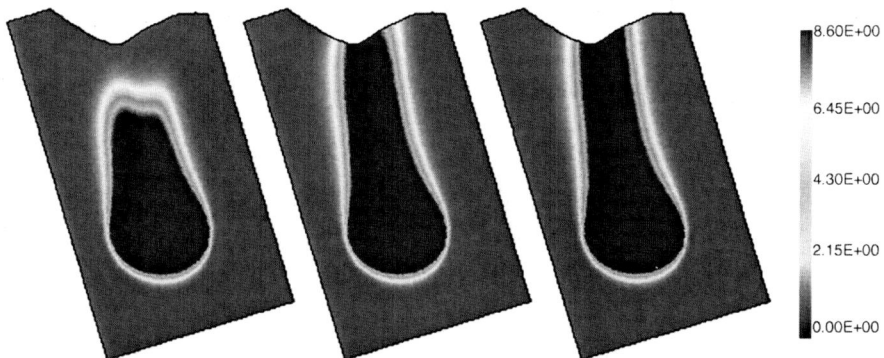

Fig. 3. Concentration of oxygen after 3, 5, 10 years. Within the plume, the oxygen is consumed.

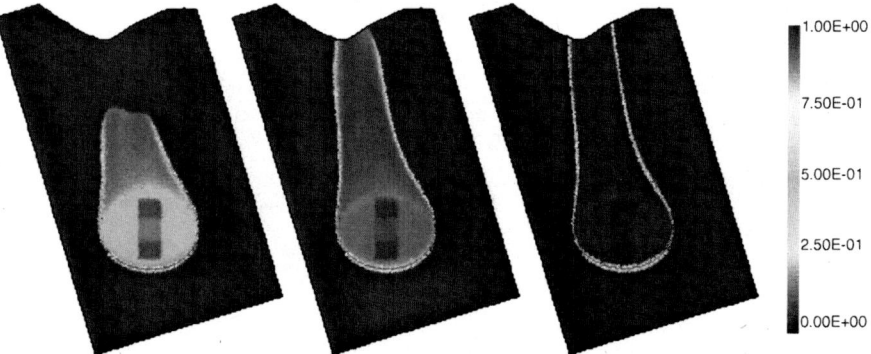

Fig. 4. Concentration of biomass after 3, 5, 10 years. The biomass is concentrated on a thin layer where oxygen and xylene mix. For a reproduction of figures Fig. 2 – Fig. 4 in colour, see Fig. A.9 – Fig. A.11

the prognosis of future behaviour under numerous scenarios (a variety of parameter sets), such a long waiting period for each simulation is of course not tolerable and would force to decrease the spatial or temporal resolution of the scheme. Three dimensional simulations of the scenario would be desirable to make allowance for a possible vertical transport of oxygen from the unsaturated zone into the contamination plume.

4 Simulation of the BTEX contamination site using M++ and higher order finite elements

It turned out that the artificial (numerical) diffusion is a crucial point in the accurate numerical solution of the Monod model: Since the reaction typically takes place in a very thin mixing zone of the reactants (cf. Figs. 2-4), any numerical diffusion leads to an artificial mixing of the compounds and, thereby, to a severe overestimation

Table 1. Upscaling for the scenario of Sect. 3.

processors	elements	elem. per proc.	comp. time [h]	comp. time × proc. / comp. time sequential
1	45523	45523	226.9	1.00
2	45523	22761	125.2	1.10
4	45523	11381	59.8	1.05
8	45523	5690	45.1	1.59
16	45523	2845	26.4	1.86

of the biodegradation process [BaK04] and an underestimation of the length of the contaminant plume. To reduce artificial diffusion, we implemented quadratic order conforming finite elements, using the M++ parallel programming model. Compared to the approach of Sect. 3, the conforming methods have the advantages of a more straight-forward solution process of the linear system and an easier implementation of stabilization techniques of SUPG (streamline-upwind-Petrov-Galerkin) type for convection dominated problems.

The temporal discretization uses an extrapolated implicit Euler method and converges formally of second order. This time stepping scheme provides an automatic error control. In future computations we will use this time stepping for a step size control (see [BaK04] for a similar step size control based on the combination of two BDF methods).

The simulation covers 7 300 macro time steps, i.e., 21 900 Euler steps of size $\Delta t = 0.05$ [days]. Besides the higher order methods, we chose a finer discretization of 197 152 triangular elements, which corresponds to 1 186 011 degrees of freedom (D.O.F.). Note that due to the size of the computational domain, even with such a large number of nodes, the average edge length of the elements is still approximately 5 metres.

At the present level of implementation, the release of xylene from an immobile oil phase is not included in the model. Instead, an initial concentration of mobile xylene is assumed, and the computation uses a homogeneous and constant flow field. The upscaling of the computation time can be found in Table 2 and Fig. 8.

Table 2. Upscaling for the scenario of Sect. 4.

processors	elements	D.O.F.	comp. time [h]	comp. time per macro time step[s]
16	197 152	1 186 011	70.60	34.8
24	197 152	1 186 011	48.48	23.9
32	197 152	1 186 011	39.18	19.3

5 Simulation of EDTA contamination using UG/DDD

5.1 Problem setting

As an example for a rather complex reactive system with a larger number of species and reactions we consider the reactions of the metal complexant EDTA (cf. Sect. 2) in the groundwater and sediments. Some of these EDTA-metal-complexes, e.g., the cobalt (Co-)EDTA, are harmful contaminants found in the groundwater near military sites. The species, reactions and rate parameters are taken from [CGS98,FYB03]. The reaction network consists of 14 species from which 6 are mobile (aqueous) und 8 are immobile, and 10 biogeochemical reactions:

Table 3. Reactions and rate parameters (units: k_f, k_b, μ: h^{-1}, K_D, K_A: [mMol/l]).

R_1:	Co(II)(aq) + S$^-$ ↔ S$^-$Co	$k_f = 1200$	$k_b = 100$
R_2:	Co(II)EDTA(aq) + S$^+$ ↔ S$^+$Co(II)EDTA	$k_f = 2500$	$k_b = 100$
R_3:	Fe(III)EDTA(aq) + S$^+$ ↔ S$^+$Fe(III)EDTA	$k_f = 900$	$k_b = 100$
R_4:	EDTA(aq) + S$^+$ ↔ S$^+$EDTA	$k_f = 2500$	$k_b = 100$
R_5:	Co(III)EDTA(aq) + S$^+$ ↔ S$^+$Co(III)EDTA	$k_f = 250$	$k_b = 100$
R_6:	S$^+$Co(II)EDTA ↔ Co(II)(aq)+S$^+$EDTA	$k_f = 1$	$k_b = 10^{-3}$
R_7:	S$^+$EDTA → Fe(III)EDTA(aq)+S$^+$	$k_f = 2.5$	$k_b = 0$
R_8:	Co(II)EDTA(aq) → Co(III)EDTA(aq)	$k_f = 10^{-3}$	$k_b = 0$
R_9:	EDTA(aq) + 6O$_2$ → 3CO$_2$ + Biomass	$\mu = 0.025$	$K_D = K_A = 10^{-5}$
R_{10}:	Fe(III)EDTA(aq) + 6O$_2$ → 3CO$_2$ + Biomass	$\mu = 0.00025$	$K_D = K_A = 10^{-5}$

The reaction network covers a large range of time scales: "fast" adsorption/desorption reactions (R_1–R_5), two biodegradation reactions of different speed (R_9, R_{10}), modelled by the Monod approach, and kinetic oxydation and dissolution processes.

While the computations in [CGS98, FYB03] consider a "batch" situation (i.e., no spatial inhomogeneities, no transport), we consider a two dimensional domain of size 100 × 100 [dm^2] and a homogeneous flux of $q = (0, -1.3)^t$ [dm/h] with diffusion/dispersion $D = 1.0$ [dm^2/h]. Oxygen is given on the whole computational domain. On a rectangular part of the computational domain, the initial concentrations from the batch simulation [CGS98, FYB03] (e.g., 0.032 mMol/l for the contaminant Co(II)EDTA) and zero concentrations outside that part are assumed.

5.2 Computational results

After a fraction of an hour, the local equilibrium of the fast reactions R_1–R_5 is attained. As a result, EDTA is produced and the concentration of the contaminant Co(II)EDTA decreases to approximately 0.025 mMol/l at time $T = 0.1$[h]. Within the first few hours, reaction R_9 causes a decay of EDTA and oxygen and a strong growth of the biomass (cf. Fig. 6) in the plume. At the same time, the "slow" reaction R_7 causes the concentration of Fe(III)EDTA to increase (cf. Fig. 6). The rising Fe(III)EDTA concentration triggers the bioreaction R_{10}, while Fe(III)EDTA

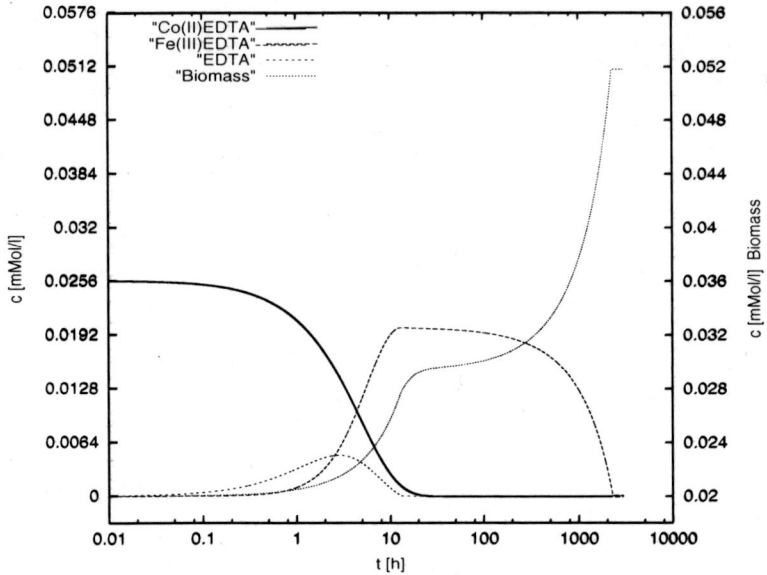

Fig. 5. Temporal development of the concentrations of Co(II)EDTA, EDTA, Fe(III)EDTA and biomass for a batch situation.

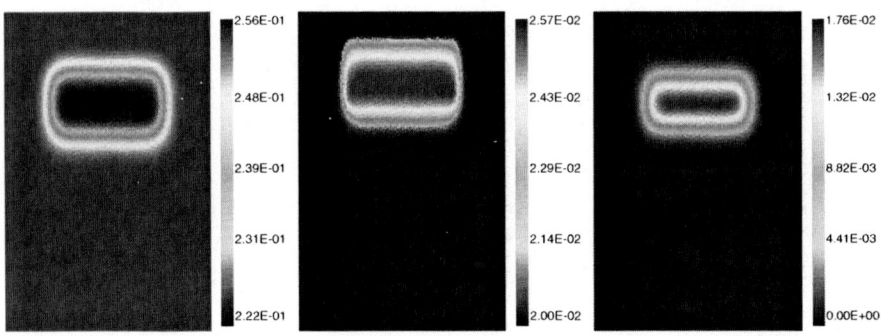

Fig. 6. Concentration of O_2, biomass and Fe(III)EDTA at time $t = 10.0[h]$ for the 2d-problem. The biodegradation of EDTA causes a strong growth of biomass and a decay of oxygen in the upper part of the domain. For a reproduction of this figure in colour, see Fig. A.12

is subject to convection, diffusion and decay (cf. Figs. 6,7). Since reaction R_{10} is very slow, only a moderate growth of the biomass can be observed.

Compared to the batch situation (cf. Fig. 5), our two dimensional simulation shows a slower biodegradation of Fe(III)EDTA, since convection transports the reactant Fe(III)EDTA from the immobile biomass population in the upper part of the computational domain (cf. Figs. 6,7).

Our simulation uses 47 396 elements, which is comparable to the simulation in Sect. 3. The computational time for 10 000 time steps with time step size $\Delta t =$

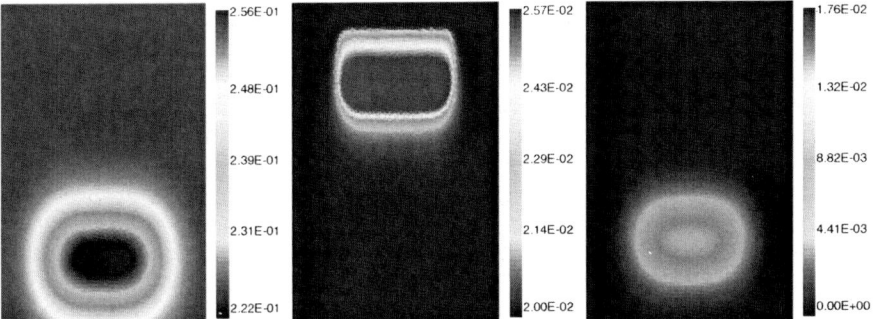

Fig. 7. Concentration of O_2, biomass and Fe(III)EDTA at time $t = 50.0[h]$ for the 2d problem. For a reproduction of this figure in colour, see Fig. A.13

$10^{-2}[h]$ is displayed in Table 4. For the upscaling we refer to Fig. 8. Each time step of the 14-species-problem takes about 4–5 times the computational time of the three-species problem of Sect. 3 if the same number of processors is used.

Table 4. Upscaling for the scenario of Sect. 5.

processors	elements	comp. time[h]	comp. time/timestep[s]
1	47396	116.72	42.02
4	47396	35.74	12.87
16	47396	10.87	3.91

6 Decoupling of multicomponent reactive transport problems

6.1 Elimination of weak couplings

For solving the nonlinear equation systems of reactive multicomponent transport we apply Newton's method, which converges locally of second order. As the size of the Jacobian grows quadratically with the number of species, we have to care for a reduction of the effort for solving the system. By an efficient implementation, e.g., with the help of pointer lists, we could exploit the fact that only a few species are directly coupled with each other via reaction terms, and thus the complexity for assembling the global sparse system only grows linearly with the number of species (cf. Tab. 5). With a growing number of species the linear solver dominates the total computing time already in the one dimensional test cases (here we use a direct sparse matrix solver). The table thus shows that here a reduction of the complexity could pay off.

This can be achieved by simplifying the Jacobian in the Newton step (cf., e.g., [KnA00]), in particular by neglecting weak couplings between single species. This

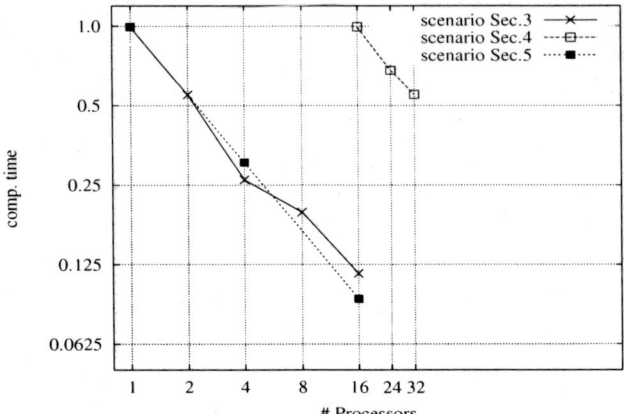

Fig. 8. Computational time for the scenarios of Sections 3,4, and 5 in dependence of the number of processors. The values are scaled with respect to the computational time for one processor (scenarios 3 and 4) or 16 processors (scenario 5).

Table 5. CPU time [s] for solving a nonlinear model problem with eight (uncoupled) species in different configurations.

Configuration	Matrix entries per node	Assembling	Solver	Total
8 single species problems	8	92.3	8.1	101.5
4 two species problems	16	73.8	14.1	89.3
2 four species problems	32	66.8	37.3	105.8
1 eight species problem	64	76.8	128.0	206.8

leads to a decoupling of parts of the global system of equations and means that these parts can be solved independently of each other. Table 5 already demonstrates that this results in a tremendous reduction of the computing time, e.g. when two "four species problems" instead of one "eight species problem" are given. The linear solver has to handle flexibly block matrices to adaptively account for different matrix sizes even during a simulation. This may become relevant when species disappear in the course of the simulation, and thus their reaction couplings can be ignored. However, this modified Newton's method can result in a deterioration of the convergence properties, and as a consequence more Newton steps become necessary. Here a trade-off has to be made between inferior convergence of Newton's method (cf. Tab. 6) and the speed up achieved by solving smaller systems (cf. Tab. 5).

For two examples, Table 6 shows that this inferior convergence behaviour does not appear immediately. The parameter δ denotes the threshold below which matrix entries are being ignored ($\delta = 0$: the exact system is solved). We see, of course, that for different problems the absolute values that let the convergence properties decrease may differ substantially (10^{-2} for the EDTA example from Tab. 3, 10^{-5} for an example of PHREEQC (No. 11 in [PaA99])). We may think of an a priori decoupling (based on the reaction network), and of an adaptive decoupling in the course of the simulation. These first results will be pursued further in other projects [Pre05].

Table 6. Neglecting weak couplings ($|$matrix entries$| < \delta$) among species in the Jacobian and effect on the convergence of Newton's method. Number of Newton steps and CPU time [s] for the EDTA example (left, cf. Table 3 and [CGS98,FYB03]) and for a PHREEQC example (right, cf. No. 11 in [PaA99]).

EDTA example					PHREEQC example				
δ	Newton steps	Assem.	Solver	Total	δ	Newton steps	Assem.	Solver	Total
0.0	210	1.34	1.86	3.20	0.0	106	2.26	2.70	4.99
10^{-9}	210	1.28	0.95	2.23	10^{-6}	106	2.20	3.33	5.59
10^{-5}	210	1.21	1.00	2.21	10^{-5}	146	2.58	4.08	6.71
10^{-2}	213	1.26	1.00	2.26	10^{-4}	372	4.94	9.37	14.45
10^{-1}	508	2.38	2.49	4.89	10^{-3}	467	5.94	11.73	17.79

6.2 Decoupling by reformulation of the given system of PDEs/ODEs

Besides the use of an approximate Jacobian proposed in Sect. 6.1, we investigated a different way to decouple the given nonlinear set of coupled partial differential equations (PDEs) and ordinary differential equations (ODEs). The new method [KrK05] is based on the choice of new variables (which are linear combinations of the entries of the local species vector) and the replacement of the given PDEs/ODEs by certain linear combinations of the equations. The algorithm leads to some decoupled scalar linear PDEs, some *local* algebraic equations at each grid point and a small remaining set of coupled nonlinear PDEs. The local equations are implicitly solved for certain variables, and these variables are substituted in the remaining coupled set of PDEs.

The reduction mechanism can handle both slow (=kinetic) reactions and fast (=equilibrium) reactions as well as mobile and immobile species. In examples, a system of 5 coupled transport equations with nonlinear reactive coupling terms could be reduced to one scalar nonlinear PDE. In a first one dimensional numerical simulation on a sequential computer, a reduction of the cpu time to 20% of the cpu time for the non-reduced scheme could be observed [KrK05].

Presently, the reduction algorithm is being implemented for 2d problems based on the parallel programming model M++. We expect that the results on the efficiency also apply for 2d problems. We think that the reduction method is especially reasonable for parallel computations, since the solution of the local problems which are generated by the reduction mechanism is trivial to distribute on several processors, while the parallel solution of systems of PDEs causes some communication overhead.

References

[Bau04] M. Bause: Computational study of field scale BTEX transport and biodegradation in the subsurface, in: Numerical mathematics and advanced applications, Proceedings of ENUMATH 2003, Prague, pp. 112–122, Springer Verlag, Berlin (2004)

[BaK04] M. Bause, P. Knabner: Numerical simulation of contaminant biodegradation by higher order methods and adaptive time stepping, Computing and Visualization in Science, **7**, pp. 61–78 (2004)

[CGS98] A. Chilakapati, T. Ginn, J. Szecsody: An analysis of complex reaction networks in groundwater modeling, Water Resour. Res., **34**, 1767–1780 (1998)

[FYB03] Fang, Y., Yeh, G.-T., Burgos, W. D.: A general paradigm to model reaction-based biochemical processes in batch systems, Water Resour. Res., **39**, 1083, doi:10.1029/2002WR001694 (2003)

[KnA00] P. Knabner, L. Angermann: Numerical methods for elliptic and parabolic partial differential equations, Springer, New York (2003)

[KrK05] S. Kräutle, P. Knabner: A new numerical reduction scheme for fully coupled multicomponent transport–reaction problems in porous media, submitted to Water Resour. Res.

[Mer04] W. Merz: Global existence result of the Monod model, accepted for publication in: Adv. Math. Sci. (2004)

[OhR02] M. Ohlberger, C. Rohde: Adaptive finite volume approximations of weakly coupled convection dominated parabolic systems, IMA J. Numer. Anal., **22**, pp. 253–280 (2002)

[PaA99] D. Parkhurst, C. Appelo: User's Guide to PHREEQC (Version 2), U.S. Geological Survey, Denver (1999)

[Pre05] A. Prechtel: On modelling and numerical solution of hydrogeochemical multicomponent transport problems in porous media, doctoral thesis, Friedrich-Alexander-Universität Erlangen, to be published in 2005.

[Rad04] F. Radu: Mixed finite element discretization of Richards' equation: error analysis and application to realistic infiltration problems, doctoral thesis, Friedrich-Alexander-Universität Erlangen (2004),
http://www.am.uni-erlangen.de/am1/publications/dipl_phd_thesis/PhD_Radu.ps.gz

[Schi01] Schirmer, A., et al.: Biodegradation modelling of a dissolved gasoline plume applying independent laboratory and field parameters, J Contam Hydrol **46**, pp. 339–374 (2001)

[Wie04] C. Wieners: Distributed point objects. A new parallel concept for parallel finite elements, to appear in: Proceedings of the 15th International Conference on Domain Decomposition, eds. R. Kornhuber, O. Pironneau, R. Hoppe, J. Périaux, D. Keyes, J. Xu.

Peridot: Towards Automated Runtime Detection of Performance Bottlenecks[*]

Karl Fürlinger, Michael Gerndt

Institut für Informatik
Lehrstuhl für Rechnertechnik und Rechnerorganisation
Technische Universität München
{fuerling, gerndt}@in.tum.de

Abstract. Performance analysis of parallel applications can be a time-consuming and daunting task, as it requires detailed a understanding of the interactions of the system's components. We present the design and the prototypical implementation of a system for the automation of the performance analysis process developed within the Peridot project. Our system is based on the notion of cooperating agents that detect performance problems automatically at runtime and in a distributed fashion, avoiding several problems of classical performance analysis techniques such as overwhelmingly large trace files.

1 Introduction

In computational science and engineering, users of high-performance computing facilities strive for the optimal use of the available resources in order to solve their research problems in the shortest possible time or to attack ever larger problem sizes.

To achieve this goal, application developers often need to spend considerable time and effort on the task of performance *tuning* or performance *debugging*. That is, after the functional correctness of the application has been established, programmers need to engage in an iterative process to bring their code to the required levels of performance. A cycle of this performance tuning process starts with measuring the performance of the application using one of the many tools that are available. This is followed by reasoning about the measurement results and drawing the necessary conclusions and finished by making the necessary corrections to the application's source code.

This process can be very daunting and time-consuming for developers as it requires very detailed knowledge of the underlying hardware, system software and

---

[*] Part of this work is funded by the Competence Network for High-Performance Computing in Bavaria KONWIHR (http://konwihr.in.tum.de) and by the European Commission via the APART working group (http://www.fz-juelich.de/apart).

communication middleware. In order to alleviate this burden, several projects in the performance tools community were initiated recently with the aim of automating the process of finding performance bottlenecks in applications.

One approach to automate performance analysis, called PERISCOPE, that offers a unique set of features has been devised within the Peridot project. This paper describes the work that has been performed within Peridot to bring the idea of *automated* detection of performance properties together with the ability to perform the analysis at *runtime* and in a *distributed* manner. We believe, that this combination will make efficient performance analysis for terascale computing feasible.

The rest of the paper is organized as follows: First, Sect. 2 discusses related approaches in the tools community that address the automation of performance analysis. Then, Sect. 3 provides an overview of the PERISCOPE approach, describing the overall structure of the analysis system, including the monitoring approach used. Sect. 4 provides a description of the design and implementation of our analysis agents, while Sect. 5 describes a concrete usage scenario to illustrate the innerworkings of the components of our system in more detail.

2 Related Work

Several projects in the performance tools community are concerned with the automation of the performance analysis process. Paradyn's [10] Performance Consultant automatically searches for performance bottlenecks in a running application by using a dynamic instrumentation approach. Based on hypotheses about potential performance problems, measurement probes are inserted into the running program. Recently MRNet [12] has been developed for the efficient collection of distributed performance data. However, the search process for performance data is still centralized.

The Expert [15] tool developed at the Forschungszentrum Jülich performs an automated post-mortem search for patterns of inefficient program execution in event traces. Potential problems with this approach are the possible large data sets and long analysis times for long-running applications that hinder the application of this approach on larger parallel machines.

Aksum [2, 8], developed at the University of Vienna, is based on a source code instrumentation to capture profile-based performance data which is stored in a relational database. The data is then analyzed by a tool implemented in Java that performs an automatic search for performance problems based on JavaPSL, a Java version of ASL.

3 Overview of the Periscope Approach

The PERISCOPE approach to performance analysis rests on three main ideas that have individually been proposed in other performance tool designs. However, to the best of our knowledge, no other approach has yet tried to reap the synergetic benefits of all three approaches taken together.

The three ideas are:

1. Automation: An automated search for performance bottlenecks is possible by formally specifying what constitutes a performance problem in the form of ASL (APART specification language) properties [5]. ASL consists of a formal static and dynamic performance data model, of performance properties and of constructs that allow the easy formulation of new properties based on existing ones (meta-properties and property templates). For details concerning ASL and for a large catalog of properties that have been specified for clusters of SMP (symmetric multiprocessor) nodes, please consult [5].

    It should be noted that ASL is not primarily intended to be used directly by performance tool users, instead it is designed to allow tool developers to easily experiment with different performance properties and thus adapt the tool to new or different programming environments or hardware platforms.

2. On-line operation: Instead of recording performance information during the execution of an application (in the form of trace files, for example), and performing a post-mortem analysis of this data, our approach is based on on-the-fly analysis of the performance data provided by our monitoring system (see Sect. 3.1). This avoids problems of overwhelmingly large trace files for long-running applications, especially on machines with large numbers of processing elements.

3. Distribution: Instead of performing a costly collection of performance data at a central location, the search for performance properties is performed in a distributed way. Low-level performance data is processed in a distributed fashion and the exchange of performance data is limited to higher-level performance data.

A design for a performance analysis framework that supports these features has been developed within the Peridot project. This approach is based on the notion of *analysis agents*, computing entities that cooperate in the process of detecting performance properties. In order to process performance data close to its place of origin in the application, analysis agents are co-located with the application subject to performance analysis on the target parallel machine (Fig. 1).

In our current implementation, the agent framework is a simple hierarchical structure where one agent is responsible for each SMP node in the system. This node-level agent detects properties for its local node and forwards the detected properties to one central master agent that reports them to the user. The agents are not currently autonomous, instead they are "programmed" by the master agent. However, we envision future research towards autonomy in our agents will prove fruitful, especially for the application to more loosely coupled systems, like computational grids. For further details on the design of our agent-based analysis framework please consult [6].

3.1 Monitoring

To provide the analysis agents with the necessary performance data, we designed a custom monitoring approach based on dynamic configuration of static instrumentation. The monitor consists of a *passive* monitoring library linked to the application and of an *active* entity called runtime information producer (RIP). The monitoring library is passive in the sense that its only function is to record program events to a shared memory buffer. In order to minimize the perturbation of the target application, all higher-level processing capability (i.e., relating the performance data to

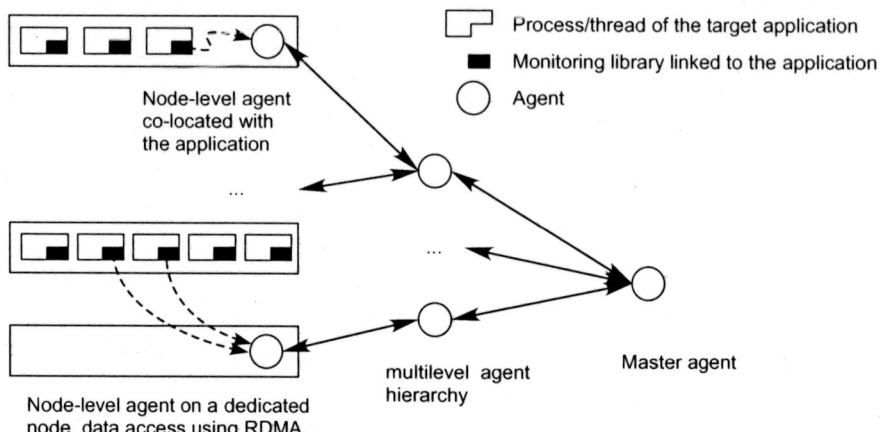

Fig. 1. The PERISCOPE agents are arranged in a hierarchy. Node-level agents are either located on dedicated nodes or co-located with the target application.

program constructs, computing statistics, ...) is handled by the RIP which itself becomes a component of the performance tool.

Depending on the hardware capabilities of the underlying platform, the RIP can be located on a different node than the corresponding instances of the monitoring library (in this case, the shared memory buffers are accessed using remote direct memory access (RDMA) mechanism, see Fig. 1) otherwise the RIP is co-located with the application on a processor set-aside for the purpose of performance analysis. For a complete description of our monitoring approach, please consult [3, 4].

4 Agent Design and Implementation

In this section we discuss the internal design of our analysis agents and describe issues concerning our prototype implementation. From the outset the system was designed with flexibility and reusability at several layers in mind:

- **Usage of the Periscope agent design and prototype implementation in other projects.** Performance tools targeting different hardware architectures or programming models or using different monitoring systems can utilize the Peridot agent technology. As an example, we are considering the usage of our approach within the EP-Cache [1, 9] project which is concerned with data-structure oriented cache performance analysis on single SMP-systems.
- **Flexibility with respect to different performance data models.** ASL properties are formulated with respect to a particular performance data model. Different programming models and new architectures might require the inclusion of new or other performance data models. An example is the support for performance counter data that varies across hardware platforms.
- **Most importantly, flexibility with respect to the formulation of new performance properties.** While an extensive set of performance properties

has already been devised [5], being able to add functionality to the performance tool without the need to change source code of the tool itself is a major improvement. With our approach, it is straightforward to add new properties to our performance tool. As long as the new property does not require a changed data model, the user simply writes the new property as a text file and runs it through the ASL compiler. The generated C++ file can then be simply added to the list of available properties for the PERISCOPE agent.

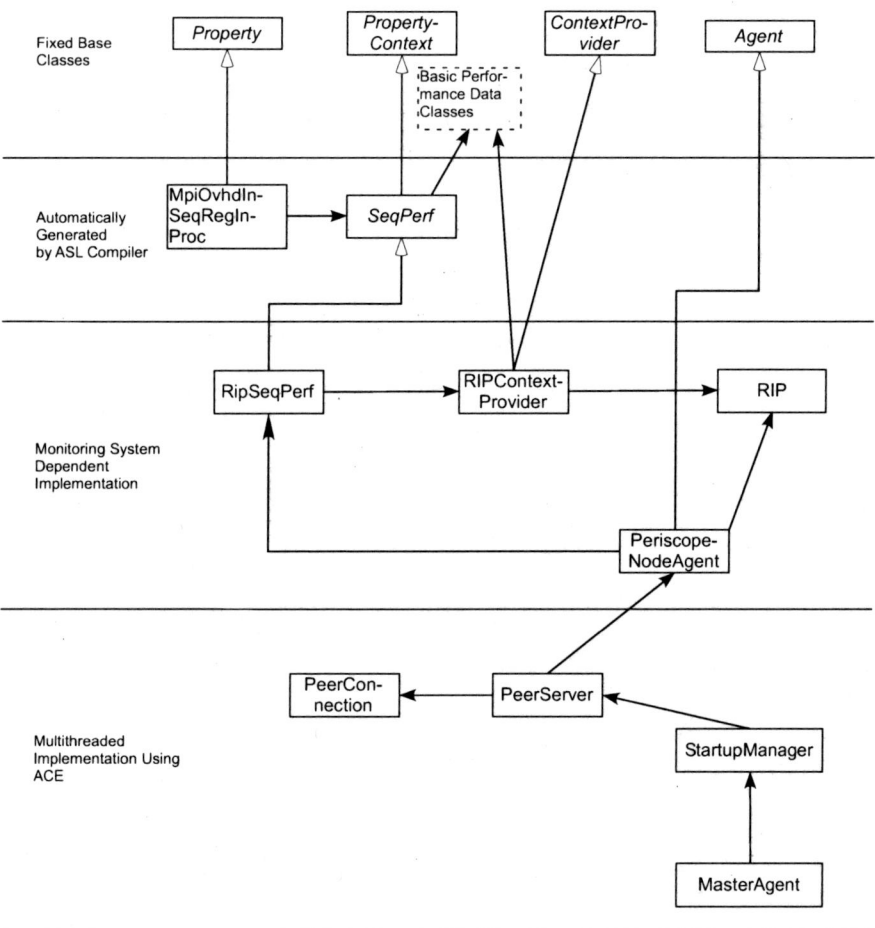

Fig. 2. Design of the PERISCOPE agents.

The prototypical implementation done in the Peridot project is outlined in an UML-like notation in Fig. 2. Diamond-shaped arrows represent an "inherits" relation, while solid arrows represent a "uses" relation between components. Names in *italics* represent abstract classes. The classes are arranged in layers, from top to bottom. This layering corresponds to the transition from most general to most implementation-specific for our PERISCOPE prototype implementation.

4.1 Fixed Base Classes

On the top layer we have three abstract base classes for the formulation and evaluation of performance properties. A `Property` describes a certain performance trait of an application, typically with an expression involving measured performance data that are formalized in an performance data model. The context (i.e., the instantiated part of the performance data model) needed to evaluate a performance property is represented by the abstract `PropertyContext` class. The entity that establishes the connection between the data model and the actual data source (i.e., the monitoring system) is represented by the `ContextProvider` class.

These abstract base classes as well as the abstract `Agent` class are intended to remain fixed for all instances of ASL-based performance analysis tools.

4.2 ASL Compiler Generated Classes

```
property MpiOvhdInSeqRegInProc (SeqPerf pD)
{
  condition: pD.mpiT > 0;
  confidence: 1;
  severity: pD.mpiT/RB(pD.exp);
}
```

Fig. 3. ASL specification of the `MpiOvhdInSeqRegInProc` property.

An ASL compiler is used for the automatic generation of C++ classes from an ASL specification of performance properties and the corresponding performance data models. The generated classes are derived form the abstract base classes already described. As an example, Figure 2 shows a particular property, `MpiOvhdInSeqRegInProc`, that represents the fact that time is spent in message passing operations in a particular sequential (i.e., non-OpenMP) region in the program. The ASL specification of the `MpiOvhdInSeqRegInProc` property is given in Fig. 3.

The context in which the `MpiOvhdInSeqRegInProc` property can be evaluated is the `SeqPerf` class of the performance data model that represents summary (i.e., profiling) information on the execution of sequential regions in the application. Among others, this class has a data member `mpiT`, representing the total time spent in MPI routines in the particular sequential region.

4.3 Monitoring System Dependent Classes

While the `MpiOvhdInSeqRegInProc` class generated by the ASL compiler was already a concrete (i.e, non-abstract) class, the `SeqPerf` class is still an abstract class, since

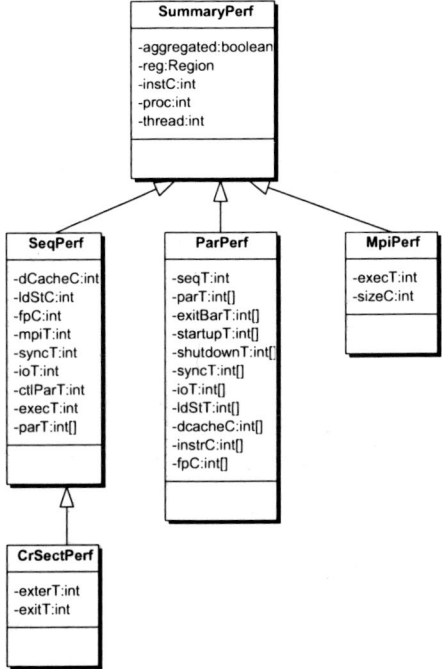

Fig. 4. Part of the performance data model used in the definition of the MpiOvhd-InSeqRegInProc property of Fig. 3

the connection to the actual monitoring system to fill in the required data cannot be provided automatically by the ASL compiler.

This connection to the particular monitoring system used has to be established manually by sub-classing the abstract ASL-compiler generated property context classes and providing a connection to a particular monitoring system. In PERISCOPE, the monitoring system is represented by the RIP ("runtime information producer") briefly described in section 3.1. Correspondingly, the RipSeqPerf class acquires its data form the RIP, indirectly, by using the RipContextProvider.

Finally, a concrete class PeriscopeNodeAgent can be derived from the abstract Agent class that constitutes the node-level agents in the PERISCOPE agent hierarchy outlined in Fig. 1. Note that up to this level, all components reside within one node of the target parallel system. Communication and coordination between the instances of the system on the indivual nodes is handled by the components in the next layer.

4.4 Multithreaded Implementation using ACE

The components on this level handle the startup, mutual discovery, communication and coordination between the constituents on the individual nodes of the hardware platform of the PERISCOPE system. For an efficient and portable implementation we decided to make use of the ACE framework for interprocess communication and multithreading. ACE is a framework for building efficient communication middleware, available for many platforms [7,13,14].

Finally, a single `MasterAgent` orchestrates the detection of performance properties on the individual nodes and reports the detected performance properties to the user.

5 Usage Scenario

To illustrate a usage of our analysis framework consider the following usage scenario, where a user wants to conduct an performance analysis of his MPI-based application. Note that this just represents one possible application scenario of our tool, a usage for OpenMP-based tools or using a batch-system instead of interactive startup is possible too, for example.

1. Preparation of the application for performance analysis. In order to enable performance analysis of the application it has to be instrumented to enable the recording of performance data by our monitoring systems. Currently we support OpenMP monitoring based on the Opari [11] source-to-source instrumenter, MPI monitoring based on the MPI performance monitoring interface and function-call monitoring based the function call instrumentation capability of GNU compilers (the `-finstrument-functions` compiler option). The instrumented application is then compiled and linked against our monitoring library.
2. Interactive startup of the application. The user starts his application using the `mpirun` command on a number of nodes of the system. The application begins to execute and on the first call into our monitoring library (typically, the `MPI_Init` call), the library performs the necessary initialization procedures like allocating buffers for the generated performance data. Furthermore the monitoring library registers itself with a registry service that is used by the components in our system to discover each other. At this point, the monitoring library halts the execution of the target application (unless a special environment variable is set) to allow the agents system to be started to monitor the execution of the application from the start.
3. The user interactively starts the performance analysis system by executing `./periscope <name of the application>`. The startup component then contacts the registry service to determine on which nodes of the systems the application is executing and starts a `NodeAgent` on each of the identified nodes.
4. After all `NodeAgent`s have been started a `MasterAgent` is started on the interactive node and a `start` command is issued by the `MasterAgent` to all `NodeAgent`s. This causes the halted application to be resumed.
5. Performance data is now generated by the target application and processed by the NodeAgents on the fly according to their programming by the MasterAgent. Specifically, property contexts are instantiated as demanded for the evaluation of the properties to be checked. The `MpiOvhdInSeqRegInProc` property, for example, is evaluted on `SeqPerf` contexts. Therefore, the NodeAgents access the `RIP` (via the `RipContextProvider`) for the list of all sequential regions in the program. The NodeAgent can then apply the `MpiOvhdInSeqRegInProc` property which in turn internally accesses the `mpiT` data member of the `SeqPerf` class.

6. The discovered performance properties are communicated to the MasterAgent that reports them to the user.

6 Summary and Future Work

We have described the design an implementation of our agent-based framework for automated performance analysis. The approach is based on a novel combination of ideas designed to make performance analysis of large systems feasible. We have presented the main components of the approach and described how we try to accomplish flexibility and reusability in our design.

We have prototypically implemented our approach in C and C++. The monitoring infrastructure is ported to the Hitachi SR8000 at LRZ Munich (with RDMA support) and Linux-based SMP clusters (RDMA support for Infiniband is planned). For the future we plan to investigate the possibilities for the collaboration and coordination between our agents beyond the current simple flat hierarchical model. This is needed for the application on large machines. Further we plan to investigate the usability of our approach for performance analysis on the Grid.

References

1. EP-Cache projekt homepage.
   http://wwwbode.cs.tum.edu/~gerndt/home/Research/EP-Cache/EPcache.htm.
2. Thomas Fahringer and Clovis Seragiotto Jr. Aksum: A performance analysis tool for parallel and distributed applications. In Vladimir Getov, Michael Gerndt, Adolfy Hoisie, Allen Malony, and Barton Miller, editors, *Performance Analysis and Grid Computing*, pages 189–208. Kluwer Academic Publishers, 2004.
3. Karl Fürlinger and Michael Gerndt. Distributed application monitoring for clustered smp architectures. In Harald Kosch, László Böszörményi, and Hermann Hellwagner, editors, *Proceedings of the 9th International Euro-Par Conference on Parallel Processing*, pages 127–134. Springer, August 2003.
4. Karl Fürlinger and Michael Gerndt. Distributed configurable application monitoring on smp clusters. In Jack Dongarra, Domenico Laforenza, and Salvatore Orlando, editors, *Proceedings of the 10th European PVM/MPI Users' Group Meeting*, pages 429–437. Springer, September 2003.
5. Michael Gerndt. Specification of performance properties of hybrid programs on hitachi SR8000. Technical report, Lehrstuhl für Rechnertechnik und Rechnerorganisation, Institut für Informatik, Technische Universität München, 2002.
6. Michael Gerndt, Karl Fürlinger, and Andreas C. Schmidt. Towards automatic performance analysis for large scale systems. In *International Workshop on Compilers for Parallel Computers*, January 2003.
7. Stephen D. Huston, James CE Johnson, and Umar Syyid. *The ACE Programmer's Guide*. Pearson Education, 2003.
8. Clovis Seragiotto Jr., Thomas Fahringer, Michael Geissler, Georg Madsen, and Hans Moritsch. On using aksum for semi-automatically searching of performance problems in parallel and distributed programs. In *11th Euromicro Conference*

on Parallel, Distributed and Network-Based Processing (Euro PDP 2003), pages 385–392. IEEE Computer Society, February 2003.
9. Edmond Kereku, Tianchao Li, Michael Gerndt, and Josef Weidendorfer. A data structure oriented monitoring environment for fortran OpenMP programs. In *Proceedings of Euro-Par 2004*, pages 133–140, 2004.
10. B.P. Miller, M.D. Callaghan, J.M. Cargille, J.K. Hollingsworth, R.B. Irvin, K.L. Karavanic, K. Kunchithapadam, and T. Newhall. The Paradyn parallel performance measurement tool. *IEEE Computer*, 28(11):37–46, 1995.
11. Bernd Mohr, Allen D. Malony, Sameer Shende, and Felix Wolf. Towards a performance tool interface for OpenMP: An approach based on directive rewriting. In *Proceedings of EWOMP'01, Third European Workshop on OpenMP*, September 2001.
12. Philip C. Roth, Dorian C. Arnold, and Barton P. Miller. MRNet: A software-based multicast/reduction network for scalable tools. In *Proc. of the 2003 ACM/IEEE Conference on Supercomputing*, 2003.
13. Douglas C. Schmidt. The ADAPTIVE communication environment (ACE) http://www.cs.wustl.edu/~schmidt/ACE.html.
14. Douglas C. Schmidt, Stephen D. Huston, and Frank Buschmann. *C++ Network Programming Vol. 1: Mastering Complexity with ACE and Patterns*. Pearson Education, 2002.
15. Felix Wolf and Bernd Mohr. Automatic performance analysis of hybrid MPI/OpenMP applications. In *11th Euromicro Conference on Parallel, Distributed and Network-Based Processing (Euro PDP 2003)*, pages 13–22. IEEE Computer Society, February 2003.

Part III

Natural Sciences

Werner Hanke

Institut für Theoretische Physik
Universität Würzburg
Am Hubland
97074 Würzburg

The topics in Natural Sciences, concerning large scale computations in physics and chemistry, are contained in part three. They range from computational seismology over modern material science and solid state physics to complex systems in chemistry. These topics are summarised in what follows:

The chemistry project of the Munich group around N. Rösch concentrates on density functional (LDA-) treatments of so-called "Complex Systems in Chemistry". The complex systems are molecules, which are embedded in some "environment" such as molecules in solution, adsorbates and solid surfaces, etc. Not surprisingly, modelling of such complex systems is computationally very demanding. To meet these challenges, the parallel density functional program ParaGauss has been developed by the Munich group. The project described in this book contains a summary of the Gaussian scheme and an overview of selected applications, which illustrate the performance of this program for large scale simulations.

Another project from chemistry by the group around J. Gasteiger (Erlangen) is the simulation of enzyme reactions. These reactions play a key role in research related to the pharmaceutical industry because of their role as targets for the design of new drugs. Important results comprise the identification of the pattern for molecules that bind to the same receptor.

In a computational approach, again from the Technical University in Munich, H. Meier and co-workers study the so-called "Maximum Likelihood Method". The aim of this project is to develop novel algorithms for computations of so-called "Phylogenetic Trees" based on sequence data from a certain data base. Simply put, the Maximum Likelihood Method tries to extract from incomplete data as much as possible reliable information. It is used in such different fields as the recognition search for identifying number plates of speeding cars from photos with low resolution up to (as much as possible) reliable evaluation of experimental data in solid state spectroscopy. The progress is summarised in the contribution of the ParBaum project, also contained in the natural science division of the present book.

The seismology project deals with recent examples in global seismology and earthquake scenario simulations. They, in particular, embrace the more detailed relation to "shaking hazard" estimation and associated problems. This is a project of the physics group of the University of Munich headed by B. Schuberth.

Gas-liquid "Foaming" Phase Transitions and the application of Lattice Boltzmann Methods are studied in the FreeWIHR project of the Erlangen group. This project is concerned with a detailed understanding of the so-called foaming process, more specifically with the internal gas-liquid interface. The Erlangen group has developed a Lattice Boltzmann model for the simulation of this foaming process which, in particular, treats the three-dimensional free surface problem more accurately than previous techniques.

In solid-state theory two projects of the Würzburg group around W. Hanke deal with the microscopic mechanism of high-temperature superconductivity. The first of the two projects concerns the question of the driving force behind the binding of electrons into Cooper-pairs in the high-temperature superconductors. In many of the previous calculations the glue for this binding was assumed to be the electron-spin or the magnetic interaction. The point of the present work is to look at this assumption with a combination of up-to-date computational physics possibilities. In particular, Quantum-Monte-Carlo techniques reveal that the spin-mediated interaction between electrons and the Cooper-pair binding is strongly reduced, due to so-called "vertex corrections". These vertex corrections are induced by the - in the high-TC cuprates - very strong on-site Coulomb correlations (Hubbard U). With these type of computer simulations it is now possible to directly shed light on the decisive microscopic question, namely what is the physical mechanism behind the Cooper-pair formation and superconductivity in these systems. The second project by the same group deals with the SO(5) theory of high temperature superconductivity, which has been proposed by the Würzburg group in close collaboration with a group (S.C. Zhang) from the Stanford University. This theory predicts the low-temperature phase diagram including the competing phases of anti-ferromagnetism and superconductivity on the basis of a general [SO(5)] symmetry principle.

Finally, the project by the group around H. Fehske (Greifswald, Erlangen) addresses a long-standing problem in solid state physics, namely the Anderson localisation in disordered electron systems. The corresponding metal-insulator transition, which is observed in many solid state systems due to disorder, is here studied for the case of three-dimensions. For this three-dimensional, physically relevant situation only preliminary numerical results are available. These previous numerical approaches are tested and largely refined concerning their localisation criteria, namely the localisation length and the inverse participation number.

In summary, the simulations in "Natural Sciences" document very convincingly the importance of large-scale computations and their profit from embedding them into the unique and extremely stimulating (and helpful) KONWIHR environment.

CUHE: Electron-Spin Interaction in High-T_c Superconductors

Zhongbing Huang, Werner Hanke, and Enrico Arrigoni

Institut für Theoretische Physik
Universität Würzburg
am Hubland
97074 Würzburg, Germany
{*zbhuang,hanke*}*@physik.uni-wuerzburg.de*

1 Abstract

In this paper, we study numerically the renormalization of the electron-spin (el-sp) interaction or vertex due to Coulomb correlations in a two-dimensional one-band Hubbard model with spin-fluctuation momentum transfer $\mathbf{q} = (\pi, \pi)$. Our simulations are based on a new numerically exact technique to extract the vertex, which is especially important for the physically relevant case, i.e., strong correlations, which cannot be controlled perturbatively. We find that the renormalized el-sp vertex decreases quite generally with increasing doping from the underdoped to the overdoped region. In the underdoped region, the corresponding effective pairing interaction increases strongly with lowering temperature in the weak- to intermediate-correlation regime. In contrast to this, it depends weakly on temperature in the strong-correlation regime. This behavior in the physically relevant strong-correlation case is due to a near cancellation between the temperature-driven enhancement of the spin susceptibility χ and the reduction of the el-sp interaction vertex. Thus, the spin-mediated d-wave attraction, which is peaked in weak coupling due to χ, is strongly reduced due to the el-sp vertex corrections for strong correlations.

2 Introduction

More than fifteen years after their discovery, the pairing mechanism of high-temperature superconductivity is still not completely resolved. One central issue is how to describe correctly the interactions of charge carriers (electrons or holes) with bosonic excitations such as phonons or collective magnetic excitations. On the one hand, anomalous magnetic and transport properties suggest that strong Coulomb interactions are dominant and the electron-phonon (el-ph) interaction plays a secondary role [1]. These anomalous normal-state properties stimulated a large effort toward an unconventional superconductivity mechanism [2,3]. On the other hand, a variety of experiments also display pronounced phonon and electron-lattice effects in these materials: superconductivity-induced phonon renormalization [4], large isotope coefficients away from optimal doping [5], tunneling phonon structures [6], etc., give evidence of strong electron-phonon coupling. Recently, photoemission data indicated a sudden change in the electron dispersion near a characteristic energy scale [7], which is possibly caused by coupling of electronic quasiparticles either to phonon modes or to the magnetic resonant mode [8,9].

To elucidate the effects of phonons and magnetic excitations on the physical properties of high-T_c superconductors, it is crucial to understand the renormalization of el-ph and el-sp interactions due to strong Coulomb correlations. In our previous work [10], we have addressed the issue of el-ph interaction in a strongly correlated system. Specifically, we applied the determinantal Monte Carlo [11] algorithm to calculate the el-ph vertex function in the one-band Hubbard model. This work showed that strong correlations induce an anomalous enhancement of the coupling between charge carriers and long-wavelength phonons as a function of the Coulomb correlation U. This is an unexpected result, which may have pronounced consequences for the d-wave microscopic pairing mechanism, for transport and superconducting properties. Here, we will employ the numerical technique to study the el-sp vertex function in the one-band Hubbard model.

3 Formalism

Our starting point is the one-band Hubbard model,

$$H = -t \sum_{\langle ij \rangle, \sigma} (c_{i\sigma}^\dagger c_{j\sigma} + c_{j\sigma}^\dagger c_{i\sigma}) + U \sum_i n_{i\uparrow} n_{i\downarrow}, \qquad (1)$$

The operators $c_{i\sigma}^\dagger$ and $c_{i\sigma}$ as usual create and destroy an electron with spin σ at site i, respectively and the sum $\langle ij \rangle$ is over nearest-neighbor lattice sites. Here, U is the on-site Coulomb interaction and we will choose the nearest-neighbor hopping t as the unit of energy.

In our simulations, we have used the linear-response technique in order to extract the el-sp vertex. In this method, one formally adds to Eq. (1) the interaction with a momentum- and (imaginary) time-dependent spin-fluctuation field $S_\mathbf{q} e^{-iq_0\tau}$ in the form

$$H_{el-sp} = \sum_{\mathbf{k}\mathbf{q}\sigma} g_{\mathbf{k}\mathbf{q}}^0 \sigma c_{\mathbf{k+q}\sigma}^\dagger c_{\mathbf{k}\sigma} S_\mathbf{q} e^{-iq_0\tau}, \qquad (2)$$

where $g_{\mathbf{k}\mathbf{q}}^0$ is the bare el-sp coupling. In the following, we will focus on the case of an el-sp coupling, in which the bare coupling $g_{\mathbf{k}\mathbf{q}}^0$ is a constant g^0. Since we will consider linear terms in g^0 only, we can set g^0 equal to 1. One then considers the "anomalous" single-particle propagator in the presence of this perturbation defined as

$$G_A(p,q) \equiv -\int_0^\beta d\tau \, e^{i(p_0+q_0)\tau} \langle T_\tau c_{\mathbf{p+q}\sigma}(\tau) c_{\mathbf{p}\sigma}^\dagger(0) \rangle_{H+H_{el-sp}}, \qquad (3)$$

where $\langle \rangle_{H+H_{el-sp}}$ is the Green's function evaluated with the Hamiltonian $H+H_{el-sp}$. Diagrammatically $G_A(p,q)$ has the structure shown in Fig. 1 so that the el-sp vertex $\Gamma(p,q)$ can be expressed quite generally in terms of G_A and of the single-particle Green's function $G(p)$ in the form

$$\Gamma(p,q) = \lim_{S_\mathbf{q}\to 0} \frac{1}{S_\mathbf{q}} \frac{1}{1+U\chi_{zz}(q)} \frac{G_A(p,q)}{G(p+q)G(p)}, \qquad (4)$$

with $\chi_{zz}(q)$ the longitudinal spin susceptibility. It is, thus, sufficient to calculate the leading linear response of G_A to $H_{\text{el-sp}}$, which is given by

$$G_A(p,q) = S_\mathbf{q} \int_0^\beta d\tau\, e^{i(p_0+q_0)\tau} \int_0^\beta d\tau'\, e^{-iq_0\tau'} \sum_{\mathbf{k}q\sigma'} g^0_{\mathbf{k}\mathbf{q}} \times$$
$$\langle T_\tau \sigma' c^\dagger_{\mathbf{k}+\mathbf{q}\sigma'}(\tau'+0^+) c_{\mathbf{k}\sigma'}(\tau') c_{\mathbf{p}+\mathbf{q}\sigma}(\tau) c^\dagger_{\mathbf{p}\sigma}(0) \rangle_H, \qquad (5)$$

where 0^+ is a positive infinitesimal. The two-particle Green's function in Eq. (5) is evaluated with respect to the pure Hubbard Hamiltonian (Eq. 1).

In terms of the el-sp vertex and the spin susceptibility, the effective pairing interaction is expressed in the form

$$V_{eff}(p,q) = (U\,\text{Re}\,\Gamma(p,q))^2 \chi_{zz}(q). \qquad (6)$$

with

$$\chi_{zz}(q) = \frac{1}{2}\int_0^\beta d\tau\, e^{-i\,q_0\tau}\, \langle T_\tau m^z_\mathbf{q}(\tau) m^z_{-\mathbf{q}}(0)\rangle,$$

and

$$m^z_\mathbf{q} = \frac{1}{\sqrt{N}}\sum_{\mathbf{k}\sigma}\sigma c^\dagger_{\mathbf{k}+\mathbf{q}\sigma} c_{\mathbf{k}\sigma}, \qquad (7)$$

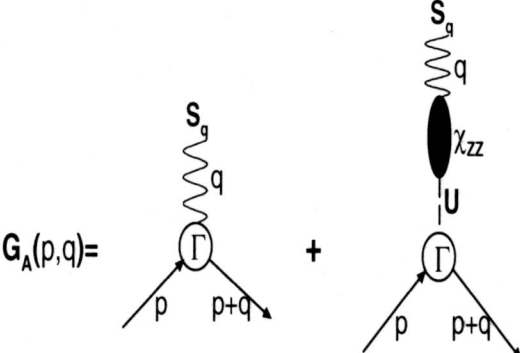

Fig. 1. Diagrammatic representation of $G_A(p,q)$ within linear response to $S_\mathbf{q}$. The thick solid lines represent dressed single-particle Green's functions of the Hubbard model. The wavy line denotes the external perturbation in Eq. (2). The dashed line represent the Hubbard interaction U and the closed ellipse stands for the longitudinal spin susceptibility $\chi_{zz}(q)$.

The low order U^2 vertex contributions to Γ are displayed in Fig. 2. The diagrams shown at the bottom of Fig. 2 are the leading terms of the random phase approximation (RPA) to the longitudinal spin susceptibility.

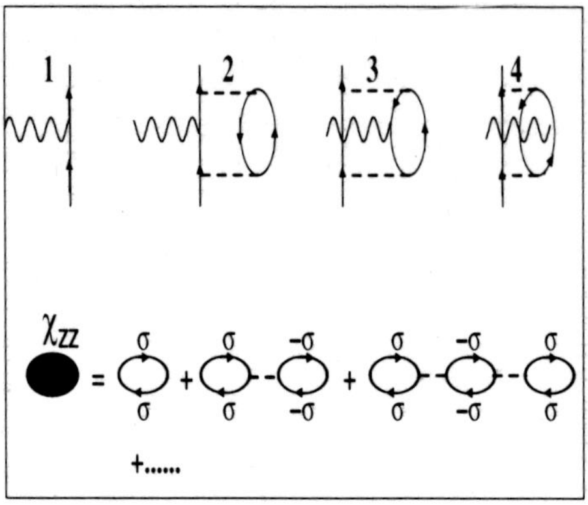

Fig. 2. Low-order Feynman diagrams for the el-sp vertex $\Gamma(p,q)$ (top) and low-order longitudinal spin susceptibility graphs (lower). The thin solid lines are the non-interacting Green's functions and the dashed lines represent the Hubbard interaction U. The wavy lines stand for the spin-fluctuation fields.

4 General Results

Our numerical Monte Carlo simulations were performed on an 8×8 lattice at different doping densities and different temperatures. We have set the frequencies to their minimum values, i.e., $p_0 = \pi T$ for fermions and $q_0 = 0$ for bosons. In high-T_c superconductors, the charge carriers near the $(\pi,0)$ region are strongly affected by antiferromagnetic spin fluctuations. Therefore, we will fix the momenta of the incoming electron and spin fluctuation at $\mathbf{p} = (-\pi, 0)$ and $\mathbf{q} = (\pi, \pi)$, respectively.

Let us first discuss the weak to intermediate ($U \leq 4$) coupling case. Figs. 3(a), 3(b), and 3(c) display the δ dependence of $Re\Gamma(p,q)$, $\chi_{zz}(q)$, and $V_{eff}(p,q)$ in the *intermediate-correlation* regime. One can see that the renormalized el-sp vertex decreases quite generally with increasing doping from the underdoped to the overdoped region except at $\beta = 4$ and large dopings. With lowering temperature, the el-sp vertex is reduced at all doping densities. As the spin susceptibility $\chi_{zz}(q)$ becomes much larger when approaching half-filling (see Fig. 3(b)), in conjunction with the behavior of $\Gamma(p,q)$, the effective pairing interaction V_{eff} in Eq. (6) is expected to dramatically increase with decreasing doping, which is clearly shown in Fig. 3(c). We also notice that, in the small doping case and for intermediate $U's$, the effective pairing interaction increases dramatically with decreasing temperature. This behavior is due to the fact that the increase of the spin susceptibility is faster than the decrease of the el-sp interaction vertex.

Figs. 3(d), 3(e), and 3(f) give the δ dependence of $Re\Gamma(p,q)$, $\chi_{zz}(q)$, and $V_{eff}(p,q)$ in the *strong-correlation* regime. Similar to the intermediate-correlation case, the renormalized el-sp vertex decreases with increasing doping from the underdoped to the overdoped region. In addition, the el-sp vertex is reduced at all doping densities when the temperature is lowered. As shown in Fig. 3(f), a *crucial difference* to *the intermediate-correlation* case is that the effective pairing interaction *depends weakly on temperature* below $T = J = 0.5$ for all doping densities. This behavior is due to the near cancellation between the temperature-driven enhancement of the

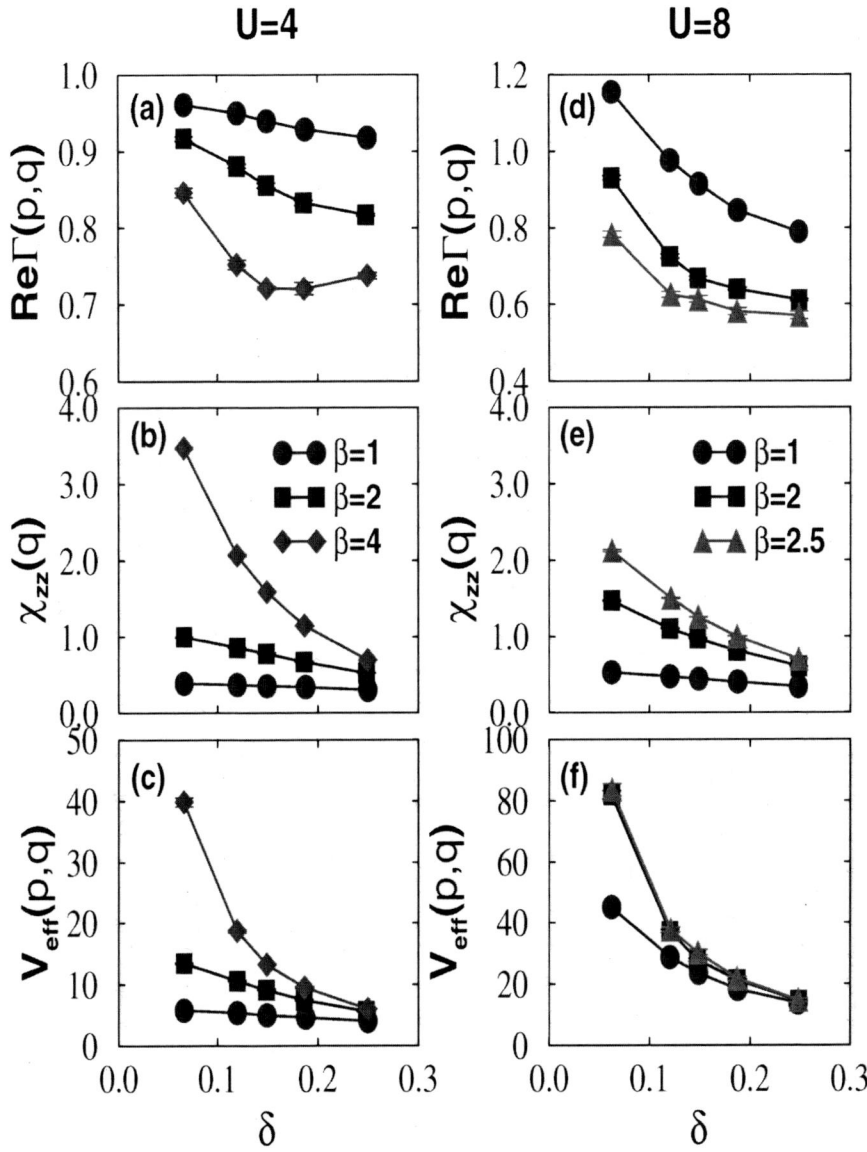

Fig. 3. (a) Real part of $\Gamma(p,q)$, (b) the spin susceptibility $\chi_{zz}(q)$, and (c) the effective pairing interaction $V_{eff}(p,q)$ as a function of doping density δ for $U=4$. (d), (e), and (f) same as (a), (b), and (c) respectively, for $U=8$. Here the results are given for inverse temperatures $\beta=1$ (closed circles), $\beta=2$ (closed squares), $\beta=4$ (closed diamonds), and $\beta=2.5$ (closed up-triangles).

spin susceptibility and the reduction of the el-sp interaction vertex. Schrieffer argued that this cancellation is valid either in the ordered spin density wave (SDW) state or in the paramagnetic state as long as spin fluctuations remain propagating excitations [12]. Our numerical results confirm Schrieffer's argument and suggest that in the physically relevant strong-correlation regime, the *spin-mediated d-wave attraction is strongly reduced due to el-sp vertex corrections.*

In order to see the temperature dependence more clearly, in Fig. 4 $Re\Gamma(p,q)$, $\chi_{zz}(q)$, and $V_{eff}(p,q)$ are plotted as a function of T at $U = 4$ and $U = 8$. From Fig. 4(b), it is evident that the spin susceptibility depends on T in a similar way for different Hubbard $U's$. On the other hand, the T dependence of the el-sp vertex and effective pairing interaction is rather different in the intermediate- and strong-correlation regimes, as shown in Figs. 4(a) and 4(c). When the temperature is lowered below $T = 0.5$, we observe that the el-sp vertex decreases much faster in the strong-correlation regime than in the intermediate-correlation regime, and that the effective pairing interaction increases with decreasing T at $U = 4$, but has very little change at $U = 8$.

5 The Need for High-Performance Computing

The massive (MPI) parallel quantum Monte Carlo program was run in the intranode mode on the Hitachi SR8000 with a sustained performance of 0.5 GFlops per processor and required a storage of 10-100 MB per processor. The total computing time of a typical run on 32 nodes (i.e. 256 processors) was about 16 hours. In general, the computing time scales with the lattice size $N = L \times L$ and the inverse temperature β in the form: $time \propto N^4 * \beta^2$, thus the computing time increases dramatically when the lattice size changes from 8×8 to 10×10 or the inverse temperature increases from 2 to 4. In order to understand how the electron-spin interaction depends on the doping density and the electron and phonon momenta in the physically relevant low energy regime, we must perform simulations on as large lattices as possible (which provides us with a dense mesh of **k**-points and a small finite-size effect) and at very low temperatures (which is required because of the low-energy scale $E \sim K_B T_c$, where T_c is the superconducting transition temperature). Therefore, we need a very large amount of computing time to obtain results at different doping densities and different temperatures, which only the Hitachi SR8000 can provide.

6 Summary

In this paper, based on quantum Monte Carlo simulations, we study the renormalization of the el-sp interaction or vertex in the one-band Hubbard model. In contrast to earlier perturbation studies, this allows for a numerically exact solution for the el-sp vertex even in the strong-correlation regime, which is very helpful for clarifying some crucial issues of high-T_c superconductivity. We find that the renormalized el-sp vertex decreases quite generally with increasing doping from the underdoped to the overdoped region. On the other hand, we find that in the underdoped case, the temperature dependence of the effective pairing interaction is rather different in

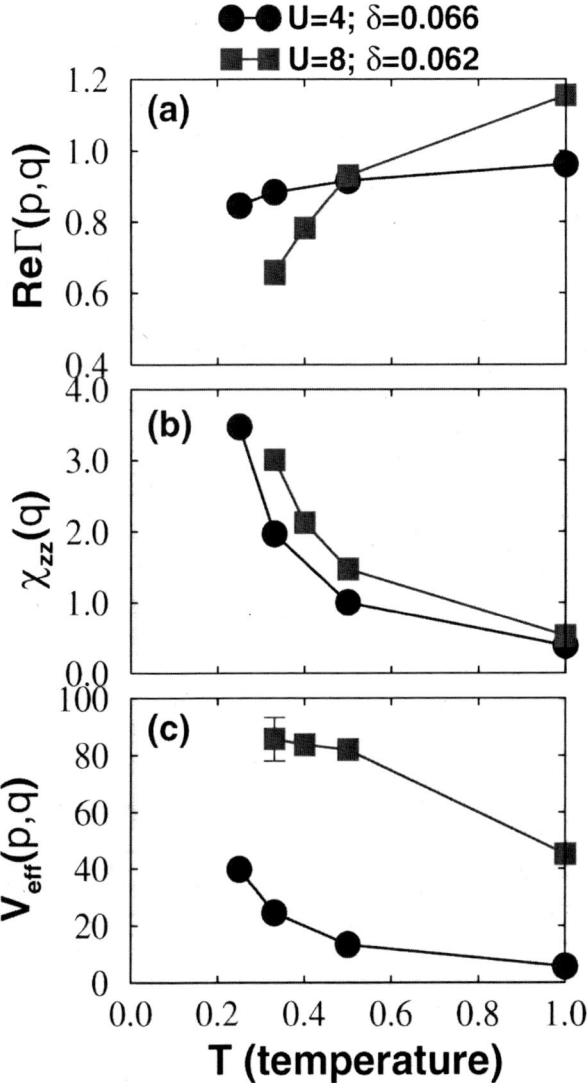

Fig. 4. (a) Real part of $\Gamma(p,q)$, (b) the spin susceptibility $\chi_{zz}(q)$, and (c) the effective pairing interaction $V_{eff}(p,q)$ as a function of temperature T. Here, in (a)-(c) the closed circles stand for the results for $U=4$ and $\delta=0.066$, and the closed squares for $U=8$ and $\delta=0.062$. T in units of t (hopping integral in Eq. (1)).

the intermediate- and strong-correlation regimes: It increases strongly with lowering temperature in the intermediate-correlation regime, but depends weakly on temperature in the strong-correlation regime. In the overdoped case, the temperature dependence of the effective pairing interaction is rather weak in both intermediate- and strong-correlation regimes.

We would like to acknowledge useful discussions with D. J. Scalapino. We also want to thank the Leibniz-Rechenzentrum (LRZ) München for computational support. This work was supported by the DFG under Grant No. Ha 1537/20-1, by a Heisenberg Grant (AR 324/3-1) and by the the KONWHIR projects OOPCV and CUHE.

References

1. M. Imada, A. Fujimori, and Y. Tokura, Rev. Mod. Phys. **70**, 1039 (1998).
2. D. J. Scalapino, Physics Reports **250**, 329–365 (1995).
3. P.W. Anderson, cond-mat/0201429.
4. V.G. Hadjiev, X.J. Zhou, T. Strohm, M. Cardona, Q.M. Lin, and C.W. Chu, Phys. Rev. B **58**, 1043 (1998); for a review, see also M.L. Kulic, Physics Reports **338**, 1–264 (2000).
5. J.P. Franck, S. Harker, and J.H. Brewer, Phys. Rev. Lett. **71**, 283 (1993).
6. D. Shimada, Y. Shiina, A. Mottate, Y. Ohyagi, and N. Tsuda, Phys. Rev. B **51**, R16495 (1995).
7. A. Lanzara, P.V. Bogdanov, X.J. Zhou, S.A. Keller, D.L. Feng, E.D. Lu, T. Yoshida, H. Eisaki, A. Fujimori, K. Kishio, J.-I. Shimoyama, T. Noda, S. Uchida, Z. Hussain, and Z.-X. Shen, Nature **412**, 510 (2001).
8. M. Eschrig and M.R. Norman, Phys. Rev. Lett. **85**, 3261 (2000).
9. M. Eschrig and M.R. Norman, Phys. Rev. B **67**, 144503 (2003).
10. Z.B. Huang, W. Hanke, E. Arrigoni, and D.J. Scalapino, Phys. Rev. B, **68**, 220507(R) (2003).
11. R. Blankenbecler, D.J. Scalapino, and R.L. Sugar, Phys. Rev. D **24**, 2278 (1981).
12. J.R. Schrieffer, J. Low Temp. Phys. **99**, 397 (1995).

ENZYMECH: Computer Simulations of Enzyme Reaction Mechanisms: Application of a Hybrid Genetic Algorithm for the Superimposition of Three-Dimensional Chemical Structures

Alexander von Homeyer and Johann Gasteiger

Computer-Chemie-Centrum
Institut für Organische Chemie
Universität Erlangen-Nürnberg
Nägelsbachstr. 25
91025 Erlangen, Germany
{alexander.von.homeyer,gasteiger}@chemie.uni-erlangen.de

Abstract. Enzymes play a key role in research of the pharmaceutical industry because they represent targets for the design of new drugs. Therefore, the determination of the mode of action of enzymes is one of the great challenges of modern chemistry and an important task in *rational drug design*. The situation is aggravated by the fact that the number of enzymes with known three-dimensional structure is small compared to the number of pharmaceutically relevant enzymes. Therefore, approaches for searching for a new *lead structure* depend on the information available about the protein structure and the ligands binding to a particular target. In this article we present a methodology based on a ligand-based approach. It can also be employed if the three-dimensional structure of the target of interest is not known. The structures of a set of molecules are superimposed based on a *parallel implementation* of a *genetic algorithm* (GA) to evaluate their *maximum common three-dimensional substructure*. This is an important step in the identification of a *pharmacophoric pattern* for molecules that bind to the same receptor. With this method it is possible to determine a complementary map of the receptor binding pocket.

Keywords:
rational drug design, lead structure, genetic algorithm, parallel computing, maximum common three-dimensional substructure, pharmacophore

1 Introduction

The determination of the mode of action of enzymes is one of the great challenges of modern chemistry. The knowledge of enzymatic mechanisms presents an excellent basis for the systematic and rational design of new drugs (*rational drug design*). With the availability of the 3D structure of a protein it is feasible to evaluate and predict the binding mode of a ligand within the active site of the receptor with *docking* methods. Unfortunately, many proteins can never be crystallized or their structure will dramatically change when taken out of their natural environment such as for membrane proteins. For many therapeutically relevant target enzymes an exact knowledge of the three-dimensional structure is not available. Depending on the situation, different strategies have to be embarked. If a set of different active ligands is at hand it is possible to draw conclusions on their binding affinities to the receptor by analyzing their similarities and dissimilarities. To this end, the ligands are superimposed to extract their *3D maximum common substructure* (3D-MCSS) [1] [2] [3] and derive from this a *pharmacophore* model. A *pharmacophore* defines the three-dimensional arrangement of substructure units such as hydrogen bonding or hydrogen accepting sites or hydrophobic areas in a molecule. It provides indications of substructures relevant for the receptor affinity of the different substrates and leads to an indirect mapping of the receptor site. By calculation of the structural requirements of the ligands it is possible to draw conclusions on the spatial requirements of the binding pocket.

One of the first programs that was able to optimize steric and physicochemical properties of two molecules simultaneously is SEAL [4]. A *Monte-Carlo* method was implemented to determine the different relative orientations of both molecules to be superimposed by rotation and translation. The program GASP [5] [6] implements a *genetic algorithm* (GA), which computes the superimposition process as well as the handling of the conformational flexibility. For the simultaneous superimposition of several ligands, one molecule is kept as a template, so that the other ligands can adjust to it with a conformation that was proofed to be optimal. Another approach is illustrated by Lemmen, Lengauer and Klebe [7]. The introduced program FLEXS follows an *iterative incremental approach*. The molecules are superimposed pairwise and one molecule is used as a reference which is kept rigid during the alignment. The second molecule is fragmented and is reconstructed in consideration of maximum and optimal superimposition. Every newly added fragment achieves conformational degrees of freedom. If conformational flexibility is taken into account during the alignment process then adequate conformations of the ligands can appear. An NP-complete search problem accrues from the large number of conformational degrees of freedom (Fig. 1). With the increase in the number of rotatable bonds, the complexity for finding an optimal superimposition grows more rapidly than exponential and can therefore not yet be enumerated by high performance computers and not be solved by exhaustive search methods. Multidimensional search spaces and problems that are NP-complete can therefore be better explored by heuristic techniques such as GAs [8] [9]. Even though GAs are able to find good solutions for a broad range of optimization problems in acceptable time scales, the computing time grows fast if they are applied to harder and larger problems. Therefore, much effort has been invested to speedup the algorithm through parallelization. The developments in parallel and distributed computing offer a means to overcome some of the limitations

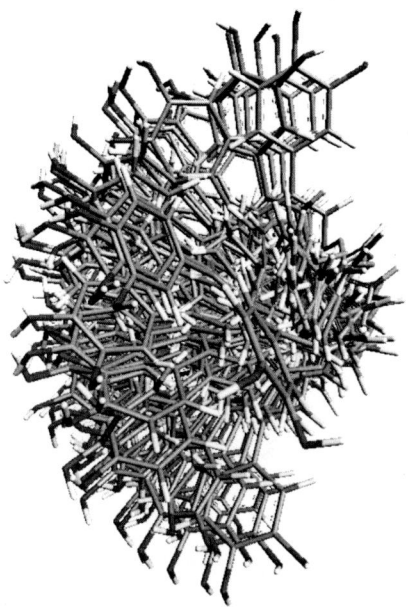

Fig. 1. Superimposition of 216 conformers of the cytochrome P450c17 inhibitor BW112 as an illustration of the search space taking into account conformational flexibility. Note that torsion angles have been restricted in this figure to certain low energy conformations. Thus, the conformational space is potentially even more extensive. For a reproduction of this figure in colour, see Fig. A.7.

of single processor machines. An overview of different implementation techniques, is given by Cantú-Paz [10].

2 Methods

2.1 Overview of Genetic Algorithms and GAMMA

GAs are stochastic search methods that are inspired by the basic mechanics of natural selection and genetics. GAs have successfully been applied to solve problems within fields that have a high dimensionality, a strong non-linearity, that are non-differential or noisy and NP-complete. A GA imitates the adaptation mechanism of a population of individuals to a changing environment (Fig. 2). At the beginning of the algorithm an initial population, $P(0)$, is usually generated randomly. These individuals represent discrete points in the search space and vary in their fitness and adaptation to the problems' solution. For each generation, t, individuals in the current population, $P(t)$, are evaluated, ranked according to their fitness and then the genetic operators *selection*, *mutation*, and *crossover* are iteratively applied. Two additional operators are implemented, called *creep* and *crunch*. *Creep* leads to a larger substructure by adding atoms to the match list taking into account restrictions imposed by the geometry of the molecules. *Crunch* acts as an antagonist to

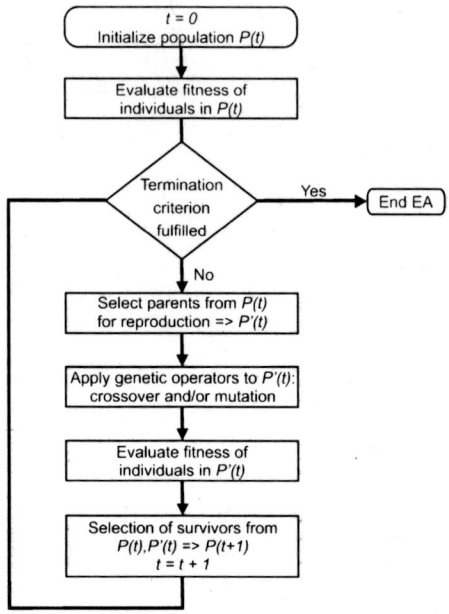

Fig. 2. Flow diagram of an evolutionary algorithm. P(0) is the initial population at the beginning of the computation. P(t) is the population at generation t. P'(t) is a sub-population whose individuals are selected from P(t) for interbreeding. P(t+1) is the population at the next generation t+1 generated from P(t) and/or P'(t). For the next iteration P(t+1) will be the new P(t).

the creep reducing the number of atom pairs in the substructure which are responsible for bad geometric distance parameters. The newly bred children represent the members of the resulting population, *P(t+1)*. The optimization process proceeds for a fixed number of iterations or until convergence is detected within the population. The method developed for the superimposition of flexible three-dimensional structures is a *hybrid genetic algorithm* implemented in the program GAMMA (*Genetic Algorithm for Multiple Molecule Alignment*) [1] [2] [3]. Because GAs are not based on a deterministic procedure the optimization by a GA does not necessarily arrive at the optimum solution. In order to alleviate this problem, an additional method, the *directed-tweak* [11] procedure was implemented to match the conformations of the molecules to be overlaid. A major goal of this hybrid procedure is to adequately address the conformational flexibility of ligand molecules. The GA optimizes in a nondeterministic process the size and the geometric fit of the overlay. The geometric fit is further improved by the directed tweak method. Two conflicting main principal parameters contribute to the fitness of a superimposition and have to be optimized: the size of the substructure, as given by the number, N, of matching atoms, and the geometric fit of the matching atoms as represented by a distance parameter. The distance parameter, D, consists of the sum of the squared differences of corresponding atom distances in the molecules k and l.

$$D = \frac{1}{4}\frac{n(n-1)}{n} \sum_{i,j, i \neq j}^{N} \sum_{k,l, k \neq l}^{n} (d_k(i,j) - d_l(i,j)) \qquad (1)$$

with $d_k(i,j)$, $d_l(i,j)$ = atom distances in molecule k and molecule l,
n = number of molecules,
i,j = indices of match tuples to be compared,
N = number of match pairs (size of the substructure).

D is related to the *root mean square* (rms) error of the distances of corresponding atoms in an optimized superimposition.

The approach for the MCSS search is based on atom mapping and the 3D substructure search starts with one conformation for each structure and investigates the conformational flexibility during the optimization process. These starting points correspond to the chromosomes or individuals of a population representing potential solutions to the search problem.

2.2 Parallelization of GAMMA

GAMMA was made parallel on a SGI ORIGIN 3400 [12] with 28 processors and 56 GBytes memory. It has a ccNUMA-architecture, that means that the whole memory can be linearly addressed from every processor, but physically it is distributed upon nodes with four CPUs. This computer is scheduled for memory-intensive, serial and moderate parallel programs. The *Message Passing Interface* (MPI) [13] [14] was chosen as the programming interface because message passing is a natural programming model for distributed-memory MIMD computers. Also because a subsequent port to workstation-clusters is planned, MPI was a convincing alternative. A complete run of the program GAMMA consists of several independent GA experiments that are consecutively executed in the serial version. The parallelization was realized on the level of the outermost program loop that enumerates the experiments of the GA. The experiments are consistently distributed upon the processes of the system. This solution was chosen because of the independent treatment of the single experiments by the algorithm. The coherence of the populations is guaranteed by making the independent experiments in parallel (Fig. 3). The mechanism is comparable to an allopatric population distribution. The individuals are separated due to a physical barrier and evolve without interaction. Resulting populations can therefore vary strongly. The processors operate asynchronously in the sense that each generation independently starts and ends at each processor. Because each of these tasks is performed independently at each processor and because the processors are not synchronized, this local search approach to parallelization efficiently uses all the processing power of each processor.

Each experiment starts with the initialization of an own separate random population of individuals per parallel process. Then, the GA loop begins with the selection based upon calculated fitness of the single individuals. After selection, the genetic and the knowledge-augmented operators are applied to the chromosomes of the populations. A new population forms the offspring generation. The presented pseudo code demonstrates the distribution of the experiments using MPI:

```
#include <mpi.h>
int ex; /* experiment */
```

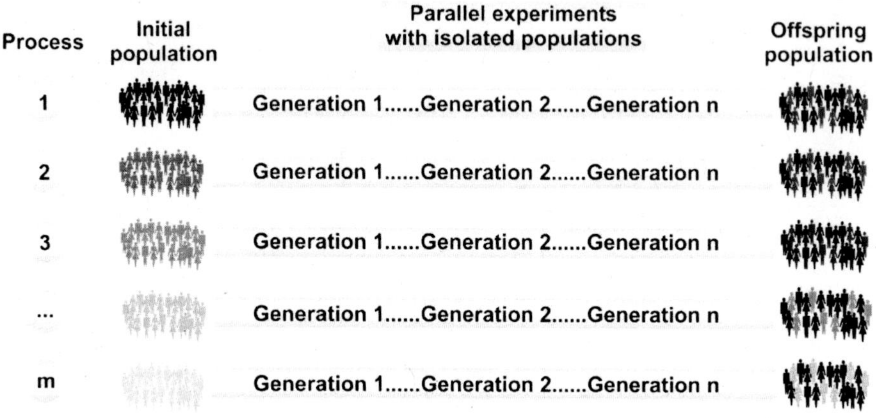

Process	Initial population	Parallel experiments with isolated populations	Offspring population
1		Generation 1......Generation 2......Generation n	
2		Generation 1......Generation 2......Generation n	
3		Generation 1......Generation 2......Generation n	
...		Generation 1......Generation 2......Generation n	
m		Generation 1......Generation 2......Generation n	

Fig. 3. Distribution of the experiments upon the different processes. The experiments are running independently in parallel per processor. This mechanism is comparable to an allopatric population distribution. The individuals are separated due to a physical barrier and evolve without interaction. Resulting populations can therefore vary strong.

```
int mpi_size, mpi_rank;

main(int argc; char *argv[]) {
  MPI_Init(&argc, &argv);
  MPI_Comm_size(MPI_COMM_WORLD, &mpi_size);
  MPI_Comm_rank(MPI_COMM_WORLD, &mpi_rank);
  nexp = nexp/mpi_size; /* distribution of experiments */
  for (ex=mpi_rank*nexp; ex<(mpi_rank+1)*nexp; ex++) {
      ... /* implementation of experiments follows here */
  }
  MPI_Finalize();
}
```

with $nexp$ = number of experiments,
mpi_size = number of processes in the group of MPI_COMM_WORLD,
mpi_rank = rank of the calling process in group of MPI_COMM_WORLD.

The distribution of the experiments onto the processors is currently managed with an integer division.

$$nexp = \frac{nexp}{mpi\_size} \tag{2}$$

The consequence of this operation is, that for the number of processes mpi_size, for which a division of the number of experiments $nexp$ through mpi_size without remainder is not possible, the remaining experiments will not be executed. Therefore, it is not possible to measure the runtime directly. To circumvent this problem an adjustment of the runtime was applied.

$$T_r = T_m \left(\frac{E}{N \lfloor \frac{E}{N} \rfloor} \right) \tag{3}$$

with T_r = real/revised runtime,
T_m = measured runtime,
E = number of experiments,
N number of processors.

This term has to has to result in 1, if the number of experiments is divisible through the number of processes without remainder.

2.3 Test Case

The described method was applied to several different data sets to measure the performance of the serial program on a single processor machine compared with the parallel version running on the SGI ORIGIN 3400. The single processor computer is equipped with an INTEL PENTIUM III Coppermine 700 MHz processor and 512 MBytes memory. The speedup, which is associated with the number of processors, and the parallel efficiency, to obtain the utilization of the processors, are determined. Five test sets with molecules of different size and different flexibility were chosen for the measurement. Every set contained three molecules to be superimposed. The data sets consisted of inhibitors and ligands of cytochrome P450c17 (Fig. 4), of HSV1 thymidine kinase, of HIV protease, of a F_{ab} fragment of a monoclonal antibody (immunoglobulin) and of the glycogen phosphorylase. For all compound sets the

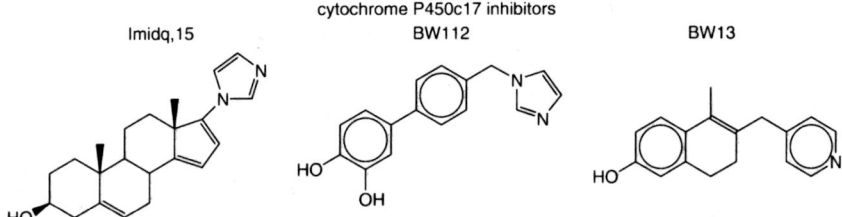

Fig. 4. Structures of cytochrome P450c17 inhibitors.

same number of experiments, generations and individuals was chosen for all program runs. In addition, the same operator probabilities were applied. The number of processors ranged from 1 to 15.

Superimpositions of these molecule sets were carried out to determine their 3D-MCSS. To do so, a universal access to the 3D structure of molecules is necessary. Such an approach can be provided by the automatic 3D structure generator CORINA [15] [16] [17]. All 3D structures investigated in this report have been obtained by CORINA which can provide a single low-energy conformation of any organic molecule. The 3D substructure search starts with one conformation for each structure and investigates the conformational flexibility during the optimization process.

3 Results and Discussion

The scalability of GAMMA is studied using both, the serial and the parallel version, and the results are shown in Fig.5. A serial GAMMA can be regarded as the program running on one processor. It can be seen from the *speedup* in Fig. 5 that the

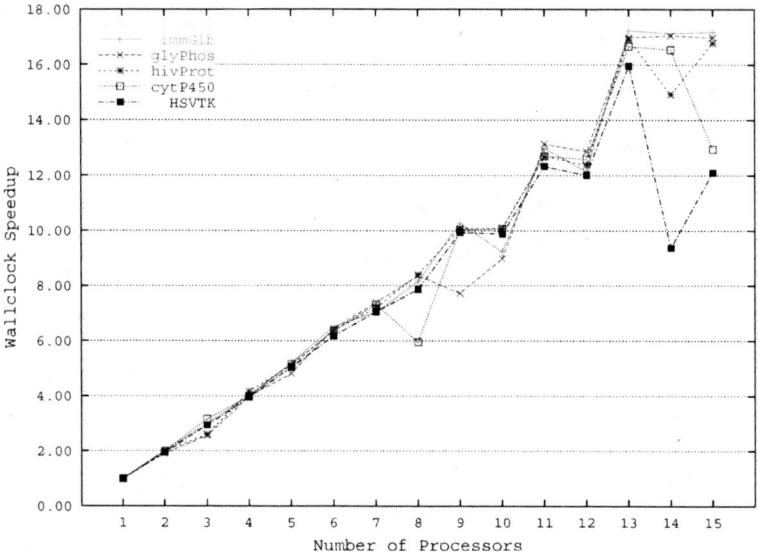

Fig. 5. Measurement of the speedup of the parallel GAMMA version. Results are shown for all five test sets: cytochrome P450c17 inhibitors (cytP450), HSV1 thymidine kinase inhibitors (HSVTK), HIV protease inhibitors (HIVProt), ligands of a Fab fragment of a monoclonal antibody (immGlb) and inhibitors of the glycogen phosphorylase (glyPhos).

performance increase of the algorithm is fairly good, although not ideal. Increasing the number of processors results in a proportional decrease of CPU time. The parallel program has a good scalability; especially when the number of processors was lower than 8. The measured *speedup* shows a quite linear curve for all five test sets. Strong downward deviations can be found for the HSVTK data set for 14 and 15 processors, for the cytP450 data set for 8 and 15 processors, for the glyPhos data set for 3 and 9 processors and for the hivProt data set for 3 and 14 processors. An inspection of the load status of the SGI ORIGIN 3400 for the time of the declining performance indicated, that the number of processes exceeded the CPU number limit of 28 by two to three processes. This leads to the conclusion that the deviations are caused by load imbalances due to the trivial parallelization technique. Some of the performance runs show a super linear speedup, especially for the immGlb and for the glyPhos data set.

For one of the five test sets the result for a superimposition is shown in Fig. 6. The three cytochrome P450c17 inhibitors, namely imidq,15, BW112 and BW13, are used for the shown alignment. Cytochrome P450c17 (17-alpha-hydroxylase/C17-20 lyase) is a key enzyme for the androgen and glucocorticoid biosynthesis. Like most cytochrome P450 isoenzymes, P450c17 also has heme as prosthetic group. Substances conjugate to this enzyme by coordinating to the central iron atom at one end and by a hydrogen bond at the other end of their skeleton. Thus, substances with a high affinity to the enzyme should have a free electron pair (e.g. a nitrogen atom) and at least one hydrogen bond acceptor or donor. It can be seen that the oxygen atoms as well as nitrogen atoms are matched on both ends of all three molecules. In a following step a van der Waals surface is constructed over the set. The surface visually conveys the steric requirements of the receptor binding site (Fig. 6 left part).

It should be pointed out that it is still necessary to do a more thorough study to determine the effects of the parameters of the GA on the evolutionary process and the superimposition results.

4 Conclusions

GAMMA overlays and aligns structures independent of the initially chosen conformation. Therefore, only one conformation per structure is necessary, and, thus, the program can work even when only one conformation of a compound is stored in a database. The automatic finding of structural similarities is based on a combination of a GA and a numerical optimization method, the *directed-tweak*. The hybrid technique drives the optimization of the geometric fit by adapting the conformations of molecules to each other. The GA is extended by two additional operators which are tailored to the superposition problem: the *creep* and *crunch* operators. GAMMA was further modified by implementing a parallel variant using MPI. The method was applied to sets of inhibitors and ligands of cytochrome P450c17, of HSV1 thymidine kinase, of HIV protease, of a Fab fragment of a monoclonal antibody (immunoglobulin) and of the glycogen phosphorylase.

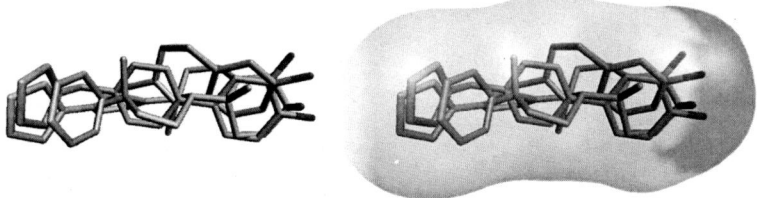

Fig. 6. Superposition of the three molecules in the cytP450 data set: imidq,15, bw112, bw13. It can be seen that the oxygen atoms as well as nitrogen atoms are matched on both ends of all three molecules. The generation of an averaged van der Waals surface around several active, superimposed molecules leads to the identification of the steric requirements of the receptor binding pocket (left part). For a reproduction of this figure in colour, see Fig. A.8.

Acknowledgement. The present work was carried out under the financial support of the "Competence Network for Technical, Scientific High Performance Computing in Bavaria" (KONWIHR). Part of this work was presented at the 6th International Conference on Chemical Structures.

References

1. Handschuh, S., Wagener, M., Gasteiger, J.: Superposition of three-dimensional chemical structures allowing for conformational flexibility by a hybrid method. J. Chem. Inf. Comput. Sci., **38**, 220–232 (1998)
2. Handschuh, S., Gasteiger, J.: Pharmacophores Derived from 3D Substructure Perception. In: Güner, O. (ed) Pharmacophore Perception, Development and Use in Drug Design. International University Line, La Jolla (2000)
3. Handschuh, S., Gasteiger, J.: The search for the spatial and electronic requirements of a drug. J. Mol. Model., **6**, 358–378 (2000)
4. Kearsley, S.K., Smith, G.M.: An alternative method for the alignment of molecular structures: maximizing electrostatic and steric overlap. Tetrahedron Computer Methodology,**3**, 615–633 (1990)
5. Jones, G., Willett, P., Glen, R.C.: A genetic algorithm for flexible molecular overlay and pharmacophore elucidation. J. Comput.-Aided Mol. Des., **9**, 532–549 (1995)
6. G. Jones, P. Willett, R. C. Glen, GASP: Genetic Algorithm Superimposition Program. In: Güner, O. (ed) Pharmacophore Perception, Development, and Use in Drug Design. International University Line, La Jolla (2000)
7. Lemmen, C., Lengauer, T., Klebe, G.: FLEXS: A method for fast flexible ligand superposition.J. Med. Chem., **41**, 4502–4520 (1998)
8. Holland, J.H.: Adaption in Natural and Artificial Aystems. The University of Michigan Press, Ann Arbor (1975)
9. Goldberg, D.E.: Genetic Algorithms in Search Optimization and Machine Learning. Addison-Wesley, New York (1989)
10. Cantú-Paz, E.: A survey of parallel genetic algorithms. Calculateurs Paralleles, Reseaux et Systems Repartis, **10**, 141–171 (1998)
11. Hurst, T.: Flexible 3D searching: the directed tweak technique. J. Chem. Inf. Comput. Sci., **34**, 190–196 (1994)
12. v. Homeyer, A., Handschuh, S., Wagener, M., Gasteiger, J.: Similarities of Inhibitors of Cytochrom P450c17 and of HSV-1 TK Derived from the Superimposition of Three-Dimensional Structures by Serial and Parallel Genetic Algorithms. Abstr. Sixth International Conference On Chemical Structures. Noordwijkerhout (2002)
13. Dongarra J.J. and Walker D.W.: MPI: a standard message passing interface. Supercomputer, **12**, 56–68 (1996)
14. Walker, D.W.: The design of a standard message-passing interface for distributed memory concurrent computers. Parallel Computing, **20**, 657–673 (1994)
15. Sadwoski, J., Gasteiger, J.: From atoms and bonds to three-dimensional atomic coordinates: automatic model builders. J. Chem. Rev., **93**, 2567-2581 (1993)
16. Sadwoski, J., Gasteiger, J., Klebe, G.: Comparison of automatic three-dimensional model builders using 639 x-ray structures. J. Chem. Inf. Sci., **34**, 1000–1008 (1994)
17. Molecular Networks GmbH: CORINA, version 2.4, http://www.mol-net.de, Erlangen (1999)

FreeWIHR: Lattice Boltzmann Methods with Free Surfaces and their Application in Material Technology

Carolin Körner[1], Thomas Pohl[2], Ulrich Rüde[2], Nils Thürey[1], Torsten Hofmann[1]

[1] Lehrstuhl für Werkstoffkunde und Technologie der Metalle, Universität Erlangen-Nürnberg, Martensstraße 5, D-91058 Erlangen, Germany
  Carolin.Koerner@ww.uni-erlangen.de
[2] Lehrstuhl für Systemsimulation (LSS), Universität Erlangen-Nürnberg, Cauerstraße 6, D-91058 Erlangen, Germany
  {Thomas.Pohl, Ulrich.Ruede}@informatik.uni-erlangen.de

Abstract. Metal foams are interesting as lightweight materials that have an excellent combination of mechanical, thermal, and acoustic properties. However, the production process is currently not fully understood. Therefore, the goal of the FreeWiHR project is the development and high performance implementation of a model for simulating the formation process of metal foams based on the lattice Boltzmann method.

1 Introduction

Foams show very interesting properties with respect to bending stiffness, energy absorption or damping behavior. Nearly every material can be foamed. However, the preparation of metal foams is a comparatively new field of research [1]. An example of an aluminum foam produced via the so-called powder metallurgical route is depicted in Fig. 1. Surprisingly, the physical understanding of foaming processes is yet very poor. This is particularly true for metal foam.

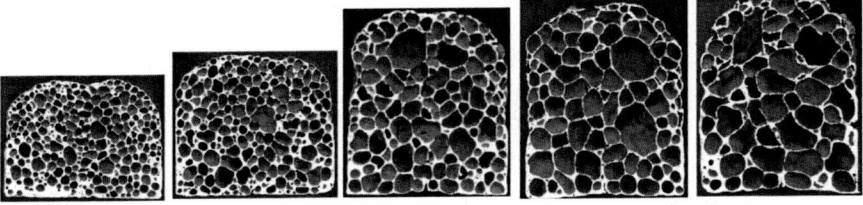

Fig. 1. Evolution of an aluminum foam structure

The main problem associated with the numerical simulation of foaming processes is the huge internal gas–liquid interface which strongly evolves with time. In addition, collapsing cell walls are able to induce avalanche-like coalescence and rearrangement processes of the whole foam structure. The time scale of these highly dynamic processes, which are governed by the Navier–Stokes equations (NSE), is typically much smaller than that of the foam expansion process itself.

A clear advantage of the LB approach compared to CFD lies in its local character, i. e. there are no global systems of equations which have to be solved. The computation time rises linearly with the system size. In addition, boundaries do not have a strong impact on the computation time. These features are essential regarding the complex internal structure of foams [2,3].

The paper is subdivided into two parts. The first part describes a Lattice Boltzmann Model (LBM) for the simulation of foaming processes. The LBM comprises the underlying physical model and a new algorithm which has to be developed to treat 3D free surface problems within the LB approach. An example demonstrates the potential of the method. The second part describes the implementation and parallelization of the code for the SR8000.

2 Physical Model

The underlying physical model describes foaming by blowing agents including nucleation, bubbles growth, bubble coalescence and eventually foam collapse. The blowing agent releases gas which diffuses to bubble nuclei and leads to foam expansion which is in all stages intimately related with cell coalescence processes. Rupture of the cell walls occurs if their thickness falls below a critical value which is characteristic for the foaming material and is for metals about 20–50 μm.

The foam is considered in the liquid state; i.e, melting and solidification are not taken into account. Due to the large density difference between gas and liquid the two phase hydrodynamic system is reduced to a one phase system which describes fluid flow with free surfaces. That is, the exact dynamics of the gas is not taken into account. At the interface the gas pressure balances the hydrodynamic pressure. Bubble nucleation is assumed to be heterogeneous and statistical. Presently, gas diffusion is simply modeled by a continuous increase of the amount of gas within each bubble which is proportional to the bubble surface.

Pure melts do not foam. Capillary forces due to the surface energy rapidly destroy cell walls by thinning. Prerequisite for the development of a polygonal cell structure is the presence of a stabilizing mechanism. Generally, metal foams get stabilized by particles. The origin of the particles is quite different. They are either deliberately added to the melt or develop during foam preparation. The effect of these particles is to generate a restoring, stabilizing pressure, the disjoining pressure Π, if they are captured within a cell wall. Both, the effect of the surface tension and the disjoining pressure are treated as a local modification of the gas pressure p^G at the gas–liquid interface $p^G \longrightarrow p^G - 2\kappa\sigma - \Pi$ where κ and σ denote the curvature and the surface energy, respectively. The disjoining pressure Π comprises the forces which stabilize the foam structure. Π is a function of the distance to the nearest neighboring interface d_{int}

$$\Pi(d_{int}) = \begin{cases} c_\Pi \, |d_{range} - d_{int}| \\ 0 \end{cases} \quad \text{for} \quad \begin{cases} d_{int} < d_{range} \\ d_{int} \geq d_{range} \end{cases} \quad (1)$$

where the magnitude and the range of the disjoining pressure are determined by the two phenomenological parameters c_Π and d_{range}.

In addition, a critical cell wall thickness is defined. If the wall thickness falls below that value, the cell wall ruptures and two bubbles merge.

3 Basic Equations

3.1 D3Q19 Model

We use the so-called D3Q19 model [4] which equilibrium distribution functions are defined as follows [5]

$$f_0^{eq}(\rho, \mathbf{v}) = \tfrac{12}{36}\rho\left[1 - \tfrac{3}{2}\mathbf{v}\cdot\mathbf{v}\right]$$
$$f_i^{eq}(\rho, \mathbf{v}) = \tfrac{2}{36}\rho\left[1 + 3(\mathbf{e}_i\cdot\mathbf{v}) + \tfrac{9}{2}(\mathbf{e}_i\cdot\mathbf{v})^2 - \tfrac{3}{2}\mathbf{v}\cdot\mathbf{v}\right], \quad |\mathbf{e}_i|^2 = 1$$
$$f_i^{eq}(\rho, \mathbf{v}) = w_2 = \tfrac{1}{36}\rho\left[1 + 3(\mathbf{e}_i\cdot\mathbf{v}) + \tfrac{9}{2}(\mathbf{e}_i\cdot\mathbf{v})^2 - \tfrac{3}{2}\mathbf{v}\cdot\mathbf{v}\right], |\mathbf{e}_i|^2 = 2. \quad (2)$$

Numerically, the LBE is solved in two steps:

Streaming $\quad f_i^{in}(\mathbf{x}, t) = f_i^{out}(\mathbf{x} - \mathbf{e}_i\cdot\Delta t, t - \Delta t) \quad (3)$

Collision $\quad f_i^{out}(\mathbf{x}, t) = f_i^{in}(\mathbf{x}, t) - \tfrac{\Delta t}{\tau}\left(f_i^{in}(\mathbf{x}, t) - f_i^{eq}(\mathbf{x}, t)\right) + \Delta t\cdot F_i \quad (4)$

where τ is the relaxation time and F_i an external force. During streaming (3) all distribution functions but f_0 are advected to their neighbor lattice site defined by their velocity. After advection the particle distribution functions approach their equilibrium distributions due to a collision step given by (4). The incoming and outgoing distribution functions; i.e., before and after collision, are denoted with f_i^{in} and f_i^{out}, respectively. The macroscopic density ρ and momentum $\rho\mathbf{v}$ in a cell are the 0-th and 1-th moments of the distribution functions: $\rho = \sum_{i=0}^{18} f_i$ and $\rho\mathbf{v} = \sum_{i=1}^{18} f_i\mathbf{e}_i$

3.2 Free Surface and Fluid Advection

The description of the liquid–gas interface is very similar to that of volume of fluid methods. An additional variable, the volume fraction of fluid ϵ, defined as the portion of the area of the cell filled with fluid, is assigned to each interface cell. The representation of liquid–gas interfaces is depicted in Fig. 2. Gas cells are separated from liquid cells by a layer of interface cells. These interface cells form a completely closed boundary in the sense that no distribution function is directly advected from fluid to gas cells and vice versa. This is a crucial point to assure mass conservation since mass coming from the liquid or mass transfered to the liquid always passes through the interface cells where the total mass is balanced. Hence, global conservation laws are fulfilled if mass and momentum conservation is ensured for interface cells. Per definition, the volume fraction ϵ of fluid and gas cells is 1 and 0, respectively. The fluid mass content of a cell is denoted with $M = M(\mathbf{x}, t)$. The mass content is a function

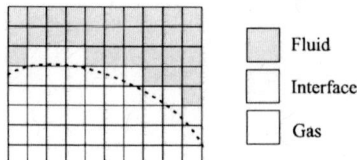

Fig. 2. 2D representation of a free liquid–gas interface by interface cells. The real interface (*dashed line*) is captured by assigning the interface cells their liquid fraction

of the volume fraction and the density. For a gas cell the fluid mass content M is zero whereas that of a fluid cell is given by its density ρ and the cell volume ΔV: $M(\mathbf{x},t) = \rho(\mathbf{x},t) \cdot \Delta V$ for $\mathbf{x} \in F$. Fluid cells gain and lose mass due to streaming of the f_i. For fluid cells M and ρ are equivalent. If interface cells are considered, M and ρ are not equivalent and we have to account for the partially filled state by introducing a second parameter, the volume fraction $\epsilon = \epsilon(\mathbf{x},t)$. The fluid mass content M, the volume fraction ϵ and the density ρ are related by $M(\mathbf{x},t) = \rho(\mathbf{x},t) \cdot \epsilon \cdot \Delta V$ for $\mathbf{x} \in I$.

All cells are able to change their state. It is important to notice that direct state changes from fluid to gas and vice versa are not possible. Hence, fluid and gas cells are only allowed to transform into interface cells whereas interface cells can be transformed into both gas and fluid cells. A fluid cell is transformed into an interface cell if a direct neighbor is transformed into a gas cell. At the moment of transformation the fluid cell contains a certain amount of fluid mass M which is stored. During further development the interface cell may gain mass from or lose mass to the neighboring cells. These mass currents are calculated and lead to a temporal change of M. If M drops below zero, the interface cell is transformed into a gas cell. It is important to pronounce that mass and density are completely decoupled for interface cells. While the density of the interface cells is given by the pressure boundary conditions and fluid dynamics, M is determined by the mass exchange ΔM with the neighboring fluid and interface cells.

The mass exchange $\Delta M_i(\mathbf{x},t)$ between an interface cell at lattice site $\mathbf{x}$ and its neighbor in $\mathbf{e}_i$-direction at $\mathbf{x} + \mathbf{e}_i$ is calculated as ($\mathbf{e}_{\bar{i}} = -\mathbf{e}_i$)

$$\Delta M_i(\mathbf{x},t) = \begin{cases} 0 \\ f_{\bar{i}}(\mathbf{x}+\mathbf{e}_i,t) - f_i(\mathbf{x},t) \\ \frac{1}{2}[\epsilon(\mathbf{x},t) + \epsilon(\mathbf{x}+\mathbf{e}_i,t)][f_{\bar{i}}(\mathbf{x}+\mathbf{e}_i,t) - f_i(\mathbf{x},t)] \end{cases} \text{ for } \mathbf{x}+\mathbf{e}_i \in \begin{cases} G \\ F \\ I \end{cases} \tag{5}$$

There is no mass transfer between gas cells and interface cells. The interchange between an interface cell and a fluid cell should be the same as that of two fluid cells since the cell boundary is completely covered with liquid. In this case, the mass exchange can be directly calculated from the particle distribution functions. The interchange between two interface cells is approximated by assuming that the mass current is weighted by the mean occupied volume fraction. It is crucial to note that mass is explicitly conserved in (5):

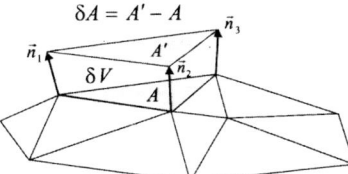 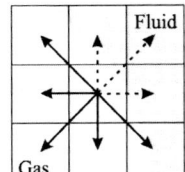 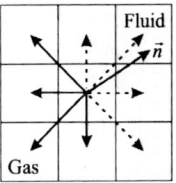

Fig. 3. Calculation of the curvature

Fig. 4. Missing distribution functions at interface cells after streaming. Left: Undefined distribution functions after streaming (*broken lines*). Right: Set of distribution functions with $\mathbf{n} \cdot \mathbf{e}_i \geq 0$ (*broken lines*)

$$\Delta M_{\bar{i}}(\mathbf{x} + \mathbf{e}_i, t) = -\Delta M_i(\mathbf{x}, t). \tag{6}$$

That is, the mass which a certain cell receives from a neighboring cell is automatically lost there and vice versa. The temporal evolution of the mass content of an interface cell is thus given by

$$M(\mathbf{x}, t + \Delta t) = M(\mathbf{x}, t) + \sum_{i=1}^{18} \Delta M_i(\mathbf{x}, t). \tag{7}$$

An interface cell is transformed into a gas or fluid cell if $M < 0$ or $M > \rho \Delta V$, respectively. At the same moment, new interface cells emerge in order to guarantee the continuity of the interface. The initial distribution functions of these new interface cells are extrapolated from the cells in normal direction towards the fluid.

The calculation of the local curvature of the interface is very complex and time consuming, see Fig. 3. In a first step, a marching cube algorithm is used to generate a triangulation of the interface. Secondly, the curvature κ belonging to each triangle is estimated by $\kappa = \frac{1}{2}\frac{\delta A}{\delta V}$ where δA denotes the alteration of the triangle area when its vertices are infinitesimally shifted in normal direction. The covered volume is denoted by δV. In the last step, the curvature of an interface cell is estimated by averaging the curvature of the triangles belonging to it.

3.3 Boundary Conditions

Interface cells separate gas cells from fluid cells. After streaming, only distribution functions from fluid and interface cells are defined. Distribution functions arriving from gas cells are not defined (see Fig. 4, left). The symmetry between known and unknown distribution functions; i.e., if f_i is known $f_{\bar{i}}$ is unknown, is essential to fulfill the boundary conditions. We demand force balance for opposite lattice directions. In addition, we make use of the fact that the forces exerted by the gas are known and are given by the gas pressure and the velocity at the interface. Hence, the missing distribution functions are reconstructed as

$$f_i^{out}(\mathbf{x} - \mathbf{e}_i, t) = f_i^{eq}(\rho^G, \mathbf{v}) + f_{\bar{i}}^{eq}(\rho^G, \mathbf{v}) - f_{\bar{i}}^{out}(\mathbf{x}, t) \qquad \forall i : \mathbf{n} \cdot \mathbf{e}_i \geq 0 \quad (8)$$

with the gas density $\rho^G = 3\,p^G$ and the velocity $\mathbf{v}$ of the interface cell.

It is important to note that not only the missing distribution functions are reconstructed but all distribution functions with $\mathbf{e}_i \cdot \mathbf{n} \geq 0$ (see Fig. 4, right). After completion of the whole set of distribution functions, the new density ρ and velocity $\mathbf{v}$ can be calculated. The outgoing distribution functions are calculated as

$$f_i^{out}(\mathbf{x}, t) = f_i^{in}(\mathbf{x}, t) - \frac{1}{\tau}\left(f_i^{in}(\mathbf{x}, t) - f_i^{eq}(\rho, \mathbf{v})\right) + \epsilon(\mathbf{x})\,w_i\,\rho\,\mathbf{e}_i\,\mathbf{g} \qquad \begin{array}{c} i = 1, \cdots, b \\ \forall \mathbf{x} \in I \end{array}$$

where $w_0 = \frac{12}{36}, w_1 = \frac{2}{36}, w_2 = \frac{1}{36}$ and g is the gravity constant.

3.4 Example: 3D foam

The development of a 3D foam is depicted in Fig. 5. A large number of bubbles starts growing within the fluid. The disjoining pressure delays bubble coalescence but does not completely prevent it. Consequently, the number of bubbles decreases with increasing gas volume. At the end, a polygonal foam structure has developed.

4 Implementation and Parallelization

For testing and performance evaluation purposes we implemented a standard LBM code (SLBM). This code was used as a base for the free surface extensions (freeLB).

4.1 Data Layout

Due to the regular grids in LBM codes there are several possibilities for choosing the most appropriate data layout. Common to all is a simple array structure. A choice had to be made about the ordering of indices: three indices representing the coordinates (X, Y, Z), one index for selecting the data item in each cell (I), and one index determining the source or destination grid (G).

Although the mixing of the source and destination grid exhibited a performance increase on commodity PCs [6, 7], the performance decreased when applying the same data layout optimizations on the SR8000. Further investigations left three alternatives, that can be described in C notation as `double data[G][Z][Y][X][I]`, `double data[G][Z][Y][I][X]`, and `double data[G][Z][I][Y][X]`.

Fig. 5. 3D foam: The bubbles grow and coalescence occurs. The disjoining pressure Π stabilizes the foam and eventually a polygonal structure develops (initial number of bubbles: 1000; system size: $120 \times 120 \times 140$; $\tau = 0.8$; g = 0; $\sigma = 0.01$; $c_\Pi = 0.006$)

For testing purposes the first data layout where all data for one cell is stored contiguously in memory has been implemented as a standard LBM code (SLBM) in C [8]. As can be seen in Fig. 6 this code exhibits a good speed-up behavior for a small number of SR8000 nodes which is close to linear scaling. When switching to a larger domain for the same number of nodes, the performance per processor increases because of the diminishing influence of the overlapping boundary interfaces which will be discussed in Sect. 4.3.

Figure 7 shows the scale-up behavior for up to 64 SR8000 nodes which amounts to 512 CPUs. For the largest simulation a total number of $1.08 \cdot 10^9$ cells have been used which require 370 GByte of memory in total. The effects of communication latency start to degrade the performance for such large-scale simulations as almost 64 MByte have to be sent to and received from each adjacent subdomain in every time step. Nevertheless, the code still achieves an efficiency of 75 % as compared to the single node performance.

The second data layout has been tested by the HPC group at the RRZE with good performance results. For various reasons which will be described in Sect. 4.3 we chose the third data layout for the LBM code with the free surface extension (freeLB) based on the SLBM code.

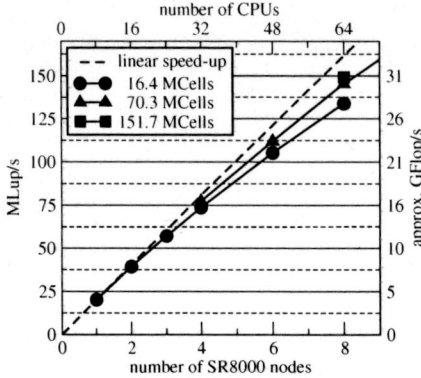 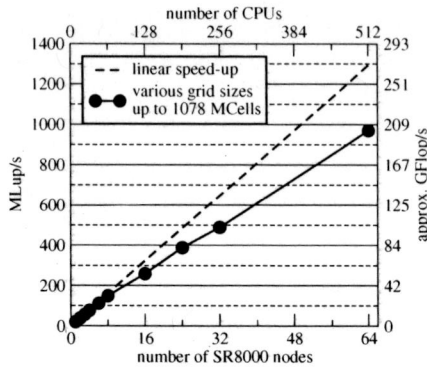

Fig. 6. Comparison of the performance for different domain sizes on the SR8000 (SLBM). The performance was measured in million lattice site updates per second (MLup/s). For our implementation a lattice site update corresponds to approximately 209 floating-point operations

Fig. 7. Scale-up performance on the SR8000 (SLBM). The grid sizes were chosen to allocate most of the available memory on all nodes

4.2 Computations

In a standard LBM code with fused stream/collide step, a time step can be performed in a single sweep over the computational domain. For the more complex handling of free surfaces with the described method, however, five sweeps are necessary:

1. The first sweep starts with the well known streaming step augmented by the reconstruction of missing distribution functions from adjacent gas cells which takes the gas pressure of the concerned bubble into account. After that, the mass exchange with neighboring cells is calculated. For interface cells the new mass M leads to an updated fill value ϵ. This value determines whether an interface cell is now completely filled or emptied resulting in a conversion to a gas or fluid cell, respectively. Finally, the collision step according to the BGK approximation with an additional force term representing gravity is performed.
2. The second sweep checks and reestablishes the strict division of fluid and gas cells by the interface cell layer. Therefore, it might be necessary to undo cell conversions that have been scheduled in the first sweep.
3. In the third sweep converted cells have to initialized. Depending on the new cell type, a set of distribution functions (only for former gas cells) and the fill ϵ and mass value M are computed.
4. Interface cells converting to gas or fluid cells are hardly ever completely filled or empty. In almost every case, they still contain or miss a certain amount of mass. This possibly negative mass is distributed among the

adjacent interface cells. Infrequently, no such cells can be found and the mass is lost. A better way to deal with this exception is to distribute the mass in a wider neighborhood.
5. During the last sweep, for all former and new interface cells, the change in the fill value ϵ is calculated and the volume of the bubble is updated.

Apart from the grid data the code also handles an array of data sets for each gas bubble in the computational domain containing the initial gas mass and the current volume of the bubble. This gas bubble data is not distributed among the involved CPUs, but stored entirely on each machine, because a single bubble can potentially span across all partitions of the computational domain.

4.3 Parallelization Technique

In general, domain decomposition is the canonical and most common way to parallelize LBM codes; i.e., the computational domain is divided up in several subdomains which are distributed to the computational units. Since information in the standard LBM can only travel one cell unit per time step, it is sufficient to have a halo (also known as ghost nodes) of one cell layer around any data adjacent to other subdomains. Due to the free surface extension with its several sweeps over the data a halo of four cell layers is necessary if the halo is updated only once per time step.

For the freeLB code we implemented a one dimensional domain partitioning; i.e., the computational domain is cut in slices parallel to the xy-plane. This allows for easy extensions as load balancing and can help to reduce the amount of data that has to be exchanged between neighboring subdomains in combination with the chosen data layout. Furthermore, the interesting domain sizes for real applications often feature a large spatial aspect ratio which attenuates the restrictions of a 1D domain partitioning.

The combination of the chosen data layout and the 1D domain partitioning features a major advantage compared to the other layouts. Without copying and/or reordering of data an entire cell layer in the xy-plane can be sent to or received from adjacent subdomains. By appropriately ordering the cell data it is even possible to communicate only a certain part of cell data; e.g, all distribution functions pointing upward. The current implementation does not yet fully exploit this feature.

In addition to the communication of the grid data, the changes in the volume of the gas bubbles have to be collected and added for each subdomain to obtain the global volume change for each bubble. Therefore, the uppermost process starts to send its volume changes to its lower neighbor which adds its changes and sends this updated data to its lower neighbor. This procedure is continued until the lowermost process receives the volume update. Meanwhile the same chain of communication has been started at the lowermost process traveling upwards. When both chains reach the opposite end of the domain

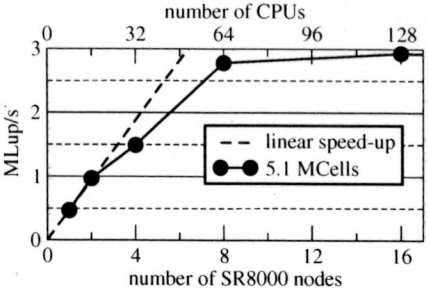

Fig. 8. Speed-up performance on the SR8000 (freeLB)

all processes have a global update of the gas bubbles' volumes. Instead of this procedure an "all to all" communication could be initiated using a dedicated MPI call. For a large number of subdomains, however, this would result in a large amount of send/receive events.

5 Performance Evaluation

In spite of the good performance of the SLBM code which was used as the base for the freeLB code with free surface handling, both the single node performance and the speed-up behavior exhibit a disappointing performance on the SR8000 as can be seen in Fig. 8.

The degradation of performance is caused by a combination of effects:

- Instead of just one layer of halo the freeLB code requires an update of four layers. The superior communication bandwidth of the SR8000 should be able to compensate the increased network traffic, but the additional workload caused by redundant cell updates degrades the performance particularly in speed-up performance tests.
- Profiling the code on an Intel Pentium 4 showed a significant increase in conditional branches. While the standard SLBM code requires only 2.9 conditionals per lattice site update, the new freeLB for handling the free surfaces needs as many as 51 conditional branches per cell update. The performance on the Pentium 4 architecture is almost unaffected by the additional branches (SLBM: 2.5 MLup/s; freeLB: 2.2 MLup/s). The modified PowerPC architecture of the SR8000, however, draws its performance from a predictable and steady instruction and data stream based on its pseudo vector processing capabilities which can no longer be applied in the more complicated freeLB code.
- The compile logs indicate that the C compiler (cc) is no longer able to apply software pipelining in the freeLB code.

This all together leads to a performance degradation on each processor by a factor of 40 of the freeLB code as compared to the SLBM version. It is unclear at this time whether improved compilers could help to recover satisfactory node performance for codes as complicated as the present freeLB implementation. Manually restructuring the codes to avoid the conditionals in the innermost loops seems to be virtually impossible due to the dynamically changing geometry of the computational domains.

As reported previously, achieving high node performance on the SR8000 architecture is already quite nontrivial for the relatively simple SLBM code. From our experience, we therefore expect that at best moderate gains will be possible for the freeLB code.

6 Conclusions

The FreeWiHR project was based on a close collaboration of the Lehrstuhl für Werkstoffkunde und Technologie der Metalle and the Lehrstuhl für Systemsimulation at University of Erlangen-Nürnberg. It has achieved a number of ambitious project goals, including the

- development, implementation, and testing of a new algorithm for calculating the surface curvature in 3D which improved the stability and accuracy of the simulation.
- development, implementation, and testing of a model for the disjoining pressure of gas bubbles.
- implementation, performance tuning, and performance evaluation of a parallel LBM code for complex geometries using different data layouts and various optimization techniques.
- re-design of the algorithms for handling free surfaces that have been used in the sequential code in order to allow for parallelization.
- extending the parallel LBM code with the handling of free surfaces.
- performance evaluation of the parallel LBM code with free surface extensions.

The standard LBM code has been implemented in C using a conservative coding design which is well adapted to the peculiarities of the SR8000 architecture. Consequently, the SLBM code exhibits an excellent single-node performance, scalability, and sustained performance on the SR8000. The more complex code with the free surface extensions, in contrast, suffers from a severe performance degradation caused by the large number of conditional branches which are unfortunately unavoidable when treating dynamically changing geometries within the LBM. At least to our current knowledge, and even assuming additional extensive profiling and optimization efforts, we cannot expect a significant improvement of the performance for this type of application on the SR8000.

A comparison with other architectures shows that the sustained performance of the freeLB code can be quite good: On an Pentium 4; e.g., freeLB runs essentially with the same performance as the SLBM code. We must therefore conclude that in the case of our algorithms for dynamically changing geometries, the SR8000 might not be the best choice of architecture.

Future work in the project will therefore concentrate on evaluating the suitability of alternative architectures, including clusters, and possibly also classical vector processors. Unfortunately, we must expect that standard clusters will suffer from insufficient communication bandwidth. Vector processors, on the other hand, may not be able to vectorize the freeLB codes easily. This is being evaluated in ongoing work.

Acknowledgements

We acknowledge the helpful cooperation with our colleagues from RRZE at the University of Erlangen, in particular Gerhard Wellein and Thomas Zeiser. This work has been financially supported by the Competence Network for Technical, Scientific High Performance Computing in Bavaria KONWIHR.

References

1. C. Körner, M. Thies, M. Arnold, and R. F. Singer. The Physics of Foaming: Structure Formation and Stability. In H. P. Degischer and B. Kriszt, editors, *Handbook of Cellular Metals*, pages 33–42. Wiley-VCH, 2002.
2. C. Körner, M. Thies, M. Arnold, and R. F. Singer. Modeling of metal foaming by in-situ gas formation. In J. Banhart et al., editor, *Cellular Metals. Manufacture, Properties, Applications*, pages 93–98. Verlag MIT Publishing, Bremen 2001.
3. C. Körner, M. Thies, and R. F. Singer. Modeling of metal foaming with Lattice Boltzmann Automata. *Advanced Engineering Materials*, 4:765–769, 2002.
4. X. He and L.-S. Luo. Theory of the lattice Boltzmann method: From the Boltzmann equation to the lattice Boltzmann equation. *Physical Review E*, 6:6811, 1997.
5. Y.H. Qian, D. d'Humières, and P. Lallemand. Lattice BGK Models for Navier-Stokes Equation. *Europhysics Letters*, 17:479–484, 1992.
6. J. Wilke, T. Pohl, M. Kowarschik, and U. Rüde. Cache Performance Optimizations for Parallel Lattice Boltzmann Codes in 2D. In *Lecture Notes in Computer Science*, volume 2790, pages 441–450. Springer, 2003.
7. T. Pohl, M. Kowarschik, J. Wilke, K. Iglberger, and U. Rüde. Optimization and Profiling of the Cache Performance of Parallel Lattice Boltzmann Codes. *Parallel Processing Letters*, 13(4):549–560, 2003.
8. T. Pohl, F. Deserno, N. Thürey, U. Rüde, P. Lammers, G. Wellein, and T. Zeiser. Performance Evaluation of Parallel Large-Scale Lattice Boltzmann Applications on Three Supercomputing Architectures. Supercomputing Conference 2004. Accepted.

HQS@HPC: Comparative numerical study of Anderson localisation in disordered electron systems

Gerald Schubert[1], Alexander Weiße[2], Gerhard Wellein[3], and Holger Fehske[1]

[1] Ernst-Moritz-Arndt-Universität Greifswald, Institut für Physik, Domstr. 10a, D-17489 Greifswald, Germany
holger.fehske@physik.uni-greifswald.de
[2] School of Physics, The University of New South Wales, Sydney, NSW 2052, Australia
[3] Regionales Rechenzentrum Erlangen (RRZE), Martensstraße 1, D-91058 Erlangen, Germany
gerhard.wellein@rrze.uni-erlangen.de

Abstract. Taking into account that a proper description of disordered systems should focus on distribution functions, the authors develop a powerful numerical scheme for the determination of the probability distribution of the local density of states (LDOS), which is based on a Chebyshev expansion with kernel polynomial refinement and allows the study of large finite clusters (up to 100^3). For the three-dimensional Anderson model it is demonstrated that the distribution of the LDOS shows a significant change at the disorder induced delocalisation-localisation transition. Consequently, the so-called typical density of states, defined as the geometric mean of the LDOS, emerges as a natural order parameter. The calculation of the phase diagram of the Anderson model proves the efficiency and reliability of the proposed approach in comparison to other localisation criteria, which rely, e.g., on the decay of the wavefunction or the inverse participation number.

1 Introduction

The localisation of quantum particles in disordered systems is one of the most intensively studied problems in condensed matter physics [1–5]. In real systems disorder can arise for a number reasons. We may think of randomly distributed impurities, vacancies or dislocations in an otherwise ideal crystal, of random arrangements of electronic or nuclear spins, etc. While the disorder appears in many forms that are sometimes difficult to characterise theoretically, the randomness in the model introduced and discussed by Anderson is simple but sufficient to capture the basic features of the disorder-induced metal insulator transition [6]. The Anderson Hamiltonian,

$$H = -t\sum_{\langle ij \rangle}\left[c_i^\dagger c_j + \text{H.c.}\right] + \sum_{j=1}^{N}\epsilon_j c_j^\dagger c_j\,, \qquad (1)$$

describes noninteracting electrons moving on a lattice with random on-site potentials (compositional disorder). The operators $c_j^\dagger$ (c_j) create (annihilate) an electron in a Wannier state centred at site j, and the local potentials ϵ_j are assumed to be independent, uniformly distributed random variables,

$$p(\epsilon_j) = \frac{1}{W}\,\theta\left(\frac{W}{2} - |\epsilon_j|\right)\,. \qquad (2)$$

The parameter W is a measure for the strength of disorder and is usually given in units of the nearest neighbour hopping matrix element t. Throughout this work we consider d-dimensional hyper-cubic lattices with $N = L^d$ sites, impose periodic boundary conditions (PBC), and set the lattice spacing equal to unity.

The spectral properties of the Anderson model (1) have been carefully analysed (see, e.g., Ref. [7]). For sufficiently large disorder or near the band tails, the spectrum consists exclusively of discrete eigenvalues, and the corresponding eigenfunctions are exponentially localised. Since localised electrons do not contribute to the transport of charge or energy, the energy that separates localised and extended eigenstates is called the mobility edge. For any finite disorder $W > 0$, on a one-dimensional (1d) lattice, all eigenstates of (1) are localised [8,9]. This is believed to hold also in 2d, where the existence of a transition from localised to delocalised states at finite W would contradict the one parameter scaling theory [10,11].

In spite of the progress made over the last decades, the Anderson metal-insulator transition is still not completely understood. There are several reasons why the existing theories remain unsatisfactory. Especially when electron-electron or electron-phonon interactions come into play, the very successful one-parameter scaling approach might be problematic, because close to the localisation transition the energy scales associated with both disorder and interactions are comparable to the Fermi energy [12]. On the other hand, the numerical study of the localisation-delocalisation transition is demanding, since the involved length scales can become extraordinary large, in particular near the critical point. Obviously, methods that are based on a full diagonalisation of the Hamiltonian and on the study of the one-particle eigenstates are restricted to rather small systems. Examples are the calculation of the localisation length from the decay of the electronic wavefunction or the evaluation of the inverse participation number. In addition, one-particle eigenstates are not defined for interacting systems. Hence, none of these criteria can easily be generalised to interacting disordered systems. To overcome these difficulties is perhaps the most challenging issue of current research on disordered materials.

Motivated by this situation, this contribution provides a (quasi approximation free) numerical analysis of the recently revived "local order parameter" approach to the Anderson transition, which, within the framework of the statistical dynamical mean field approximation, has been successfully applied also to correlated electron (phonon) systems [12,14]. Adopting a local point of view and focusing on the distribution of the physically interesting quantities, the method follows the original route to the localisation problem established by Abou-Chacra et al. [16]. In particular, we demonstrate that for Anderson type models the distribution of the local density of states (LDOS) can be determined very easily by the kernel polynomial method (KPM) [17], a refined Chebyshev expansion technique. Based on the distribution of the LDOS, localised states are distinguished from extended states by a vanishing geometrical average, which is usually called the "typical DOS". In addition, it turns out that this quantity characterises the disorder-induced metal-insulator transition also in more complex systems [12, 14, 19–22].

To examine the efficiency and accuracy of the proposed LDOS-KPM approach, we carry out a comparative numerical study of the localisation-delocalisation transition. In view of the wealth of known results, the 3d Anderson model seems to be best suited for this kind of investigation. The results we obtain for the mobility edge from different methods allow for a detailed understanding of the typical DOS concept and open the road towards an application to more complex situations.

2 Anderson transition

As the Anderson transition is expected only for $d > 2$, in this chapter we focus on the 3d case, for which the lack of successful analytical approaches necessitates a numerical treatment. In contrast to the widely used numerical transfer matrix method [23–25], which describes the 3d system as a quasi-1d system of finite cross section, below we stress the bulk properties of the system and consider cubic clusters which extend equally in all spatial directions. Due to the large length scales that emerge in the critical region, it is generally a difficult task to interpret the results of such finite cluster calculations.

As a kind of benchmarking, we review and compare established localisation criteria, namely, the localisation length (Sec. 2.1) and the inverse participation number (Sec. 2.2). Both can be extracted from the one-particle wavefunctions, which, however, requires the complete numerical diagonalisation of the Hamiltonian (1). In Sec. 2.3, we present the new approach that is based on the distribution of the LDOS. Since the calculation of the LDOS via the KPM requires only sparse matrix vector multiplications, this technique scales linearly with the number of lattice sites and permits the study of significantly larger systems.

2.1 Decay of the wavefunction

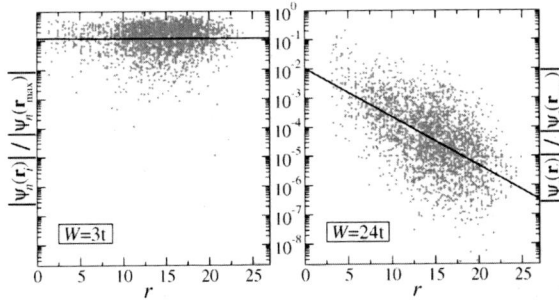

Fig. 1 Decay of an electronic wavefunction ψ_n in the band centre as a function of the distance $r = |\mathbf{r}_i - \mathbf{r}_{\max}|$ to the site with maximum amplitude, $\mathbf{r}_{\max}$. Results are given each for one fixed energy and realisation of disorder, $N = 30^3$, PBC.

The most obvious but costly way to access the localisation properties of an electronic wavefunction is the direct calculation of the localisation length λ, which is infinite for extended states and finite otherwise. For localised states, the envelope of the wavefunction decays exponentially from some point $\mathbf{r}_{\max}$ in the crystal.

$$|\psi_n(\mathbf{r}_i)| \sim f(\mathbf{r}_i) \exp\left(-\frac{|\mathbf{r}_i - \mathbf{r}_{\max}|}{\lambda}\right). \tag{3}$$

The random function $f(\mathbf{r}_i)$ describes the statistical fluctuations of the amplitudes $\psi_n(\mathbf{r}_i)$ of the eigenfunction ψ_n at energy E_n. Given ψ_n, the localisation length $\lambda(E_n)$ is obtained by locating the site of maximum amplitude, $\mathbf{r}_{\max}$, and fitting Eq. (3) to the data. In contrast to the case of weak disorder, where the amplitude is essentially independent of the distance from $\mathbf{r}_{\max}$, at higher values of W a clear exponential decay is observed (see Fig. 1). Note, that besides the direct fit with equal weight for the amplitudes of all sites, λ can also be determined using the method of asymptotic slope [26]. Here the data is first averaged within shells of fixed distance from $\mathbf{r}_{\max}$ and fitted thereafter. However, using this second approach the detection of the Anderson transition is not as robust and more sensitive to the fluctuations of the data. We therefore refrain from considering corresponding results.

2.2 Inverse participation number

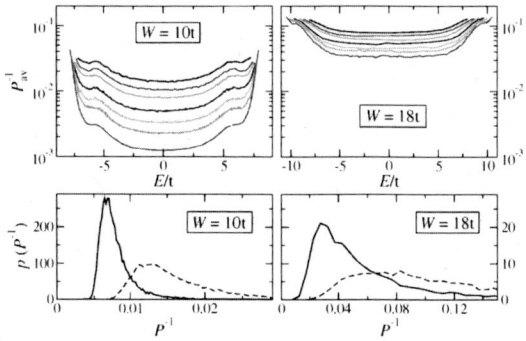

Fig. 2 Upper part: Averaged inverse participation number P_{av}^{-1} of 1000 systems with L^3 sites and PBC (top to bottom: $L = 8, 9, 10, 12, 14, 16, 20$). Lower part: Probability density of P^{-1} for the 10^3 system in the band centre (solid line) and near the band edges (dashed line). Note the different scales in the lower panels.

Yet another quantity that measures the Anderson transition is the inverse participation number [27],

$$P^{-1}(E_n) = \sum_{i=1}^{N} |\psi_n(\mathbf{r}_i)|^4, \tag{4}$$

which is proportional to the inverse number of sites that contribute to a given one-particle wavefunction ψ_n. For delocalised states we find $P^{-1} \sim 1/N$, which vanishes in the thermodynamic limit. Localised states, on the other hand, approximately extend over a finite volume N_0, yielding $P^{-1?} \sim 1/N_0$ independent of the system size N. In Fig. 2 this behaviour is demonstrated for small and large disorder W. While in the localised case ($W = 18t$) P^{-1} is almost independent of L, apparently it decreases with L for extended states ($W = 10t$). Apart from the different scaling, the distribution of P^{-1} changes at the transition, being sharply peaked for extended states and rather broad for localised ones (lower part of Fig. 2).

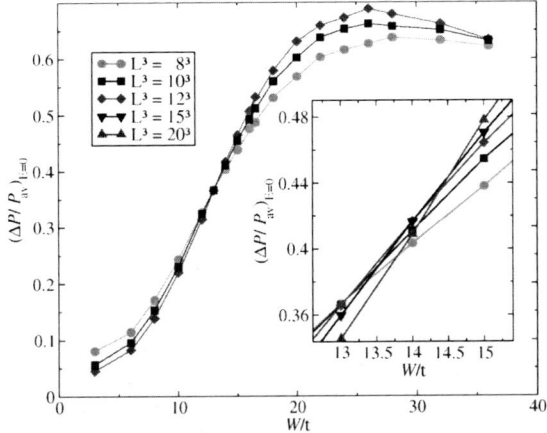

Fig. 3 Normalised standard deviation of the participation number in the band centre as a function of disorder for different system sizes using PBC. The obtained results were averaged over 1000 realisations of disorder.

Based on the distribution of the participation number P recently an alternative numerical approach for monitoring the Anderson transition was proposed [28]. In analogy to results for a certain class of power-law random banded matrices, which indicate the scale invariance of the distribution of P at the Anderson transition, Malyshev et al. [28] suggest to detect the transition by studying the ratio of the standard deviation of P, ΔP, to the mean participation number P_{av}, which should be independent of the system size at the critical disorder.

So far this approach has only been tested for a one-dimensional model with diagonal disorder and power-law long-range hopping [28], which shows a transition at the band edge and can thus be tackled with the Lanczos method. In Fig. 3 we present first data for the band centre of the standard Anderson model. While for small disorder the ratio $\Delta P/P_{\mathrm{av}}$ decreases with increasing system size, at large disorder the opposite happens. The intersection of the curves is not completely independent of the system size (see inset of Fig. 3), and a precise determination of the transition requires some finite size scaling of the data. Performing a finite-size scaling [28] our data is consistent with a critical disorder strength of $W_c \approx 16.1 \pm 0.8$ in the thermodynamic limit.

2.3 Local density of states

Already in his seminal paper [6] Anderson pointed out that in order to describe the transition from delocalised to localised states it is very instructive to discuss the distribution of local quantities of interest, such as the escape rate or recurrence probability from or to a given site. Another suitable quantity that becomes critical at the Anderson transition is the LDOS [12, 29],

$$\rho_i(E) = \sum_{n=1}^{N} |\psi_n(\mathbf{r}_i)|^2 \delta(E - E_n), \qquad (5)$$

which for a given energy directly measures the local amplitude of the wavefunction at site $\mathbf{r}_i$. So far the LDOS has been considered mainly within analytical approaches or by the Lanczos recursion method [29]. Typically the latter suffers from severe

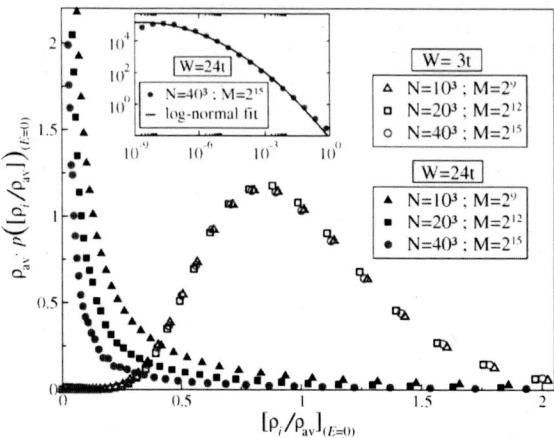

Fig. 4 General shape and finite size scaling of the LDOS distribution $p(\rho_i/\rho_{\text{av}})$. Keeping the ratio $N/M = 1.95$ fixed for $N = 10^3, 20^3, 40^3$ and $K_r \times K_s = 10^4 \times 100, 100 \times 100, 32 \times 32$ respectively, we calculated histograms for $E \in [-0.1t, 0.1t]$. Inset: Double logarithmic plot of $p(\rho_i/\rho_{\text{av}})$ for the localised case together with a log-normal fit to the data.

stability problems at high expansion order and conclusive results for the Anderson transition are difficult to obtain [30]. Fortunately the KPM technique [17] described in Appendix A is a very efficient way to circumvent these difficulties and allows the calculation of high-resolution LDOS data for very large systems. In a nutshell, within this approach the function of interest is expanded in a finite series of Chebyshev polynomials. To weaken the effects of the truncation and ensure properties such as positivity and normalisation, the function is convoluted with an appropriate integral kernel. The resolution of the method is inversely proportional to the order of the expansion M (the number of so-called Chebyshev moments).

Adopting Anderson's original point of view that a proper description of disordered systems should focus on distribution functions, we calculated $\rho_i(E)$ for a large number of samples, K_r, and sites, K_s, and studied its statistical properties. In Fig. 4 we show the resulting distribution of $\rho_i(E = 0)$, normalised by its mean value ρ_{av}, for two characteristic values of disorder. As ρ_{av} is a function of disorder, this normalisation ensures $\langle \rho_i/\rho_{\text{av}} \rangle = 1$ independent of W, allowing thus an appropriate comparison. In the delocalised phase, $W = 3t$, the distribution is rather symmetric and peaked close to its mean value. Note that increasing the system size and the expansion order, such that the ratio of mean level spacing and KPM resolution is fixed, does not change the distribution. This is in strong contrast to the localised phase, e.g., $W = 24t$, where the distribution of $\rho_i(E)$ is extremely asymmetric. Although most of the weight is now concentrated close to zero, the distribution extends to very large values of ρ_i, causing the mean value to be much larger than the most probable value. In addition, a similar finite size scaling increases the asymmetry and underlines the singular behaviour expected in the thermodynamic limit and at infinite resolution. Note also, that the distribution of the LDOS is well approximated by a log-normal distribution [31],

$$p(x) = \frac{1}{\sqrt{2\pi\sigma^2}} \frac{1}{x} \exp\left(-\frac{(\ln(x/x_0))^2}{2\sigma^2}\right), \qquad (6)$$

as illustrated in the inset of Fig. 4.

Of course, the study of entire distributions is a bit inconvenient, and for practical calculations, instead, we will prefer an appropriate statistics that uniquely charac-

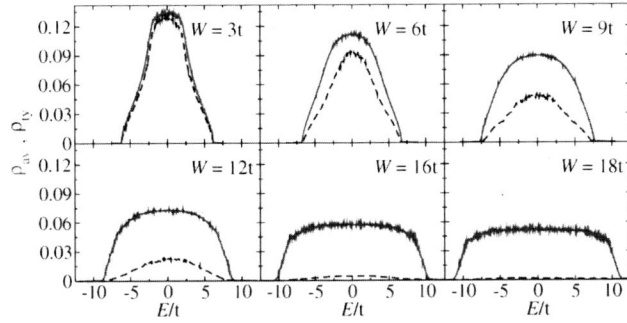

Fig. 5 Average (solid line) and typical (dashed line) DOS for a 50^3 lattice with PBC. $K_s \times K_r = 32 \times 32$, $M = 8192$.

terises the distribution. The above findings suggest, that such a statistics is given by the arithmetic and geometric averages of $\rho_i(E)$,

$$\rho_{\mathrm{av}}(E) = \frac{1}{K_r K_s} \sum_{k=1}^{K_r} \sum_{i=1}^{K_s} \rho_i(E), \qquad (7)$$

$$\rho_{\mathrm{ty}}(E) = \exp\left(\frac{1}{K_r K_s} \sum_{k=1}^{K_r} \sum_{i=1}^{K_s} \ln(\rho_i(E))\right). \qquad (8)$$

On the one hand, the arithmetic mean for large enough K_r and K_s converges to the standard density of states $\rho(E) = \sum_{n=1}^{N} \delta(E - E_n)$, which is not critical at the Anderson transition. The geometric mean, on the other hand, represents the typical value of the distribution, which, as shown above, is finite in the delocalised phase, but goes to zero in the localised phase. As can be seen from Fig. 5, $\rho_{\mathrm{av}}(E)$ and $\rho_{\mathrm{ty}}(E)$ are almost equal for extended states, whereas for localised states $\rho_{\mathrm{ty}}(E)$ vanishes and $\rho_{\mathrm{av}}(E)$ remains finite. This implies, that the ratio of these two quantities, the normalised typical density of states

$$R(E) := \frac{\rho_{\mathrm{ty}}(E)}{\rho_{\mathrm{av}}(E)}, \qquad (9)$$

can serve as an order parameter for the Anderson transition with $R > 0$ for extended states and $R = 0$ for localised ones. As for most numerical calculations the transition is slightly washed out by the finite size of the considered cluster, and by the KPM resolution. However, for large clusters and increasing M a plot of the ratio R versus disorder strength W (see Fig. 6) allows for a reliable determination of the critical disorder W_c, and, e.g., in the band centre we obtain $W_c(E=0) \simeq 16.5t$ in accordance with other numerical results for the 3d Anderson model [5, 32, 33]. The quality of this criterion is underlined also by our data for a 1d system shown in the inset of Fig. 6. Here, as mentioned above, arbitrarily small disorder leads to localisation of the entire spectrum. Clearly, in our approach this is reflected by a typical DOS that vanishes for large M.

2.4 Comparison of the different methods

Comparing the value of the critical disorder obtained by the various methods discussed in the previous sections, the two main results are the following: (i) As can

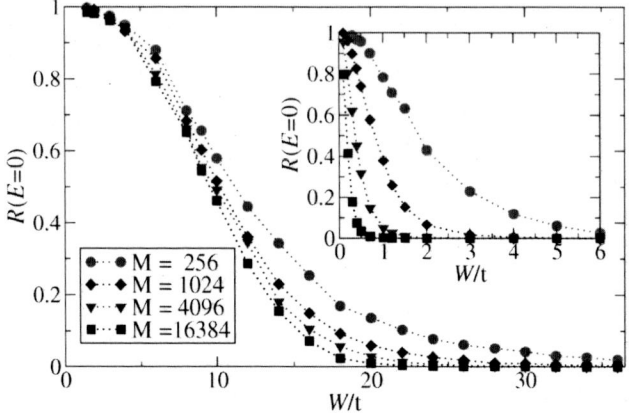

Fig. 6 Normalised typical DOS as a function of disorder calculated with increasing expansion order on a 50^3 lattice. The inset shows the corresponding behaviour of the 1d system with 125000 sites. $K_s \times K_r = 32 \times 32$.

be seen from Fig. 7, the established criteria and methods show an uncertainty of the critical value W_c in the order of $\pm 0.5t$, which is mainly due to the finite system sizes accessible to the numerical calculations. Note that our data widely agrees with the results in the literature [5, 34]. (ii) The value $W_c \simeq 16.5t$ can be reproduced with the same accuracy using a vanishing typical DOS as an indicator for localisation. An improvement of the accuracy of this result can in principle be obtained by extending the numerical effort (larger systems, higher resolution, high-performance computers), which is facilitated by the straightforward parallelisability of the KPM

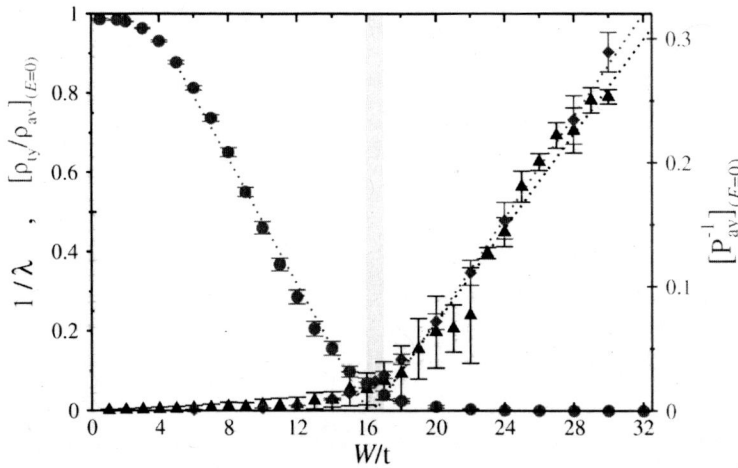

Fig. 7 Comparison of the critical values of disorder, W_c, obtained by the methods outlined in Secs. 2.1-2.3. Decay of the wavefunction (triangles): $N = 20^3$ and 30^3, $E_n \in [-0.01, 0.01]$, $K_r = 10$. Average inverse participation number (diamonds): $N = 16^3$, $E_n \in [-0.1, 0.1]$, $K_r = 100$. Normalised typical DOS (circles): $N = 50^3$, $M = 16384$, $K_r \times K_s = 32 \times 32$.

algorithm. On the other hand, an appropriate scaling ansatz may improve the estimate of W_c on the basis of the presented data.

Using the well-established value $W_c(E = 0) \simeq 16.5t$ as a calibration of the critical R, required to distinguish localised from extended states for the used values of N and M, we reproduce the mobility edge in the energy-disorder plane [5, 33] using $R_c \simeq 0.05$ (see Fig. 8). We also find the well-known reentrant behaviour near the unperturbed band edges [32,35]: Varying W for some fixed values of E ($6t < E \leq 7.6t$) a region of extended states separates two regions of localised states. The Lifshitz boundaries, shown as dashed lines, indicate the energy range, where eigenstates are in principle allowed. As the probability of reaching the Lifshitz boundaries is exponentially small, we cannot expect to find states near these boundaries for the finite ensembles considered in any numerical calculation.

With respect to numerical resources, clearly, the methods that are based on a complete diagonalisation of the Hamiltonian (decay of the wavefunction or participation number) are the least favourable ones, since their CPU requirements scale as N^3 and the memory as N^2. Using LAPACK [36] routines for dense matrices on a standard PC-system diagonalisations are feasible for systems up to 21^3. Banded matrix routines together with the bandwidth reduction described in Appendix B allow to increase this size to 30^3.

The calculation of the LDOS via KPM is based on sparse matrix vector multiplications, whose CPU and memory requirements scale only linearly in N. Hence, systems up to 100^3 can be easily handled with desktop computers, and the use of high-performance environments permits the study of even larger ensembles and systems. We conclude that the new method substantially increases the size of numerically accessible systems, which may lead to a more thorough understanding of the Anderson transition.

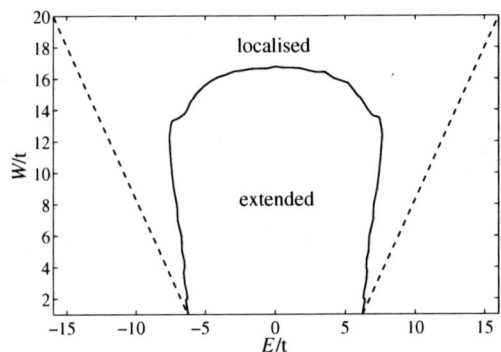

Fig. 8 Phase diagram of the Anderson model on a 3d cubic lattice. Shown are the mobility edge (solid curve) as well as the Lifshitz boundaries (dashed lines).

3 Conclusions

With this contribution we aimed to compare well-established numerical localisation criteria for electrons in disordered systems with a recently proposed approach that is based on the evaluation of the typical density of states.

Considering the 3d cubic Anderson model we proved that the local DOS can be very efficiently calculated using a Chebyshev expansion with kernel polynomial refinement. Given the numerically obtained distribution of the LDOS, the corresponding typical DOS allows for the detection of the delocalisation-localisation transition with a precision that is comparable to results known from other methods. Like for all numerical schemes, the method is restricted to finite systems, and the obtained critical values are subjected to finite-size effects. However, due to the low computational demands the accessible system sizes increase.

Finally we established the use of the typical DOS as a kind of order parameter, which is important in view of its application to interacting disordered systems.

The authors greatly acknowledge support from the Competence Network for Technical/Scientific High-Performance Computing in Bavaria (KONWIHR). Special thanks go to LRZ München, NIC Jülich and HLRN (Zuse-Institut Berlin) for granting resources on their computing facilities. Discussions with A. Alvermann, F.X. Bronold, S.A. Trugman and W. Weller were greatly appreciated.

A Calculation of the LDOS via the kernel polynomial method

At first glance, Eq. (5) suggests that the calculation of the LDOS could require a complete diagonalisation of H. It turns out, however, that an expansion of ρ_i in terms of Chebyshev polynomials $T_n(x) = \cos(n \arccos x)$ allows for an incredibly precise approximation. Since the Chebyshev polynomials form an orthogonal set on the interval $[-1, 1]$, prior to an expansion the Hamiltonian H needs to be rescaled,

$$\mathfrak{X} = \frac{H}{W/2 + 2dt + 0.01t}. \tag{10}$$

Here $W/2 + 2dt$ reflects half the bandwidth of the Anderson model and $0.01t$ is an additional offset that ensures numerical stability of the expansion. In terms of the coefficients

$$\mu_m = \int_{-1}^{1} \rho_i(x) T_m(x)\, dx = \sum_{n=1}^{N} \langle i|n\rangle \langle n|T_m(x_n)|i\rangle = \langle i|T_m(\mathfrak{X})|i\rangle \tag{11}$$

the approximate LDOS $\tilde{\rho}_i(x)$ reads

$$\tilde{\rho}_i(x) = \frac{1}{\pi\sqrt{1-x^2}} \left(g_0 \mu_0 + 2 \sum_{m=1}^{M} g_m \mu_m T_m(x) \right). \tag{12}$$

The factors

$$g_m = \frac{1}{M+1} \left((M - m + 1) \cos(m\phi) + \frac{\sin(m\phi)}{\tan(\phi)} \right), \tag{13}$$

where $\phi = \pi/(M+1)$, result from a convolution of the finite series with the so-called Jackson kernel [17], which mainly damps out the Gibbs oscillations known

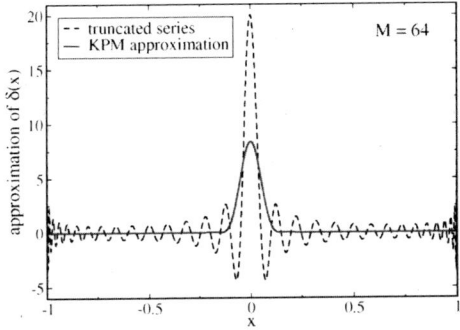

Fig. 9 Chebyshev expansion of a δ-peak: The plain truncated series of order $M = 64$ is a strongly oscillating curve (dashed). By convolution with the Jackson kernel it transforms into a strictly positive, well localised peak at $x = 0$ (solid), which is much closer to our usual notion of $\delta(x)$.

from polynomial approximations (cf. Fig. 9). The width of the kernel, $\Delta x = \pi/M$, scales inversely with the order of the expansion M and defines the resolution of the method.

Using the recursion relations of the Chebyshev polynomials,

$$T_{m+1}(x) = 2xT_m(x) - T_{m-1}(x), \qquad (14)$$

the moments μ_m can be calculated iteratively. An additional trick allows for the generation of two moments with each matrix vector multiplication by $\mathcal{X}$,

$$\mu_{2m-1} = \sum_{i=1}^{N} 2\langle i|T_m(\mathcal{X})T_{m-1}(\mathcal{X})|i\rangle - \mu_1,$$

$$\mu_{2m} = \sum_{i=1}^{N} 2\langle i|T_m(\mathcal{X})T_m(\mathcal{X})|i\rangle - \mu_0, \qquad (15)$$

reducing the numerical effort by another factor $1/2$. Note that the algorithm requires storage only for the sparse matrix $\mathcal{X}$ and two vectors of the corresponding dimension.

B Reduction of the bandwidth of the Anderson matrix

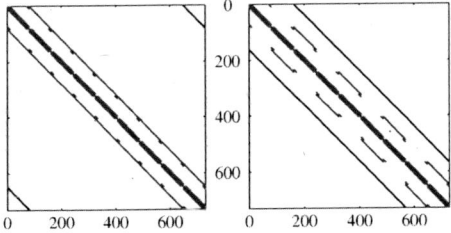

Fig. 10 Sparsity pattern of the Anderson matrix. Left: Standard Anderson matrix for a 9^3 system with PBC. The use of DSBEV would require the storage of 648 off-diagonals. Right: Reduced matrix after transformation (16). Only 162 off-diagonals have to be stored.

While for Krylov sub-space methods [37] (like Lanczos or Jacobi-Davidson) we can take advantage of the sparsity of the tight-binding type matrices, the full diagonalisation with LAPACK routines requires their complete storage. Unfortunately, for periodic boundary conditions the cyclic tridiagonal structure of the matrices spoils the use of band matrix routines like DSBEV. For a L^3-system there are non-vanishing matrix elements in a distance of $L^3 - L^2$ from the diagonal (Fig. 10). Thus almost all matrix elements (most of them zero) need to be stored, giving no advantage compared to full matrix diagonalisation routines like DSYEV.

For linear systems there are tricks to use tridiagonal matrices instead of the corresponding cyclic tridiagonal matrices, which are based on the use of the Sherman-Morrison formula [38]. We are not aware of similar ideas for eigenvalue problems. It turns out, however, that an appropriate sequence of Givens rotations [38] allows the transformation of the cyclic tight-binding matrix (with quadratic blocks close to the outer edges) onto a matrix with blocks only along five diagonals. The corresponding transformation reads

$$H_{\text{red}} = T^T H T, \qquad (16)$$

where $T = P \otimes \mathbb{1}_{L^2 \times L^2}$ and for odd L the $L \times L$ matrix P is given by

$$P = \frac{1}{\sqrt{2}} \begin{pmatrix} \sqrt{2} & 0 & \cdots & \cdots & \cdots & 0 \\ 0 & -1 & 1 & 0 & \cdots & 0 \\ 0 & \cdots & 0 & -1 & 1 & 0 & \cdots & 0 \\ \cdots & \cdots & \cdots & \cdots & \cdots & \cdots \\ 0 & \cdots & \cdots & \cdots & 0 & -1 & 1 \\ 0 & \cdots & \cdots & \cdots & 0 & 1 & 1 \\ \cdots & \cdots & \cdots & \cdots & \cdots & \cdots \\ 0 & \cdots & 0 & 1 & 1 & 0 & \cdots & 0 \\ 0 & 1 & 1 & 0 & \cdots & \cdots & 0 \end{pmatrix}. \qquad (17)$$

For even L the first row and column are absent.

The bandwidth of H can thus be substantially reduced (to $2L^2$, see Fig. 10), which allows for the full diagonalisation of systems up to 30^3 sites on PC-systems with a memory of 512 MB. Furthermore, the sparsity of the transformation (16) can be used to avoid an explicit matrix-matrix multiplication [39]. Hence the change from H to H_{red} is not time consuming. The advantage of the transformation is not primarily a gain in CPU time but storage.

References

1. D. J. Thouless, Physics Reports **13**, 93 (1974).
2. F. J. Wegner, Z. Phys. B **25**, 327 (1976).
3. P. A. Lee and T. V. Ramakrishnan, Rev. Mod. Phys. **57**, 287 (1985).
4. D. Vollhardt and P. Wölfle, in *Electronic Phase Transitions*, edited by W. Hanke and Y. V. Kopaev, North Holland, Amsterdam, (1992), p. 1.
5. B. Kramer and A. Mac Kinnon, Rep. Prog. Phys. **56**, 1469 (1993).
6. P. W. Anderson, Phys. Rev. **109**, 1492 (1958).
7. J. Fröhlich, F. Martinelli, E. Scoppola and T. Spencer, Commun. Math. Phys. **101**, 21 (1985).
8. R. E. Borland, Proc. Roy. Soc. London, Ser. A **274**, 529 (1963).
9. N. F. Mott and W. D. Twose, Adv. Phys. **10**, 107 (1961).
10. E. Abrahams, P. W. Anderson, D. C. Licciardello and T. V. Ramakrishnan, Phys. Rev. Lett. **42**, 673 (1979).

11. M. Janssen, Physics Reports **295**, 1 (1998).
12. V. Dobrosavljević, A. A. Pastor and B. K. Nikolić, Europhys. Lett. **62**, 76 (2003);
13. V. Dobrosavljević and G. Kotliar, Phys. Rev. Lett. **78**, 3943 (1997).
14. F. X. Bronold and H. Fehske, Phys. Rev. B **66**, 073102 (2002);
15. F. X. Bronold, A. Alvermann, and H. Fehske, Philos. Mag. **84**, 673 (2004).
16. R. Abou-Chacra, P. W. Anderson and D. J. Thouless, J. Phys. C **6**, 1734 (1973).
17. R. N. Silver, H. Röder, A. F. Voter and D. J. Kress, J. of Comp. Phys. **124**, 115 (1996);
18. R. N. Silver and H. Röder, Phys. Rev. E **56**, 4822 (1997).
19. K. Byczuk, W. Hofstetter, and D. Vollhardt, URLhttp://arXiv.org/abs/cond-mat/0403765.
20. G. Schubert, A. Weiße and H. Fehske, URL http://arXiv.org/abs/cond-mat/0406212, to appear in Physica B (2005).
21. G. Schubert, A. Weiße and H. Fehske, URLhttp://arXiv.org/abs/cond-mat/0406750.
22. A. Alvermann, G. Schubert, A. Weße, F. X. Bronold and H. Fehske, URL http://arXiv.org/abs/cond-mat/0406051, to appear in Physica B (2005).
23. A. Mac Kinnon and B. Kramer, Z. Phys. B **53**, 1 (1983).
24. J. L. Pichard and G. Sarma, J. Phys. C **14**, L127 (1981).
25. I. V. Plyushchay, R. A. Römer, and M. Schreiber, Phys. Rev. B **68**, 064201 (2003).
26. B. J. Last and D. J. Thouless, J. Phys. C **7**, 699 (1974).
27. F. Wegner, Z. Phys. B **36**, 209 (1980).
28. A. V. Malyshev, V. A. Malyshev and F. Domínguez-Adame, URL http://arXiv.org/abs/cond-mat/0303092.
29. R. Haydock and R. L. Te, Phys. Rev. B **49**, 10845 (1994).
30. W. T. Arnold and R. Haydock, Phys. Rev. B **66**, 155121 (2002).
31. E. W. Montroll and M. F. Shlesinger, J. Stat. Phys. **32**, 209 (1983).
32. B. Bulka, M. Schreiber and B. Kramer, Z. Phys. B **66**, 21 (1987).
33. H. Grussbach and M. Schreiber, Phys. Rev. B **51**, 663 (1995).
34. B. Kramer, K. Broderix, A. Mac Kinnon and M. Schreiber, Physica A **167**, 163 (1990).
35. S. L. A. de Queiroz, Phys. Rev. B **653**, 214202 (2001).
36. *The Linear Algebra PACKage*, URL http://www.netlib.org.
37. Y. Saad, *Numerical Methods for Large Eigenvalue Problems* (University Press, Manchester, 1992).
38. W. H. Press, B. P. Flannery, S. A. Teukolsky and W. T. Vetterling, *Numerical Recipes* (Cambridge University Press, Cambridge, 1986).
39. G. Schubert, diploma thesis, Universität Bayreuth (2003).

NBW: Computational Seismology: Narrowing the Gap Between Theory and Observations

Bernhard Schuberth, Michael Ewald, Heiner Igel, Markus Treml, Haijiang Wang, Gilbert Brietzke

Department für Geo- und Umweltwissenschaften, Bereich Geophysik,
Ludwig-Maximilians-Universität München,
Theresienstrasse 41,
80333 München, Germany,
bernhard@geophysik.uni-muenchen.de

Abstract. Numerical solutions to the problem of seismic wave propagation, that allow simulations of complete wave fields through 3D structures, are currently revolutionizing seismology and related fields. So far - in order to calculate theoretical seismograms in the observed frequency bands - one had to resort to solution methods with severe limitations (e.g., ray theoretical approximations, one-dimensional structures, perturbation theory, etc.). Only in the past few years, computational power has allowed us to simulate wave fields that can be directly compared to observations. Even though the computations still require substantial resources, the methodologies developed in the past decade are beginning to enter routine processing steps in all branches ranging from exploration seismics to global seismology. Here we present recent examples in global seismology (spectral element modeling of global wave propagation) and earthquake scenario simulations, their relation to shaking hazard estimation, and associated problems. The next decade will see fundamental changes in the way data fitting (inverse problem, parameter estimation) is done in seismology with the potential of advances in several fields of Earth Sciences.

1 Introduction

Many phenomena of (visco-)elastic (acoustic) wave propagation are the basis for diagnostic tools in fields such as medicine, meteorology, seismology, exploration geophysics, engineering, material sciences and others. Because of the increasing computational power, the methodologies used in these fields have dramatically converged in the past decade. Numerical solutions to wave propagation problems - using finite differences (FD), finite volumes (FV), finite and spectral element methods (FEM and SEM respectively) and others - are now common tools in most disciplines. While the algorithm development is now at a fairly advanced stage, the routine application of those tools with the associated potential large impact is just at the beginning. The KONWIHR project described here (numerical wave propagation) focused in the past years on the program development and verification. Below, we will link these developments to other ongoing projects and demonstrate that KONWIHR was fundamental in providing the more technical progress, thereby serving these related projects.

1.1 Current Issues in Computational Seismology

Several of the technical and scientific objectives and results were presented in the two previous project reports [4, 8]. These focused on the development of 3D wave propagation tools for media with strong topography [23,24], the calculation of earthquake scenarios and the resulting ground motions [5, 31–35], the simulation of wave propagation in seismically active fault zones [6, 7, 12], the simulation of the actual rupture process during earthquakes [1–3] and finally the emerging field of numerical wave propagation on a global scale [9, 11, 13, 19, 20, 25, 27–30]. It is beyond the scope of this paper to describe the results in detail. The main scientific advances are summarized here:

Earthquake Scenarios. 3D simulations of earthquake scenarios in seismically active regions (e.g., Cologne Basin) show that the 3D structure, source location, source mechanism, near-surface structure, etc. strongly influence the peak ground motion observed at a particular point in the region of interest (e.g., [5]). This has tremendous implications for seismic hazard studies, as many more aspects need to be taken into account for reliable estimations than was previously thought. The most important factor is the 3D seismic velocity structure, that is often not sufficiently well known. Studies are being undertaken (see Sect. 2.4) to incorporate uncertainties into the estimation of shaking hazard.

Dynamic Rupture. The physical processes that happen at the fault during rupture are still poorly understood. Recently it was discovered that - if across the fault the material properties change, a likely feature at many large deformation-rate strike slip faults such as the San Andreas in California - rupture at bi-material interfaces may dramatically influence the rupture behavior. A large number of simulations carried out as a parameter study illustrated the behavior of ruptures in such circumstances. It was shown that if rupture initializes in the vicinity of such material interfaces, the rupture is likely to migrate to this interface and continue to break that particular fault [1–3].

Global Seismology. Wave propagation on a planetary scale was so far predominantly carried out using quasi-analytical approaches (e.g., spherical harmonics) and perturbation theory. Only recently the impact of 3D structures on the observed wave field is being appreciated. A large part of the information contained in the recorded seismograms is still not used to understand the structure of the Earth's deep interior. With the tools developed in the Munich group (axi-symmetric approach [11] and spherical sections [9, 19, 20]) as well as a spectral element approach [14–16] that was developed elsewhere but was extended and installed on the Munich supercomputer facilities, a new era of global seismic data modeling is just beginning.

After putting the KONWIHR project in the context of other activities, we will present recent results from three ongoing topics.

1.2 KONWIHR and Related Projects

There was strong beneficial interaction between the KONWIHR project and other studies. The technical lessons learned through the code development (debugging, profiling, parallelization) within KONWIHR lead to successful applications of the algorithms in projects like the **International Quality Network: Georisk** (www.iqn-georisk.de, funded by the German Academic Exchange Service); the **Geosensor Project** (BMBF, Geotechnologies program) that aims at understanding earthquake induced rotational motions [10]; a study that aims at understanding **the seismic signature of plumes** (German Research Foundation); and others. The experiences within the KONWIHR project were instrumental in the preparation for a large European network in computational seismology that was funded by the European Union in 2003. This project called **SPICE** (Seismic wave propagation and imaging in complex media: a European network, www.spice-rtn.org) is the first Research and Training Network in computational seismology connecting 14 European institutions and aims at developing a digital library with verified wave propagation codes for the Earth Science community. The project will also train young researchers in this field through workshops and online training material.

2 Recent Scientific Results

2.1 Spectral Element Modeling of Global Wave Propagation

In the last years the spectral element method became one of the most important tools in computational seismology. Being a modified finite element method, its name derives from the convergence behavior of the method with increasing order, which is the same as in the spectral methods.
First introduced for fluid dynamics [21], it was further developed in the 1990's for seismological applications ([22] and [17, 18]). Today the SEM can be used to simulate the wave propagation in global spherical Earth models including various features such as topography/bathymetry, laterally heterogenous velocity structures in the crust and the mantle, attenuation, anisotropy and also second order effects as for example Earth's rotation, gravity or the influence of ocean water on the wave field [15, 16]. The advantages of the method are not only the capability of dealing with the complex problems mentioned above, but also its high accuracy and the ease of implementing free boundary conditions. Especially for global Earth models, the latter is very appreciable.

The model for global wave propagation is built using the "cubed sphere" approach. This is illustrated in Fig. 1, where the initial cube is gradually distorted from left to right, until its six faces match the surface of the sphere. In the lower part of the picture the procedure is shown for one of the six "chunks" comprising the cubed sphere. The clue in this procedure is to keep a small cube in the interior of the mesh undistorted thus avoiding singularities in the center at $r = 0$.
In the current implementation of the SEM in a parallel MPI-FORTRAN code, the model space is decomposed for each chunk separately, but in the same manner. The number of divisions in both "horizontal" directions of the chunks has to be equal. In radial direction, the chunks are not split up. Thus, for given integer values n of

possible divisions (1,2,3,...), the resulting number of processes for the whole sphere gets $6 \times n^2$ (= 6, 24, 54, 96 etc.).

As first tests on the SR8000 showed minor performance with auto-parallelization, we switched to intra-node MPI parallelization. The reasons for that are numerous indirect addressings in most of the loops, and therefore auto-parallelization of the loops fails. In addition, the communication between the processes is only at around 10% of the CPU time. Because of those reasons the number of nodes used on the HITACHI is very unusual compared to widely used 2^n values.

At the moment we use a standard setup for various calculations using 5×5 processors per chunk resulting in a total number of 150 processes. These are distributed on 19 nodes. The typical memory needed by our models, which are accurate for periods greater than 20 s, is between 60 to 90 GB, depending whether attenuation is incorporated or not. The typical runtime is 3.4 seconds per time step leading to a total runtime of 19 hours for the calculation of a 90 minute seismogram. This setup already allows us to enter a new realm of data modeling with dramatic improvement of misfit between observations and theory as shown in Sect. 2.2.

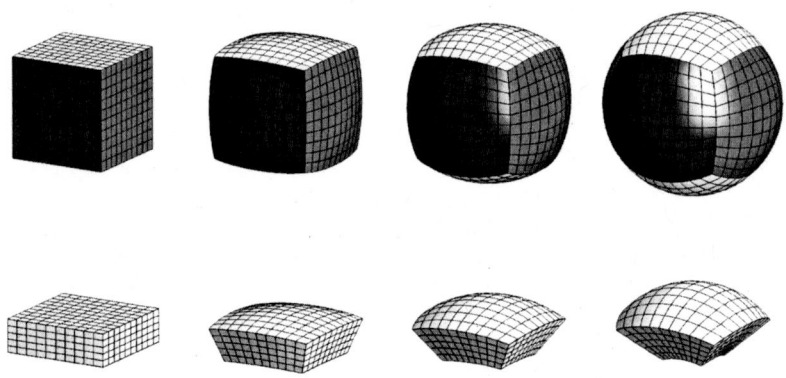

Fig. 1. Upper part: creation of a global Earth model by expanding an initial cube to the sphere (i.e., cubed sphere mesh). Keeping a small central cube undistorted avoids singularities at r=0. Lower part: same as above for only one of the six chunks of the cubed sphere mesh. (Picture courtesy of Peter Danecek).

2.2 Convergence of Observations and Theory - Examples from Spectral Element Modeling

The SEM described above, was used in this study to show the decreasing misfit of theoretical simulations compared to real data with increasing complexity of the applied models. We simulated the M8.1 Tokachi-oki earthquake that happened on September 25, 2003. As a starting model we used a purely elastic (i.e., no attenuation), isotropic, spherically symmetric Earth model. The source itself was described as a point source double couple. In a second step we extended the simulation using a finite source model (Ji Chen, personal communication). The next step was using the same radially symmetric velocity structure but this time including attenuation for both source models. Finally, we included all effects, that are up to date incorporated in the SEM code, only excluding anisotropy. Thus, the final model consisted of a laterally heterogenous, attenuating 3D velocity structure for the Earth's crust and mantle together with a $2' \times 2'$ topography/bathymetry grid, as well as rotation and gravity of the Earth and the effects of ocean water. Comparisons between respective simulations and observations are shown in Fig. 2. The bottom seismogram shows the transverse component (w.r.t. the great circle path from the source to the receiver) of the translational ground velocity recorded with a broad-band seismometer at the Geodetic Observatory Wettzell/Germany. The first two seismograms from top show the results for the radially symmetric Earth model without attenuation for point source (topmost trace) and finite source (second from top) simulations. The amplitudes were scaled to the S-wave amplitude of the real data which shows up at around 1350 s. The ratio of synthetic S-wave amplitude to the real amplitude (here called amp) is given in Fig. 2 for every seismogram. The third (point source) and fourth (finite source) traces from top show the results for the complex 3D Earth model described above. The simulations for the radially symmetric model with attenuation are not presented here, as they are very similar to the ones without attenuation. This illustrates that attenuation has much less effect on the amplitudes than the representation of the source by an extended fault plane (difference of amp of a factor of 10 for the two simulations using the radially symmetric model in Fig. 2). The azimuthal dependence of the amplitude is called the "directivity effect", which is a result of the anisotropic radiation from the large source area.

The results show a huge increase of fit to the real data with increasing incorporation of the various effects. The synthetic seismograms of the both source-type simulations for the 3D model are already quite close to reality, but show still too small amplitudes for the first part of the surface waves (between 2000 s and 2600 s). Nevertheless, the amplitudes of the finite source seismogram are closer to the observed ones. Thus, the complex 3D model with a finite source representation is the best fitting one. To better illustrate this, Fig. 3 shows both seismograms plotted on top of each other. Only for displaying purposes, the synthetic seismograms (dark grey) have been offset by 5% of the maximum amplitude.

These comparisons show that with increasing effort and complexity of the models we are able to fit observations quite well. Thus, by studying even more complex aspects in future and continuously comparing the results to real data, the models of Earth's structure can be improved step by step with implications for other fields like geodynamics, tectonics and geodesy.

2.3 Effects of Plume Structures on the Wave Field

Apart from studying the effects of Earth's structure at a global scale, it is useful to examine the effects of prominent regional features in the Earth isolated in order to improve understanding of their specific influence. A prominent structure on a regional scale would be for example subduction zones. However, in recent years the hot topic of the scientific community has been mantle plumes. From earlier studies using a regional, Cartesian 3D code [26], the development headed for examination of the influence of plumes on incoming teleseismic wavefields. For this purpose a hybrid 2D axi-symmetric/regional spherical 3D code was developed, allowing for an effective computation of the wavefield by concentrating computational power on the

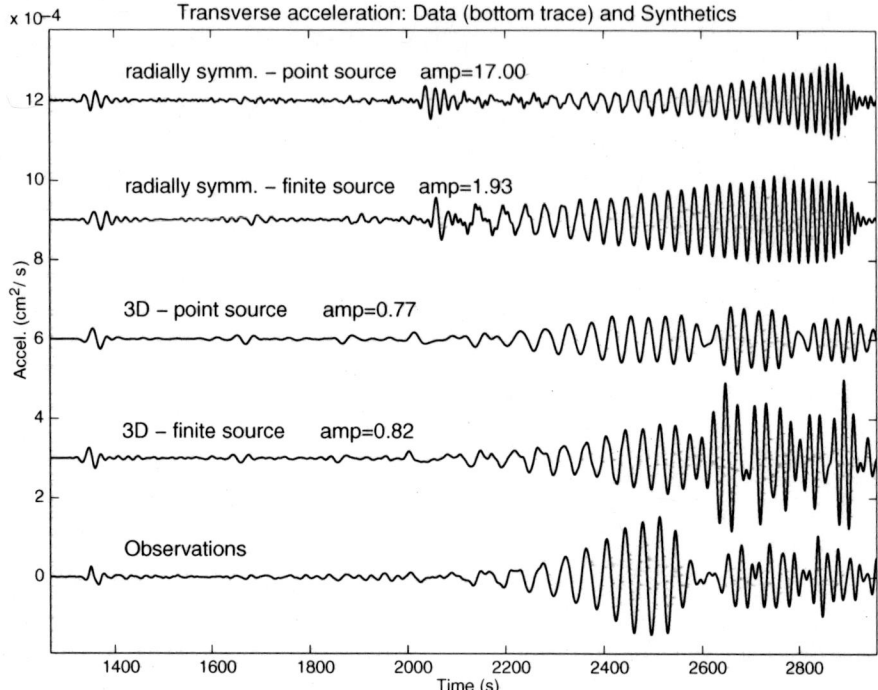

Fig. 2. Simulations using different Earth and source models compared to real data of the M8.1 Tokachi-oki earthquake, September 25 2003: The bottom trace shows the broad-band seismogram recorded at Wettzell/Germany. The two top traces show seismograms for a radially symmetric Earth model using a point source (uppermost trace) and a finite source model. The two middle traces show simulations using a 3D velocity model incorporating attenuation (anelasticity), topography/bathymetry, Earth's rotation and gravity and ocean water together with finite and point source representations (second and third trace from below respectively). All synthetic seismograms have been scaled to the S-wave amplitude of the observations. The given factor (amp) is the ratio of the synthetic S-wave amplitude and the S-wave amplitude of the real data.

area around the structure of interest [4]. Meanwhile, the focus is on a full 3D spectral element approach because it allows for a wider range of studies. Its drawback are the enormous requirements in computational power.

Using the same code as described in Sect. 2.1, we additionally implemented plume models and are now beginning to systematically study effects of plume geometry and the nature of the spatial perturbation pattern of plumes. In this first plume parameter study to date we aim at finding a typical "signature" of a plume in the wavefield. A further goal is to be able in future to suggest optimal experiment configuration in order to detect this signature with the least effort possible. This has become necessary since experiments carried out to image plume structures became

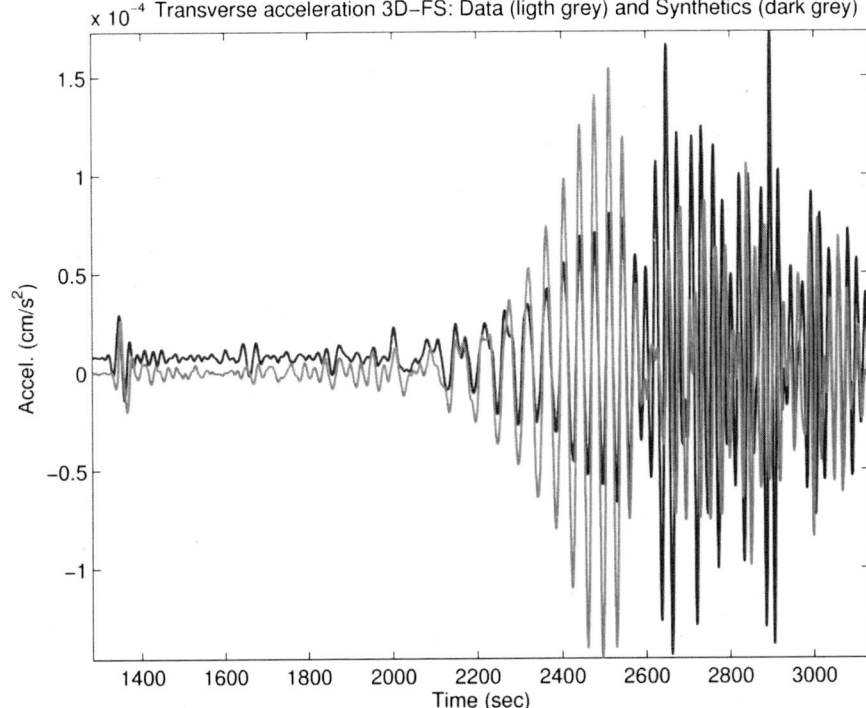

Fig. 3. Comparison of the best fitting simulation (3D – finite source; plotted in dark grey) to the real observations (light grey). The seismograms are offset in vertical direction by 5% of the maximum amplitude for better distinction. Especially the first part of the surface waves between 2000 s and 2600 s fit very well in waveform and phase.

more and more expensive, especially when employing ocean bottom seismometers, which are furthermore very difficult to install and maintain. Nevertheless, they turn out to be mandatory for studying important ocean island plumes such as Hawaii.

2.4 Earthquake Scenarios: Uncertainties in Ground Motion Estimation

In this study 3D FD techniques are applied to simulate wave propagation of earthquake scenarios in seismically active regions and thus assess the seismic hazard of such areas. Amplifications of ground motion due to low velocity structures such as sedimentary basins is of special interest in these investigations. Our studies of earthquake scenario simulations concentrate on basins near the city of Cologne and the Beijing metropolitan area [4, 8].

In FD ground motion simulations the velocity model is discretized onto a spatial grid on which the wave equations are solved. This solution provides the complete 3D wave field over the model space. Information on the ground motion in terms of velocity, acceleration or displacement according to the purposes of the simulation can be determined at any point (typically on the surface). Quantities relevant for engineering purposes (e.g., intensity or shaking duration) are then derived from these values.

Results from FD simulations are afflicted with errors caused by two major sources: 1) approximation errors, due to uncertainties of the input data (e.g., velocities, q-values, densities, source location especially source depth etc.) and 2) modeling errors, due to natural imperfections of mathematical abstractions of real physical events depending on the algorithms used in the simulations.

Whereas modeling errors can be minimized by properly chosen operator accuracy and simulation parameters, the influence of uncertainties in the input data are hard to quantify. To investigate such effects multiple simulations were performed using varying models within realistic error margins on the input data sets. We chose basin depth and hypocenter depth as parameter axes along which the model is modified. Figure 4 shows five synthetic seismograms for the same earthquake scenario at the same receiver location using velocity models varying in terms of basin depth by -10, -5, 0, 5 and 10% respectively relative to the original model. Notable effects on waveforms can be observed in the later arrivals which are caused by surface waves multiply scattered within the sedimentary basin. These phases are most sensitive to slight variations on the basin shape as resonant amplification occurs to different extents and at different points in time. A variation of ±30% in peak ground velocity like in this example would result in a difference in predicted seismic intensity of one unit on the Mercalli scale. The expected shaking level would change for example from so-called "severe" to "violent" which has distinct implications on local building codes. Source depth is another input parameter in earthquake scenario modeling afflicted with noticeable error. Besides the expected effects of earlier arrival times and higher amplitudes with declining source depth our investigations show a strong impact on resonant wave trains that can account for quite the opposite behavior. Such effects are due to the interaction of source location, velocity model and receiver location and can be quantified only by 3D simulations. Our goal is to pin down the most critical model parameters and quantify their influence on the simulation results.

3 General Conclusions and Outlook

In the field of seismology the calculation of synthetic (theoretical) seismograms for generally heterogeneous 3D media by numerical means is now beginning to complement (maybe soon in large part to replace) the previous tools (ray theory, 1D-, 2D-approximations) used to process, analyse and explain observations. For realistic problems, these calculations will remain large-scale computing problems with the necessity of parallel programming for some time. However, with the more and more common cluster facilities and the exciting prospects of GRID computing, the developed codes are likely to become routine tools for the seismological community. Providing these new facilities to the non-specialist Earth scientist (i.e., non programdeveloper) is one of the most important tasks for the coming years and will require substantial software engineering particularly in the domain of www-interfaces and data bases.

The results that are beginning to appear (e.g., [10]), demonstrate that the scientific value of synthetic seismograms (i.e., the results of large scale simulations) is

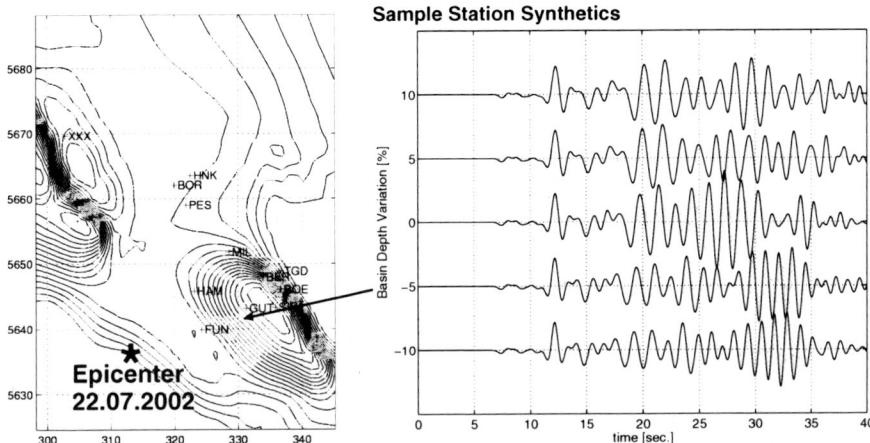

Fig. 4. Left part: contour map of the sedimentary basin depth in the Cologne area. Right part: synthetic seismograms for the same earthquake scenario (the Alsdorf, July 22 2002 earthquake) and receiver location using velocity models with varying basin depth by -10, -5, 0, +5 and +10% respectively. Note the strong influence of the model variation on the later phases of the individual traces.

approaching that of observations. This has tremendous consequences as it suggests that not only the observations (seismograms measured around the world) should be archived and made publicly available but also the synthetic seismograms. This problem is taken up within the European SPICE project (see Sect. 1.2) and has led to an international working group of Earth scientists and seismic network managers with the aim of defining common data formats to store synthetic data in the same way as observations, also using the same infrastructure (e.g., international seismic data centers like www.iris.org). As an example, this will imply that in the not so distant future, scientists will be able to download not only observed seismograms for specific sites, but also theoretical seismograms computed using the latest 3D reference Earth model. It is likely that this will allow new ways of rapidly interpreting seismograms and progress in the associated fields such as (1) the determination of earth structure, (2) the recovery of earthquake rupture properties, and (3) the reliable estimation of shaking hazards.

Acknowledgement. We like to acknowledge funding through the KONWIHR Project, and the Leibniz-Rechenzentrum and its steering committees for providing access to the Hitachi SR8000. We also want to thank the LRZ supporting staff for their help. These projects where partly supported through the DAAD (IQN-Georisk) and the German Research Foundation.

References

1. G. Brietzke and Y. Ben-Zion. Examining tendencies of in-plane rupture to migrate to material interfaces. In *Eos Trans. AGU, Fall Meet. Suppl., Abstract S42C-0190*, volume 84(46), 2003.
2. G. Brietzke, Y. Ben-Zion, H. Igel, and A. Cochard. Simulation of wavefields emanating from finite sources in anisotropic media. In *Eos Trans. AGU, Fall Meet. Suppl., Abstract S61B-1143*, volume 83(47), 2002.
3. G. Brietzke, H. Igel, and Y. Ben-Zion. 3D modeling of kinematic and dynamic in anisotropic media. In *DGG-Tagung Hannover, Germany*, 2002.
4. G. Brietzke, H. Igel, G. Jahnke, M. Treml, M. Ewald, H. Wang, A. Cochard, and G. Wang. *High Performance Computing in Science and Engineering*, chapter Computational elastic wave propagation: advances in global and regional seismology, pages 444–458. Springer, Heidelberg, 2004.
5. M. Ewald, H. Igel, F. Scherbaum, and K. Hinzen. 3D simulation of earthquake ground motion in the Cologne basin, Germany. In *Eos Trans. AGU*, volume 82(47), 2002.
6. M. Fohrmann, H. Igel, G. Jahnke, and Y. Ben-Zion. Guided waves from sources outside faults: an indication for shallow fault zone structure? In *Proceedings of 3rd ACES Workshop 2002*. Geoprint ISBN 0-9750394-0-7, 2002.
7. M. Fohrmann, G. Jahnke, H. Igel, and Y. Ben-Zion. Guided waves generated by sources outside a low velocity fault zone. In *Eos Trans. AGU, Fall Meet. Suppl.*, volume 82(47), 2001.
8. H. Igel, G. Brietzke, M. Ewald, M. Fohrmann, G. Jahnke, T. Nissen-Meyer, J. Ripperger, M. Strasser, M. Treml, and G. Wang. *High Performance Computing in Science and Engineering, ISBN 3-540-00474-2*, chapter 3-D Seismic Wave Propagation on a Global and Regional Scale: Earthquakes, Fault Zones, Volcanoes, pages 353–362. Springer, Heidelberg, 2002.
9. H. Igel, T. Nissen-Meyer, and G. Jahnke. Wave propagation in 3D spherical sections. Effects of subduction zones. *Phys. Earth Planet. Int.*, 132:219–234, 2002.
10. H. Igel, U. Schreiber, A. Flaws, B. Schuberth, A. Velikoseltsev, and A. Cochard. Rotational motions induced by the M8.1 Tokachi-oki earthquake, September 25, 2003. *submitted to Nature*, 2004.
11. G. Jahnke and H. Igel. High resolution global wave propagation through the whole Earth: the axi-symmetric PSV and SH case. In *EGS General Assembly, Nice, France*, 2003.
12. G. Jahnke, H. Igel, and Y. Ben-Zion. Three-dimensional calculations of fault zone guided waves in various irregular structures. *Geophys. J. Int.*, 151:416–426, 2002.
13. G. Jahnke, T. Nissen-Meyer, and M. Treml. High resolution global wave propagation for axisymmetric and additional 3D geometries. In *Poster, presented at the EGS General Assembly, Nice, France*, 2002.
14. D. Komatitsch and J. Tromp. Introduction to the spectral-element method for 3-D seismic wave propagation. *Geophys. J. Int.*, 139:806–822, 1999.
15. D. Komatitsch and J. Tromp. Spectral-element simulations of global seismic wave propagation-I. Validation. *Geophys. J. Int.*, 149:390–412, 2002.
16. D. Komatitsch and J. Tromp. Spectral-element simulations of global seismic wave propagation-II. 3-D models, oceans, rotation, and self-gravitation. *Geophys. J. Int.*, 150:303–318, 2002.

17. D. Komatitsch and J-P Vilotte. The spectral-element method: an efficient tool to simulate the seismic response of 2D and 3D geological structures. *Bull. Seismol. Soc. Am.*, 88(2):368–392, 1998.
18. Dimitri Komatitsch. *Méthodes spectrales et éléments spectraux pour l'équation de l'élastodynamique 2D et 3D en milieu hétérogène (Spectral and spectral-element methods for the 2D and 3D elastodynamics equations in heterogeneous media)*. PhD thesis, Institut de Physique du Globe, Paris, France, 1997.
19. T. Nissen-Meyer, G. Brietzke, and H. Igel. Numerische Simulationen von Wellenausbreitung in seismisch aktiven Zonen. *KONWIHR Quarterly*, 31, 2001.
20. T. Nissen-Meyer and H. Igel. 3D wave effects of sources inside subduction zone. In *Eos Trans. AGU, Fall Meet. Suppl., Abstract T41C-0903,*, volume 82 (47), 2001.
21. A. T. Patera. A spectral element method for fluid dynamics: laminar flow in a channel expansion. *J. Comput. Phys.*, 54:468–488, 1984.
22. E. Priolo and G. Seriani. A numerical investigation of Chebyshev spectral element method for acoustic wave propagation. In *Proc. 13th IMACS Conf. on Comp. Appl. Math.,v. 2*, pages 551–556, Dublin, Ireland, 1991.
23. J. Ripperger, H. Igel, and J. Wassermann. Simulation of 3D seismic wave propagation with volcano topography. In *Eos Trans. AGU, Fall Meet. Suppl.*, volume 82(47), 2001.
24. J. Ripperger, H. Igel, and J. Wassermann. Seismic wave simulation in the presence of real volcano topography. *J. Volcanol. Geotherm. Res.*, 128:31–44, 2003.
25. N. Schmerr, E. Garnero, H. Igel, M. Treml, and G. Jahnke. Probing the nature of 410- and 660km discontinuities beneath hotspots using SS precursors. In *poster presented at the AGU fall meeting, San Francisco, CA, USA*, 2003.
26. M. Strasser. Numerical modeling of 3D wave effects of plumes. Master's thesis, Universität München, 2001. unpublished.
27. M. Thorne, E. Garnero, G. Jahnke, M. Treml, and H. Igel. Investigating the core-mantle boundary and ULVZ topography with synthetic FD seismograms for 3D axi-symmetric geometries: predictions and data. In *poster presented at the AGU fall meeting, San Francisco, CA, USA*, 2003.
28. M. Treml, S. Goes, G. Jahnke, T. Nissen-Meyer, and H. Igel. Synthetic seismic wave propagation through thermal mantle plumes. In *poster presented at the EGS General Assembly, Nice, France*, 2003.
29. M. Treml, H. Igel, F. Tilmann, J. Phipps Morgan, and M. Strasser. Towards modelling 3D wavefield effects of plumes. In *talk and poster, presented at the Meeting of Geoscientific Iceland Research, Frankfurt, Germany*, 2002.
30. W. Treml, G. Jahnke, T. Nissen-Meyer, H. Igel, and E. Garnero. A hybrid finite-difference method for global wave propagation. In *Eos Trans. AGU, Fall Meet. Suppl.*, volume 82(47), 2002.
31. G. Wang, H. Igel, and H. Wang. Three-dimensional finite-difference simulations of strong ground motions during the 1720 Shacheng earthquake (Mw = 7.0) of Yanhuai area, Beijing, China using a stochastic finite-fault approach. In *Eos Trans. AGU, Fall Meet. Suppl.*, volume 84(46), 2003.
32. G. Wang, H. Igel, X. Zhou, and H. Wang. Three-dimensional finite-difference simulations of strong ground motions in the beijing metropolitan area. In *Eos Trans. AGU*, volume 82(47), 2001.

33. G.-Q Wang, H. Boore, D. M.and Igel, and Zhou X. Y. Some observations on colocated and closely-spaced strong ground motion records of the 1999, Chi-Chi, Taiwan earthquake. *Bull. Seism. Soc. Am.*, 93:674–693, 2003.
34. G.-Q. Wang, H. Igel, H.-J. Wang, and X.-Y. Zhou. Simulations of strong ground motion in the Beijing metropolitan area using the finite difference and stochastic combined method,. In *poster presented at the EGS General Assembly, Nice, France*, 2003.
35. H. Wang, H. Igel, and G. Wang. The effect of 3D structure and finite-source on ground motion in the Yanhuai basin. In *AGU 2003 Fall Meeting, San Francisco, CA, USA*, 2003.

OOPCV: Phasediagram and Scaling Properties of the Projected SO(5) Model in Three Dimensions

Martin Jöstingmeier[1], Ansgar Dorneich[1], Enrico Arrigoni[2], Werner Hanke[1], and S.C. Zhang[3]

[1] Institute for Theoretical Physics and Astrophysics, University of Würzburg, Am Hubland, D-97074 Würzburg
joesel@physik.uni-wuerzburg.de
[2] Enrico Arrigoni, Institute for Theoretical Physics, Technical University of Graz, Petersgasse 16, A-8010 Graz
arrigoni@itp.tu-graz.ac.at
[3] Department of Physics, Stanford University, Stanford, 94305 California, USA
sczhang@stanford.edu

Abstract. We study the scaling properties of the quantum projected $SO(5)$ model in three dimensions by means of a highly accurate Quantum-Monte-Carlo analysis. Within the parameter regime studied (temperature and system size), we show that the scaling behavior is consistent with a $SO(5)$-symmetric critical behavior in the numerically accessible region. This holds both when the symmetry breaking is caused by quantum fluctuations only as well as when also the static (mean-field) symmetry is moderately broken. We argue that possible departure away from the $SO(5)$ - symmetric scaling occurs only in an extremely narrow parameter regime, which is inaccessible both experimentally and numerically.

1 Introduction

High-Temperature Superconductivity is one of the most fascinating phenomena of modern solid state physics. This fascination is motivated on the one hand by the possible technical innovations connected with high-temperature superconductivity, such as loss free energy storing, faster computer chips or simply loss free energy transport. On the other hand a consistent theoretical description of high-temperature superconductivity is still not available. The difficulty of a microscopic understanding of this phenomenon, which would allow to synthesize high-temperature superconductors with even enhanced material properties, is caused by an unusual strong entanglement of the many body wave function. This strong entanglement of about 10^{23} electrons within typically a cubic centimeter, is the reason why one can observe "quantum mechanical behavior" on a macroscopic level, but it is also responsible for the failure of the standard analytical approach in theoretical solid state physics, where one attempts to describe the interaction between two particles by a small perturbation of the non interacting system. Obviously this attempt fails, if the interaction plays a major role and affects the physics of the system under consideration substantially. Therefore numerical simulations provide a very powerful tool to achieve a detailed understanding of the microscopic physics of high-temperature superconductors.

In this paper we will first (Sec. 2) introduce very briefly the idea of the $SO(5)$ theory of high-temperature superconductivity. In Sec. 3 we present the phase diagram for the 3 dimensional projected $SO(5)$ model and address the question of symmetry restoration. A conclusion and the discussion of our obtained results can be found in Sec. 4.

2 The $SO(5)$ - model

A common feature of the phase diagram of most high-temperature superconductors (HTSC) is the close proximity of the superconducting (SC) and the antiferromagnetic (AF) phases. The $SO(5)$ theory of High-Temperature Superconductivity de-

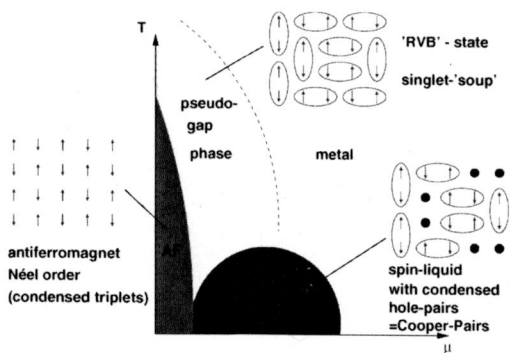

Fig. 1. Generic Temperature versus phase diagram of the cuprate HTSC. In real HTSC crystals, the chemical potential μ can be varied by various hole doping concentrations.

scribes the transition between these two phases by an effective quantum non-linear σ model with approximate $SO(5)$ symmetry, which unifies the antiferromagnetic and superconducting order parameters into a five dimensional superspin [1]. The basic idea has been explained in [2] for a more detailed review we refer to [3]. Several microscopic $SO(5)$ -symmetric models have been proposed which succesfully describe many features of the cuprate physics [4–8].

In the p$SO(5)$ model each coarse-grained lattice site represents a plaquette of the original lattice model, and the lowest energy state on the plaquette is a spin singlet state at half-filling. There are four types of excitations, namely, three magnon modes and a hole-pair mode. Their dynamics are described by the following Hamiltonian:

$$\hat{H} = \Delta_s \sum_{x,\alpha=2,3,4} t_\alpha^\dagger(x) t_\alpha(x) + (\Delta_c - 2\mu) \sum_x t_h^\dagger(x) t_h(x) \qquad (1)$$

$$- J_s \sum_{<xx'>,\alpha=2,3,4} n_\alpha(x) n_\alpha(x') - J_c \sum_{<xx'>} (t_h^\dagger(x) t_h(x') + \text{h.c.}),$$

Here $t_{\alpha=2,3,4}^{(\dagger)}$ anihilates (creates) a triplet state, $t_h^{(\dagger)}$ anihilates (creates) a hole pair state and $n_\alpha = (t_\alpha + t_\alpha^\dagger)/\sqrt{2}$ are the three components of the Néel order parameter. Δ_s and $\Delta_c \sim U$ are the energies to create a magnon and a hole-pair excitation, respectively, at vanishing chemical potential $\mu = 0$. This model can also be effectively obtained by a coarse-grained reduction of more common models such as $t - J$ or Hubbard [9]. In order to study the effect of symmetry breaking we consider different situations associated with different sets of parameters. First, we consider the case where $J_s = J = J_c/2$ (our zero of the chemical potential is such that $\Delta_s = \Delta_c$). It has been shown [10] that this model has a static $SO(5)$ symmetry at the meanfield level and that the symmetry is only broken by quantum fluctuations [11]. Since we want to carry out our analysis also for a more realistic model in which also the static $SO(5)$ symmetry is broken, we also consider a system with a different ratio J_s/J_c. In particular, one would like to reproduce the order of mangitude of T_c/T_N observerved in the cuprates, where T_c (T_N) denominates the SC critical temperature (Néel Temperature). However, this behavior is obtained for $J_s/J_c \sim 2$, for which the numerical simulation is rather unstable, making it impossible to determine the critical exponents with sufficient accuracy. For this reason, we choose a value of the parameter "in between" ($J_c = J_h = J$), for which also the static $SO(5)$ symmetry is broken. The phase diagram of this model in two dimensions has been analyzed in detail by a numerical Quantum-Monte-Carlo approach in Ref. [12]. In particular, the model has been shown to provide a semiquantitative description of many properties of the HTSC in a consistent way. In Ref. [12], the SC transition has been identified as a Kosterlitz-Thouless phase in which the SC correlations decay algebraically. Unfortunately, there is no such transition for the AF phase in two dimensions, as all AF correlations decay exponentially at finite temperatures. Therefore, in order to analyze the multicritical point where the AF and SC critical lines meet, it is necessary to work in three dimensions, which is what we investigate in the present paper. The calculations presented here have been performed using the object-oriented C++ class library descibed in Ref [2].

3 Results

3.1 Case $J_s = J_c/2$

We start by presenting the phase diagram of the 3D p$SO(5)$ model for the "symmetric" case $J_s = J_c/2$. Figure 2 shows an AF and a SC phase extending to finite temperatures as expected. Furthermore, the two phase transition lines merge into a multicritical point (at $T_b = 0.960 \pm 0.005$ and $\mu_b = -0.098 \pm 0.001$). The line of equal correlation decay of hole-pairs and triplet bosons also merges into this multicritical point P. Unlike the corresponding phase in the classical model, the SC phase extends only over a finite μ range; this is due to the hardcore constraint of the hole-pair bosons and agrees with experimentally determined phase diagrams of the cuprates. In this sense, the quantum mechanical p$SO(5)$ model is more physical than the classical $SO(5)$ model. However, in real cuprates the ratio between the max-

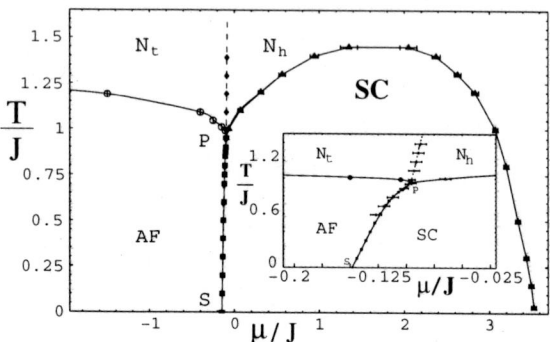

Fig. 2. Phase diagram $T(\mu)$ of the three-dimensional projected $SO(5)$ model with $J = J_s = J_c/2$ and $\Delta_s = \Delta_c = J$. N_h and N_t are, respectively, the hole-pair and the magnon-dominated regions of the disordered phase. The separation line between N_h and N_t is the line of equal spatial correlation decay of hole-pairs and bosons. The inset shows a detailed view of the region near the multicritical point.

imum SC temperature T_c and Néel temperature T_N is about 0.17 to 0.25, whereas in the p$SO(5)$ model we obtain the values $T_c/J = 1.465 \pm 0.008$ at $\mu_{opt}/J \approx 1.7$ and $T_N/J = 1.29 \pm 0.01$, hence T_c is slightly larger than T_N. In order to obtain realistic values for the transition temperatures, it is necessary to relax the *static* $SO(5)$ condition and take a smaller value for the ratio $J_c/(2J_s)$, which breaks $SO(5)$ symmetry even on a mean field level. The phase diagram with $J_c/(2J_s) = 0.225$ is plotted in Fig. 3. As one can see, this gives a more realistic ratio of $T_N/T_c \approx 0.2$. However, it should be pointed out that the numerical effort to treat such different values of J is order of magnitudes larger than considering J_c and J_s of the same order of magnitude, as we have done in Fig. 2. Therefore, we will also consider a system with $J_c = J_s = 1$ for which also the static $SO(5)$ symmetry is broken. For the same reason, we neglect here the c-axis anisotropy and consider an isotropic 3D model.

We first carry out an analysis of the critical properties for $J_c/(2J_s) = 1$ A closer look to the phase transition line between the points S and P reveals (inset of Fig. 2) that this line is not vertical as in the classical $SO(5)$ model but slightly inclined. This indicates that a finite latent heat is connected with the AF-SC phase transition. Moreover, this means that in contrast to the classical model, μ is not a scaling variable for the bicritical point P.

3.2 Scaling analysis

We now perform a scaling analysis similar to the one performed by Hu [6] in a classical $SO(5)$ system. The most important outcome of this analysis will be the strong numerical indication that in a large region around the multicritical point the full $SO(5)$ symmetry is approximately restored. This is non trivial for a system

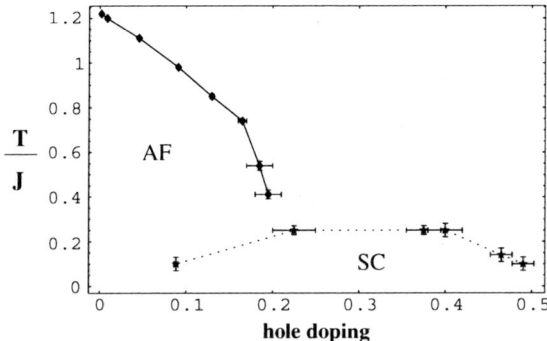

Fig. 3. Phase diagram for $J_c/(2J_s) = 0.225$ as function of the hole doping δ.

whose $SO(5)$-symmetry has manifestly been broken by projecting out all doubly-occupied states. First we want to determine the form of the $T_N(\mu)$ and $T_c(\mu)$ curves in the vicinity of the bicritical point. For crossover behavior with an exponent $\phi > 1$ one would generally expect that the two curves merge tangentially into the first-order line. However, this holds for the scaling variables, therefore, one should first perform a transformation from the old μ axis to a new μ' axis defined by $\mu'(T) = \mu - (T - T_b)/m$, where $m \approx 0.11$ is the slope of the first order line below T_b.

After this transformation, the transition curves $T_N(\mu')$ and $T_c(\mu')$ are quite well described by the crossover behavior (we now drop the prime for convenience)

$$\frac{T_c(\mu)}{T_b} - 1 = B_2 \cdot (\mu - \mu_b)^{1/\phi}$$

and
$$\frac{T_N(\mu)}{T_b} - 1 = B_3 \cdot (\mu_b - \mu)^{1/\phi} \qquad (2)$$

The fit to this behavior is shown in more detail in Fig. 4. However, the value of ϕ we obtain ($\phi \approx 2.35$) is considerably larger than the value expected form the ϵ-expansion. It should be noted that the above determination of ϕ is not very accurate: the data points in Fig. 4 are the result of a delicate finite-size scaling for lattice sites up to 18^3, followed by the transformation from μ to μ' which again increases the numerical error bars. For this reason it cannot be excluded that the difference in the ϕ values is mainly due to statistical and finite-size scaling errors. In fact, a more accurate evaluation of ϕ will be provided below.

On the SC side, the finite-size scaling carried out in order to extract the order parameter and the transition temperature turns out to be quite reliable. On the other hand, on the AF side, the fluctuations in the particle numbers of the three triplet bosons slightly increase the statistical errors of the SSE results and make the finite-size scaling more difficult.

The critical exponents for the onset of AF and SC order as a function of temperature for various chemical potentials can be extracted from Fig. 4. Far into the SC range, at $\mu = 1.5$, we find for the SC helicity modulus [13]

$$\Upsilon \propto (1 - T/T_c)^\nu \quad \text{with} \quad \nu = 0.66 \pm 0.02,$$

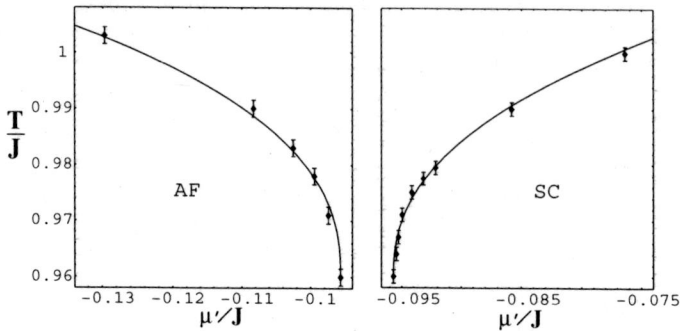

Fig. 4. Plot of the AF (left) and SC (right) critical lines in the vicinity of the multicritical point.

which matches very well the values obtained by the ϵ-expansion and by numerical analyses of a 3D XY model. On the AF side, error bars are larger, as discussed above. We obtain for the AF order parameter

$$C_{AF}(\infty) \propto (1 - T/T_c)^{\beta_3} \quad \text{with} \quad \beta_3 = 0.35 \pm 0.03,$$

for $\mu = -2.25$, also in accordance with the value expected for a 3D classical Heisenberg model.

In order to determine ν and ϕ more accurately in the crossover regime, we use two expressions derived from the scaling behavior (cf. Ref. [6])

$$\Upsilon(T_b, \mu)/\Upsilon(T_b, \mu'') = ((\mu - \mu_b)/(\mu'' - \mu_b))^{\nu_5/\phi}. \quad (3)$$

and

$$\phi = \frac{\ln\left(\frac{\mu_2 - \mu_b}{\mu_1 - \mu_b}\right)}{\ln\left(\frac{\partial}{\partial T}\frac{\Upsilon(T,\mu_1)}{\Upsilon(T,\mu_1')}\Big|_{T=T_b} \Big/ \frac{\partial}{\partial T}\frac{\Upsilon(T,\mu_2)}{\Upsilon(T,\mu_2')}\Big|_{T=T_b}\right)} \quad (4)$$

where μ_1, μ_1', μ_2, and μ_2' are related by $(\mu_1 - \mu_b)/(\mu_1' - \mu_b) = (\mu_2 - \mu_b)/(\mu_2' - \mu_b) > 0$.

The result is shown in Fig. 5: we obtain the ratio

$$\nu_5/\phi = 0.52 \pm 0.01,$$

which is in excellent accordance with the results of the ϵ-expansion and other numerical analyses [6]. ϕ is then obtained by using 4. We have applied 4 onto 9 different combinations of $(\mu_1, \mu_1' = \mu_2, \mu_2')$ values with $\mu_1/\mu_1' = \mu_2/\mu_2' = 0.5$. The result is

$$\phi = 1.43 \pm 0.05,$$

which is again in good agreement with the ϵ-expansion for a $SO(5)$ bicritical point and with the results of Ref. [6].

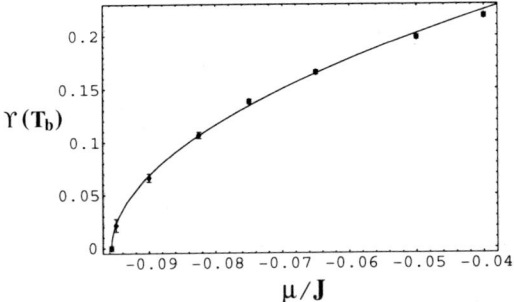

Fig. 5. Helicity Υ as a function of the chemical potential μ at $T = T_b$. From this function, the value of ν_5/ϕ can be extracted via equation (3).

3.3 Case $J_s = J_c$

This agreement between the critical exponents obtained in the previsous section may not come completely as a surprise, since the $SO(5)$ symmetry is only broken by quantum fluctuations for the parameter we have taken. The question we want to adress now is wether $SO(5)$ symmetry is also asymptotically restored for a more realistic set of parameters for which the static $SO(5)$ symmetry is broken as well. As already mentioned above, the case, where the phase diagram of the cuprates is qualitatively well reproduced ($J_c/(2J_s) = 0.225$, see Fig. 3), is too difficult to address numerically, so that the critical exponents cannot be determined with sufficient precision in this case. Therefore, we repeat our analysis for the model in an intermediate regime ($J_c = J_h$), which is not so realistic but for which the static $SO(5)$ symmetry is broken as well. One could hope that if $SO(5)$ symmetry is restored for here, then it might be also restored for the case $J_c/(2J_s) = 0.225$, although one may expect that the asymptotic region in which this occurs will be less extended. We stress again the fact that eventually one should expect the system to flow away from the $SO(5)$ fixed point, although in a very small critical region [14]. The phase diagram for $J_c = J_h$ is presented in Fig. 6 and a detailed view of the region close to the bicritical point is plotted in Fig. 7. Here, the points in the plots were obtained by a finite-size scaling with lattices up to 5032 (18^3) sites. In some cases, we were able to simulate lattices up to 10648 (22^3) sites. An example of the finite-size scaling is shown in Fig. 8. Our analysis yields $T_b = 0.682 \pm 0.005$ and $\mu_b = 0.548 \pm 0.0005$. Here the line of equal correlation decay is vertical within the error bars, so the transformation from μ to μ' is not necessary and the error bars are not increased by the transformation. This allows to determine the critical exponents by fitting the data points visible in Fig. 7 against $T(\mu) = T_b * \left(1 + (B_2 + B_3 * Sign[\mu_b - \mu]) * \mid x - \mu_b \mid^{\frac{1}{\phi}}\right)$. We obtain:

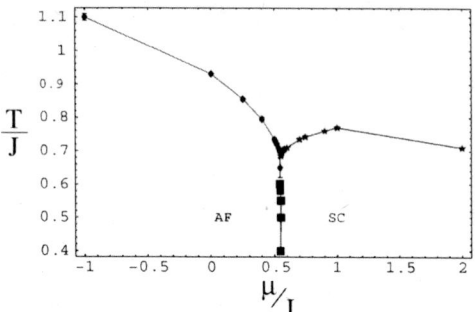

Fig. 6. Phase diagram as a function of the chemical potential for $J_c = J_h = 1$, the lines are guides to the eyes.

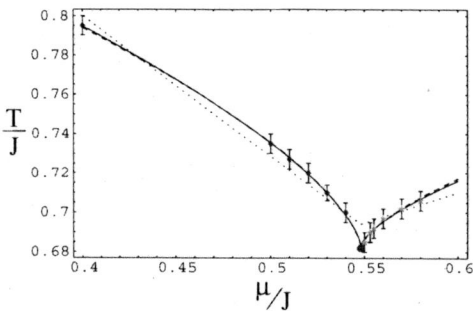

Fig. 7. Detailed view of the phase diagram as a function μ ($J_c = J_h = 1$), the two lines have been obtained by fits to $T(\mu) = T_b * \left(1 + (B_2 + B_3 * Sign[\mu_b - \mu]) * \mid \mu - \mu_b \mid^{\frac{1}{\phi}}\right)$. The continuous (dashed) line is the 'normal' ('weighted') fit. The decoupled fixpoint case is plotted as a dashed-dotted line.

$$B_2 = 0.47 \pm 0.07, \tag{5}$$
$$B_3 = 0.11 \pm 0.04,, \tag{6}$$
$$\phi = 1.49 \pm 0.18, \tag{7}$$
$$Tb = 0.683 \pm 0.004, \tag{8}$$
$$\frac{B_2}{B_3} = 1.67, \pm 0.36 \tag{9}$$

Since points further away from the bicritical point are expected to show a larger deviation from the bicritical behavior, we also performed a weighted fit, which takes this fact into account. Here, data points closer to the bicritical point are weighted more than the ones further away. Specifically, in both the SC and the AF phase, the point closest to the bicritical point is weighted six times the one with the largest distance to the bicritical point. The second closest is weighted 5 times and so on. The results are, within the error bars, quite similiar to the ones obtained without this different weighting procedure:

$$B_2 = 0.46 \pm 0.05, \quad (10)$$
$$B_3 = 0.11 \pm 0.03, \quad (11)$$
$$\phi = 1.53 \pm 0.12 \quad (12)$$
$$T_b = 0.682 \pm 0.003 \quad (13)$$
$$\frac{B_2}{B_3} = 1.61 \pm 0.23 \quad (14)$$

The agreement between Eqs. 5-9 and Eqs. 11-14 suggests that the data we have considered are still controlled by the bicritical point, In order to test whether alternativly proposed fixed points may be excluded, we carried out a least-square fit of our data to the decoupled fixpoint behavior ($\phi = 1, B_2, B_3$ and T_b arbitrary). The results are shown in Fig. 7 (dashed-dotted line). As one can see form the curve, our data do not support this hypothesis in the numerically accessible region.

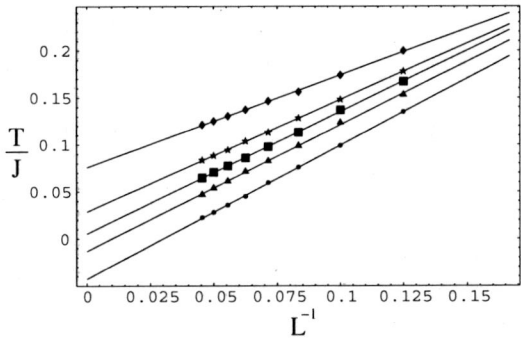

Fig. 8. Finite size scaling of the antiferromagnetic order parameter for $\mu = 0.5$, the temperatures cover 0.72J (lozenge), 0.73J (star), 0.735 (square), 0.74J (triangle) and 0.75 (cirle). The lattice size was varied from 216 (8^3) upto 10648 (22^3) sites, scanning all cubes with even edge length (L).

4 Discussion and Conlusions

Within this work we have shown that the projected $SO(5)$ model not only reproduces salient features of the high-temperature superconductors phase diagram but also that the scaling analysis of the 3D p$SO(5)$ model has produced a crossover exponent which matches quite well with the corresponding value obtained from a classical $SO(5)$ model and from the ϵ-expansion. This gives convincing evidence that the static correlation functions at the p$SO(5)$ multicritical point is controlled by a fully $SO(5)$ symmetric point in a large parameter region which is relevant experimentally and in the numerically accessible region. However, one should point out that within the statistical and finite-size error, as well as within the error due to the extrapolation of the ϵ-expansion value to $\epsilon = 1$ one cannot exclude that the actual fixed point one approaches is the biconical one, which has very similar exponents to the isotropic $SO(5)$ one. On the other hand, the biconical fixed point should be accompanied by a AF+SC coexistence region (as a function of chemical potential), which we do not observe. As discussed above we can certainly exclude in this transient region the *decoupled* fixed point for which $\phi = 1$ [14]. Of course, our limited system sizes cannot tell which fixed point would be ultimately stable in the deep asymptotic region. Here, Aharony's exact statement shows that the decoupled fixed point should be ultimately the stable one in the deep asymptotic region [14].

We argue that the resolution between this exact result and the numerically observed $SO(5)$ critical behavior lies in the size of the critical region [14]. We now give an estimate, based on ϵ expansion, for the scale at which the instability of the $SO(5)$ fixed point could be detectable. This estimate holds for the case in which one has a "static" $SO(5)$ symmetry at the mean-field level. The symmetry-breaking effects due to quantum fluctuations have been estimated in Ref. [11] and are given by Eq. (36) there. By replacing the initial conditions for the bare couplings in terms of the microscopic parameters of the Hamiltonian (cf. Eq. 26 of Ref. [11]), and projecting along the different scaling variables around the $SO(5)$ fixed point, one obtains a quite small projection along the variable that scales away from the fixed point. Combined with the fact that the exponent for this scaling variables is quite small ($\lambda = 1/13$ at the lowest-order in the ϵ expansion, although more accurate estimates [15–17] give a somewhat larger value of $\lambda \approx 0.3$), we obtain an estimate for the scaling region in which the $SO(5)$ fixed point is replaced by another – e.g. the biconical or the decoupled – fixed point at $t \equiv (T_b - T)/T_b \sim 10^{-10}$ if one takes the $O(\epsilon)$ result for the exponent. Notice that taking the result of Ref. [16] for the exponent, one obtains a quite larger value $t \sim 2.10^{-3}$. However, since the multi-critical temperatures of relevant materials (organic conductors, and, more recently, $YBa_2Cu_3O_{6.35}$) are around 10 K, the critical region is still basically unaccessible experimentally as well as with our *quantum* simulation. On the other hand, the other scaling variables, although being initially of the order of 1, rapidly scale to zero due to the large, negative, exponents. Therefore, the $SO(5)$ regime starts to become important as soon as the AF and SC correlation lengths become large and continues to affect the scaling behavior of the system basically in the whole accessible region. Possible flow away from the symmetric fix point occurs only within an extremely narrow region in reduced temperature, making it impossible to observe both experimentally and numerically. We would like to point out that this situation is very similar to many other examples in condensed-matter physics. The ubiquitous Fermi-liquid fix point is strictly speaking always unstable because of the Kohn-Luttinger effect [18]. But for most metals this instability occurs only at extremely low temperatures, and is practically irrelevant. Another example is the "ordinary" superconductor to normal-state transition at T_c. Strictly speaking, coupling to the fluctuating electromagnetic field renders this fix point unstable [19]. However, this effect has never been observed experimentally, since the associated critical region is too small. Therefore, irrespective of the question of ultimate stability, we argue that the $SO(5)$ fix point is a robust one in a similar sense, and it controls the physics near the AF and SC transitions.

In conclusion, we applied the software package developped within the KONWIHR OOPCV project, to a very demanding actual solid state physics problem. A detailed description of the library layout and the algorithms provided within this project can be found in [2]. This package in combination with the computational power of the Hitachi SR8000 in Munich allowed not only to calculate the phase diagram of the projected $SO(5)$ model in three dimensions, but also to achieve an accuracy that permitted us to extract even the critical exponents. This accuracy could only be obtained with a detailed finite size study of the order parameters of systems with, at least to our knowledge, unprecedented sizes of the order of 10^4.

Acknowledgments

This work is supported by the DFG via a Heisenberg grant (AR 324/3-1), as well as by KONWIHR (OOPCV and CUHE). The calculations were carried out at the high-performance computing centers HLRZ (Jülich) and LRZ (München).

References

1. S.-C. Zhang, Science **275**, 1089 (1997).
2. A. Dorneich, M. Jöstingmeier, E. Arrigoni, C. Dahnken, T. Eckl, W. Hanke, S. C. Zhang, and M. Troyer, in *Proceedings of the First Joint HLRB and KONWIHR Result and Reviewing Workshop, Garching, Oct. 2002*, edited by S. Wagner, W. Hanke, A. Bode, and F.Durst (Springer, Berlin, Heidelberg, New York, 2003).
3. E. Demler, W. Hanke, and S. C. Zhang, to appear in Rev. Mod. Phys. (unpublished).
4. R. Eder, A. Dorneich, M. G. Zacher, W. Hanke, and S.-C. Zhang, Phys. Rev. B **59**, 561 (1999).
5. E. Demler, H. Kohno, and S.-C. Zhang, Phys. Rev. B **58**, 5719 (1998).
6. X. Hu, Phys. Rev. Lett. **87**, 057004 (2001).
7. D. P. Arovas, A. J. Berlinsky, C. Kallin, and S.-C. Zhang, Phys. Rev. Lett. **79**, 2871 (1997).
8. E. Arrigoni and W. Hanke, Phys. Rev. Lett. **82**, 2115 (1999).
9. E. Altman and A. Auerbach, Phys. Rev. B **65**, 104508 (2002).
10. S.-C. Zhang, J.-P. Hu, E. Arrigoni, W. Hanke, and A. Auerbach, Phys. Rev. B **60**, 13070 (1999).
11. E. Arrigoni and W. Hanke, Phys. Rev. B **62**, 11770 (2000).
12. A. Dorneich, W. Hanke, E. Arrigoni, M. Troyer, and S. C. Zhang, Phys. Rev. Lett. **88**, 057003 (2002).
13. M. E. Fisher, M. N. Barber, and D. Jasnow, Phys. Rev. A **8**, 1111 (1973).
14. A. Aharony, Phys. Rev. Lett. **88**, 059703 (2002).
15. P. Calabrese, A. Pelissetto, and E. Vicari, cond-mat/0203533 (unpublished).
16. P. Calabrese, A. Pelissetto, and E. Vicari, Phys. Rev. B **67**, 054505 (2002).
17. A. Pelissetto and E. Vicari, Phys. Rep. **368**, 549 (2000).
18. W. Kohn and J. M. Luttinger, Phys. Rev. Lett. **15**, 524 (1965).
19. B. I. Halperin, T. C. Lubensky, and S.-K. Ma, Phys. Rev. Lett. **32**, 292 (1974).

ParBaum: A Fast Program for Phylogenetic Tree Inference with Maximum Likelihood *

Alexandros P. Stamatakis[1], Thomas Ludwig[2] and Harald Meier[1]

[1] Technische Universität München, Department of Computer Science, Boltzmannstr. 3 D-85748 Garching b. München Germany
{*Alexandros.Stamatakis,Harald.Meier*}*@in.tum.de*
[2] Ruprecht-Karls-Universität, Department of Computer Science, Im Neuenheimer Feld 348 D-69120 Heidelberg Germany
thomas.ludwig@informatik.uni-heidelberg.de

Abstract. Inference of large phylogenetic trees using elaborate statistical models is computationally extremely intensive. Thus, progress is primarily achieved via algorithmic innovation rather than by brute-force allocation of all available computational resources. We present simple heuristics which yield accurate trees for synthetic (simulated) as well as real data and improve execution time compared to the currently fastest programs. The new heuristics are implemented in a sequential program (RAxML) which is available as open source code. Furthermore, we present a non-deterministic parallel version of our algorithm which in some cases yielded super-linear speedups for computations with 1000 organisms. We compare sequential RAxML performance with the currently fastest and most accurate programs for phylogenetic tree inference based on statistical methods using 50 synthetic alignments and 9 real-world alignments comprising up to 1000 sequences. RAxML outperforms those programs for real-world data in terms of speed and final likelihood values.

1 Introduction

Within the ParBaum project at the Technische Universität München, we work on phylogenetic tree inference based on the maximum likelihood method by J. Felsenstein [F81]. The aim of the project is to develop novel systems and algorithms for the computation of huge phylogenetic trees based on sequence data from the ARB [L03] database. In previous work [S02] we have introduced a mechanism to reduce topology evaluation time which represents the most cost-intensive part of maximum likelihood-based phylogeny programs. We implemented our concept in

---

* This work is sponsored under the project ID **ParBaum**, within the framework of the "Competence Network for Technical, Scientific High Performance Computing in Bavaria" (KONWIHR).

parallel fastDNAml [O94, S01] and named the resulting program PAxML (Parallel Axelerated Maximum Likelihood). In tests with large alignments, we achieved global run time improvements of 26% up to 65% compared to parallel fastDNAml. In this paper we describe simple heuristics which accelerate the tree optimization process, yield accurate results, and outperform the currently fastest and most accurate programs on real data. This algorithm has firstly been introduced in [SML04]. We also present a new scalable non-deterministic parallel version of our program. The new RAxML-release presented here is a significantly changed and improved version of the initial RAxML algorithm we describe in [S04].

Related Work: An excellent comparison of popular phylogeny programs using statistical approaches such as fastDNAml, MrBayes [HR01], treepuzzle [SH96], and PAUP [PAUP03] based on synthetic data may be found in [WM03]. MrBayes carries out bayesian phylogenetic inference and outperforms all other phylogeny programs in terms of speed and tree quality in this survey. More recently, Guidon and Gascuel published an interesting paper about their new "traditional" maximum likelihood program PHYML [GG03], which seems to be able to compete with MrBayes. To our best knowledge MrBayes and PHYML are currently the fastest and most accurate programs for phylogenetic tree inference.

2 Heuristics

Sequential Algorithm: "Traditional" maximum likelihood searches can be implemented in two ways: On the one hand they can start from scratch and insert organisms progressively into the tree such as the stepwise addition algorithm (implemented e.g. in [F81, O94, W00]). On the other hand they can start with an initial tree already containing all organisms built by a simpler method such as Neighbor Joining (NJ) or by random (implemented in [GG03, HR01]). The likelihood of such a starting tree is then progressively optimized by application of minor topological changes. RAxML belongs to this second class of algorithms. The first part of our heuristics consists in building a starting tree using dnapars from PHYLIP [PHYLIP03] for two reasons:

Firstly, parsimony is related to maximum likelihood under simple evolutionary models [TS97], such that we can expect to obtain a starting tree with a relatively good likelihood value compared to random or NJ starting trees.

Secondly, dnapars uses stepwise addition for tree building and is relatively fast. This enables the construction of different starting trees by using a randomized input sequence ordering, since distinct input orderings produce distinct final trees. Thus, RAxML can be run several times with different starting trees and the set of final trees may be used to build a consensus tree and augment confidence in the final result. We removed however some optimization steps from the dnapars algorithm to accelerate computations. The second and most important part of our heuristics is the tree optimization process. RAxML performs simple tree rearrangements by subsequently removing all subtrees from the present tree and inserting them into neighboring branches up to a specified distance of nodes. RAxML inherited this optimization strategy from fastDNAml. One rearrangement step in fastDNAml consists of moving all subtrees within the currently best tree by the minimum up to the maximum distance of nodes specified (rearrangement setting). This process is outlined for a single subtree (ST5) and a distance of 1 in Figure 1 (not all possible moves are shown). The likelihood values of the such generated topologies are evaluated and the best tree is kept. If one alternative topology improves the likelihood the process is repeated with the new tree until no better topology is found. The rearrangement process of RAxML differs in two major points: In fastDNAml after each insertion of a subtree into an alternative branch the branch lengths of the entire tree are optimized. As depicted in Figure 1 with bold lines RAxML only optimizes the three branches adjacent to the insertion point before computing its likelihood value. Since the likelihood of the tree strongly depends on the topology per se this fast pre-scoring can be used to establish a small list of good alternative trees. RAxML uses a list of only size 20 to store the best trees obtained during one rearrangement step. This list size proved to be a practical value in terms of speed and thoroughness of the search. The algorithm performs global branch length optimizations only on those 20 best trees after completion of each rearrangement step. Due to the capability to analyze many more alternative topologies in less time higher rearrangements settings can be used e.g. 1–5 or 1–10 which results in significantly improved final trees.

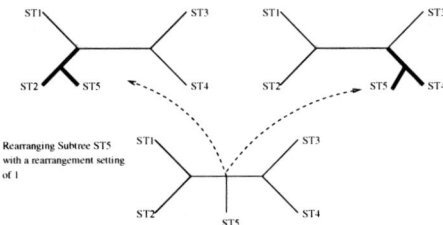

Fig. 1. Rearrangements traversing one node for subtree ST5, branches which are optimized are indicated by bold lines

Another important change especially for the initial optimization phase, i.e the first 3-4 rearrangement steps, consists in the subsequent application of topological improvements during one rearrangement step. If during the insertion of one specific subtree into an alternative branch a topology with a better likelihood is found this tree is kept immediately and all subsequent subtree rearrangements are performed on the improved topology. This enables rapid optimization of random starting trees as depicted e.g. for an alignment containing 150 taxa in Figure 5.

Parallel Algorithm: Our parallel implementation is based on a simple master-worker architecture and consists of two phases.

In **phase I** the master distributes the alignment file to all worker processes if no common file system is available, otherwise it is read directly from the file. Thereafter, each worker independently computes a randomized parsimony starting tree and sends it to the master process.

In **phase II** the master initiates the optimization process for the best parsimony starting tree. Due to the high speed of a single topology evaluation it is not feasible to distribute work by topologies as e.g. in parallel fastDNAml. Instead, we distribute work by sending a span of subtree node numbers i.e. IDs for the subtrees which shall be moved, along with the currently best topology ct, to each worker. Since the subsequent application of topological improvements during 1 rearrangement step is closely coupled we slightly modify the algorithm according to the following observation: Our experiments have shown that subsequent improved topologies occur only during the first rearrangement steps (initial optimization phase). Thereafter, only few alternative topologies per rearrangement step which are drawn form the list containing the 20 best trees improve the likelihood. This behavior is illustrated in Figure 2 where we plot the number of improved topologies per rearrangement step for a phylogenetic reconstruction of a 150 taxa tree with a random and a parsimony starting tree. When the number of improved topologies is zero the improved tree has been obtained by optimizing a topology of the best tree list (final optimization phase). This phase requires the largest amount of computation time, especially with big alignments ($\geq$ 500 organisms, $\approx$ 70% of execution time).

Thus, during the initial optimization phase we send only one single subtree ID $i, i = 2...\#species * 2 - 1$ along with the currently best tree ct to each worker for rearrangements. The worker returns the best tree wt_i obtained by rearranging subtree i to the master. If wt_i has a better likelihood than ct at the master, ct is set to wt_i and distributed to each worker along with the subsequent work (subtree ID) requests.

In the final optimization phase we reduce communication costs by generating only $5 * \#workers$ jobs (subtree ID spans) and entirely avoiding the transfer of tree topologies.

Finally, irrespective of the current optimization phase the best 20 topologies (or $\#workers$ topologies if $\#workers > 20$) computed by each worker during one rearrangement step are stored in a local worker tree list. When all subtree rearrangements i of one rearrangement step have been completed, each worker sends its tree list to the master. The master process merges the lists and redistributes the 20 ($\#workers$) best tree topologies to the workers for branch length optimization. When all topologies have been branch-length optimized the master initiates the next rearrangement step until no better tree is found. Due to the required changes to the algorithm the parallel program is non-deterministic, since final output depends on the number of workers and on the arrival sequence of results for runs with equal numbers of workers.

This is due to the changed implementation of the subsequent application of topological improvements during the initial rearrangement steps which leads to a traversal of search space on different paths.

3 Results

For our experiments we extracted alignments comprising 150, 200, 250, 500 and 1000 taxa (150_ARB,...,1000_ARB) from the ARB [L03] small subunit ribosomal ribonucleic acid (ssu rRNA) database containing organisms from the domains Eucarya, Bacteria and Archaea. In addition, we used the 101 and 150 sequence data sets (101_SC, 150_SC [S01]) which can be downloaded at superconducting www.indiana.edu/ rac/hpc/fastDNAml and have proved to be very hard to compute, especially for MrBayes. In addition we used two well-known real data sets of 218 and 500 sequences (218_RDPII, 500_ZILLA). Finally, we used 50 synthetic 100-taxon alignments with 500bp each and the respective true reference trees which are available at superconducting www.lirmm.fr/w3ifa/MAAS. Details on the generation

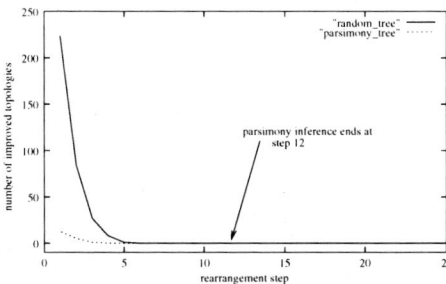

Fig. 2. Number of improved topologies per rearrangement step for a SC_150 random and parsimony starting tree

of those data sets can be found in [GG03]. To facilitate and accelerate testing we used the HKY (Hasegawa et al. 1985) model of sequence evolution and a transition/transversion (Tr/Tv) ratio of 2.0 except for 150_SC (1.24) and 101_SC (1.45). All alignments including the best topologies are available together with the RAxML source code at superconducting wwwbode.cs.tum.edu/stamatak.

Since the transition/transversion ratio is defined differently in PHYML we scaled it accordingly for the test runs (the PAML-manual [PAML03] contains a nice description of differences in the Tr/Tv ratio definitions).

For real data MrBayes was executed for 2.000.000 generations using 4 Metropolis-Coupled MCMC (MC^3) chains and the recommended random starting trees. Furthermore we used a sample and print frequency of 5000. To enable a fair comparison we evaluated all 400 output trees with fastDNAml and we report the value of the topology with the best likelihood and the execution time at that point. For synthetic data we executed MrBayes for 100.000 generations using 4 MCMC chains and random starting trees. We used sample and print frequencies of 500 and built a majority-rule consensus tree from the last 50 trees. Those significantly faster settings proved to be sufficient since trees for synthetic data converged much faster than trees for real data in our experiments.

We decided to assess performance only for those three programs since results in [WM03] and [GG03] indicate that MrBayes and PHYML are the fastest and most accurate methods for phylogenetic tree reconstruction, i.e. the methods to beat.

Sequential Tests: All sequential tests were performed on an Intel Xeon 2.4 GHz Processor. In Table 1 we summarize the final likelihood values and execution times in seconds obtained with PHYML, MrBayes, and RAxML. The results listed for RAxML correspond to the best of 10 runs. In addition, since execution times of RAxML might seem long compared to PHYML in column $R > PHY$ we indicate the likelihood and the time at which RAxML passed the final likelihood obtained by PHYML for a separate series of RAxML runs.

The long overall execution times of RAxML compared to PHYML are due to the asymptotic convergence of likelihood over time which is typical for the tree optimization process. Therefore, the comparatively small differences in final likelihood values (usually <1%) should not be underestimated, in terms of the computational effort required to obtain those values.

Table 1. PHYML, MrBayes, RAxML execution times and likelihood values for real data sets

data	PHYML	secs	MrBayes	secs	RAxML	secs	R > PHY	secs
101_SC	-74097.6	153	-77191.5	40527	-73919.3	617	-74046.9	31
150_SC	-44298.1	158	-52028.4	49427	-44142.6	390	-44262.9	33
150_ARB	-77219.7	313	-77196.7	29383	-77189.7	178	-77197.6	67
200_ARB	-104826.5	477	-104856.4	156419	-104742.6	272	-104809.0	99
250_ARB	-131560.3	787	-133238.3	158418	-131468.0	1067	-131549.4	249
500_ARB	-253354.2	2235	-263217.8	366496	-252499.4	26124	-252986.4	493
1000_ARB	-402215.0	16594	-459392.4	509148	-400925.3	50729	-401571.9	1893
218_RDPII	-157923.1	403	-158911.6	138453	-157526.0	6774	-157807.9	244
500_ZILLA	-22186.8	2400	-22259.0	96557	-21033.9	29916	-22036.9	67

Finally, in Figure 3 we plot the topological accuracy (Robinson-Foulds rate) of PHYML, RAxML, and MrBayes for 50 100-taxon trees which are enumerated on the x-axis. The average R-F rate for PHYML is 0.0796, 0.0808 for RAxML, 0.0818 for RAxML with a less exhaustive search and 0.0741 for MrBayes. The average execution time of RAxML was 131.05 seconds and 29.27 seconds for the less exhaustive search. PHYML required an average of 35.21 seconds and MrBayes 945.32 seconds.

The experiments illustrate that there seems to be no significant difference between PHYML and RAxML for synthetic data in contrast to the results obtained with real data. Thus, real as well as synthetic data should be used to perform comparative analysis of phylogeny programs.

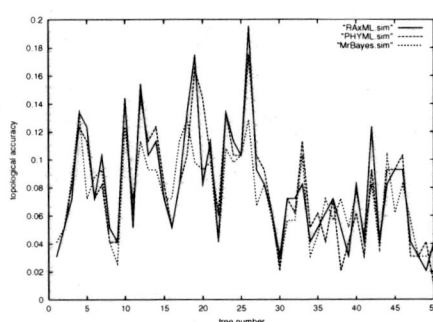

Fig. 3. Topological accuracy of PHYML, RAxML and MrBayes for 50 100-taxon trees

An example which underlines this argument is depicted in Figure 4 for the 101_SC alignment. We plot the MrBayes likelihood values over generation numbers for runs with RAxML and random starting trees. Figure 4 also illustrates the main problem of MCMC analysis pointed out by Huelsenbeck in [H02]: When to stop the chain?

In this example the run with a random starting tree seems to have reached apparent stationarity. Furthermore, it shows that "good" user trees can be useful both as reference as well as starting trees and significantly accelerate computations. This justifies the work on fast "traditional" maximum likelihood methods after the emergence and great impact of bayesian methods. Thus, we do not see RAxML as concurrence to MrBayes, but rather as useful tool to improve bayesian inference and vice versa. Therefore, RAxML produces an output file containing the alignment and the final tree in MrBayes input format.

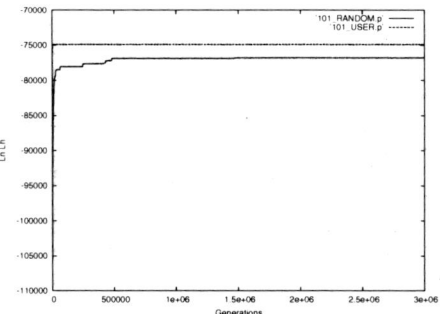

Fig. 4. Convergence behavior of MrBayes for 101_SC with user and random starting trees

Finally, in order to demonstrate the rapid tree optimization capabilities of RAxML in Figure 5 we plot the likelihood improvement over time of RAxML and MrBayes for the same 150_SC random starting tree (the final likelihood for this RAxML run was -44149.18).

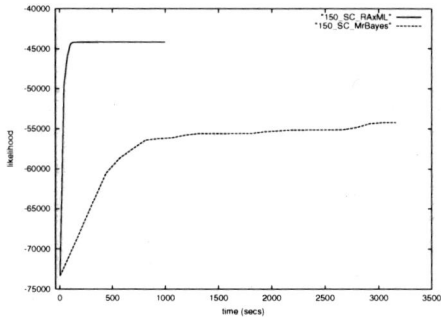

Fig. 5. 150_SC likelihood improvement over time of RAxML and MrBayes for the same random starting tree

Parallel Tests: We conducted parallel tests using a fixed starting tree for 1000_ARB. The program was executed on the Hitachi SR8000-F1 using 8, 32, and 64 processors (1, 4 and 8 nodes), as well as on the 2.66GHz Xeon cluster at the RRZE [RRZE03]

on 1,4,8,16, and 32 processors. For calculating the speedup values we only count the number of workers, since the master process hardly produces any load. In Figure 6 we plot "fair" and "normal" speedup values obtained for the experiments with the 1000_ARB data set at the RRZE PC-cluster. "Fair" speedup values take into account the first point of time at which the parallel code encountered a tree with a better likelihood than the final tree of the sequential run or vice versa (also indicated in column "P > S" of Table 2). These "fair" values correspond better to real program performance. Furthermore, we also indicate "normal" speedup values which are based on the entire execution time of the parallel program, irrespective of final likelihood values. Due to the non-determinism of the program we executed the parallel code 4 times for each job-size and calculated average "normal"/"fair" execution times and likelihood values. On the Hitachi SR8000-F1 we executed 1 run with 8 processors (1 node, 6 workers), 3 runs with 32 processors (4 nodes, 27 workers), and 2 runs with 64 processors (8 nodes, 57 workers) in intra-node MPI mode to assess performance. According to their Spec data the Intel processors should roughly be 3-4 times faster than the Hitachi CPUs. A comparison of execution times shows that the acceleration factor is > 6. We will make an effort to tune our program for the Hitachi SR8000-F1 which has been compiled with -O3 and -model=F1 only. The data from those test runs is also summarized in Table 2.

Table 2. RAxML execution times and final likelihood values for 1000_ARB

#workers	Average Likelihood	Average Execution Time (secs)	Platform	P > S
1	-400964.07	67828	Intel	void
3	-401025.23	23006	Intel	20117
7	-400917.95	11359	Intel	9233
15	-400951.36	5920	Intel	4779
31	-400942.26	3021	Intel	2199
6	-400911.91	72889	Hitachi	void
27	-400953.24	24883	Hitachi	void
57	-400912.86	17676	Hitachi	void

4 Conclusion, Current & Future Work

We presented heuristics for phylogenetic inference which outperform the currently fastest and most accurate programs on real-world data. Furthermore, we have shown that for some real data sets MrBayes does not converge in reasonable times or has reached apparent stationarity while the likelihood values of the chain are significantly inferior to those obtained by "traditional" maximum likelihood searches. Currently, we are implementing a distributed version of our code which is based on an appropriately adapted parallel algorithm.

Future work will mainly cover the execution of large production runs to compute a first small "tree of life" containing about 10.000 representative organisms of all three domains.

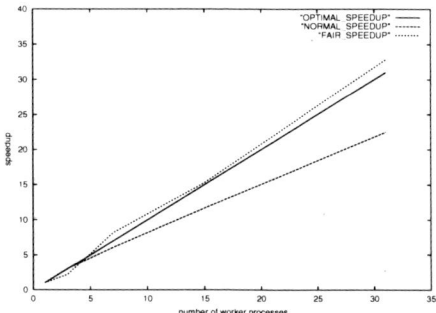

Fig. 6. Normal, fair, and optimal speedup values for 1000_ARB with 3,7,15, and 31 worker processes on the RRZE PC Cluster

References

[F81] Felsenstein, J.: Evolutionary Trees from DNA Sequences: A Maximum Likelihood Approach. In: J. Mol. Evol., 17:368–376, 1981.
[GG03] Guindon, S., and Gascuel, O.: A Simple, Fast, and Accurate Algorithm to Estimate Large Phylogenies by Maximum Likelihood. In: Syst. Biol., 52(5):696–704, 2003.
[HL03] Holder, M.T., and Lewis, P.O.: Phylogeny Estimation: Traditional and Bayesian Approaches. In: Nat. Rev. Gen., 4:275–284, 2003.
[HR01] Huelsenbeck, J.P., and Ronquist, F.: MRBAYES: Bayesian inference of phylogenetic trees. In: Bioinf., 17(8):754-5, 2001.
[H02] Huelsenbeck, J.P., et al.: Potential Applications and Pitfalls of Bayesian Inference of Phylogeny. In: Syst. Biol., 51(5):673–688, 2002.
[L03] Ludwig, W. et al.: ARB: A Software Environment for Sequence Data. In: Nucl. Acids Res., in press, 2003.
[O94] Olsen, G., et al.: fastdnaml: A Tool for Construction of Phylogenetic Trees of DNA Sequences using Maximum Likelihood. In: Comput. Appl. Biosci., 10:41–48, 1994.
[PAML03] PAML Manual: superconducting bcr.musc.edu/manuals, visited Nov 2003.
[PAUP03] PAUP: superconducting paup.csit.fsu.edu, visited May 2003.
[PHYLIP03] PHYLIP: superconducting evolution.genetics.washington.edu, visited Nov 2003.
[RRZE03] RRZE: superconducting www.rrze.uni-erlangen.de, visited Oct 2003.
[SML04] Stamatakis, A.P., et al: New Fast and Accurate Heuristics for Inference of Large Phylogenetic Trees. In: Proc. of IPDPS2004, to be published.
[S04] Stamatakis, A.P., et al: A Fast Program for Maximum Likelihood-based Inference of Large Phylogenetic Trees. In: Proc. of SAC'04, to be published.
[S02] Stamatakis, A.P., et al.: Accelerating Parallel Maximum Likelihood-based Phylogenetic Tree Computations using Subtree Equality Vectors. In: Proc. of SC2002, 2002.
[S01] Stewart, C. et al.: Parallel Implementation and Performance of fastdnaml - a Program for Maximum Likelihood Phylogenetic Inference. In: Proc. of SC2001, 2001.

[SH96] Strimmer, K., Haeseler, A.v.: Quartet Puzzling: A Maximum-Likelihood Method for Reconstructing Tree Topologies. In: Mol. Biol. Evol., 13:964–969, 1996.

[WM03] Williams, T.L., Moret, B.M.E.: An Investigation of Phylogenetic Likelihood Methods. In: Proc. of BIBE'03, 2003.

[TS97] Tuffley, C., Steel, M.: Links between Maximum Likelihood and Maximum Parsimony under a Simple Sodel of Site Substitution. In: Bull. Math. Biol., 59(3):581–607, 1997.

[W00] Wolf, M.J., et al.: TrExML: A Maximum Likelihood Program for Extensive Tree-space Exploration. In: Bioinf., 16(4):383-394, 2000.

ParaGauss: The Density Functional Program ParaGauss for Complex Systems in Chemistry

Notker Rösch[1], Sven Krüger[1], Vladimir A. Nasluzov[1,2], and Alexei V. Matveev[1]

[1] Department Chemie, Technische Universität München, 85747 Garching, Germany,
roesch@ch.tum.de
[2] Institute of Chemistry and Chemical Technology, Russian Academy of Sciences, 660049 Krasnoyarsk, Russian Federation

Abstract. The quantum chemistry software PARAGAUSS, which implements various density functional methods to determine the electronic structure of molecular systems, has been ported to and optimized for the use on the Hitachi SR8000 supercomputer platform at Leibniz Rechenzentrum München. The effort focused on tuning the code and extending it by methods that allow the simulation of molecules in an environment, e.g., in solution or adsorbed at a solid surface or in a zeolite cavity.

1 Introduction

Today, quantum chemistry is no longer restricted to determining properties of single molecules in the gas phase at an ever increasing accuracy. Rather, molecules in some environment, a much more demanding problem, attract growing interest. Examples of this type of systems are molecules in solution, adsorbates at solid surfaces, or molecules interacting with supported metal particles of a catalyst. These type of problems are often solved by a divide-and-conquer strategy where only the relevant part of the system is treated at an accurate level while the remaining systems, the "environment", is described with a less accurate method. With such an embedding approach, one is able to apply quantum chemistry methods in a variety of fields, also those of technological importance like materials science, catalysis, environmental and geochemistry, as well as to nanoscience. Not unexpectedly, modeling of such complex systems is computationally very demanding. However, also requirements regarding the accuracy of the chosen approach have to be met to allow simulations which can compete with experiments at the atomic level. As an example, we mention relativistic quantum chemistry methods which permit an accurate treatment of compounds of heavy elements. In all-electron calculations, a relativistic approach is mandatory for elements starting with xenon, but relativistic effects can no longer be neglected for elements beyond krypton if one desires highly accurate results.

To meet these various challenges, the parallel density functional program PARAGAUSS [1, 2] has been developed during the last decade. The KONWIHR project

ParaGauss aimed at porting this software package to the Hitachi SR8000 high-performance architecture, but also at tuning as well as extending and redesigning algorithms of various modules to increase the program efficiency. In that way, density functional methods as well as various embedding techniques have been made available on a powerful computing platform which allows simulations in various application fields that otherwise would be too demanding or rather time consuming. In the following, we will sketch the density functional approach to electronic structure calculations, focusing on recent improvements of PARAGAUSS. Finally, we will present selected applications to illustrate the performance of PARAGAUSS for large scale simulations as well as the type of problems that can be treated on a high-performance platform.

2 The Kohn-Sham Problem and its Extensions

Similar to other quantum chemistry methods, the central equation of a density functional approach is an eigenvalue equation for the spin orbitals ψ_i [3]. In fact, one solves the Kohn-Sham equation:

$$H\psi_k = \varepsilon_k \psi_k$$

The Kohn-Sham (Hamilton) operator

$$H = -\frac{1}{2}\nabla^2 - \sum_A Z_A/|\mathbf{r}-\mathbf{r}_A| + \int \rho(\mathbf{r}')/|\mathbf{r}-\mathbf{r}'|\,dr' + v_{xc}[\rho](\mathbf{r})$$

includes the terms of the kinetic energy of an electron, its electrostatic potential in the field of all nuclei A, the Coulomb potential of electron density ρ as well as the exchange-correlation potential v_{xc}. This latter potential is specific for the Kohn-Sham approach to density functional theory and accounts for quantum effects of the electron-electron interaction, namely exchange and correlation. Because the Kohn-Sham operator H, which determines the electronic orbitals ψ_i, depends via the electron density ρ on these spin orbitals,

$$\rho = \sum_{k=occ} |\psi_k|^2$$

the Kohn-Sham equation has to be solved iteratively via a self-consistency procedure (SCF). For computational convenience, one represents the spin orbitals as a finite linear combination of Gaussian-type "basis" functions χ_k:

$$\psi_k = \sum_j \chi_j c_{jk}$$

One maps the Kohn-Sham equation onto a generalized matrix eigenvalue problem by forming the scalar product with functions χ_i. Due to the Gaussian-type character of these functions, most resulting matrix elements can be efficiently evaluated as analytic integrals; however, the matrix elements of v_{xc} require a numerical integration.

A typical density functional calculation involves several steps. One starts by processing the information on the point group symmetry of a system. Then various "integrals", required during the SCF procedure, are precalculated. As main steps, the iterations of the SCF procedure comprise the construction of the Kohn-Sham Hamilton matrix, the solution of the resulting generalized eigenvalue problem, the selection of the occupied orbitals according to the Aufbau principle, and the construction of electron (and spin) density from the spin orbitals. After the electronic density has converged, forces on atoms are calculated as displacement derivatives of the total energy. These forces are used in a separate optimization procedure to search for an equilibrium or transition state structure in an automatic fashion. Normally, of all these tasks, the numerical integration of the exchange-correlation potential v_{xc} and the corresponding contribution to the forces, are the most time consuming steps.

Including the effect of an environment in the simulation of a molecular system implies different procedures depending on the type and model of the surrounding. Long-range solvation effects are commonly accounted for via a polarizable continuum model (PCM) where one places the molecule in a suitably shaped cavity inside a dielectric medium [4]. The solvent-induced electric field acting on the solute is modeled by a charge density on the cavity surface, represented by a set of point charges which have to be determined selfconsistently in response to the polarization of the molecule (selfconsistent reaction field – SCRF). Adsorption complexes at oxide surfaces are modeled by a density functional treatment of a cutout of a solid surface. The surrounding environment is described by a hierarchical model which accounts for the long-range electrostatic (Madelung) field of the extended crystalline solid surface as well as for the relaxation of the atomic structure in response to the perturbation exerted by the adsorbate. For this purpose, the cutout is embedded in a large set of polarizable charged centers (shell model) which are allowed to shift in response to the perturbation. Ions further away are kept at fixed positions according to the unperturbed surface structure which, without any perturbation, is relaxed compared to the terminated bulk. To prevent an artificial polarization of the atoms treated quantum mechanically by nearby (positive) point charges, these latter model cations are endowed with a repulsive potential which mimics the Pauli repulsion between the ions. For simple ionic substrates, this embedding model, which is based on a fully variational total energy of the entire system, has been termed elastic polarization environment (EPE) [5]. For materials featuring directional bonds of covalent ionic character, the model is augmented by a force field treatment of atoms at the border of the quantum mechanical and the molecular mechanical partitions (covEPE) [6]. As in the PCM treatment of solvation, both embedding variants, EPE and covEPE, require that one calculates a large number of integrals involving point charge centers.

Finally, we mention the recent success with modeling adsorption complexes on metal surfaces. For these substrates, nanocrystalline clusters of 80 and more metal atoms, terminated by low-index crystal planes, featured properties of adsorption complexes that were essentially converged with cluster size [7]. These converged properties also comprise the very sensitive adsorption energies that had long escaped reliable computation via cluster models. This success of nanocluster models relied in an essential fashion on the efficient symmetry treatment afforded by PARA-GAUSS which allows one to treat such large metal clusters if the overall point group symmetry of the model is sufficiently large (D_{4h} and higher).

For the treatment of systems with heavy elements, a relativistic variant of the density functional approach is required, either a scalar relativistic method or one that includes spin-orbit interaction. The relativistic variants implemented in PARAGAUSS [8, 9] are based on the Douglas-Kroll-Hess procedure to decouple electronic and positronic degrees of freedom in the relativistic analogue of the Kohn-Sham equation, the Dirac-Kohn-Sham equation. The resulting formalism leads to new terms in an effective Kohn-Sham operator which imply new computational tasks in the program, e.g., the transformation of all operators to their relativistic form. The relativistic approaches available in PARAGAUSS demand computational resources which increase with increasing sophistication of the method, especially when spin orbit interaction is taken into account.

Straightforward parallelization of a code relies on data which are available throughout the calculation and exhibit a structure that permits their independent treatment. This is not the case for a density functional program like PARAGAUSS because the various computational steps involve different quantities which can not be treated independently of each other. Thus, each of several tasks requires a different parallelization strategy and a specific distribution of data structures. Analytical integrals are bundled according to the atoms and the angular momenta of the orbitals involved; they are treated according to the farming concept to achieve load balancing. Numerical integration is parallelized by distributing the grid points over available processors. Algorithms involving matrix algebra, like the construction of the Kohn-Sham matrix, exploit the block-diagonal structure of the matrices when the space spanned by the "basis" functions is separated according to the irreducible representations of the pertinent point group. Overall, coarse grain parallelization with specialized algorithms for each major task has been constructed in PARAGAUSS, based on the MPI communication library [1]. To achieve a reasonable efficiency, every task consuming typically more than 1 % of the overall computation time has been parallelized. Taking into account the size of the program, which comprises about 400.000 lines of Fortran95 code, it is evident that porting and optimization for a new platform is a substantial task.

3 Optimization and Performance

Porting the program PARAGAUSS to the supercomputer platform Hitachi SR8000 at Leibniz Rechenzentrum München and optimizing the various density functional methods implemented in the software package implied that several modules had to be adapted to this specific architecture. In addition, new algorithms have been developed aiming at a more efficient use of a larger number of processors. In a first step, basic optimization was carried out by invoking BLAS and LAPACK libraries as optimized for Hitachi SR8000. In addition, larger subroutines had to be partitioned so that the optimization capabilities of the Fortran compiler could be exploited.

First test calculations revealed a large difference between real and user timings, which could be traced back to the slow operations of opening and closing of external files as well as to slow processing of system calls. The resulting overhead amounted to up to 100% of the actual CPU time, especially for smaller systems and calculations with a larger number of processors. In particular for larger calculations, this behavior prevented efficient applications as not all integrals required in the SCF procedure could be stored in memory. By restructuring the file handling, employing predominately direct access files, and reducing the number of file handling operations, the overhead was reduced to less than 20% for the program version with integrals on file and to less than 10% for the version with integrals in memory.

Out of the various algorithmic improvements achieved in this project, we shall describe three instructive examples in more detail. While some developments, like the parallel diagonalization of blocks of the Kohn-Sham matrix, relied exclusively on novel algorithms, the most successful reconstructions combined algorithmic and methodic improvements.

In an earlier version of PARAGAUSS, the diagonalization of the Kohn-Sham matrix had been parallelized according to the blocks of the irreducible representations of the point group. Thus, depending on the symmetry group, a number of these sub-matrices of different size was diagonalized in parallel. For common point groups, this strategy is limited to systems with symmetry and to less than about 20 processors. Applying in addition a parallel eigen solver from ScaLAPACK, it became possible to diagonalize in parallel a Kohn-Sham matrix that features only one block. To construct an efficient scheme for the general case, one has to take into account that only matrices of order 500 and higher can be diagonalized faster in parallel than in serial mode. Based on one-time test calculations, for each matrix the optimum number of processors for the diagonalization is estimated; then, matrices are distributed over available processors in such a way as to minimize the overall time for diagonalizing all of them. Especially for problems of low symmetry, with only a few matrix blocks, the corresponding part of the SCF problem was accelerated by factors from 2 to 6. A better speed-up is achieved for larger problems due to the increasing efficiency of the parallel solver.

Relativistic calculations including spin orbit interaction are rather demanding because the coupling of spin and angular momenta prevents a representation of the orbitals by real functions; rather, the resulting orbital functions are inherently complex-valued. Especially for the evaluation of the electron density and the determination of the exchange-correlation potential, this entails a considerably larger computational effort. By separate symmetrization of spin and spatial parts of the orbitals, a tremendous saving is possible for operators that are not explicitly spin-dependent if one couples both parts at a very late stage in the algorithm [10]. In this way, the time for evaluating matrix elements of the exchange-correlation potential in serial fashion was reduced by a factor of up to 100. For a complete electronic structure calculation, the savings still amounted to an overall speed-up by a factor of 4 in real time.

Solvation models as well as EPE embedding techniques for cluster models of solids and surfaces demand the analytical calculation of a large number of integrals involving the potential of point charges. This part of the calculations originally amounted to 15 to 30 % of the total CPU time of these overall rather demanding calculations. In addition, the corresponding part of the program PARAGAUSS did not exhibit a satisfactory parallel efficiency. As shown in Figure 1 for actinide complexes in aqueous solution, parallel acceleration originally was below 2 for the diuranyl complex $[(UO_2)_2(OH)_2(H_2O)_6]^{2+}$ and about 4 for the diuranyl carbonate complex $(UO_2)_2(OH)_2(H_2O)_4CO_3$. The former system features D_{2h} symmetry, the latter only a single reflection plane. A significant speed-up of these solvent-model calculations was achieved by introducing a novel strategy for integral evaluation. It relies on a careful separation of angular and radial integrations and a reusage of data. This strategy is also beneficial for the evaluation of other types of three-center integrals. For serial runs, acceleration values from 5 to 10 have been obtained and even the parallel performance was notably improved. For the smaller diuranyl complex (Fig. 1), the scaling still levels off at four processors, but the speed-up was increased by a factor of two. Also for the more demanding problem of a diuranyl carbonate complex, the parallel speed-up is about doubled. Overall, the effort to calculate point-charge

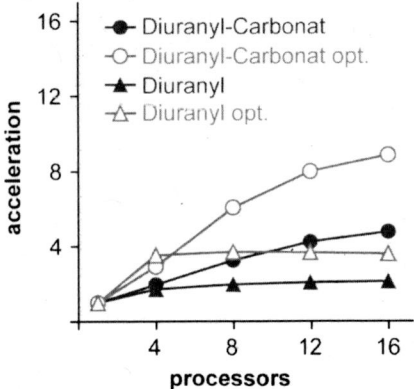

Fig. 1. Improvement of parallel performance of point charge related integrals in solvation models of actinide complexes. Filled symbols mark results before, open symbols after optimization.

related integrals was reduced to only about 2 % of the total time of the electronic structure calculation.

4 Applications

4.1 Transition metal clusters

Metal clusters form a state of matter between molecules and solids. Due to their size at the nanometer scale, they exhibit quantum effects. Their properties vary with size, which offers an additional degree of freedom for tailoring their characteristics. These features lead to a considerable interest in metal clusters with respect to applications in nano electronics, heterogeneous catalysis, sensors, and biophysical chemistry [11]. Calculational modeling of transition metal clusters is one of the challenges of quantum chemistry because a large number of atoms with many electrons have to be treated. Moreover, these species commonly exhibit very many low lying states; as a result, the SCF procedure is difficult to converge.

Synthesis of transition metal cluster compounds normally results in samples containing a distribution of cluster sizes, except for small species with only few transition metal atoms. $Au_{55}Cl_6PPh_{12}$ is one of the rare cases where synthesis of a well defined larger species is claimed [12]. Although this cluster has been examined in quite a number of experimental studies, its structure is still controversially discussed. The Au_{55} cluster core is assumed to have a cuboctahedral structure (Fig. 2) with chlorine atoms singly coordinated in the center of the six square (100) surface facets, while phosphine ligands occupy the twelve corners. Computational modeling of chlorine adsorption on the bare cluster revealed (Fig. 2) that threefold coordinated sites are preferred by about 40 kJ/mol. While this result agrees with experimental determinations of the chlorine position on gold surfaces, the phosphine ligands may well

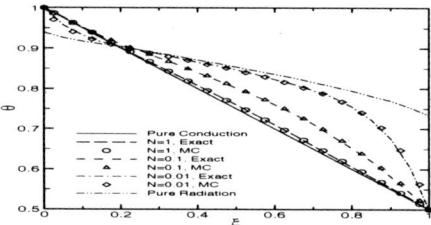

Fig. 2. Chlorine coordination to the cuboctahedral cluster Au_{55}: singly coordinated to the (100) facet (left) and threefold coordinated on the (111) facet (right).

affect the attachment of chlorine to Au_{55}. This latter question is currently investigated with a scalar relativistic density functional approach. While highly symmetric clusters of this size are easily modeled by PARAGAUSS [1,2] due to its efficient exploitation of symmetry, studies of ligand shells are more complicated due to lower symmetry. Calculations of gold clusters are well performing on Hitachi SR8000. For Au_{55} in D_{4h} and D_{3d} symmetry, up to 300 MFLOPS per processor were achieved for up to 24 processors.

Another important area of application for nanoclusters is their use as models of transition metal surfaces. Commonly, metal surfaces are modeled either with a periodic slab approach, exploiting periodicity by means of a plane wave expansion of the orbital functions, or by "two-dimensional" cut-outs of the ideal surface structure, called surface cluster models, which are treated by a conventional quantum chemistry method. Whereas periodic models allow only the treatment of periodic arrays of adsorbates, normal (two-dimensional) surface cluster models of isolated adsorbates suffer from unphysical boundary conditions on their bulk side. In a study of CO adsorption at (111) surface facets of symmetric nanocrystalline Pd clusters with 55 to 146 atoms, we showed that the CO binding energy is converged for clusters with about 80 atoms and more [7]. Also structural features and vibrational frequencies of the adsorbate are converged for nanoclusters of this size. The success of these models is due to the fact that all surface facets are formed by low-index crystal planes. In further studies, this type of models has been also used to inspect the effect of impurity atoms like H, C, N, and O on the properties of Pd surfaces [13]. For carbon, adsorption on the surface as well as subsurface sites are energetically accessible; however, surface adsorption was found to be preferred for the other atoms. As a result of subsurface carbon contamination, a reduced CO adsorption energy was calculated. For these large Pd clusters, an even better performance than for Au_{55} was obtained on the Hitachi SR8000. For $Pd_{116}C_6$, PARAGAUSS reached up to 460 MFLOPS per processor in single-point calculations on 16 processors; for 32 processors, the performance dropped to 380 MFLOPS. The latter example represents the highest overall performance achieved with PARAGAUSS until now, about 12 GFLOPS.

4.2 Zeolites

Zeolites, nanoporous silicate cyrstals, are applied as molecular sieves and catalyst, in pure as well as in chemically modified form. The acidity of these lattices can be

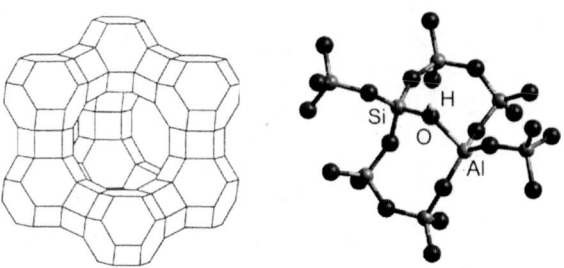

Fig. 3. Stucture of the zeolite fausjasite (left) and atomic structure model of an acidic OH site (right).

varied by incorporating Al centers which require nearby compensating OH groups. Shape and size of the pores, as well as the chemically variable character of the active sites or transition metal species in the cages allow a specific and tunable steering of chemical reactions inside these materials.

This type of systems can be studied with a cluster approach together with the covalent variant of the EPE embedding technique. Small parts of a zeolite, typically 20-50 atoms (Fig. 3) around the active site, are treated quantum mechanically, while the surrounding crystal is modelled by a force-field based approach. This procedure was successfully applied on the Hitachi SR8000 after integral evaluation had been optimized (see above). To exemplify the achievements for that type of calculations, we quote the performance for the cluster model $AlSi_7O_{23}H$ embedded in a faujasite lattice (Fig. 3). The GFLOP rates for this benchmark system increased due to optimization from 2.5 to 3.1 for 16 processors and from 4.7 to 5.6 on 32 processors.

One of the zeolite studies performed on the Hitachi SR8000 aimed at a characterization of acidic OH sites in faujasite (Fig. 3) and ZSM-1 zeolites, focusing on the effect of the Al content of the zeolite lattice [14]. Experimentally observed shifts of the OH vibrational stretching frequency had been tentatively assigned to OH groups in cavities of different size. This interpretation was confirmed by our calculations, which yielded frequency shifts in qualitative agreement with experiment. Also experimental hints for a decreasing deprotonation propensity of OH groups with increasing Al content have been supported by the model calculations. Moreover, it has been demonstrated that for a given Al concentration also the relative position of Al centers with respect to the OH groups affects their acidity. Thus, in combination with experimental information on systems, where detailed knowledge about the atomistic structure is missing, computational results on well defined model systems result in a consistent and considerably refined picture.

Another type of zeolite lattice, MFI, has been shown to be not very reactive, while substitution of some of its Si centers by titanium Ti yields a remarkable catalyst for selective oxidation. Due to the low concentrations of Ti centers, X-ray as well as neutron diffraction experiments lead to different conclusions about the preferred location for Ti incorporation among the 12 cyrstallographically distinguishable sites. Previously, various simulations at different level of sophistication produced contradictory results about the nature of Ti sites. An EPE embedded cluster model study systematically inspected all 12 possible sites and compared their stability by means of a model substitution reaction $Si_{zeo} + TiO_4 \longrightarrow SiO_4 + Ti_{zeo}$, where the subscript "zeo" refers to an atom incorporated in the zeolite lattice [15]. The substitution energies for Si by Ti determined in that way varied in the narrow range from 88 to 107 kJ/mol. Energy differences of this small size require an accurate modelling including long-range polarization effects as well as lattice relaxation, as was done here by means of the covalent EPE technique. Favorable sites have been determined. From this study, we conclude that a careful examination of this material is necessary to distinguish the various possibilities for locating intra-lattice Ti centers. The variation of the relative energies of these sites clearly indicates preferences, at variance with the hypothesis of a random distribution of Ti centers. This study also exemplified that typical problems in computational chemistry are not solved by a single very large and highly accurate calculation; rather, a number of quite demanding computations on similar systems are needed to solve a given problem.

4.3 Large actinide complexes

The quantum chemical treatment of small actinide complexes poses no severe computational demand nowadays, if the conventional approach of approximating the large number of core electrons of the actinide atoms by means of a pseudopotential is applied. On the other hand, accurate treatments involving all electrons with a first principles relativistic method are still rare. If in addition larger complexes involving more than a single actinide atom are considered, than high performance computational resources are necessary.

As already mentioned above, a part of this project was devoted to the optimization of solvation models to allow the efficient treatment of actinide complexes in solution. Knowledge of the detailed properties of these species is important for understanding the chemical speciation and the transport behavior of actinides in the environment. The study of actinide species in a natural aqueous environment is considerably complicated by the complex hydrolysis behavior of these elements. At lower pH values, they are present as mononuclear hydroxides or carbonates, which show a strong tendency for some elements to associate as dimers and larger multinuclear complexes at about neutral pH. In the regime of basic pH this leads to colloid formation and precipitation.

Test calculations for the solvation modeling of such species have been carried out on the Hitachi SR8000 for dimers of uranyl UO_2^{2+} connected by bridging OH ligands and surrounded by water molecules. If a carbonate ligand is attached to this type of complexes, various structures can arise, all complying with the formula $(UO_2)_2(OH)_2(H_2O)_4CO_3$. In agreement with experimental suggestions, it was shown that there is no preferred structure, but that the energetic differences between various geometries are so small that isomers can coexist. Whereas mononuclear uranyl carbonates like $UO_2(CO_3)_n^{2(n-1)-}$ are well established in solution, the existence of the trimeric species $(UO_2)_3(CO_3)_6^{6-}$ (Fig. 4) is less certain. Inspection of the calculated structures revealed rather long U-O bond distances between uranyl and bridging carbonate ligands. This finding points towards a reduced stability. In the gas phase, a decay reaction to uranyl tricarbonat $UO_2(CO_3)_3^{4-}$ was estimated to be slightly endothermic by 46 kJ/mol. Addition of some water ligands of the first solvation shell changed the reaction energetics: it became exothermic by 96 kJ/mol. From the computational aspect these calculations are remarkable because, to our knowledge,

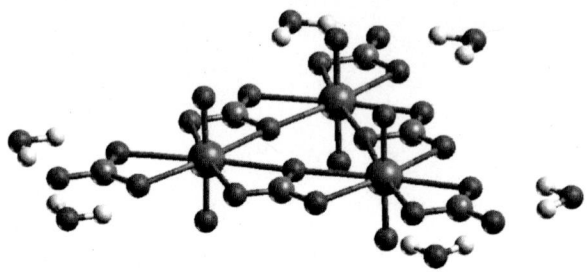

Fig. 4. Uranyl carbonate trimer $(UO_2)_3(CO_3)_6^{6-}$ with some water ligands for modeling solvation effects.

$(UO_2)_3(CO_3)_6^{6-}$ is the largest actinide complex modeled so far with a first principles method. Further studies will be devoted to the interaction of actinide species in solution with humic acids, which are modelled by larger organic acids. Here, the experience gained already for larger hydrolysis species will be very helpful.

5 Summary

The parallel density functional software suite PARAGAUSS for quantum chemistry modeling was ported to and optimized for the high-performance computing platform Hitachi SR8000 at Leibniz Rechenzentrum München. While in the beginning the work concentrated on modifications specific to the Hitachi architecture, algorithmic as well as methodological developments to increase serial as well as parallel performance dominated the second phase of the project. Remarkable progress has been achieved especially for embedding methods, to simulate the solvation of molecules and to employ embedded local models of solids and surfaces in their extended crystalline environment. Also, accurate relativistic density functional methods including spin-orbit interaction have been improved; this lead to a considerable extension of their applicability for problems of practical interest. While PARAGAUSS has been originally designed for moderately parallel applications with up to 24 processors, the present, improved version is able to run efficiently on up to twice as many processors. This parallel efficiency is not common for quantum chemistry codes which use a Gaussian-type (local) representation of wave functions. The progress due to this project made it possible to release version 3.0 of PARAGAUSS [2]. Besides the features mentioned above, this new version includes a module to calculate hyperfine coupling constants and g-tensor elements, a time-dependent DFT approach for determining excitation energies, a model density approach that also encompassed the exchange-correlation contributions, a traditional embedding approach by means of coupling of a quantum mechanical system with a surrounding modelled by molecular mechanics (IMOMM), as well as new variants for SCF convergence acceleration and an improved optimizer for molecular structures.

The achievements of the project have been validated and tested in practice for several classes of demanding systems. Large transition metal clusters like Au_{55} have been modelled with relativistic methods. Even larger nanoparticles of Pd with up to 146 atoms have been successfully applied as a new type of models for adsorption at transition metal surfaces, showing superior convergence properties for calculating binding energies compared to other model classes. For this type of systems also the best performance has been achieved with about 12 GFLOPS for 32 processors, opening a wide area of application. Another field of application has been zeolites and zeolite supported metal particles, where considerably new insight into the structure and properties at the atomistic level has been gained due to a large series of demanding model calculations. Calculations on large actinide complexes, which are found at typical environmental conditions in natural waters, have been notably facilitated by the new version of PARAGAUSS on Hitachi SR8000. Among these novel applications, the uranyl trimer carbonate $(UO_2)_3(CO_3)_6^{6-}$ is probably the largest actinide complex treated until now by a first principles all-electron quantum chemistry method. Studies in this field are currently extended to carboxylate complexes, which serve as models to understand the interaction of actinide species with humic acids in the environment. By porting the newly developed software modules to other platforms, the experience and achievements gained on Hitachi SR8000 will also be available for other computing environments.

References

1. Belling, Th., Grauschopf, T., Krüger, S., Mayer, M., Nörtemann, F., Staufer, M., Zenger, C., Rösch, N., in Bungartz, H.-J., Durst, F., Zenger, C. (eds): High Performance Scientific and Engineering Computing, Proceedings of the First International FORTWIHR Conference 1998. Lecture Notes in Computational Science and Engineering, Vol. 8, p. 439–453, Springer, Heidelberg (1999).
2. Belling, Th., Grauschopf, T., Krüger, S., Nörtemann, F., Staufer, M., Mayer, M., Nasluzov, V. A., Birkenheuer, U., Shor, A., Matveev, A., Hu, A., Fuchs-Rohr, M. S. K., Neyman, K. M., Ganyushin, D. I., Kerdcharoen, T., Woiterski, A., Gordienko, A. B., Majumder, S., Rösch, N., PARAGAUSS, Version 3.0, Technische Universität München (2004).
3. Koch, W., Holthausen, M. C., A Chemist's Guide to Denisty Functional Theory, Wiley VCH, Weinheim, 2002.
4. Fuchs, M., Shor, A., Rösch, N., Int. J. Quant. Chem., **86**, 487–501 (2002).
5. Nasluzov, V. A., Rivanenkov, V. V., Gordienko, A. B., Neyman, K. M., Birkenheuer, U., Rösch, N., J. Chem. Phys., **115**, 8157–8171 (2001).
6. Nasluzov, V. A., Ivanova, E. A., Shor, A. M., Vayssilov, G. N., Birkenheuer, U., Rösch, N., J. Phys. Chem. B, **107**, 2228–2241 (2003).
7. Yudanov, I. V., Sahnoun, R., Neyman, K. M., Rösch, N., J. Chem. Phys., **117**, 9887–9896 (2002).
8. Rösch, N., Krüger, S., Mayer, M., Nasluzov, V. A., in Seminario, J. (ed): Recent Developments and Applications of Modern Density Functional Theory, Theoretical and Computational Chemistry Series, Vol. 4, p. 497–566, Elsevier, Amsterdam (1996).
9. Rösch, N., Matveev, A. V., Nasluzov, V. A., Neyman, K. M., Moskaleva, L., Krüger, S., in Schwerdtfeger, P. (ed): Relativistic Electronic Structure Theory

- Applications, Theoretical and Computational Chemistry Series, Vol. 14, p. 656–722, Elsevier, Amsterdam (2004).
10. Matveev, A. V., Mayer, M., Rösch, N., Comp. Phys. Comm., **160**, 91–119 (2004).
11. Daniel, M.-C., Astruc, D., Chem. Rev., **104**, 293–346 (2004).
12. Schmid, G., Boese, R., Pfeil, R., Bandermann, F., Meyer, G., Calis, G. H. M., van der Velden, J. W. A., Chem. Ber., **114**, 3634–3642 (1981).
13. Yudanov, I. V., Neyman, K. M., Rösch, N., Phys. Chem. Chem. Phys., **6**, 116–123 (2004).
14. Ivanova Shor, E. A., Shor, A. M., Nasluzov, V. A., Vayssilov, G. N., Rösch, N., to be published.
15. Deka, R. C., Nasluzov, V. A., Ivanova Shor, E. A., Shor, A. M., Vayssilov, G. N., Rösch, N., to be published.

Appendix

Color figures

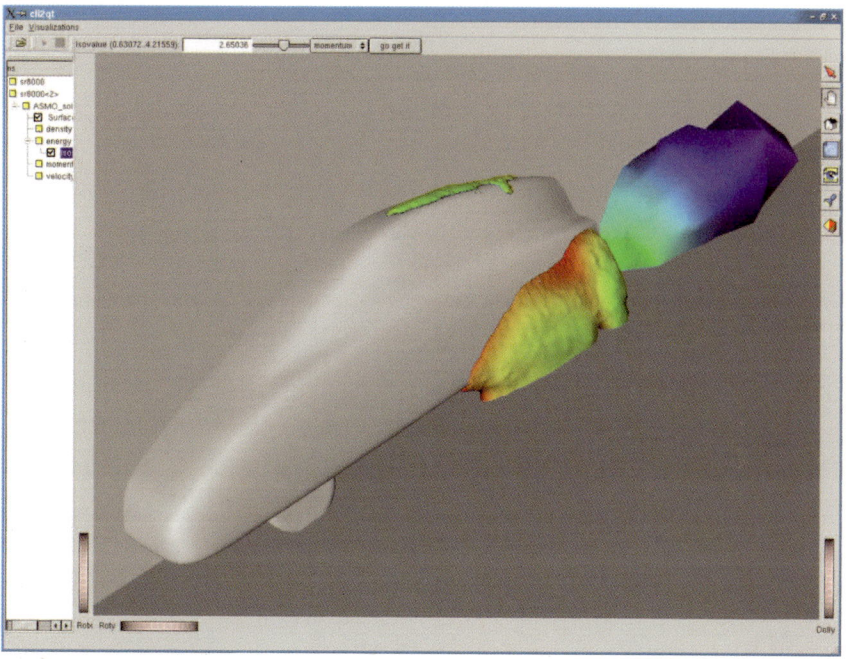

Fig. A.1. [F. Hülsemann, S. Meinlschmidt, B. Bergen, G. Greiner, U. Rüde] Flow solution around ASMO – energy isosurface with color mapped momentum. The data set resides and is post processed on the Hitachi SR8000 in Munich while being viewed on a PC in Erlangen

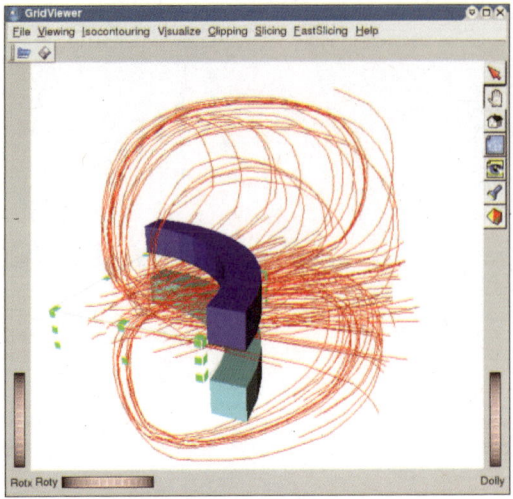

Fig. A.2. [F. Hülsemann, S. Meinlschmidt, B. Bergen, G. Greiner, U. Rüde] Magnetic field around a coil visualized by locally exact stream lines

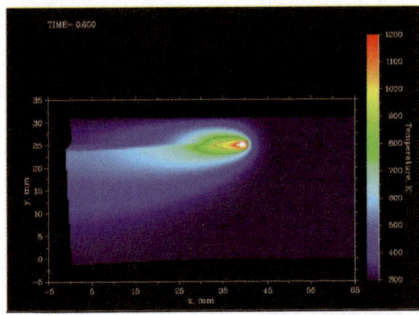 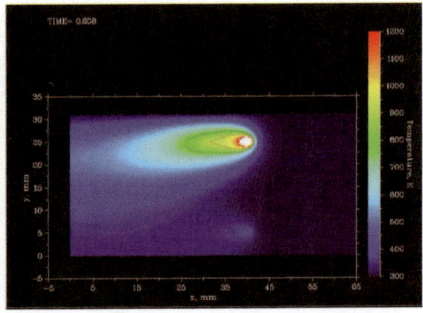

Fig. A.3. Simulation of single welding: a crack arises.

Fig. A.4. Simulation of two-beam welding: no crack arises.

 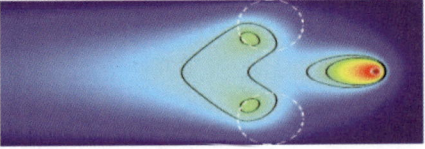

Fig. A.5. Temperature field generated by conventional laser beam welding, i.e. without additional laser beams.

Fig. A.6. Temperature field generated by multi-beam welding, i.e. with additional laser beams.

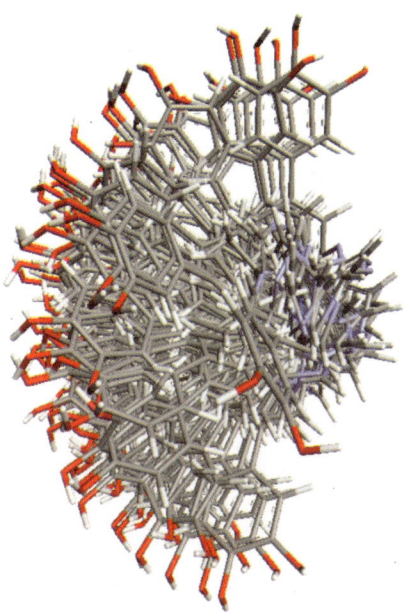

Fig. A.7. [A. von Homeyer, J. Gasteiger] Superimposition of 216 conformers of the cytochrome P450c17 inhibitor BW112 as an illustration of the search space taking into account conformational flexibility. Note that torsion angles have been restricted in this figure to certain low energy conformations. Thus, the conformational space is potentially even more extensive.

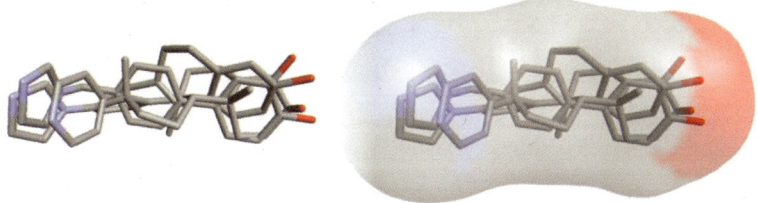

Fig. A.8. [A. von Homeyer, J. Gasteiger] Superposition of the three molecules in the cytP450 data set: imidq,15, bw112, bw13. It can be seen that the oxygen atoms as well as nitrogen atoms are matched on both ends of all three molecules. The generation of an averaged van der Waals surface around several active, superimposed molecules leads to the identification of the steric requirements of the receptor binding pocket (left part).

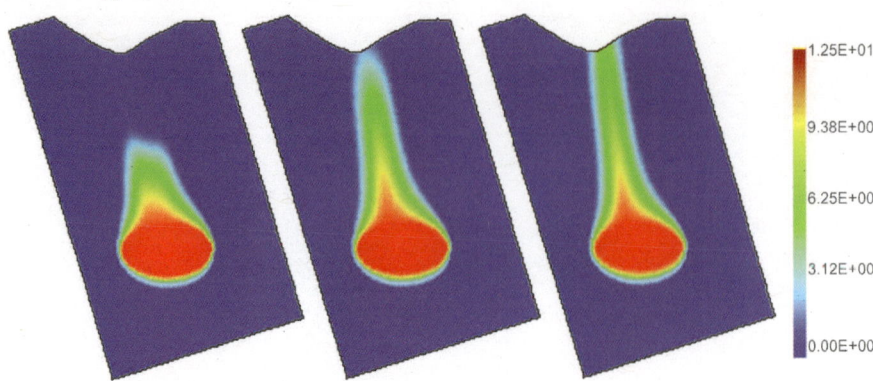

Fig. A.9. Concentration of xylene after 3, 5, 10 years.

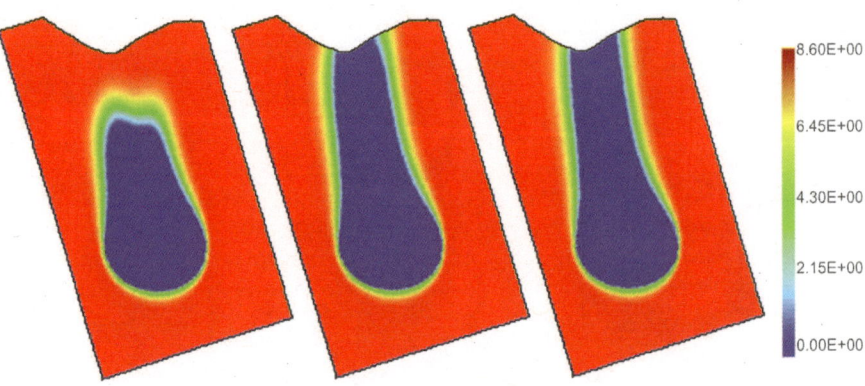

Fig. A.10. Concentration of oxygen after 3, 5, 10 years. Within the plume, the oxygen is consumed.

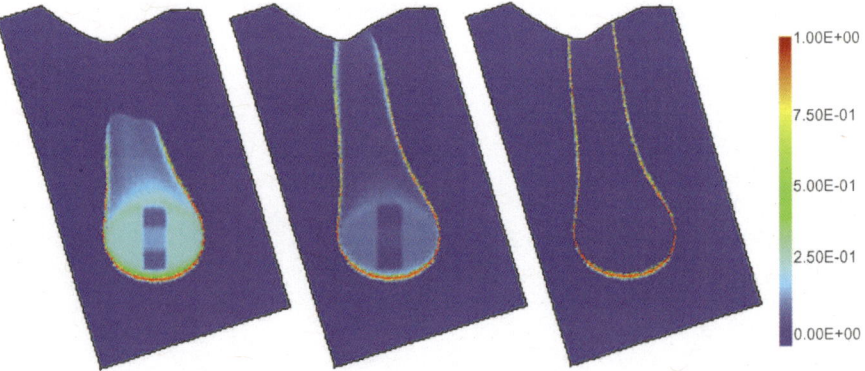

Fig. A.11. Concentration of biomass after 3, 5, 10 years. The biomass is concentrated on a thin layer where oxygen and xylene mix.

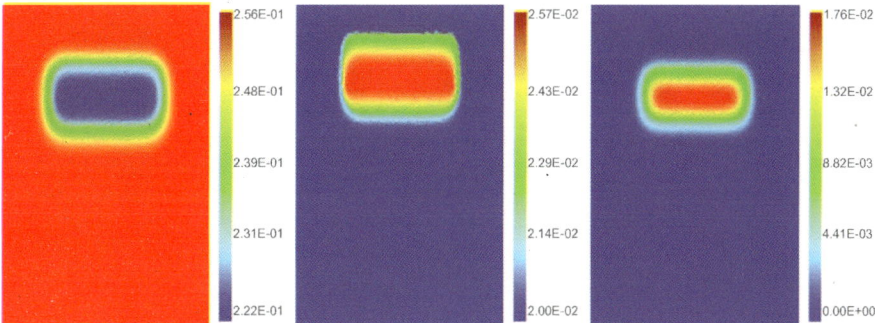

Fig. A.12. Concentration of O_2, biomass and Fe(III)EDTA at time $t = 10.0$[h] for the 2d-problem. The biodegradation of EDTA causes a strong growth of biomass and a decay of oxygen in the upper part of the domain.

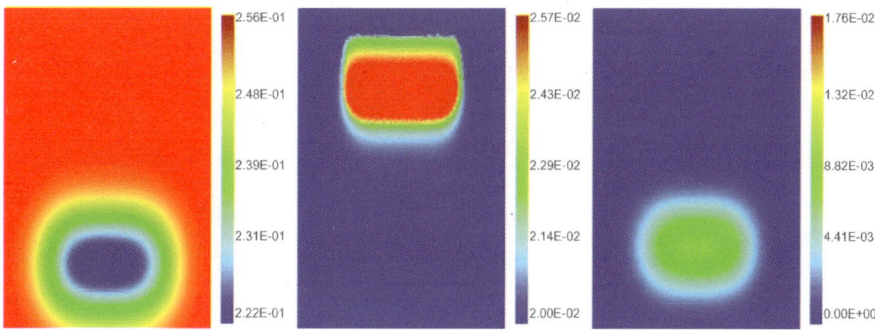

Fig. A.13. Concentration of O_2, biomass and Fe(III)EDTA at time $t = 50.0$[h] for the 2d problem.